Molecular Biology of Fungal Development

MYCOLOGY SERIES

Editor

J. W. Bennett
Professor
Department of Cell and Molecular Biology
Tulane University
New Orleans, Louisiana

Founding Editor

Paul A. Lemke

1. *Viruses and Plasmids in Fungi*, edited by Paul A. Lemke
2. *The Fungal Community: Its Organization and Role in the Ecosystem*, edited by Donald T. Wicklow and George C. Carroll
3. *Fungi Pathogenic for Humans and Animals* (in three parts), edited by Dexter H. Howard
4. *Fungal Differentiation: A Contemporary Synthesis*, edited by John E. Smith
5. *Secondary Metabolism and Differentiation in Fungi*, edited by Joan W. Bennett and Alex Ciegler
6. *Fungal Protoplasts*, edited by John F. Peberdy and Lajos Ferenczy
7. *Viruses of Fungi and Simple Eukaryotes*, edited by Yigal Koltin and Michael J. Leibowitz
8. *Molecular Industrial Mycology: Systems and Applications for Filamentous Fungi*, edited by Sally A. Leong and Randy M. Berka
9. *The Fungal Community: Its Organization and Role in the Ecosystem, Second Edition*, edited by George C. Carroll and Donald T. Wicklow
10. *Stress Tolerance of Fungi*, edited by D. H. Jennings
11. *Metal Ions in Fungi*, edited by Günther Winkelmann and Dennis R. Winge
12. *Anaerobic Fungi: Biology, Ecology, and Function*, edited by Douglas O. Mountfort and Colin G. Orpin
13. *Fungal Genetics: Principles and Practice*, edited by Cees J. Bos
14. *Fungal Pathogenesis: Principles and Clinical Applications*, edited by Richard A. Calderone and Ronald L. Cihlar
15. *Molecular Biology of Fungal Development*, edited by Heinz D. Osiewacz

Additional Volumes in Preparation

Pathogenic Fungi in Humans and Animals: Second Edition, edited by Dexter Howard

Molecular Biology of Fungal Development

edited by
Heinz D. Osiewacz
Johann Wolfgang Goethe-Universität
Frankfurt, Germany

MARCEL DEKKER, INC.   NEW YORK • BASEL

ISBN: 0-8247-0744-3

This book is printed on acid-free paper.

Headquarters
Marcel Dekker, Inc.
270 Madison Avenue, New York, NY 10016
tel: 212-696-9000; fax: 212-685-4540

Eastern Hemisphere Distribution
Marcel Dekker AG
Hutgasse 4, Postfach 812, CH-4001 Basel, Switzerland
tel: 41-61-261-8482; fax: 41-61-261-8896

World Wide Web
http://www.dekker.com

The publisher offers discounts on this book when ordered in bulk quantities. For more information, write to Special Sales/Professional Marketing at the headquarters address above.

Current printing (last digit):
10 9 8 7 6 5 4 3 2 1

PRINTED IN THE UNITED STATES OF AMERICA

Preface

Developmental biology, which deals with the various aspects of species-specific changes during the lifetime of individuals, is central to the biological sciences and connects the traditional disciplines—zoology, botany, and microbiology. In the past two decades, research in the field has strongly benefited from approaches aimed at the elucidation of the molecular mechanisms of developmental processes. In particular, in animal systems including vertebrates, nematodes, and insects—and more recently also in higher plants—key concepts and principles have been elaborated. These are documented in books on developmental biology and in more general biological textbooks. In contrast, although fungi have been actively investigated and they do play a major role in fundamental as well as applied fields, molecular developmental biology of this group of lower eukaryotes is either excluded completely from such books or dealt with marginally. In order to create an overview of this field, one is forced to consult a number of books dealing with specific topics (e.g., biotechnology, pathology). The aim of this book is to cover different aspects of molecular fungal development in one volume. Special emphasis has been put on mycelial fungi. The book, designed to present an up-to-date overview of the field, is written for students at the upper-undergraduate and graduate levels as well as for researchers in various disciplines of the life sciences.

In the first part of the book, basic developmental processes as they occur during vegetative growth, reproduction, and sexual propagation are addressed. Apart from true mycelial fungi, it also deals with the pseudohyphal growth of yeast as it is observed under certain conditions. The other chapters deal with hyphal tip growth, the genetic and molecular control of hyphae fusion, degenerative processes (senescence), and the control of the generation of vegetative and sexual reproduction units.

The second part is devoted to another level of complexity: interactions of fungi with different hosts. Symbiotic interactions of fungi with the roots of higher plants (mycorrhiza), although rather difficult to understand, are getting more and more attention. The same is true for parasitic interactions of fungi with both plants and animals. This is not surprising since the latter interactions lead to a number of important diseases and have a clear application. The elucidation of the various molecular pathways, in both the parasite and the host, will certainly help us to develop strategies for intervention against a variety of diseases.

Heinz D. Osiewacz

Contents

Contributors

Markus Aebi, Prof.Dr. Institut für Mikrobiologie, ETH Zurich, Zurich, Switzerland

Dipnath Baidyaroy, Ph.D. Department of Energy Plant Research Laboratory, Michigan State University, East Lansing, Michigan

Flora Banuett, Ph.D. Department of Biochemistry and Biophysics, University of California–San Francisco, San Francisco, California

Salomón Bartnicki-García, Ph.D. Unidad de Biología Experimental y Aplicada, Centro de Investigación Científica y de Educación Superior de Ensenada (CICESE), Ensenada, Mexico, and Department of Plant Pathology, University of California, Riverside, California

Deborah Bell-Pedersen, Ph.D. Department of Biology, Texas A&M University, College Station, Texas

Helmut Bertrand, Ph.D. Department of Microbiology and Molecular Genetics, Michigan State University, East Lansing, Michigan

Alan P. F. Bottoli, Ph.D.* Institut für Mikrobiologie, ETH Zurich, Zurich, Switzerland

Robert P. Boulianne, Ph.D.† Institut für Mikrobiologie, ETH Zurich, Zurich, Switzerland

Axel A. Brakhage, Ph.D. Institute of Microbiology, University of Hannover, Hannover, Germany

Gerhard H. Braus, Prof.Dr. Department of Molecular Biology, Institute of Microbiology and Genetics, Georg-August-Universität, Göttingen, Germany

Richard Calderone, Ph.D. Department of Microbiology and Immunology, Georgetown University Medical Center, Washington, D.C.

Sébastien Duplessis, Ph.D. Unité Mixte de Recherche INRA-UHP ''Interactions Arbres/Micro-Organismes,'' Centre INRA de Nancy, Champenoux, France

Sabine E. Eckert, Ph.D.‡ Department of Molecular Biology, Institute of Microbiology and Genetics, Georg-August-Universität, Göttingen, Germany

Reinhard Fischer, Ph.D. Philipps-Universität Marburg and Max-Planck-Institut für Terrestrische Mikrobiologie, Marburg/Lahn, Germany

Philipp Franken, Ph.D. Department of Biochemistry, Max-Planck-Institut für Terrestrische Mikrobiologie, Marburg/Lahn, Germany

Vivienne Gianinazzi-Pearson, Ph.D. UMR INRA/Université de Bourgogne, Dijon, France

N. Louise Glass, Ph.D. Department of Plant and Microbial Biology, University of California–Berkeley, Berkeley, California

John E. Hamer, Ph.D. Research Department, Paradigm Genetics, Inc., Research Triangle Park, North Carolina

Patrick M. Heidrich Institut für Botanik, Westfälische Wilhelms-Universität, Münster, Germany

Joseph Heitman, M.D., Ph.D. Department of Genetics, Howard Hughes Medical Institute, Duke University Medical Center, Durham, North Carolina

---

* *Current affiliation*: Credit Suisse, Zurich, Switzerland

† *Current affiliation*: National Institute of Diabetes and Digestive Kidney Diseases, National Institutes of Health, Bethesda, Maryland

‡ *Current affiliation*: Imperial College of Science, Technology and Medicine, London, England

Daniel Herman, Ph.D. Biomedical Health Science Department, Grand Valley State University, Allendale, Michigan

René Hermann, Ph.D. Labor für Elektronenmikroskopie, ETH Zurich, Zurich, Switzerland

Marcel Hollenstein* Institut für Mikrobiologie, ETH Zurich, Zurich, Switzerland

Bernhard Jahn, M.D. Institute of Medical Microbiology and Hygiene, Johannes Gutenberg-Universität Mainz, Mainz, Germany

Erika Kothe, Prof.Dr. Institut für Mikrobiologie, Friedrich-Schiller-Universität, Jena, Germany

Sven Krappmann, Ph.D. Department of Molecular Microbiology, Institute of Microbiology and Genetics, Georg-August-Universität, Göttingen, Germany

Gerrit Kuhn Institut d'Ecologie, University of Lausanne, Lausanne, Switzerland

Ursula Kües, Prof.Dr. Institut für Mikrobiologie, ETH Zurich, Zurich, Switzerland, and Institut für Forstbotanik, Georg-August-Universität Göttingen, Göttingen, Germany

Klaus B. Lengeler, Ph.D. Department of Genetics, Howard Hughes Medical Institute, Duke University Medical Center, Durham, North Carolina

Hartmut Linden, Ph.D. Lehrstuhl für Physiologie und Biochemie der Pflanzen, University of Konstanz, Konstanz, Germany

Francis Martin, Ph.D. Unité Mixte de Recherche INRA-UHP "Interactions Arbres/Micro-Organismes," Centre INRA de Nancy, Champenoux, France

Martina Mihlan Institut für Botanik, Westfälische Wilhelms-Universität, Münster, Germany

Sabine M. Moore, Dipl.Bio. Institut für Botanik, Westfälische Wilhelms-Universität, Münster, Germany

Hans-Ulrich Mösch, Ph.D. Department of Molecular Microbiology, Institute for Microbiology and Genetics, Georg-August-Universität, Göttingen, Germany

Birgitt Oeser, Ph.D. Institut für Botanik, Westfälische Wilhelms-Universität, Münster, Germany

---

* *Current affiliation*: PFC Pharma Focus Consultants AG, Volketswil, Switzerland

Heinz D. Osiewacz, Ph.D. Botanisches Institut, Johann Wolfgang Goethe-Universität, Frankfurt, Germany

Eline Polak, Ph.D.* Institut für Mikrobiologie, ETH Zurich, Zurich, Switzerland

Namboori B. Raju, Ph.D. Department of Biological Sciences, Stanford University, Stanford, California

Sven J. Saupe, Ph.D. Institut de Biochimie et de Génétique Cellulaire, Centre National de la Recherche Scientifique, Bordeaux, France

Christian Q. Scheckhuber Botanisches Institut, Johann Wolfgang Goethe-Universität, Frankfurt, Germany

Denis Tagu, Ph.D. Unité Mixte de Recherche INRA-UHP ''Interactions Arbres/Micro-Organismes,'' Centre INRA de Nancy, Champenoux, France

Klaus B. Tenberge, Ph.D. Institut für Botanik, Westfälische Wilhelms-Universität, Münster, Germany

Fernando A. Tenjo, Ph.D. Department of Microbiology, Duke University Medical Center, Durham, North Carolina

Paul Tudzynski, Ph.D. Institut für Botanik, Westfälische Wilhelms-Universität, Münster, Germany

Piers J. Walser Institut für Mikrobiologie, ETH Zurich, Zurich, Switzerland

---

* *Current affiliation*: Tecan Schweiz AG, Männedorf, Switzerland

1

Pseudohyphal Growth in Yeast

Hans-Ulrich Mösch
Institute for Microbiology and Genetics, Georg-August-Universität, Göttingen, Germany

1 INTRODUCTION

1.1 Definition and Importance of Yeasts

Yeasts comprise a group of ~700 ascomycetous or basidiomycetous fungi, whose predominant mode of vegetative reproduction is growth as single cells [1]. In other words, growth as unicellular organism is the main criterion that unifies taxonomically diverse fungi as yeasts. In addition, many molds that predominantly grow as hyphal filaments can adopt the yeast form as part of their life cycle. One of the prominent genera among yeasts is *Saccharomyces*, which has been used in making bread and brewing liquor throughout human history. The literary meaning of the terms used for yeasts in different languages can often be associated with the property of fermentation [2]. The English word *yeast* is related to the Dutch word *gist* or the German word *Gischt* which both mean foam. The French expression *levure* is derived from *lever* (= to rise) and the Latin term *levere*, both of which refer to the evolution of CO_2 that pushes up solid substances during fermentation. Apart from *Saccharomyces*, many other yeast genera are of central importance in today's biotechnology and medicine. Among others, strains of *Schwanniomyces occidentalis*, *Kluyveromyces lactis*, *Pichia pastoris*, *Hansenula polymorpha*, *Yarrowia lipolytica*, and *Candida maltosa* belong to the so-called nonconventional yeasts that are used in a wide variety of biotechnological processes of great economical importance [3].

A number of human pathogenic yeasts and medically important fungi with a yeast phase during infection have been described and include *Candida albicans* and other *Candida* spp. (causative agents for candidosis), *Cryptococcus neoformans* (cryptococcosis), *Histoplasma capsulatum* (histoplasmosis), and *Blastomyces dermatitidis* (blastomycosis) [4]. In basic biological research, *Saccharomyces cerevisiae* and *Schizosaccharomyces pombe* are among the molecularly best-characterized organisms and have become model systems for the eucaryotic cell in general [5].

1.2 Dimorphism

The unicellular growth form is the preferred mode of reproduction of yeasts. However, many yeasts are able to grow as multicellular filaments that consist of either pseudohyphae or hyphae (Fig. 1). Each of the yeast form (YF), pseudohyphal (PH), and hyphal (H) growth forms represents a distinct fungal cell type that is characterized by its cellular attributes and the mode of cell division (see below) [6]. The ability of yeasts to change between unicellular and filamentous growth phases is referred to as yeast-mycelial dimorphism and has long been used as an important morphological criterion for taxonomic and systematic classification of yeasts [1,7,8]. For the medical mycologist, dimorphism is a significant virulence factor characteristic for human pathogenic yeasts [4]. Numerous exogenous factors have been described that control interconversion among the different growth forms of yeasts and include specific nutrients, peptides, certain sugars,

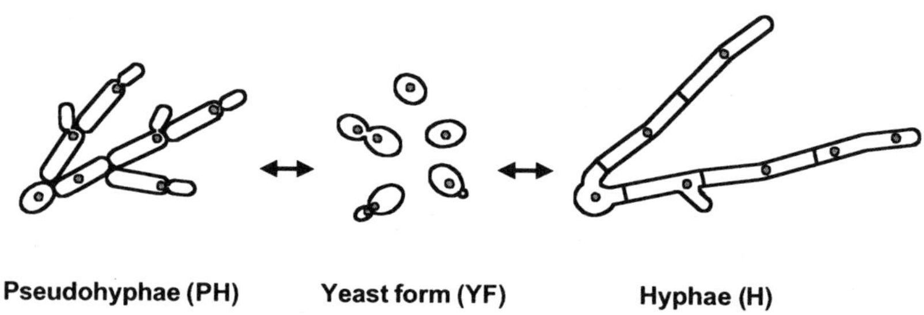

FIGURE 1 Vegetative growth forms of budding yeasts. In the yeast form (YF), cells divide by budding, followed by complete separation of the mature bud from the mother cell. Pseudohyphae (PH) originate from the budding of elongated cells that remain attached to each other, resulting in the formation of pseudohyphal filaments or pseudomycelium. Hyphae (H) and hyphal mycelium develop by continuous tip growth of hyphal cells, followed by fission of cells through the formation of septa.

metal ions, oxygen, pH, and temperature [4,9]. Therefore, dimorphism of yeasts also serves as a general model for fungal development in response to external signals.

The molecular mechanisms underlying dimorphism are far from being understood in detail, despite the fact that a wealth of information on the cellular and molecular biology of model yeasts like *S. cerevisiae* has become available [5]. A profound understanding of yeast-mycelial dimorphism requires the molecular analysis of all growth forms of yeasts and of the signals and regulatory networks that control interconversion among them. In a first step, easily tractable model organisms serve as initial source of molecular information. At a later stage, however, molecular studies must include analysis of dimorphic yeast species from a broad range of taxa.

2 GROWTH FORMS OF YEAST

2.1 The Yeast Form

By definition, yeasts prefer unicellular growth as their favorite mode of vegetative reproduction. The typical yeast cell reproduces by either budding, exemplified by *S. cerevisiae*, or fission, as in *S. pombe* [5]. Budding and fission both confer a unique relationship between cell growth and the sequence of events that constitute the cell division cycle. Budding is initiated by the emergence of a bud as a small protuberance from the cell surface (Fig. 1). Until cytokinesis, further growth is restricted to the bud. The bud lengthens in parallel to DNA synthesis (during S phase) and then swells to achieve its characteristic form during mitosis (M phase). Once the bud receives a daughter nucleus, it separates from the mother cell via septation and enters G1 as an independent daughter cell. Reiteration of this process rapidly produces a rounded pile of single cells, or colony, on the surface of solid growth media. Cell division by fission starts by cellular growth at both tips which leads to cell elongation and stops when mitosis is initiated. Upon completion of mitosis a septum is centrally placed in the dividing cell, and cell division occurs by binary fission. However, it is not the mode of cell division (budding or fission) that determines whether yeasts reproduce as single cells or build filaments. The crucial step that confers unicellular growth is complete separation of mother and daughter at the end of each round of cell division, whether cells divide by budding or by fission.

The yeast form has several advantages when compared to more complex growth forms such as branched hyphae [10]. Morphologically, the yeast form is a sphere. Although few yeasts are exactly spherical, most are spheroids with axes that differ only little in their length. Owing to their small surface area per unit volume, spherical yeast cells are very economical with cell wall materials. Because a single mother cell survives many cell divisions and produces up to 25 daughter cells, further wall material is saved. Yeast form cells are very resistant to osmotic rupture throughout cell division, especially when compared to the osmotically more fragile tips of hyphal cells. This attribute makes yeasts particularly suited for growth on their preferred natural habitat, the surfaces of plants, where they may be exposed to sudden changes in osmolarity owing to sudden rain. A further advantage of the unicellular nature of yeasts is efficient dispersal by water, air, or insects. In the case of pathogenic yeasts, the unicellular yeast form is probably transported more efficiently in the vascular and lymphatic systems of warm-blooded animals than the filamentous growth forms. In industrial processes, the yeast form of producing strains is often preferred over filamentous growth forms owing to easier handling of single cells.

2.2 Pseudohyphae

The development of pseudohyphae was observed and described in industrial yeast strains >100 years ago by Hansen [11]. Since then, the ability of yeasts to form pseudomycelia has been used as an important taxonomic criterion [1,8]. Pseudohyphal growth was accurately defined by Lodder [8], who wrote in 1934: "By pseudomycelium I understand septate, frequently branched filaments, which have arisen in such a way that the filament-producing, mostly longer cells are formed from one another by budding." Thus, filament-building PH cells differ from single YF cells by an elongated morphology and the incomplete separation of mothers from daughters. Nevertheless, it is important to point out that both PH and YF cells reproduce by budding. A further difference between YF and PH cells was found in the early 1940s, when the microstructures of colonies of various yeast strains were analyzed in detail [12]. This study showed that certain yeast colonies consisted of a pile of single YF cells on top of the agar medium but that beneath the surface, pseudohyphal mycelium had developed and grown invasively deep into the agar substrate. Therefore, PH cells further differ from YF cells by their ability to grow invasively into substrates, suggesting that the development of pseudomycelia reflects a foraging mechanism that allows yeasts to explore new habitats. In pathogenic yeasts, pseudomycelium (along with true mycelium) is thought to be important for penetrating barriers of the host organism [4].

Pseudohyphal development is observed in a wide variety of yeasts including the genera *Candida*, *Endomyces*, *Pichia*, *Saccharomyces*, and *Yarrowia* [1]. However, pseudohyphal growth depends not only on the yeast species, but also on the growth conditions. In general, formation of pseudomycelium seems to be favored by poor conditions. Early observations showed that elongated cells and pseudomycelia are formed in cultures which have been grown at temperatures markedly below the optimum [11,13]. Partial anaerobiosis is another factor that stimulates the formation of pseudomycelium. This factor has found its application in the so-called Dalmau plate technique, where part of the surface of an agar streak is covered with a coverslip and where the portion of the culture beneath the glass produces pseudomycelium more readily than the other part of the culture [14]. This standard procedure for observing pseudohyphal growth on solid medium is still used for taxonomical classification of yeasts [1].

Filamentous forms are also frequently observed in older cultures, and fusel oils and higher alcohols appear to be at least in part responsible for pseudohyphal development [15–18]. A further classical way to stimulate pseudomycelia is growth in media that are relatively poor sources of nutrients and include potato water or cornmeal agar [1,8,19]. In the case of *S. cerevisiae*, starvation for nitrogen is known to stimulate pseudohyphal growth of diploid strains [20]. Apart from natural conditions or routine laboratory practice, several specific substances have been described that stimulate the development of pseudomycelium and include camphor, penicillin, auxin, and certain salts of cobalt or boron [21]. Nevertheless, the molecular mechanisms by which any of the above factors stimulate pseudohyphal growth are not known.

2.3 Hyphae

Hyphal development is a further growth option of many dimorphic yeasts and is an important criterion in taxonomic studies [1]. Hyphal mycelium is different from pseudomycelium, as has been pointed out by Lodder [8]: ''True mycelium arises either as non-septate, long, thread-like, eventually branched cells, or from septate and often branched filaments, in which the single limbs have been built by the formation of cross-walls in the filaments.'' The main difference between the filamentous pseudohyphal and hyphal growth forms is the mode of origin and not the end product, which in both cases is a mycelium. As described above, pseudomycelium is produced by budding cells. In contrast, hyphal mycelia develop by continuous tip growth of hyphal cells, followed by fission of cells through the formation of septa (Fig. 1). The existence of different types of septa is widely used for taxonomical classification of yeasts and filamentous fungi, because septal ontogeny and ultrastructure are robust indicators of taxonomic relationships [1]. For instance, ascomycetous and basidiomycetous hyphae are easily distinguishable by their types of septa.

Hyphal development is characteristic of many medically important yeasts, such as *Candida albicans*, *Cryptococcus neoformans*, and *Blastomyces dermatitidis*. The main advantage of the hyphal over the yeast form is the ability of hyphae to attack and penetrate barriers of host organism, to grow through tissues, and to escape host defense mechanisms such as phagocytosis [10,22]. Factors that favor hyphal growth in dimorphic yeasts include growth on poor nitrogen sources or on specific carbon sources, but also changes in temperature or pH [4]. However, the molecular nature of the signals that trigger the transition from yeast to hyphal growth and back remains obscure. In addition, molecular analysis of hyphal development has been hampered by the fact that many of the medically important dimorphic yeasts are experimentally not easily tractable. Therefore, model systems have been developed with the goal to elucidate the molecular biology of dimorphism. Among the most prominent of these systems are the pathogenic yeasts *Candida albicans* (a deuteromycete) and *Cryptococcus neoformans* (a basidiomycete) and the apathogenic hemiascomycetes *Yarrowia lipolytica* and *Saccharomyces cerevisiae* [4]. During the last decade, studies with these model organisms have greatly contributed to get first insights into the molecular mechanisms underlying dimorphism [6,23,24].

3 REGULATION OF PSEUDOHYPHAL GROWTH IN *SACCHAROMYCES CEREVISIAE*

3.1 *Saccharomyces cerevisiae* as Model for Molecular Analysis of Pseudohyphal Growth

Dimorphism in yeasts has been observed and studied for >100 years. Yet, little was known about the molecular mechanisms that control filamentous growth in *S. cerevisiae* before 1992, when the baker's yeast *Saccharomyces cerevisiae* was "rediscovered" to be dimorphic [20]. The unawareness among yeast geneticists about the dimorphic potential of *S. cerevisiae* may be due to the fact that the most commonly used laboratory strains are unable to form pseudohyphae. Most of the strains used for genetic studies derive from the wild isolate EM93, a strain that readily forms pseudohyphae and consequently grows as highly adherent colonies that interfere with replica plating and easy dispersion [25,26]. Derivatives of EM93, such as the widely used laboratory strain S288C, were selected for nonadherence and easy dispersal in liquid cultures. Unfortunately, selection for easy dispersion also selected against dimorphism in these strains [26,27]. As a consequence, dimorphism was not studied on a molecular level in baker's yeast until 1992, although half a century of genetics and molecular biology of *S. cerevisiae* has defined the pathways of almost every aspect of its physiology.

In some laboratories, *S. cerevisiae* strains of distinct genetic background were developed, which had retained the ability to form pseudohyphae. These include strains of the Σ1278b background that led to the "rediscovery" of dimorphism in *S. cerevisiae* [20]. Pseudohyphal growth of *S. cerevisiae* is initiated by nitrogen starvation. The switch from yeast form to pseudohyphal growth is accompanied by changes in several distinct cellular processes. The budding pattern of cells changes, resulting in linear filamentous chains of cells. Cell morphogenesis is altered from ellipsoidal-shaped yeast form cells to long, thin pseudohyphal cells. Pseudohyphal cells, in contrast to yeast form cells, exhibit invasive growth behavior, resulting in direct substrate invasion. Cell separation switches from complete to incomplete scission, leading to multicellular growth, where cells remain attached to each other. Therefore, yeast and pseudohyphal forms of diploid *S. cerevisiae* are thought to be distinct cell types each with a unique budding pattern, cell shape, invasive growth behavior, and cell cycle.

The possibility for molecular biological studies of pseudohyphal development in baker's yeast is of profound importance, because *S. cerevisiae* is one of the best-studied model systems for molecular genetic analysis and genomics. Its power lies in being amenable to genetic manipulation, having a completely characterized genomic sequence, and the evolutionary conservation of the most basic biological processes common to all eukaryotic organisms. Indeed, numerous studies have led to great progress in understanding pseudohyphal growth in *S. cerevisiae* in a relatively short period of time. Because pseudohyphal growth of *S. cerevisiae* has become one of the best-understood models for dimorphism in yeasts, the following sections will mainly focus on *S. cerevisiae* and review the current knowledge of the molecular mechanisms underlying pseudohyphal development in this organism. The genes and gene products are described that constitute the signaling pathways transducing environmental stimuli, control the cell division cycle in pseudohyphae, establish and regulate pseudohyphal cell polarity, are required for pseudohyphal cell morphogenesis, and induce substrate adhesion and invasive growth.

3.2 Signals and Sensors

The developmental options in the life cycle of *S. cerevisiae* are largely controlled by nutrient availability (Fig. 2). In laboratory routine, *S. cerevisiae* is often cultured on media containing ammonium and glucose, conditions that favor growth in the yeast form. When deprived of either of these nutrients, diploid cells exhibit distinct responses. Glucose depletion induces growth arrest, but ammonium depletion favors pseudohyphal growth. In the case of ammonium and glucose depletion, diploids will undergo meiosis. Thus, pseudohyphal growth requires at least two nutritional stimuli: starvation for nitrogen, and availability of a fermentable carbon source (Fig. 3). Haploid cells of strains capable of pseudohyphal development are highly adherent and form colonies with granular consistency. This phenomenon is called haploid invasive growth, because cells in these colonies form short filaments that will grow into the agar substrate (Fig. 2). In contrast to diploid pseudohyphae, haploid filaments lack a markedly elongated cell shape and fail to grow out from the colony border.

Pseudohyphal growth of *S. cerevisiae* is suppressed on media that contain standard amounts of ammonium, arginine, glutamine, glutamate, or a mixture of proline and histidine as sole nitrogen sources [20]. In contrast, low ammonium levels or supply with standard concentrations of histidine, proline or uracil alone is permissive for pseudohyphal development. The sensor systems that differenti-

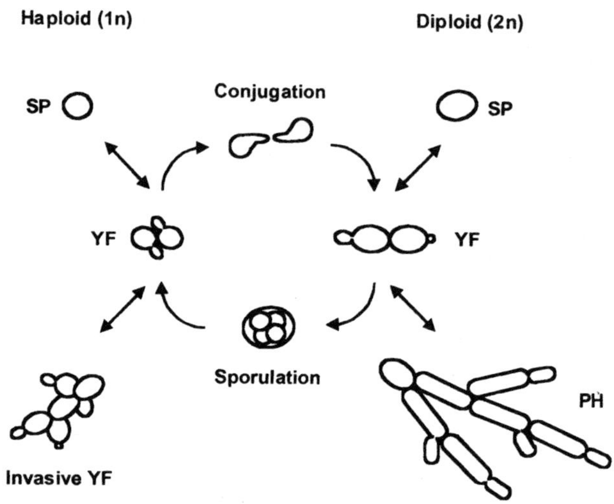

Figure 2 Life cycle of *S. cerevisiae*. Haploids have four options: (1) reproduction as single YF cells; (2) reproduction as invasive YF; (3) stationary phase (SP); (4) conjugation with haploids of opposite mating type to form diploids. Diploids can choose among: (1) reproduction as single YF cells; (2) reproduction as PH filaments; (3) stationary phase (SP); (4) sporulation to form haploids.

ate between diverse nitrogen compounds and control pseudohyphal dimorphism are not understood in complete detail. Ammonium availability is thought to be sensed by the high-affinity ammonium permease Mep2p [28]. Mep2p is localized at the plasma membrane and has been proposed to transmit a signal to intracellular signaling pathways (Fig. 4). Regulation of pseudohyphal development by amino acids might involve intracellular and extracellular sensing systems. Intracellular sensing of amino acids is thought to require the function of a glutamine tRNA [29]. Extracellular sensing of amino acids requires a separate system, components of which are the plasma membrane proteins Ssy1p and Ptr3p [30].

The fermentable sugars that promote pseudohyphal growth under nitrogen starvation conditions include glucose, galactose, sucrose, maltose, and raffinose [20,31,32]. The nonfermentable carbon source acetate prevents the formation of pseudohyphae, but promotes sporulation. Several sensing systems for sugars are known in *S. cerevisiae*, some of which are involved in regulation of pseudohyphal growth [33,34]. The G-protein coupled receptor Gpr1p is part of a sensing system for glucose and other structurally related sugars. Gpr1p regulates pseudohyphal growth together with the G-alpha protein Gpa2p and with Plc1p, a phosphatidylinositol-specific phospholipase C [32,35–37]. The Gpr1p–Gpa2p–Plc1p com-

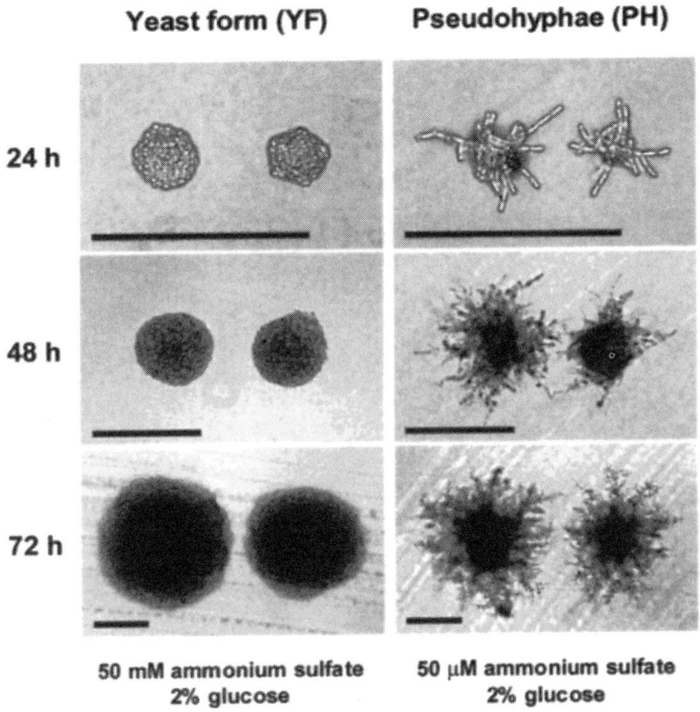

FIGURE 3 Yeast form and pseudohyphal growth of *S. cerevisiae*. Diploid *S. cerevisiae* strains were streaked for single colonies on either nitrogen-rich medium (50 mM ammonium sulfate) or nitrogen-starvation medium (50 μM ammonium sulfate). Development of YF and PH colonies was photographed at different time points (24 h, 48 h, and 72 h). Scale bars all equal 100 μm.

plex is activated by glucose and regulates pseudohyphal growth via the cAMP pathway (Fig. 4). A second system, which is responsive to changes in the availability of glucose, is the Ras-cAMP signaling pathway of *S. cerevisiae* [38,39]. The central player of this system is the small GTP-binding protein Ras2p together with its guanine nucleotide exchange factor Cdc25p and the GTPase-activating proteins Ira1p and Ira2p [40]. Activation of Ras2p induces hyperfilamentous growth, but only under nitrogen starvation conditions. This indicates that Ras2p may be a transmitter that regulates pseudohyphal development in response to glucose availability. Glucose sensor proteins that activate Ras2p, however, have not been identified.

Two further mechanisms are known in yeast that play an important role in glucose sensing [33,41]. One mechanism involves the Snf1p–Snf4p protein ki-

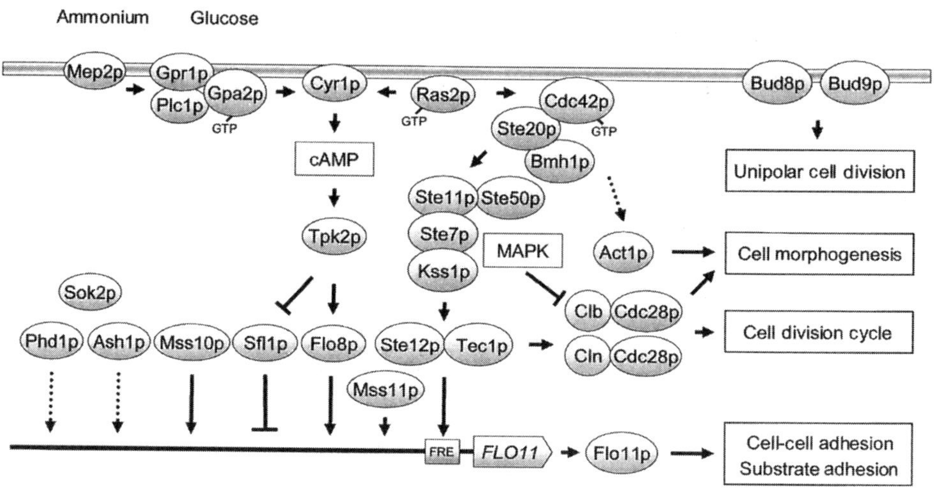

FIGURE 4 Model of signaling pathways regulating pseudohyphal growth in *S. cerevisiae* (see text for details).

nase and Reg1p–Glc7p protein phosphatase, and another pathway employs the Rgt1p transcriptional repressor, the ubiquitin ligase protein complex SCF(Grr1p), and two glucose sensors in the membrane, Snf3p and Rgt2p. The role of these pathways in pseudohyphal growth, however, has not been investigated.

Certain "fusel" alcohols have been described as stimuli of pseudohyphal growth not only in *S. cerevisiae*, but also in certain species of *Candida* and *Brettanomyces* [17]. Fusel alcohols are products of amino acid catabolism and include isoamyl alcohol and 1-butanol, both of which induce pseudohyphal development independently of the Gpr1p glucose-sensing system, but dependent on the pseudohyphal MAPK cascade [18]. A further stimulus known to control dimorphism in yeasts is high osmolarity, which has been found to negatively regulate pseudohyphal growth of *S. cerevisiae* [42]. Further factors that regulate dimorphism such as temperature and pH have not been systematically tested in *S. cerevisiae*.

3.3 Signal Transduction Pathways

Pseudohyphal growth of *S. cerevisiae* is under the control of a complex regulatory network (Fig. 4). Two main signal transduction pathways are essential for regulation, the pseudohyphal MAPK cascade and the cAMP pathway. The pseudohyphal MAPK cascade is an evolutionary, highly conserved signaling module composed of four protein kinases that act in a sequence and that have been named MAPK (mitogen-activated protein kinase), MEK (MAPK kinase), MEKK (MEK kinase), and MEKKK (MEKK kinase) [43–45]. In the pseudohyphal regulatory network, the MAPK module is composed of the protein kinases Ste20p (MEKKK), Ste11p (MEKK), Ste7p (MEK), and Kss1p (MAPK) [46–48]. Activity of the pseudohyphal MAPK cascade can be modulated by the 14-3-3 proteins Bmh1p and Bmh2p at the level of Ste20p, and by the Ste50p protein that controls activity of Ste11p [49,50]. The function of the MAPK cascade is phosphorylation of target proteins via the Kss1p MAPK in response to nutrient signals that stimulate pseudohyphal growth. A crucial target of Kss1p is the transcription factor Ste12p, whose activity is negatively regulated by the nuclear proteins Dig1p and Dig2p [51]. Upon activation by the Kss1p MAPK cascade, Ste12p promotes transcription of target genes required for pseudohyphal growth. For this function, Ste12p acts in combination with Tec1p, a regulator originally identified to control expression of yeast transposons [52]. Tec1p is required for pseudohyphal growth [53,54] and contains the TEA/ATTS DNA binding domain, which is shared by several eukaryotic transcription factors, including *Aspergillus nidulans* AbaAp [55,56]. Ste12p and Tec1p regulate gene transcription by cooperative binding to specific filamentation response elements (FREs), which are present in the promoter regions of target genes including *TEC1* [57], *FLO11* [58], and many others [59]. A central target gene of the Kss1p MAPK cascade and the Ste12p–Tec1p transcription factor is *FLO11*, which encodes a cell surface flocculin required for pseudohyphal growth [58]. The *FLO11* promoter is large and complex, and is an important integration site for multiple regulatory pathways that control pseudohyphal growth [60,61].

The cAMP pathway is a further, well-characterized signaling route that regulates pseudohyphal growth via expression of *FLO11* [61–63]. Central components of the cAMP pathway are the adenylyl cyclase Cyr1p, the phosphodiesterases Pde1p and Pde2p, and the cAMP-dependent protein kinase or A kinase that is composed of an inhibitory subunit Bcy1p and any of the catalytic subunits Tpk1p, Tpk2p, or Tpk3p [38]. All three Tpk proteins are redundant for viability [64]. However, only Tpk2p positively regulates pseudohyphal development, whereas Tpk1p and Tpk3p appear to have negative functions [62,63]. Putative targets of the A kinase are the two transcription factors Sfl1p and Flo8p. Sfl1p is a negative regulator of *FLO11* expression and itself is thought to be negatively regulated by Tpk2p [62]. Flo8p is a positive regulator of *FLO11* expression and is required for regulation of pseudohyphal growth by Tpk2p [61,63]. Therefore, Flo8p is thought to be a target of the cAMP pathway with positive function in pseudohyphal growth control. Interestingly, the *FLO8* gene has been inactivated in strains of the S288C background that fail to form pseudohyphae when starved for nitrogen [27]. In contrast, dimorphic strains of the Σ1278b background or EM93 (the progenitor of S288C) have a functional *FLO8* gene. This suggests that the *flo8* mutation in S288C was selected during laboratory cultivation.

How are the cAMP pathway and Kss1p MAPK cascade connected to the sensor systems that regulate pseudohyphal growth? In *S. cerevisiae*, activated Ras proteins have long been known to bind and stimulate the adenylyl cyclase

Cyr1p, thereby causing elevated cAMP levels and activating the A kinase [40]. Indeed, the small GTP-binding protein Ras2p was early found to stimulate pseudohyphal growth when expressed as the dominant activated $Ras2^{Val19}$p form [20]. Stimulation of pseudohyphal growth by active Ras2p is suppressed by increased activity of the cAMP-hydrolyzing phosphodiesterase Pde2p [65]. Deletion of Ras2p causes pseudohyphal growth defects that can be rescued by activation of the A kinase [66,67]. Thus, Ras2p has been proposed to control pseudohyphal growth by activating the cAMP pathway. A further system that controls pseudohyphal development via the cAMP pathway is the Gpr1p–Gpa2p–Plc1p sensor module, as has been demonstrated in several studies [32,35–37,63,66,68–71]. Upon glucose stimulus, Gpr1p and Plc1p are thought to activate Gpa2p, which in turn stimulates the adenylyl cyclase Cyr1p. The small GTP-binding proteins Ras2p and Cdc42p have been identified as regulators of the pseudohyphal MAPK cascade. When activated, both Ras2p and Cdc42p stimulate pseudohyphal growth and FRE-dependent gene expression, but require the pseudohyphal MAPK cascade for these functions [67,72,73]. The double functions of Ras2p in stimulating both the cAMP pathway and the pseudohyphal MAPK cascade are separable by mutations in Ras2p that block binding of the adenylyl cyclase Cyr1p [67]. The osmosensor protein Sho1p is a further regulator of pseudohyphal development that is thought to feed into the pseudohyphal MAPK cascade [42].

A number of further regulators of pseudohyphal growth have been identified including the transcription factors Phd1p, Sok2p, Mss10p, Mss11p, and Ash1p [60,65,74,75]. Phd1p promotes pseudohyphal growth, whereas Sok2p appears to antagonize it. Phd1p and Sok2p may either act in a linear pathway or converge independently on the same target, e.g., *FLO11*. Mss10p and Mss11p were identified as regulators of both pseudohyphal growth and expression of *FLO11* [60,76]. Mss10p acts independently of the pseudohyphal MAPK cascade, whereas Mss11p appears to act downstream of Ste12p. Ash1p is a daughter cell–specific transcription factor that was originally identified as regulator of mating type switching in haploid cells of *S. cerevisiae* [77]. In diploid cells, Ash1p is required for pseudohyphal growth and is also asymmetrically localized in the nuclei of daughter cells [75]. Ash1p is thought to act independently of the pseudohyphal MAPK cascade and might regulate expression of *FLO11*.

3.4 Cell Cycle Regulation

The molecular machinery that constitutes and regulates the cell division cycle of *S. cerevisiae* is understood in great detail, although most studies have addressed analysis of the YF [5]. Many studies have shown that a central regulatory protein kinase, Cdc28p, undergoes changes in activity through the cell cycle by associating with distinct groups of cyclins that accumulate at different times. The various cyclin/Cdc28p complexes control different aspects of cell cycle progression, including the commitment step known as START, and mitosis. However, activity of Cdc28p also affects morphogenesis during the yeast cell cycle. Activation of Cdc28p by G1 cyclins (Cln1p, Cln2p, or Cln3p) during G1 triggers polarization of the cortical actin cytoskeleton, while activation of Cdc28p by mitotic cyclins (Clb1p or Clb2p) in G2 cells causes depolarization of the cortical actin cytoskeleton [78].

The ''rediscovery'' of dimorphism in *S. cerevisiae* has allowed to compare the division cycles of yeast form cells and PH cells. A striking difference between YF cells and PH cells is the timing of the different steps of their division cycles [31]. The YF cell cycle is controlled at the G1/S transition by the cell-size checkpoint START. YF cells divide asymmetrically, producing small daughters from full-size mothers. As a result, mothers and daughters bud asynchronously. Mothers bud immediately, but daughters grow in G1 until they achieve a critical cell size to pass START. In contrast, PH cells divide symmetrically, restricting mitosis until the bud grows to the size of the mother. As a consequence, mother and daughter bud synchronously in the next cycle, without a G1 delay before START. Hence, PH cells are thought to delay mitosis at a G2 cell-size checkpoint until the size of the bud reaches that of the mother cell.

Several studies indicate that the G2/M progression delay is under control of the pseudohyphal MAPK cascade [6,31,79]. These reports suggest that the MAPK pathway promotes pseudohyphal growth by a mechanism that inhibits mitotic cyclin/CDK complexes consisting of Cdc28p and the mitotic cyclins Clb1p and Clb2p. In addition, several lines of evidence suggest that pseudohyphal growth is also regulated by G1/CDK complexes composed of Cdc28p and the G1 cyclins Cln1p and Cln2p [80]. However, whereas mitotic cyclin/CDK activity is downregulated in PH cells, active G1 cyclin/CDK complexes are essential for efficient pseudohyphal growth. The positive role of G1 cyclin function in pseudohyphal growth is further emphasized by the finding that the *CLN1* gene is transcriptionally upregulated by Ste12p and Tec1p via FREs present in the promoter of *CLN1* [59]. In the human pathogenic yeast *Candida albicans*, G1 cyclin function is necessary for maintenance of filamentous growth [81].

Activity of the cyclin-dependent kinase Cdc28p toward pseudohyphal growth is further regulated by the protein kinase Elm1p. The normal function of Elm1p appears to be inhibition of pseudohyphae formation under optimal growth conditions, because absence of Elm1p leads to constitutive pseudohyphal morphology [82,83]. Genetic evidence suggests that Elm1p acts in a protein kinase cascade that regulates pseudohyphal growth by modulating activity of Cdc28p [84].

3.5 Regulation of PH Cell Polarity

S. cerevisiae cells divide by budding and choose cell division sites in different spatial patterns that are under genetic control of their cell type [85–87]. Haploid **a** or α cells bud in an axial pattern, where mother and daughter cells bud adjacent to their cell pole that defined the previous mother–daughter junction. This region of the yeast cell surface is also referred to as the proximal pole or the birth end of the cell. Diploid **a**/α YF cells bud in a bipolar pattern, where buds form equally either at the proximal pole or at the site opposite to it, called the distal pole (Fig. 5). Yeast cell polarity and accordingly budding patterns are affected by extracelluar stimuli, such as nutrients. Upon nitrogen starvation, diploid cells that have switched to growth as pseudohyphal filaments preferentially bud in a unipolar distal pattern, where most of the buds emerge at the distal pole [20,31] (Fig. 5). The unipolar distal budding program is a prerequisite for the establishment of filamentous structures and therefore can be viewed as a process regulated by nutritional signals and guiding the direction of the growing PH filaments.

In yeast, selection of cell division sites is regulated by at least three different classes of genes and corresponding proteins [88,89]. One class of genes is required for axial and bipolar budding and includes *RSR1/BUD1*, *BUD2*, and *BUD5* [90–92]. Mutations in these genes cause random budding patterns in haploid and diploid YF cells. Rsr1p/Bud1p, Bud2p, and Bud5p constitute a GTPase signaling module that is thought to help to direct bud formation components to the selected cell division site. A second class of genes is required specifically for axial budding of haploids without affecting the bipolar pattern of diploids. Genes of this class include *AXL1*, *BUD10/AXL2*, *BUD3*, and *BUD4* [91,93–95]. A third class of genes is required for the bipolar budding pattern of diploid yeast cells but not for haploid axial budding. Many genes of this class have been identified by genetic screens and include *AIP3/BUD6*, *BUD7*, *BUD8*, *BUD9*, *BNI1*, *PEA2*, and *SPA2* [96–98]. Mutations in most of these genes cause a random budding pattern specifically only in diploids without affecting axial budding in haploids. Only two genes of this class, *BUD8* and *BUD9*, have been described to shift the bipolar pattern to a unipolar pattern and therefore appear to have the

Figure 5 Regulation of cell polarity in *S. cerevisiae*. (a) Budding patterns of YF and PH cells. YF cells bud in a bipolar pattern, where buds form with equal (50%:50%) probability at either the proximal cell pole (birth end) or the distal cell pole (site opposite to birth end). PH cells prefer the unipolar distal budding pattern, where most buds (90%) emerge at the distal pole. Photographs below show distribution of bud scars of YF and PH cells that were stained with calcofluor and visualized by fluorescence microscopy. (b) Unipolar distal budding of PH requires the *BUD8* gene. Wild-type and *bud8/bud8* mutant strains were analyzed for selection patterns of first buds of virgin PH cells. Numbers indicate the percentage of virgin PH cells that produced their first bud at the proximal or the distal pole, respectively. After 3 days of growth, PH development of cells at the edges of the colonies was visualized under the microscope using Nomarski optics.

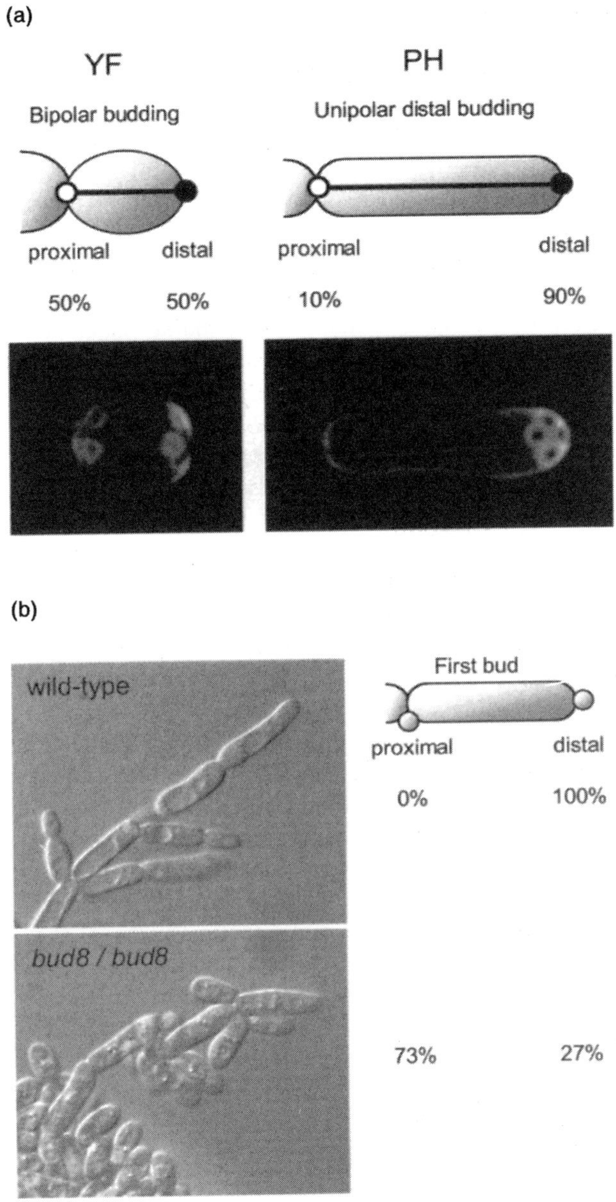
(a)
YF
PH
Bipolar budding
Unipolar distal budding
proximal
distal
proximal
distal
50%
50%
10%
90%
(b)
wild-type
First bud
proximal
distal
0%
100%
bud8 / bud8
73%
27%

most specific effects on bipolar budding. Mutations in *BUD8* cause a unipolar proximal budding pattern in diploids, whereas *bud9* mutants bud with high frequency from the distal cell pole [98]. Therefore, Bud8p and Bud9p have been proposed to act as bipolar landmarks that might recruit components of the common budding factors, e.g., Bud2p, Bud5p, or Rsr1p/Bud1p, to either of the cell poles [89].

Most studies that have addressed the function of genes controlling bud site selection were performed under nutrient-rich conditions, where *S. cerevisiae* will grow and divide as single YF cells. Only little is known about the molecular mechanisms that control changes in cell polarity in response to nutritional starvation. Because nitrogen starvation causes a switch in the budding pattern from bipolar to unipolar distal in diploid cells, pseudohyphal development is an ideal model to study factors that control oriented cell division in response to external signals. To date, no class of genes has been identified that is specifically required for the unipolar distal pattern of PH cells without affecting bipolar budding of YF cells. An initial study has identified Rsr1p/Bud1p to be required for pseudohyphal development, because expression of a dominant negative form of *RSR1/BUD1*, $RSR1^{Asn16}$, suppresses filament formation in response to nitrogen starvation [20]. A genetic screen directed at the identification of genes specifically required for pseudohyphal development has uncovered several of the bipolar specific bud site selection genes including *BUD8*, *BNI1*, *PEA2/DFG9*, and *SPA2* [54] (Fig. 5). This suggests that the pseudohyphal polarity switch might be achieved by alteration of components that control bipolar budding. The function of the Bud8p and Bud9p proteins in YF and PH cells has been investigated [99]. This study shows that Bud8p and Bud9p are asymmetrically localized at the distal cell pole, where they regulate bud initiation. Bud8p acts as a landmark for bud initiation at the distal cell pole, whereas Bud9p is an inhibitor that might interfere with Bud8p functions in YF cells. In response to nitrogen starvation, Bud9p is prevented from being localized to the distal cell pole, causing a switch in cell polarity from bipolar to unipolar budding. Whether any of the known signaling pathways mediates the nitrogen starvation signal to Bud8p or Bud9p remains to be determined.

3.6 PH Cell Morphogenesis

The morphology of yeast form and pseudohyphal cells is markedly different. YF cells are spheres of oval shape with axes that differ by a factor of not more than two. In contrast, PH cells are tubular and have a length-to-width ratio of up to 5 (Fig. 5). The long and thin morphology of PH cells has several advantages over the round YF, especially with regard to the purpose of the organism to build a filamentous structure that allows escaping unfavorable nutrient conditions. The elongated morphology provides more surface area than the shape of an oval YF cell and allows PH cells a more efficient absorption of nutrients. A second consequence of changing the cell shape is polarization of both cell growth and division along the same axis, an effect that enhances the ability of the growing chain of cells to escape the colony. The velocity of PH filament elongation is proportional to the length of the cells that comprise the chain.

A key factor in regulating polarized cell growth and cell morphology is the actin cytoskeleton [88,100–102]. In yeast cells, the actin cytoskeleton consists of two major structures, patches and cables. Actin patches are punctate ''dots'' of filamentous actin present at the cell cortex, and actin cables are oriented along the mother–bud axis. Patches are highly mobile and are localized to areas of new cell growth throughout the cell cycle. YF and PH cells have distinct arrangements of the actin cytoskeleton [31]. In YF cells, a distinct period between bud emergence and cytokinesis is observed, during which actin patches are distributed around the entire cortex of the emerging bud. This leads to isotropic growth of the daughter cell and the typical oval shape of YF cells. In PH cells, actin patches remain polarized at the distal tip of the cell throughout bud emergence. Actin patches play an important role in directing secretion of new membrane and cell wall material in the emerging bud [102]. Thus, polarization of actin patches to the distal pole of PH cells throughout bud emergence promotes the long and thin cell shape in PH.

Several cytoskeletal proteins have been found to be required for PH cell morphogenesis including the actin-binding protein Tpm1p (tropomyosin), the formin homology (FH) domain protein Bni1p, the actin filament-bundling protein Sac6p (fimbrin), and the cyclase-associated protein Srv2p [54,103]. An important issue in understanding PH cell morphogenesis is the molecular connection between the PH signaling pathways and regulators of the actin cytoskeleton. A central link is the Rho-type GTPase Cdc42p together with several of its effector proteins [101]. Cdc42p not only is an activator of the pseudohyphal MAPK cascade, but also forms complexes with the FH domain protein Bni1p and actin [72,104]. Two further Rho-type GTPases that interact with Bni1p and are involved in reorganization of the actin cytoskeleton are Rho1p and Rho3p [105–107]. Interestingly, Rho1p also associates with the 1,3-β-D-glucan synthase Fks1p [108]. Both proteins are part of a multienzyme complex that catalyzes the synthesis of 1,3-β-linked glucan, a major structural component of the yeast cell wall. Thus, cell morphology might be regulated by modeling of the cell wall during cell growth through the action of Rho1p and the actin cytoskeleton. Whether Rho1p directly affects PH cell morphogenesis, however, has not been tested.

3.7 Substrate Adhesion and Invasion

An important property of PH cells is marked adhesion to other cells and to solid substrates. This feature of PH cells is reflected by the fact that mother and daughter cells remain attached to each other at the end of cell division, a prerequisite for filament formation. In contrast, YF mothers and daughters completely fall apart, a requirement for growth as single cells. A further consequence of enhanced adhesiveness is the observation that PH cells stick to the surface of the agar plate and grow down into the medium, whereas YF cells grow by spreading out on the surface of the agar. In the laboratory, adhesion and invasive growth can be assayed by a simple wash test, in which the surface of the agar plate is rinsed under running water (Fig. 6). PH cells invade the agar and cannot be washed away from the plate, even when the surface of the plate is rubbed with a hand. Pseudohyphal cells that have invaded the agar remain as visible colonies beneath the surface of the washed plate and can be reached only by piercing a microneedle into the agar. In contrast, YF cells are nonadhesive and can be easily washed away from the surface of the plate (Fig. 6). How does adhesion contribute to substrate invasive growth? PH filaments are linked structures consisting of many adhesive cells, in which the previous generations act as an anchor for the cell at the apex. Anchored filaments combined with the force of unipolar cell division by thin pseudohyphal cells might generate more force than a single cell and could be sufficient to propel a column through the agar.

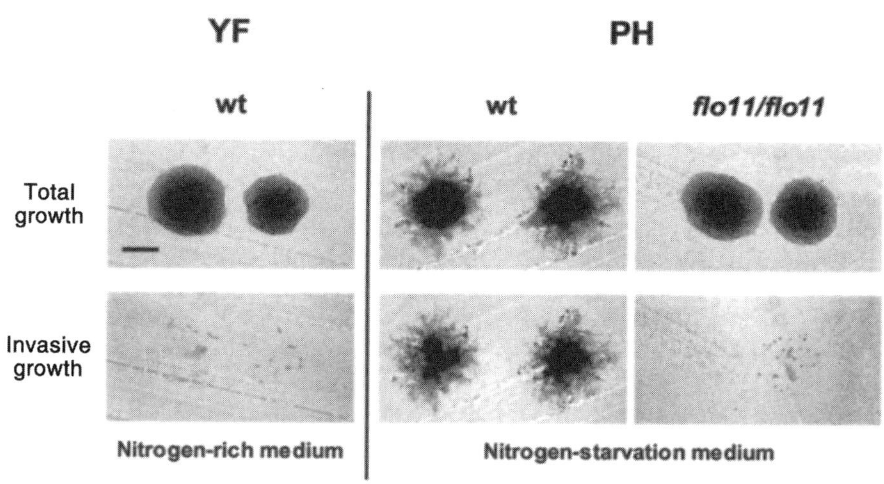

FIGURE 6 Induction of substrate adhesion and invasive growth by nitrogen starvation. Diploid wild-type (wt) and *flo11/flo11* mutant strains were grown on nitrogen-rich medium to form YF colonies and on nitrogen starvation medium for PH growth. After 3 days, colonies were photographed before (total growth) and after (invasive growth) a wash assay. Scale bar = 100 μm.

Adhesion of PH cells to each other and to solid substrates requires Flo11p, a cell surface flocculin with a structure similar to yeast serine/threonine-rich GPI-anchored cell wall proteins [58]. Mutant strains lacking the cell wall protein Flo11p are unable to develop invasively growing pseudohyphae (Fig. 6). Flo11p was initially identified as a cell surface molecule responsible for the calcium-dependent nonsexual aggregation known as flocculation [109]. However, flocculation per se does not cause invasiveness or filamentation, because other flocculent strains have been described that do not grow invasively [58]. In addition, other flocculins that are related to Flo11p, e.g., Flo1p, are not required for PH growth [61]. Expression of Flo11p is induced by nitrogen starvation and is under the control of both the pseudohyphal MAPK cascade and the cAMP pathway [58,61,62]. With a spanning of at least 2.8 kb, the *FLO11* promoter is unusually large and contains many upstream activation sequences (UASs) and repression elements [61]. Like the promoters of *HO* and *IME1*, other key developmental genes in yeast, the *FLO11* promoter has the ability to integrate multiple inputs and has become a key transcriptional reporter for pseudohyphal signaling pathways.

Apart from adhesion, further factors are likely to promote invasiveness of PH cells. Many invasive fungi, including *Candida albicans*, have been shown to secrete proteases and other enzymes believed to facilitate their invasion. In *S. cerevisiae*, secretion of lytic enzymes capable of hydrolyzing polysaccharides may enhance substrate invasive growth. This assumption is based on the finding that expression of the *PGU1* gene is upregulated by an activated pseudohyphal MAPK cascade [59]. *PGU1* encodes a secreted enzyme that hydrolyzes polygalacturonic acid, which is a structural barrier to microbial invasion and is present in the natural plant substrate of *S. cerevisiae*.

4 CONCLUSIONS

Pseudohyphal growth of yeasts is the result of a dimorphic transition that enables certain fungi to change from unicellular growth to reproduction as filaments. Pseudohyphal growth allows yeasts to escape from unfavorable growth conditions and to penetrate natural barriers. The baker's yeast *S. cerevisiae* is a genetically highly tractable organism and represents a simple model for pseudohyphal growth with the potential for a profound characterization of the pathways and targets involved. Studies in *S. cerevisiae* have provided a framework of genes and gene products that constitute the sensors and signaling pathways that transduce environmental signals, control the pseudohyphal cell division cycle, establish pseudohyphal cell polarity and morphogenesis, and induce substrate invasive growth. Pseudohyphal growth of *S. cerevisiae* has common features to dimorphism of pathogenic fungi that often experimentally are less accessible. Lessons learned from molecular analysis of dimorphism in *S. cerevisiae* have turned out to be true for human and plant pathogens as well. Homologs of G-proteins, protein kinases, transcription factors, and further elements required for pseudohyphal growth in *S. cerevisiae* have been found to control dimorphism of *Candida albicans*, *Cryptococcus neoformans*, and *Ustilago maydis* [44,110–121]. In conclusion, study of pseudohyphal growth in yeast not only offers important insights into mechanism and control of fungal dimorphism and disease, but also serves as an excellent model for differentiation in more complex organisms.

ACKNOWLEDGMENTS

I thank Gerhard Braus, Tim Köhler, and Naimeh Taheri for critically reading the manuscript. This work was supported by grants of the Deutsche Forschungsgemeinschaft and the Volkswagenstiftung.

REFERENCES

1. CP Kurtzman, JW Fell. The Yeasts, A Taxonomic Study. 4th ed. Amsterdam: Elsevier Science, 1998.
2. J Lodder, WC Slooff, JW Kreger–Van Rij. The classification of yeasts. In: AH Cook, ed. The Chemistry and Biology of Yeasts. New York: Academic Press, 1958, pp 1–62.
3. K Wolf. Nonconventional Yeasts in Biotechnology. Heidelberg: Springer-Verlag, 1996.
4. JF Ernst, A Schmidt. Dimorphism in Human Pathogenic and Apathogenic Yeasts. Basel: Karger, 2000.
5. JR Pringle, RB Broach, EW Jones. The Molecular and Cellular Biology of the Yeast *Saccharomyces*. Cold Spring Harbor, NY: Cold Spring Harbor Laboratory Press, 1997.
6. SJ Kron, NA Gow. Budding yeast morphogenesis: signalling, cytoskeleton and cell cycle. Curr Opin Cell Biol 7:845–855, 1995.
7. NM Stelling-Dekker. Die sporogenen Hefen. Amsterdam: Koninklijke Akademie van Wetenschappen, 1931.
8. J Lodder. Die anaskosporogenen Hefen, erste Hälfte. ed. Amsterdam: Noord-Hollandsche Uitgevers Maatschappij, 1934.
9. EO Morris. Yeast growth. In: AH Cook, ed. The Chemistry and Biology of Yeasts. New York: Academic Press, 1958, pp 251–321.
10. MJ Carlile. The success of the hypha and mycelium. In: NAR Gow, GM Gadd, eds. The Growing Fungus. London: Chapman & Hall, 1995, pp 3–19.
11. EC Hansen. Recherches sur la physiologie et la morphologie des fermentes alcooliques. VI. Les voiles chez le genre *Saccharomyces*. C R Trav Laborat Carlsberg T II:106–147, 1886.
12. CC Lindegren, E Hamilton. Bot Gaz 105:316–321, 1944.
13. S Levine, ZJ Ordal. Factors influencing the morphology of *Blastomyces dermatitidis*. J Bacteriol 52:687–694, 1946.

14. HA Diddens, J Lodder. Die anaskosporogenen Hefen, zweite Hälfte. ed. Amsterdam: Noord-Hollandsche Uitgevers Maatschappij, 1942.
15. RB Segal. The effect of fusel oil and its components on yeast (in Russian). Mikrobiologiya 7:93–105, 1938.
16. RB Segal. The effect of higher alcohols on yeast (in Russian). Mikrobiologiya 8: 466–470, 1939.
17. JR Dickinson. 'Fusel' alcohols induce hyphal-like extensions and pseudohyphal formation in yeasts. Microbiology 142:1391–1397, 1996.
18. MC Lorenz, NS Cutler, J Heitman. Characterization of alcohol-induced filamentous growth in *Saccharomyces cerevisiae*. Mol Biol Cell 11:183–199, 2000.
19. J Lodder, NJW Kreger–Van Rij. The Yeasts, A Taxonomic Study. 1st ed. Amsterdam: North-Holland Publishing Company, 1952.
20. CJ Gimeno, PO Ljungdahl, CA Styles, GR Fink. Unipolar cell divisions in the yeast *S. cerevisiae* lead to filamentous growth: regulation by starvation and *RAS*. Cell 68:1077–1090, 1992.
21. GH Scherr, RH Weaver. The dimorphism phenomenon in yeasts. Bacteriol Rev 17:51–92, 1953.
22. NAR Gow. Yeast-hyphal dimorphism. In: NAR Gow, GM Gadd, ed. The Growing Fungus. London: Chapman & Hall, 1995, pp 403–422.
23. AJ Brown, NA Gow. Regulatory networks controlling *Candida albicans* morphogenesis. Trends Microbiol 7:333–338, 1999.
24. H-U Mösch. Pseudohyphal development of *Saccharomyces cerevisiae*. In: JF Ernst, A Schmidt, eds. Dimorphism in Human Pathogenic and Apathogenic Yeasts. Basel: Karger, 2000, pp 185–200.
25. MN Hall, P Linder. The Early Days of Yeast Genetics. Cold Spring Harbor, NY: Cold Spring Harbor Laboratory Press, 1993.
26. SJ Kron. Filamentous growth in budding yeast. Trends Microbiol 5:450–454, 1997.
27. H Liu, CA Styles, GR Fink. *Saccharomyces cerevisiae* S288C has a mutation in *FLO8*, a gene required for filamentous growth. Genetics 144:967–978, 1996.
28. MC Lorenz, J Heitman. The MEP2 ammonium permease regulates pseudohyphal differentiation in *Saccharomyces cerevisiae*. EMBO J 17:1236–1247, 1998.
29. LE Murray, N Rowley, IW Dawes, GC Johnston, RA Singer. A yeast glutamine tRNA signals nitrogen status for regulation of dimorphic growth and sporulation. Proc Natl Acad Sci USA 95:8619–8624, 1998.
30. H Klasson, GR Fink, PO Ljungdahl. Ssy1p and Ptr3p are plasma membrane components of a yeast system that senses extracellular amino acids. Mol Cell Biol 19: 5405–5416, 1999.
31. SJ Kron, CA Styles, GR Fink. Symmetric cell division in pseudohyphae of the yeast *Saccharomyces cerevisiae*. Mol Biol Cell 5:1003–1022, 1994.
32. MC Lorenz, X Pan, T Harashima, ME Cardenas, Y Xue, JP Hirsch, J Heitman. The G protein–coupled receptor gpr1 is a nutrient sensor that regulates pseudohyphal differentiation in *Saccharomyces cerevisiae*. Genetics 154:609–622, 2000.
33. JM Gancedo. Yeast carbon catabolite repression. Microbiol Mol Biol Rev 62:334–361, 1998.
34. JM Thevelein, JH de Winde. Novel sensing mechanisms and targets for the cAMP-

protein kinase A pathway in the yeast *Saccharomyces cerevisiae*. Mol Microbiol 33:904–918, 1999.
35. Y Xue, M Batlle, JP Hirsch. GPR1 encodes a putative G protein–coupled receptor that associates with the Gpa2p Galpha subunit and functions in a Ras-independent pathway. EMBO J 17:1996–2007, 1998.
36. K Ansari, S Martin, M Farkasovsky, IM Ehbrecht, H Küntzel. Phospholipase C binds to the receptor-like GPR1 protein and controls pseudohyphal differentiation in *Saccharomyces cerevisiae*. J Biol Chem 274:30052–30058, 1999.
37. L Kraakman, K Lemaire, P Ma, AW Teunissen, MC Donaton, P Van Dijck, J Winderickx, JH de Winde, JM Thevelein. A *Saccharomyces cerevisiae* G-protein coupled receptor, Gpr1, is specifically required for glucose activation of the cAMP pathway during the transition to growth on glucose. Mol Microbiol 32:1002–1012, 1999.
38. JR Broach. *RAS* genes in *Saccharomyces cerevisiae*: signal transduction in search of a pathway. Trends Genet 7:28–33, 1991.
39. Y Jiang, C Davis, JR Broach. Efficient transition to growth on fermentable carbon sources in *Saccharomyces cerevisiae* requires signaling through the Ras pathway. Embo J 17:6942–6951, 1998.
40. JR Broach, RJ Deschennes. The function of *RAS* genes in *Saccharomyces cerevisiae*. Adv Cancer Res 54:79–139, 1990.
41. S Ozcan, M Johnston. Function and regulation of yeast hexose transporters. Microbiol Mol Biol Rev 63:554–569, 1999.
42. SM O'Rourke, I Herskowitz. The Hog1 MAPK prevents cross talk between the HOG and pheromone response MAPK pathways in *Saccharomyces cerevisiae*. Genes Dev 12:2874–2886, 1998.
43. I Herskowitz. MAP kinase pathways in yeast: for mating and more. Cell 80:187–197, 1995.
44. F Banuett. Signalling in the yeasts: an informational cascade with links to the filamentous fungi. Microbiol Mol Biol Rev 62:249–274, 1998.
45. F Posas, M Takekawa, H Saito. Signal transduction by MAP kinase cascades in budding yeast. Curr Opin Microbiol 1:175–182, 1998.
46. H Liu, CA Styles, GR Fink. Elements of the yeast pheromone response pathway required for filamentous growth of diploids. Science 262:1741–1744, 1993.
47. JG Cook, L Bardwell, J Thorner. Inhibitory and activating functions for MAPK Kss1 in the *S. cerevisiae* filamentous-growth signalling pathway. Nature 390:85–88, 1997.
48. HD Madhani, CA Styles, GR Fink. MAP kinases with distinct inhibitory functions impart signaling specificity during yeast differentiation. Cell 91:673–684, 1997.
49. RL Roberts, H-U Mösch, GR Fink. 14-3-3 proteins are essential for RAS/MAPK cascade signaling during pseudohyphal development in *S. cerevisiae*. Cell 89: 1055–1065, 1997.
50. M Ramezani Rad, G Jansen, F Buhring, CP Hollenberg. Ste50p is involved in regulating filamentous growth in the yeast *Saccharomyces cerevisiae* and associates with Ste11p. Mol Gen Genet 259:29–38, 1998.
51. JG Cook, L Bardwell, SJ Kron, J Thorner. Two novel targets of the MAP kinase

Kss1 are negative regulators of invasive growth in the yeast *Saccharomyces cerevisiae*. Genes Dev 10:2831–2848, 1996.
52. I Laloux, E Dubois, M Dewerchin, E Jacobs. *TEC1*, a gene involved in the activation of Ty1 and Ty1-mediated gene expression in *Saccharomyces cerevisiae*: cloning and molecular analysis. Mol Cell Biol 10:3541–3550, 1990.
53. V Gavrias, A Andrianopoulos, CJ Gimeno, WE Timberlake. *Saccharomyces cerevisiae TEC1* is required for pseudohyphal growth. Mol Microbiol 19:1255–1263, 1996.
54. H-U Mösch, GR Fink. Dissection of filamentous growth by transposon mutagenesis in *Saccharomyces cerevisiae*. Genetics 145:671–684, 1997.
55. A Andrianopoulos, WE Timberlake. ATTS, a new and conserved DNA binding domain. Plant Cell 3:747–748, 1991.
56. TR Burglin. The TEA domain: a novel, highly conserved DNA-binding motif. Cell 66:11–12, 1991.
57. HD Madhani, GR Fink. Combinatorial control required for the specificity of yeast MAPK signaling. Science 275:1314–1317, 1997.
58. WS Lo, AM Dranginis. The cell surface flocculin Flo11 is required for pseudohyphae formation and invasion by *Saccharomyces cerevisiae*. Mol Biol Cell 9:161–171, 1998.
59. HD Madhani, T Galitski, ES Lander, GR Fink. Effectors of a developmental mitogen-activated protein kinase cascade revealed by expression signatures of signaling mutants. Proc Natl Acad Sci USA 96:12530–12535, 1999.
60. M Gagiano, D van Dyk, FF Bauer, MG Lambrechts, IS Pretorius. Msn1p/Mss10p, Mss11p and Muc1p/Flo11p are part of a signal transduction pathway downstream of Mep2p regulating invasive growth and pseudohyphal differentiation in *Saccharomyces cerevisiae*. Mol Microbiol 31:103–116, 1999.
61. S Rupp, E Summers, HJ Lo, H Madhani, GR Fink. MAP kinase and cAMP filamentation signaling pathways converge on the unusually large promoter of the yeast *FLO11* gene. EMBO J 18:1257–1269, 1999.
62. LS Robertson, GR Fink. The three yeast A kinases have specific signaling functions in pseudohyphal growth. Proc Natl Acad Sci USA 95:13783–13787, 1998.
63. X Pan, J Heitman. Cyclic AMP–dependent protein kinase regulates pseudohyphal differentiation in *Saccharomyces cerevisiae*. Mol Cell Biol 19:4874–4887, 1999.
64. T Toda, S Cameron, P Sass, M Zoller, M Wigler. Three different genes in *S. cerevisiae* encode the catalytic subunits of the cAMP-dependent protein kinase. Cell 50:277–287, 1987.
65. MP Ward, CJ Gimeno, GR Fink, S Garrett. *SOK2* may regulate cyclic AMP–dependent protein kinase-stimulated growth and pseudohyphal development by repressing transcription. Mol Cell Biol 15:6854–6863, 1995.
66. E Kübler, H-U Mösch, S Rupp, MP Lisanti. Gpa2p, a G-protein alpha-subunit, regulates growth and pseudohyphal development in *Saccharomyces cerevisiae* via a cAMP-dependent mechanism. J Biol Chem 272:20321–20323, 1997.
67. H-U Mösch, E Kübler, S Krappmann, GR Fink, GH Braus. Crosstalk between the Ras2p-controlled mitogen-activated protein kinase and cAMP pathways during invasive growth of *Saccharomyces cerevisiae*. Mol Biol Cell 10:1325–1335, 1999.

68. MC Lorenz, J Heitman. Yeast pseudohyphal growth is regulated by GPA2, a G protein alpha homolog. EMBO J 16:7008–7018, 1997.
69. S Colombo, P Ma, L Cauwenberg, J Winderickx, M Crauwels, A Teunissen, D Nauwelaers, JH de Winde, MF Gorwa, D Colavizza, JM Thevelein. Involvement of distinct G-proteins, Gpa2 and Ras, in glucose- and intracellular acidification-induced cAMP signalling in the yeast *Saccharomyces cerevisiae*. EMBO J 17: 3326–3341, 1998.
70. CW Yun, H Tamaki, R Nakayama, K Yamamoto, H Kumagai. Gpr1p, a putative G-protein coupled receptor, regulates glucose-dependent cellular cAMP level in yeast *Saccharomyces cerevisiae*. Biochem Biophys Res Commun 252:29–33, 1998.
71. H Tamaki, T Miwa, M Shinozaki, M Saito, CW Yun, K Yamamoto, H Kumagai. GPR1 regulates filamentous growth through *FLO11* in yeast *Saccharomyces cerevisiae*. Biochem Biophys Res Commun 267:164–168, 2000.
72. H-U Mösch, RL Roberts, GR Fink. Ras2 signals via the Cdc42/Ste20/mitogen-activated protein kinase module to induce filamentous growth in *Saccharomyces cerevisiae*. Proc Natl Acad Sci USA 93:5352–5356, 1996.
73. H-U Mösch, T Köhler, GH Braus. Different domains of the essential GTPase Cdc42p required for growth and development of *S. cerevisiae*. Mol Cell Biol 21: 235–248, 2001.
74. CJ Gimeno, GR Fink. Induction of pseudohyphal growth by overexpression of *PHD1*, a *Saccharomyces cerevisiae* gene related to transcriptional regulators of fungal development. Mol Cell Biol 14:2100–2112, 1994.
75. S Chandarlapaty, B Errede. Ash1, a daughter cell-specific protein, is required for pseudohyphal growth of *Saccharomyces cerevisiae*. Mol Cell Biol 18:2884–2891, 1998.
76. MG Lambrechts, FF Bauer, J Marmur, IS Pretorius. Muc1, a mucin-like protein that is regulated by Mss10, is critical for pseudohyphal differentiation in yeast. Proc Natl Acad Sci USA 93:8419–8424, 1996.
77. N Bobola, RP Jansen, TH Shin, K Nasmyth. Asymmetric accumulation of Ash1p in postanaphase nuclei depends on a myosin and restricts yeast mating-type switching to mother cells. Cell 84:699–709, 1996.
78. DJ Lew, SI Reed. Morphogenesis in the yeast cell cycle: regulation by Cdc28 and cyclins. J Cell Biol 120:1305–1320, 1993.
79. SH Ahn, A Acurio, SJ Kron. Regulation of G2/M progression by the *STE* mitogen-activated protein kinase pathway in budding yeast filamentous growth. Mol Biol Cell 10:3301–3316, 1999.
80. JD Loeb, TA Kerentseva, T Pan, M Sepulveda-Becerra, H Liu. *Saccharomyces cerevisiae* G1 cyclins are differentially involved in invasive and pseudohyphal growth independent of the filamentation mitogen-activated protein kinase pathway. Genetics 153:1535–1546, 1999.
81. JD Loeb, M Sepulveda-Becerra, I Hazan, H Liu. A G1 cyclin is necessary for maintenance of filamentous growth in *Candida albicans*. Mol Cell Biol 19:4019–4027, 1999.
82. MJ Blacketer, CM Koehler, SG Coats, AM Myers, P Madaule. Regulation of dimorphism in *Saccharomyces cerevisiae*: involvement of the novel protein kinase

homolog Elm1p and protein phosphatase 2A. Mol Cell Biol 13:5567–5581, 1993.

83. MJ Blacketer, P Madaule, AM Myers. Mutational analysis of morphologic differentiation in *Saccharomyces cerevisiae*. Genetics 140:1259–1275, 1995.
84. NP Edgington, MJ Blacketer, TA Bierwagen, AM Myers. Control of *Saccharomyces cerevisiae* filamentous growth by cyclin-dependent kinase Cdc28. Mol Cell Biol 19:1369–1380, 1999.
85. D Freifelder. Bud position in *Saccharomyces cerevisiae*. J Bacteriol 80:567–568, 1960.
86. JB Hicks, JN Strathern, I Herskowitz. Interconversion of yeast mating types III. Action of the homothallism (HO) gene in cells homozygous for the mating type locus. Genetics 85:395–405, 1977.
87. J Chant, JR Pringle. Patterns of bud-site selection in the yeast *Saccharomyces cerevisiae*. J Cell Biol 129:751–765, 1995.
88. K Madden, M Snyder. Cell polarity and morphogenesis in budding yeast. Annu Rev Microbiol 52:687–744, 1998.
89. J Chant. Cell polarity in yeast. Annu Rev Cell Dev Biol 15:365–391, 1999.
90. A Bender, JR Pringle. Multicopy suppression of the *cdc24* budding defect in yeast by *CDC42* and three newly identified genes including the ras-related gene *RSR1*. Proc Natl Acad Sci USA 86:9976–9980, 1989.
91. J Chant, I Herskowitz. Genetic control of bud site selection in yeast by a set of gene products that constitute a morphogenetic pathway. Cell 65:1203–1212, 1991.
92. J Chant, K Corrado, JR Pringle, I Herskowitz. Yeast *BUD5*, encoding a putative GDP-GTP exchange factor, is necessary for bud site selection and interacts with bud formation gene *BEM1*. Cell 65:1213–1224, 1991.
93. A Fujita, C Oka, Y Arikawa, T Katagai, A Tonouchi, S Kuhara, Y Misumi. A yeast gene necessary for bud-site selection encodes a protein similar to insulin-degrading enzymes. Nature 372:567–570, 1994.
94. A Halme, M Michelitch, EL Mitchell, J Chant. Bud10p directs axial cell polarization in budding yeast and resembles a transmembrane receptor. Curr Biol 6:570–579, 1996.
95. T Roemer, K Madden, J Chang, M Snyder. Selection of axial growth sites in yeast requires Axl2p, a novel plasma membrane glycoprotein. Genes Dev 10:777–793, 1996.
96. M Snyder. The SPA2 protein of yeast localizes to sites of cell growth. J Cell Biol 108:1419–1429, 1989.
97. N Valtz, I Herskowitz. Pea2 protein of yeast is localized to sites of polarized growth and is required for efficient mating and bipolar budding. J Cell Biol 135:725–739, 1996.
98. JE Zahner, HA Harkins, JR Pringle. Genetic analysis of the bipolar pattern of bud site selection in the yeast *Saccharomyces cerevisiae*. Mol Cell Biol 16:1857–1870, 1996.
99. N Taheri, T Köhler, GH Braus, H-U Mösch. Asymmetrically localized Bud8p and Bud9p proteins control yeast cell polarity and development. EMBO J 19:6686–6696, 2000.

100. E Cabib, J Drgonova, T Drgon. Role of small G proteins in yeast cell polarization and wall biosynthesis. Annu Rev Biochem 67:307–333, 1998.
101. A Hall. Rho GTPases and the actin cytoskeleton. Science 279:509–514, 1998.
102. D Pruyne, A Bretscher. Polarization of cell growth in yeast. J Cell Sci 113:571–585, 2000.
103. BM Cali, TC Doyle, D Botstein, GR Fink. Multiple functions for actin during filamentous growth of *Saccharomyces cerevisiae*. Mol Biol Cell 9:1873–1889, 1998.
104. M Evangelista, K Blundell, MS Longtine, CJ Chow, N Adames, JR Pringle, M Peter, C Boone. Bni1p, a yeast formin linking Cdc42p and the actin cytoskeleton during polarized morphogenesis. Science 276:118–122, 1997.
105. H Imamura, K Tanaka, T Hihara, M Umikawa, T Kamei, K Takahashi, T Sasaki, Y Takai. Bni1p and Bnr1p: downstream targets of the Rho family small G-proteins which interact with profilin and regulate actin cytoskeleton in *Saccharomyces cerevisiae*. EMBO J 16:2745–2755, 1997.
106. T Fujiwara, K Tanaka, A Mino, M Kikyo, K Takahashi, K Shimizu, Y Takai. Rho1p–Bni1p–Spa2p interactions: implication in localization of Bni1p at the bud site and regulation of the actin cytoskeleton in *Saccharomyces cerevisiae*. Mol Biol Cell 9:1221–1233, 1998.
107. T Fujiwara, K Tanaka, E Inoue, M Kikyo, Y Takai. Bni1p regulates microtubule-dependent nuclear migration through the actin cytoskeleton in *Saccharomyces cerevisiae*. Mol Cell Biol 19:8016–8027, 1999.
108. H Qadota, CP Python, SB Inoue, M Arisawa, Y Anraku, Y Zheng, T Watanabe, DE Levin, Y Ohya. Identification of yeast Rho1p GTPase as a regulatory subunit of 1,3-beta-glucan synthase. Science 272:279–281, 1996.
109. WS Lo, AM Dranginis. *FLO11*, a yeast gene related to the *STA* genes, encodes a novel cell surface flocculin. J Bacteriol 178:7144–7151, 1996.
110. F Banuett, I Herskowitz. Identification of fuz7, a *Ustilago maydis* MEK/MAPKK homolog required for *a*-locus-dependent and -independent steps in the fungal life cycle. Genes Dev 8:1367–1378, 1994.
111. H Liu, J Köhler, GR Fink. Suppression of hyphal formation in *Candida albicans* by mutation of a *STE12* homolog. Science 266:1723–1726, 1994.
112. JR Köhler, GR Fink. *Candida albicans* strains heterozygous and homozygous for mutations in mitogen-activated protein kinase signaling components have defects in hyphal development. Proc Natl Acad Sci USA 93:13223–13228, 1996.
113. JA Alspaugh, JR Perfect, J Heitman. *Cryptococcus neoformans* mating and virulence are regulated by the G-protein alpha subunit GPA1 and cAMP. Genes Dev 11:3206–3217, 1997.
114. HJ Lo, JR Köhler, B DiDomenico, D Loebenberg, A Cacciapuoti, GR Fink. Nonfilamentous *C. albicans* mutants are avirulent. Cell 90:939–949, 1997.
115. VR Stoldt, A Sonneborn, CE Leuker, JF Ernst. Efg1p, an essential regulator of morphogenesis of the human pathogen *Candida albicans*, is a member of a conserved class of bHLH proteins regulating morphogenetic processes in fungi. EMBO J 16:1982–1991, 1997.
116. JA Alspaugh, JR Perfect, J Heitman. Signal transduction pathways regulating differentiation and pathogenicity of *Cryptococcus neoformans*. Fungal Genet Biol 25: 1–14, 1998.

117. Q Feng, E Summers, B Guo, G Fink. Ras signaling is required for serum-induced hyphal differentiation in *Candida albicans*. J Bacteriol 181:6339–6346, 1999.
118. R Kahmann, C Basse, M Feldbrugge. Fungal-plant signalling in the *Ustilago maydis-maize* pathosystem. Curr Opin Microbiol 2:647–650, 1999.
119. JA Alspaugh, LM Cavallo, JR Perfect, J Heitman. *RAS1* regulates filamentation, mating and growth at high temperature of *Cryptococcus neoformans*. Mol Microbiol 36:352–365, 2000.
120. JF Ernst. Transcription factors in *Candida albicans*—environmental control of morphogenesis. Microbiology 146:1763–1774, 2000.
121. A Sonneborn, DP Bockmuhl, M Gerads, K Kurpanek, D Sanglard, JF Ernst. Protein kinase A encoded by *TPK2* regulates dimorphism of *Candida albicans*. Mol Microbiol 35:386–396, 2000.

2

Hyphal Tip Growth
Outstanding Questions

Salomón Bartnicki-García
Centro de Investigación Científica y de Educación Superior de Ensenada (CICESE), Ensenada, Mexico, and University of California, Riverside, California

1 GENERAL

The remarkable ability of fungi to make tubular cells or hyphae is an exquisite case of polarized growth and has been the subject of intense attention and experimentation for more than 100 years. Much has been learned about many facets of hyphal morphogenesis, but a precise understanding of the structural, biochemical, and genetic basis of apical growth remains to be attained. In this chapter I will summarize some new developments in hyphal morphogenesis and address some of the outstanding questions in this field. Hyphal morphogenesis or tip growth of fungi has been the subject of recent reviews, each with a somewhat different perspective [1–8].

2 KEY STRUCTURES AND PROCESSES IN TIP GROWTH

Myriad genetic and biochemical processes are involved in the growth of a hypha. It is therefore not surprising that a wide variety of genetic, biochemical, and environmental alterations have been reported to have a controlling impact on hyphal growth and morphogenesis. The challenge is to narrow the search to those events that are immediately or directly involved in the production of a tubular cell wall by tip growth. Understanding how a fungus establishes a polarized gradient of wall formation is clearly the ultimate objective in the search for the basis of tip growth. Different researchers have employed different experimental tools and applied widely different emphasis. The end result is a number of seemingly divergent models to explain tip growth in fungi (see Sec. 3).

2.1 The Central Question—Polarized Secretion

Since cell wall formation in fungi is the result of a secretory process, it follows that tip growth is basically a polarization of the secretory apparatus of the cell. A most vivid example of the subtlety of the polarization mechanism can be seen during germ tube emergence in *Mucor rouxii* (Fig. 1). Without any detectable change in the overall rates of cell growth or wall synthesis, or in any other major metabolic parameter, the pattern of wall deposition in the germ sphere switched from uniformly dispersed (isotropic) to highly polarized at the site of germ tube emergence. We concluded that a corresponding spatial reorganization of the underlying secretory apparatus took place with no other perceptible change in cell physiology or metabolism [9]. This morphogenetic transition seems ideal to identify genes selectively expressed during the switch from isotropic to polarized growth, an approach recently adopted (see Sec. 3.4.3).

2.2 The Spitzenkörper

Because of its location in the apical dome of growing hyphae, the Spitzenkörper has attracted much attention from those interested in understanding the mechanism of apical growth in fungi. Brunswik's [10] original discovery of an iron-hematoxylin–staining body in the hyphal tips of *Coprinus* spp. went largely unnoticed until Girbardt's studies confirmed the existence of a Spitzenkörper in living hyphae by phase contrast microscopy [11]. Phase contrast microscopy provides the best optics to study the structure and behavior of the Spitzenkörper in living specimens of fungal hyphae [12–18]. The video microscopic surveys made by Lopez-Franco and Bracker [13], on >30 different fungal species, recognized eight different patterns of Spitzenkörper organization in higher fungi; the large Spitzenkörper found by Vargas et al. [19] in the lower fungus *Allomyces macrogynous* constitutes a ninth unique type. This morphological variability has yet to be reconciled with the proposed function of the Spitzenkörper.

2.2.1 Organizer of Vesicle Traffic

Although the exact function of the Spitzenkörper has not been well established, there is strong reason to believe, because of its position, composition, and behavior, that it serves to organize or direct vesicle traffic for hyphal elongation. This

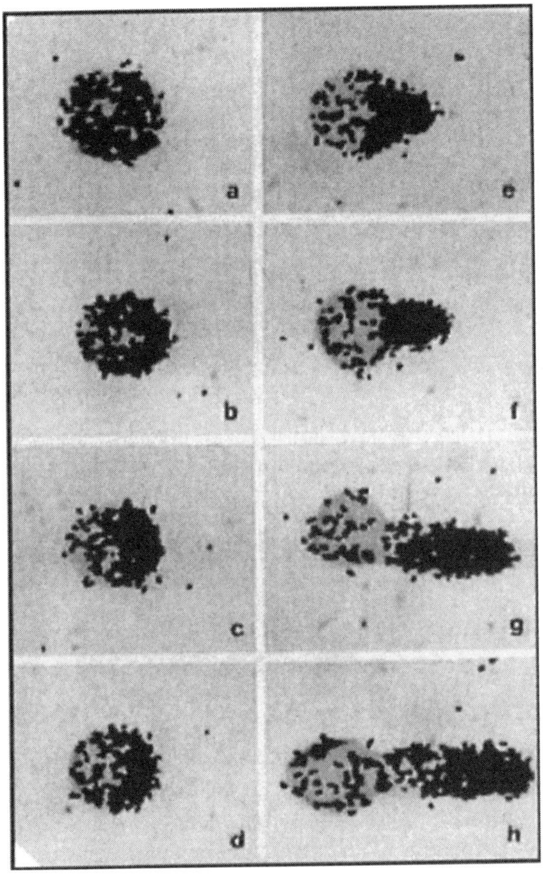

FIGURE 1 Autoradiographs of germinating spores of *Mucor rouxii* assembled to show the progressive polarization of cell wall deposition (chitin and chitosan) in the germ sphere prior to (a–d), during (e and f), and after germ tube emergence (g and h). (From Ref. 9.)

characteristic accumulation of wall-building vesicles in the hyphal apex is much more than an simple case of traffic congestion. The Spitzenkörper appears to be a highly evolved distribution center for collecting vesicles and delivering them to the cell surface. The cluster of Spitzenkörper vesicles accumulates around a core region, whose biochemical identity remains a mystery. Microtubules [20–22], actin microfilaments [23,24], and other, unrecognized granular material have been found in the Spitzenkörper region. More recently, McDaniel and Roberson [25] discovered that γ-tubulin is a component of the central region of the Spit-

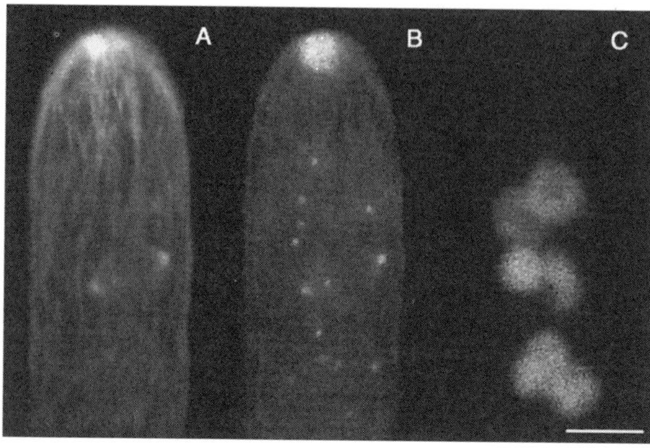

Figure 2 Confocal microscopy of the microtubular cytoskeleton in a hyphal tip of *Allomyces macrogynous*. (A) Immunofluorescence labeling of α-tubulin shows cytoplasmic microtubules converging at the apex. A single mitotic spindle is seen in the subapex. (B) Immunofluorescence labeling of γ-tubulin shows a large accumulation in the Spitzenkörper region. The small discrete spots in the subapex correspond to centrosomes. (C) DAPI staining of nuclei in the same hypha. (Micrographs courtesy of R. Roberson. B is previously unpublished image from Ref. 22; A and C are from Ref. 25.) Scale bar = 5 μm.

zenkörper of *Allomyces macrogynous* (Fig. 2), a finding that corroborated earlier suspicions that the Spitzenkörper functions as an MTOC [26]. Since microtubules are considered to be cellular tracks for the long-range transport of vesicles [26–29], a system of microtubules rooted in the MTOC of the Spitzenkörper would explain the vesicle-gathering role of the Spitzenkörper.

2.2.2 The Spitzenkörper as a Vesicle Supply Center

A vesicle-based computer simulation of fungal morphogenesis led us to the realization that different cell shapes could be generated by displacing the immediate source of wall-making vesicles, called the vesicle supply center (VSC) [30]. We showed that linear displacement of the VSC generated hyphal shapes. This shape-building process was described by a simple equation

$$y = x \cot(XV/N)$$

named the hyphoid, that relates the number of wall-building vesicles released from the VSC per unit time with the rate of advancement of the VSC. The 2D shape described by the plotted hyphoid was almost identical to the shape of well-preserved hyphae in longitudinal median section [30].

In view of the remarkable coincidence between the position of the VSC in the hyphoid and the position of the Spitzenkörper in a real hypha, we proposed that the key function of the Spitzenkörper was to serve as a VSC, i.e., a movable distribution center for wall-building vesicles [30,31]. Accordingly, morphogenesis would be primarily determined by the interplay between movement of the Spitzenkörper and the amount of wall-building vesicles emanating from the Spitzenkörper.

Several examples of hyphal morphogenesis in different fungi were analyzed by computer simulation (Fungus Simulator) [32] including normal hyphal growth, spontaneous hyphal bulging [14,33], induced apical branching in a temperature-sensitive mutant [15], hyphal meandering [17], and deformed growth of mutated hyphae [18]. In each morphogenetic example, the VSC of the Simulator was programmed to follow the actual trajectories of the Spitzenkörper. In all cases, morphology could be explained by assuming that the Spitzenkörper functions as a VSC.

2.2.3 Spitzenkörper Trajectory

Video microscopy and image analysis of living hyphae of *Neurospora crassa* showed a close correlation among Spitzenkörper position, trajectory, and the growth direction of a hypha [18]. A permanent change in growth direction, i.e., the establishment of a new growth axis, was correlated with a sustained shift in Spitzenkörper trajectory away from the existing cell axis. This study confirmed Girbardt's finding [11] that an off-center displacement of the Spitzenkörper precedes a change in growth direction of the hypha.

Spitzenkörper trajectory determines not only growth directionality [17] but also the overall appearance of a hypha [14,18]. In the straightest hyphae, the Spitzenkörper advances along a straight path with frequent but minute transverse oscillations. In hyphae with highly distorted morphology, e.g., in the *ropy* mutants of *N. crassa*, the trajectory of the Spitzenkörper became erratic; sustained departures of the Spitzenkörper from the hyphal growth axis produced corresponding distortions in hyphal morphology [18]. The growing scaffolding of cytoplasmic microtubules in a hypha was proposed as the mechanism that maintained the Spitzenkörper on a rather fixed trajectory [17]. An alternative explanation [34] invoking inhibitory substances secreted by advancing hyphae as the primary determinants of growth directionality was considered unlikely [17]. A future challenge would be to find out how external factors such as light, chemicals, etc., affect the intrinsically fixed directionality of hyphae to bring about the well known tropic responses of fungal hyphae.

Perhaps the most compelling case for believing that the Spitzenkörper functions as a ''steering wheel'' responsible for growth directionality and morphogenesis of a hypha comes from the laser experiments done by Bracker and coworkers [35]. By manipulating (chasing) the Spitzenkörper with laser tweezers,

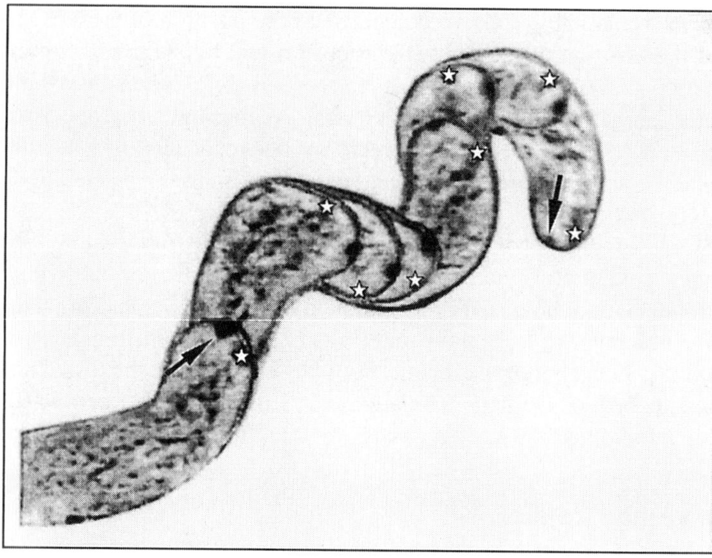

Figure 3 Abrupt changes in growth directionality of a hypha of *Trichoderma viride* during a Spitzenkörper chase lasting 9 min. The Spitzenkörper was forced to move in different directions by shifting the position of the laser trap (white stars). Black arrows show the Spitzenkörper at the beginning and the end of the sequence. Reconstruction made from a videotaped sequence supplied by C. E. Bracker and R. Lopez Franco. The montage contains eight images captured at 0, 2.0, 2.6, 3.0, 4.7, 5.8, 6.9, and 8.9 min. (From Ref. 35.)

they discovered that it could be displaced at will to produce corresponding changes in the direction of hyphal elongation, and other morphological alterations (Fig. 3). Their findings support the hypothesis that the patterns of apical exocytosis are governed by the position of the Spitzenkörper.

2.2.4 Growth Pulses and Satellite Spitzenkörper

Two recent findings by Lopez-Franco and coworkers [12,36] have modified our basic understanding of the physiology of hyphal tip growth. First, the well-established notion that hyphae elongate at a steady rate when grown under constant environmental conditions is, strictly speaking, incorrect. When analyzed with the high precision made practical by computer-enhanced video microscopy, i.e., measuring elongation rates at high magnification and at 1- to 5-sec intervals (rather than minutes or hours as is the usual practice), elongation was found to be always a pulsatile phenomenon [36]. In the seven fungal species examined, representing major taxonomic groups, periods of slow and fast growth alternated at a somewhat regular frequency. Pulsation varied from species to species and ranged from 2.7 to 14 pulses/min on the average. The other related finding was the discovery of satellite Spitzenkörper, i.e., small packages of vesicles that arise a few micrometers behind the apical pole, migrate rapidly to the hyphal apex and merge with the main Spitzenkörper and thus appear to contribute to the growth of the hyphal apex [12]. Satellites were frequently detected and their occurrence may be related to growth pulsation. In two fungi, the fast phase of the growth pulses was correlated with the merger of satellite Spitzenkörper with the main Spitzenkörper. It remains to be determined whether satellite Spitzenkörper are mainly responsible for pulsation or whether they simply exacerbate a presumed intrinsic pulsation of the secretory apparatus. Alternative interpretations have been suggested by Johns et al. [37] who measured the force exerted by the tips of *Achlya bisexualis* on a miniature strain gauge, and found that the force fluctuated with a periodicity comparable to that of growth pulses. They speculated that such fluctuations may reflect minute changes in turgor pressure that are beyond measurement, or pulses in wall hardening and wall loosening.

2.2.5 Spitzenkörper Origin

Studies by Reynaga-Peña and coworkers [15,16] on an apical-branching, temperature-sensitive mutant of *Aspergillus niger* (*ramosa-1*) addressed the question of Spitzenkörper biogenesis. Basically, at the restrictive temperature, single growing hyphal tips split into two tips. The original Spitzenkörper did not divide; instead, it retracted from its polar position and disappeared. A few minutes later two new Spitzenkörper appeared, each giving rise to an apical branch. The two new Spitzenkörper arose seemingly de novo from vesicle clouds that formed in the apical region next to the future site of branch emergence. It remains to be seen whether the invisible core of the original Spitzenkörper may have divided to serve as nucleation site for the formation of the two new Spitzenkörper. The behavior of satellite Spitzenkörper [12] and observations on the origin of lateral branches [38] suggest that nucleation sites for new Spitzenkörper can appear repeatedly in the subapical region independently of the main Spitzenkörper.

Regalado [39] proposed a mathematical model to explain the accumulation of vesicles in the Spitzenkörper through changes in the rheological properties of the cytoskeleton conditioned by the Ca^{2+} gradient in the hyphal tip.

2.2.6 Questions

1. *If a Spitzenkörper is so essential for tip growth, why don't we see one in oomycetous hyphae*? This question was often a reason for skepticism about earlier claims on the significance of the Spitzenkörper. It is important to keep in mind that although no Spitzenkörper can be seen with the optical microscope, transmission electron microscopy provides ample proof that Oomycetes have a similar cluster of vesicles in their hyphal apices [40]; an excellent example can be seen in the hyphal tip of *Saprolegnia ferax* [Fig. 1 in Ref. 2].

Consequently, there is no reason to think that the mechanism of tip growth would be radically different between cellulosic and chitinous fungi, despite the entirely different evolutionary history of these two groups of fungi [4]. As we postulated earlier [30], fungi lacking a visible Spitzenkörper must have its functional equivalent—namely, a VSC from which vesicles start on the final leg of their journey to the cell surface. Presumably, in Oomycetes the cluster of apical vesicles does not have sufficient density and/or refractivity to be visible by light microscopy. On the other hand, Harold [5] is of the opinion that the diversity of apical organization bespeaks the existence of alternative mechanisms for shaping a hypha.

2. *Is γ-tubulin, ergo an MTOC, present in the Spitzenkörper of higher fungi?* Attempts to extend the finding of γ-tubulin to the hyphal apices of higher fungi have been unsuccessful (R. Roberson, private communication). The failure to detect γ-tubulin in the apical region was not due to staining problems since the antibody did stain the γ-tubulin present in basal bodies. Given the functional significance of γ-tubulin presence in a Spitzenkörper, it is worth exhausting alternative technical reasons for the lack of γ-tubulin staining. But if γ-tubulin is truly absent in the hyphal apices of septate fungi, what replaces it? Is there a major difference in hyphal organization and function that obviates the need for an apical MTOC between the lower and the higher members of the same phylogenetic trunk line? Note that the arrangement of microtubules in the hyphal apex also differs. In *Allomyces* [20,22], the microtubules are sharply focused on the Spitzenkörper, in higher fungi, they are not [41]. Perhaps higher fungi have evolved a modified MTOC that does not depend on γ-tubulin.

3. *Is the Spitzenkörper a maturation site for vesicles?* Given that secretory vesicles undergo extensive biochemical modification in their transit from ER to plasma membrane, the Spitzenkörper may be a final station mediating or regulating a final modification that prepares the vesicles for the final leg of their exocytotic journey.

4. *Is the Spitzenkörper a transfer station where vesicles that arrive from the subapex on microtubular tracks switch to actin tracks?* The Spitzenkörper may be the place where vesicles switch motors and shift from incoming longitudinal travel on microtubules to outgoing travel on actin filaments.

5. *Is the Spitzenkörper a vesicle-recycling center?* Contrary to earlier studies, investigators from several laboratories (R. Lopez-Franco [42,43]; and C.E. Bracker, personal communication) found that endocytosis does take place in fungal hyphae and that this process of membrane internalization contributes material to the Spitzenkörper. The fluorescent dye FM4-64 has been especially useful. The dye was taken up and internalized from the plasma membrane appearing progressively in structures corresponding to presumed endosomes, the Spitzenkörper, the vacuolar membrane, and mitochondria [43]. The pattern of stain distribution was broadly similar in a wide range of fungal species. Accord-

ingly, the Spitzenkörper would be a site not only for collecting exocytotic vesicles but also for recycling internalized membranes.

The preceding questions highlight the need for biochemical studies to identify the core components of the Spitzenkörper and cytological studies to label vesicles and cytoskeleton in living cells to clarify their dynamics and mutual interaction. The application of molecular genetics could be of great help in overcoming technical obstacles that heretofore have limited progress in elucidating details on the structure and function of the secretory apparatus of a fungus. Such new quest is already under way [44–46].

2.3 The Cytoskeleton

A good number of studies in recent years have reaffirmed the central importance of the cytoskeleton in hyphal morphogenesis. Both the F-actin and the microtubular skeletons, plus their associated proteins, have been implicated but some of the findings, and conclusions have been somewhat divergent, particularly those based on inhibitor experiments. Given the vagaries of negative results, it may be safer not to regard them as final and conclude that *both* F-actin and microtubules are of critical but different importance in hyphal morphogenesis.

2.3.1 Actin Cytoskeleton

There seems to be general agreement that actin is abundantly present in growing hyphal tips though its organization varies. In Oomycetes, actin forms a cap next to the apical plasma membrane [47], while in the rest of the mycelial fungi actin is more likely to appear in small plaques [48–50]. Intriguingly, in some fungi actin is sometimes more abundant in the subapical than in the apical region [51–53]; since the functional significance of this arrangement is not obvious, the possibility of it being a fixation artifact has been advanced [7]. Inhibitor experiments with cytochalasin A by Torralba et al. [53] confirmed that a polymerized actin cytoskeleton is required for normal apical growth, hyphal tip shape, and polarized enzyme secretion in *Aspergillus nidulans*. Likewise, actin inhibition by Latrunculin B disrupted tip growth in *Saprolegnia ferax* hyphae [54,55]. Bachewich and Heath [55] reported that radial arrays of F-actin precede new hypha formation in *Saprolegnia ferax* and suggested that F-actin participates in establishing polar growth. Heath et al. [41] concluded that there was an obligatory role for F-actin in hyphal polarization and tip morphogenesis. By a different, molecular route, Harris et al. [56] determined that the normal pattern of germ tube emergence in *Aspergillus nidulans* is dependent on the integrity of the actin cytoskeleton.

2.3.2 Microtubular Cytoskeleton

Although past studies questioned the importance of microtubules in hyphal growth and morphogenesis [2,57], new studies with mutants deficient in motor proteins have shown that a fully functional microtubular cytoskeleton is necessary to maintain normal growth rates, normal nuclear distribution, and regular hyphal morphology [18,58–63]. Inoue et al. [62] reported that the heavy chain of dynein was required for normal secretory vesicle transport to the hyphal apex and normal hyphal tip cell morphogenesis in *Nectria haematococca*. Riquelme et al. [18] concluded that dynein and dynactin deficiencies of two *ropy* mutants of *Neurospora crassa* distorted hyphal morphogenesis by destabilizing the Spitzenkörper and causing it to deviate widely from an axial trajectory. Kinesin deficiency in *Nectria haematococca* also caused various effects including severe reduction in colony growth rate, helical or wavy hyphae with reduced diameter, and reduction in Spitzenkörper size. Wu et al. [63] noted that these effects were not due to altered microtubule distribution, as microtubules were abundant throughout the length of hyphal tip cells of the mutant. These studies suggest that both the anterograde and retrograde movement of vesicles on microtubules are important in maintaining the large Spitzenkörper size and high growth rate of the wild-type strains. Findings implicating dynein in apical transport [18,62] differ from the negative conclusion reached by Seiler et al. [61] but are in accord with the video microscopy observations of McDaniel and Roberson [25] who showed that microtubules were required for vesicle movement in hyphae of *Allomyces macrogynous* and noted that movement can occur in both directions along a common path.

The involvement of opposite motors in apical transport suggests that cytoplasmic microtubules in a hypha may not all have the same orientation. Thus the finding of a bona fide MTOC in hyphal apices [22] and the accumulation of dynein at the tip [61] indicate that the minus ends of microtubules are at the apex. On the other hand, a reverse orientation is indicated by the observation that kinesin, a plus-end motor, accumulates at growing tips [60]. A similar discrepancy in microtubule orientation, found in nuclear migration, prompted Xiang and Morris [8] to conclude that fungi may have a "need for different microtubule polarities" and may even result in "different mechanisms to move nuclei in the hyphae." The same could be said for vesicle traffic in the growing tips.

2.3.3 Questions

1. *Are actin and microtubules required for tip growth*? Based on inhibitor experiments with Nocodazole, MBC, and Latrunculin B on *N. crassa* and *Saprolegnia ferax*, Heath et al. [41] concluded that there was an obligatory role for F-actin in hyphal polarization and tip morphogenesis but only an indirect role for microtubules. A reevaluation of their evidence leads to a modified conclusion: inhibition of either actin or microtubular functions has a direct impact on tip growth albeit with different consequences on hyphal morphogenesis. Actin inhibition by cytochalasin often [54,64,65] but not always [17] stops elongation and leads to formation of bulbous tips; microtubule inhibition by nocodazole [41] or MBC [17] also inhibits growth but causes morphological distortions that may extend over a long stretch of the hyphal tube. It is essential to keep in mind that

most of the distortions in the shape of a hypha originate when the distorted or deviated portion was part of the growing tip; i.e., the observed alterations in the cylindrical portion of the hypha were direct effects on tip growth. Actin inhibition causes a more drastic loss of polarized growth, but the effect does not usually persist, either because the hypha stops growing completely or because it ceases to respond to the inhibitor. Microtubule inhibition tends to produce a less severe but more persistent disruption or disorientation of tip growth. The different effects are probably a reflection of the different localization and different roles that actin and microtubules play in vesicle traffic.

2. *Is tip growth initiated by a localized site of actin polymerization*? The strong correlation between actin presence and tip growth poses the question of whether a localized site of actin polymerization at the plasma membrane may be the initiator of tip growth. Such view, however, needs to be reconciled with other findings showing no changes in levels of actin or gene expression during extensive induction of growing points. Thus, in *Achlya ambisexualis*, Brunt et al. [66] reported that antheridiol increased transcription of the heat shock protein chaperones (Hsp90 and Hsp70 family), but there was no similar increase in the level of transcripts encoding actin even though 90% of hyphae in the hormone-treated thalli were undergoing antheridial branching. Likewise, Tinsley et al. [50] used the temperature-sensitive mutants *cot-1* and *mcb* of *N. crassa* to show that there was no increase in actin following a >20-fold increase in the number of hyphal tips. They suggested that the level of actin monomers within *N. crassa* hyphae is sufficient to accommodate the need for additional actin in the new tips. These two different studies provide evidence that actin reorganization, but not necessarily new synthesis, is needed to establish new growth zones. In turn, this is evidence that actin per se is not the initiator of hyphal growth but that something else in the incipient or future apex provides the signal for actin polymerization at the growing tip. Rho GTPases have been implicated in establishing cell polarity by controlling the localization of F-actin in budding of *Saccharomyces cerevisiae* [67] and in the apical growth of pollen tubes [68] and fungal hyphae [69].

3. *Do nuclear events compete with apical growth for microtubule resources*? Since microtubules are involved in both mitosis and tip growth, the question arose as to whether the two processes compete for the same pool of tubulin. This was found no to be the case [70]. Presumably, the population of cytoplasmic microtubules involved in apical growth operates independently of those involved in mitosis.

2.4 Turgor

2.4.1 Evolutionary and Ecological Significance

Thanks to the work and writings of Money and coworkers, the role of turgor in fungal biology has received a great deal of attention in recent years [37, 71–77].

The importance of turgor in growth, morphogenesis, pathogenesis, and ecology of fungi has been extensively discussed. As I noted earlier [78,79], the development of high turgor concomitant with the formation of microfibrillar cell walls of high tensile strength has been a central factor in the evolution of the fungal kingdom. In the same vein, Johns et al. [37] remarked that the evolution of the fungal cell wall and the generation of turgor pressure afforded by this structure are at the root of the ecological diversity of the filamentous fungi [37]. The enormous turgor that fungal cells can generate explains the penetrating physical power of fungal hyphae [80,81].

2.4.2 Role of Turgor in Hyphal Growth and Morphogenesis

Money and coworkers [37,72] are probably correct in challenging claims for turgor having a *controlling* role in hyphal growth and morphogenesis. The conclusions reached by Eamus and Jennings [82] and Gervais et al. [83] linking growth rate with internal pressure have been disputed [37]. Previously, Kaminskyj et al. [84] had also questioned that turgor was a determinant factor in growth rate. Probably, growth rate has far more to do with metabolic activity, wall loosening, and exocytosis than the degree of internal osmotic pressure [37].

Putting aside the controversy of turgor being a controlling agent of growth rate, a more basic question emerges. Is turgor essential for hyphal growth, i.e., does turgor provide the physical force needed to expand the cell wall? Here, Money is probably incorrect in doubting whether turgor fulfills this role and limiting the role of turgor to invasive growth [37,72]. His argument is based on some unique observations made by Harold et al. [71] on the ability of hyphae of *Saprolegnia ferax* to grow at high concentration of osmolytes without adjusting their turgor. The turgor of such hyphae fell below the limit of detection of the micropipet pressure probe used (0.02 MPa), i.e., 5% or less of the turgor in control hyphae. Despite the fact that the actual pressure of osmotically stressed hyphae could not be determined, these hyphae were prominently regarded as "turgorless" and held as reason to contradict the common belief that turgor supplies the driving force for hyphal extension. But the same article contains more measured conclusions that I believe paint the true picture. The role of turgor was not completely ruled out but rather conditioned: "If hydrostatic pressure plays a role it is one that can be filled by 3% of the normal turgor." Since hyphae grown at such high concentration of osmolytes are known to have weak cell walls [85], such low value of turgor may have been sufficient to expand the debilitated wall. In a parting answer to the question Is turgor required for extension of *Saprolegnia* hyphae?, they concluded: "In the extreme case when turgor is essentially 'zero' and the wall most plastic, the answer is no. But when turgor is high and the wall rigid, hydrostatic pressure may well be required to stress the wall allowing it to expand and admit new wall material." This last statement addresses the normal condition of fungal hyphae. But it is no longer necessary to dispute the validity

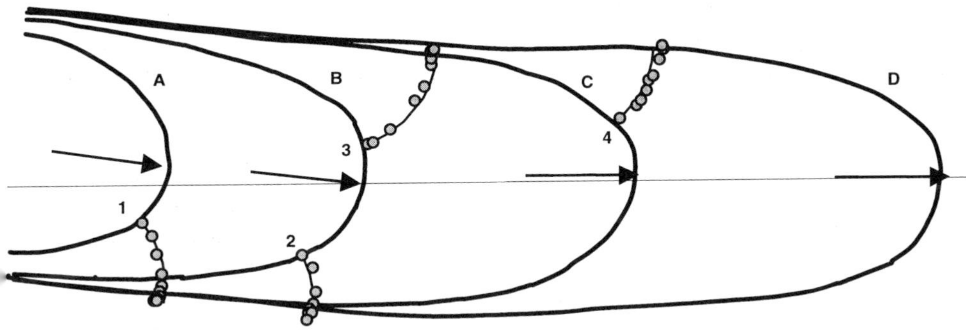

FIGURE 4 Orthogonal pattern of surface expansion in hyphal tip growth. Circles depict the trajectories of four carbon particles that became attached to the surface of an elongating hypha of *Rhizoctonia solani*. Growth was followed for 219 sec (D). Particle 1 was attached at 0 time (A), particles 2 and 3 after 58 sec (B), and particle 4 at 136 sec (C). The thin solid lines are the theoretical curves calculated for orthogonal displacement of each particle. Arrows show the growth axis for each tip profile. (From Ref. 86.)

of the claims on logic alone; experimental evidence showing that the wall expands orthogonally over the entire growth zone of a hyphal tip [86] (Fig. 4) leaves turgor as the only viable candidate to provide the physical force for wall expansion. But beyond a minimum threshold value to permit expansion, the magnitude of turgor has no role in the 3D model of hyphal morphogenesis [87]. Presumably, as Harold et al. [71] hinted, most of the turgor pressure in a growing fungus is in excess of that needed for normal hyphal growth.

As to hyphal morphology, until recently there was no experimental basis to suspect that turgor would have a morphogenetic role [72]. By forcing cell wall expansion to follow an orthogonal pattern, turgor modulates the 3D shape of a hypha [86]. However, turgor does not dictate the tubular shape of a hypha, and its overall effect on hyphal shape is relatively minor [87] compared to the major factor in hyphal morphogenesis, namely, the highly polarized distribution of cell wall building vesicles created by the Spitzenkörper (see Sec. 2.1).

2.4.3 Question

1. *If not turgor, what, then, controls cell wall extensibility?* This is one of the most fundamental questions in cell wall growth, particularly tip growth, for which there is no decisive answer. The extension of an inelastic cell wall of high tensile strength requires a plasticizing action that would yield to turgor to increase the surface area of the cell. Accordingly, the crucial issue is the need for a mechanism to coordinate synthetic and lytic processes so that the wall attains

an exact measure of controlled plasticity necessary for cell wall extension [79]. As I suggested earlier [88], the proper balance of wall synthesis and lysis in the growing regions of the wall may be established through a coordinated discharge of different types of vesicles carrying different ingredients for wall extension. In other words, the well-proven gradient of wall synthesis [89–91] needs to be accompanied by a parallel gradient of wall softening. So far, all evidence for the involvement of a softening, lysing, or plasticizing action is circumstantial and qualitative [75,92–94]. Here, molecular genetics could have a crucial impact by helping to elucidate and quantitate the fleeting but crucial biochemical process that softens the fungal cell wall to permit extensibility.

An alternative explanation for the control of wall extensibility assigns this role to the F-actin membrane skeleton underlying the apical wall (see Sec. 3.2). This explanation, however, fails to take into account that wall extensibility requires autolysis [92].

3 MODELS OF HYPHAL TIP GROWTH

Different models have been proposed to explain how the tubular cell of a hypha is generated by apical growth. Each model focuses on a different aspect of hyphal biology and invokes or emphasizes different morphogenetic criteria. Although each of the models or approaches listed below is usually presented to the exclusion of the others, in reality no single model provides a satisfactory answer. Each model explains but a portion of the mystery of the deceivingly simple process of hyphal tip growth.

3.1 The Hyphal Apex Is Shaped by Gradual Rigidification of the Cell Wall

The steady-state model of tip growth [3,95] is a refinement of previous ideas invoking changes in the physical properties of the wall to explain tip growth [96–98]. Accordingly, the hyphal tube is manufactured by a steady-state process that transforms and expands the newly deposited plastic wall of the apical dome into a rigid cylindrical wall at the base of the dome. The newly added wall material is plastic owing to the presence of individual polymer chains and the lack of crosslinkages. Vermeulen and Wessels [99,100] showed that the newly synthesized chitin at the hyphal tips was not microfibrillar and was highly susceptible to chitinase; this susceptibility diminished as the wall progressed from apex to subapex, and the chitin chains crystallized into microfibrils. As the nascent wall becomes progressively crosslinked by covalent bonds between β-1,3-glucan and chitin [101–102], it develops a greater resistance to turgor pressure.

3.1.1 Comments

Even though there is much merit in the discovery that the overall physical properties of the apical wall may change significantly because of crosslinks between β-1,3-glucan and chitin polymers, it is questionable whether any changes in rigidification of the apical wall would regulate its shape. The wall of the apical dome, even at its most plastic point, must be strong enough to resist deformation by the high turgor pressure of the cytoplasm; otherwise, the cell would explode under normal growth conditions. The steady-state model assumes that the apical dome adopts a hemiellipsoidal shape [95], but it is not evident why a plastic dome adopts this particular shape, and there is no quantitative formulation to correlate the gradual change in wall rigidity with the shape of the hyphal apex.

3.2 The Expanding Cytoplasm Molds the Hyphal Apex

In this model, favored by Heath and coworkers [2,7,57], the fungus is viewed primarily as a protozoon living inside a rigid tubular casing. Morphogenesis is thought to originate from the same elements that mold the shape of a wall-less protozoon, namely the underlying cytoskeleton. Specifically, in its more recent version, a scaffolding of F-actin [104] in conjunction with spectrin- and integrin-like components [105,106] forms a membrane skeleton associated with the inside of the apical plasma membrane. This membrane skeleton is believed to regulate tip extensibility and thus hyphal morphology [106]. There is some circumstantial evidence to support these ideas. The high concentration of actin found in hyphal tips of many fungi [47,48,50,52,107,108] indicates that actin must play a major role in apical growth. The evidence gathered by Degousee et al. [106] showed that F-actin is attached to a membrane skeleton that is rich in spectrinlike protein and also contains an integrinlike protein. This complex is concentrated in growing hyphal tips of *Neurospora crassa*.

3.2.1 Comments

The idea that actin has a direct physical role in shaping a hypha [41,57,104] runs contrary to new experimental findings showing that the apical wall of a hypha expands orthogonally, i.e., expansion is always perpendicular to the cell surface [86] (Fig. 4). There is no evidence that actin is deployed perpendicular to the cell surface over the entire growing region, which includes the apex and the neighboring subapex. The ever-present high force generated by the turgor of the cytoplasm does have the orientation, and the strength, to explain expansion of a walled cell. More likely, the high concentration of actin in the apex is an indicator of intense exocytosis and plays a crucial role in the transport of vesicles to the cell wall. Inhibition experiments with Latrunculin B are consistent with F-actin regulating polar vesicle delivery and controlling vesicle fusion at the plasma membrane [55].

Studies by Heath and coworkers [see reviews 2,7,57] have contributed greatly to our understanding of the fungal cytoskeleton and its role in fungal growth. Although knowledge is still too fragmented, there is little doubt that the cytoskeleton is intimately involved in the elongation of a hyphal cell. But the intense focus on the cytoskeleton relegates the cell wall to a secondary role, and downplays the importance of the ultimate defining process of fungal morphogenesis: the making of the cell wall. The importance of cell wall biogenesis cannot be underestimated. The evolution of fungi and their extraordinary ecological role were attained largely because eons ago fungi learned to construct microfibrillar cell walls to satisfy innumerable ecological challenges [73,78,79]. The conceptual approach to tip growth adopted in the membrane skeleton model and its earlier versions [2,57], was not strengthened by reviving 19th-century ideas equating a fungus with a tube-dwelling amoeba [7]. Such analogy may have been appropriate for the days when the cell wall was regarded merely as a hardened exudate of the cell, and when nothing was known about the chemical and structural complexity of the cell wall or the existence of a highly sophisticated secretory apparatus to construct the cell wall. By equating tip growth with pseudopodium formation [7], the membrane skeleton model disregards the crucial difference between an amoeba and a hypha, namely, the existence a polarized secretion process for building and shaping a tubular cell wall.

3.3 The Hyphal Apex Is Shaped by a Moving VSC

The VSC model described in Section 2.2.2 was based on an earlier qualitative model that postulated that hyphal shape is determined by a polarized pattern of distribution of vesicles involved in cell wall construction [88]. The model assumed that extension growth was the result of three concomitant actions: synthesis and deposition of new cell wall polymers, enzymic plasticizing action to loosen a basically rigid wall structure, and turgor to force cell wall expansion [88]. The basic premise of the qualitative model stipulates that the biochemical gradients needed for hyphal wall construction are created by a gradient of exocytosis. From these qualitative assumptions, we developed a mathematical model [30] that postulated that the spatial discharge of wall-building vesicles was centrally controlled by a VSC. The VSC model provides a plausible mechanism to explain how a continuous sharp gradient of vesicle-discharge can be generated by a growing hypha. By the simple action of moving forward while continuously releasing wall-destined vesicles, the Spitzenkörper could generate such gradient [30].

3.3.1 Comments

Aside from a VSC solution, Harold [5] considered two other alternative explanations for shaping the hyphal tip through exocytosis. One was the calcium hypothesis in which secretory vesicles carrying calcium channels merge with the plasma membrane creating an influx of calcium ions. The calcium gradient, known to be present in hyphal tips [109], would promote actin polymerization, vesicle fusion, and localized cell wall deposition. The other one was targeted exocytosis where vesicles are carried to marked fusion sites, an idea explored by Gupta and Heath [110]. Both alternatives represent viable mechanisms to support exocytosis, but neither one can explain the origin of the exocytosis gradient since they both depend on molecules that have to be carried to the surface by the vesicles themselves. One virtue of the VSC model is that it does not need preexisting signals or targets on the cell surface to initiate or maintain its operation. The control of morphology obtained by laser manipulation of the Spitzenkörper (Fig. 3; Sec. 2.2.3) supports an internal origin for the gradient of wall-building vesicles.

The VSC model was a deliberate exercise in physiological reductionism that attempted to extract the essence of cell wall biogenesis. Cell wall construction was reduced to its minimum expression: one vesicle discharge equals one unit of cell surface. The model does not address the complex interaction between wall synthesis and wall softening and, in its original 2D version, it did not need to address the role of turgor pressure (see below). Despite supporting circumstantial evidence (Sec. 2.2.2), the ultimate validity of the VSC hypothesis depends on demonstrating that the flow of wall-building vesicles passes through a Spitzenkörper control gate. Such traffic of vesicles in/out of the Spitzenkörper is yet to be demonstrated and measured. If the Spitzenkörper is proven unequivocally to be a VSC, it follows that the mechanism that advances the Spitzenkörper would be a key regulator of hyphal morphogenesis. Circumstantial evidence suggests that the microtubular cytoskeleton is involved [17], but one cannot rule out a complex interplay between various components of the cytoskeleton providing the propelling force [30].

A common criticism of the VSC model, variously voiced by Koch [111], Green [112], and Harold [5], was that it "dealt with the two-dimensional analogue of the hypha" and as such it accounted for cell substance and not for surface area. We had initially assumed that mere rotation of the 2D hyphoid model, along its longitudinal axis, would automatically supply a 3D model of hyphal morphogenesis [30], but Koch predicted correctly that the 2D VSC model would be indetermined in three dimensions and concluded that "the velocity of the VSC is not a sufficient condition to define the shape." Indeed, when an attempt was made to derive a 3D model based on the VSC concept, an indetermination was encountered whose solution required defining a priori the pattern of expansion of the wall, i.e., defining the overall spatial movement of the wall as the newly inserted wall elements displace the existing wall fabric [87]. The actual mode of wall expansion was determined experimentally and found to follow orthogonal trajectories [86] (Fig. 4), as depicted by Reinhardt in [113] 1892! Fortunately, the close similarity in profile between the 2D hyphoid and the 3D orthogonal hyphoid [87] validates the use of the simpler 2D model in morphogenetic studies described in Section 2.2.2. Although not claimed by the authors, the data presented by Shaw et al. [114] on the mode of surface growth of root hairs show clearly that orthogonal expansion applies to tip-growing cells of the plant world.

Harold [5] pointed out that the VSC model lacked a mechanism for the performance of physical work. Although we did not set out to deal with this issue, the mathematical impasse mentioned above compelled us to consider wall expansion patterns and, in turn, the physical forces behind them. We now have experimental reason to invoke turgor as the physical force that expands the wall of a hypha (see Sec. 2.4.2).

3.4 The Molecular Genetics Approach

A number of laboratories have taken advantage of the methodology of molecular genetics to unravel the genetic components involved in fungal tip growth. No major breakthrough has emerged, but steady progress has been made in detecting genes mainly on three several fronts: cytoskeleton, signal transduction, and polarized cell wall formation.

3.4.1 Cytoskeleton Genes

That many of the mutated genes responsible for distortions in hyphal morphogenesis belong to the cytoskeleton and associated molecules underscores the importance of the cytoskeleton. As already described in Section 2.3, mutants deficient in microtubule motor proteins (dynein and kinesin) show overall growth rate reduction, a smaller and often unstable Spitzenkörper, and distorted morphogenesis. Harris et al. [56] identified the genes *podB* and *sepA* as necessary for the organization of the actin cytoskeleton at sites of polarized growth in *Aspergillus nidulans*. With a conditionally null *myoA* strain of *Aspergillus nidulans*, McGoldrick et al. [115] showed that MYOA, a gene encoding an essential myosin I, is required for secretion and polarized growth.

Much work has focused on mutations disrupting nuclear distribution, mainly *nud* in *Aspergillus nidulans* [116] and *ro* in *Neurospora crassa* [49], which led to the identification of dynein as the major motor for nuclear migration in hyphae. The subject has been reviewed by Xiang and Morris [8], who conclude that the models proposed to explain nuclear migration are still controversial.

3.4.2 Signal Transduction

A variety of mutations point to the involvement of signal transduction pathways in hyphal morphogenesis. For example, studies by Feng et al. [117] with the *ras1-2/ras1-3* mutant of *Candida albicans* indicated that low-molecular-weight molecules in serum induce hyphal differentiation in *C. albicans* through a Ras-mediated pathway. Similarly, Truesdell et al. [118] found that a mutationally activated Ras homolog (*CT-Ras*) induced abnormal hyphal proliferation and defects in polarized growth, an indication that proper regulation of Ras is required for normal growth in *Colletotrichum trifolii*. Disruption of another RAS-related gene (*CaRSR1*) in *C. albicans* indicated its involvement in polarized growth: initiation of budding in yeast, germ tube emergence, and pseudohyphal elongation [119]. Alex et al. [120] found that *COS1*, a gene encoding a two-component histidine kinase, is involved in hyphal but not yeast morphogenesis. Mutants of *C. albicans* lacking both copies of *COS1* produced normal yeast cells but showed defective hyphal formation in response to nutrient deprivation or serum. Yarden and coworkers [121] reported that *cot-1*, a kinase-encoding gene required for hyphal cell elongation in *Neurospora crassa* hyphae, was photoregulated by blue light, an effect blocked by L-sorbose. These interactions indicate the involvement of alternative and potentially interdependent signaling pathways for the regulation of hyphal elongation/branching [122].

Examples of induction of polarized (hyphal and pseudohyphal) growth abound in the dimorphic and mating responses of fungi. The signaling pathways play a fundamental role connecting external signal with some internal metabolic or regulatory response [see review by Banuett, 123]. But in dissecting the ultimate basis of tip growth, it is important to keep in mind that the polarization of wall growth that leads to hyphal development is usually initiated internally with no obvious external input (Fig. 1).

3.4.3 Polarized Cell Wall Construction

Although the basic structure and major wall components of fungi are known, much remains to be learned about minor components of the wall, about the linkages among different components, and about the entire process of cell wall assembly. Molecular studies may help characterize proteinaceous components of the wall that play a key role in hyphal morphogenesis. For instance, Staab and Sundstrom [124] have cloned the complete hyphal wall protein 1 gene (*HWP1*) of *C. albicans*. The Hwp1 is a glucan-linked protein with serine/threonine-rich regions that are predicted to function in extending a ligand-binding domain into the extracellular space.

In search of genes involved directly in conversion from isotropic to tropic cell wall growth, i.e., the establishment of polarity, Momany's group [125] screened for temperature-sensitive swollen-cell mutants (*swo*) of *Aspergillus nidulans*. The screen yielded eight genes involved in polarity establishment, polarity maintenance, and hyphal morphogenesis. They concluded that *swo C, D,* and *F* are required to establish polarity and that *swoA* is required to maintain polarity. *swo B, E, G,* and *H* are involved later in hyphal morphogenesis. Their results suggest that polarity establishment and polarity maintenance are genetically separate events and that a persistent signal is required for apical extension in *A. nidulans*. Wendland and Philippsen [69] searched for molecular similarities between the onset of polarized growth in germinating spheres of *Ashbya gossypii* and bud emergence in unicellular yeastlike fungi. They found a common requirement of rho-GTPase modules for the establishment and maintenance of polarized hyphal growth. Most recently, Knechtle et al. [126] investigated the dynamics of polarized growth of *A. gossypii* with a polarity marker (AgSpa2p) homologous to the one in *Saccharomyces cerevisiae* (Spa2p). GFP-labeled AgSpa2p was sharply localized in hyphal tips of *A. gossypii* (The video microscopy images shown at the conference gave the distinct impression that the fluorescence of this polarity marker coincided with the usual position of the Spitzenkörper in a hyphal tip.)

Much of our knowledge on the cytology [127,128], biochemistry [129,130], and genetics of cell wall formation in fungi pertains to chitin synthesis. Chitin synthases are coded by a multigene family divided into at least five classes. This genetic multiplicity may be related to morphogenesis and pathogenesis as there is ample evidence for differential gene expression [131–135]; however, the redundancy of function complicates interpretation. Thus, two different *CHS* genes, *Umchs1* and *Umchs2*, were identified in *Ustilago maydis*. Transcripts of both genes appeared more abundant in the mycelial form, but both genes were found not to be essential [136]. Single-gene disruption and replacements of class I, II, or IV enzymes of various fungi including *Neurospora crassa* [137,138], *A. nidulans* [139–141], and *Ustilago maydis* [136,142] did not affect development. On the other hand, mutations affecting chitin synthases of class III of *Neurospora crassa* [143], *A. nidulans* [140,144], and *Aspergillus fumigatus* [145] did affect hyphal growth. Also, the chitin synthases with a myosin domain (class V) found in *A. nidulans* [146] and *Pyricularia oryzae* [147] appear to be important for the maintenance of hyphal wall integrity and the polarized synthesis of the cell wall.

To resolve the issue of whether the polarized growth of a hypha requires a specific kind of chitin synthase, it would be helpful to find out if the spatial redirection of chitin synthesis that occurs during polarization of wall growth (Fig. 1) requires expression of a specific chitin synthase gene or if the same population of chitosomes [148,149] involved in isotropic growth are simply rerouted to a polar destination.

3.4.4 Comment

The molecular approach is largely an attempt to identify individual genes and gene products that play a role in hyphal morphogenesis. A good number have already been implicated, but their relative relevance, hierarchy, and chronology are far from clear. Will molecular reductionism eventually succeed in explaining the basis of hyphal morphogenesis? Beyond identifying specific molecules affecting apical growth, can reductionism provide a coherent scheme that explains tip growth? Is polarized growth triggered by the action of a single gene setting up a cascade of metabolic events that establishes a polarized gradient of exocytosis?

Or do we need a holistic approach [150,151] to understand how a complex interplay of simultaneous events coalesce to trigger the onset of polarized growth?

3.5 Model Reconciliation

The aforementioned models of hyphal morphogenesis deal with structures and processes that occur in living hyphae, and all, or substantial parts of them, need to be considered in any final picture of hyphal development. Thus, a comprehensive model would include changes in physical properties of the wall (Sec. 3.1), the cytoskeleton to provide the scaffolding for vesicle dynamics (Sec. 3.2), and a VSC to orchestrate the exocytosis gradient (Sec. 3.3). Ultimately, none of the current models define the precise biochemical reaction(s) that confer(s) polarity to a hypha. It remains to be seen whether any of the genes so far identified (Sec. 3.4) would provide a biochemical answer to the ultimate question—the origin of polarity.

Despite firmly entrenched past postures, model reconciliation is possible and desirable. For instance, our VSC model and Wessels' steady-state model are not necessarily incompatible [152]; they account for different features of the wall-building process. A key conceptual difference between these models was in the role of lysins (i.e., wall-softening or wall-plasticizing enzymes). In the VSC model, lysin action is implicitly needed to permit wall extension. In the steady-state model, lysins were deliberately excluded; although Wessels [103] agreed that lysins were needed for wall growth, he limited the need for lytic action to the initiation of growth from a rigidified wall, such as in branching or spore germination. I contend that there is no basic difference between these processes: both initiation and continuation of apical growth require a steady supply of plasticizing agents. This contention is perhaps most vividly affirmed by the findings of Bracker et al. [35] during manipulation of the Spitzenkörper with a laser beam (Fig. 3). When a Spitzenkörper that was actively engaged in the elongation of a growing tip was forced away from its usual position in the apex, it would start deforming any wall adjacent to it. Clearly, the vesicles emanating from the Spitzenkörper have the power to render any wall region plastic. The inclusion of lysins in the steady-state model strengthens its key premise, namely, the progressive change in physical properties of the expanding wall. Lysin action would allow the growing wall, which is always a mixture of preexisting plus new wall, to be plasticized adequately. The newly deposited wall would be plastic by virtue of the nascent nature of the polymer chains and the lack of crosslinks, while the existing rigid wall would be rendered plastic by lytic action. Once plasticized, the wall would expand under the force of turgor. But such expansion needs to be discrete, or the wall would bulge out and eventually break. The expansion-limiting factors would be the rigidifying processes invoked in the steady-state model plus rapid inactivation of both lytic and synthesizing enzymes. The latter is supported by autoradiographic evidence showing that the apical accumulation of chitin synthetase disappears rapidly in the subapical region [153]. Similarly, lysin action must have an intrinsic short half-life to limit its plasticizing effect. A tandem VSC–steady-state model would embody the spatial and temporal controls needed for wall biogenesis. The VSC model would explain the *spatial* control of wall synthesis, while the steady-state model would account for the *temporal* control of wall extensibility.

REFERENCES

1. NAR Gow. Tip growth and polarity. In: NAR Gow, GM Gadd, eds. The Growing Fungus. London: Chapman & Hall, 1994, pp 277–297.
2. IB Heath. The cytoskeleton in hyphal growth, organelle movements, and mitosis. In: JGH Wessels, F Meinhardt, eds. The Mycota, Vol 1. Berlin: Springer-Verlag, 1994, pp 43–65.
3. JH Sietsma, JGH Wessels. Apical wall biogenesis. In: JGH Wessels, F Meinhardt, eds. The Mycota, Vol 1. Berlin: Springer-Verlag, 1994, pp 126–141.
4. S Bartnicki-García. The hypha: unifying thread of the fungal kingdom. In: B Sutton, ed. A Century of Mycology. Cambridge: Cambridge University Press, 1996, pp 105–133.
5. FM Harold. How hyphae grow: morphogenesis explained? Protoplasma 197:137–147, 1997.
6. FM Harold. In pursuit of the whole hypha. Fungal Genet Biol 27:128–133, 1999.
7. IB Heath, G Steinberg. Mechanisms of hyphal tip growth: tube dwelling amebae revisited. Fungal Genet Biol 28:79–93, 1999.
8. X Xiang, NR Morris. Hyphal tip growth and nuclear migration. Curr Opin Microbiol 2:636–640, 1999.
9. S Bartnicki-García, E Lippman. Polarization of cell wall synthesis during spore germination of *Mucor rouxii*. Exp Mycol 1:230–240, 1977.
10. H Brunswik 1924. Untersuchungen über Geschlechts und Kernverhältnisse bei der Hymenomyzetengattung *Coprinus*. In: K Goebel, ed. Botanische Abhandlungen. Jena: Gustav Fischer, pp 1–152.
11. M Girbardt. Der Spitzenkörper von *Polystictus versicolor*. Planta 50:47–59, 1957.
12. R Lopez-Franco, RJ Howard, CE Bracker. Satellite Spitzenkörper in growing hyphal tips. Protoplasma 188:85–103, 1995.
13. R Lopez-Franco, CE Bracker. Diversity and dynamics of the Spitzenkörper in growing hyphal tips of higher fungi. Protoplasma 195:90–111, 1996.
14. S Bartnicki-García, DD Bartnicki, G Gierz, R López-Franco, CE Bracker. Evidence that Spitzenkörper behavior determines the shape of a fungal hypha: a test of the hyphoid model. Exp Mycol 19:153–159, 1995.
15. CG Reynaga-Peña, G Gierz, S Bartnicki-García. Analysis of the role of the Spitzenkörper in fungal morphogenesis by computer simulation of apical branching in *Aspergillus niger*. Proc Natl Acad Sci USA 94:9096–9101, 1997.

16. CG Reynaga-Peña, S Bartnicki-García. Apical branching in a temperature sensitive mutant of *Aspergillus niger*. Fungal Genet Biol 22:153–167, 1997.
17. M Riquelme, CG Reynaga-Peña, G Gierz, S Bartnicki-García. What determines growth direction in fungal hyphae? Fungal Genet Biol 24:101–109, 1998.
18. M Riquelme, G Gierz, S Bartnicki-García. Dynein and dynactin deficiencies affect the formation and function of the Spitzenkörper and distort hyphal morphogenesis of *Neurospora crassa*. Microbiology UK 146:1743–1752, 2000.
19. MM Vargas, JM Aronson, RW Roberson. The cytoplasmic organization of hyphal tip cells in the fungus *Allomyces macrogynus*. Protoplasma 176:43–52, 1993.
20. UP Roos, G Turian. Hyphal tip organization in *Allomyces arbuscula*. Protoplasma 93:231–247, 1977.
21. YH Kwon, HC Hoch, JR Aist. Initiation of appressorium formation in *Uromyces appendiculatus*: organization of the apex, and the responses involving microtubules and apical vesicles. Can J Bot 69:2560–2573, 1991.
22. DP McDaniel, RW Roberson. Gamma tubulin is a component of the Spitzenkörper and centrosomes in hyphal tip cells of *Allomyces macrogynus*. Protoplasma 203: 118–123, 1998.
23. RJ Howard. Ultrastructural analysis of hyphal tip cell growth in fungi: Spitzenkörper, cytoskeleton and endomembranes after freeze-substitution. J Cell Sci 48: 89–103, 1981.
24. TM Bourett, RJ Howard. Ultrastructural immunolocalization of actin in a fungus. Protoplasma 163:199–202, 1991.
25. DP McDaniel, RW Roberson. Microtubules are required for vesicle and mitochondria motility and positioning in hyphal tip cells of *Allomyces macrogynus*. Fungal Genet Biol 31:233–244, 2000.
26. RJ Howard, JR Aist. Effects of MBC on hyphal tip organization, growth and mitosis of *Fusarium acuminatum*, and their antagonism by D2O. Protoplasma 92:195–210, 1977.
27. RJ Howard, JR Aist. Cytoplasmic microtubules and fungal morphogenesis: ultrastructural effects of methyl benzimidazole-2-ylcarbamate determined by freeze-substitution of hyphal tip cells. J Cell Biol 87:55–64, 1980.
28. LJ McKerracher, IB Heath. Cytoplasmic migration and intracellular organelle movements during tip growth of fungal hyphae. Exp Mycol 11:79–100, 1987.
29. G Steinberg. Organelle transport and molecular motors in fungi. Fungal Genet Biol 24:161–177, 1998.
30. S Bartnicki-García, F Hergert, G Gierz. Computer simulation of fungal morphogenesis and the mathematical basis for hyphal tip growth. Protoplasma 153:46–57, 1989.
31. S Bartnicki-García. Role of vesicles in apical growth and a new mathematical model of hyphal morphogenesis. In: IB Heath, ed. Tip Growth in Plant and Fungal Cells. San Diego: Academic Press, 1990, pp 211–232.
32. DD Bartnicki, G Gierz, S Bartnicki-García. ''Fungus Simulator'': a Windows application to model fungal morphogenesis. Abstracts of the 5th Int Mycol Congress, Vancouver, Canada, 1994, p 12.
33. S Bartnicki-García, DD Bartnicki, G Gierz. Determinants of fungal cell wall morphology: the vesicle supply center. Can J Bot 73:S372–S378, 1995.

34. EJ Bottone, N Nagarsheth, K Chiu. Evidence of self-inhibition by filamentous fungi accounts for unidirectional hyphal growth in colonies. Can J Microbiol 44:390–393, 1998.
35. CE Bracker, DJ Murphy, R López-Franco. Laser microbeam manipulation of cell morphogenesis in growing fungal hyphae. In: DL Farkas BJ Tromberg, eds. Functional Imaging and Optical Manipulation of Living Cells. Bellingham, WA: International Society for Optical Engineering (Proceedings of SPIE Vol 2983), 1997, pp 67–80.
36. R Lopez-Franco, S Bartnicki-García, CE Bracker. Pulsed growth of fungal hyphal tips. Proc Natl Acad Sci USA 91:12228–12232, 1994.
37. S Johns, CM Davis, NP Money. Pulses in turgor pressure and water potential: resolving the mechanics of hyphal growth. Microbiol Res 154:225–231, 1999.
38. M Riquelme, G Gierz, S Bartnicki-García. Comparative morphometric analysis of lateral and apical branching in hyphae of *Neurospora crassa*. Proceedings of the XXI Fungal Genetics Conference, Asilomar, California, 2001, p 46.
39. CM Regalado. Roles of calcium gradients in hyphal tip growth: a mathematical model. Microbiology UK 144:2771–2782, 1998.
40. SN Grove, CE Bracker. Protoplasmic organization of hyphal tips among fungi: vesicles and Spitzenkörper. J Bacteriol 104:989–1009, 1970.
41. IB Heath, G Gupta, S Bai. Plasma membrane-adjacent actin filaments, but not microtubules, are essential for both polarization and hyphal tip morphogenesis in *Saprolegnia ferax* and *Neurospora crassa*. Fungal Genet Biol 30:45–62, 2000.
42. J Hoffmann, K Mendgen. Endocytosis and membrane turnover in the germ tube of *Uromyces fabae*. 24:77–85, 1998.
43. S Fischer-Parton, RM Parton, PC Hickey, J Dijksterhuis, HA Atkinson, ND Read. Confocal microscopy of FM4-64 as a tool for analysing endocytosis and vesicle trafficking in living fungal hyphae. J Microsc (Oxford) 198:246–259, 2000.
44. CL Gordon, DB Archer, DJ Jeenes, JH Doonan, B Wells, APJ Trinci, GD Robson. A Glucoamylase :: GFP gene fusion to study protein secretion by individual hyphae of *Aspergillus niger*. J Microbiol Methods 42:39–48, 2000.
45. CL Gordon, V Khalaj, AFJ Ram, DB Archer, JL Brookman, APJ Trinci, D Jeenes J, JH Doonan, B Wells, PJ Punt, CAMJJ van den Hondel, GD Robson. Glucoamylase :: green fluorescent protein fusions to monitor protein secretion in *Aspergillus niger*. Microbiology UK 146:415–426, 2000.
46. X Xiang, GS Han, DA Winkelmann, WQ Zuo, NR Morris. Dynamics of cytoplasmic dynein in living cells and the effect of a mutation in the dynactin complex actin-related protein Arp1. Curr Biol 10:603–606, 2000.
47. IB Heath. Preservation of a labile cortical array of actin filaments in growing hyphal tips of the fungus *Saprolegnia ferax*. Eur J Cell Biol 44:10–16, 1987.
48. J Hasek, S Bartnicki-García. The arrangement of F-actin and microtubules during germination of *Mucor rouxii* sporangiospores. Arch Microbiol 161:363–369, 1994.
49. KS Bruno, JH Tinsley, PF Minke, M Plamann. Genetic interactions among cytoplasmic dynein, dynactin, and nuclear distribution mutants of *Neurospora crassa*. Proc Natl Acad Sci USA 93:4775–4780, 1996.
50. JH Tinsley, IH Lee, PF Minke, M Plamann. Analysis of actin and actin-related

protein 3 (ARP3) gene expression following induction of hyphal tip formation and apolar growth in *Neurospora*. Mol Gen Genet 259:601–609, 1998.
51. HC Hoch, RC Staples. The microtubule cytoskeleton in hyphae of *Uromyces phaseoli* germlings: its relationship to the region of nucleation and to the F-actin cytoskeleton. Protoplasma 124:112–122, 1985.
52. RW Roberson. The actin cytoskeleton in hyphal cells of *Sclerotium rolfsii*. Mycologia 84:41–51, 1992.
53. S Torralba, M Raudaskoski, AM Pedregosa, F Laborda. Effect of cytochalasin A on apical growth, actin cytoskeleton organization and enzyme secretion in *Aspergillus nidulans*. Microbiology UK 144:45–53, 1998.
54. GD Gupta, IB Heath. Actin disruption by latrunculin B causes turgor-related changes in tip growth of *Saprolegnia ferax* hyphae. Fungal Genet Biol 21:64–75, 1997.
55. C Bachewich, IB Heath. Radial F-actin arrays precede new hypha formation in *Saprolegnia*. Implications for establishing polar growth and regulating tip morphogenesis. J Cell Sci 111:2005–2016, 1998.
56. SD Harris, AF Hofmann, HW Tedford, MP Lee. Identification and characterization of genes required for hyphal morphogenesis in the filamentous fungus *Aspergillus nidulans*. Genetics 151:1015–1025, 1999.
57. IB Heath. The roles of actin in tip growth of fungi. Int Rev Cytol 123:95–127, 1990.
58. M Plamann, PE Minke, JH Tinsley, KS Bruno. Cytoplasmic dynein and actin-related protein arp1 are required for normal nuclear distribution in filamentous fungi. J Cell Biol 127:139–149, 1994.
59. C Lehmler, G Steinberg, KM Snetselaar, M Schliwa, R Kahmann, M Bolker. Identification of a motor protein required for filamentous growth in *Ustilago maydis*. EMBO J 16:3464–3473, 1997.
60. S Seiler, FE Nargang, G Steinberg, M Schliwa. Kinesin is essential for cell morphogenesis and polarized secretion in *Neurospora crassa*. EMBO J 16:3025–3034, 1997.
61. S Seiler, M Plamann, M Schliwa. Kinesin and dynein mutants provide novel insights into the roles of vesicle traffic during cell morphogenesis in *Neurospora*. Curr Biol 9:779–785, 1999.
62. S Inoue, BG Turgeon, OC Yoder, JR Aist. Role of fungal dynein in hyphal growth, microtubule organization, spindle pole body motility and nuclear migration. J Cell Sci 111:1555–1566, 1998.
63. Q Wu, TM Sandrock, BG Turgeon, OC Yoder, SG Wirsel, JR Aist. A fungal kinesin required for organelle motility, hyphal growth, and morphogenesis. Mol Biol Cell 9:89–101, 1998.
64. SN Grove, JA Sweigard. Cytochalasin A inhibits spore germination and hyphal tip growth in *Gilbertella persicaria*. Exp Mycol 4:239–250, 1980.
65. SN Grove, JA Sweigard. Apical cytoplasmic organization and polar growth in *Rhizoctonia solani* are modified by cytochalasins. http://epress.com/w3jbio/growve/grove.htm, 1996.
66. SA Brunt, M Borkar, JC Silver. Regulation of hsp90 and hsp70 genes during antheridiol-induced hyphal branching in the oomycete *Achlya ambisexualis*. Fungal Genet Biol 24:310–324, 1998.

67. M Ziman, JM Obrien, LA Ouellette, WR Church, DI Johnson. Mutational analysis of CDC42Sc, a *Saccharomyces* gene that encodes a putative GTP-binding protein involved in the control of cell polarity. Mol Cell Biol 11:3537–3544, 1991.
68. YK Lin, YL Wang, JK Zhu, ZB Yang. Localization of a Rho GTPase implies a role in tip growth and movement of the generative cell in pollen tubes. Plant Cell 8:293–303, 1996.
69. J Wendland, P Philippsen. Determination of cell polarity in germinated spores and hyphal tips of the filamentous ascomycete *Ashbya gossypii* requires a rhoGAP homolog. J Cell Sci 113:1611–1621, 2000.
70. M Riquelme, R Fischer, S Bartnicki-García. Apical growth and mitosis do not compete for the same microtubule resources in *Aspergillus nidulans*. Proceedings of the XXI Fungal Genetics Conference, Asilomar, California, 2001, p 46.
71. RL Harold, NP Money, FM Harold. Growth and morphogenesis in *Saprolegnia ferax*: is turgor required? Protoplasma 191:105–114, 1996.
72. NP Money. Wishful thinking of turgor revisited: the mechanics of fungal growth. Fungal Genet Biol 21:173–187, 1997.
73. NP Money. On the origin and functions of hyphal walls and turgor pressure. Mycol Res 103:1360, 1999.
74. NP Money. To perforate a leaf of grass. Fungal Genet Biol 28:146–147, 1999.
75. NP Money, TW Hill. Correlation between endoglucanase secretion and cell wall strength in oomycete hyphae: implications for growth and morphogenesis. Mycologia 89:777–785, 1997.
76. NP Money, TC Caesar-TonThat, B Frederick, JM Henson. Melanin synthesis is associated with changes in hyphopodial turgor, permeability, and wall rigidity in *Gaeumannomyces graminis* var. *graminis*. Fungal Genet Biol 24:240–251, 1998.
77. L Brush, NP Money. Invasive hyphal growth in *Wangiella dermatitidis* is induced by stab inoculation and shows dependence upon melanin biosynthesis. Fungal Genet Biol 28:190–200, 1999.
78. S Bartnicki-García. Kingdoms with walls. In: WM Dugger, S Bartnicki-García, eds. Structure, Function and Biosynthesis of Plant Cell Walls. Rockville, MD: American Society of Plant Physiologists, 1984, pp 1–18.
79. S Bartnicki-García. The cell wall: a crucial structure in fungal evolution. In: ADM Rayner, CM Brasier, D Moore, eds. Evolutionary Biology of the Fungi. Cambridge: Cambridge University Press, 1987, pp 389–403.
80. RJ Howard, MA Ferrari, DH Roach, NP Money. Penetration of hard substrates by a fungus employing enormous turgor pressures. Proc Natl Acad Sci USA 88:11281–11284, 1991.
81. NP Money. Turgor pressure and the mechanics of fungal penetration. Can J Bot 73(suppl 1):S96–S102, 1995.
82. D Eamus, DH Jennings. Turgor and fungal growth: studies on water relations of mycelium of *Serpula lacrimans* and *Phallus impudicus*. Trans Br Mycol Soc 86: 527–535, 1986.
83. P Gervais, C Abadie, P Molin. Fungal turgor pressure is directly involved in the hyphal growth rate. Microbiol Res 154:81–87, 1999.
84. SGW Kaminskyj, A Garrill, IB Heath. The relation between turgor and tip growth

in *Saprolegnia ferax*—turgor is necessary, but not sufficient to explain apical extension rates. Exp Mycol 16:64–75, 1992.
85. NP Money, FM Harold. Extension growth of the water mold *Achlya*—interplay of turgor and wall strength. Proc Natl Acad Sci USA 89:4245–4249, 1992.
86. S Bartnicki-García, CE Bracker, G Gierz, R López-Franco, H Lu. Mapping the growth of fungal hyphae: orthogonal cell wall expansion during tip growth and the role of turgor. Biophys J 79:2382–2390, 2000.
87. G Gierz, S Bartnicki-García. A three-dimensional model of fungal morphogenesis based on the vesicle supply center concept. J Theor Biol 208:151–164, 2001.
88. S Bartnicki-García. Fundamental aspects of hyphal morphogenesis. In: JM Ashworth, JE Smith, eds. Microbial Differentiation. Cambridge, UK: Cambridge University Press, 1973, pp 245–267.
89. S Bartnicki-García, E Lippman. Fungal morphogenesis: cell wall construction in *Mucor rouxii*. Science 165:302–304, 1969.
90. GW Gooday. An autoradiographic study of hyphal growth of some fungi. J Gen Microbiol 67:125–133, 1971.
91. M Fèvre, M Rougier. Autoradiographic study of hyphal cell wall synthesis of *Saprolegnia*. Arch Microbiol 131:212–215, 1982.
92. S Bartnicki-García, E Lippman. The bursting tendency of hyphal tips of fungi: presumptive evidence for a delicate balance between wall synthesis and wall lysis in apical growth. J Gen Microbiol 73:487–500, 1972.
93. GW Gooday, NAR Gow. Enzymology of tip growth in fungi. In: IB Heath, ed. Tip Growth in Plant and Fungal Cells. San Diego: Academic Press, 1990, pp 31–58.
94. DM Rast, M Horsch, R Furter, GW Gooday. A complex chitinolytic system in exponentially growing mycelium of *Mucor rouxii*: properties and function. J Gen Microbiol 137:2797–2810, 1991.
95. JGH Wessels. Cell wall synthesis in apical hyphal growth. Int Rev Cytol 104:37–79, 1986.
96. NF Robertson. Presidential address: the fungal hypha. Trans Br Mycol Soc 48:1–8, 1965.
97. D Park, PM Robinson. Aspects of hyphal morphogenesis in fungi. In: EG Cutter, ed. Trends in Plant Morphogenesis. London: Longmans, Green & Co., 1966, pp 27–44.
98. PT Saunders, APJ Trinci. Determination of tip shape in fungal hyphae. J Gen Microbiol 110:469–473, 1979.
99. CA Vermeulen, JGH Wessels. Ultrastructural differences between wall apices of growing and non-growing hyphae of *Schizophyllum commune*. Protoplasma 120: 123–131, 1984.
100. CA Vermeulen, JGH Wessels. Chitin biosynthesis by a fungal membrane preparation: evidence for a transient noncrystalline state of chitin. Eur J Biochem 158: 411–415, 1986.
101. JH Sietsma, JGH Wessels. Evidence for covalent linkages between chitin and β-glucan in a fungal wall. J Gen Microbiol 114:99–108, 1979.
102. JH Sietsma, JGH Wessels. Solubility of (1→3)-β-D(1→6)-β-D-glucan in fungal walls: importance of presumed linkage between glucan and chitin. J Gen Microbiol 125:209–212, 1981.

103. JGH Wessels, Apical hyphal wall extension. Do lytic enzymes play a role? In: C Nombela, ed. Microbial Cell Wall Synthesis and Autolysis. Amsterdam: Elsevier Science Publishers, 1984, pp 31–42.
104. SL Jackson, IB Heath. Evidence that actin reinforces the extensible hyphal apex of the oomycete *Saprolegnia ferax*. Protoplasma 157:144–153, 1990.
105. SGW Kaminskyj, IB Heath. Integrin and spectrin homologues, and cytoplasm-wall adhesion in tip growth. J Cell Sci 108:849–856, 1995.
106. N Degousee, GD Gupta, RR Lew, IB Heath. A putative spectrin-containing membrane skeleton in hyphal tips of *Neurospora crassa*. Fungal Genet Biol 30:33–44, 2000.
107. E Temperli, U Roos, HR Hohl. Actin and tubulin cytoskeletons in germlings of the oomycete fungus *Phytophthora infestans*. Eur J Cell Biol 53:75–88, 1990.
108. K Yokoyama, H Kaji, K Nishimura, M Miyaji. The role of microfilaments and microtubules in apical growth and dimorphism of *Candida albicans*. J Gen Microbiol 136:1067–1075, 1990.
109. GJ Hyde, IB Heath. Ca^{+2} gradients in hyphae and branches of *Saprolegnia ferax*. Fungal Genet Biol 21:238–251, 1997.
110. GD Gupta, IB Heath. A tip-high gradient of a putative plasma membrane SNARE approximates the exocytotic gradient in hyphal apices of the fungus *Neurospora crassa*. Fungal Genet Biol 29:187–199, 2000.
111. AL Koch. The problem of hyphal growth in streptomycetes and fungi. J Theor Biol 171:137–150, 1994.
112. PB Green. Morphogenesis In: FC Bidwell, RGS Steward, eds. Plant Physiology. Growth and Development, Vol X. San Diego: Academic Press, 1991, pp 1–64.
113. MO Reinhardt. Das Wachsthum der Pilzhyphen. Jahrb Wissenschaft Bot 23:479–566, 1892.
114. SL Shaw, J Dumais, SR Long. Cell surface expansion in polarly growing root hairs of *Medicago truncatula*. Plant Physiol 124:959–970, 2000.
115. CA McGoldrick, C Gruver, GS May. *myoA* of *Aspergillus nidulans* encodes an essential myosin I required for secretion and polarized growth. J Cell Biol 128: 577–587, 1995.
116. YH Chiu, X Xiang, AL Dawe, NR Morris. Deletion of nudC, a nuclear migration gene of *Aspergillus nidulans*, causes morphological and cell wall abnormalities and is lethal. Mol Biol Cell 8:1735–1749, 1997.
117. Q Feng, E Summers, B Guo, G Fink. Ras signaling is required for serum-induced hyphal differentiation in *Candida albicans*. J Bacteriol 181:6339–6346, 1999.
118. GM Truesdell, C Jones, T Holt, G Henderson, MB Dickman. A Ras protein from a phytopathogenic fungus causes defects in hyphal growth polarity, and induces tumors in mice. Mol Gen Genet 262:46–54, 1999.
119. L Yaar, M Mevarech, Y Koltin. A *Candida albicans* RAS-related gene (CaRSR1) is involved in budding, cell morphogenesis and hypha development. Microbiology UK 143:3033–3044, 1997.
120. LA Alex, C Korch, CP Selitrennikoff, MI Simon. COS1, a two-component histidine kinase that is involved in hyphal development in the opportunistic pathogen *Candida albicans*. Proc Natl Acad Sci USA 95:7069–7073, 1998.

121. O Yarden, M Plamann, DJ Ebbole, C. Yanofsky. *cot-1*, a gene required for hyphal elongation in *Neurospora crassa*, encodes a protein kinase. EMBO J 11:2159–2166, 1992.
122. FR Lauter, U Marchfelder, VEA Russo, CT Yamashiro, E Yatzkan, O Yarden. Photoregulation of cot-1, a kinase-encoding gene involved in hyphal growth in *Neurospora crassa*. Fungal Genet Biol 23:300–310, 1998.
123. F Banuett. Signalling in the yeasts: an informational cascade with links to the filamentous fungi. Microbiol Mol Biol Rev 62:249–274, 1998.
124. JF Staab, P Sundstrom. Genetic organization and sequence analysis of the hypha-specific cell wall protein gene HWP1 of *Candida albicans*. Yeast 14:681–686, 1998.
125. M Momany, PJ Westfall, G Abramowsky. *Aspergillus nidulans* swo mutants show defects in polarity establishment, polarity maintenance and hyphal morphogenesis. Genetics 151:557–567, 1999.
126. P Knechtle, J Wendland, P Philippsen. Control of polarity in the filamentous fungus *Ashbya gossypii*. Proceedings of the XXI Fungal Genetics Conference, Asilomar, California, 2001, p 58.
127. CA Leal-Morales, CE Bracker, S Bartnicki-García. Localization of chitin synthetase in cell-free homogenates of *Saccharomyces cerevisiae*: chitosomes and plasma membrane. Proc Natl Acad Sci USA 85:8516–8520, 1988.
128. CA Leal-Morales, CE Bracker, S Bartnicki-García. Distribution of chitin synthetase and various membrane marker enzymes in chitosomes and other organelles of the slime mutant of *Neurospora crassa*. Exp Mycol 18:168–179, 1994.
129. R Sentandreu, S Mormeneo, J Ruiz-Herrera. Biogenesis of the fungal cell wall. In: JGH Wessels, F Meinhardt, eds. The Mycota, Vol 1. Berlin: Springer-Verlag, 1994, pp 111–124.
130. E Cabib, JA Shaw, PC Mol, B Bowers, WJ Choi. Chitin biosynthesis and morphogenetic processes. In: R Brambl, GA Marzluf, eds. The Mycota, Vol 3. Berlin: Springer-Verlag, 1996, pp 243–267.
131. L Lanfranco, M Vallino, P Bonfante. Expression of chitin synthase genes in the arbuscular mycorrhizal fungus *Gigaspora margarita*. New Phytol 142:347–354, 1999.
132. M Fujiwara, M Ichinomiya, T Motoyama, H Horiuchi, A Ohta, M Takagi. Evidence that the *Aspergillus nidulans* class I and class II chitin synthase genes, *chsC* and *chsA*, share critical roles in hyphal wall integrity and conidiophore development. J Biochem 127:359–366, 2000.
133. MA Lopez-Matas, AP Eslava, JM Diaz-Minguez. Mcchs1, a member of a chitin synthase gene family in *Mucor circinelloides*, is differentially expressed during dimorphism. Curr Microbiol 40:169–175, 2000.
134. GA Nino-Vega, CA Munro, G San-Blas, GW Gooday, NAR Gow. Differential expression of chitin synthase genes during temperature-induced dimorphic transitions in *Paracoccidioides brasiliensis*. Med Mycol 38:31–39, 2000.
135. Z Wang, PJ Szaniszlo. WdCHS3, a gene that encodes a class III chitin synthase in *Wangiella (Exophiala) dermatitidis*, is expressed differentially under stress conditions. J Bacteriol 182:874–881, 2000.
136. B Xoconostle-Cazares, C Leon-Ramirez, J Ruiz-Herrera. Two chitin synthase genes from *Ustilago maydis*. Microbiology UK 142:377–387, 1996.

137. AB Din, O Yarden. The *Neurospora crassa chs-2* gene encodes a non-essential chitin synthase. Microbiology UK 140:2189–2197, 1994.
138. AB Din, CA Specht, PW Robbins, O Yarden. *chs 4*, a class IV chitin synthase gene from *Neurospora crassa*. Mol Gen Genet 250:214–222, 1996.
139. CA Specht, YL Liu, PW Robbins, CE Bulawa, N Iartchouk, KR Winter, PJ Riggle, JC Rhodes, CL Dodge, DW Culp, PT Borgia,. The *chsD* and *chsE* genes of *Aspergillus nidulans* and their roles in chitin synthesis. Fungal Genet Biol 20:153–167, 1996.
140. K Yanai, N Kojima, N Takaya, H Horiuchi, A Ohta, M Takagi. Isolation and characterization of two chitin synthase genes from *Aspergillus nidulans*. Biosci Biotechnol Biochem 58:1828–1835, 1994.
141. T Motoyama, N Kojima, H Horiuchi, A Ohta, M Takagi. Isolation of a chitin synthase gene (chsC) of *Aspergillus nidulans*. Biosci Biotechnol Biochem 58:2254–2257, 1994.
142. SE Gold, JW Kronstad. Disruption of two genes for chitin synthase in the phytopathogenic fungus *Ustilago maydis*. Mol Microbiol 11:897–902, 1994.
143. Yarden, C Yanofsky. Chitin synthase 1 plays a major role in cell wall biogenesis in *Neurospora crassa*. Genes Dev 5:2420–2430, 1991.
144. PT Borgia, N Iartchouk, PJ Riggle, KR Winter, Y Koltin, CE Bulawa. The *chsb* gene of *Aspergillus nidulans* is necessary for normal hyphal growth and development. Fungal Genet Biol 20:193–203, 1996.
145. E Mellado, A Aufauvre-Brown, NAR Gow, DW Holden. The *Aspergillus fumigatus chsC* and *chsG* genes encode class III chitin synthases with different functions. Mol Microbiol 20:667–679, 1996.
146. M Fujiwara, H Horiuchi, A Ohta, M Takagi. A novel fungal gene encoding chitin synthase with a myosin motor-like domain. Biochem Biophys Res Commun 236: 75–78, 1997.
147. IC Park, H Horiuchi, CW Hwang, WH Yeh, A Ohta, JC Ryu, M Takagi. Isolation of *csm1* encoding a class V chitin synthase with a myosin motor-like domain from the rice blast fungus, *Pyricularia oryzae*. FEMS Microbiol Lett 170:131–139, 1999.
148. CE Bracker, J Ruiz-Herrera, S Bartnicki-García. Structure and transformation of chitin synthetase particles (chitosomes) during microfibril synthesis in vitro. Proc Natl Acad Sci USA 73:4570–4574, 1976.
149. JH Sietsma, AB Din, V Ziv, KA Sjollema, O Yarden. The localization of chitin synthase in membranous vesicles (chitosomes) in *Neurospora crassa*. Microbiology UK 142:1591–1596, 1996.
150. S Bartnicki-García. Biochemical events and controls: discussant's introduction. In: G Turian, HR Hohl, eds. The Fungal Spore: Morphogenetic Controls. London: Academic Press, 1981, pp 457–461.
151. FM Harold. To shape a cell: an inquiry into the causes of morphogenesis of microorganisms. Microbiol Rev 54:381–431, 1990.
152. S Bartnicki-García. Glucans, walls, and morphogenesis: on the contributions of JGH Wessels to the golden decades of fungal physiology and beyond. Fungal Genet Biol 27:119–127, 1999.
153. I McMurrough, A Flores-Carreon, S Bartnicki-García. Pathway of chitin synthesis and cellular localization of chitin synthetase in *Mucor rouxii*. J Biol Chem 246: 3999–4007, 1971.

3

Conidiation in *Aspergillus nidulans*

Reinhard Fischer
Philipps-Universität Marburg and Max-Planck-Institut für Terrestrische Mikrobiologie, Marburg/Lahn, Germany

1 INTRODUCTION

Spore formation is a common mechanism among most fungi to reproduce, to spread in the environment, or to survive unfavorable conditions. For many pathogenic fungi, such as *Aspergillus fumigatus* or rust fungi spores are the propagule to infect host organisms. In addition, every year thousands of tons of stored agricultural products are contaminated with spores of saprophytic fungi, which cause dramatic losses of feed and food. Therefore, the understanding of sporulation is an important issue with the aim to control spreading of fungi and to prevent infections. In addition, differentiation of reproductive structures in filamentous fungi is a fascinating process. The principal question is how a complex structure develops from an omnipotent vegetative cell. Since the end product of *A. nidulans* asexual development is rather simple, this mold serves as an excellent model to unravel the basic processes of temporal and spatial differential gene expression.

Conidiation in *A. nidulans* has been studied for many years, and knowledge of the molecular regulation has greatly extended throughout the last decade. The power of mutant analysis in combination with sophisticated molecular biological methods has led to good progress in understanding the signals leading to the initiation of the differentiation process, the transmittance of the signals into cellular actions, and the understanding of the cellular processes underlying development.

A. nidulans was first described by Eidam 1883 and the strain that is used today in most laboratories was isolated in Glasgow by Pontecorvo [35,89]. From the Glasgow laboratory this mold started its successful travel into many laboratories worldwide. *A. nidulans* grows well on artificial media at 37°C. Developmental processes are easily visible on an agar plate and can be analyzed with the light microscope but also with the electron microscope for more details [77,86]. From the genetic point of view, *A. nidulans* is amenable to mutagenesis experiments because the conidiospores contain only one haploid nucleus. However, diploid strains can be generated and also propagated through conidiospores [50]. This allows the study of essential genes and dominance assays. In contrast to many other molds, *A. nidulans* is able to reproduce in a sexual manner (Fig. 1). Although it is a homothallic ascomycetous fungus, the sexual cycle allows to set up crosses among different strains. This is especially convenient because *A. nidulans* does not possess any true mating type. Hundreds of different mutants

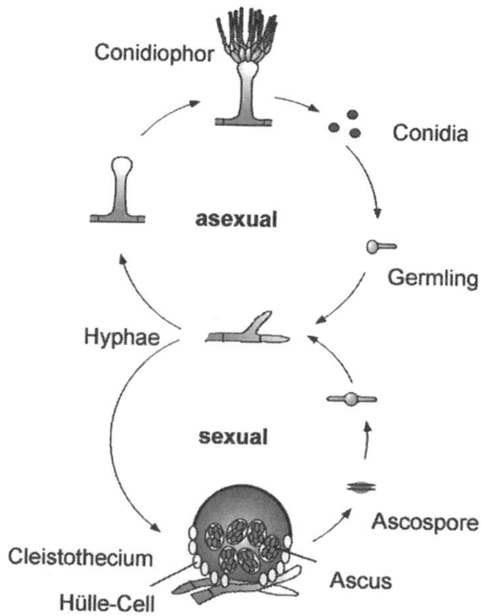

FIGURE 1 Life cycle of *Aspergillus nidulans*. A conidium germinates and produces the vegetative mycelium. Competent mycelium can undergo the asexual developmental cycle and generate conidiophores or enter the sexual cycle in which cleistothecia are formed.

have been generated over time and subsequently been mapped to one of the eight linkage groups (http://aspergillus-genomics.org) [91].

One great advantage of *A. nidulans* as a model for developmental processes is that conidiation can be synchronized. If conidia are germinated and grown in liquid culture, they do not produce conidiophores until they are exposed to an air interphase. Experimentally one can filter vegetative hyphae and place the filter on top of an agar surface. This will lead to an immediate, synchronous initiation of development, and mycelium of identical developmental stage can thus be harvested and analyzed, e.g., for gene expression. The process is called induction. For detailed analyses of cellular processes in *A. nidulans* it is important that since almost 20 years it is possible to introduce DNA into the mold to construct genetically engineered strains [111,120]. The combination of different classical and modern molecular biological methods allows an analysis of the differentiation process at all levels.

2 THE CONIDIOPHORE—A SIMPLE SPORE-FORMING STRUCTURE?

Conidiation in *A. nidulans* is initiated from a thick-walled hyphal cell, the foot. This cell extends into the air and produces a stalk 50–70 μm in length before it swells terminally to form the vesicle (Fig. 2). In a buddinglike process up to 70 metulae are formed on top of the vesicle. Those cells themselves generate two to three phialides, the spore-forming cells. The phialides repeatedly undergo mitosis and provide the emerging conidia with a single nucleus each. Therefore the youngest conidium is located close to the sterigmatum and the oldest at the tip of a conidia chain. Gradually conidia become dark green in color. The conidio-

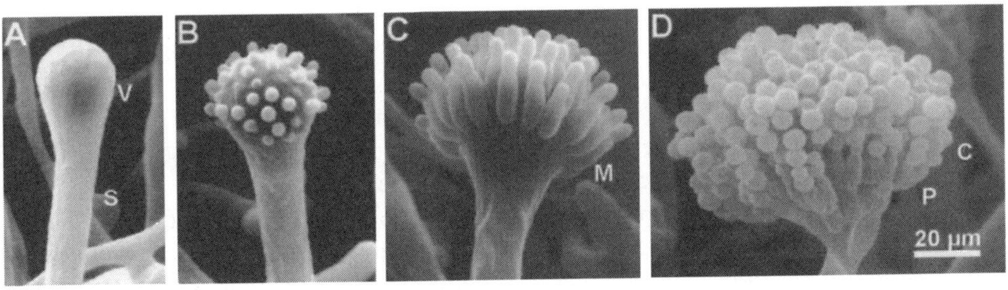

FIGURE 2 Conidiophore development as observed in the scanning electron microscope. An aerial hypha (S = stalk) swells terminally to a vesicle (V) (**A**), which nearly synchronously form metulae (M) in a buddinglike process (**B, C**). Metulae produce two to three phialides (P), which continuously generate conidia (C) (**D**). (From Refs. 39, 51.)

phore, a relatively simple, reproductive structure, consists only of four different cell types—the foot with the stalk, the metulae, the phialides, and the conidia. They are all easily distinguishable in the light microscope and thus abnormal morphologies are easy to detect.

Despite the simple organization, many interesting aspects can be considered. How is development intiated and how are environmental signals processed and transduced into cellular actions? What determines the height of the conidiophore stalk or the diameter of the vesicle? How is the switch between polarized hyphal growth and buddinglike growth in the conidiophore regulated? What determines the number of metulae, the cell volume of metulae and phialides, and the number of spores? How is the cell cycle coordinated to developmental requirements? All these questions touch basic cell biological problems, and thus the study of conidiation in *A. nidulans* might help to understand some of the questions, which are of general importance.

3 SIGNALS, SIGNAL TRANSDUCTION, AND DEVELOPMENTAL DECISIONS

The first step for asexual development is the transformance of an undifferentiated vegetative cell into a cell committed to enter the differentiation program. Several signals have been described which trigger the developmental decisions. One essential condition is the exposure to a water–air interface. One can grow *A. nidulans* in liquid culture, which only allows vegetative growth. Exposure of the mycelium to the air-exposed surface of a medium induces synchronous induction of asexual development. However, *A. nidulans* requires ~18 h of vegetative growth until induction of development is possible. This period was defined as the time to acquire developmental competence [14]. Although not much is known of what happens to the mycelium during this process, mutants were isolated with different time length of competence. One gene, the developmental modifier *stuA*, is transcriptionally induced when developmental competence is achieved [75]. This might indicate a role of STUA in this process (see below).

Filamentous fungi are sessile organisms which live in the soil. Since organic nutrients are not evenly distributed, fungi need to explore the terrestrial environment to find and colonize new substrates. One important strategy to achieve this is the production of airborne spores. It is therefore not surprising that differentiation processes are triggered by the nutritional status of the mycelium. Although it was long believed that conidiation of *A. nidulans* was programmed into the life cycle rather than strictly dependent on unfavorable conditions, Scromne et al. [101] showed that carbon and nitrogen starvation leads to conidiophore development under noninducing conditions, namely in liquid culture. The complexity of the conidiophores was bigger in strains starved for nitrogen than starved for glucose, although the number of conidia produced was higher in glucose-starved mycelium. How starvation induces development is not known. However, in *Saccharomyces cerevisiae* carbon starvation involves the actions of a ras protein and adenylate cyclase [61]. In *A. nidulans* a *ras* homolog has been isolated and dominant active alleles were created. It appears that different threshold concentrations of active RAS must exist, which allow development to proceed to certain points [102]. In addition, it was found recently that RAS signaling is required during conidia germination [87]. In addition to RAS, the level of cAMP, synthesized by adenylate cyclase, has been shown to trigger developmental processes in *Schizosaccharomyces pombe* and *Neurospora crassa* [53,56,79]. Whether the cAMP level changes during different developmental stages in *A. nidulans* is not clearly shown. Early experiments related the cAMP level to the nutritional status of the mycelium and induction of sexual development [124]. Recent molecular analyses of adenylate cyclase function suggest several important roles in the life cycle of *A. nidulans* (d'Enfert, personal communication). Whether under normal conditions, surface growth on an agar plate, nutritional limitation is important for induction of conidiation has not been demonstrated.

Besides the exposure to an air interface and nitrogen and carbon starvation, conidiation is only effectively initiated in cultures exposed to light. When *A. nidulans* mycelium is induced for development it takes 6 h until the first vesicles appear and another 6 h for the production of mature conidia. When wild-type *A. nidulans* mycelium is induced for development in the dark, aerial hyphae are produced but they do not differentiate mature conidiophores. However, after exposure to light, development occurs. Mooney and Yager [80] defined a time of 6 h after induction in which *A. nidulans* is susceptible to light. Within this critical period a light pulse of 15–30 min is sufficient to elicit vesicle formation and subsequently conidiation. If the light pulse was given after 6 h, it had no effect. Determination of the light quality revealed that red light of 650–700 nm is very effective [80]. Interestingly, the effect of red light can be reversed by far red light (720 nm), a phenomenon which resembles the phytochrome system of higher plants. Besides the action of red light, in certain *A. nidulans* mutant strains also blue light (436 nm) appears to be effective, suggesting that both red and blue light are important triggers for developmental processes in *A. nidulans* [117].

Similarly to the question of how developmental competence is achieved, the question also arises how the physical parameter of light quality is being detected and transduced to initiate the developmental program. In *A. nidulans* one gene has been known for many years which is believed to play a central role in light sensing or transduction, the *velvet* (*veA*) gene, located on chromosome VIII [49]. Mutation of this gene causes light-independent asexual development. Since asexually derived conidiospores are very useful for many experiments and since *velvet* mutant strains conidiate well in the dark, most of the common laboratory strains harbor this mutation. In contrast to the wide use of the mutation, little is known about the molecular function of the gene. It has been cloned recently independently in two different laboratories, and the results were somewhat contradictory. Whereas one group claimed that deletion of the gene was lethal, the other group did not. The gene encodes a protein with a nuclear localization signal and might act as a repressor of asexual and an inducer of sexual development. Induced expression of *veA* in liquid culture led to the formation of sexual structures (Chae and Yager, personal communications). Some evidence for blue light regulation of conidiation in *A. nidulans* came from comparisons with the related fungus *N. crassa*, in which many different cellular processes depend on light. Although the light receptor has still not been identified, recent findings suggest a very interesting regulatory system [15,62]. Two transcription factors have been identified, WC-1 (white collar) and WC-2, which contain a PAS dimerization and a LOV domain [16]. The latter was shown to be crucial for blue light responses. There is evidence that a flavin is the photoactive component and that this is associated with one of the proteins. This means that the photoreceptor would be part of one of the regulatory proteins [109] (Macino, personal communication). The activity of WC-1 might be modified through phosphorylation by protein kinase C [13]. Interestingly, WC-1 and WC-2 are also part of the circadian clock of *N. crassa* [31]. In agreement with a general role of WC-1 and WC-2 in fungi is the recent identification of homologous genes in *A. nidulans* (H. Haas, Innsbruck, personal communication). Vice versa, a gene with very high homology to *A. nidulans velvet* appeared in the *N. crassa* sequencing project. The sensors for red and blue light in fungi are still unknown, as is the signaling cascade transmitting the light to cellular actions. It will be very interesting to analyze the relation between the *velvet* light-sensing system and the blue light response mediated through WC-1 and WC-2. Taken together, sensing of an air interface and light guarantees that the soilborne fungus *A. nidulans* undergoes asexual sporulation when it reaches the soil surface. This enables the mold to most efficiently disperse the conidiospores into the air or into free water.

In addition to developmental competence, nutritional status, and light, asexual development appears to be triggered by a pheromone system. Although mating in *A. nidulans* is not dependent on a pheromone-based partner selection, a pheromone system was described several years ago [26]. Ethylacetate extraction of the mycelium of a certain aconidial *A. nidulans* strain allowed the characterization of a compound which had severe effects on development. Applied to a lawn of growing *A. nidulans* on a plate caused a block of asexual development and precocious sexual differentiation. It thus behaved much like a pheromone and was named PSI (*p*recocious *s*exual *i*nduction). The molecule can exist in mainly four different but related forms. Each of them had slightly different agonistic or antagonistic signaling properties [25]. The structures were solved as unsaturated C-18 fatty acid derivatives [70,71]. The question of how PSI factor is recognized by the fungus, whether a membrane-bound receptor is required, how the signaling occurs, and whether a G-protein and a MAP-kinase cascade are involved in signal transduction, as is the case in *S. cerevisiae* pheromone signaling, remains to be determined. In *N. crassa* a homolog of the *S. cerevisiae* Ste11 MAPKK kinase has been recognized recently to be involved in the repression of the onset of conidiation [55]. In our laboratory a homolog has been identified in *A. nidulans*, and studies of its function are under way (unpublished results).

A protein, which might be directly or indirectly involved in early signal integration and processing, could be PHOA. The *phoA* gene encodes a cyclin-dependent kinase, which is a homolog of *S. cerevisiae* Pho85. In yeast, this kinase is involved in the regulation of the cell cycle but serves many additional functions, such as in glycogen metabolism, phosphate acquisition, or morphogenesis [6, 72,73]. In *A. nidulans*, deletion of the gene had an effect on early decisions between asexual and sexual development [24]. The reaction was dependent on pH, phosphate concentration, and cell density; this last condition resembles the effects with the PSI factor (see above). Other insights into the early events of conidiation came from the analysis of a class of mutants named ''fluffy.'' The current knowledge has been excellently reviewed recently [4] and will therefore be only briefly summarized here.

4 THE FLUFFY GENES

Another candidate for a signaling molecule came from the analysis of the *fluG* mutant, which produces masses of aerial mycelium but no mature conidiophores. The developmental defect could be overcome by growing the strain close to wild type, suggesting the lack of a signaling molecule to be responsible for the differentiation block. This cross-feeding also worked when the two strains were separated through a dialysis membrane with a 6000- to 8000-dalton pore size, suggesting that the molecule is a low-molecular-weight, soluble, and diffusible compound. The corresponding gene was cloned through complementation of the recessive mutation [3,116]. Sequence comparisons of the deduced polypeptide identified homology to bacterial glutamine synthetase I [59]. The biochemical nature of the compound is unknown. Interestingly, there a genetic link between the *fluG* and the *veA* gene has been discovered [117]. The extracellular rescue of *fluG* mutant strains was dependent on the *velvet* gene. Furthermore, *fluG* was isolated as an extragenic suppressor of the *velvet* mutation. Although FLUG appears to be involved in the regulation of a specific step in development, the expression is only slightly upregulated upon induction of development. The protein is abundant in the cytoplasm of hyphae grown in liquid culture as well as in developing conidiophores. However, the protein itself negatively regulates its expression [59].

Like the *fluG* mutant, all fluffy mutants are characterized by the proliferation of undifferentiated aerial hyphae, which give the colony a white, cottonlike or fluffy appearance. However, different types can be distinguished. Whereas some mutants overcome the defect upon longer incubation, other strains undergo lysis of the hyphae (Fig. 3). Mutants were generated over the years in numerous intelligent genetic screening approaches. Early screening methods based on the mutagenesis of a wild-type strain and visual inspection of recovered strains for a fluffylike phenotype [113,116]. The mutations were recessive and allowed epistasis analyses among them and determination of the induction of the central regulatory cascade through the activation of the *bristle* transcription factor (see below).

In another approach genes were classified according to their potential to induce *brlA* (measured as a *brlA(p)::lacZ* gene fusion) and named *flb* (=*f*luffy *l*ow b*rlA* expression), *fmb* (=*f*luffy *m*oderate b*rlA* expression), and *fhb* (=*f*luffy *h*igh *b*rlA expression). One interesting gene obtained in this screen was *flbA* [60]. The mutant belongs to the class, whose hyphae autolyse as the colonies mature. The gene encodes a protein with similarity to *S. cerevisiae* SST2, a protein involved in the pheromone response pathway [32]. The proteins share a 120 amino acids long domain, defining a large protein family of G-protein interacting regulators [33]. The domain was therefore named RGS (*r*egulator of *G*-protein *s*ignaling). Forced expression of FLBA induced expression of BRLA and the formation of simple conidiophores in liquid culture, similar to the structures obtained through overexpression of BRLA in submersed culture (see below). These results suggest an important regulatory role during early stages of induction.

Novel insights into the function of FLBA came from a mutant screen for dominant mutations. A diploid *A. nidulans* wild-type strain was mutagenized and fluffy autolytic strains have been isolated. The mutants were named *fad* (=*f*luffy *a*utolytic *d*ominant) and *fadA* has been characterized in detail. The FADA protein shares high homology to α-subunits of heterotrimeric G proteins. The dominant active mutant allele caused proliferation and inhibited conidiation. Since the gain-of-function mutation of *fadA* led to a phenotype similar to that of a loss-of-function mutation of *flbA*, Adams and coworkers suggested that the role of FLBA is in controlling growth and activating sporulation by negatively affecting FADA signaling [121]. Furthermore, a suppressor analysis of *flbA* identified *sfaB*, which encodes the β-subunit of a heterotrimeric G-protein [95]. Taken together, these experiments show that G-protein signaling is of crucial importance for the switch between vegetative growth and initiation of development. This raises the question of the signal feeding into this cascade. Since other heterotrimeric G-proteins interact with seven-transmembrane receptor molecules, it is likely that such a receptor type exists also in *A. nidulans* and is involved in signal perception [19]. It will be the challenge for future research to identify this receptor and the corresponding signal.

Other "*f*luffy-*l*ow expression of b*rlA*" (*flb*) genes, such as *flbD*, encode nucleic acid-binding proteins, whose function could be directly in activating downstream developmental genes such as *brlA* [112]. However, target sequences

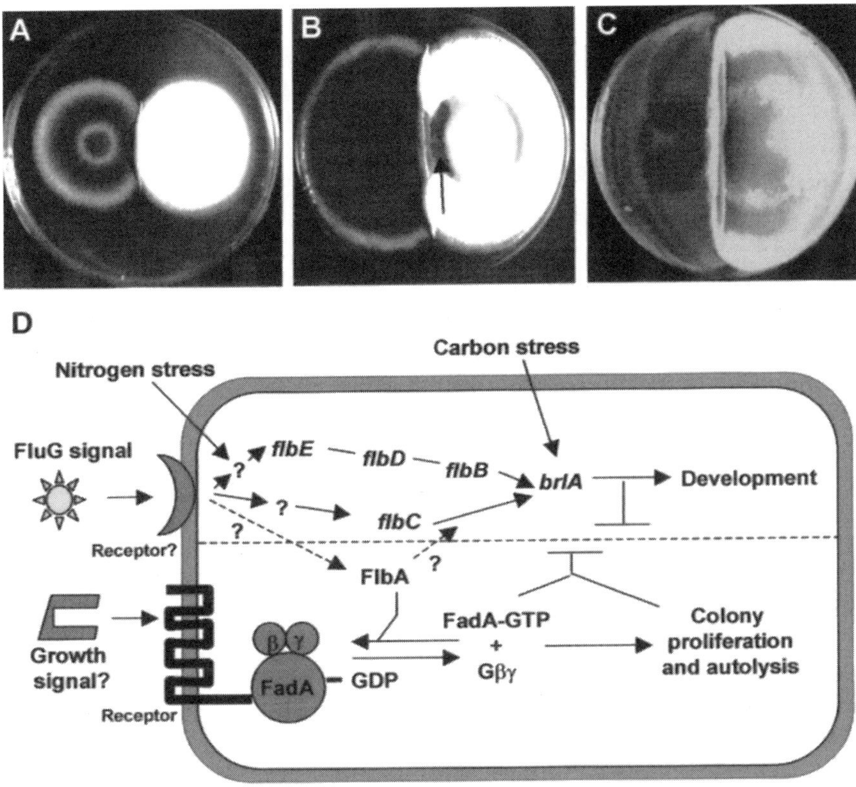

FIGURE 3 Fluffy mutants are characterized by a cottonlike appearance on agar plates. (A) A wild-type (left colony) and a fluffy mutant (right colony) were inoculated on an agar plate and incubated for three days at 37°C. (B) One class of fluffy mutants undergoes lysis of the hyphae after prolonged incubation. After 5–6 days, lysis is obvious in the middle of the colony (arrow), and (C) after 8 days almost all mycelium of the right colony disappeared. (D) Scheme showing the regulatory interactions of some of the fluffy genes. Two antagonistic signaling pathways appear to regulate *A. nidulans* growth and development. Growth signaling is mediated by the FadA G protein α-subunit. Activation of FADA by exchange of GDP for GTP results in a proliferative phenotype and blocks sporulation. To induce conidiation, FLBA has to be activated, which in turn deactivates the growth signaling and favors the differentiation pathway. (D from Ref. 4.)

have not been determined, and the exact roles of FLBD and the other potential regulators (FLBB and FLBC) in conidiation remain to be determined.

Another approach for isolation of genes responsible for activating the central regulatory pathway took advantage of the vegetative growth inhibition after activation of the pathway. Hence, genes upstream of *brlA*, which lead to its activation, should result in a block of hyphal growth. Therefore, Marhoul and Adams [64,65] constructed a DNA library under the control of the carbon source regulated *alcA* promoter. Transformants could be isolated which did not form colonies under inducing conditions. They were named *fig* (=*f*orced expression *i*nhibition of *g*rowth) or *fab* (*f*orced expression *a*ctivation of *b*rlA). One of them, *figA*, was analyzed and the corresponding protein displayed homology to BNI1 form *S. cerevisiae*, a protein required for polarized growth [36,64]. One member of the *fab* class of mutants, *fabM*, encodes a poly(A)-binding protein which is essential for viability. Since it activates development when overexpressed, it defines a class of genes that are required for vegetative functions but are in addition necessary for certain development-specific steps [65].

5 BRISTLE AND ABACUS ARE DEVELOPMENT-SPECIFIC TRANSCRIPTION FACTORS

In the late 1960s J. Clutterbuck mutagenized a wild-type strain and identified strains with abnormal conidiophore morphology or different color of the conidia [28,67]. Among those mutants were strains with defects in the genes *brl*, *aba* and *wet*, which were later shown to be central regulators.

BrlA mutants initiate conidiophore formation, but after elongation of the stalk, swelling of the stalk does not occur and the mutants fail to elaborate the next cell generation, the metulae (Fig. 4). Since the colonies are characterized by those elongated bristlelike structures on the surface, the mutant was named *brlA*. The severity of the phenotype is dependent on the corresponding mutant allele suggesting a rather complex interaction of this gene with the developmental program [44]. Cloning of the corresponding gene, by complementation of the recessive mutation, proved that *brlA* encodes a regulatory protein, a transcriptional activator with a typical TFIII Zn finger DNA-binding domain [1,2,22,48]. Although the BRLA protein has not been purified and tested for specific DNA binding in vitro, an in vivo assay in *S. cerevisiae* was done [27]. This assay allowed demonstration that BRLA is sufficient for specific gene activation. Furthermore, the system has been used to define consensus target sequences 5′-(C/A)(G/A)AGGG(G/A)-3′ in the promoter of *brlA*-dependent genes. Detailed analysis of the *brlA* locus revealed that the locus consists of two overlapping transcription units, α and β [90]. The initiation of the α-transcript is located in a region, which is spliced out in the β-transcript. The transcripts are 2.1 and 2.5 kb in size, respectively. The β-transcript initiates ~850 bp upstream of the initia-

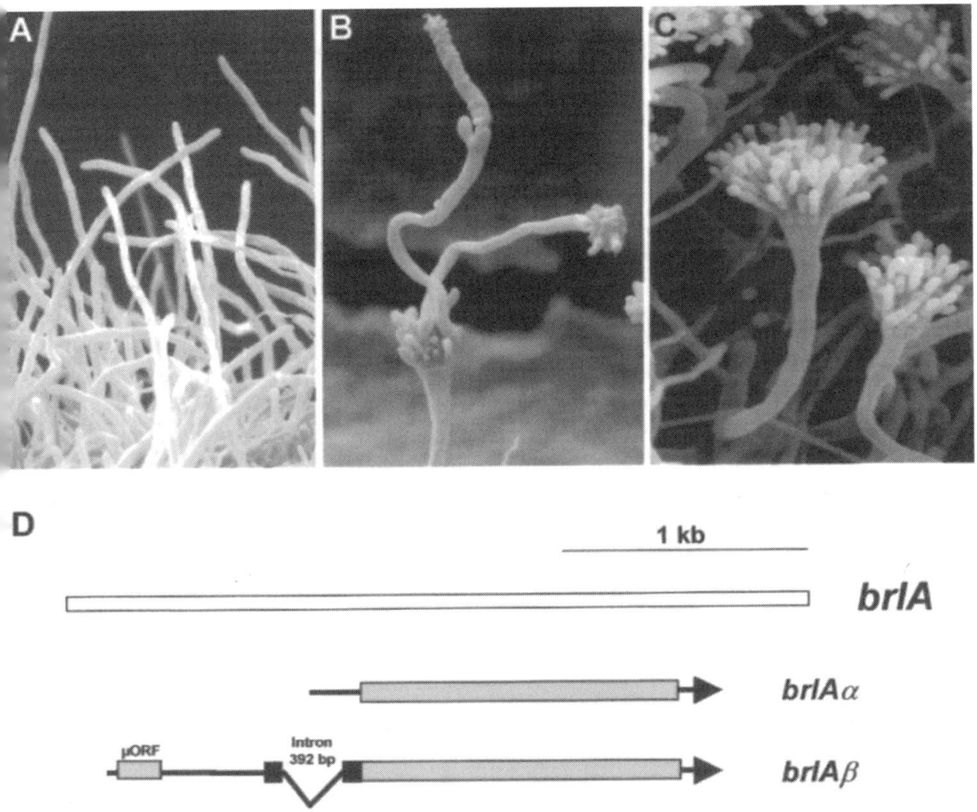

FIGURE 4 The *brlA* mutation. The phenotype of *brlA* mutants depends on the molecular defect. In complete deletion mutants, only elongated stalks are formed (A), whereas in *brlA*β mutants secondary conidiophores may arise from the vesicles of aberrant primary conidiophores (B). In brlAα mutants development proceeds further, but conidia are not produced (C). (D) Scheme of the *brlA* locus. The transcripts are indicated with an arrow and the open reading frames are shown by gray boxes. The N-terminal extension of the BRLAβ protein, disrupted by the intron, is drawn with a black box. (B from Ref. 90; C from R. Prade, Oklahoma City.)

tion site of the α-transcript and is characterized by an unusually large intron of 392 bp. The two corresponding proteins are mostly identical, but the one derived from the β-transcript has a 23-amino-acid-long extension at the N-terminus. Although the two proteins are almost identical, each of them appears to fulfill specific functions during development. Deletion of either one caused abnormal conidiophore morphology, which was different from the phenotype of *brlA* null mutants. Both individual mutants developed further than the null mutant and formed metulae and abnormal sterigmata or secondary stalks and vesicles (Fig. 4).

Interestingly, the α- and β-transcripts could substitute for each other when overexpressed. This means the two transcripts serve specific but overlapping functions. Besides the complex organization of the gene locus, the regulation appears also to be complex. The promoter region is with 2.9 kb quite long. Besides the activation of development-specific genes, BRLA activates its own expression. This is true for the α-transcript. In contrast, the β-transcript is present at any time, in hyphae and during development. However, translation of the β-transcript is suppressed in hyphae and induced upon induction of conidiophore formation. The regulation of translation occurs through a 41-amino-acid small open reading frame in the 5′-leader of the β-transcript [46]. Removal of the initiation codon of the μORF results in inappropriate induction of development. This demonstrates that the expression of BRLA is very complex and fine-tuned, and suggests very distinct roles of the proteins in cellular processes. The importance of BRLA for asexual development is demonstrated not only by mutagenesis experiments but also by overexpression in liquid culture where normal development is suppressed (see above). When BRLA is overexpressed in submerged culture, vegetative growth ceases and conidia differentiate directly from hyphal tips [1].

One of the target genes of BRLA is another transcriptional activator encoding gene, *abaA*. When *abaA* is mutated, conidiophores resemble a mechanical calculator—an abacus (Fig. 5). The gene was therefore named *abaA*. Genetic evidence suggested already *abaA* to be a major regulator, which was proven through many molecular biological experiments [28]. Finally, with this protein direct DNA sequence-specific protein–DNA interaction was shown [7]. The protein harbors an ATTS/TEA DNA-binding motif, which is also present in a number of other transcription factors such as the human TEF-1 or the *S. cerevisiae* Ty1 regulator TEC1 [58,115]. Direct targets for ABAA are *brlA*α, *abaA* itself, *wetA*, and several structural genes, such as *yA*, *wA*, and *rodA*. All target genes share multiple elements with the consensus sequence 5′-CATTCY-3′ in their promoter. Overexpression of *abaA* in liquid cultures causes cessation of vegetative growth and extensive vacuolization, but no conidial differentiation [78]. This demonstrates that earlier genes are required throughout development to ensure the correct temporal and spatial expression of later genes.

A third gene placed into the central transcriptional cascade together with *brlA* and *abaA* is *wetA*. As for *brlA* and *abaA*, mutation of the gene causes a well-defined developmental phenotype. Conidiospores are generated as in the wild type but they lyse during the final stages of differentiation. This leaves a droplet in the conidiophore head and makes them look wet-white [22]. Ultrastructural analyses revealed that the cell walls of *wetA* mutant strains are different

Figure 5 Conidiophore of an *abaA* mutant. (From Ref. 51.)

from wild type, and thus one function of WETA is the modification of the wall to gain the stability of mature conidia [99]. The gene is expressed in mature conidia, in contrast to *brlA* and *abaA*, whose transcripts are only detectable in earlier stages, but not in conidia [22]. WETA reinforces its own expression. Forced expression of the gene in hyphae leads to highly branched cells, again pointing to a possible effect on the remodeling of cell walls. Expression led to the activation of several spore-specific genes as well as *wA*, whose mRNA does not occur in spores but rather in phialides. Although these experiments demonstrate that WETA induces expression of development-specific genes, it is not yet clear how this is achieved. The WETA protein does not contain any motifs or homologs, which would allow to assign a direct DNA interaction and thus direct gene activation activity [66]. However, homologous proteins have been found in *Penicillium chrysogenum* and recently in the sequencing project in *N. crassa* [92]. One interesting aspect of the central regulatory pathway is its reinforcement once development is induced. Two of the three regulators, ABAA and WETA, act in a positive feedback loop and thus guarantee the high and quick expression required for efficient downstream gene activation (Fig. 6).

There is evidence for a novel spore-specific regulator, which came from the analysis of a gene expressed in mature conidia [104]. The gene *spoC1-C1C* is a member of a cluster of 14 genes, spanning 38 kb [45,110]. All genes are coordinately regulated, in part, by a regional, position-dependent regulatory

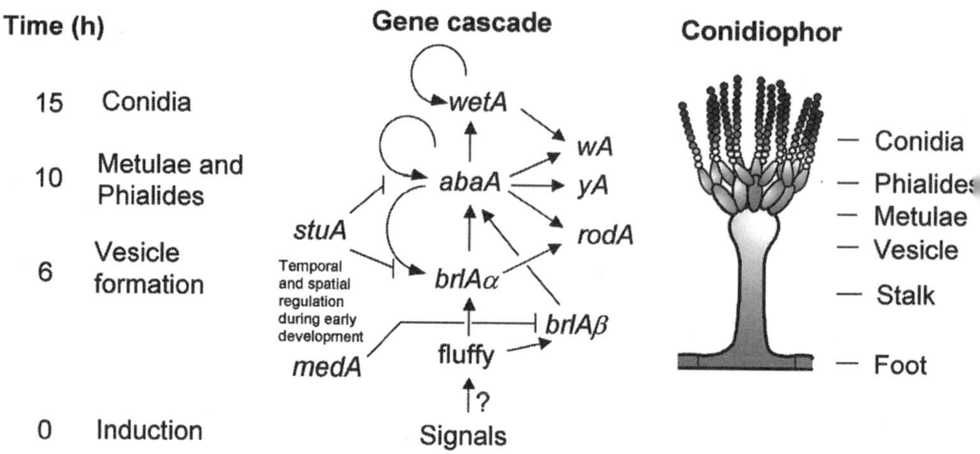

Figure 6 Regulatory circuits of asexual reproduction. (From Refs. 7, 23.)

mechanism that represses expression in undifferentiated hyphae and that may involve developmentally altered changes in chromatin conformation within the *spoC1* cluster [74]. The biochemical function of the proteins encoded by the genes is not known, because deletion of the entire cluster had no discernible phenotype [9]. Although the developmental upregulation of transcription is dependent on functional BRLA and ABAA proteins, the 5′-region of the *spoC1-C1* gene lacks response elements for either transcriptional regulator, suggesting a novel regulator downstream of the two central regulators.

Besides the transcriptional cascade, several genes have been identified which are targets of the regulators. Examples are the spore-specific genes *wA* and *yA*, which encode a polyketide synthase and a laccase, respectively, and which are both involved in synthesis of the green pigment in the conidiospores [11,12,63,68,69]. Other examples are two genes involved in conidiophore pigmentation, *ivoA* and *ivoB*. They encode proteins required for the synthesis of a melaninlike molecule, which colors the conidiophore brown [17,18,29]. In the wild type this color is hidden, because of the highly pigmented conidia. Other examples are *rodA* and *dewA*, which encode highly hydrophobic proteins (hydrophobins) and which are involved in spore wall formation [105,106].

6 STUNTED AND MEDUSA GENES

With the transcriptional cascade of the fluffy genes, the *brl*, *aba*, and *wet* genes, a first concept has been established to explain the differentiation process. Two genes, which might explain some of the open questions, are *stunted* and *medusa*.

Both mutants were isolated as strains with an aberrant conidiophore morphology, but they are still able to produce some viable conidiospores [28]. *Stunted* conidiophores are shorter than wild type and fail to produce metulae or phialides but instead generate viable conidia directly from the vesicle. Early expression studies showed that in *stuA* mutants, expression of several other developmental genes is altered [122]. The *stuA* transcript appears in hyphae and is significantly induced at the time when *A. nidulans* acquires developmental competence [75]. The *stuA* gene locus is as complex as the *brlA* locus [76]. It consists of two overlapping transcripts, *stuA*α and *stuA*β, which are initiated from different promoters and which are characterized by three common introns. Both transcripts contain a long (>1 kb) nontranslated leader, which harbors another, relatively large intron of 497 bp in the case of the β-transcript. The encoded STUA proteins are identical, but several small, open reading frames in the nontranslated leaders of the transcripts suggest translational control [114]. The protein contains a bipartite nuclear localization signal whose functionality was shown by fusion to GFP [107]. The homologous protein in *N. crassa* was also localized to the nucleus [10]. STUA also contains a basic helix–loop–helix DNA binding domain, which was found in several related proteins and therefore named APSES (*A*SM-1, *P*HD1, *S*TUA, *E*FGTF-1, and *S*OK2) [10,34].

Taken together, these features plus the developmental phenotype of *stuA* mutants suggest that STUA is a regulatory protein, which directly acts on the expression of target genes. The complex regulation of STUA expression and its modulation of expression of *brlA* and *abaA* was nicely shown with the help of reporter constructs. STUAα and -β are expressed in hyphae after acquisition of developmental competence, but *stuA*α then is further induced transcriptionally through BRLA. In addition, *stuA* translation is stimulated through a micro open reading frame located in the 5′ nontranslated leader of the *stuA*α transcript [114]. Expression of STUA is also feedback regulated. During conidiation, expression is restricted to the periphery of the conidiophore vesicle, metulae, and phialides [76]. STUA binds to MCB-like boxes in the promoters of target genes (STUA response elements) and is able to activate transcription from MCB elements in yeast. STUA response elements are found in the promoters of developmental genes such as *brlAa* or *abaA*, but also in the promoters of genes involved in cell cycle regulation, such as *nimE* or *nimO*. The potential regulation of the latter class of genes suggests a coordination of cell cycle events with developmental processes (see below). Despite the activating capacity of STUA in yeast, in *A. nidulans* it represses transcription of the *abaA* gene [34]. In summary, STUA expression leads to a restriction of *brlA* expression to the periphery of the conidiophore vesicle, metulae, phialides, and immature conidia and *abaA* expression to metulae, phialides, and immature conidia [5,76].

Conidiophores of *medusa* are characterized by a delay of differentiation of phialides and conidia, which leads to branching chains of sterigmata. Frequently, secondary conidiophores are also produced. A detailed molecular analysis has not been done, but genetic interactions among *medA*, *brlA*, and *abaA* have been studied [23]. The *medA* gene is induced upon induction of development of the conidiophore and regulates the expression of *brlA* and *abaA*. MEDA represses premature expression of both *brlA* transcripts during early development and downregulates *brlA*β during later stages. On the other hand, it is required for sufficient *abaA* expression in phialides. Interestingly, the *medA* defect can be suppressed by extra copies of *brlA*. This and other results led Busby et al. [23] to suggest that MEDA and BRLA might form a heterodimeric protein complex and together regulate gene activities. Both modifiers, STUA and MEDA, are also required for sexual development, which is in contrast to other regulators of the central cascade. Whether they directly regulate sexual-development-specific genes or the effects are rather indirect, is not yet known.

Recently, a novel regulator of the leucine zipper protein family has been discovered. The gene, *dopA*, was identified by complementation of aconidial mutants described earlier [14,116]. Mutant strains display a somewhat pleiotropic phenotype and produce 2.5-fold fewer conidiophores than wild type. Furthermore, conidiophore morphology is rather rudimentary. Deletion of the gene also affects the sexual cycle and therefore resembles the *stuA* function. Interestingly, the protein is conserved from yeast to man and has been shown to be essential for viability in *S. cerevisiae* [88]. The exact interaction with other regulators is not completely understood.

7 COORDINATION OF DEVELOPMENT AND CELL BIOLOGY IN THE CONIDIOPHORE

So far we have analyzed the conidiophore with respect to the morphological changes and the genetic program underlying these changes. However, besides the transcriptional program and the target genes, which are specifically required for conidiation, e.g., the pigment synthesis genes, many other changes are occurring as the different cell types of the conidiophore elaborate. These changes are cytoskeletal rearrangements, changes in the growth mode, etc., and these do require slight modifications or modulations of gene or protein activity rather than the expression of conidiophore-specific genes. One phenomenon that has been studied to some extent is the provision of conidiophore cell types with nuclei. Hyphal cell compartments are multinucleate, whereas metulae, phialides, and conidia all contain a single nucleus, which implies that nuclear division is strictly coupled to cell division (Fig. 7). Two genes that are involved in nuclear distribution from the vesicle to the metulae, the initial step before the coupling of cell cycle and cytokinesis occurs, are *apsA* and *apsB*. In corresponding mutant strains, nuclei fail to migrate into metulae, and development is therefore blocked at this developmental stage [30] (Fig. 8). Both genes were analyzed at the molecular level [96,108]. It was found that both were required for nuclear migration in

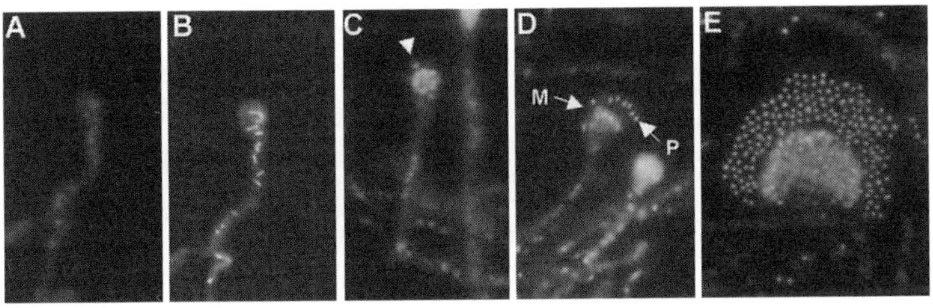

Figure 7 Nuclear distribution during conidiation. Several nuclei are distributed in the stalk and the vesicle, where they undergo synchronized mitoses. In (A) nuclei are stained with DAPI, and in (B) microtubules of the same conidiophore were stained by secondary immunofluorescence. Mitotic spindles are visible. (C) Immediately after the appearance of young metulae, nuclei move into these cells (arrowhead). Metulae (M), phialides (P), and conidia contain a single nucleus per cell (D, E). (From Ref. 39.)

hyphae and in the conidiophore. Homologs of *apsA* are found in *S. cerevisiae* and *Podospora anserina* [37,43]. Since Num1 affects microtubule stability in yeast, nuclear distribution in *A. nidulans* could also be disturbed because of altered microtubule dynamics [38,94].

Besides genes required for nuclear migration, recent evidence has accumulated that genes with a well-characterized role in cell cycle regulation are also

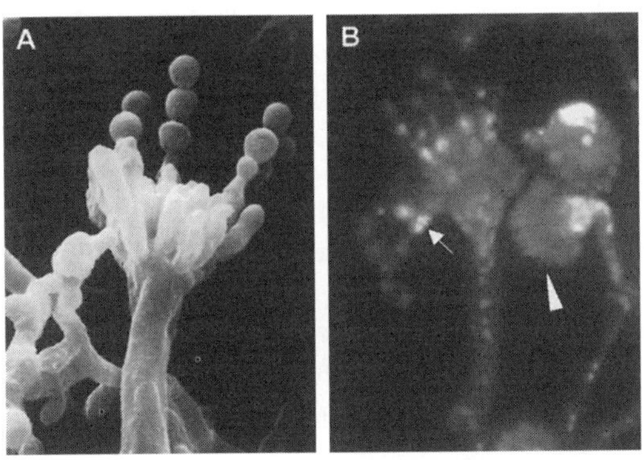

Figure 8 Phenotype of *aps* mutants. (A) Scanning electron microscopic picture of a conidiophore of an *apsA* mutant. Some metulae proceeded with development and produced single chains of conidia. (B) Nuclear distribution in an *aps* mutant. Nuclei were stained with DAPI. Metulae do not contain nuclei (arrowhead). However, sometimes several nuclei moved into a metula and development continued (arrow). (From Ref. 39.)

transcriptionally upregulated during conidiation and thus are likely to be specifically required during conidiation. Among those genes are the NIMX^{cdc2}-interacting cyclin NIME^{cyclinB} and PCLA, a member of the pcl cyclin family in *S. cerevisiae* [85,97]. If and how the cell cycle is adjusted to the proliferation in the conidiophore is an interesting question and has not been solved. In addition to cell cycle regulation, NIMX appeared also to be involved in correct cell pattern formation in the conidiophore and the suppression of septation in the stalk and vesicle [119]. Since the coupling of the cell cycle to morphogenetic changes is a common process from yeast to man, the analysis of this phenomenon in *A. nidulans* might help to understand these basic questions.

8 CELL WALL SYNTHESIS AND OXIDATIVE STRESS

Fungal cell walls are rigid structures composed mainly of polymeric sugars, such as glucans and mannans, chitin, and proteins [42,123]. Since the chemical composition of the wall determines its plasticity, it is not surprising that remodeling of the cell wall is a prerequisite for morphogenetic processes during conidiation. Chitin synthases exist in *A. nidulans* as at least five different isoenzymes with redundant functions. Deletion of *chsB* had a severe effect on hyphal growth, suggesting a role in chitin synthesis in hyphae. Whereas deletion of *chsE* had no discernible phenotye, mutation of *chsD* resulted in a swelling of conidia and subsequent lysis. Conidia, which did not lyse, formed hyphae and, on osmotically stabilized media, also conidiophores. However, many conidiophores swelled subapically and lysed before producing metulae [20,103]. In comparison, Motoyama and colleagues [40,81] reported that mutation of *chsA* only in combination with *chsD* or *chsC* had a remarkable effect on conidiation. *chsA* was mainly expressed in metulae, phialides, and conidia. Since the promoter regions of *chsA* and *chsD* harbor putative BRLA- and ABAA-binding sites, it is likely that these chitin synthases together with *chsC* serve conidiophore-specific functions.

Differentiation is often accompanied by different stress conditions, and many genes related to stress response are therefore expressed upon induction of development [47]. Two catalases were described in *A. nidulans*, one of which appears to have a function during conidiation. *catA* is transcriptionally upregulated during development and the enzyme accumulated in conidia. Interestingly, gene expression was independent of *brlA*, and expression of enzyme activity was in addition posttranscriptionally regulated [52,83,84].

9 GENOMEWIDE APPROACHES TO UNDERSTAND DEVELOPMENT

In the last few years genomic sequencing was greatly improved and enabled determination of the primary DNA sequence of entire organisms. Since the *S. cerevisiae* genomic DNA sequence was completed and released, in 1996, several attempts were undertaken to sequence filamentous fungi. As a first step to genomewide expression studies, several expressed sequence tag projects were launched. In *A. nidulans* >12,000 EST sequences of a cDNA library obtained from asexual developed cultures are available and provide a great tool for gene identification (http://www.genome.ou.edu/fungal.html). In addition, sequencing of chromosome IV is completed (http://bioinfo.okstate.edu/PipeOnline_db/ancivquery.html). EST sequences in addition to chromosome IV sequences will allow now to design DNA arrays with >4000 different genes. Using this tool, the expression pattern of those thousands of genes can be studied simultaneously and will give new insights into the fascinating developmental process of *A. nidulans*. Parallel to the publicly funded EST and genome-sequencing attempts, most of the *A. nidulans* genomic DNA sequence was resolved commercially, and thus is available only to the scientific community. DNA high-density array experiments with >70% of the estimated 8000–10,000 genes were already performed with respect to developmentally induced genes [56] but the data are not yet available (Kellner et al., Cereon Genomics, personal communication and ECFG5 abstract) [57].

10 FUTURE PERSPECTIVES

Analysis of conidiation has revealed a basic knowledge of the regulatory circuits underlying the developmental process. However, many interesting questions need to be solved—e.g., which other components are required, and how gene activity is modulated and fine-tuned during differentiation to achieve the structural and physiological changes in the conidiophore. Several regulatory principles well established in other systems have not yet been studied during development. During the preparation of this manuscript an excellent comparative review of signal transduction cascades appeared and is highly recommended for further reading [61]. One example of a regulatory mechanism not discovered yet in asexual development in *A. nidulans* is the modulation of gene activity through chromatin changes, as they were observed in the nitrogen metabolism regulatory system of *A. nidulans* [82] or in the regulation of pathogenicity of *U. maydis* [93].

First indications for such a regulatory mechanism during conidiation of *A. nidulans* came from the analysis of the *spoC1* cluster, which is positionally regulated [74]. Whether chromatin remodeling is required for stage-specific gene activation in this case and how it is achieved has not been solved. Another example is the modulation of the activity of regulatory proteins through their subcellular localization. A shuttle of a transcription factor between the nucleus and the cytoplasm has been established for the regulation of acetate utilization in *A. nidulans* and for several other systems in a variety of eukaryotes [54]. Most likely this regulatory principle also plays a role during conidiation of *A. nidulans*, and it will be the challenge of future research to identify those regulatory circuits.

There is also evidence that "two-component" systems, which are well-known signal transduction pathways in prokaryotes, also operate in *A. nidulans* and are involved in conidiation. However, a detailed analysis of the link to existing regulatory pathways has not been established, and the exact role of this two-component system is not clear [8].

There are mainly two rationales for studying conidiation in *A. nidulans* in such detail. One is to understand the process of conidium formation in this mold and thus broaden our knowledge of fungal biology. A second reason is the possible discovery and/or better understanding of general regulatory principles and their contribution to morphological changes. Whereas the interaction of, e.g., transcriptional activators with their target DNA sequence and the principles of gene activation are evolutionarily conserved and thus the study of *A. nidulans* helps to understand these processes, the question arises how widely distributed the developmental regulation revealed in *A. nidulans* will be among other fungi. Studies are rather limited and the few examples that have been published do not answer this question satisfactorily. The *brlA* gene, which does not have a homolog in *S. cerevisiae*, was identified in *A. oryzae*, and functional analysis revealed a conserved function [118]. The same holds true for the *wetA* gene. It appears to be functionally conserved in *P. chrysogenum* [92]. The *abaA* gene has been identified in the pathogenic and dimorphic growing *Penicillium marneffei* [21]. This fungus grows filamentous at 25°C, where it reproduces with conidiophoreborne asexual spores, and it grows fission yeastlike at 37°C. The phenotype of an *P. marneffei abaA* mutant strain is similar to the phenotype described in *A. nidulans*. The two proteins were interchangeable between the organisms, suggesting conserved functions (Andrianopoulos, personal communication). The ABAA homolog in *S. cerevisiae*, TEC1, regulates pseudohyphal development [41] and in *C. albicans* triggers pathogenicity [98]. These results indicate that the central regulatory pathway seems to be conserved among filamentous fungi and that some regulators are also involved in developmental processes in yeast. However, the example of *P. marneffei* nicely demonstrates that different fungi need to be studied, since in this pathogenic fungus sensing of the temperature or other host-specific environmental conditions must trigger the developmental cascade defined by the regulators. Another example for a species-specific modulation of the regulatory cascade is *flbD*. This gene is well characterized in *A. nidulans* as a typical ''fluffy'' gene (see above), but deletion of the gene in *N. crassa* revealed no evidence for an important developmental role, although the gene could complement the defect in *A. nidulans* [100]. Since the fluffy genes are required early during development, perhaps for the integration of environmental signals, this example might indicate the differences in the processing of these initial signals.

These last examples show that although our knowledge about fungal development has improved in recent years, much more research is required to fully understand conidiation in *A. nidulans* and other filamentous fungi before specific approaches can be applied to control fungal growth and development.

REFERENCES

1. TH Adams, MT Boylan, WE Timberlake. *brlA* is necessary and sufficient to direct conidiophore development in *Aspergillus nidulans*. Cell 54:353–362, 1988.
2. TH Adams, H Deising, WE Timberlake. *brlA* requires both zinc fingers to induce development. Mol Cell Biol 10(4):1815–1817, 1990.
3. TH Adams, WA Hide, LN Yager, BN Lee. Isolation of a gene required for programmed initiation of development by *Aspergillus nidulans*. Mol Cell Biol 12(9): 3827–3833, 1992.
4. TH Adams, JK Wieser, J-H Yu. Asexual sporulation in *Aspergillus nidulans*. Microbiol Mol Biol Rev 62(1):35–54, 1998.
5. J Aguirre. Spatial and temporal controls of the *Aspergillus brlA* developmental regulatory gene. Mol Microbiol 8(2):211–218, 1993.
6. B Andrews, V Measday. The cyclin family of budding yeast: abundant use of a good idea. Trends Genet 14(2):66–72, 1998.
7. A Andrianopoulos, WE Timberlake. The *Aspergillus nidulans abaA* gene encodes a transcriptional activator that acts as a genetic switch to control development. Mol Cell Biol 14(4):2503–2515, 1994.
8. MVCL Appleyard, WL Mcpheat, MJR Stark. A novel ''two-component'' protein containing histidine kinase and response regulator domains required for sporulation in *Aspergillus nidulans*. Curr Genet 37:364–372, 2000.
9. R Aramayo, TH Adams, WE Timberlake. A large cluster of highly expressed genes is dispensable for growth and development in *Aspergillus nidulans*. Genetics 122: 65–71, 1989.
10. R Aramayo, Y Peleg, R Addison, R Metzenberg. *Asm-1*$^+$, a *Neurospora crassa* gene related to transcriptional regulators of fungal development. Genetics 144:991–1003, 1996.
11. R Aramayo, WE Timberlake. Sequence and molecular structure of the *Aspergillus nidulans yA* (laccase I) gene. Nucleic Acids Res 18(11):3415, 1990.
12. R Aramayo, WE Timberlake. The *Aspergillus nidulans yA* gene is regulated by *abaA*. EMBO J 12(5):2039–2048, 1993.
13. G Arpaia, F Cerri, S Baima, G Macino. Involvement of protein kinase C in the repsonse of *Neurospora crassa* to blue light. Mol Gen Genet 262:314–322, 1999.
14. DE Axelrod, M Grealt, M Pastushok. Gene control of developmental competence in *Aspergillus nidulans*. Dev Biol 34(9):9–15, 1973.
15. P Ballario MG. White collar proteins: PASsing the light signal in *Neurospora crassa*. Trends Microbiol 5(11):458, 1997.
16. P Ballario, C Talora, D Galli, H Linden, G Macino. Roles in dimerization and blue light photoresponse of the PAS and LOV domains of *Neurospora crassa* white collar proteins. Mol Microbiol 29(3):719–729, 1998.
17. CE Birse, AJ Clutterbuck. N-Acetyl-6-hydroxytryptophan oxidase, a developmentally controlled phenol oxidase from *Aspergillus nidulans*. J Gen Microbiol 136: 1725–1730, 1990.
18. CE Birse, AJ Clutterbuck. Isolation and developmentally regulated expression of an *Aspergillus nidulans* phenol oxidase–encoding gene, *ivoB*. Gene 98:69–76, 1991.

19. M Bölker. Sex and crime: heterotrimeric G-proteins in fungal mating and pathogenesis. Fungal Genet Biol 25:143–156, 1998.
20. PT Borgia, N Iartchouk, PJ Riggle, KR Winter, Y Koltin, CE Bulawa. The *chsB* gene of *Aspergillus nidulans* is necessary for normal hyphal growth and development. Fungal Genet Biol 20:193–203, 1996.
21. AR Borneman, MJ Hynes, A Andrianopoulos. The *abaA* homologue of *Penicillium marneffei* participates in two developmental programmes: conidiation and dimorphic growth. Mol Microbiol 38(5):1034–1047, 2000.
22. MT Boylan, PM Mirabito, CE Willett, CR Zimmerman, WE Timberlake. Isolation and physical characterization of three essential conidiation genes from *Aspergillus nidulans*. Mol Cell Biol 7(9):3113–3118, 1987.
23. TM Busby, KY Miller, BL Miller. Suppression and enhancement of the *Aspergillus nidulans medusa* mutation by altered dosage of the *bristle* and *stunted* genes. Genetics 143:155–163, 1996.
24. HJ Bussink, SA Osmani. A cyclin-dependent kinase family member (PHOA) is required to link developmental fate to environmental conditions in *Aspergillus nidulans*. EMBO J 17(14):3990–4003, 1998.
25. SP Champe, AE El-Zayat. Isolation of a sexual sporulation hormone from *Aspergillus nidulans*. J Bacteriol 171(7):3982–3988, 1989.
26. SP Champe, P Rao, A Chang. An endogenous inducer of sexual development in *Aspergillus nidulans*. J Gen Microbiol 133:1383–1387, 1987.
27. YC Chang, WE Timberlake. Identification of Aspergillus *brlA* response elements (BREs) by genetic selection in yeast. Genetics 133:29–38, 1992.
28. AJ Clutterbuck. A mutational analysis of conidial development in *Aspergillus nidulans*. Genetics 63:317–327, 1969.
29. AJ Clutterbuck. The genetics of conidiophore pigmentation in *Aspergillus nidulans*. J Gen Microbiol 136:1731–1738, 1990.
30. AJ Clutterbuck. Mutants of *Aspergillus nidulans* deficient in nuclear migration during hyphal growth and conidiation. Microbiology 140:1169–1174, 1994.
31. SK Crosthwaite, JC Dunlap, JJ Loros. *Neurospora wc-1* and *wc-2*: transcription, photoresponses, and the origins of circadian rhythmicity. Science 276:763–769, 1997.
32. C Dietzel, J Kurjan. Pheromonal regulation and sequence of the *Sacchaomyces cerevisiae SST2* gene: a model for desensitization to pheromone. Mol Cell Biol 7: 4169–4177, 1987.
33. KM Druey, KJ Blumer, VH Kang, JH Kehrl. Inhibition of G-protein mediated MAP kinase activation by a new mammalian gene family. Nature 379:742–746, 1996.
34. JR Dutton, S Johns, BL Miller. StuAp is a sequence-specific transcription factor that regulates developmental complexity in *Aspergillus nidulans*. EMBO J 16(18): 5710–5721, 1997.
35. E Eidam. Zur Kenntniss der Entwicklung bei den Ascomyceten. III. *Sterigmatocystis nidulans*. Beitr Biol Pfl 3:392–411, 1883.
36. M Evangelista, K Blundell, MS Longtine, CJ Chow, N Adames, M Peter, C Boone. Bni1p, a yeast formin linking cdc42p and the actin cytoskeleton during polarized morphogenesis. Science 276:118–122, 1997.

37. M Farkasovsky, H Küntzel. Yeast Num1p associates with the mother cell cortex during S/G2 phase and affects microtubular functions. J Cell Biol 131(4):1003–1014, 1995.
38. R Fischer. Nuclear migration in fungi—different motors at work. Res Microbiol 151:247–254, 2000.
39. R Fischer, WE Timberlake. *Aspergillus nidulans apsA* (anucleate primary sterigmata) encodes a coiled-coil protein necessary for nuclear positioning and completion of asexual development. J Cell Biol 128(4):485–498, 1995.
40. M Fujiwara, M Ichinomiya, T Motoyama, H Horiuchi, A Ohta, M Takagi. Evidence that the *Aspergillus nidulans* class I and class II chitin synthase genes, *chsC* and *chsA*, share critical roles in hyphal wall integrity and conidiophore development. J Biochem 127(3):359–366, 2000.
41. V Gavrias, A Andrianopoulos, CJ Gimeno, WE Timberlake. *Sacchromyces cerevisiae TEC1* is required for pseudohyphal growth. Mol Microbiol 19(6):1255–1263, 1996.
42. NH Georgopapadakou, JS Tkacz. The fungal cell wall as a drug target. Trends Microbiol 3(3):98–104, 1995.
43. F Graia, V Berteaux-Lecellier, D Zickler, M Picard. *ami1*, an orthologue of the *Aspergillus niudlans apsA* gene, is involved in nuclear migration events throughout the life cycle of *Podospora anserina*. Genetics 155:633–646, 2000.
44. GW Griffith, MS Stark, AJ Clutterbuck. Wild-type and mutant alleles of the *Aspergillus nidulans* developmental regulator gene *brlA*: correlation of variant sites with protein function. Mol Gen Genet 262:892–897, 1999.
45. DI Gwynne, BL Miller, KY Miller, WE Timberlake. Structure and regulated expression of the *SpoC1* gene cluster from *Aspergillus nidulans*. J Mol Biol 180:91–109, 1984.
46. S Han, J Navarro, RA Greve, TH Adams. Translational repression of *brlA* expression prevents premature development in *Aspergillus*. EMBO J 12(6):2449–2457, 1993.
47. W Hansberg, H de Groot, H Sies. Reactive oxygen species associated with cell differentiation in *Neurospora crassa*. Free Radic Biol Med 14(3):287–293, 1993.
48. IL Johnstone, SG Hughes, AJ Clutterbuck. Cloning an *Aspergillus nidulans* developmental gene by transformation. EMBO J 4(5):1307–1311, 1985.
49. E Käfer. Origins of translocations in *Aspergillus nidulans*. Genetics 52:217–232, 1965.
50. E Käfer. Meiotic and mitotic recombination in *Aspergillus* and its chromosomal aberrations. Adv Genet 19:33–131, 1977.
51. M Karos, R Fischer. *hymA* (hypha-like metulae), a new developmental mutant of *Aspergillus nidulans*. Microbiology 142(11):3211–3218, 1996.
52. L Kawasaki, D Wysong, R Diamond, J Aguirre. Two divergent catalase genes are differentially regulated during *Aspergillus nidulans* development and oxidative stress. J Bacteriol 179(10):3284–3292, 1997.
53. AM Kays, PS Rowlry, RA Baasiri, KA Borkovich. Regulation of conidiation and adenyly cyclase levels by the Gα protein GNA-3 in *Neurospora crassa*. Mol Cell Biol 20(20):7693–7705, 2000.

54. A Komeili, EK O'Shea. Nuclear transport and transcription. Curr Opin Cell Biol 12(3):355–360, 2000.
55. GO Kothe, SJ Free. The isolation and characterization of *nrc-1* and *nrc-2*, two genes encoding protein kinases that control growth and development in *Neurospora crassa*. Genetics 149:117–130, 1998.
56. H Kunitomo, T Higuchi, Y Iino, M Yamamoto. A zinc-finger protein, rst2p, regulates transcription of the fission yeast ste11(+) gene, which encodes a pivotal transcription factor for sexual development. Mol Biol Chem 11(9):3205–3217, 2000.
57. DM Kupfer, CA Reece, SW Clifton, BA Roe, RA Prade. Multicellular ascomycetous fungal genomes contain more than 8000 genes. Fungal Genet Biol 21(3):364–372, 1997.
58. I Laloux, E Dubois, M Dewerchin, E Jacobs. TEC1, a gene involved in the activation of Ty1 and Ty1-mediated gene expression in *Saccharomyces cerevisiae*: cloning and molecular analysis. Mol Cell Biol 10:3541–3550, 1990.
59. BN Lee, TH Adams. The *Aspergillus nidulans fluG* gene is required for production of an extracellular developmental signal and is related to prokaryotic glutamine synthetase I. Genes Dev 8:641–651, 1994.
60. BN Lee, TH Adams. Overexpression of *flbA*, an early regulator of *Aspergillus* asexual sporulation, leads to activation of *brlA* and premature initiation of development. Mol Microbiol 14(2):323–334, 1994.
61. KB Lengeler, RC Davidson, C D'Souza, T Harashima, W-C Shen, P Wang, X Pan, M Waugh, J Heitman. Signal transduction cascades regulating fungal development and virulence. Microbiol Mol Biol Rev 64(4):746–785, 2000.
62. H Linden, P Ballario, G Macino. Blue light regulation in *Neurospora crassa*. Fungal Genet Biol 22:141–150, 1997.
63. S Machida, Y Itoh, H Kishida, T Higasa, M Saito. Localization of chitin synthase in *Absidia glauca* studied by immunoelectron microscopy: application of cryo-ultramicrotomy. Biosci Biotechnol Biochem 58(11):1983–1989, 1994.
64. JF Marhoul, TH Adams. Identification of developmental regulatory genes in *Aspergillus nidulans* by overexpression. Genetics 139:537–547, 1995.
65. JF Marhoul, TH Adams. Aspergillus *fabM* encodes an essential product that is related to poly(A)-binding proteins and activates development when overexpressed. Genetics 144:1463–1470, 1996.
66. MA Marshall, WE Timberlake. *Aspergillus nidulans wetA* activates spore-specific gene expression. Mol Cell Biol 11(1):55–62, 1991.
67. SD Martinelli, AJ Clutterbuck. A quantitative survey of conidiation mutants in *Aspergillus nidulans*. J Gen Microbiol 69:261–268, 1971.
68. ME Mayorga, WE Timberlake. Isolation and molecular characterization of the *Aspergillus nidulans wA* gene. Genetics 126:73–79, 1990.
69. ME Mayorga, WE Timberlake. The developmentally regulated *Aspergillus nidulans wA* gene encodes a polypeptide homologous to polyketide and fatty acid synthases. Mol Gen Genet 235:205–212, 1992.
70. P Mazur, HV Meyers, K Nakanishi. Structural elucidation of sporogenic fatty acid metabolites from *Aspergillus nidulans*. Tetrahedron Lett 31(27):3837–3840, 1990.
71. P Mazur, K Nakanishi. Enantioselective synthesis of PsiAβ, a sporogenic metabolite of *Aspergillus nidulans*. J Org Chem 57:1047–1051, 1992.

72. V Measday, H McBride, J Moffat, D Stillman, B Andrews. Interactions between Pho85 cyclin-dependent kinase complexes and the Swi5 transcription factor in budding yeast. Mol Microbiol 35(4):825–834, 2000.
73. V Measday, L Moore, R Retnakaran, J Lee, M Donoviel, AM Neiman, B Andrews. A family of cyclin-like proteins that interact with the Pho85 cyclin-dependent kinase. Mol Cell Biol 17(3):1212–1223, 1997.
74. BL Miller, KA Roberti, WE Timberlake. Position-dependent and independent mechanisms regulate cell-specific expression of the *SpoC1* gene cluster of *A. nidulans*. Mol Cell Biol 7:427–434, 1987.
75. KY Miller, TM Toennis, TH Adams, BL Miller. Isolation and transcriptional characterization of a morphological modifier: The *Aspergillus nidulans* stunted (*stuA*) gene. Mol Gen Genet 227:285–292, 1991.
76. KY Miller, J Wu, BL Miller. *StuA* is required for cell pattern formation in *Aspergillus*. Genes Dev 6:1770–1782, 1992.
77. CW Mims, EA Richardson, WE Timberlake. Ultrastructural analysis of conidiophore development in the fungus *Aspergillus nidulans* using freeze-substitution. Protoplasma 44:132–141, 1988.
78. PM Mirabito, TH Adams, WE Timberlake. Interactions of three sequentially expressed genes control temporal and spatial specificity in *Aspergillus* development. Cell 57:859–868, 1989.
79. N Mochizuki, M Yamamoto. Reduction in the intracellular cAMP level triggers initiation of sexual development in fission yeast. Mol Gen Genet 233:17–24, 1992.
80. JL Mooney, LN Yager. Light is required for conidiation in *Aspergillus nidulans*. Genes Dev 4:1473–1482, 1990.
81. T Motoyama, M Fujiwara, N Komjima, H Horiuchi, A Ohta, M Takagi. The *Aspergillus nidulans* genes *chsA* and *chsD* encode chitin synthases which have redundant functions in conidia formation. Mol Gen Genet 253:520–528, 1997.
82. MI Muro-Pator, R Gonzalez, J Strauss, F Narendja, C Scazzocchio. The GATA factor AreA is essential for chromatin remodelling in a eukaryotic bidirectional promoter. EMBO J 15(6):1584–1597, 1999.
83. RE Navarro, J Aguirre. Posttranscriptional control mediates cell type-specific localization of catalase A during *Aspergillus nidulans* development. J Bacteriol 180(21): 5733–5738, 1998.
84. RE Navarro, MA Stringer, W Hansberg, WE Timberlake, J Aguirre. *catA*, a new *Aspergillus nidulans* gene encoding a developmentally regulated catalase. Curr Genet 29:352–359, 1996.
85. MJ O'Connell, AH Osmani, NR Morris, SA Osmani. An extra copy of *nim*E^{cyclinB} elevates pre-MPF levels and partially suppresses mutation of *nim*T^{cdc25} in *Aspergillus nidulans*. EMBO J 11(6):2139–2149, 1992.
86. PTP Oliver. Conidiophore and spore development in *Aspergillus nidulans*. J Gen Microbiol 73:45–54, 1972.
87. N Osherov, G May. Conidial germination in *Aspergillus nidulans* requires RAS signaling and protein synthesis. Genetics 155(2):647–656, 2000.
88. RC Pascon, BL Miller. Morphogenesis in *Aspergillus nidlans* requires Dopey (DopA), a member of a novel family of leucine zipper-like proteins conserved from yeast to humans. Mol Microbiol 36(6):1250–1264, 2000.

89. G Pontecorvo, JA Roper, LM Hemmons, KD MacDonald, AWJ Bufton. The genetics of *Aspergillus nidulans*. Adv Genet 5:141–238, 1953.
90. R Prade, WE Timberlake. The *Aspergillus nidulans brlA* regulatory locus consists of two overlapping transcription units that are individually required for conidiophore development. EMBO J 12(6):2439–2447, 1993.
91. RA Prade. The reliability of the *Aspergillus nidulans* physical map. Fungal Genet Biol 29(3):175–185, 2000.
92. RA Prade, WE Timberlake. The *Penicillium chrysogenum* and *Aspergillus nidulans wetA* developmental regulatory genes are functionally equivalent. Mol Gen Genet 244:539–547, 1994.
93. C Quadbeck-Seeger, G Wanner, S Huber, R Kahmann, J Kämper. A protein with similarity to the human retinoblastoma binding protein 2 acts specifically as a repressor for genes regulated by the b mating type locus in *Ustilago maydis*. Mol Microbiol 38(1):154–166, 2000.
94. N Requena, C Alberti-Segui, E Winzenburg, C Horn, M Schliwa, P Philippsen, R Liese, R Fischer. Genetic evidence for a microtubule-destabilizing effect of conventional kinesin and analysis of its consequences for the control of nuclear distribution in *Aspergillus nidulans*. Mol Microbiol 42:121–132, 2001. See also: http://www.blackwell-science.com/products/journals/suppmat/mole2609/mmi2609sm.htm.
95. S Rosén, J-H Yu, TH Adams. The *Aspergillus nidulans sfaD* gene encodes a G protein β subunit that is required for normal growth and repression of sporulation. EMBO J 18(20):5592–5600, 1999.
96. EP Sablin, RB Case, SC Cai, CL Hart, A Ruby, RD Vale, RJ Fletterick. Direction determination in the minus-end-directed kinesin motor nice. Nature 395:813–816, 1998.
97. N Schier, R Liese, R Fischer. A pcl-like cyclin of *Aspergillus nidulans* is transcriptionally activated by developmental regulators and is involved in sporulation. Mol Cell Biol 21:4075–4088, 2001.
98. A Schweizer, S Rupp, BN Taylor, M Röllinghoff, K Schröppel. The TEA/ATTS transcription factor CaTec1p regulates hyphal development and virulence in *Candida albicans*. Mol Microbiol 38(3):435–445, 2000.
99. TC Sewall, CW Mims, WE Timberlake. Conidium differentiation in *Aspergillus nidulans* wild-type and wet-white (*wet*) mutant strains. Dev Biol 138:499–508, 1990.
100. WC Shen, J Wieser, TH Adams, DJ Ebbole. The Neurospora *rca-1* gene complements an Aspergillus *flbD* sporulation mutant but has no identifiable role in *Neurospora* sporulation. Genetics 148:1031–1041, 1998.
101. I Skromne, O Sánchez, J Aguirre. Starvation stress modulates the expression of the *Aspergillus nidulans brlA* regulatory gene. Microbiology 141:21–28, 1995.
102. T Som, VSR Kolaparthi. Developmental decisions in *Aspergillus nidulans* are modulated by ras activity. Mol Cell Biol 14(8):5333–5348, 1994.
103. CA Specht. The *chsD* and *chsE* genes of *Aspergillus nidulans* and their roles in chitin synthesis. Fungal Genet Biol 20:153–167, 1996.
104. KE Stephens, KY Miller, BL Miller. Functional analysis of DNA sequences re-

quired for conidium-specific expression of the SpoC1–C1C gene of *Aspergillus nidulans*. Fungal Genet Biol 2–3:231–242, 1999.

105. MA Stringer, RA Dean, TC Sewall, WE Timberlake. *Rodletless*, a new *Aspergillus* developmental mutant induced by directed gene inactivation. Genes Dev 5:1161–1171, 1991.
106. MA Stringer, WE Timberlake. *dewA* encodes a fungal hydrophobin component of the *Aspergillus* spore wall. Mol Microbiol 16(1):33–44, 1994.
107. R Suelmann, N Sievers, R Fischer. Nuclear traffic in fungal hyphae: in vivo study of nuclear migration and positioning in *Aspergillus nidulans*. Mol Microbiol 25(4):757–769 (http://www.blacksci.co.uk/products/elecpro/anims/mole254.gif), 1997.
108. R Suelmann, N Sievers, D Galetzka, L Robertson, WE Timberlake, R Fischer. Increased nuclear traffic chaos in hyphae of *apsB* mutants of *Aspergillus nidulans*: molecular characterization of *apsB* and in vivo observation of nuclear behaviour. Mol Microbiol 30(4):831–842 (http://www.uni-marburg.de/mpi/movies/apsbnew.mov) 1998.
109. C Talora, L Franchi, H Linden, P Ballario, G Macino. Role of a white collar-1–white collar-2 complex in blue-light signal transduction. EMBO J 18:4961–4968, 1999.
110. WE Timberlake, EC Barnard. Organization of a gene cluster expressed specifically in the asexual spores of *A. nidulans*. Cell 26:29–37, 1981.
111. WE Timberlake, MA Marshall. Genetic engineering of filamentous fungi. Science 244:1313–1317, 1989.
112. J Wieser, TH Adams. *flbD* encodes a myb-like DNA-binding protein that coordinates initiation of *Aspergillus nidulans* conidiophore development. Genes Dev 9: 491–502, 1995.
113. J Wieser, BN Lee, JW Fondon III, TH Adams. Genetic requirements for initiating asexual development in *Aspergillus nidulans*. Curr Genet 27:62–69, 1994.
114. J Wu, BL Miller. *Aspergillus* asexual reproduction are differentially affected by transcriptional and translational mechanisms regulating *stunted* gene expression. Mol Cell Biol 17(10):6191–6201, 1997.
115. JH Xiao, I Davidson, H Matthes, J-M Garnier, P Chambon. Cloning, expression, and transcriptional properties of the human enhancer factor TEF-1. Cell 65:551–568, 1991.
116. LN Yager, MB Kurtz, S Champe. Temperature-shift analysis of conidial development in *Aspergillus nidulans*. Dev Biol 93:92–103, 1982.
117. LN Yager, HO Lee, DL Nagle, JE Zimmermann. Analysis of *fluG* mutations that affect light-dependent conidiation in *Aspergillus nidulans*. Genetics 149:1777–1786, 1998.
118. O Yamada, BR Lee, K Bomi, Y Iimura. Cloning and functional analysis of the *Aspergillus oryzae* conidiation regulator gene *brlA* by its disruption and misscheduled expression. J Biosci Bioeng 87:424–429, 1999.
119. XS Ye, S-L Lee, TD Wolkow, S-L McGuire, JE Hamer, GC Wood, SA Osmani. Interaction between developmental and cell cycle regulators is required for morphogenesis in *Aspergillus nidulans*. EMBO J 18(24):6994–7001, 1999.

120. MM Yelton, JE Hamer, WE Timberlake. Transformation of *Aspergillus nidulans* by using a *trpC* plasmid. Proc Natl Acad Sci USA 81:1470–1474, 1984.
121. J-H Yu, J Wieser, TH Adams. The *Aspergillus* FlbA RGS domain protein antagonizes G-protein signaling to block proliferation and allow development. EMBO J 15(19):5184–5190, 1996.
122. CR Zimmermann, A molecular genetic analysis of developmental gene regulation in *Aspergillus nidulans*. University of California, Davis: PhD thesis, 1986, p 267.
123. BJM Zonneveld. Biochemical analysis of the cell wall of *Aspergillus nidulans*. Biochim Biophys Acta 249:506–514, 1971.
124. BJM Zonneveld. The effect of glucose and manganese on adenosine-3′,5′-monophosphate levels during growth and differentiation of *Aspergillus nidulans*. Arch Microbiol 108:41–44, 1976.

4

Senescence in *Podospora anserina*

Heinz D. Osiewacz and
Christian Q. Scheckhuber
Johann Wolfgang Goethe-Universität, Frankfurt, Germany

1 INTRODUCTION

Podospora anserina is a filamentous ascomycete naturally growing on the dung of herbivores. This substrate dries out very fast and is therefore a suitable nutritional resource for only a very limited time. To survive, *P. anserina* is forced to propagate efficiently in a short time. Propagation proceeds via a sexual cycle (Fig. 1) giving rise to the formation of meiospores, the so-called ascospores. These spores are generated in a special sporangium, the ascus. Many asci are produced in one fruiting body, the perithecium. Mature ascospores are shot out of the ascus and the perithecium and adhere to surrounding herbage. After ingestion by a herbivore and after passing through its intestine, they germinate on dung and give rise to the formation of a new vegetation body, a mycelium. Under natural conditions senescence of the mycelium does not play a role because it is destroyed as the result of the unfavorable changes of the habitat before reaching the senescent phase. It appears that, in accordance with the ''disposable soma theory of aging'' [1–3], the adaptation to the special ecological niche has led to the evolution of a life cycle in which most of the available energy is expended into an efficient and fast reproduction and only little energy into cellular maintenance functions (e.g., defense and repair).

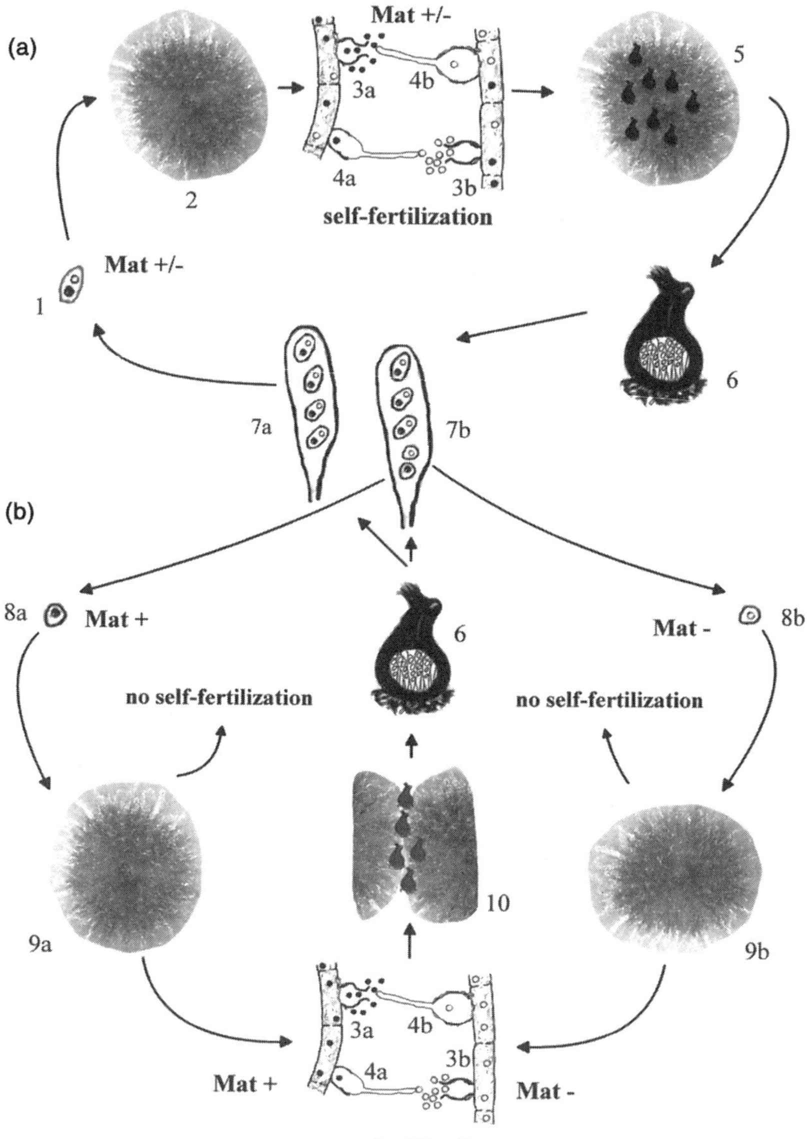
(a)
Mat +/-
3a
4b
5
2
4a
3b
self-fertilization
1
Mat +/-
6
7a
7b
(b)
8a
Mat +
6
Mat -
8b
no self-fertilization
no self-fertilization
10
9a
9b
3a
4b
3b
4a
Mat +
Mat -
cross-fertilization

Only under well-defined and constant laboratory conditions do wild-type strains display the senescence phenotype first reported >50 years ago [4]. Under such conditions, the progression through a complete life cycle takes ~2 weeks (Fig. 1). After ascospore germination, a fast-growing, highly branched mycelium develops. On this mycelium two types of gametangia are produced: the protoperithecia, representing the female gametangia, and the spermogonia, in which the male gametes, the spermatia, develop (Fig. 1a). After fertilization of a protoperithecium by a spermatium and after a subsequent karyogamy and meiosis, many asci are formed in a single perithecium. The vast majority of asci contains four dicaryotic ascospores. Most of them, owing to the presence of two different mating-type idiomorphs (Mat+/−) give rise to self-fertile mycelia (Fig. 1a) which produce a new generation of progeny. Thus, like *Neurospora tetrasperma*, *P. anserina* is secondary homothallic [5,6]. However, in ~1–2% of all asci, two smaller, monocaryotic ascospores (Fig. 1) are produced instead of one dinucleate, larger one. These ascospores and the derived mycelia are of particular significance for genetic investigations because the mycelia are self-incompatible although they produce both types of gametangia (Fig. 1b). Completion of the sexual cycle takes only place when two homocaryotic mycelia of the opposite mating type contact each other and the protoperithecia of one mating type are fertilized by spermatia of the opposite mating type (cross-fertilization). Thus, in principle, the use of homocaryotic mycelia derived from mononucleate ascospores allows us to experimentally perform well-defined genetic crosses between selected strains (e.g., mutants, different geographical races) and to analyze the inheritance

Figure 1 Life cycle of *Podospora anserina*. (a) Since the vast majority of all ascospores [1] produced during sexual reproduction contain two nuclei and both mating type idiomorphs (Mat+ and Mat−), the developing mycelia are self-fertile. At these mycelia, male gametangia, the spermogonia (3a, 3b), and female gametangia, the protoperithecia (4a, 4b) of both mating types (indicated by black-and-white nuclei), develop. Protoperithecia of one mating type are fertilized by the male gametes, the spermatia, of the opposite mating type, and vice versa. After karyogamy and meiosis meiosporangia, the asci develop in special fruiting bodies, the perithecia [6]. Most asci contain four linearly ordered ascospores with two nuclei. These dicaryotic ascospores give rise to the next sexual cycle. One percent to 2% of all asci are irregular, containing 5–8 ascospores. (b) In a five-spored ascus two ascospores are smaller and contain only one nucleus and only one mating type idiomorph (8a, 8b). The resulting mycelium growth is normal but is unable to proceed via the complete sexual cycle. Only if two mycelia of the opposite mating type contact each other does a cross-fertilization lead to the production of fruiting bodies in the contact zone of both individual mycelia. (From Ref. 10.)

of any character of interest expressed by these strains (e.g., spore color, life span). Moreover, since the cytoplasm of a mating product is almost completely derived from the female gametangium and not from the male gamete, the analysis of reciprocal crosses allows us to discriminate between mendelian and maternal inheritance and thus the type of genetic trait (nuclear or cytoplasmic) controlling a given phenotype [7–10]. Because of these characteristics *P. anserina* has been used extensively to unravel the genetic control of different biological processes. In particular, the basis of senescence has been investigated in great detail. This is because, in contrast to many other systems, the life span of wild-type strains is short and specific for different geographical isolates (different wild-type isolates) and mutants, allowing the formal and molecular genetic analysis of the mechanisms giving rise to differences in the onset of senescence. This type of analysis resulted in the elucidation of the first clear molecular pathways involved in life span control and has inspired experimental aging research in general. Today, as in other systems [11], it is clear that aging of *P. anserina* cultures is under the control of a complex molecular network. Individual components and branches of this network have been identified and characterized, and interactions among different components are emerging. In this chapter, after referring to the most important data of earlier investigations, the current view of the molecular basis of aging in this model system is summarized, and open questions and future directions are discussed. This chapter is not aimed at providing a complete overview. For more details about earlier investigations, the reader is referred to previous reviews in which different aspects of fungal senescence have been addressed [8,12–25].

2 THE SENESCENCE PHENOTYPE IS CONTROLLED BY ENVIRONMENTAL AND GENETIC FACTORS

2.1 The Senescence Syndrome

In the early 1950s, George Rizet described for the first time that all wild isolates of *P. anserina* senesce when cultivated under vegetative conditions [4]. This phenotype, the so-called senescence syndrome, is characterized by an age-related decrease of the growth rate of a mycelium, a reduction in the formation of aerial hyphae, and an increase in the pigmentation of an aging colony (Fig. 2). Finally, the growth of a culture ceases completely and the peripheral hyphae die. At the microscopic level, the peripheral hyphae show abnormal branching and swellings [26].

2.2 Onset of Senescence Depends on Environmental and Genetic Conditions

2.2.1 Environmental Control

Early investigations demonstrated that life span in *P. anserina* is clearly depending on different environmental conditions. In general, conditions leading to

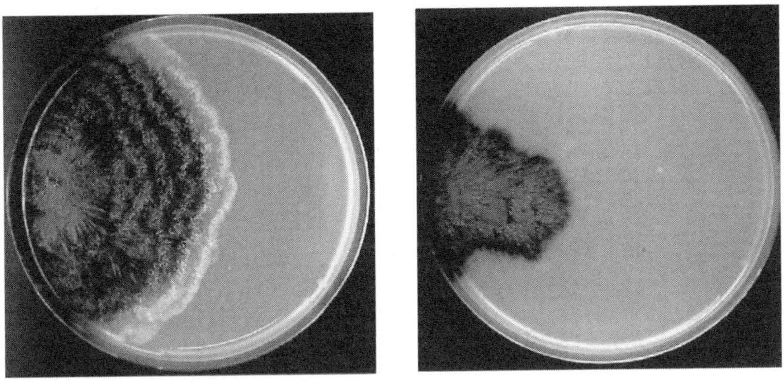

FIGURE 2 Comparison of a juvenile (left) and a senescent (right) wild-type culture of *P. anserina* grown on agar plates. The senescent culture stopped growing at the hyphal tips. The culture is further characterized by a darker pigmentation and a reduced production of aerial hyphae.

a reduction of the metabolism (e.g., low temperature, growth on nutritionally "poor media") were found to result in an increased life span.

Temperature. The impact of the temperature under which strains are cultivated was carefully investigated in the past [27]. Generally cultures are grown at temperatures between 25°C and 27°C. Constant growth at higher temperature was found to significantly shorten life span whereas growth of senescent cultures for a short period of time at 36°C postpones senescence [28]. Growth at temperatures <27°C slows the process of senescence. Below 16°C strains do not senesce. Moreover, incubation at 4°C was found to lead to the rejuvenation of senescent cultures [27]. Although some of the reported data are suggestive and link aging to different aging theories (e.g., rate of living theory), the underlying molecular mechanisms remain to be elucidated in detail.

Metabolic Inhibitors. In early investigations, different metabolic inhibitors added to the growth medium were found to significantly affect the life span of strains [29]. At concentrations that do not or only slightly affect the growth rate of the mycelia, life span is clearly extended. Inhibitors of mitochondrial ribosomes like kanamycin, neomycin, streptomycin, puromycin, and tiamulin were reported to be effective [26,30]. Two other classes of compounds leading to an increase in life span are also linked to mitochondrial functions. The first class are intercalating substances like ethidium bromide, acridine, and acriflavine which, in eukaryotes, preferentially act on mitochondrial DNA (mtDNA). As a second class, inhibitors of the mitochondrial respiratory chain like mucidin and potassium cyanide were found to be effective [29,31,32]. Moreover, at higher concentrations, cycloheximide, an inhibitor of cytoplasmic ribosomes, also increases life span [32]. Recently, molecular data suggested that an age-dependent change in the specificity of cytoplasmic translation appears to affect mitochondrial functions via an age-dependent alteration in the spectrum of nuclear-encoded mitochondrial proteins [25]. Taken together, the data are highly indicative and emphasize the importance of metabolism and in particular of mitochondrial functions in life span control and senescence.

Carbon Sources and Carbon Repression. Since probably the most important function of mitochondria is their role in energy transduction, it is not surprising that environmental interferences in the energy metabolism also have a strong impact on life span in *P. anserina*. A first clear indication was the observation that cultures grown in nutritionally ''rich media'' are characterized by an early onset of senescence [27]. Previously, a detailed analysis was performed using different carbon sources. Cultivation of strains on media containing acetate, glycerol, melibiose, and raffinose led to a significant life span increase in comparison to growth on cornmeal medium, a commonly used, undefined ''rich medium.'' Life span extension was not observed when cultures were grown on media containing glucose, lactose, or galactose as the sole carbon source. Interestingly, these carbon sources are effective in carbon catabolite repression. However, although showing a repressing effect at concentrations >5 g/L, glucose concentrations of 0.5 g/L did not lead to this effect. At this low glucose concentration, life span was greatly increased to 95 days in comparison to 24 days on medium containing the higher glucose concentration. Evidently, and most significantly, this type of caloric restriction extends life span enormously. Moreover, under these growth conditions, the growth rate of the mycelium was faster than on medium containing 5 g/L glucose [29,33], indicating that the caloric restriction had no negative impact on the general performance of the culture.

From studies with different yeasts it is known that cAMP levels are inversely proportional to the concentration of glucose in the growth medium [34] and that carbon catabolite repression can be relieved by extracellular cAMP addition [35]. Taking these observations into account, the effect of increasing cAMP levels on the life span of *P. anserina* was analyzed [33]. In the corresponding experiments, intracellular cAMP levels were increased either directly by the addition of cAMP or indirectly by inhibition of the enzyme cAMP phosphodiesterase via theophylline and caffeine. All three additives led to a significant increase in life span without affecting the morphology or the growth rate of the strain. At defined concentrations (20 μM/L cAMP, 0.5 mM/L theophylline, 1.0 mM/L caffeine) the life span was doubled in comparison to the control.

At the molecular level, another link between life span and carbon catabolite repression emerged recently. A reverse transcriptase differential display (RT-PCR) analysis of the wild-type strain of *P. anserina* and the long-lived mutant grisea resulted in the identification and a subsequent characterization of *PaGrg1*, a homolog of the glucose-repressible gene of *Neurospora crassa* [36]. The impact of glucose on the expression of this gene was verified by Northern blot analysis. Significantly, glucose repression was found to decrease during aging. In both analyzed strains, *PaGrg1* transcript levels were found to increase during senescence. Interestingly, in the long-lived mutant *PaGrg1*, transcripts were found to be much lower than in the wild-type strain. This appears to be due to the control of the expression of *PaGrg1* by the copper-modulated transcription factor GRISEA. In the grisea mutant, this transcription factor is not available and thus not involved in the control of *PaGrg1* expression [21,37–39]. Interestingly, the upstream sequence of *PaGrg1* contains two CREA (cyclic AMP–responsive elements) consensus sequences [40]. In *Aspergillus nidulans*, this sequence is known to bind a negative regulator involved in carbon catabolite repression [41,42]. Furthermore, a CREB consensus sequence was identified in the 5′ untranslated region of *PaGrg1*. In mammalian cells, CREB is a nuclear transcription factor that, after phosphorylation via a cAMP-dependent proteine kinase A, activates the transcription of the corresponding target genes [43,44].

Taken together, it is evident that, although the underlying mechanisms of carbon catabolite repression in *P. anserina* and those involved in caloric restriction are not clear at the moment, there is good evidence indicating that these types of metabolic response are effective in *P. anserina* and are of significance in the control of life span.

2.2.2 Genetic Control

In the past, one main focus of aging research in *P. anserina* was the elucidation of molecular mechanisms involved in life span control. The direction from which this question was addressed was mainly from genetics. In particular, the demonstration of age-related changes occurring in the mtDNA, the characterization of specific life span mutants, the cloning and characterization of specific genes, and the construction of transgenic strains revealed important clues. A complex network emerged in which the semiautonomous mitochondria and the energy metabolism play a major role.

Age-Related mtDNA Reorganizations. From earlier genetic investigations it was clear that the onset of senescence in *P. anserina* is under the control of specific nuclear genes as well as extranuclear genetic traits [27,29,45]. A covalently closed circular DNA species, termed plDNA or α-senDNA, was demonstrated to accumulate in mitochondria of senescent cultures [46,47]. In juvenile cultures, plDNA is an integral part of the high molecular weight mtDNA and represents the first intron (pl-intron) of the gene coding for the largest subunit of the cytochrome oxidase (COX) [26,48–51]. During aging of wild-type cultures, the pl-intron becomes systematically liberated and amplified. In parallel, large parts of the mtDNA are deleted (Fig. 3).

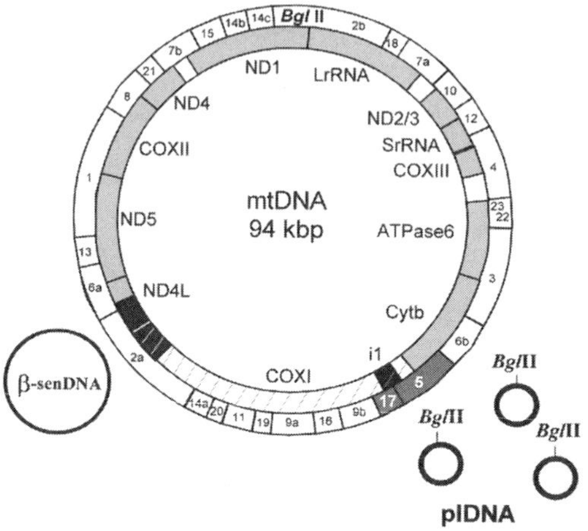

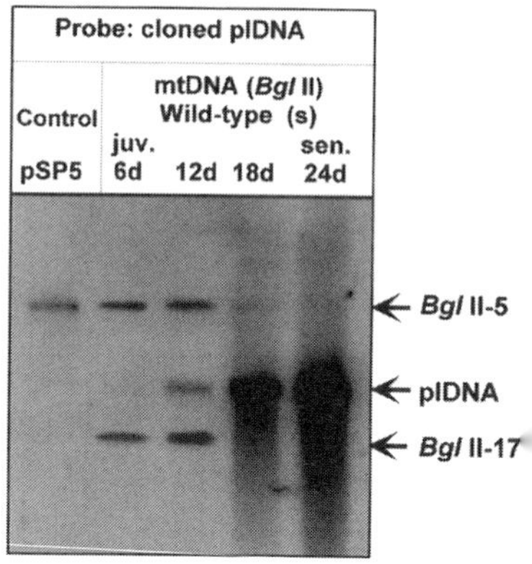

FIGURE 3 Age-related mtDNA reorganizations in *P. anserina.* The mtDNA of juvenile cultures is a circular molecule of 94 kbp and codes for different proteins of the mitochondrial respiratory chain (abbreviations of the different components are indicated in the inner part of the restriction map of the mtDNA) and of the mitochondrial ATP-synthase. The approximate position of these genes and of two genes coding for the mitochondrial rRNA (LrRNA, SrRNA) is indicated in the inner circle of the mtDNA map. The first intron of the *CoxI* gene (il) giving rise to the formation of the circular plDNA is indicated in black. The approximate region giving rise to the generation of another circular molecule, the β-senDNA, is also indicated in black. The recognition sites for the restriction endonuclease *Bgl*II are indicated on the outer circle. The right part of the figure shows a Southern blot of mtDNA from *P. anserina* cultures of different ages from juvenile (juv.) to senescent (sen.). The DNA was cut with *Bgl*II and hybridized to the cloned plDNA. In juvenile cultures only two fragments (*Bgl*II-5 and -17) are recognized. These fragments indicate a functional mtDNA. During aging, as a result of rearrangements, an additional DNA band of 2.5 kbp shows up. In senescent cultures this is the only detectable band hybridizing to the cloned plDNA. The 2.5-kbp *Bgl*II fragment corresponds to the autonomous circular plDNA molecule that systematically accumulates during aging of *P. anserina* wild-type strains.

Since the age-related reorganization of the mtDNA is almost quantitative, the vast majority of mtDNA molecules in senescent cultures are extensively rearranged [49,52]. In addition, reorganization processes in which the pl-intron is involved, pl-intron-independent mtDNA rearrangements between short, dispersed, direct repeats occur frequently, but not systematically, during aging, and resemble mtDNA reorganizations found in different biological system occurring in pathological situations and during aging [53–57]. In *P. anserina*, it was demonstrated that this type of recombination leads to the deletion of one part of the mtDNA. Significantly, since the resulting mtDNA subcircle, termed β-senDNA (Fig. 3), is able to replicate, it is retained in senescent cultures [58,59].

For a long time, the age-related mtDNA reorganizations giving rise to the amplification of the plDNA were thought to be a prerequisite for aging of *Podospora* cultures. This conclusion was based on different pieces of evidence. Most importantly, plDNA was systematically demonstrated to accumulate during aging of different wild-type isolates, strains that all senesce after a specific growth period. Moreover, various long-lived mutants were reported that either do not contain this autonomous genetic element in higher amounts or are characterized by a delayed amplification [60–65]. Surprisingly, recently, different strains with an increased but still finite life span were found to contain no amplified plDNA in the senescent stage [38,66]. It thus became clear that the amplification of plDNA is not a prerequisite for senescence although, under natural conditions, it plays an important role in accelerating senescence.

The molecular mechanism by which plDNA is involved in life span control has been elucidated in detail. It is dependent on the transposition of the pl-intron leading to an age-related increase in repeated mtDNA sequences and the subsequent recombination between these sequences (Fig. 4). Intron transposition occurs either to a position directly downstream of the first *CoxI* exon (''homing-like'' transposition) or to other acceptor sites in the mtDNA (''ectopic'' transposition). As a consequence, two or more copies of the 2.5-kbp intron sequence are found either in tandem or dispersed in the same mtDNA molecule [67,68]. Subsequent homologous recombination processes between these duplicated sequences appear to account for the formation of circular plDNA molecules or to other mtDNA subcircles of different size. Depending on whether or not these circles contain an ''origin of replication,'' they are retained or become lost during subsequent growth. It thus appears that the occurrence of the amplified plDNA is a good marker of transposition processes and of subsequent homologous recombination greatly contributing to the characteristic age-related mtDNA reorganizations observed during senescence of *P. anserina*. These processes depend on different factors. Intron transposition appears to be proceeded by a reverse-transcriptase step depending on the activity of a protein encoded by an open reading frame on the pl-intron [51,52,69,70]. Homologous recombination between the duplicated sequences was recently demonstrated to be dependent on the availability of copper [68].

There are different life span mutants in which, for different reasons, the amplification of the plDNA is either completely affected or delayed. From the considerations above, it is trivial that a complete or partial deletion of the pl-intron has the first consequence [64,71]. However, other cases are not so clear. For example, in long-lived mutant AL2, a delayed amplification of plDNA was

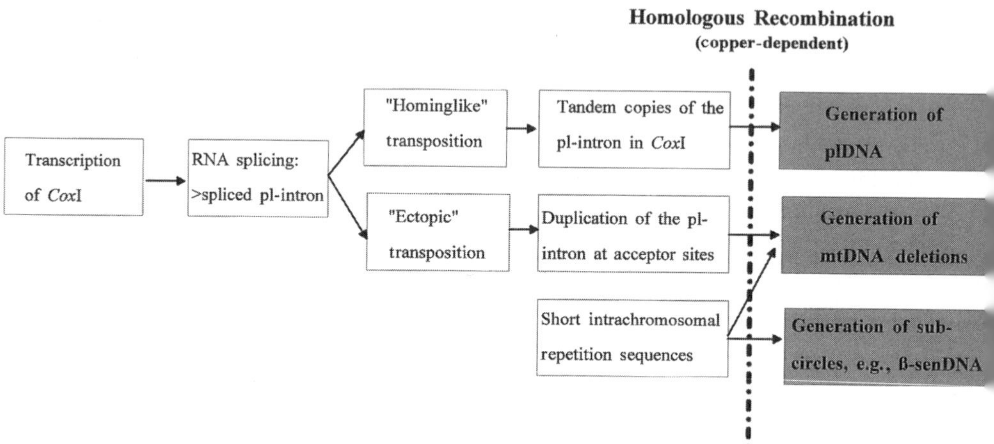

FIGURE 4 Model explaining the different age-related mtDNA rearrangements as the result of homologous recombination between direct repeats. The upper two lanes explain the generation of the autonomous circular plDNA and of gross mtDNA deletions as the result of homologous recombination between duplicated copies of the pl-intron. This intron is able to transpose either to the position in the *CoxI* gene where one copy is already located ("hominglike" transposition) or into different acceptor sites in the mtDNA ("ectopic" transposition). Transposition of the intron is dependent on *CoxI* expression and on the activity of an intron-encoded reverse transcriptase. In addition to the pl-intron-dependent homologous recombination processes, homologous recombination between short repeats scattered throughout the mtDNA can also lead to mtDNA reorganizations. One prominent process is the generation of the circular β-senDNA which is often but not systematically found in senescent *P. anserina* cultures.

correlated with the occurrence of a linear plasmid, pAL2-1 [62,63,72,73]. In the mutant, the plasmid was demonstrated to be present in an autonomous and an integrated stage. Plasmid integration occurred primarily in the third intron of the apocytochrome b gene [74]. In addition, a deletion of the mtDNA was identified in the mutant. The deletion corresponded to a wild-type mtDNA stretch that appears to be derived from a linear plasmid with homology to pAL2-1 [75,76]. The significance of the linear plasmid in longevity is demonstrated by the selection of the offspring of defined crosses. Progeny containing the mutant-specific mtDNA but no pAL2-1-specific sequences were characterized by a short, wild-type-specific life span [77]. Cytoplasmic transfer of the autonomous linear plasmid to a short-lived strain containing no pAL2-1 sequences resulted in long-lived strains containing both autonomous and integrated pAL2-1 sequences. These data revealed that in *P. anserina* the linear plasmid is indeed responsible for the life span–extending effect [78]. This finding is in marked contrast to a number of senescing *Neurospora* strains in which linear plasmids similar to pAL2-1 act as mtDNA mutators leading to mitochondrial deficiency and senescence [79–81].

Taken together, it is clear that there are a number of different mechanisms that affect the stability of the mtDNA and consequently the onset of senescence. In *P. anserina*, the mobile pl-intron plays an important role in accelerating mtDNA instabilities. Since the underlying processes take place in all wild-type strains and only some laboratory mutants were found not to follow this route of aging, it seems to be likely that the mechanism has evolved due to some specific constraints under which the life cycle of *P. anserina* is completed in nature (see Sec. 1).

Mitochondrial Dysfunction, Retrograde Regulation, and Life Span Control. What is the significance of the above-mentioned findings in respect to aging of *P. anserina* cultures? To answer this question, data from the characterization of different long-lived strains have to be considered. In particular, two mutants in which mitochondrial energy transduction processes are affected provide significant clues. In the nuclear grisea mutant [82], respiratory electron transfer via the cyanide-sensitive COX is severely affected [83]. The mutant phenotype is due to a loss-of-function mutation in nuclear gene *Grisea* [38]. The gene codes for the copper-modulated transcription factor GRISEA which is involved in tight control of copper homeostasis [37,84]. Owing to the mutation, high-affinity copper uptake is defective, leading to a cellular copper deficiency (Fig. 5). Since copper is a cofactor of COX, electron transport via complex IV is affected. In principle, such a defect should be lethal since *P. anserina*, as an obligate aerobe, depends on mitochondrial ATP generation. However, it appears that partially dysfunctional mitochondria of the grisea mutant are able to signal to the nucleus and lead to the expression of *PaAox*, a gene coding for an alternative oxidase (AOX). This enzyme contains iron instead of copper and branches at the ubiquinone pool. Like COX, it transfers electrons to oxygen, giving rise to the formation of water (Fig. 6). Since the AOX is located upstream of complex III, the formation of the electron motive force is almost completely restricted to complex I. Consequently, the production of ATP is reduced. However, since in the grisea mutant copper deficiency is not complete and low amounts of copper enter the cell via a low-affinity uptake system, the mitochondrial respiratory chain of the grisea mutant respires via both a copper-dependent COX and the iron-dependent AOX [83]. The Δ life span of this mutant is increased ~60% in comparison to the wild-type strain.

The extranuclear ex mutant is characterized by a deletion of large parts of the *CoxI* gene coding for the largest subunit of complex IV [64]. In this mutant, the block of the COX-dependent respiratory chain is complete. *PaAox* is induced.

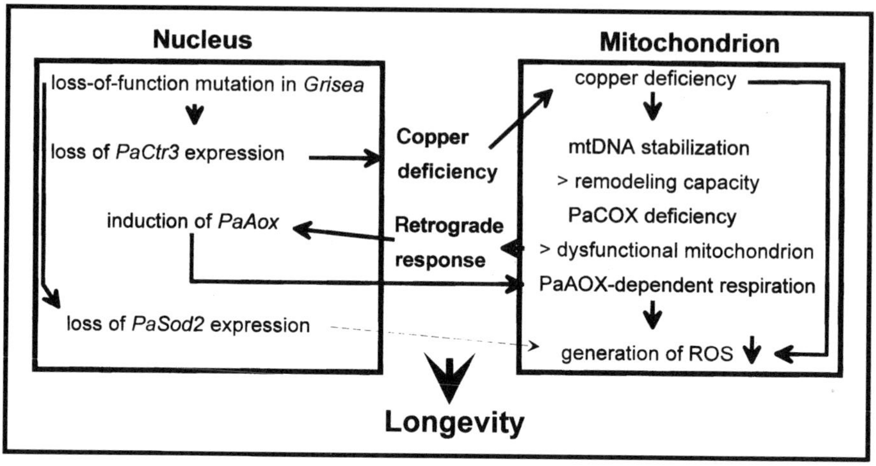

FIGURE 5 Mitochondrial–nuclear interactions in the long-lived grisea mutant of *P. anserina*. A loss-of-function mutation in *Grisea* coding for a copper-modulated transcription factor leads to cellular copper deficiency since *PaCtr3*, the gene coding for a high-affinity copper transporter, is not expressed in the mutant. In mitochondria this leads to a stabilization of the mtDNA which remains available for remodeling processes. On the other hand, copper deficiency leads to deficiency in COX-dependent respiration. The resulting dysfunctional mitochondria signal to the nucleus and induce the copper-independent alternative oxidase (PaAOX). The PaAOX-dependent respiration rescues the mutant. Moreover, since this respiratory pathway produces less ROS than the COX-dependent respiratory chain, the mutant is characterized by an increased life span.

Protein levels of the AOX are much higher than in the grisea mutant. Significantly, the ex mutant does not show any symptoms of senescence even after >10 years of continuous growth.

Collectively, the data show that in *P. anserina*, dysfunctional mitochondria can compensate for specific defects by the induction of certain genes which, under normal conditions, are not expressed. This type of response was first demonstrated in yeast and was named the ''retrograde response'' [85–87]. Interestingly, the induction of the retrograde response also in yeast was found to lead to an increase in life span [88,89]. At this time, we have only demonstrated that a deficiency in COX can be compensated for by the induction of *PaAox* encoding a backup system of a terminal oxidase in the mitochondria respiratory chain. We do not know whether, as in yeast, the expression of other genes routing

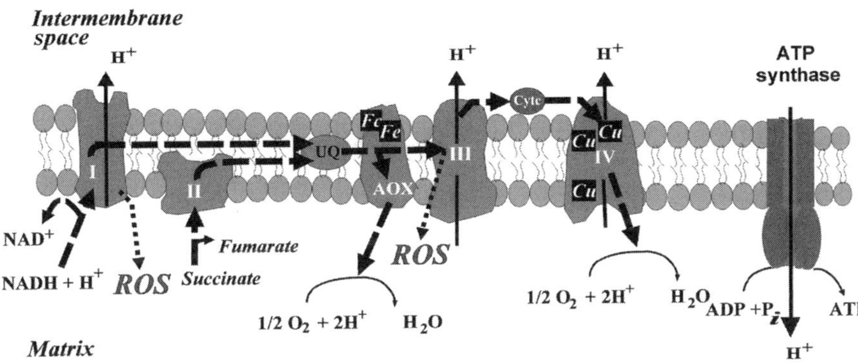

FIGURE 6 A branched respiratory chain in the copper-deficient grisea mutant. Because copper is limited, most electrons funneled into the respiratory chain at complexes I and II are not transferred to oxygen at complex IV but via the di-iron alternative oxidase. Following this route one prominent site of ROS generation at complex III is bypassed.

the metabolism to completely different pathways can be generally induced by dysfunctional mitochondria.

Although the induction of a retrograde response in *P. anserina* is intriguing, the question of how it can lead to an extension of life span remains open. In *P. anserina*, the answer appears to be strongly related to the characteristics of the two types of respiratory pathways. In one recent study in which *PaAox* was induced in a transgenic COX deficiency strain, it was shown that the generation of reactive oxygen species is lower than in the wild-type strain respiring via the standard COX-dependent oxidase [90]. These data are in agreement with data from higher plants demonstrating a reduced generation of ROS via an AOX-dependent respiratory chain [91]. It appears that a reduction of ROS significantly affects the life span of *P. anserina* and links the mitochondrial energy metabolism to the refined free radical theory of aging [92–94].

Oxidative Stress and Senescence. The refined free radical theory of aging is one of only a few aging theories of a more general applicability [reviewed in 95]. The theory is strongly linked to mitochondrial energy transduction. As byproducts of this metabolic pathway, ROS are generated that can damage all types of biomolecules including nucleic acids, lipids, and proteins. In *P. anserina*, as mentioned above, it has been shown that ROS formation in a COX-deficient transgenic strain respiring via the alternative pathway is strongly reduced [90]. This appears also to be the case in the two long-lived mutants grisea and ex. The differences in life span of these mutants, 39 days vs >10 years, is due to different reasons. First, in the grisea mutant, owing to respiration via both the COX-dependent and the AOX-dependent pathway ROS production appears to be higher than in the immortal ex mutant. Second, because of the reduced copper levels in the grisea mutant, the activity of the Cu/ZnSOD in the cytoplasm is lower than in the ex mutant (unpublished). In addition, since the gene coding for the MnSOD, as part of the mitochondrial enzymatic systems protecting against ROS, is a target gene of transcription factor GRISEA, mitochondria of the grisea mutant are MnSOD deficient and consequently less protected than in the ex mutant (Fig. 5). There are certainly other, additional factors contributing to the observed life span differences but it appears to be clear that differences in ROS generation and in the protection system against oxidative stress have an important impact on life span. A part of this protection system is the recently identified metallothionein and a putative *o*-methyltransferase of *P. anserina* [96,97]. Whereas the role of the metallothionein in protecting against copper-derived oxidative stress is rather clear, the function of the *o*-methyltransferase is still speculative. However, from its biochemical characteristics the protein may catalyze methylation reactions of hydroxyl groups which, if not modified, may be converted to a free radical and participate in radical chain reactions leading to increased levels of ROS.

As mentioned, life span in *P. anserina* is strongly affected by the stability of the mtDNA. Given that during growth of the peripheral hyphae proteins of the respiratory chain become progressively damaged (e.g., via ROS), a replacement of these components by newly synthesized proteins is only possible if the genes encoding these proteins are accessible. Since mitochondrial proteins are encoded by both the mtDNA and the nuclear DNA, remodeling is an integrated action of the nucleus, the cytoplasm, and the mitochondrion (Fig. 7). Moreover, in the actively growing peripheral hyphal tips, the propagation of mitochondria by division is also dependent on the integrity of the mtDNA. However, in wild-type strains, the mtDNA becomes extensively reorganized in a rather short period of time and consequently the synthesis of mtDNA encoded proteins of the respiratory chain is time limited. In contrast, in different mutants, the mtDNA is stabilized and the encoded genes are longer available to be expressed in order to remodel affected respiratory chains. Of course, since the vast majority of genes coding for different components of functional mitochondria are encoded by the nucleus, the impact of nuclear genes is obvious. These genes code for various functions including parts of the respiratory chain, the whole set of enzymes of the citric acid cycle, the components of the protein import machinery (TIM and TOM) as well as the enzymes involved in mtDNA replication, and the expression of mitochondrial genes. The transport of the various gene products and of cofactors like copper into the different compartments of the organelle and the correct assembly of supramolecular complexes appears to be of prime significance for the remodeling of existing mitochondria and for division of mitochondria in actively

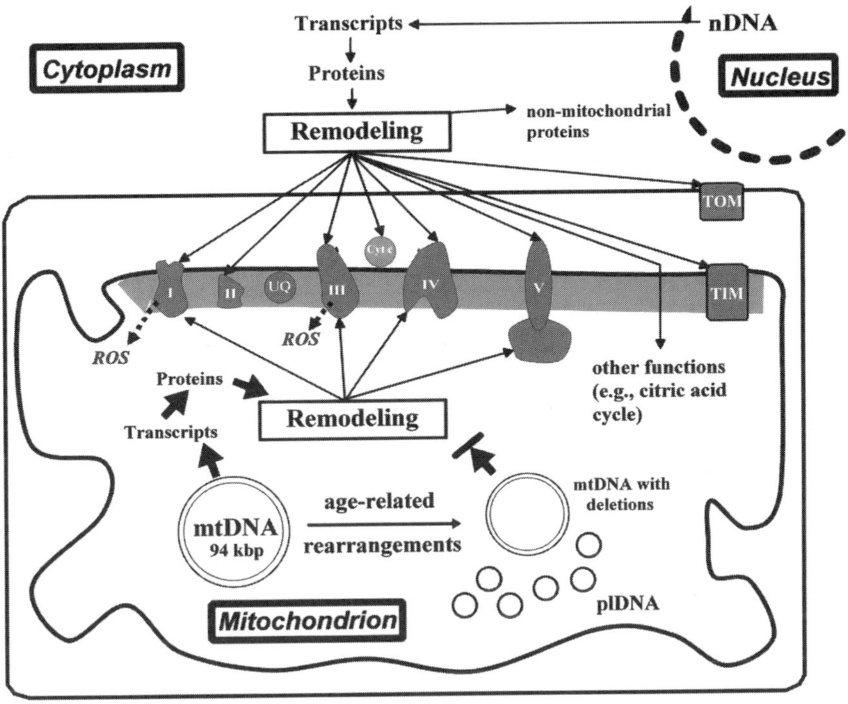

FIGURE 7 Nuclear, cytoplasmic, and mitochondrial interactions involved in remodeling of impaired mitochondria. In particular, the components of the respiratory chain are prone to damage via ROS generated at complexes I and III. As the respiratory chain becomes impaired, more ROS are formed until this type of a vicious cycle leads to nonfunctional organelles. However, within certain limits, impaired pathways can be remodeled. Remodeling of the respiratory chain requires the expression of both mitochondrial genes and nuclear genes. In wild-type strains, mtDNA-encoded proteins are synthesized only for a short period of time since the mtDNA becomes heavily rearranged during aging. Mutants in which the mtDNA is stabilized are characterized by a longer remodeling capacity. In addition to this mitochondrial basis, there are many other factors contributing to mitochondrial remodeling functions. Various nuclear genes need to be expressed in the cytoplasm, and the resulting proteins need to be transported to the correct compartment via the mitochondrial translocators TIM and TOM. Finally, a correct assembly of individual compounds has to occur. The basis of many of these processes is poorly understood but appears to be important also for aging processes.

growing parts of a mycelium. At this time only very limited data are available about the age-related mechanisms of these basic processes. As mentioned above, the delivery of copper to the respiratory chain is one specific example. Another one emerged from the analysis of a long-lived mutant in which the mutated gene was demonstrated to code for PaTOM70, a component of the mitochondrial protein import machinery. Finally, it has been speculated that a number of nuclear mutations modifying translational accuracy also affect mitochondrial functions and life span via a modification of the mitochondrial protein spectrum [25].

3 CONCLUSIONS AND PERSPECTIVES

In *P. anserina*, a large body of data accumulated demonstrating a central role of the energy metabolism and of mitochondria. In respect to aging, these semiautonomous organelles can be viewed as a kind of a cellular ''Achilles' heel.'' Since the energy metabolism and also the function of mitochondria depend on a large number of individual components, it is no surprise that senescence is controlled by a complex network of interacting branches of individual molecular pathways. Many different components and conditions have an effect on the performance of these organelles. As a side effect of a decreased performance of the respiratory chain, an increased generation of ROS as byproducts of oxygenic energy transduction processes may occur. Within certain limits, cells are able to cope with these harmful products since they contain specific defense systems directed against ROS. Moreover, once damage has been manifested, repair and remodeling systems can deal with the problem. These ''caretaker'' systems are genetically controlled and are part of the complex network affecting life span in all biological systems.

In *P. anserina* it is clear that there are a variety of pathways affecting longevity. First, the generation of ROS appears to be of prime significance. Reducing the production of ROS, as it occurs in different mutants and specific transgenic strains, leads to increased life span. In this context metabolic processes are of special significance. Second, the instability of the mtDNA is greatly responsible for the short life span observed in the different wild-type strains, and appears to have evolved in order to adapt the organism to the specific biological niche. Under laboratory conditions, any modification resulting in a stabilization of the mtDNA was found to significantly increase life span. The instability of the mtDNA of *P. anserina* is the result of different mechanisms. Consequently, there are different ways of increasing the stability of the mtDNA. Third, the impact of the cellular defense system against different types of damage is beginning to emerge. In recent years, a number of individual genes coding for components of this system have been cloned and characterized [96,97]. However, more work is needed to obtain a solid picture of the contribution of these factors to the complete molecular network. At this time, although certainly important, the role of repair systems in *P. anserina* is completely unexplored. In the future, it will be of great interest to elucidate the detailed role not only of this system but also of additional pathways (e.g., details of the significance of cytoplasmic translational accuracy). Such an integrative analysis of the molecular basis of aging in one specific biological system can be expected to provide important clues of general importance to unravel the corresponding mechanisms of aging and of age-related diseases also in complex long-lived species including humans.

ACKNOWLEDGMENTS

The experimental work of the authors was supported by a grant of the Deutsche Forschungsgemeinschaft (Bonn, Germany) to H.D.O.

REFERENCES

1. TBL Kirkwood. Evolution of ageing. Nature 270:301–304, 1977.
2. TBL Kirkwood. Human senescence. BioEssays 18:1009–1016, 1996.
3. TBL Kirkwood, SN Austad. Why do we age? Nature 408:233–238, 2000.
4. G Rizet. Sur l'impossibilite d'obtenir la multiplication vegetative ininterrompue et illemite de l'ascomycete *Podospora anserina*. C R Acad Sci Paris 237:838–855, 1953.
5. HLK Whitehouse. Heterothallism and sex in fungi. Biol Rev 24:411–447, 1949.
6. HLK Whitehouse. Multiple-allelomorph heterothallism in the fungi. New Phytol 48: 212–224, 1949.
7. K Esser, R Kuenen. Genetik der Pilze. Heidelberg: Springer-Verlag, 1967.
8. K Esser. *Podospora anserina*. In: RC King, ed. Handbook of Genetics, Vol I. New York: Plenum Press, 1974, pp 531–551.
9. HD Osiewacz. Genetic analysis of senescence in *Podospora anserina*. In: CJ Bos, ed. Fungal Genetics. New York: Marcel Dekker, 1996, pp 317–335.
10. K Esser. Kryptogamen. Heidelberg: Springer-Verlag, 2000.
11. TBL Kirkwood. Molecular gerontology. Bridging the simple and the complex. In: O Toussaint, H Osiewacz, G Lithgow, C Brack, eds. Molecular and Cellular Gerontology, Vol 908. New York: New York Academy of Sciences, 2000, pp 14–18.
12. K Esser, P Tudzynski. Senescence in fungi. In: KV Thimann, ed. Senescence in Plants. Boca Raton: CRC Press, 1980, pp 67–83.
13. K Esser, HD Osiewacz, U Stahl, P Tudzynski. Nuclear mitochondrial interactions cause senescence in the filamentous fungus *Podospora anserina*. In: RJ Schweyen, K Wolf, F Kaudewitz, eds. Mitochondria 1983. Berlin: Walter de Gruyter & Co, 1983, pp 251–257.
14. K Esser. Genetic control of aging: the mobile intron model. In: M Bergener, M Ermini, HB Stähelin, eds. The 1984 Sandoz Lectures in Gerontology. London: Academic Press, 1984, pp 3–20.

15. K Esser, U Kück, C Lang-Hinrichs, PA Lemke, HD Osiewacz, U Stahl, P Tudzynski. Plasmids of eukaryotes. Fundamentals and applications. Heidelberg: Springer-Verlag, 1986.
16. HD Osiewacz. Molecular analysis of aging processes in fungi. Mutat Res 237:1–8, 1990.
17. AJF Griffiths. Fungal senescence. Annu Rev Genet 26:351–372, 1992.
18. HD Osiewacz. Genetic Control of Aging in the Ascomycete *Podospora anserina*. Berlin: Springer-Verlag, 1992.
19. HD Osiewacz. Aging and genetic instabilities. In: K Esser, GM Martin, eds. Molecular Aspects of Aging. New York: Wiley & Sons, 1995, pp 29–44.
20. HD Osiewacz, A Hamann. DNA reorganization and biological aging. A review. Biochemistry (Moscow) 62:1275–1284, 1997.
21. HD Osiewacz, E Kimpel. Mitochondrial–nuclear interactions and lifespan control in fungi. Exp Gerontol 34:901–909, 1999.
22. HD Osiewacz, C Borghouts. Cellular copper homeostasis, mitochondrial DNA instabilities, and lifespan control in the filamentous fungus *Podospora anserina*. Exp Gerontol 35:677–686, 2000.
23. HD Osiewacz, C Borghouts. Mitochondrial oxidative stress and aging in the filamentous fungus *Podospora anserina*. Ann NY Acad Sci 908:31–39, 2000.
24. HD Osiewacz, S Stumpferl. Metabolism and aging in the filamentous fungus *Podospora anserina*. Arch Gerontol Geriatr 32:185–197, 2001.
25. P Silar, H Lakucque, C Vierny. Cell degeneration in the model system *Podospora anserina*. Biogerontology 2:1–17, 2001.
26. K Esser, P Tudzynski. Prevention of senescence in the ascomycete *Podospora anserina* by the antibiotic tiamulin. Nature 265:454–456, 1977.
27. D Marcou. Notion de longevite et nature cytoplasmique du determinant de la senescence chez quelques champignons. Ann Sci Nat Bot Ser 12:653–764, 1961.
28. G Rizet, D Marcou. Longevite et senecence chez *Podospora anserina*. VIII Congr Intern Bot, 1954, pp 121–128.
29. P Tudzynski, K Esser. Chromosomal and extrachromosomal control of senescence in the ascomycete *Podospora anserina*. Mol Gen Genet 173:71–84, 1979.
30. P Tudzynski, K Esser. Inhibitors of mitochondrial function prevent senescence in the ascomycete *Podospora anserina*. Mol Gen Genet 153:111–113, 1977.
31. J Subik, M Behun, P Smigan, V Musilek. Mode of action of mucidin, a new antifungal antibiotic produced by the basidiomycete *Oudemansiella mucida*. Biochim Biophys Acta 343:363–370, 1974.
32. JR Smith, I Rubenstein. The development of senescence in *Podospora anserina*. J Gen Microbiol 76:283–296, 1973.
33. P Tudzynski. Untersuchungen zur Genetik und Physiologie der Seneszenz des Ascomyceten *Podospora anserina*. Dissertation, Ruhr-Universität, Bochum, Germany, 1978.
34. J Sy, D Richter. Content of cyclic 3′,5′-adenosine monophoshate and adenylyl cyclase in yeast at various growth conditions. Biochemistry 11:2788–2791, 1972.
35. M Fang, RA Butow. Nucleotide reversal of mitochondrial repression in *Saccharomyces cerevisiae*. Biochem Biophys Res Commun 41:1579–1583, 1970.

36. MT McNally, SJ Free. Isolation and characterization of a *Neurospora* glucose-repressible gene. Curr Genet 14:545–551, 1998.
37. HD Osiewacz, U Nuber. GRISEA, a putative copper-activated transcription factor from *Podospora anserina* involved in differentiation and senescence. Mol Gen Genet 252:115–124, 1996.
38. C Borghouts, E Kimpel, HD Osiewacz. Mitochondrial DNA rearrangements of *Podospora anserina* are under the control of the nuclear gene grisea. Proc Natl Acad Sci USA 94:10768–10773, 1997.
39. E Kimpel, HD Osiewacz. *PaGrgl*, a glucose-repressible gene of *Podospora anserina* that is differentially expressed during lifespan. Curr Genet 35:557–563, 1999.
40. B Cubero, C Scazzocchio. Two different, adjacent and divergent zinc-finger binding sites are necessary for CREA-mediated carbon catabolite repression in the proline gene cluster of *Aspergillus nidulans*. EMBO J 13:407–415, 2001.
41. C Bailey, HN Arst. Carbon catabolite repression in *Aspergillus nidulans*. Eur J Biochem 51:573–577, 1975.
42. CE Dowzer, JM Kelly. Analysis of the *creA* gene, a regulator of carbon catabolite repression in *Aspergillus nidulans*. Mol Cell Biol 11:5701–5709, 1991.
43. DOM Andrisani, TE Hayes, B Roos, JE Dixon. Identification of the promoter sequences involved in the cell-specific expression of the rat somatostatin gene. Nucleic Acids Res 15:5715–5728, 1987.
44. GA Gonzalez, P Menzel, J Leonard, WH Fisher, MR Montminy. Characterization of motifs which are critical for activity of the cyclic AMP-responsive transcription factor CREB. Mol Cell Biol 11:306–312, 2001.
45. K Esser, W Keller. Genes inhibiting senescence in the ascomycete *Podospora anserina*. Mol Gen Genet 144:107–110, 1976.
46. U Stahl, PA Lemke, P Tudzynski, U Kück, K Esser. Evidence for plasmid-like DNA in a filamentous fungus, the ascomycete *Podospora anserina*. Mol Gen Genet 162: 341–343, 1978.
47. DJ Cummings, L Belcour, C Grandchamp. Mitochondrial DNA from *Podospora anserina*. II. Properties of mutant DNA and multimeric circular DNA from senescent cultures. Mol Gen Genet 171:239–250, 1979.
48. U Stahl, U Kück, P Tudzynski, K Esser. Characterization and cloning of plasmid-like DNA of the ascomycete *Podospora anserina*. Mol Gen Genet 178:639–646, 1980.
49. U Kück, K Esser, U Stahl. Plasmid-like DNA is part of mitochondrial DNA in *Podospora anserina*. Curr Genet 9:373–382, 1981.
50. HD Osiewacz. Mitochondriale Mosaikgene: Strukturaufklärung eines mobilen Introns aus dem Ascomyceten *Podospora anserina*. Vaduz, Liechtenstein: J Cramer, 1984.
51. HD Osiewacz, K Esser. The mitochondrial plasmid of *Podospora anserina*: a mobile intron of a mitochondrial gene. Curr Genet 8:299–305, 1984.
52. U Kück, HD Osiewacz, U Schmidt, B Kappelhoff, E Schulte, U Stahl, K Esser. The onset of senescence is affected by DNA rearrangements of a discontinuous mitochondrial gene in *Podospora anserina*. Curr Genet 9:373–382, 1985.
53. AW Linnane, S Marzuki, T Ozawa, M Tanaka. Mitochondrial DNA mutations as an important contributor to ageing and degenerative disease. Lancet 1:642–645, 1989.

54. HD Osiewacz, J Hermanns. The role of mitochondrial DNA rearrangements in aging and human diseases. Aging (Milan) 4:273–286, 1992.
55. DC Wallace. Diseases of mitochondrial DNA. Annu Rev Biochem 6:1175–1212, 1992.
56. DA Cottrell, EL Blakeley, GM Borthwick, MA Johnson, GA Taylor, EJ Brierley, PG Ince, DM Turnbull. Role of mitochondrial DNA mutations in disease and aging. In: O Toussaint, H Osiewacz, G Lithgow, C Brack, eds. Molecular and Cellular Gerontology, Vol 908. New York: New York Academy of Sciences, 2000, pp 199–207.
57. DC Wallace. Mitochondrial diseases in man and mouse. Science 283:1482–1488, 2001.
58. C Jamet-Vierny, J Boulay, O Begel, P Silar. Contribution of various classes of defective mitochondrial DNA molecules to senescence in *Podospora anserina*. Curr Genet 31:171–178, 1997.
59. C Jamet-Vierny, J Boulay, JF Briand. Intramolecular cross-overs generate deleted mitochondrial DNA molecules in *Podospora anserina*. Curr Genet 31:162–170, 1997.
60. U Stahl, P Tudzynski, U Kück, K Esser. Replication and expression of a bacterial–mitochondrial hybrid plasmid in the fungus *Podospora anserina*. Proc Natl Acad Sci USA 79:3641–3645, 1982.
61. C Vierny, AM Keller, O Begel, L Belcour. A sequence of mitochondrial DNA is associated with the onset of senescence in a fungus. Nature 297:157–159, 1982.
62. HD Osiewacz, J Hermanns, D Marcou, M Triffi, K Esser. Mitochondrial DNA rearrangements are correlated with a delayed amplification of the mobile intron (plDNA) in a long-lived mutant of *Podospora anserina*. Mutat Res 219:9–15, 1989.
63. J Hermanns. Mitochondriale Genomveränderungen und Altern: Struktur and Funktion eines linearen Plasmides einer langlebigen Mutante von *Podospora anserina*. Berlin: J Cramer, 1992.
64. E Schulte, U Kück, K Esser. Extrachromosomal mutants from *Podospora anserina*: permanent vegetative growth in spite of multiple recombination events in the mitochondrial genome. Mol Gen Genet 211:342–349, 1988.
65. E Schulte. Einfluß struktureller Umordnungen des Chondrioms auf die Seneszenz bei *Podospora anserina*. Berlin: J Cramer, 1988.
66. P Silar, F Koll, M Rossignol. Cytosolic ribosomal mutations that abolish accumulation of circular intron in the mitochondria without preventing senescence of *Podospora anserina*. Genetics 145:697–705, 1997.
67. CH Sellem, G Lecellier, L Belcour. Transposition of a group II intron. Nature 366: 176–178, 1993.
68. C Borghouts, S Kerschner, HD Osiewacz. Copper-dependence of mitochondrial DNA rearrangements in *Podospora anserina*. Curr Genet 37:268–275, 2000.
69. F Michel, B Lang. Mitochondrial class I introns encode proteins related to the reverse transcriptases of retroviruses. Nature 316:641–643, 1985.
70. S Fassbender, KH Bruhl, M Ciriacy, U Kück. Reverse transcriptase activity of an intron encoded polypeptide. EMBO J 13:2075–2083, 1994.
71. F Koll. A 1100-bp sequence of mitochondria DNA is involved in senescence process in *Podospora*: study of senescent and mutant cultures. Plasmid 14:106–117, 1995.

72. J Hermanns, HD Osiewacz. The linear mitochondrial plasmid pAL2-1 of a long-lived *Podospora anserina* mutant is an invertron encoding a DNA and RNA polymerase. Curr Genet 22:491–500, 1992.
73. F Kempken, J Hermanns, HD Osiewacz. Evolution of linear plasmids. J Mol Evol 35:502–513, 1992.
74. J Hermanns, A Asseburg, HD Osiewacz. Evidence for giant linear plasmids in the ascomycete *Podospora anserina*. Curr Genet 27:379–386, 1995.
75. HD Osiewacz, J Hermanns, D Marcou, M Triffi, K Esser. Mitochondrial DNA rearrangements are correlated with a delayed amplification of the mobile intron (plDNA) in a long-lived mutant of *Podospora anserina*. Mutat Res 219:9–15, 1989.
76. J Hermanns, HD Osiewacz. Three mitochondrial unassigned open reading frames of *Podospora anserina* represent remnants of a viral-type RNA polymerase gene. Curr Genet 25:150–157, 1994.
77. J Hermanns, A Asseburg, HD Osiewacz. Evidence for a life span–prolonging effect of a linear plasmid in a longevity mutant of *Podospora anserina*. Mol Gen Genet 243:297–307, 1994.
78. J Hermanns, HD Osiewacz. Induction of longevity by cytoplasmic transfer of a linear plasmid in *Podospora anserina*. Curr Genet 29:250–256, 1996.
79. H Bertrand, BSS Chan, AJF Griffiths. Insertion of a foreign nucleotide sequence into mitochondrial DNA causes senescence in *Neurospora intermedia*. Cell 41:877–884, 1985.
80. BSS Chan, DA Court, PJ Vierula, H Bertrand. The *kalilo* linear senescence-inducing plasmid of *Neurospora* is an invertron and encodes DNA and RNA polymerases. Curr Genet 20:225–237, 1991.
81. DA Court, AJF Griffiths, SR Kraus, PJ Russell, H Bertrand. A new senescence-inducing mitochondrial linear plasmid in field-isolated *Neurospora crassa* strains from India. Curr Genet 19:129–137, 1991.
82. H Prillinger, K Esser. The phenoloxidases of the ascomycete *Podospora anserina*. XIII. Action and interaction of genes controlling the formation of laccase. Mol Gen Genet 156:333–345, 1977.
83. C Borghouts, A Werner, T Elthon, HD Osiewacz. Copper-modulated gene expression and senescence in the filamentous fungus *Podospora anserina*. Mol Cell Biol 21:390–399, 2001.
84. C Borghouts, HD Osiewacz. GRISEA, a copper-modulated transcription factor from *Podospora anserina* involved in senescence and morphogenesis, is an ortholog of MAC1 in *Saccharomyces cerevisiae*. Mol Gen Genet 260:492–502, 1998.
85. XS Liao, WC Small, PA Srere, RA Butow. Intramitochondrial functions regulate nonmitochondrial citrate synthase (CIT2) expression in *Saccharomyces cerevisiae*. Mol Cell Biol 11:38–46, 1991.
86. X Liao, RA Butow. RTG1 and RTG2: two yeast genes required for a novel path of communication from mitochondria to the nucleus. Cell 72:61–71, 1993.
87. T Sekito, J Thornton, RA Butow. Mitochondria-to-nuclear signaling is regulated by the subcellular localization of the transcription factors rtg1p and rtg3p. Mol Cell Biol 11:2103–2115, 2000.
88. PA Kirchman, S Kim, CY Lai, SM Jazwinski. Interorganelle signaling is a determinant of longevity in *Saccharomyces cerevisiae*. Genetics 152:179–190, 1999.

89. SM Jazwinski. Metabolic control and gene dysregulation in yeast aging. In: O Toussaint, H Osiewacz, G Lithgow, C Brack, eds. Molecular and Cellular Gerontology, Vol 908. New York: New York Academy of Sciences, 2000, pp 21–30.
90. E Dufour, J Boulay, V Rincheval, A Sainsard-Chanet. A causal link between respiration and senescence in *Podospora anserina*. Proc Natl Acad Sci USA 97:4138–4143, 2000.
91. AM Wagner, AL Moore. Structure and function of the plant alternative oxidase: its putative role in the oxygen defense mechanism. Biosci Rep 17:319–333, 1997.
92. D Harman. A theory based on free radical and radiation chemistry. J Gerontol 11: 298–300, 1956.
93. D Harman. The aging process. Proc Natl Acad Sci USA 78:7124–7128, 1981.
94. D Harman. Aging and oxidative stress. J Int Fed Clin Chem 10:24–27, 1998.
95. ADNJ de Grey. The mitochondrial free radical theory of aging. Austin: R.G. Landes, 1999.
96. NB Averbeck, ON Jensen, M Mann, H Schägger, HD Osiewacz. Identification and characterization of PaMTH1, a putative *O*-methyltransferase accumulating during senescence of *Podospora anserina* cultures. Curr Genet 37:200–208, 2000.
97. NB Averbeck, C Borghouts, A Hamann, V Specke, HD Osiewacz. Molecular control of copper homeostasis in filamentous fungi: increased expression of a metallothionein gene during aging of *Podospora anserina*. Mol Gen Genet 264:604–612, 2001.

5

Vegetative Incompatibility in Filamentous Ascomycetes

N. Louise Glass
University of California–Berkeley, Berkeley, California
Sven J. Saupe
Centre National de la Recherche Scientifique, Bordeaux, France

1 INTRODUCTION

Filamentous fungi grow by hyphal tip extension, branching, and hyphal fusion (anastomoses [1]) to form a complex tridimensional hyphal network. As an individual colony grows in nature, it will often come into contact with other individuals from the same or related species. Interactions with other individuals can have a number of consequences. In certain cases, anastomoses among different individuals can lead to the formation of a vegetative heterokaryon, where genetically distinct nuclei occupy a common cytoplasm. Two possible outcomes to such a vegetative cell fusion event are possible. A vigorous heterokaryon may be established (the two involved strains are termed "compatible," or the heterokaryon is inviable (the two strains are "incompatible"). Vegetative incompatibility (also referred to as heterokaryon or somatic incompatibility) is a common phenomenon in filamentous ascomycetes and basidiomycetes [2–4]. Vegetative compatibility between isolates is genetically determined by specific loci termed *het* (for *het*erokaryon incompatibility) or *vic* (*v*egestative *inc*ompatibility) loci [1,3,5]. Two strains are compatible and capable of forming vigorous heterokaryons if they have the same *het* (*vic*) genotype. It has been proposed that this phenomenon constitutes a non-self-recognition system that operates during vegetative growth. Although the selective mechanisms operating to maintain vegetative incompatibility are unclear, it has been proposed that preventing heterokaryon formation between unlike individuals may limit horizontal transfer of cytoplasmic infectious elements such as mycoviruses [6,7] and/or resource plundering between individuals [8].

This chapter focuses on a description of vegetative incompatibility in filamentous ascomycetes and summarizes the work performed on the characterization of genes involved in vegetative incompatibility in two model systems for this phenomenon, *Podospora anserina* and *Neurospora crassa*.

2 MANIFESTATIONS OF VEGETATIVE INCOMPATIBILITY

The lack of vigourous heterokaryon formation between two individuals can be due to a variety of mechanisms. Different individuals may fail to undergo hyphal fusion to establish a heterokaryon [9,10]. If individuals undergo hyphal fusion but have genetic differences at *het* loci, most frequently, the heterokaryotic fusion cell is compartmentalized and dies or is inhibited in its growth [11–13]. Vegetative incompatibility may also lead to the progressive loss of one of the nuclear types of the heterokaryon [14] without growth inhibition, hyphal compartmentation, or death.

Several techniques are used to visualize vegetative incompatibility at the macroscopic level. Compatibility can be determined using forced heterokaryons (Fig. 1) using strains that differ for an auxotrophic marker. Establishment of a vigorous heterokaryon indicates that the strains have undergone hyphal fusion, have no *het* differences and are therefore compatible. Nitrate non-utilizing mutants have been selected based on chlorate resistance (see for example [15,16]) and have been used in heterokaryon tests to assay for compatibility groups among naturally occurring isolates in a number of different fungi. In numerous species, including *P. anserina* and *Cryphonectria parasitica*, vegetative incompatibility can be visualized by the formation of an abnormal contact zone termed ‘‘barrage’’ when strains are confronted on solid medium [11,17] (Fig. 2). The barrage corresponds to the accumulation of dead hyphal fusion cells in the confrontation zone between two incompatible isolates.

Microscopic manifestations of the cell death reaction due to vegetative incompatibility have also been analyzed. In most cases, the hyphal fusion event per se is normal, and the first microscopic manifestation of the incompatibility reaction occurs ~15 min after anastomosis [12,13,18]. Cytoplasmic granules appear which show agitation and the septal pores which bracket the heterokaryotic

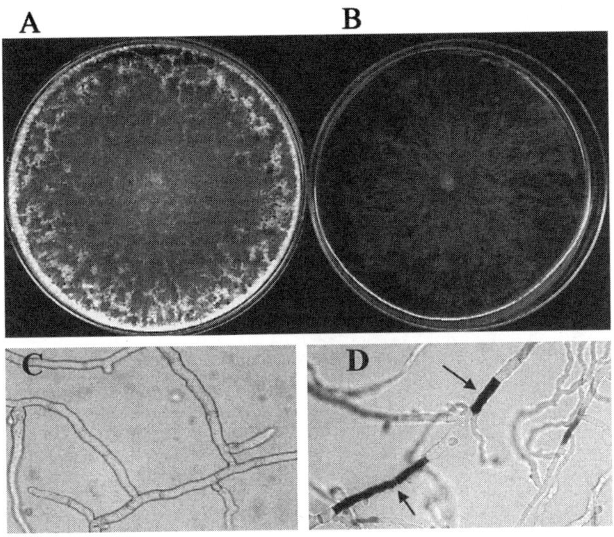

Figure 1 Macroscopic and microscopic features of vegetative incompatibility. (A) A *Neurospora crassa*–compatible heterokaryon made by using two strains that contain different auxotrophic markers grown on minimal medium. The strains are identical in *het* genotype. (B) An *N. crassa*–incompatible heterokaryon made by growing strains containing different auxotrophic markers on minimal medium. The strains contain alternative specificity at *het-c*, but are otherwise isogenic. Note growth inhibition and lack of conidiation. (C) Microscopic features associated with the compatible heterokaryon shown in (A). (D) Microscopic features associated with the *het-c*-incompatible heterokaryon shown in (B). In both (C) and (D), the hyphae have been stained using the vital dye Evan's Blue [18]; the dye is excluded from hyphae that contain intact and functional plasma membranes. Note the dead hyphal compartments in the *het-c*-incompatible heterokaryon that have taken up Evan's Blue as shown by the arrows in (D).

cell become occluded. Large vacuoles develop and eventually the cyploplasm retracts and completely disappears. Destruction of the heterokaryotic cell is complete within 1 h after anastomosis. Surrounding hyphae regenerate around or even within the destroyed hyphal compartment. The phenotype of fusion cells undergoing hyphal compartmentation and death show similarities among different fungal species, suggesting common cellular mechanisms may be involved. Ultrastructural studies of incompatible partial diploids in *N. crassa* show organelle degeneration, shrinkage of the plasma membrane, and septal plugging [18] (Fig. 3).

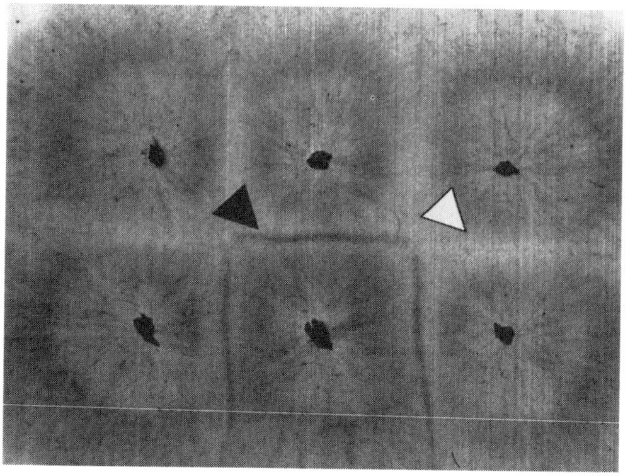

FIGURE 2 Barrage reaction in *Podospora anserina* caused by the *het-b1/het-b2* interaction. The white arrowhead shows a normal contact line between compatible strains. The black arrowhead shows an abnormal contact line or barrage between incompatible strains. The strain in the middle of the bottom line is of the *het-b2* genotype and thus incompatible with the other five strains which are of the *het-b1* genotype.

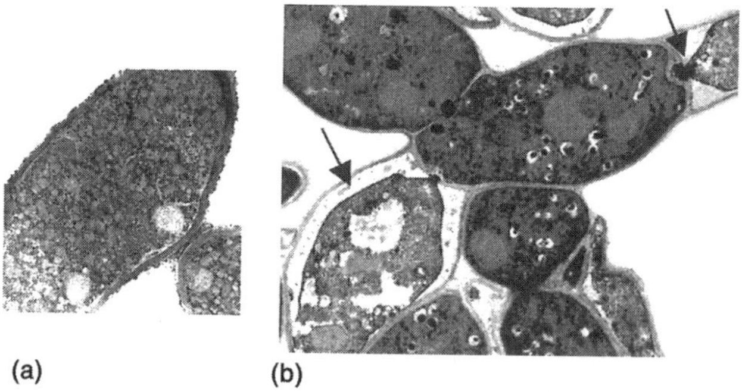

FIGURE 3 Transmission electron micrograph of sections of hyphae from *Neurospora crassa*. (a) Wild-type hyphae. (b) An incompatible transformant containing alleles of alternative *het-c* specificity. Note the vacuolization of the cytoplasm and shrinkage of the plasma membrane from the cell wall in the dead hyphal compartments (as shown by arrow at bottom left). Septal plugs compartmentalize the dead hyphal segments (arrow, top right).

Biochemical modifications associated with vegetative incompatibility have been studied in *P. anserina*. Upon the induction of vegetative incompatibility, the production of many cellular proteins stops while at least 20 new polypeptides are synthesized [19]. New enzymatic activities appear, including laccases, dehydrogenases, an amino acid oxidase, and two specific proteases [20,21]. DNA fragmentation has also been associated with vegetative incompatibility triggered by the *het-c* locus in *N. crassa*, suggesting that some biochemical similarities may exist between apoptosis or programmed cell death in multicellular eucaryotes and vegetative incompatibility in filamentous fungi [22].

3 GENETICS OF VEGETATIVE INCOMPATIBILITY

Vegetative incompatibility is induced if hyphal fusion occurs between individuals that differ in *het* specificity. Although this phenomenon is widespread in filamentous fungi, the genetic determinants of vegetative incompatibility have been examined in only a few ascomycete species. In *P. anserina*, the systematic genetic analysis of 17 wild-type strains led to the identification of nine *het* loci [23,24]. Two types of vegetative incompatibility systems were distinguished. The *het-b*, *het-q*, *het-s*, *het-v*, and *het-z* loci function as allelic incompatibility systems; individuals are incompatible if they have different allelic specificity at the same *het* locus. The *het-c*, *het-d*, *het-e*, *het-r*, and *het-v* loci define three nonallelic incompatibility systems (*het-c/het-d*, *het-c/het-e*, and *het-r/het-v*). In nonallelic *het* systems, vegetative incompatibility is triggered by the interaction of specific alleles at distinct loci. The *het-v* locus is the only locus that is involved in both allelic and nonallelic incompatibility. The nonallelic incompatibility systems in *P. anserina* also function as sexual incompatibility systems. A cross between incompatible strains can be partially or even completely sterile [23], presumably because vegetative incompatibility is induced when fertilization occurs. For example, a *het-c/het-E* × *het-C/het-e* cross is totally sterile when the *het-c/het-E* strain is the female parent of the cross. Thus, *het* loci may participate in speciation by conferring reproductive isolation. Allelic differences at the *het-s* locus also affect sexual reproduction. In a *het-s* × *het-S* cross in which *het-s* is the maternal parent, a proportion of the *het-S* spores are abortive [23]. Such *het-s* × *het-S* crosses produce an excess of *het-s* offspring; *het-s* thus behaves as a spore-killer locus.

In *N. crassa*, 11 *het* loci have been identified. All function as allelic incompatibility systems. Among them is the mating-type locus (*mat*) which, in addition to its role in the sexual cycle, functions as a *het* locus [25]. The *mat* locus and the *het-c*, *d*, *e*, and *i* loci were identified using forced heterokaryons between near isogenic, inbred strains [12,14,26]. The other *het* loci, *het-5*, through *het-10*, were identified by using translocation strains [27] that generate partial duplication progeny when crossed with normal sequence strains [28,29]. Partial diploid progeny that are heterozygous at a *het* locus display phenotypic aspects of vegetative incompatibility—e.g., suppression of conidiation, growth inhibition, hyphal compartmentation, and death. Translocation strains have been used to analyze *het* genotype in natural isolates [30] because they dispense with the construction of near-isogenic strains required for *het* genotyping using forced heterokaryons. Unlike other *het* loci in *N. crassa*, the *het-I/het-i* interaction does not lead to an immediate destruction of the heterokaryotic cell but to the loss of one of the nuclear components of the heterokaryon [14].

Vegetative incompatibility has been genetically defined in only two other ascomycete species. In *Aspergillus nidulans*, at least eight different allelic *het* loci have been identified [31–33]. Heterokaryon compatibility is determined by using an assay based on complementation of conidial color markers. In a compatible heterokaryon constructed from homokaryons differing for conidiospore color, the sporeheads are striped. In *A. nidulans*, vegetative incompatibility can be overcome by protoplast fusion and monochromosomic somatic hybrids can be obtained in that manner [34]. Such somatic hybrids can be used to demonstrate parasexual linkage between a *het* locus and a standard genetic marker. A hybrid is then backcrossed to the tester strains from which it differs only by one linkage group. This method is somewhat analogous to the use of translocation strains as *het* testers in *Neurospora*, as it simplifies subsequent genetic analyses by reducing the number of *het* loci that segregate in each cross.

In *C. parasitica*, six *vic* loci (*vic* 1, 2, 3, 4, 6, 7) have been genetically identified by crossing European isolates defining 31 different vegetative compatibility types; compatibility of progeny was assessed by barrage tests until a *het* (*vic*) genotype could be assigned [35,36]. All six *vic* loci are allelic systems with apparently only two allelic specificities at each locus. A seventh *vic* locus, *vic* 5, does not lead to a typical barrage reaction.

These genetic analyses show that the number of *het* loci is relatively high (between six and 11) in a given species. With few exceptions, there are generally only two or three allelic specificities at each *het* locus. Even though the number of allelic specificites is low, the fact that a particular species possesses a number of unlinked *het* loci greatly increases the possible number of different vegetative compatibility groups (vcg). An analysis of 128 isolates of *Fusarium oxysporum* f. sp. *phaseoli* revealed 96 different vcg [37]. However, vcg diversity in fungal populations can vary. In *C. parasitica*, analysis of more than 1000 isolates from Italy and Switzerland identified 31 vcg; subpopulations usually had 10 or fewer vcg [38]. In fungi that lack a sexual stage, have infrequent sexual recombination, or primarily in-breed, it is often inferred that strains belonging to the same vcg represent clonal derivatives from a common ancestor. Analysis of vcg in populations is often used by plant pathologists to characterize the structure and dynamics of fungal populations (reviewed in [2,3]).

4 MOLECULAR CHARACTERIZATION OF GENES INVOLVED IN VEGETATIVE INCOMPATIBILITY

4.1 *Neurospora crassa*

Among the 11 *het* loci that have been genetically characterized in *N. crassa*, three (the *mat* and *het-c* loci and the *het-6* region) have been cloned (Table 1). A mediator of mating type–associated incompatibility (*tol* for *tol*erant) has also been characterized.

4.1.1 The *mat* Locus and *tol*

During sexual reproduction in *N. crassa*, the fusion of opposite-mating-type reproductive structures places opposite-mating-type nuclei in a common cytoplasm within the developing fruit body (perithecium). Proliferation of opposite-mating-type nuclei occurs within reproductive hyphae prior to karyogamy. However, during vegetative growth, hyphal fusion between *mat A* and *mat a* strains results in hyphal compartmentation and death of heterokaryotic cells [25,39,40]. The fact that opposite-mating-type nuclei are in close proximity during sexual reproduction has led to the hypothesis that mating-type vegetative incompatibility is

TABLE 1 Cloned *het* Genes of *N. crassa* and *P. anserina*

	Number of alleles	Size of encoded polypeptide	Protein feature	Ref.
N. crassa				
mat A-1	1	293 aa	region of similarity to *mat* αl of *S. cerevisiae*	42
mat a-1	1	381 aa	HMG box	43
het-C	3	966 aa	signal peptide, glycine-rich repeats	62
het-6	2	680 aa	region of similarity to TOL and HET-E	56
un-24	2	929 aa	large subunit of type I ribonucleotide reductase	69
P. anserina				
het-s	3	289 aa	prionlike protein	70
het-c	4	208 aa	similarity to glycolipid transfer proteins	80
het-e	4	1056 aa	WD-40 repeats, GTP-binding site, region of similarity to TOL and HET-6	83

suppressed during the sexual cycle. Crosses with translocation strains that generate *mat A/mat a* partial diploid progeny are completely normal; upon ascospore germination, the *mat A/mat a* partial diploid progeny exhibit the phenotypic aspects of vegetative incompatibility. Suppression of vegetative incompatibility during sexual reproduction is not restricted to the *mat* locus because strains that differ not only in mating type but in specificity at all other known *het* loci are completely fertile. This is in contrast to nonallelic *het* loci in *P. anserina*, which also function as sexual incompatibility factors. The differences between the two species may be partly due to reproductive behavior; *P. anserina* is a pseudohomothallic species (an inbreeding species), while *N. crassa* is a heterothallic species (an obligately outbreeding species).

The *mat a* and *mat A* mating-type alleles have been termed idiomorphs as they completely differ in DNA sequence [41–44]. Three different genes have been identified in the *mat A* idiomorph. *mat A-1* specifies mating identity and is required for postfertilization functions whereas *mat A-2* and *mat A-3* have only postfertilization functions [42,45,46]. The *a* idiomorph has a single gene, *mat a-1*, which specifies all functions of the *a* mating type [43,47]. The vegetative incompatibility function of the *mat* locus is conferred by *mat A-1* and *mat a-1*; mutations in *mat A-1* or *mat a-1* result in strains that are sterile but are able to form vigorous heterokaryons with either mating type [48]. *mat A-1* encodes a putative transcriptional regulator that contains an α-box (a domain conserved in various mating-type polypeptides including MATα1 of *S. cerevisiae*) [42]. *mat a-1* encodes a putative transcriptional regulator containing a HMG box DNA binding domain [43]. Additional mutants of *mat a-1* and *mat A-1* have been isolated that have lost vegetative incompatibility but have retained mating function [48–51]. A *mat A-1* mutant which has a stop-codon at amino acid 86 forms heterokaryons with both *mat A* and *mat a* strains, but is still fertile as a female [51]. An arginine-to-serine substitution at amino acid position 258 in *mat a-1* inactivates vegetative incompatibility function, but does not affect mating or fertility function of MAT a-1. In vitro, DNA binding activity of this mutant MAT a-1 is unaffected [50]. Conversely, a deletion within the HMG domain alleviates DNA binding of MAT a-1, but does not affect vegetative incompatibility function. These data suggest that MAT A-1 and MAT a-1 mediate vegetative incompatibility and mating identity functions by distinct mechanisms.

Mutations at the *tol* locus do not affect mating capacity, but suppress mating type–associated incompatibility such that *tol mat A* and *tol mat a* strains form vigorous heterokaryons [52,53]. The *tol* mutations do not suppress vegetative incompatibility triggered by allelic differences at other *het* loci [54]. The *tol* gene was isolated by a chromosome walk from a linked marker (*trp-4*) and encodes a 1011 amino acid polypeptide [53]. TOL possesses a putative LRR motif (leucine-rich repeat motif), a motif that has been implicated in protein–protein interactions in other systems [55]. TOL also displays a 140 aa region of similarity to HET-E (involved in vegetative incompatibility in *P. anserina*) and HET-6 (involved in vegetative incompatibility in *N. crassa*; see below [56]). In addition, numerous other TOL-like proteins have been identified by genome sequencing efforts (http://www.mips.biochem.mpg.de/proj/neurospora/). The significance of these amino acid similarities and the relationship of these proteins to TOL function are unknown.

Mating type–associated incompatibility is not unique to *N. crassa*, but has been described in other species, such as *Ascobolus stercorarius* [57], *Aspergillus heterothallicus* [58], and *Sordaria brevicollis* [53]. However, other *Neurospora* species, such as *N. sitophila* and *N. tetrasperma*, do not display mating type–associated incompatibility. Introgression studies showed that the lack of mating-type-associated incompatibility in *N. sitophila* and *N. tetrasperma* is not due to differences at the *mat* locus, but to the presence of *tol*-like mutant alleles in species that do not mediate vegetative incompatibility [59–61].

The enigma of how mating type–associated incompatibility is suppressed during the sexual cycle was addressed by examining *tol* expression during vegetative growth and sexual reproduction. Although *tol* cDNAs were detected during vegetative growth in a *mat A*, *mat a*, a *mat A/mat a* partial diploid and even in a Δ*mat* mutant, expression of *tol* was not detected in developing perithecia, suggesting that transcriptional repression mechanisms directed at *tol* play a role in the suppression of mating type–associated incompatibility during sexual reproduction [53].

4.1.2 The *het-c* Locus

Forced heterokaryons or partial diploids heterozygous for *het-c* are slow growing and are aconidial with a slight budding morphology [26,28,62]. Microscopic examination of hyphae showed that ~20–30% of the hyphal segments are compartmentalized and are dead or dying [18]. The distribution of dead hyphal compartments in the hyphae is apparently random.

Genetic analysis using partial duplication strains indicated that at least three allelic specificities are encoded by the *het-c* locus [63,64], which have been termed Oak Ridge–compatible, Panama-compatible, and Groveland-compatible, based on the *het-c* allelic specificities of translocation tester strains. The *het-c* allele from the Oak Ridge strain (*het-c*^{OR}) was cloned by locating crossover points between *het-c* and flanking markers in a cosmid contig [62] combined with a functional transformation assay. The *het-c*^{OR} allele encodes a 966 amino acid polypeptide with a putative signal peptide and a C-terminal glycine-rich domain. Glycine-rich domains are found in a large number of structural proteins, such as RNA-binding proteins and keratin. These glycine-rich domains contain regularly spaced aromatic residues that have been proposed to allow tension-adaptable protein–protein interactions [65]. The HET-C protein is predicted to have two transmembrane helices and reside in the plasma membrane; the glycine-rich domain is predicted to be extracellular. Inactivation of *het-c* does not lead to any detectable vegetative or sexual phenotype [62], other than the loss of *het-c*-mediated vegetative incompatibility; *het-c* null mutants form vigorous heterokaryons with strains containing alternative *het-c* specificities.

Representatives of all three mutually incompatible *het-c* allelic specificities, $het\text{-}c^{OR}$, $het\text{-}c^{PA}$, and $het\text{-}c^{GR}$, have been molecularly characterized [64]. Overall, HET-C^{OR}, HET-C^{PA}, and HET-C^{GR} display 86% identity and show both variable and conserved regions. Three highly polymorphic regions were distinguished; outside of these regions amino acid conservation is 99%. Analysis by chimeric allele construction showed that one polymorphic region (102–144 bp in length) controls *het-c* allelic specificity. The $het\text{-}c^{OR}$, $het\text{-}c^{PA}$, and $het\text{-}c^{GR}$ alleles differ both by point mutations and in insertion/deletion (indel) pattern within this region. A perfect correlation was observed between indel pattern type and *het-c* specificity of strains determined by genetic analysis [64]. Further analysis of the *het-c* specificity domain showed that variations in the indel pattern conferred allelic specificity; new *het-c* allelic specificities were generated by altering either indel size or amino acid composition of the indel motif [66].

4.1.3 The *het-6* Region

Heterozygosity at *het-6* causes a severe growth inhibition in partial diploids or heterokaryons [28,29,67,68] with a growth rate ~100 times slower than wild type. Microscopic examination revealed that, as with *het-c* incompatibility, ~20–30% of the hyphal segments are compartmentalized and dead in incompatible *het-6* partial diploids [18]. Therefore, the phenotypic differences between *het-c* and *het-6* incompatibility lie in differences in the rate of growth inhibition, not the percentage of dead hyphal segments. Partial diploid strains that are heterozygous at *het-6* eventually ''escape'' to near wild-type growth. This escape event is associated with deletions that remove one of the *het-6* loci. By probing genomic DNA with a cosmid walk that spanned *het-6*, Smith et al. [68] determined that the *het-6* locus resided on a 35-kbp segment. Two different loci within this region exhibit *het-6* incompatibility activity based on functional transformation experiments [56]. One of these loci encodes the large subunit of type I ribonucleotide reductase and has been named *un-24* owing to a temperature-sensitive mutant isolated by a different mutagenic screen [69]. The allele that confers vegetative incompatibility function is derived from an Oak Ridge strain and is therefore termed $un\text{-}24^{OR}$ The alternative allelic specificity is termed $un\text{-}24^{PA}$. UN-24^{OR} and UN-24^{PA} allele products differ in the C-terminal end of the ribonucleotide reductase in a region that has an insertion that is unique to *N. crassa*. The second locus, termed $het\text{-}6^{OR}$, encodes a 680 aa polypeptide that has similarity to TOL and to HET-E [56]. The alternative allele, $het\text{-}6^{PA}$, encodes a polypeptide that displays only 68% identity to HET-6^{OR} [56], the lowest identity level reported between alternative alleles of a *het* locus.

Although *un-24*^{OR} and *het-6*^{OR} have vegetative incompatibility activity based on functional transformation assays into a strain containing alternative *het-6* specificity (a *het-6*^{PA} strain), the alternative alleles, *un-24*^{PA} and *het-6*^{PA}, do not reduce the frequency of recoverable transformants when introduced into a *het-6*^{OR} strain; transformants were identical in growth phenotype to vector controls [56]. These data suggest that *het-6* incompatibility is not reciprocal and instead is mediated by nonallelic interactions between *un-24*^{OR} and/or *het-6*^{OR} and another locus (or loci) in the *het-6* region of the *het-6*^{PA} strain. Initial molecular characterization of the *het-6* locus, though incomplete, reveals an unexpected genetic complexity. Multiple genes in the *het-6* region are involved, and although they are genetically located at the same locus [67], their interaction is apparently nonallelic in nature.

4.2 *Podospora anserina*

In *P. anserina*, the genes belonging to an allelic system (*het-s*/*het-S*) and to one nonallelic system (*het-c*/*het-e*) have been molecularly characterized (Table 1). The isolation of *het* genes was based on the acquisition of a vegetative incompatibility phenotype via transformation that was detected by a barrage tests between transformants and specific tester strains.

4.2.1 The *het-s* Locus

Two alternative allelic specificities, *het-s* and *het-S*, occur at the *het-s* locus. Confrontation of a *het-s* strain with a *het-S* strain leads to a barrage reaction [11]. The *het-s* and *het-S* alleles both encode proteins of 289 amino acids that differ at 13 amino acid positions [70,71]. Analyses of *het-s* sequences from natural isolates revealed that all *het-s* strains contain a 354-bp-long transposon LTR inserted in the promoter sequence and a 2-kbp insertion downstream of the ORF. Although polymorphism exists between *het-s* and *het-S* alleles, variability was not observed between *het-s* alleles from different isolates or when alleles from different isolates conferring *het-S* specificity were compared [72].

Inactivation of *het-s* by gene replacement did not result in a detectable phenotype other than the loss of the *het-s*-mediated vegetative incompatibility [71]. Chimeric allele analysis and site-directed mutagenesis of *het-s* and *het-S* alleles showed that mutations resulting in a single amino acid replacement (histidine to proline at amino acid position 33) in HET-S was sufficient to switch allelic specificity to *het-s* [72]. All HET-S/HET-S, HET-s/HET-s homomeric, and HET-S/HET-s heteromeric complexes were detected by the yeast two-hybrid system [73], suggesting that vegetative incompatibility may be triggered by the formation of HET-s/HET-S heterocomplex.

The *het-s* allele shows non-Mendelian segregation; in a *het-s* $\times$ *het-S* cross, 50% of the progeny are *het-S*, but a proportion of the *het-s* progeny fail to induce a barrage reaction with either *het-S* or *het-s* tester strains [11]. These progeny are of neutral *het-s* [Het-s*] phenotype. The proportion of [Het-s*] progeny is low if the female parent is *het-s*, but reaches 50% when the *het-s* strain is the male parent (which contributes little cytoplasm). The maternal inheritance of [Het-s] is also illustrated by the fact that in a [Het-s] × [Het-s*] cross, all progeny have the phenotype of the female parent. [Het-s*] strains can be propagated vegetatively over an extended period of time but eventually spontaneously acquire the reactive [Het-s] phenotype; a phenotypic conversion can occur in any spot of the mycelium and propagates by an infectious process [13]. This conversion is also systematically induced after a cytoplasmic contact (anastomosis) with a [Het-s] strain. Conversely, a [Het-s] strain can return to [Het-s*] phenotype when protoplasts are generated from a [Het-s] strain or after fragmentation of the mycelium [13,74]. In this case, a proportion of strains regenerated from protoplasts or mycelial fragments display the neutral [Het-s*] phenotype.

It was determined that the *het-s*-encoded protein is present in both [Het-s*] and [Het-s] strains. It is not the absence of the *het-s* encoded protein that accounts for the nonreactive [HET-s*] phenotype [73], and therefore it was proposed that the [Het-s] cytoplasmic element is a prion. Prions are ''infectious proteins,'' a term introduced to define the infectious agent of spongiform encephalopathies [75,76]. Transmission of this infectious agent occurs as a transmissible conformational modification in a cellular protein called Prp. Two non-Mendelian elements of *Saccharomyces cerevisiae*, the [URE3] and [PSI] elements, are prions [77]. In the prion model, [Het-s*] strains have HET-s in a nonreactive conformation, HET-s*. In [Het-s] strains, the protein is in the reactive HET-s conformation; HET-s can catalyze the conversion of HET-s* into HET-s. This model readily explains the unusual genetic and physiological properties of the [HET-s] element; namely, it accounts for the cytoplasmic inheritance of the [Het-s] element, the rapid and infectious transmission of [Het-s], and the reversible curing of [Het-s] (loss of the [Het-s] character and spontaneous reappearance). The [Het-s] system displays additional properties that are common to all prion systems. First, as expected in an autocatalytic system, overexpression of the *het-s*-encoded protein increases the frequency of spontaneous appearance of the [Het-s] phenotype [73]. Second, the HET-s protein is more resistant to proteinase K digestion than HET-s*, which indicates the occurrence of some sort of modification of the protein [73].

A genetic screen based on the escape from *het-s/het-S* self-incompatibility resulted in the isolation of a *het-s* mutant that contained a mutation resulting in a stop codon at amino acid position 26 of *het-s* [78]. This strain lost vegetative incompatibility function but was capable of converting a [Het-s*] strain to the [Het-s] phenotype. The prion aspects and vegetative incompatibility function of [Het-s] are separable, and a short N-terminal peptide from HET-s is sufficient to propagate [Het-s] character [78].

For the two yeast prions, transition to the prion form leads to a loss of function of the protein. It is the opposite case for [Het-s]. Here the prion state corresponds to the reactive form of the *het-s*-encoded protein, the one that triggers the vegetative incompatibility reaction. This conclusion is only valid, however, if one assumes that the only biological function of the *het-s* is to control heterokaryon incompatibility. It remains possible that a yet-unknown cellular function of the HET-s protein is lost during the transition to the prion state as for the yeast prions.

4.2.2 The *het-c* and *het-e* Nonallelic Incompatibility Loci

The *P. anserina het-c* and *het-e* loci are both multiallelic [79]. Four *het-c*, four *het-e* and three *het-d* alleles have been described in wild-type isolates. Vegetative incompatibility is mediated by interactions between a specific set of *het-c* alleles with specific set of *het-e* and *het-d* alleles. The *het-c* locus encodes a 208 amino acid protein which displays amino acid similarity to a glycolipid transfer protein (GLTP) isolated from from pig brain [80]. The precise cellular function of the glycolipid transfer proteins is not known, although in vitro, GLTPs bind several glycolipids and mediates their transfer between donor and acceptor liposomes [81]. Inactivation of *het-c* in *P. anserina* does not affect vegetative growth, but in a cross homozygous for the *het-c* deletion, ascospore maturation is drastically impaired [80]. It is unknown how the putative glycolipid transfer activity of HET-c is related to the vegetative incompatibility reaction and to spore maturation defect observed in *het-c* mutants.

The four allelic forms of *het-c* are 92% identical in amino acid sequence [82]. Chimeric alleles constructed among the four wild-type alleles did not result in the identification of a specificity domain, as many polymorphic positions in *het-c* participated in affecting allelic specificity. Some of the chimeric alleles displayed a novel specificity different from that of all known wild-type *het-c* alleles.

One allele of *het-e* has been isolated using a functional approach [83] and encodes a 1056 amino acid polypeptide. The N-terminal region shows similarity with TOL and HET-6 and is followed by a GTP binding site (P-loop motif) and a C-terminal WD repeat domain. The HET-e protein binds GTP in vitro [84], and mutations that abolish binding also abolish vegetative incompatibility activity [83]. The WD repeat domain was first described in the α-subunits of heterotrimeric G-proteins and is a common protein structural motif believed to provide interaction sites for other protein partners. The HET-el protein has 10 WD repeats; allelic specificity is not dependent on the number of repeats but alleles displaying fewer than 10 repeats were inactive [83]. The number of WD repeats ranges from three to 10 in other wild-type *het-e* alleles. Amino acid conservation between WD repeats ranges from 81% to 98%. Size polymorphisms in the WD repeats of HET-e may be generated by unequal crossovers in *het-e* that lead to the expansion or reduction of the number of repeats, a mechanism that can generate allelic polymorphisms.

Inactivation of *het-e* by gene replacement has no detectable effect during vegetative growth or sexual reproduction [83]. In particular, the defect in ascospore maturation seen in *het-c* mutants was not observed in *het-e* mutants. Sequences similar to *het-e* exist in the *P. anserina* genome [84], suggesting that paralogs of *het-e* exist which may complement its function when inactivated. Based on its genetic interaction with *het-c*, *het-d* is an obvious candidate for being such a paralog.

4.2.3 The *mod-A* Locus

Crosses between strains that have genetic differences at nonallelic *het* loci generate progeny that display vegetative incompatibility due to independent assortment of the unlinked *het* loci. Such self-incompatible strains often ''escape'' from vegetative incompatibility, as observed by the emergence of growing sectors [79,85]. Most often, mutations at nonallelic *het* loci relieve vegetative incompatibility. However, a number of mutations at unlinked loci lead to a modified self-incompatible (MSI) phenotype; most MSI strains contain mutations at the *mod-A* locus. The *mod-A1* mutation suppresses self-incompatibility mediated by differences at three nonallelic interactions: *het-C/het-E*, *het-C/het-D*, and *het-R/het-V* [85]. The *mod-A1* mutation prevents growth arrest owing to the nonallelic *het* interactions [85], but does not suppress cell lysis. Complete suppression of vegetative incompatibility is mediated by mutations at the unlinked *mod-B* locus [86]. In a *mod-A1/mod-B1* double mutant, all three nonallelic *het* vegetative incompatibility interactions are fully suppressed. The *mod-A* gene has been isolated [87] and encodes a 687 amino acid polypeptide with a proline-rich region, but otherwise is without significant similarity to known proteins.

The *mod-A1/mod-B1* double mutant displays developmental defects [88] and is defective in the development of the female reproductive organs—the protoperithecia. In *P. anserina*, development of protoperithecia occurs once the culture medium is exhausted and is accompanied by the death of surrounding vegetative hyphae. This autolysis presumably provides nutrients required for differentiation of the female reproductive structures. It has been suggested that the *mod-A* and *mod-B* and nonallelic *het* genes control this limited self-lysis process during sexual reproduction [89].

Mutations at *mod-C* specifically suppress *het-R/het-V* vegetative incompatibility but do not affect *het-C/het-D* or *het-C/het-E* incompatible interactions. This observation indicates that the cell death pathways induced by the *het-R/het-V* and *het-C/het-E* interactions are at least partially distinct. Similar to *mod-A1/mod-B1* double mutants, the *mod-C1* mutant is female sterile [90].

4.2.4 Genes Induced During Vegetative Incompatibility

The *het-R/het-V* vegetative incompatibility reaction is temperature conditional; a strain containing alternative *het-V* and *het-R* alleles displays normal growth at 32°C, but undergoes a cell death reaction when transferred to 26°C [24]. When the vegetative incompatibility reaction is induced in a *het-R/het-V* strain, at least a 20 new polypeptides appear. The cloning of genes induced by *het-R/het-V* incompatibility was undertaken by a differential hybridization approach [91]. Three *idi* genes (*idi-1*, *2*, and *3* for *i*nduced *d*uring *i*ncompatibility) are characterized by a very high expression level during vegetative incompatibility (each one represents ~1% of the mRNA). All three *idi* genes encode small proteins with putative signal peptides. The IDI-2 protein is cysteine rich and displays a region of similarity with IDI-3. These proteins may mediate the cell death reaction or, alternatively, be induced as a consequence of the vegetative incompatibility reaction.

The production of a protease that is specifically induced when a *het-R/het-V* strain is shifted from 32°C to 26°C has been purified. The corresponding gene for this protease (*pspA*) encodes a 532 amino acid serine protease precursor of the subtilisin type (Matthieu Paoletti and Corinne Clavé, unpublished results). Subtilisins are broad-specificity endopeptidases with an S–H–D catalytic triad. PSPA shows amino acid similarity with prB, a vacuolar protease from *S. cerevisiae* [92]. The induction of *pspA* during vegetative incompatibility might be related to the intensive vacuolization observed during hyphal compartmentation and death.

5 BIOLOGICAL SIGNIFICANCE OF VEGETATIVE INCOMPATIBILITY

One of the central questions about vegetative incompatibility is the biological significance of the phenomenon in filamentous fungi. There are clear benefits associated with heterokaryon formation; growth as a heterokaryon provides some of the avantages of diploidy, and there are examples of heterosis associated with heterokaryon formation [93]. Also, in certain species, fertilization begins with formation of a vegetative heterokaryon between strains of opposite mating type [94]. But what are the possible disadvantages of heterokaryon formation? Heterokaryon formation permits the transmission of cytoplasmic elements such as transposons, mitochondrial senescence plasmids, and RNA mycoviruses [6,35]. The prevention of productive heterokaryon formation by vegetative incompatibility has been proposed to limit the horizontal transfer of such infectious elements [95]. This aspect has been investigated in the phytopathogenic species *C. parasitica* and *Ophiostoma ulmi*, in which dsRNA viruses that cause hypovirulence have been described [96]. Vegetative incompatibility provides an advantage for the fungal plant pathogen by restricting transmission of these hypovirulence factors. Under laboratory conditions, it has been observed that different *het* loci vary in their efficiency to restrict cytoplasmic transmission of such deleterious genetic elements [6,16,97,98]. Additional hypotheses attributing a biological function to vegetative incompatibility have been proposed, such as the maintenance of genetic individuality and the prevention of nuclear parasitism in an heterokaryon [8,99]. A alternative perspective is to consider the existence of *het* genes (genes that are deleterious at the heteroallelic state) as analagous to loci involved in hybrid sterility, in which polymorphisms accumulate in genetically separated individuals. Heterokaryon formation among such individuals results in presence of such polymorphic alleles in a common cytoplasm, resulting in a deleterious effect on the hyphal fusion cell [100]. This aspect may be particularly relevant in species undergoing speciation (and thus genetically separated) or in species that primarily inbreed (such as homothallic and pseudohomothallic species) and which may be clonal in population structure. Whether or not polymorphisms at these loci would be under selection for non-self-recognition function in fungal populations is an unanswered question.

If *het* loci function as non-self-recognition systems, it is inferred that selective pressures favor the accumulation and maintenance of polymorphisms at these loci. The best-studied examples of non-self-recognition systems are the *S* (self-incompatibility) locus in plants and the class I and class II MHC loci in mammals. In these two examples, balancing selection favors the maintenance of allelic polymorphisms. Balancing selection can be assessed based on the increased frequency of nonsynonymous compared to synonymous substitutions in regions that confer allelic specificity, the existence of large allelic series, and the maintenance of allelic lineages through multiple speciation events [101].

At present, the strongest evidence that balancing selection may be acting on a *het* locus comes from a study of the *het-c* locus of *N. crassa*. The three *het-c* allelic specificities ($het\text{-}c^{OR}$, $het\text{-}c^{PA}$, and $het\text{-}c^{GR}$) had an equal frequency of distribution, both in a field population of about 40 individuals and in a sample of 15 geographic isolates from various subtropical locations [102]. Polymorphisms associated with all three *het-c* specificities were present not only in different species of *Neurospora*, but even in different genera in the Sordariacae [102]. These data indicate that *het-c* polymorphisms were generated in an ancestral species and have been maintained through multiple speciation events. *Sordaria* and *Neurospora* diverged at least 36 million years ago and thus *het-c* polymorphisms have been maintained for an evolutionarily long period. This evolutionary characteristic has also been observed for alleles at the *S* locus and loci in the MHC and has been termed ‘‘transspecies polymorphisms’’ [101]. An analysis of the frequency of synonymous to nonsynonymous replacements in the *het-c* specificity domain showed an increase in nonsynonomous substitutions, suggesting that selection is favoring diversity in this region [102]. Together these results strongly suggest that balancing selection is acting to maintain polymorphism and allele frequencies at *het-c*. However, whether it is the restriction of the transfer of deleterious elements, the prevention of resource plundering, the maintenance of genetic individuality or another undiscovered function of *het-c* that results in the retention of allelic polymorphisms is unclear.

Molecular and genetic analyses suggest that other *het* loci may also be subject to balancing selection. Analysis of *P. anserina het-c* showed an excess of nonsynonymous substitutions and more DNA sequence variability in coding regions than in noncoding regions [82]. Analysis of a limited number of *P. anserina* strains from different geographic locations showed that all known *het* loci had a relatively equal frequency of allele distribution (Léon Belcour, personal communication). The most unequal allele frequency was ~1:2. In *N. crassa*, *het-6* haplotypes had equal frequency of distribution within populations [56]. However, not all alleles at *het* loci show equal frequency of distribution within populations, as shown by an analysis of *vic* allele frequency in *C. parasitica*. In numerous cases, *vic* allele frequency deviated significantly from 1:1 [38]. It is premature to conclude that balancing selection is a force acting on all *het* loci and that these systems all function as non-self-recognition mechanisms. Additional molecular analysis of *het* loci and an analysis of *het* allele frequencies in fungal populations are necessary to resolve this question. In addition, the selective pressures that maintain polymorphisms at *het* loci have not been well delineated. Experimental approaches that address not only the mechanism of vegetative incompatibility but also its function in filamentous fungal populations, are required to gain a fuller understanding of the role of this phenomenon in filamentous fungi.

6 CONCLUSION

The phenomenon of vegetative incompatibility has intrigued many fungal biologists since its discovery. It is a ubiquitous phenomenon in filamentous fungi, yet its significance remains unresolved. None of the hypotheses that have been proposed to explain the existence of vegetative incompatibility are adequate to explain the number of *het* loci within a species or the selective pressures that result in the maintanance of polymorphisms at *het* loci [103].

Elucidation of the genetic mechanisms that mediate vegetative incompatibility, which represents an impressive task, has opened the way to molecular studies. Now a number of *het* loci involved in vegetative incompatibility have been characterized, and the mechanisms of allelic specificity have been determined. At least some *het* loci are not solely involved in vegetative incompatibility, but appear to have additional, cellular functions. Although the molecular differences that define allelic specificity at several *het* loci have been identified, the molecular mechanism by which distinction between self and nonself is achieved is not known for any *het* system, and the biochemical or genetic mechanism of cell death has not been elucidated. It is unclear whether each *het* locus mediates vegetative incompatibility by common or different genetic and biochemical mechanisms. We hope that some of these questions will be resolved in the near future, so that a better understanding of this intriguing phenomenon and its role in the biology of filamentous fungi can be elucidated.

REFERENCES

1. NL Glass, DJ Jacobson, KT Shiu. The genetics of hyphal fusion and vegetative incompatibility in filamentous ascomycetes. Annu Rev Genet 34:165–186, 2000.
2. NL Glass, GA Kuldau. Mating type and vegetative incompatibility in filamentous ascomycetes. Annu Rev Phytopathol 30:201–224, 1992.
3. JF Leslie. Fungal vegetative compatibility. Annu Rev Phytopathol 31:127–150, 1993.
4. JJ Worrall. Somatic incompatibility in basidiomycetes. Mycologia 89:24–36, 1997.
5. SJ Saupe. Molecular genetics of heterokaryon incompatibility in filamentous ascomycetes. Microbiol Mol Biol Rev 64:489–502, 2000.
6. F Debets, X Yang, AJF Griffiths. Vegetative incompatibility in *Neurospora*—its effect on horizontal transfer of mitochondrial plasmids and senescence in natural populations. Curr Genet 26:113–119, 1994.
7. AD van Diepeningen, AJM Debets, RF Hoekstra. Heterokaryon incompatibility blocks virus transfer among natural isolates of black Aspergilli. Curr Genet 32: 209–217, 1997.
8. AJM Debets, AJF Griffiths. Polymorphism of *het*-genes prevents resource plundering in *Neurospora crassa*. Mycol Res 102:1343–1349, 1998.
9. JF Wilson, JA Dempsey. A hyphal fusion mutant in *Neurospora crassa*. Fungal Genet Newsl 46:31, 1999.
10. JC Correll, CJR Klittich, JF Leslie. Heterokaryon self-incompatibility in *Gibberella fujikuroi (Fusarium moniliforme)*. Mycol Res 93:21–27, 1989.
11. G Rizet. Les phenomenes de barrage chez *Podospora anserina*. I. Analyse genetique des barrages entre le souches *S* et *s*. Rev Cytol Biol Veg 13:51–92, 1952.
12. L Garnjobst, JF Wilson. Heterocaryosis and protoplasmic incompatibility in *Neurospora crassa*. Proc Natl Acad Sci USA 42:613–618, 1956.
13. J Beisson-Schecroun. Incompatibilité cellulaire et interactions nucleo-cytoplasmiques dans les phenomènes de barrage chez *Podospora anserina*. Ann Genet 4:3–50, 1962.
14. TH Pittinger, TG Browner. Genetic control of nuclear selection in *Neurospora* heterokaryons. Genetics 46:1645–1663, 1961.
15. JE Puhalla, PT Speith. A comparison of heterokaryosis and vegetative incompatibility among varieties of *Gibberella fujikuroi (Fusarium moniliforme)*. Exp Mycol 9: 39–47, 1985.
16. A Coenen, F Debets, R Hoekstra. Additive action of partial heterokaryon incompatibility (partial-het) genes in *Aspergillus nidulans*. Curr Genet 26:233–237, 1994.

17. SL Anagnostakis. Vegetative incompatibility in *Endothia parasitica*. Exp Mycol 1:306–316, 1977.
18. DJ Jacobson, K Beurkens, KL Klomparens. Microscopic and ultrastructural examination of vegetative incompatibility in partial diploids heterozygous at *het* loci in *Neurospora crassa*. Fungal Genet Biol 23:45–56, 1998.
19. H Boucherie, CH Dupont, J Bernet. Polypeptide synthesis during protoplasmic incompatibility in the fungus *Podospora anserina*. Biochim Biophys Acta 653:18–26, 1981.
20. J Bégueret, J Bernet. Proteolytic enzymes and protoplasmic incompatibility in *Podospora anserina*. Nat New Biol 243:94–96, 1973.
21. H Boucherie, J Bernet. Protoplasmic incompatibility and self-lysis in *Podospora anserina*: enzyme activities associated with cell destruction. Can J Bot 56:2171–2176, 1978.
22. S Marek, J Wu, NL Glass, DG Gilchrist, R Bostock. 1999. Programmed cell death in fungi: heterokaryon incompatibility involves nuclear DNA degradation (abstr). 20th Fungal Genetics Conference, # 335 (http://www.fgsc.net/asilo99/posterabs3.htm).
23. J Bernet. Mode d'action des genes de barrage et relation entre l'incompatibilite cellulaire et l'incompatibilite sexuelle chez le *Podospora anserina*. Ann Sci Nat Bot 6:611–768, 1965.
24. J Labarère. Propriété d'un système d'incompatibilité chez le champignon *Podospora anserina* et interêt de ce système pour l'étude de l'incompatibilité. C R Acad Sci Paris 276:1301–1304, 1973.
25. GW Beadle, VL Coonradt. Heterokaryosis in *Neurospora crassa*. Genetics 29:291–308, 1944.
26. L Garnjobst. Genetic control of heterocaryosis in *Neurospora crassa*. Am J Bot 40:607–614, 1953.
27. DD Perkins. Chromosome rearrangements in *Neurospora* and other filamentous fungi. Adv Genet 36:239–398, 1977.
28. DD Perkins. The use of duplication-generating rearrangements for studying heterokaryon incompatibility genes in *Neurospora*. Genetics 80:87–105, 1975.
29. OM Mylyk. Heterokaryon incompatibility genes in *Neurospora crassa* detected using duplication-producing chromosome rearrangements. Genetics 80:107–124, 1975.
30. OM Mylyk. Heteromorphism for heterokaryon incompatibility genes in natural populations of *Neurospora crassa*. Genetics 83:275–284, 1976.
31. RBG Dales, JH Croft. Investigation of the *het* genes that control heterokaryon incompatibility between members of heterokaryon-compatibility (h-c) group-A and group-G1 of *Aspergillus nidulans*. J Gen Microbiol 136:1717–1724, 1990.
32. RBG Dales, J Moorhouse, JH Croft. Evidence for a multi-allelic heterokaryon incompatibility (*het*) locus detected by hybridization among 3 heterokaryon-compatibility (h-c) groups of *Aspergillus nidulans*. Heredity 70:537–543, 1993.
33. MM Anwar, JH Croft, RBG Dales. Analysis of heterokaryon incompatibility between heterokaryon-compatibility (h-c) groups R and G1 provides evidence that at least 8 *het* loci control somatic incompatibility in *Aspergillus nidulans*. J Gen Microbiol 139:1599–1603, 1993.

34. RBG Dales, JH Croft. A chromosome assay method for the detection of heterokaryon incompatibility (*het*) genes operating between members of different heterokaryon compatibility (h-c) groups in *Aspergillus nidulans*. J Gen Microbiol 129:3643–3649, 1983.
35. SL Anagnostakis. Genetic analysis of *Endothia parasitica*: linkage data for four single genes and three vegetative incompatibility types. Genetics 102:25–58, 1982.
36. P Cortesi, MG Milgroom. Genetics of vegetative incompatibility in *Cryphonectria parasitica*. Appl Environ Microbiol 64:2988–2994, 1998.
37. FM Alves-Santos, EP Benito, AP Eslava, JM Diaz-Minguez. Genetic diversity of *Fusarium oxysporum* strains from common bean fields in Spain. Appl Environ Microbiol 65:3335–3340, 1999.
38. MG Milgroom, P Cortesi. Analysis of population structure of the chestnut blight fungus based on vegetative incompatibility genotypes. Proc Natl Acad Sci USA 96:10518–10523, 1999.
39. ER Sansome. Heterokaryosis, mating-type factors, and sexual reproduction in *Neurospora*. Bull Torrey Bot Club 73:397–409, 1946.
40. TH Pittenger. The mating type alleles and heterokaryon formation in *Neurospora crassa*. Microb Genet Bull 15:21–22, 1957.
41. NL Glass, SJ Vollmer, C Staben, RL Metzenberg, C Yanofsky. DNAs of the two mating type alleles of *Neurospora crassa* are highly dissimilar. Science 241:570–573, 1988.
42. NL Glass, J Grotelueschen, RL Metzenberg. *Neurospora crassa A* mating-type region. Proc Natl Acad Sci USA 87:4912–4916, 1990.
43. C Staben, C Yanofsky. *Neurospora crassa a* mating-type region. Proc Natl Acad Sci USA 87:4917–4921, 1990.
44. RL Metzenberg, NL Glass. Mating type and mating strategies in *Neurospora*. BioEssays 12:53–59, 1990.
45. AV Ferreira, SJ Saupe, NL Glass. Transcriptional analysis of the *mt A* idiomorph of *Neurospora crassa* identifies two genes in addition to *mt A-1*. Mol Gen Genet 250:767–74, 1996.
46. AV Ferreira, Z An, RL Metzenberg, NL Glass. Characterization of *mat A-2*, *mat A-3* and Δ*matA* mating-type mutants of *Neurospora crassa*. Genetics 148:1069–1079, 1998.
47. S Chang, C Staben. Directed replacement of *mt A* by *mt a* effects a mating type switch in *Neurospora crassa*. Genetics 138:75–81, 1994.
48. AJF Griffiths. Null mutants of the *A* and *a* mating-type alleles of *Neurospora crassa*. Can J Genet Cytol 24:167–176, 1982.
49. AJF Griffiths, AM Delange. Mutations of the *a* mating-type gene in *Neurospora crassa*. Genetics 88:239–254, 1978.
50. ML Philley, C Staben. Functional analysis of the *Neurospora crassa* MT a-1 mating type polypeptide. Genetics 137:715–722, 1994.
51. S Saupe, L Stenberg, KT Shiu, AJF Griffiths, NL Glass. The molecular nature of mutations in the *mt A-1* gene of the *Neurospora crassa A* idiomorph and their relation to mating-type function. Mol Gen Genet 250:115–122, 1996.
52. D Newmeyer. A suppressor of the heterokaryon-incompatibility associated with mating type in *Neurospora crassa*. Can J Genet Cytol 12:914–926, 1970.

53. PKT Shiu, NL Glass. Molecular characterization of *tol*, a mediator of mating-type-associated vegetative incompatibility in *Neurospora crassa*. Genetics 151:545–555, 1999.
54. JF Leslie, CT Yamashiro. Effects of the *tol* mutation on allelic interactions at *het* loci in *Neurospora crassa*. Genome 40:834–840, 1997.
55. B Kobe, J Deisenhofer. The leucine-rich repeat: a versatile binding motif. Trends Biochem Sci 19:415–421, 1994.
56. ML Smith, OC Micali, SP Hubbard, N Mir-Rashed, DJ Jacobson, NL Glass. Vegetative incompatibility in the *het-6* region of *Neurospora crassa* is mediated by two linked genes. Genetics 155:1094–1104, 2000.
57. GN Bistis. Retardation of the growth of transplanted apothecia: a manifestation of vegetative incompatibility in *Ascobolus stercorarius* (Bull.). Schrot Exp Mycol 18: 103–110, 1994.
58. K Kwon, KB Raper. Heterokaryon formation and genetic analyses of color mutants in *Aspergillus heterothallicus*. Am J Bot 54:49–60, 1967.
59. RL Metzenberg, SK Ahlgren. Behaviour of *Neurospora tetrasperma* mating-type genes introgressed into *Neurospora crassa*. Can J Genet Cytol 15:571–576, 1973.
60. DD Perkins. Behavior of *Neurospora sitophila* mating-type alleles in heterozygous duplications after introgression into *Neurospora crassa*. Exp Mycol 1:166–172, 1977.
61. DJ Jacobson. Control of mating type heterokaryon incompatibility by the *tol* gene in *Neurospora crassa* and *N. tetrasperma*. Genome 35:347–353, 1992.
62. SJ Saupe, GA Kuldau, ML Smith, NL Glass NL. The product of the *het-C* heterokaryon incompatibility gene of *Neurospora crassa* has characteristics of a glycine-rich cell wall protein. Genetics 143:1589–1600, 1996.
63. B Howlett, JF Leslie, DD Perkins. Putative multiple alleles at the vegetative (heterokaryon) incompatibility loci *het-c* and *het-8* in *Neurospora crassa*. Fungal Genet Newsl 40:40–42, 1993.
64. SJ Saupe, NL Glass. Allelic specificity at the *het-c* heterokaryon incompatibility locus of *Neurospora crassa* is determined by a highly variable domain. Genetics 146:1299–1309, 1997.
65. PM Steinert, JW Mack, BP Korge, SQ Gan, SR Haynes, AC Steven. Glycine loops in proteins: their occurrence in certain intermediate filament chains, loricrins and single-stranded *RNA* binding proteins. Int J Biol Macromol 13:130–139, 1991.
66. J Wu, NL Glass. Identification of specificity determinants and the generation of alleles with novel specificity at the *het-c* heterokaryon incompatibility locus of *Neurospora crassa*. Mol Cell Biol 21:1045–1057, 2001.
67. N Mir-Rashed, DJ Jacobson, RM Dehghany, OC Micali, ML Smith. Molecular and functional analysis of incompatibility genes at *het-6* in a population of *Neurospora crassa*. Fungal Genet Biol (in press).
68. ML Smith, CJ Yang, RL Metzenberg, NL Glass. Escape from *het-6* incompatibility in *Neurospora crassa* partial diploids involves preferential deletion within the ectopic segment. Genetics 144:523–531, 1996.
69. ML Smith, SP Hubbard, DJ Jacobson, OC Micali, NL Glass. An osmotic-remedial, temperature-sensitive mutation in the allosteric activity site of ribonucleotide reductase in *Neurospora crassa*. Mol Gen Genet 262:1022–1035, 2000.

70. B Turcq, M Denayrolles, J Bégueret. Isolation of the two allelic incompatibility gene-*S* and gene-*s* of the fungus *Podospora anserina*. Curr Genet 17:297–303, 1990.
71. B Turcq, C Deleu, M Denayrolles, J Bégueret. Two allelic genes responsible for vegetative incompatibility in the fungus *Podospora anserina* are not essential for cell viability. Mol Gen Genet 228:265–269, 1991.
72. C Deleu, C Clavé, J Bégueret. A single amino acid difference is sufficient to elicit vegetative incompatibility in the fungus *Podospora anserina*. Genetics 135:45–52, 1993.
73. V Coustou, C Deleu, SJ Saupe, J Bégueret. The protein product of the *het-s* heterokaryon incompatibility gene of the fungus *Podospora anserina* behaves as a prion analog. Proc Natl Acad Sci USA 94:9773–9778, 1997.
74. L Belcour. Loss of a cytoplasmic determinant through formation of protoplasts in *Podospora*. Neurospora Newsl 23:26–27, 1976.
75. SB Prusiner. Prions. Proc Natl Acad Sci USA 95:13363–13383, 1998.
76. G Pontecorvo. The parasexual cycle in fungi. Annu Rev Microbiol 10:393–400, 1956.
77. RB Wickner, KL Taylor, HK Edskes, ML Maddelein, H Moriyama, BT Roberts. Prions in *Saccharomyces* and *Podospora* spp.: protein-based inheritance. Microbiol Mol Biol Rev 63:844–861, 1999.
78. V Coustou, C Deleu, SJ Saupe, J Bégueret. Mutational analysis of the [Het-s] prion analog of *Podospora anserina*: a short N-terminal peptide allows prion propagation. Genetics 153:1629–1640, 1999.
79. J Bernet, J Bégueret, J Labarère. Incompatibility in the fungus *Podospora anserina*. Are the mutations abolishing the incompatibility reaction ribosomal mutations? Mol Gen Genet 124:35–50, 1973.
80. S Saupe, C Descamps, B Turcq, J Bégueret. Inactivation of the *Podospora anserina* vegetative incompatibility locus *het-c*, whose product resembles a glycolipid transfer protein, drastically impairs ascospore production. Proc Natl Acad Sci USA 91: 5927–5931, 1994.
81. A Abe. Primary structure of glycolipid transfer protein from pig brain. J Biol Chem 265:9634–9637, 1990.
82. S Saupe, B Turcq, J Bégueret. Sequence diversity and unusual variability at the *het-c* locus involved in vegetative incompatibility in the fungus *Podospora anserina*. Curr Genet 27:466–471, 1995.
83. S Saupe, B Turcq, J Bégueret J. A gene responsible for vegetative incompatibility in the fungus *Podospora anserina* encodes a protein with a GTP-binding motif and G beta homologous domain. Gene 162:135–139, 1995.
84. E Espagne, P Balhadère, J Bégueret, B Turcq. Reactivity in vegetative incompatibility of the HET-E protein of the fungus *Podospora anserina* is dependent on GTP-binding activity and a WD40 repeated domain. Mol Gen Genet 256:620–627, 1997.
85. L Belcour, J Bernet. Sur la mise en évidence d'un gène dont la mutation supprime spécifiquement certaines manifestations d'incompatibilité chez le *Podospora anserina*. C R Acad Sci Paris 269:712–714, 1969.

86. J Bernet. Sur un cas de suppression de l'incompatibilite cellulaire chez le champignon filamenteux *Podospora anserina*. C R Acad Sci Paris 273:120–122, 1971.
87. C Barreau, M Iskandar, G Loubradou, V Levallois, J Bégueret. The *mod-A* suppressor of nonallelic heterokaryon incompatibility in *Podospora anserina* encodes a proline-rich polypeptide involved in female organ formation. Genetics 149:915–926, 1998.
88. H Boucherie, J Bernet. Protoplasmic incompatibility in *Podospora anserina*: a possible function for incompatibility genes. Genetics 96:399–411, 1980.
89. J Bernet. In *Podospora anserina*, protoplasmic incompatibility genes are involved in cell death control via multiple gene interactions. Heredity 68:79–87, 1992.
90. J Labarère, J Bernet. Protoplasmic incompatibility and cell lysis in *Podospora anserina*. I. Genetic investigations on mutations of a novel modifier gene that suppresses cell destruction. Genetics 87:249–257, 1977.
91. N Bourges, A Groppi, C Barreau, C Clavé, J Bégueret. Regulation of gene expression during the vegetative incompatibility reaction in *Podospora anserina*. Characterization of three induced genes. Genetics 150:633–641, 1998.
92. CM Moehle, R Tizard, SK Lemmon, J Smart, EW Jones. Protease B of the lysosomelike vacuole of the yeast *Saccharomyces cerevisiae* is homologous to the subtilisin family of serine proteases. Mol Cell Biol 7:4390–4399, 1987.
93. EW Buxton. Heterokaryosis and parasexual recombination in pathogenic strains of *Fusarium oxysporum*. J Gen Microbiol 15:133–139, 1956.
94. LS Olive. Genetics of homothallic fungi. Mycologia 55:93–103, 1963.
95. CE Caten. Vegetative incompatibility and cytoplasmic infection in fungi. J Gen Microbiol 72:221–229, 1972.
96. DL Nuss, Y Koltin. Significance of dsRNA genetics elements in plant pathogenic fungi. Annu Rev Phytopathol 28:37–58, 1990.
97. D Baidyaroy, JM Glynn, H Bertrand. Dynamics of asexual transmission of a mitochondrial plasmid in *Cryphonectria parasitica*. Curr Genet 37:257–267, 2000.
98. SL Anagnostakis. Conversion to curative morphology in *Endothia parasitica* and its restriction by vegetative compatibility. Mycologia 75:777–780, 1983.
99. DL Hartl, ER Dempster, SW Brown. Adaptive significance of vegetative incompatibility in *Neurospora crassa*. Genetics 81:553–569, 1975.
100. J Bégueret, B Turcq, C Clavé. Vegetative incompatibility in filamentous fungi—*het* genes begin to talk. Trends Genet 10:441–446, 1994.
101. J Klein, A Sato, S Nagl, CO Uigin. Molecular trans-species polymorphism. Annu Rev Ecol Syst 29:1–21, 1998.
102. J Wu, SJ Saupe, NL Glass. Evidence for balancing selection operating at the *het-c* heterokaryon incompatibility locus in a group of filamentous fungi. Proc Natl Acad Sci USA 95:12398–12403, 1998.
103. RF Hoekstra. Population genetics of filamentous fungi. Antonie Van Leeuwenhoek Int J Gen Mol Microbiol 65:199–204, 1994.

6

Vegetative Development in *Coprinus cinereus*

Ursula Kües
ETH Zurich, Zurich, Switzerland, and Georg-August-Universität Göttingen, Göttingen, Germany

Eline Polak,* Alan P. F. Bottoli,† Marcel Hollenstein,‡ Piers J. Walser, Robert P. Boulianne,* René Hermann, and Markus Aebi
ETH Zurich, Zurich, Switzerland

1 INTRODUCTION

Fungal organisms may produce several types of spores, at the same or different time periods, and either mitotically or, after nuclear fusion, meiotically. Fungi that have more than one independent form or spore stage in their life cycle are called pleomorphic. The whole fungus in all its forms (morphs) and phases is called the holomorph. The term teleomorph is used to designate the sexual or ''perfect'' state of a fungus and is characterized by the sexual spores resulting from meiosis (meiospores). In contrast, the term anamorph denotes the asexual or ''imperfect'' stage or stages exhibiting only mitotic divisions. An anamorph may produce asexual spores by means of mitosis (mitospores) [1]. Mitosporic fungi (formerly classified as Deuteromycetes or Fungi Imperfecti) are those that

* *Current affiliation*: PFC Pharma Focus Consultants AG, Volketswil, Switzerland
† *Current affiliation*: Credit Suisse, Zurich, Switzerland
‡ *Current affiliation*: National Institute of Diabetes and Digestive Kidney Diseases, National Institutes of Health, Bethesda, Maryland

are missing a teleomorphic form and exist only in an anamorphic stage(s). As deduced from the shape of the anamorph(s) and, more significantly, from DNA sequence analysis, many of the mitosporic fungi are closely related to the ascomycetous fungi. These have a sexual cycle and form their meiotic spores (ascospores) endogenously within asci, which are either naked cells or are present within fruiting bodies. Much attention has thus been given to the anamorphs of Ascomycetidae that often occur well separated in either time or space from their teleomorphs [2–4; this volume, Chapter 3 by Fischer].

In the basidiomyceteous fungi, karyogamy and meiosis occur within the specialized basidia that can be localized within fruiting bodies. Upon meiosis, the exogenously formed sexual spores (basidiospores) bud off the basidium [5; this volume, Chapters 14 by Banuett and 10 by Kothe]. Among the yeastlike hemibasidiomycetes, anamorphs are frequently distinguished from their teleomorphs [6]. Anamorphs are recognized in the group of the plant pathogenic rust fungi (Uredinales) that produce various types of mitotic spores, some of which are accompanied by a change of host [7,8]. More commonly, different life forms of Basidiomycetes are not strictly separated. Anamorphs often appear side by side with the teleomorphs and are usually considered an integral part of the teleomorphs. For this reason, anamorphs are rarely defined by a separate generic name [9].

Within the homobasidiomycetes, the occurrence of asexual spores might be thought to be the exception. However, mitotic spores of homobasidiomycetes are typically not very conspicuous and are therefore frequently overlooked. In their literature survey, Kendrick and Watling [9] list ~100 species of Agaricales and several hundreds of Aphyllophorales where the occurrence of one or more types of mitotic spores have been noted. Brodie [10,11] initiated studies on asexual spore production in *Coprinus cinereus* (formerly referred to as *lagopus*) and *Flammulina velutipes* (formerly called *Collybia velutipes*) in the 1930s. However, it took several decades before studies were taken up again first in *F. velutipes* [12,13] and later in *C. cinereus* [14,15], whose anamorph is *Hormographiella aspergillata* [16,17]. *C. cinereus*, an Inky Cap, is a well-employed model organism for studies on fruiting body (mushroom) development [5,18–20], meiosis [21–23], and mating types [5,24,25] [this volume, Chapter 10 by Kothe].

2 TWO FORMS OF VEGETATIVE MYCELIUM: THE MONOKARYON AND THE DIKARYON

Coprinus cinereus is coprophilous. In nature, it grows and fruits on horse dung and has the typical life cycle of a homobasidiomycete (Fig. 1). The meiotic basidiospores germinate into a primary vegetative mycelium (equivalent to the anamorph *H. aspergillata*) [17,27] that is characterized by hypha with simple dolipore-septa (Fig. 2) and one type of haploid nuclei in its cells. Therefore, this mycelium is classified as a homokaryon. Traditionally, the primary mycelium of

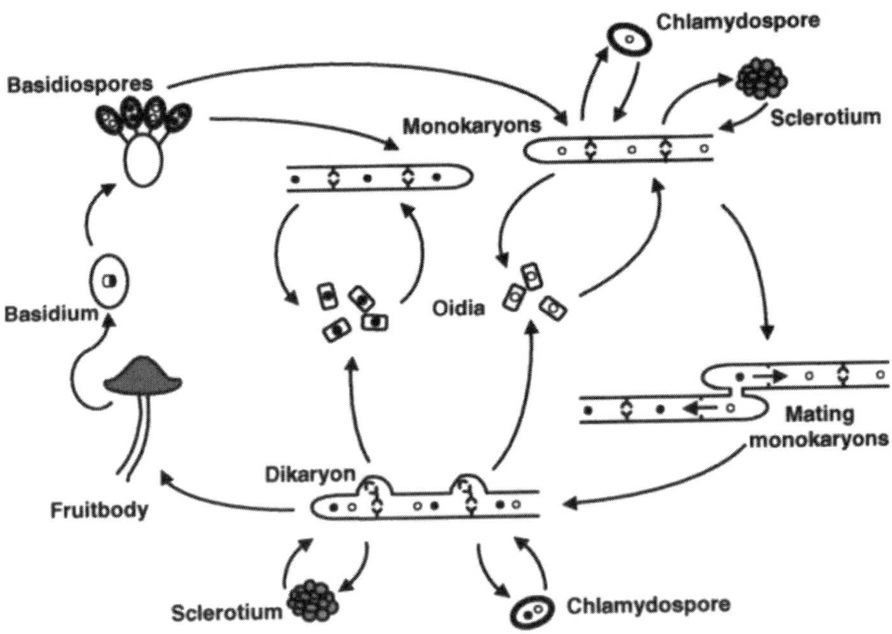

FIGURE 1 Life cycle of *Coprinus cinereus*. (From Ref. 5.)

C. cinereus is called a monokaryon [5] which in its strictest sense implies that it is a homokaryon that has one nucleus per hyphal segment [28]. However, nuclear staining of various *C. cinereus* strains revealed that the primary mycelium has typically one or two, occasionally three or four nuclei in its cells [14] (Table 1, Fig. 2). Usually, the hyphae of monokaryons are thin with a diameter of ~3 μm [18,29,30]. Genetically distinct monokaryons differ in growth speed and also in colony appearance, primarily owing to variations in the amount and structure of the aerial mycelium produced [30,31]. Distances between hyphal branches may vary between different monokaryons but branches mostly protrude in broad angles of ~70–75° from the parental monokaryotic hypha [18,29,30].

The secondary mycelium of *C. cinereus* is a specific heterokaryon called a dikaryon and arises upon hyphal fusion (anastomosis) of two mating-compatible monokaryons—i.e., two primary mycelia with different mating types. Following fusion of hyphal tips (tip-to-tip fusion), fusion of a hyphal tip and a lateral hyphal wall (tip-to-tip fusion), fusion between a hyphal tip and a lateral swelling of a hyphae (tip-to-peg fusion), or fusion between two lateral swellings of two neighboring hyphae (peg-to-peg fusion), nuclei exchange between the two monokaryotic mycelia and migrate with a speed of 1–3 mm h^{-1} through the mycelium of opposite mating type until they reach a hyphal tip cell. Importantly, the two differ-

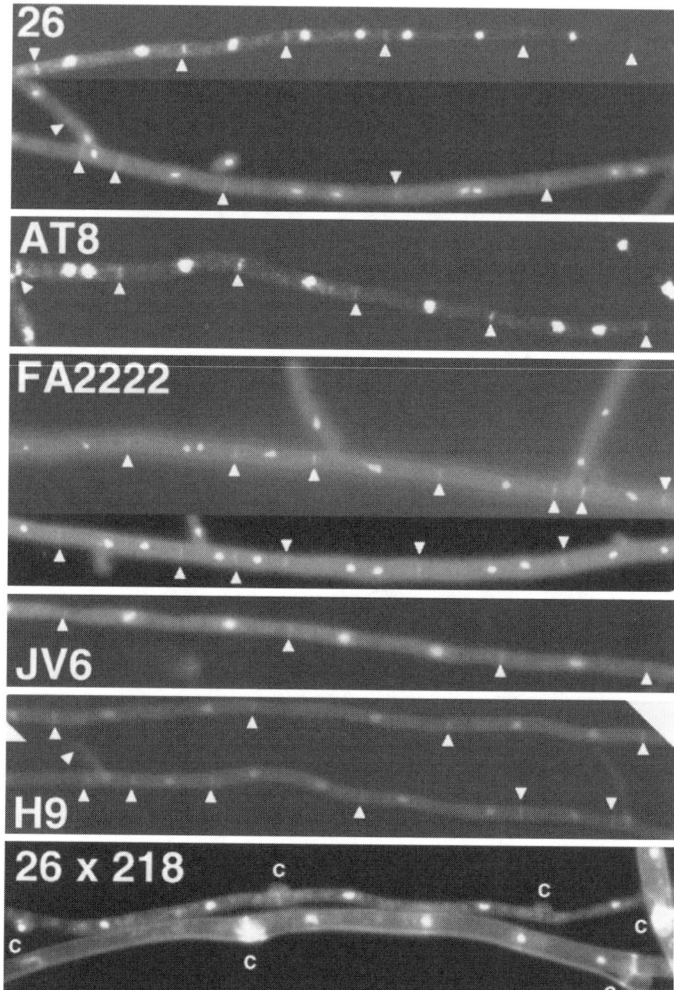

Figure 2 Morphology of aerial hyphae in various monokaryons (26, AT8, FA2222, JV6, H9) and a dikaryon (26 × 218) of *Coprinus cinereus*. Nuclei were stained with DAPI (4′,6-diamidine-2-phenylindole dichloride) [14]. Arrows mark the positions of septa in the hyphae of monokaryons, showing that one or two nuclei are present in the hyphal cells. For the dikaryon, the letter "c" marks septa with clamp cells. Usually, two nuclei are found in the hyphal cells of dikaryons as shown by the upper hypha in the picture, but occasionally four nuclei were observed (hypha at the bottom). (For origin and genetic characterization of these and other strains in this paper, see Ref. 35.)

TABLE 1 Nuclear Distribution Within Aerial Hyphae of Various *C. cinereus* Monokaryons and Dikaryons[a]

C. cinereus strains[b]	Total cells analyzed	Number of cells with: 1 Nucleus	2 Nuclei	3 Nuclei	4 Nuclei
Monokaryon					
5026	81	41	40	—	—
26	81	40	37	2	2
218	45	23	22	—	—
LT2	52	24	26	—	2
AT8	40	24	16	—	—
PG78	51	30	21	—	—
H5	61	28	30	3	—
H9	48	26	22	—	—
FA2222	44	24	19	1	—
JV6	52	26	26	—	—
LN118	47	19	22	4	2
301	49	29	17	3	—
306	53	26	27	—	—
Dikaryon					
26 × 218	120	1	107	2	10
26 × LT2	124	6	116	—	2
26 × AT8	107	2	101	2	2
26 × PG78	105	1	101	—	3
AT8 × PG78	55	—	55	—	—

[a] Strains were grown at 37°C in light on complete medium on microslides, nuclei stained with DAPI, and septa stained with Blankophor BA 267% (Bayer Leverkusen, Germany) as described in Ref. 14.
[b] For strain origins and genetic characteristics see Ref. 35.
Source: Ref. 14.

ent haploid nuclei do not fuse at this stage. In the hyphal tip cell, the two distinct nuclei pair and divide in a synchronous manner. For this a clamp cell develops laterally on the hyphal tip cell at the place where a new septum will be laid following mitosis [5,24; this volume, Chapter 10 by Kothe] (Fig. 3). The nucleus closer to the hyphal tip enters the clamp cell and divides with a relatively short spindle. One of the dividing daughter nuclei migrates back into the parental hyphal cell, and a dolipore septum is laid between the clamp and hyphal cell, trapping one of the newly formed nuclei within the clamp cell. Simultaneously to nuclear division in the clamp cell, the more subapical nucleus in the parental hyphal cell divides with a long spindle. In consequence, one of the dividing daughter nuclei passes the other pair of dividing nuclei in the clamp and becomes

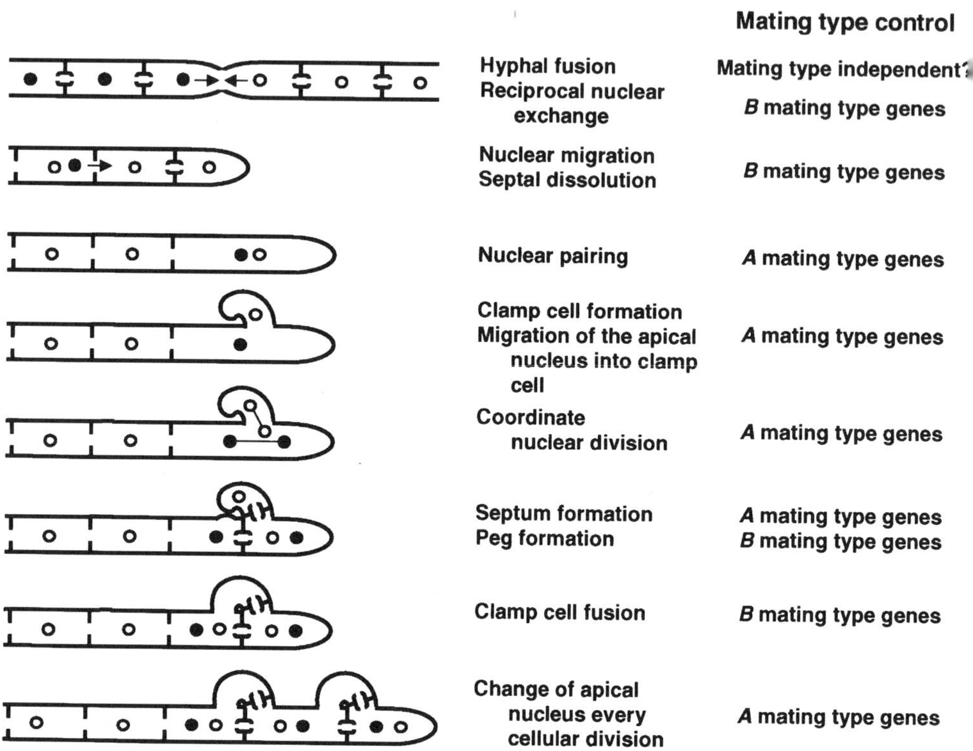

FIGURE 3 Formation of a dikaryotic mycelium following fusion of two compatible monokaryons and control of development by the two mating-type loci *A* and *B*. Open and filled circles indicate the two distinct haploid nuclei. Note that it is not known whether nuclei divide while migrating to a monokaryotic mycelium of opposite mating type [5]. (For detailed explanations on the *A* and *B* mating-type control, see Refs. 5 and 24; Sec. 4.1; and this volume, Chapter 10 by Kothe.)

the apically localized nucleus. Formation of a dolipore septum between the second pair of dividing nuclei locks one of the nuclei into the newly formed subapical cell and combines a pair of two distinct haploid nuclei in the newly generated apical cell. Subsequently, the clamp cell fuses with the subapical cell, reestablishing the presence of two genetically distinct haploid nuclei in this cell. Each synchronized nuclear division changes the relative position of nuclei in the apical cell, which is an interesting effect of the differences in spindle length between the pairs of dividing nuclei [5,24,32] (Fig. 3). Occasionally, within dikaryotic cells a further mitotic division occurs, giving rise to four nuclei within a single cell (Table 1, Fig. 2).

Formation of clamp cells at the hyphal septum ensures the even distribution of the two genetically distinct haploid nuclei in the dikaryon, but it also gives the secondary mycelium its typical hyphal appearance (Figs. 1–3). Dikaryotic hyphae are usually broader than those of monokaryons and measure ~7 μm in width. Branches on the dikaryon arise in relatively acute angles of 10–45° [18,29]. Dikaryons tend to grow faster than monokaryons, with a dense, more protruding, and conspicious aerial mycelium [18,33,34]. Under defined environmental conditions, fruiting bodies develop on the sexually fertile dikaryon, in contrast to the sexually sterile monokaryon. At the lower surface of the cap of the fruiting body, within the hymenium that lines the gills, karyogamy, meiosis, and spore formation take place in the basidia, which completes the sexual life cycle [5,18,19,22; this volume, Chapter 10 by Kothe] (Fig. 1).

3 SUBSIDIARY REPRODUCTIVE CYCLES

In addition to the sexual reproductive cycle, *Coprinus* has three vegetative, i.e., asexual reproductive, cycles that occur in the monokaryotic and dikaryotic states. Both types of mycelia may form specialized structures (oidiophores) that produce small, rod-shaped, uninucleate, haploid, hyaline spores termed oidia [10,14,15,35]. They also produce large, thick-walled chlamydospores [28,33, 36,37] and multicellular resting bodies called sclerotia [37–40].

3.1 Oidiophores and Oidia

Oidiophores, with their attached oidia, mainly form in the aerial mycelium of the fungus [10,14,15,33] but occasionally also within the agar phase on artificial medium [11,15]. The whole process of oidiophore formation and spore production (Fig. 4) takes between 12 and 24 h at 37°C (the favored growth temperature of *C. cinereus*) and has recently been defined in its consecutive steps [14] (Fig. 4). Oidiophore formation initiates at a hyphal cell with a localized lateral bulging. A nucleus divides in the hyphal cell (now the oidiophore foot cell), and one of the daughter nuclei migrates laterally into the young oidiophore that protrudes with an angle of ±80–85° from the parental hypha [14,30]. A septum is laid that separates the oidiophore stem cell from the foot cell. The stem cell, measuring 3–7 μm at its base, is broader than an average undifferentiated monokaryotic hyphal cell and elongates and tapers with length. Oidial hyphae, one after the other, bud from the tip of the mature stem cell and consecutive nuclear divisions within the stem cell provide each oidial hypha with a nucleus. An oidial hypha separates from the stem cell by septation, another nuclear division follows, and the oidial hypha eventually breaks up into two or rarely three equal-size uninucleate oidia. Separation is achieved through splitting of the delimiting septum (schizolysis) between an oidium and the oidiophore or between two oidia [14]. Upon release from the oidial hyphae, oidia collect in a sticky liquid extruded

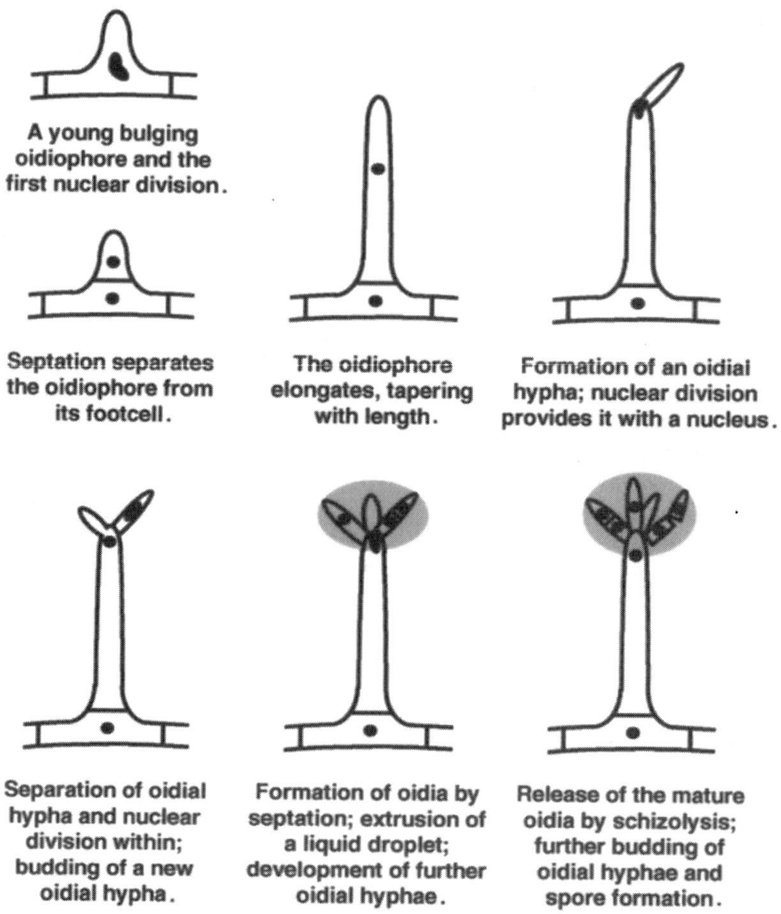

FIGURE 4 Model for development of oidiophore and oidia in *C. cinereus* [14]. For purposes of simplification, the oidiophore is shown unbranched with only one stem cell.

from the tip of the oidiophore (Figs. 5, 6), which gives the aerial mycelium an appearance as if speckled with dewdrops [10,14] (Fig. 5). Production of oidial hyphae and oidia continues until up to 200 oidia are formed at the tip of a single oidiophore [14].

An average oidiophore, as described above, measures about 20–30 μm [14]. Usually broader and morphologically different from their generating vegetative hyphae, *C. cinereus* oidiophores are considered macronematous. The basic

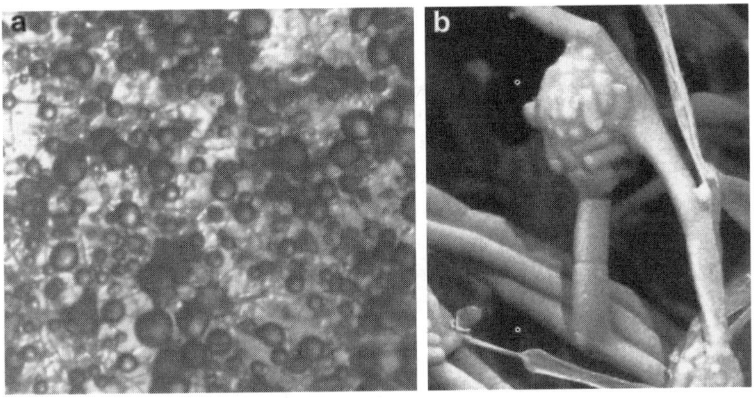

FIGURE 5 Oidia are collected in liquid droplets at the tip of oidiophores. (a) Mycelial overview of monokaryon H5 grown on microslides on complete medium at 37°C in the dark. (b) Low-temperature scanning electron microscopy image of a freeze-etched oidiophore in the aerial mycelium of monokaryon FA2222. (Experimental details as in Ref. 14.)

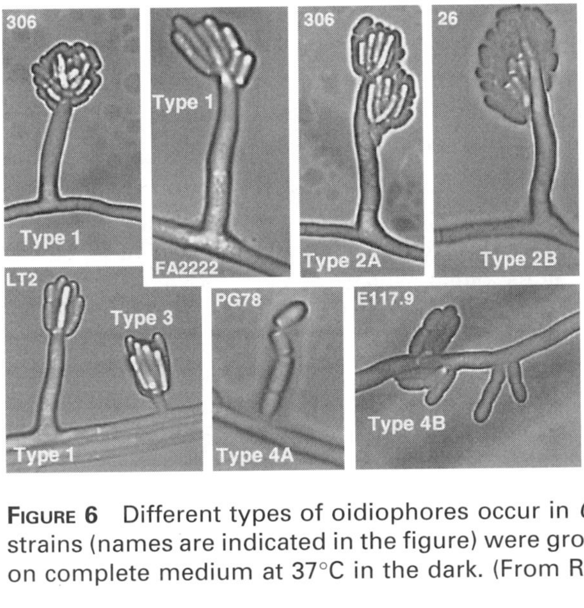

FIGURE 6 Different types of oidiophores occur in *C. cinereus*. Monokaryotic strains (names are indicated in the figure) were grown on microslide cultures on complete medium at 37°C in the dark. (From Ref. 14.)

type of macronematous oidiophores of comparable size (8–55 μm in length, 5–8 μm broad at the base, and 2.5–5.5 μm in the middle) have also been observed in *H. aspergillata* and are used for identification of the anamorph [16,27]. As for the first two *C. cinereus* strains described in detail [14], simple erect oidiophores are found next to oidiophores that have one or two, rarely three, lateral branches [16]. A more extended survey of several *C. cinereus* strains revealed that the developmental pathway of oidiation is not a strictly fixed process and that the overall oidiophore structure can be much more variable [15]. An oidiophore may form one, two, or three stem cells before oidial hyphae are formed at the tip cell (Fig. 6). However, the stem may form side branches whose tip cells will also give rise to oidial hyphae; alternatively, oidial hyphae can also form from lateral bulges on the tips of the stem cells beneath (Fig. 6). These advanced types of oidiophores are the most complex structures classified as type 1 and type 2A and type 2B oidiophores, respectively [15]. Other types of oidiophores are more simple because some steps of the developmental process of oidiophore formation are missing [15]. Type 3 oidiophores, for example, have a stem cell that does not elongate. These oidiophores are thus quite short (Fig. 6). Type 4 oidiophores have no stem cells and are subdivided into two groups: oidial hyphae bud off directly from a hyphal foot cell either singly (type 4A; Fig. 6) or in bunches (type 4B; Fig. 6). Analysis of >20 different strains revealed that each strain is able to form all types of oidiophores, but usually, one or two types are preferentially formed [15]. In the more advanced oidiophores, oidial hyphae typically emerge from the stem cells in a pattern resembling the inflorescence of an umbel (Fig. 6). In rare cases, however, oidial hyphae themselves branch before breaking up into single spores, which gives an impression of a panicle. Also rare, oidial hyphae form one behind the other at only one side of the oidiophore tip cell in a cymelike pattern [15].

Oidia in *Coprinus* measure about 2 × 4–6 μm although in some strains their size is much more variable than in others [10,14,15]. Independent of whether they form on a monokaryon or a dikaryon, oidia are nearly exclusively uninucleate [14,15,41] (Fig. 7). Since oidia formation involves both budding (of the oidial hypha) and septum schizolysis, oidia are classified as arthroconidia [9,14,42]. Owing to their position within the oidial hyphae, two types of oidia are distinguished [14,43,44]. Oidia coming from the apical ends of the oidial hyphae have one truncated end as a consequence of schizolysis and one round end corresponding to the tip of the oidial hypha (Fig. 6). Intercalarily formed oidia have two flat ends owing to splitting of the septa between two chained oidia and between oidia and oidiophore stemcells (Fig. 6). The truncated ends of oidia therefore represent a single-layered secondary cell wall [14,45], and these will be used to form germ tubes when the spores start to grow again [44]. In contrast to the limited areas of secondary cell wall, the double-layered primary cell wall of oidia is covered with hairlike structures (Fig. 6) that possibly help to keep a mucilagenous layer surrounding the oidium [14,42]. Unlike many airborne mitospores

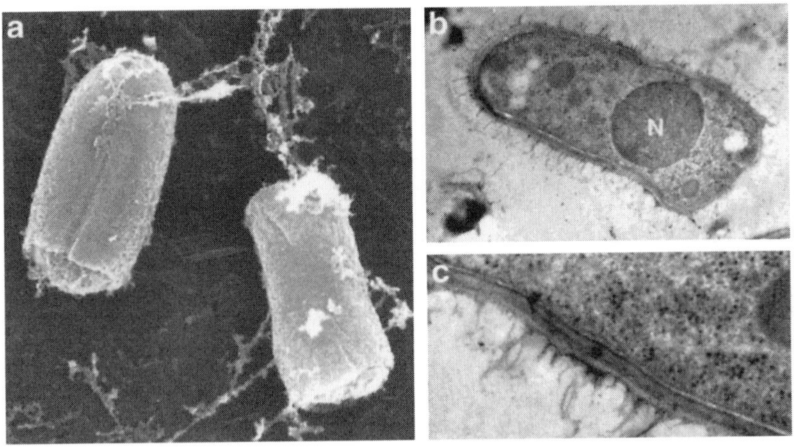

FIGURE 7 (a) Scanning electron microscopic, and (b) and (c) transmission electron microscopic photographs of oidia from homokaryon AmutBmut. An apical formed oidium with one flat and one round end is shown in (a), at the right, an intercalary formed oidium with two truncated ends at the left. Note the hairlike structures at the outer primary cell wall of the oidium shown in (b) and enlarged in (c). [Experimental details as in Refs. 14 (b and c) and 115 (a).]

of ascomycetes [46–48], oidia of *C. cinereus* are not coated by a layer of cysteinerich hydrophobic secreted proteins called hydrophobins [49]. Oidia of *C. cinereus* are strongly hydrophilic (''wet''), and because of the surrounding gelatinous layer they are also sticky [14,50,51]. They are not windborne but will attach to flies and other insects that distribute them from horse dung to horse dung [10]. Although oidia loose the ability to germinate relatively fast [41], this type of spore has two biological roles. They attract and fuse to monokaryotic hyphae of different mating type (a process known as ''oidial homing'') to give rise to a fertile dikaryon [33,52]. They also attract and fuse to hyphae of other *Coprinus* spp. competing for the substrate horse dung. Upon fusion, somatic incompatibility reactions occur, leading to the death of the foreign hyphae [52]. Thus, oidia act as spermatia toward hyphae of their own species and as killing agents toward hyphae of other species.

3.2 Chlamydospores

Chalamydospores are large, thick-walled mitospores with condensed cytoplasm [28,29,36,37]. They are variable in form and may be round, oval, or irregularly shaped (Fig. 8). Chlamydospores are found in brown-colored patches at the agar–air interface in the mycelial matting of aging cultures and permit long-term sur-

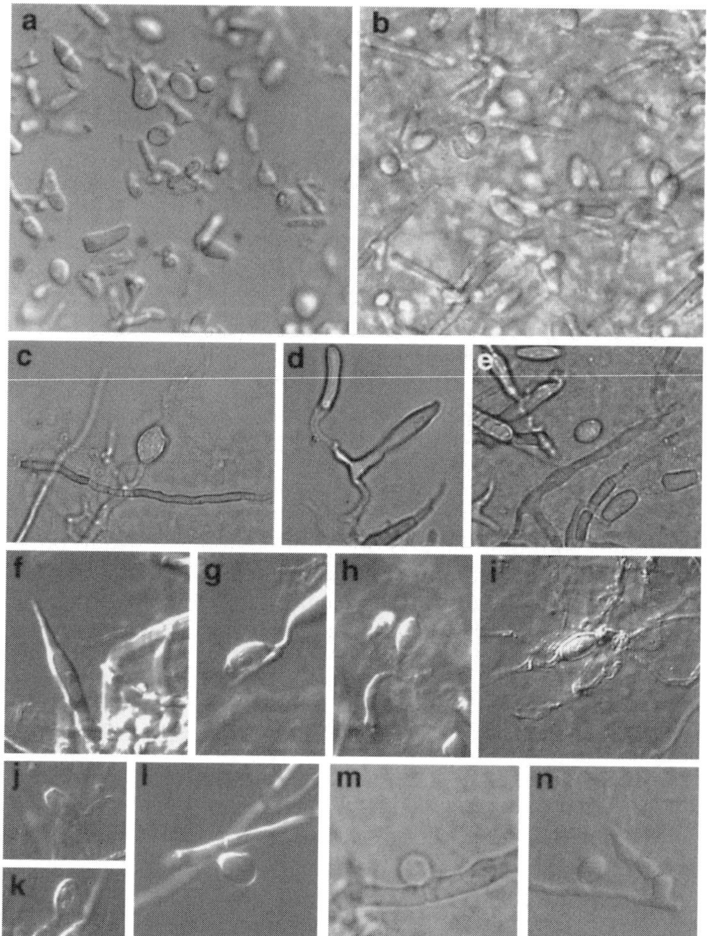

FIGURE 8 Chlamydospores can by found in squeezed preparations of the mycelial matting from aging cultures of monokaryons (a, strain 218) and dikaryons (b, strain 218 × LN118). (c–k) Chlamydospores might be formed endogeneously within vegetative hyphal cells (chlamydospores in the strictest sense). First, the cytoplasm breaks up in portions and condenses within hyphal cells (c–e). Cell walls are formed around the ellipsoidal condensed protoplasts, giving rise to the chlamydospores (f–k) with a double-layered cell wall, of which the inner is newly formed and the outer is from the original hyphal cell [53]. Alternatively, chlamydospores arise as blastocysts from hyphal cells by budding (l–n). Owing to the mode of generation, blastocysts are expected to have only a single-layered cell wall [53]. (c–l) Homokaryon AmutBmut. (m) Monokaryon 218. (n) Monokaryon AT8.

vival of the fungus [29,36]. The mode of chlamydospore production is not well understood. It appears that *C. cinereus* produces such inflated spores in at least two ways. Chlamydospores may develop endogenously within terminal or intercalary cells of the vegetative hypha, solitary or in chains, following compression of the cytoplasm (Fig. 8). This type of spore formed by modification of a preexisting cell are chlamydospores in the strictest sense [53]. We also observed spore production by formation of an inflated bud followed by transfer of the compressed cytoplasm from the hyphal cell into the bud (Fig. 8). This type of chlamydospore is classified as blastocyst [53]. Swollen, thick-walled hyphal segments resembling chlamydospores have been described in the mycelial matting of mono- and dikaryotic strains and connected with glycogen storage in preparation for fruiting [39,54]. These inflated cells also correlate with the appearance of the multicellular sclerotia that harbor similar chlamydosporelike cells [38,39,55] (Sec. 3.3, Fig. 9). If just repositories for food until needed in another developmental pathway and never to germinate, inflated cells in the mycelial matting may not be considered spores [9]. However, it is a matter for discussion and further study whether such a sharp distinction by function between a persistent chlamydospore that eventually will germinate under better environmental conditions into a new mycelium and a chlamydosporelike cell that will not germinate in favor of supplying nutrients for fruiting body formation is necessary.

Although occurrence of chlamydospores on monokaryons had been originally noted by Bensaude [33], but not rediscovered until recently [37], the nuclear conditions of chlamydospores from monokaryons and their mode of germination have yet to be clarified. In contrast, chlamydospores from dikaryons were shown to be binucleate [56] and to germinate with either one or two germ tubes [28,36]. In the case of a single germ tube, the nuclei will enter it together and a dikaryotic hyphae will arise. With two germ tubes, the distinct haploid nuclei will migrate into different germ tubes and two monokaryotic hyphae will form. When separated by microsurgery before they dikaryotize each other, it is possible to isolate the component monokaryons of the former dikaryon [28,36].

3.3 Sclerotia

Sclerotia are multicellular, oval or globular symmetrical resting bodies that develop in aerial and submerged mycelium in aging cultures of *C. cinereus* [37–40] (Fig. 9) and its anamorph *H. aspergillata* [16]. Mature sclerotia have a single- or multilayered melanized outer rind (cortex) of small, thick-walled cells and are surrounded by an outer layer of dead and moribund hyphae. The internal pseudoparenchymatous medulla is filled with large, ovate, and globular cells that are generally thick-walled and resemble chlamydospores [38,40,55,57] (Sec. 3.2, Fig. 9). Mature submerged sclerotia are usually larger (0.5–1.0 mm) than aerial sclerotia (0.1–0.5 mm), less regular in shape, and paler. In contrast to aerial

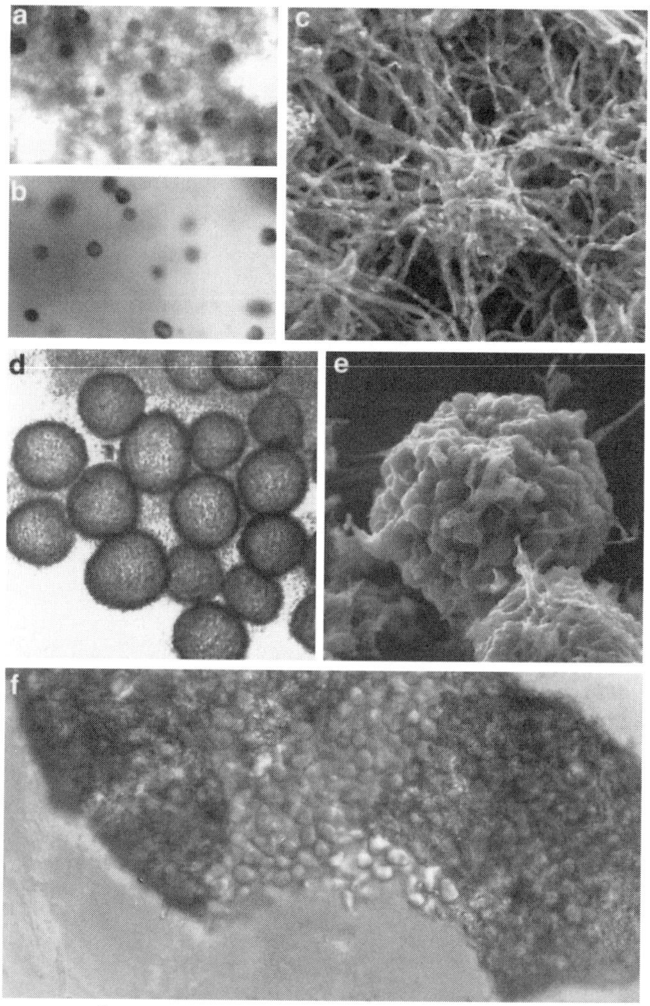

FIGURE 9 Sclerotia formation in *C. cinereus.* (a) Aerial and (b) submerged sclerotia of monokaryon FA2222 transformed with a compatible *A* mating-type gene [37]. Note the hyphal knots (shown as scanning electron microscopic overview in (c) within the aerial mycelium. (d) Mature aerial sclerotia isolated from aerial mycelium of homokaryon AmutBmut. (e) Scanning electron microscopic overview of a mature sclerotium from monokaryon FA2222 formed after transformation with a compatible *A* gene [37]. (f) Following squeezing of an aerial sclerotium of dikaryon LN118 × AT8, the internal chlamydosporelike cells become obvious.

sclerotia, their medulla is loosely organized and the thickness of the layer of rind cells is uneven. Mature aerial sclerotia are covered by an outer layer of dead and dying hyphae [39].

Sclerotia development is centrifugal. Growth initiates at a central point and continues outward, finishing with the formation of the melanized rind. Genetic evidence [37,58] suggests that sclerotia develop from hyphal knots (Fig. 6), which are areas of localized intense hyphal branching [5,40,59,60]. In the aerial mycelium, hyphal knot development may start from one or a few intercalary undifferentiated hyphal cells. Most of the abundantly formed short hyphal branches in these knots will swell [40,59] and probably give rise to the bulbous chlamydosporelike cells found within the medulla of the immature sclerotia while further short branches are still forming. As the medulla continues to develop, the rind appears as areas of pigmentation in the margin of the immature sclerotium. Unlike cells in the medulla, those in the rind do not inflate, but their cell walls thicken and become pigmented [40]. Typically, aerial sclerotia have a multilayered rind; however, in exceptional cases aerial sclerotia may have a single-layered rind [39,40,57]. The cortex in aerial sclerotia centripetally increases in width by incorporating cell layers from an intermediate layer that otherwise gives rise to additional medullary cells in the final stages of maturation [40]. Differences in the pattern of development have been encountered between aerial and submerged sclerotia [38,40]. In the submerged mycelium, individual swellings and bulbous extensions of hyphal strands initiate sclerotia formation. Hyphal branching, bulbous hyphal tip growth, and delineation of septa contribute to the centrifugal growth of the developing sclerotia [38]. For maturation, peripheral cells undergo wall deposition, always ending in a single-layered rind of thick-walled pigmented cells of the small lumen [38,39]. Hyphal knots formed in the submerged mycelium can also turn into complete nests of chlamydospores without a surrounding melanized cortex, suggesting again a close link between the two types of resting bodies [5,38]. Occurrence of sclerotia and free chlamydospores correlate with each other spatially and temporally and also with the formation of the abundant inflated hyphal cells in the mycelial matting shown to accumulate glycogen [37–39] (Sec. 3.2). If not relocated to the fruiting body [5,19,54,61], polysaccharides seem to be transported from these large inflated cells to the sclerotia to build up the glycogen storage found in the bulbous chlamydosporelike cells of the medulla [19,39,40,62,63].

4 REGULATION OF VEGETATIVE DEVELOPMENT

4.1 Genetic Regulation

Most of our knowledge on genetic regulation of vegetative development in *C. cinereus* is restricted to the actions of the mating-type genes, which are the master genes in control of all developmental processes [5,37]. *C. cinereus* has two mating-type loci, called *A* and *B*. For dikaryon formation, mating monokaryons have to be distinct at both loci; if fusing monokaryons differ only at one of the two mating-type loci, unstable common *A* or common *B* heterokaryons might be formed that exhibit only parts of the morphological characteristics of the dikaryotic mycelium [5,24,25; this volume, Chapter 10 by Kothe]. Such heterokaryons have been very useful, however, in deducing the functions of the mating-type loci in dikaryon formation. The *A* mating-type genes are thus known to control synchronized nuclear division and clamp cell production and septum formation; the *B* genes are responsible for peg formation at the subapical cell and for clamp cell fusion with the subapical cell [5,64,65; this volume, Chapter 10 by Kothe] (Fig. 3). In the monokaryon, mating-type genes are expected to have no essential function [66,67; this volume, Chapter 10 by Kothe]. Mutations in the mating-type genes selected in monokaryons never inactivated the loci but always led to a constitutive activation of the regulated developmental pathways—i.e., to expression of morphological characteristics of the dikaryon [68,69]. So far, a knockout by transformation has only been generated from the *A* locus of *C. cinereus* [66]. Monokaryotic knockout strains are viable and have no altered morphology in vegetative hyphae and oidiophores, and constitutively produce abundant oidia [30]. On wild-type monokaryons, oidia production is also constitutive in high numbers ($\sim 10^9$ spores/9 cm Petri dish culture or $1–2 \times 10^7$ spores/cm^2) [35,41]. Oidiation on the dikaryon (Fig. 10) was overlooked for nearly a century [24,70,71] before a blue light control of this process was discovered [35,37]. Blue light illumination induces oidiation at the stage of oidiophore initiation [14], but oidia are never formed in such levels as obtained with monokaryons [35]. The oidia yield on dikaryons ranges $\sim 10^7$, in rare cases $\sim 10^8$, spores per plate compared to the 10^9 spores produced by monokaryons [35]. Since dikaryon formation is under control of the mating-type loci, the products of the mating-type genes have been suspected to contribute to this phenomenon [35,72].

The *A* mating-type locus has been shown encode two types of homeodomain transcription factors (HD1 and HD2), the *B* mating type locus pheromones, and pheromone receptor [5,24,25; this volume, Chapter 10 by Kothe]. Transformation studies with cloned genes revealed that the product of an *HD1* gene of one mating type needs to interact with the product of a compatible *HD2* gene from another to induce expression of *A* mating type–regulated functions in the dikaryon. These functions include clamp cell formation (Fig. 3), chlamydospore production, hyphal knot development, and fruiting body initiation [37,73–75]. Likewise, it needs pheromones from one mating type to interact with compatible pheromone receptors of another mating type to induce *B* mating type–regulated development—for example, clamp cell fusion [76,77] (Fig. 3). Recently, in a transformant of *C. cinereus*, compatible *B* genes were also shown to trigger septal dissolution and nuclear migration (Fig. 11), repress formation of aerial mycelium

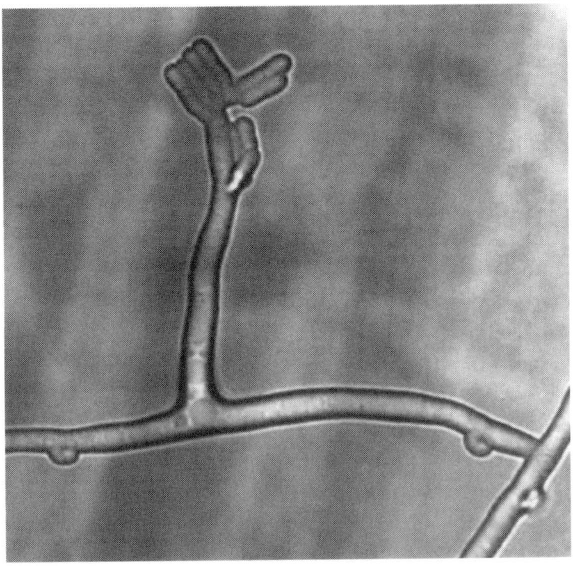

FIGURE 10 Oidiophore formed in the aerial mycelium of dikaryon FA2222 × 306. Note the clamp cells at the septa of the oidiophore clamp cell. (From Ref. 5.)

FIGURE 11 Septal dissolution and nuclear migration in a hyphae of strain NA2 transformed with a cosmid carrying a compatible *B* mating-type DNA [29]. As observed in strains with a solely activated *B* pathway of *S. commune* [113,114], nuclei can pass hyphal septa (white arrow), thereby leaving other cellular compartments within the hyphae nuclei free. Note that the transformant shown is from a monokaryon whose *A* mating-type locus has been inactivated through gene knockout [66]. It is not known whether nuclear migration and septal dissolution can also be observed in *B*-activated transformants of normal monokaryons.

(comparable to the ''flat'' phenotype of *Schizophyllum commune* [78]), and cause chlamydospore and hyphal knot formation as well as nuclear fusion in the basidia [30] (U. Kües and P.J. Walser, unpublished results).

When transforming cloned mating types solely into compatible *C. cinereus* monokaryons, *A* mating-type genes were found to repress oidia formation in the dark in all tested strains [37,74,75], whereas *B* genes, in most but not all strains, did not render constitutive oidiation a light controlled process (U. Kües, unpublished). Likewise, mutants with a constitutively activated *A* mating type pathway (*Amut* homokaryons) produce no or few oidia in the dark and abundant numbers in light. In contrast, mutants with a constitutively activated *B* mating-type pathway (*Bmut* homokaryons) behave like monokaryons [35]. Since *A*-mediated repression of oidiation in *A*-activated transformants and *Amut* homokaryons is usually fully released in light [35,37], the *B* mating-type products were thought to have a modifying role on the action of *A* mating-type genes [35,72]. Indeed, when both types of mating-type genes were introduced into compatible monokaryons, oidiation was repressed in the dark owing to the action of the heterologous *A* gene products, and repression was only partially released in light, owing to the action of the heterologous *B* genes (U. Kües, unpublished).

A and *B* mating-type genes also act together in other light-controlled processes. For example, sclerotia development is enhanced in *A*-activated transformants of certain monokaryons by induction of hyphal knot formation [37]. Whereas hyphal knot formation and sclerotia development are repressed in light in the *A*-activated transformants [37], those transformants also carrying heterologous *B* genes may still produce hyphal knots and sclerotia when incubated in light (U. Kües, unpublished). Similarly, chlamydospore production in these *A* transformants is induced in cultures kept in dark [37] whereas in light, heterologous *B* genes are also required for abundant chlamydospore production (U. Kües, unpublished). When cultures of a specific monokaryon are kept under fruiting conditions [25–28°C, 12 h dark–12 h light rhythm, 90% relative humidity], *B* genes help to convert hyphal knots into fruiting body initials (U. Kües, unpublished), a reaction that occurs less efficiently when a strain is transformed solely with heterologous *A* mating-type genes [37,74]. As indicated in this discussion, some monokaryotic strains are more susceptible to the actions of the mating-type gene products than others, showing how variable the natural pool is in *C. cinereus* of genes influencing developmental processes [37] (U. Kües, unpublished).

Apart from the mating-type genes, little is known about genetic determinants controlling hyphal morphology and asexual development. Monokaryons with defects in gene *pcc1* exhibit clamp cell and fruiting body formation uncoupled from control by the *A* and *B* mating-type genes. Such strains form unfused clamp cells at hyphal septa and develop fruiting bodies with basidiospores under the correct environmental conditions without the need to mate with another strain [79,80]. Not surprisingly, since also influenced by compatible *A* mating-type genes, mutations in *pcc1* render oidiation a light-controlled process. The *pcc1* mutant Fis[c], for example, produces 40-fold fewer oidia at 37°C in dark than in light [5]. The gene *pcc1* encodes a transcription factor with an HMG (high-mobility group) box as potential DNA binding domain. The HMG box in protein Pcc1 has greatest similarity to those present in the mating-type proteins FPR1 and MAT-Mc of the ascomycetes *Podospora anserina* and *Schizosaccharomyces pombe*, respectively, but also to that found in the transcriptional regulator Prf1 of the hemibasidiomycete *Ustilago maydis* [79] which controls expression of mating-type loci in a nutritional dependent manner [81,82]. Another, recently identified *C. cinereus* gene acting in clamp cell formation and repression of oidiation is *clp1*. Unfortunately, the *clp1* gene product has no known orthologs in the databases, making it difficult to speculate about its function [83].

Whereas genes *pcc1* and *ccf1* may act in dikaryons within the *A* mating-type pathway (and *pcc1* possibly in the *B* mating-type pathway as well, since *pcc1* mutants form basidiospores [79,84]) downstream of the mating type proteins, so far no gene has been identified that plays a direct role in the formation of oidiophores and oidia in either monokaryons or dikaryons. Defects in oidiation have been identified in a mutant collection of the mating type–defective homokaryon AmutBmut made by classical UV and by modern REMI (restriction enzyme–mediated integration) mutagenesis ([85]; U. Kües et al., in preparation). These included blocks in oidiophore development, malformations in oidiophore morphology, alterations in the type of oidiophores produced, changes in the mode of oidia release, modifications in oidia shape and cell wall structure, loss of repression, and light regulation [30]. Homokaryon AmutBmut preferentially produces type 1, type 2A, and type 2B oidiophores [14,15]. Among the characterized mutants were some producing type 3 and type 4 oidiophores instead [30]. Mutant generation therefore suggests that formation of these different types of oidiophores is indeed a genetic trait whose variable occurrence in *C. cinereus* wild-type strains [15] is manifested by different genetic backgrounds. It is hoped that by cloning the mutated genes in homokaryon AmutBmut we will learn in the future as much about oidiation in *C. cinereus* as in conidiation in the ascomycete *Aspergillus nidulans* where a number of specific genetic determinants for conidiophore development and spore production have been identified ([86]; this volume, Chapter 3 by Fischer].

4.2 Regulation by Environmental and Physiological Signals

Vegetative development in *C. cinereus* and its anamorph *H. aspergillata* is affected by environmental parameters such as temperature and light [16,35,37,63], although it is not as strictly controlled as fruiting body initiation and maturation [5,19,20]. Environmental conditions do influence growth speed and the overall mycelial appearance of a culture [30] as well as the number of spores and sclerotia produced in a culture [72]. The conditions favoring fruiting body development (25–28°C, 12 h dark–12 h light) do allow asexual development, but usually it will be restricted in favor of fruiting. In competing between asexual and sexual development, both temperature and light play decisive roles in dikaryons and homokaryons with mutated mating-type loci. According to our observations on homokaryon AmutBmut and *A*-activated monokaryotic transformants, hyphal knots, sclerotia, and chlamydospores form in the dark preferentially at 37°C and less abundantly at 25°C [37,60] (unpublished results). Incubation in constant light represses hyphal knot and chlamydospore formation as well as maturation of sclerotia from hyphal knots [37,63] (Sec. 4.1). Like sclerotia, the development of hyphal knots occurs at the beginning of fruiting body development in *C. cinereus* [37,58–60]. Generally, a light signal is required to progress from the hyphal knot to fruiting body initials [5,37,88–90]. Some initials might appear at 37°C following a short light signal but, usually many more form at 25–28°C [88,89] (unpublished observations). Once initiated, fruiting body development up to its maturation will take place only at lower temperatures [5,88,89]. Fruiting body maturation is only effective when additional defined and alternating light and dark signals are present [87–91]. In terms of oidia formation, analysis of homokaryon AmutBmut showed that oidia will be formed at 25°C upon light incubation but that spore numbers rise with increasing temperature, giving a maximum yield at 42°C [35,72]. The light intensity needed for oidiation induction is low (0.1 μE m^2 s^{-1}) [35], but that for switching to the fruiting pathway is even lower (0.98 lx) [88].

Light regulation in development in the dikaryon is clearly linked to regulation by mating-type genes [35,37] (Sec. 4.1), but light also acts independently of mating-type regulation, as for example in monokaryons repressed for hyphal knot formation and sclerotia maturation [37,63]. As in other basidiomycetes [92], light in *Coprinus* is effective in the blue wavelength range (400–500 nm) of the spectrum [35,87–91]. The light receptor has therefore been predicted to be a flavoprotein [89,92]. Unfortunately, neither a candidate for the light receptor nor any downstream elements in the light signaling pathway have been identified.

Similarly, nothing is known regarding how temperature is perceived by the fungus and how the information is translated into an output signal in the fungal cell. Just one gene (*hyt1*) has been cloned that, when mutated, disturbs tip growth of the vegetative hyphae at higher temperatures. Furthermore, this gene encodes a novel protein of unknown function [93]. In a basidiomycetous yeast, *Cryptococcus neoformans*, RAS1, a member of the small G-protein superfamily, has recently been shown to be essential for growth at higher temperatures (37°C). *RAS1* of *C. neoformans* shares a high degree of sequence identity (77%) to *CcRas* of *C. cinereus* [94]. Effects on development of *ras* mutant alleles giving rise to a constitutively activated and a constitutively inactivated Ras protein (Ras^{Val19} and

Ras^{Asn24}, respectively) are under investigation in *C. cinereus*. At least the Ras^{Val19} form has an influence on the development of aerial mycelium in monokaryons. Interestingly, when grown at high temperature (37°C), Ras^{Val19} and Ras^{Asn24} transformants of homokaryon AmutBmut are often unstable (A. P. F. Bottoli, unpublished).

In *C. neoformans*, RAS1 appears to play a central role regulating the MAP (mitogen-activated protein) kinase pathway but seems to be less important in regulation of production of cAMP, cyclic 3′–5′ adenosine monophosphate [95]. In the heterobasidiomycetous pathogen *U. maydis*, interaction between cAMP and MAP kinases pathways and the HMG box transcription factor Prf1 has been demonstrated [81,82,95,96; this volume, Chapter 14 by Banuett]. Exogeneous cAMP induces pheromone expression showing a link to the mating type loci in this fungus [96]. In *C. cinereus*, nearly all the components of these signaling pathways have yet to be identified and connections between different pathways have yet to be established. As a first step in this direction, the structural gene *cac* for the cAMP-producing adenylate cyclase has recently been cloned [97]. cAMP has long been known to take part in induction of fruiting body development [5,19,98]. Addition of cAMP to receptive monokaryotic strains (*fis*$^+$ mutants) induces fruiting body development at 28°C in a day/night light regimen in the absence of compatible *A* and compatible *B* mating-type products [98]. In the dikaryon, cAMP production is controlled by the *A* and *B* mating-type pathways together, and in addition, a light signal is required. In strains where only one or neither of the two the mating-type pathways is activated, light has no effect on induction of cAMP production [99]. In contrast, the monokaryotic fruiting strain Fis^c carrying a mutation in gene *pcc1* [80] produces increased levels of cAMP, but this is still dependent on illumination [98]. Unlike fruiting, it remains to be shown whether cAMP has a function in the vegetative reproduction of dikaryons and *pcc1* mutants—for example, in light-induced production of oidiophores. However, high levels of cAMP are clearly not essential for light-induced oidiation, since the process is also observed in *Amut* homokaryons and *A*-activated transformants. In fact, considering it needs an activated *B* mating-type pathway in addition to an activated *A* mating-type pathway and light for high-level cAMP production, it is possible that cAMP helps to repress oidia formation in favor of fruiting [35]. cAMP production and in consequence fruiting body initiation is repressed by high glucose levels (2–5% w/v) in the substrate [98]. In spite of this, cultures grown on high glucose also repress hyphal knot formation prior to fruiting body initiation and reduce the numbers of oidia formed on a culture (A. P. F. Bottoli and P. J. Walser, unpublished), indicating that a lack of cAMP does not necessarily induce asexual reproduction. Nitrogen sources available in the growth medium also take part in regulation of development. For example, high ammonium levels were shown to inhibit sclerotia maturation [63], and high asparagine concentrations negatively influence hyphal knot formation (A. P. F.

Bottoli, unpublished). Most interestingly, increasing C and N levels counteract the negative effect of the activated *B* mating-type pathway on formation of aerial mycelium (''flat'' phenotype), suggesting a link between nutritional control of development and the *B* pheromone pathway (U. Kües and A. P. F. Bottoli, unpublished).

To clearly understand the complex relationship among environmental signals, nutritional and genetic controls, and the different modes of reproduction, we need to identify more internal components of the different signaling cascades. At present we only know that *pcc1* (and likewise *clp1*?) has a central role in regulation of development. Judging by the phenotypes of *pcc1* mutants (formation of unfused clamp cells and fruit bodies, light-dependent oidiation), Pcc1 might be expected to localize in the *A* mating-type pathway downstream to the *A*-encoded homeodomain transcription factors. The analyzed mutations in Pcc1 cluster in the putative HMG box and are thus expected to cause a failure in DNA binding of the transcription factor. Since *pcc1* mutations lead to gain of dikaryon-specific functions, Pcc1 possibly acts as a repressor in monokaryons [79,80,83]. If this is true, the *A* mating-type proteins in turn have to be postulated to be repressors of *pcc1* function. Pcc1 is expressed in mono- and dikaryons [79], demanding a posttranscriptional regulation of the protein. How can this relate to the observed dark repression of oidiation in *pcc1* mutants and strains with an activated *A* mating-type pathway? If Pcc1 is not a light-independent activator of oidiation in the monokaryon but generally a repressor, another negative regulator has to be postulated to be downstream of Pcc1 that normally is only active in inhibition of oidiation within the dikaryon and is put out of action by light.

5 THE RELEVANCE OF ASEXUAL DEVELOPMENT FOR THE FUNGUS—CONCLUSIONS

C. cinereus is able to react to environmental signals with various ways of asexual as well as sexual reproduction. Sexual spore formation following karyogamy and meiosis is under the strictest environmental control and likely underlies the most complex genetic and physiological determination [5,19] (Sec. 3.2). Sexual reproduction gives rise to new combinations from the gene pools available; 10^7–10^8 of recombinant haploid basidiospores are formed on a dikaryon per single fruiting body [100] and released by autolysis of the fruiting body in droplets that fall to the ground [5,18]. The high number of basidiospores should ensure the survival and contribute to the distribution of the fungus, especially since basidiospores are long-lasting, endure adverse environmental conditions, and maintain their ability to germinate for several years [101].

Although sexual reproduction is sufficient to secure existence of the species, the various modes of vegetative reproduction that the fungus can undergo have given the organism much flexibility. Monokaryotic and dikaryotic mycelia of *C. cinereus* in principle show indefinite growth. However, horse dung, the organism's substrate, will usually be limited in natural environments. When spent, formation of the haploid oidia on either type of mycelium offers an opportunity for the fungus to conquer new substrates. Spread by insects, haploid oidia promote dispersal over relatively long distances. When landing on fresh horse dung, the substrate might even be inoculated for the first time with the species by the insects. In contrast to basidiospores, oidia quickly lose their germination ability [41], indicating that they are formed for distribution but not for persistence at a given location. However, the fungus developed other long-lasting asexual structures, the chlamydospores and the multicellular sclerotia, which overcome long hostile periods eliminating the need for the fungus to undergo meiosis. Not surprisingly, these types of resting structures therefore occur in aging cultures at the end of vegetative growth periods [36–40] in contrast to oidia that arise abundantly in growing cultures unrestricted by the substrate [11,14,35].

Since oidia act as spermatia, a major function of this type of spore will be in mixing different genomic populations in the wild. *C. cinereus* has multiple mating specificities—12,000 are estimated to exist in nature! Following meiosis, basidiospores of four different mating-type specificities arise from each fruiting body, each with 25% likelihood to find a compatible mating partner within the same sexual progeny [5,24,25; this volume, Chapter 10 by Kothe]. Multiple mating types promote outbreeding [25]; to maximize outbreeding, mixing among the various mating type populations is vital. Once entering an existing population, frequency-dependent selection can help the newcomer of different mating type to quickly establish itself in this population [102].

Formation of the dikaryon might have advantages over the simple monokaryon since the dikaryotic mycelium contains two different nuclei with the potential to complement each other in weak genetic characters. In this respect, repression of oidia development that would otherwise give rise to monokaryons with minor fitness might seem favorable. As in many other fungi, nuclear fusion in *C. cinereus* is spatially and temporally well separated from cellular fusion (Sec. 2). Keeping the two parental haploid nuclei together as separate units in the dikaryotic hyphal cells still has all the advantages of two different genomes as found in situations of diploid nuclei. In terms of genetic complementation, a dikaryon acts virtually as a diploid. The binucleate haploid state of the dikaryon offers, however, a unique opportunity to the individual nuclei. Nuclei maintain their identity although they can benefit from the presence of partner nuclei and multiply together with these, thereby overcoming possible limitations in their own fitness. It is intriguing that nuclei can leave their dikaryotic partnership upon external stimuli. Occasionally, monokaryotic hyphae occur on a dikaryotic colony and might give rise to a monokaryotic colony containing only one type of nucleus [103]. Dikaryotic chlamydospores and also the large dikaryotic veil cells covering the caps of fruiting bodies might give up their dikaryotic state when germinating

[28,36,104]. Such monokaryotic outgrowth usually will only be transient since nuclei of the opposite mating type will quickly invade the monokaryotic hyphae from adjacent cells, uniting the two types of nuclei again within the same cellular compartments. In contrast to these mycelial types of dedikaryotization, nuclear escape from the dikaryon through oidiation is much more effective since the oidia are completely detached from their generative hyphal cells and, moreover, the spores might be taken away by insects into other environments. In a new location, an oidium might germinate into an isolated monokaryon, meet other monokaryons of compatible mating types to fuse with, or encounter established dikaryons that can deliver mating type–compatible nuclei into the monokaryotic germling. This last phenomenon is known as the Buller effect [105]. In contrast to nuclear migration from a dikaryon into monokaryon, nuclei from monokaryons never invade dikaryotic mycelia. However, a monokaryotic mycelium might be entered from one or both types of the alternate nuclei of the dikaryon. The newly dikaryotized mycelium therefore can be chimerical, with some parts harboring one and other sectors containing the second nucleus of the donating dikaryon, before somatic incompatibility reactions start the disintegration of the mycelium into separate individuals. However, it is also possible that the resident nucleus in the acceptor monokaryon will be lost completely in favor of the two invading nuclei from the donor dikaryon [106–109]. These different situations indicate that haploid nuclei in *C. cinereus* are not generally equal, possibly differing in their individual fitness.

A similar conclusion can be drawn from the finding that oidia formation on the dikaryon is strongly biased. One type of nuclei usually prevails over the other in entering the oidiophore and thus the oidia formed on a dikaryon [30,41] (U. Kües, unpublished), an observation that also holds true for other basidiomycetes forming haploid mitotic spores on dikaryotic mycelium [13,110]. In *C. cinereus*, dominance of one nucleus leaving the dikaryotic state links to the mating-type genes [30] (U. Kües, unpublished). Control of biased oidiation on the dikaryon by the mating-type genes and in addition by the environmental factor light indicates that the process is not spontaneous but likely has a deeper consequence for the organism, both on the individual level and for the whole fungal population. For the individual, nuclei of high fitness, possibly well adapted to their environment, are kept unchanged through dikaryotic growth and can go unrecombined into new nuclear cooperations after leaving a former dikaryotic partnership. Generally, sexual recombination is expected to lead to a reduced fitness in many individual nuclei within the resulting progeny but on the whole, recombination tends to increases the variance of fitness, and by response to selection the mean fitness of whole populations [111]. In spite of the advantages of sexual reproduction, facultative asexuality is found stable especially within the fungi, for example, because of ecological differences between the sexual and asexual propagules [111]. Sexual reproduction is costly for performing organisms [111]; for example, in *C. cinereus*, formation of a large and complex fruiting body adds to these costs [5,18,19]. Sexual development in *C. cinereus* is at the end of a vegetative growth phase [5,19,20,112], in accordance with the presumption that the timing of sexual development is restricted and most likely set to minimize the opportunity costs of sex [111]. When, as in *C. cinereus*, bouts of sexual reproduction are interspersed with cycles of asexual reproduction, the most genetic advantage of sex and recombination is provided while greatly reducing the costs [111]. The possibility of a haploid nucleus to change their partner nuclei in the dikaryotic situation adds twist to the story. Partner-swapping of haploid nuclei in vegetative mycelium achieves a mixing of the genetic pools within populations which, in diploid systems, is usually only provided following sexual recombination and meiotic reductive divisions.

ACKNOWLEDGMENTS

We are very grateful to Takashi Kamada and Ben Lu for communicating results prior to publication, and to Erika Kothe for comments on the manuscript. Work by our group was supported by the Swiss National Science Foundation (grants 31-46′940.96, 31-46′940.96/2, and 31-59157.99) and by the ETH Zürich. Ursula Kües acknowledges support by the DBU (Deutsche Bundesstiftung Umwelt).

REFERENCES

1. DL Hawksworth, PM Kirk, BC Sutton, DN Pegler. Dictionary of the Fungi. 8th ed. Wallingford, UK: International Mycological Institute, CAB International, 1995.
2. B Kendrick. The Fifth Kingdom. 2nd ed. Newburyport, MA: Focus Texts, Focus Information Group, 1992.
3. DR Reynolds, JW Taylor, eds. The Fungal Holomorph: Mitotic, Meiotic and Pleomorphic Speciation in Fungal Systematics. Wallingford, UK: International Mycological Institute, CAB International, 1993.
4. KA Seifert, W Gams, PW Crous, GJ Samuels, eds. Molecules, Morphology and Classification: Towards Monophyletic Genera in the Ascomycetes. Baarn, Netherlands: Centraalbureau voor Schimmelcultures, 2000.
5. U Kües. Life history and developmental processes in the basidiomycete *Coprinus cinereus*. Microbiol Mol Biol Rev 64:316–353, 2000.
6. JW Fell, T Boekhout, A Fonseca, G Scorzetti, A Statzell-Tallman. Biodiversity and systematics of basidiomycetous yeasts as determined by large-subunit rDNA D1/D2 domain sequence analysis. Int J Syst Evol Microbiol 50:1351–1371, 2000.
7. DBO Saville. The evolution of anamorphs in the Uredinales. In: B Kendrick, ed. The Whole Fungus, Vol 2. Ottawa: National Museum of Natural Sciences, 1979, pp 547–554.
8. GN Agrios. Plant Pathology. 4th ed. San Diego, CA: Academic Press, 1997.
9. B Kendrick, R Watling. Mitospores in basidiomycetes. In: B Kendrick, ed. The

Whole Fungus, Vol 2. Ottawa: National Museum of Natural Sciences, 1979, pp 473–545.

10. HJ Brodie. The oidia of *Coprinus lagopus* and their relation with insects. Ann Bot 45:315–344, 1931.
11. HJ Brodie. The occurrence and function of oidia in the hymenomycetes. Am J Bot 23:309–327, 1936.
12. CT Ingold. Mycelium, oidia and sporophore initials in *Flammulina velutipes*. Trans Br Mycol Soc 75:107–116, 1980.
13. ROF Kemp. Production of oidia by dikaryons of *Flammulina velutipes*. Trans Br Mycol Soc 74:557–560, 1980.
14. E Polak, R Hermann, U Kües, M Aebi. Asexual sporulation in *Coprinus cinereus*: structure and development of oidiophores and oidia in an *Amut Bmut* homokaryon. Fungal Genet Biol 22:112–126, 1997.
15. E Polak, M Aebi, U Kües. Morphological variations in oidia formation in the basidiomycete *Coprinus cinereus*. Mycol Res 105:603–610, 2001.
16. J Guarro, J Gene, de Vroey C, E Guého. *Hormographiella*, a new genus of hyphomycetes from clinical sources. Mycotaxon 45:179–190, 1992.
17. J Gene, GJ Guillamon, J Guarro, I Pujol, K Ulfig. Molecular characterization, relatedness and antifungal susceptibility of the basidiomycetous *Hormographiella* species and *Coprinus cinereus* from clinical and environmental sources. Antonie van Leeuwenhoek Int J Gen Mol Microbiol 70:49–57, 1996.
18. AHR Buller. Researches on Fungi. IV. Further Observations on the *Coprini* Together with Some Investigations on Social Organisation and Sex in Hymenomycetes. New York: Hafner, 1931.
19. D Moore. Fungal Morphogenesis. Cambridge, UK: Cambridge University Press, 1998.
20. U Kües, Y Liu. Fruiting body production in basidiomycetes. Appl Microbiol Biotechnol 54:141–152, 2000.
21. PJ Pukkila. Meiosis in mycelial fungi. In: JGH Wessels, F Meinhardt, eds. The Mycota, Vol I. Growth, Differentiation and Sexuality. Berlin: Springer-Verlag, 1994, pp 267–282.
22. BCK Lu. Chromosomes, mitosis, and meiosis. In: CJ Bos, ed. Fungal Genetics. Principles and Practice. New York: Marcel Dekker, 1996, pp 119–176.
23. WJ Cummings, ME Zolan. Functions of DNA repair genes during meiosis. Curr Top Dev Biol 37:117–140, 1998.
24. LA Casselton, NS Olesnicky. Molecular genetics of mating recognition in basidiomycete fungi. Microbiol Mol Biol Rev 62:55–70, 1998.
25. SJ Hiscock, U Kües. Cellular and molecular mechanisms of sexual incompatibility in plants and fungi. Int Rev Cytol 193:165–295, 1999.
26. H Brunswick. Untersuchungen über die Geschlechts- und Kernverhältnisse bei der Hymenomycetengattung *Coprinus*. Botanische Abhandlungen, Heft 5. Jena, Germany: Gustav Fischer, 1924.
27. PE Verweij, M van Kasteren, J van de Nes, GS de Hoog, BE de Pauw, JFGM Meis. Fatal pulmonary infection caused by the basidiomycete *Hormographiella aspergillata*. J Clin Microbiol 35:2675–2578, 1997.
28. K Esser. Kryptogamen. Praktikum und Lehrbuch. 1. Cyanobakterien, Algen, Pilze, Flechten. 3rd ed. Berlin: Springer-Verlag, 2000.

29. D Lewis. Genetic analysis of methionine suppressors in *Coprinus*. Genet Res 2: 141–155, 1961.
30. E Polak. Asexual sporulation in the basidiomycete *Coprinus cinereus*. PhD dissertation, ETH Zürich, Zurich, Switzerland, 1999.
31. U Mittwoch. The Genetics of *Coprinus lagopus*. II. Growth rates of wild-type strains. J Genet 52:551–554, 1953.
32. M Iwasa, S Tanabe, T Kamada. The two nuclei in the dikaryon of the homobasidiomycete *Coprinus cinereus* change position after each conjugate division. Fungal Genet Biol 23:110–116, 1998.
33. M Bensaude. Recherches sur le cycle évolutif et la sexualité chez les Basidiomycetes. PhD dissertation, Nemours, France, 1918.
34. PJ Walser, M Hollenstein, MJ Klaus, U Kües. Genetic analysis of basidiomycete fungi. In: NJ Talbot, ed. Molecular and Cell Biology of Filamentous Fungi: A Practical Approach. Oxford, UK: IRL Practical Approach Series, Oxford University Press, 2001, pp 59–90.
35. K Kertesz-Chaloupková, PJ Walser, JD Granado, M Aebi, U Kües. Blue light overrides repression of asexual sporulation by mating type genes in the basidiomycete *Coprinus cinereus*. Fungal Genet Biol 23:95–109, 1998.
36. GE Anderson. The Life History and Genetics of *Coprinus lagopus*. Weston-super-Mare, UK: Phillip Harris Biological Supplies, 1971.
37. U Kües, JD Granado, R Hermann, RP Boulianne, K Kertesz-Chaloupková, M Aebi. The *A* mating type and blue light regulate all known differentiation processes in the basidiomycete *Coprinus cinereus*. Mol Gen Genet 260:81–91, 1998.
38. PA Volz, DJ Niederpruem. The sclerotia of *Coprinus lagopus*. Arch Microbiol 70: 369–377, 1970.
39. H Waters, D Moore, RD Butler. Structure of aerial and submerged sclerotia of *Coprinus lagopus*. New Phytol 174:199–205, 1975.
40. H Waters, D Moore, RD Butler. Morphogenesis of aerial sclerotia of *Coprinus lagopus*. New Phytol 174:207–213, 1975.
41. M Hollenstein. Aspekte zur genotypischen Determination der Oidienverteilung in Dikaryen von *Coprinus cinereus*. Diploma thesis, Institut für Mikrobiologie, ETH Zürich, Zurich, Switzerland, 1997.
42. R Watling. The morphology, variation and ecological significance of anamorphs in the Agaricales. In: B Kendrick, ed. The Whole Fungus, Vol 2. Ottawa: National Museums of Natural Sciences, 1979, pp 453–472.
43. CE Heintz, DJ Niederpruem. Ultrastructure and respiration of oidia and basidiospores of *Coprinus lagopus* (sensu Buller). Can J Microbiol 16:481–484, 1970.
44. CE Heintz, DJ Niederpruem. Ultrastructure of quiescent and germinated basidiospores and oidia of *Coprinus lagopus*. Mycologia 63:745–766, 1971.
45. GW Gooday. Cell walls. In: NAR Gow, GM Gadd, eds. The Growing Fungus. London: Chapman & Hall, 1994, pp 43–59.
46. HAB Wösten, JGH Wessels. Hydrophobins, from molecular structure to multiple functions in fungal development. Mycoscience 38:363–374, 1997.
47. JGH Wessels. Hydrophobins: proteins that change the nature of the fungal surface. Adv Microbiol Physiol 38:1–45, 1997.

48. MJ Kershaw, NJ Talbot. Hydrophobins and repellents: proteins with fundamental roles in fungal morphogenesis. Fungal Genet Biol 23:18–23, 1998.
49. SAM Ásgeirsdottir, JR Halsall, LA Casselton. Expression of two closely linked hydrophobin genes of *Coprinus cinereus* is monokaryon-specific and down-regulated by the *oid-1* mutation. Fungal Genet Biol 22:54–63, 1997.
50. RFO Kemp. Breeding biology of *Coprinus* species in the section Lanatuli. Trans Br Mycol Soc 65:375–388, 1975.
51. E Polak, U Kües, M Aebi. Replica plating of *Coprinus cinereus* colonies using asexual spores. Fungal Genet Newsl 44:45–46, 1997.
52. RFO Kemp. Oidial homing and the taxonomy and speciation of basidiomycetes with special reference to the genus *Coprinus*. In: H Clémençon, ed. The Species Concept in Hymenomycetes. Vaduz, Switzerland: Cramer, 1977, pp 259–276.
53. H Clémençon. Anatomie der Hymenomyceten. Eine Einführung in die Cytologie und Plectologie der Krustenpilze, Porlinge, Keulenpilze, Leistlinge, Blätterpilze und Röhrlinge. (Anatomy of the Hymenomycetes. An Introduction to the Cytology and Plectology of the Crust Fungi, Bracket Fungi, Club Fungi, Chantarelles, Agarics and Boletes). Teufen, Switzerland: F. Flück-Wirth, 1997.
54. MF Madelin. Visible changes in the vegetative mycelium of *Coprinus lagopus* Fr. at the time of fruiting. Trans Br Mycol Soc 43:105–110, 1960.
55. H Waters, RD Butler, D Moore. Thick-walled sclerotial medullary cells in *Coprinus lagopus*. Trans Br Mycol Soc 59:167–169, 1972.
56. PR Day. A cytoplasmically controlled abnormality of the tetrads of *Coprinus lagopus*. Heredity 13:81–87, 1959.
57. FV Hereward, D Moore. Polymorphic variation in the structure of aerial sclerotia of *Coprinus cinereus*. J Gen Microbiol 113:13–18, 1979.
58. D Moore. Developmental genetics of *Coprinus cinereus*: genetic evidence that carpophores and sclerotia share a common pathway of initiation. Curr Genet 3:145–150, 1981.
59. TR Matthews, DJ Niederpruem. Differentiation in *Coprinus lagopus*. I. Control of fruiting and cytology of initial events. Arch Microbiol 87:257–268, 1972.
60. RP Boulianne, Y Liu, M Aebi, BC Lu, U Kües. Fruiting body development in *Coprinus cinereus*: regulated expression of two galectins secreted by a non-classical pathway. Microbiology 146:1841–1853, 2000.
61. J Ji, D Moore. Glycogen metabolism in relation to fruit body maturation in *Coprinus cinereus*. Mycol Res 97:283–289, 1993.
62. RI Jirjis, D Moore. Involvement of glycogen in morphogenesis of *Coprinus cinereus*. J Gen Microbiol 95:348–352, 1976.
63. D Moore, RI Jirjis. Regulation of sclerotium production by primary metabolites in *Coprinus cinereus* (=*C. lagopus* sensu Lewis). Trans Br Mycol Soc 66:377–382, 1976.
64. KM Swiezynski, PR Day. Heterokaryon formation in *Coprinus lagopus*. Genet Res 1:114–128, 1960.
65. KM Swiezynski, PR Day. Migration of nuclei in *Coprinus lagopus*. Genet Res 1: 129–139, 1960.
66. EH Pardo. Organisation of the *A* mating type locus of *Coprinus cinereus*. PhD dissertation, University of Oxford, Oxford, UK, 1995.

67. TJ Fowler, MF Mitton, CA Raper. Gene mutations affecting specificity of pheromone/receptor mating interactions in *Schizophyllum commune*. In: LJLD Van Griensven, J Visser, eds. Proceedings of the Fourth Meeting on the Genetics and Cellular Biology of Basidiomycetes. Horst, Netherlands: Mushroom Experimental Station, 1998, pp 130–134.
68. S Swamy, I Uno, T Ishikawa. Morphogenetic effects of mutations at the *A* and *B* incompatibility factors in *Coprinus cinereus*. J Gen Microbiol 130:3219–3224, 1984.
69. RW Haylock, A Economou, LA Casselton. Dikaryon formation in *Coprinus cinereus*: selection and identification of the *B* factor mutants. J Gen Microbiol 121:17–26, 1980.
70. HJ Brodie. Oidial mycelia and the diploidization process in *Coprinus lagopus*. Ann Bot 46:727–732, 1932.
71. K Kimura. Studies on the sex of *Coprinus macrorhizus* Rea f. *microsporous* Hongo. I. Introductory experiments. Biol J Okayama Univ 1:72–79, 1952.
72. U Kües, JD Granado, K Kertesz-Chaloupková, JP Walser, M Hollenstein, E Polak, Y Liu, RP Boulianne, APF Bottoli, M Aebi. Mating types and light are major regulators of development in *Coprinus cinereus*. In: LJLD Van Griensven, J Visser, eds. Proceedings of the Fourth Meeting on the Genetics and Cellular Biology of Basidiomycetes. Horst, Netherlands: Mushroom Experimental Station, 1998, pp 111–118.
73. U Kües, WVJ Richardson, ES Mutasa, AM Tymon, EH Pardo, SF O'Shea, LA Casselton. The combination of dissimilar alleles of the *A*α and *A*β gene complexes, whose proteins contain homeodomain motifs, determine sexual development in the mushroom *Coprinus cinereus*. Genes Dev 6:568–577, 1992.
74. AM Tymon, U Kües, WVJ Richardson, LA Casselton. A fungal mating type protein that regulates sexual and asexual development contains a POU-related domain. EMBO J 11:164–178, 1992.
75. U Kües, RN Asante-Owusu, ES Mutasa, AM Tymon, EH Pardo, SF O'Shea, LA Casselton. Two classes of homeodomain proteins specify the multiple *A* mating types of the mushroom *Coprinus cinereus*. Plant Cell 6:1467–1475, 1994.
76. JR Halsall, MJ Milner, LA Casselton. Three subfamilies of pheromone and receptor genes generate multiple *B* mating specificities in the mushroom *Coprinus cinereus*. Genetics 154:1115–1123, 2000.
77. SF O'Shea, PT Chaure, JR Halsall, NS Olesnicky, A Leibbrandt, IF Connerton, LA Casselton. A large pheromone and receptor gene complex determines multiple *B* mating type specificities in *Coprinus cinereus*. Genetics 148:1081–1090, 1998.
78. J Wendland, LJ Vaillancourt, J Hegner, KB Lengeler, KJ Laddison, CA Specht, CA Raper, E Kothe. The mating-type locus *B*α1 of *Schizophyllum commune* contains a pheromone receptor gene and putative pheromone genes. EMBO J 14:5271–5278, 1995.
79. Y Murata, M Fujii, ME Zolan, T Kamada. Molecular analysis of *pcc1*, a gene that leads to *A*-regulated sexual morphogenesis in *Coprinus cinereus*. Genetics 149: 1753–1761, 1998.
80. Y Murata, M Fujii, ME Zolan, T Kamada. Mutations in clamp-cell formation, the *A*-regulated sexual morphogenesis, in homobasidiomycetes. In: LJLD Van Grien-

sven, J Visser, eds. Proceedings of the Fourth Meeting on the Genetics and Cellular Biology of Basidiomycetes. Horst, Netherlands: Mushroom Experimental Station, 1998, pp 143–146.
81. R Kahmann, C Basse, M Feldbrügge. Fungal-plant signalling in the *Ustilago maydis*-maize pathosystem. Curr Opin Microbiol 2:647–650, 1999.
82. HA Hartmann, J Krüger, F Lottspeich, R Kahmann. Environmental signals controlling sexual development of the corn smut fungus *Ustilago maydis* through the transcriptional regulator Prf1. Plant Cell 11:1293–1305, 1999.
83. K Inada, Y Morimoto, T Arima, Y Murata, T Kamada. The *clp1* gene of the mushroom *Coprinus cinereus* is essential for *A*-regulated sexual development. Genetics 157:133–140, 2001.
84. I Uno, T Ishikawa. Chemical and genetical control of induction of monokaryotic fruiting bodies in *Coprinus macrorhizus*. Mol Gen Genet 113:228–239, 1971.
85. J Granado, K Kertesz-Chaloupková, M Aebi, U Kües. Restriction enzyme–mediated DNA integration in *Coprinus cinereus*. Mol Gen Genet 256:28–36, 1997.
86. TH Adams, JK Wieser, J-H Yu. Asexual sporulation in *Aspergillus nidulans*. Microbiol Mol Biol Rev 62:35–41, 1998.
87. N Morimoto, Y Oda. Effects of light on fruitbody formation in a basidiomycete, *Coprinus macrorhizus*. Plant Cell Physiol 14:217–225, 1973.
88. BC Lu. Meiosis in *Coprinus*. V. The role of light on basidiocarp initiation, mitosis and hymenium differentiation in *Coprinus lagopus*. Can J Bot 52:299–305, 1974.
89. CG Elliott. Reproduction in Fungi. Genetical and Physiological Aspects. London: Chapman & Hall, 1994.
90. T Kamada, R Kurita, T Takemaru. Effects of light on basidiocarp maturation in *Coprinus macrorhizus*. Plant Cell Physiol 19:263–275, 1978.
91. BC Lu. The control of meiosis progression in *Coprinus cinereus* by light/dark cycles. Fungal Genet Biol 31:33–41, 2000.
92. T Yli-Mattila. Photobiology of fruit body formation in *Schizophyllum commune*. Luonnon Tutkija 95:116–123, 1991.
93. S Maida, M Fujii, C Skrzynia, PJ Pukkila, T Kamada. A temperature-sensitive mutation of *Coprinus cinereus*, *hyt1-1*, that causes swelling of hyphal tips. Curr Genet 32:231–236, 1997.
94. JA Alspaugh, LM Cavallo, JR Perfect, J Heitman. *RAS1* regulates filamentation, mating and growth at high temperature of *Cryptococcus neoformans*. Mol Microbiol 36:352–365, 2000.
95. P Müller, C Aichinger, M Feldbrügge, R Kahmann. The MAP kinase Kpp2 regulates mating and pathogenic development in *Ustilago maydis*. Mol Microbiol 34:1007–1017, 1999.
96. J Krüger, G Loubradou, E Regenfelder, A Hartmann, R Kahmann. Crosstalk between cAMP and pheromone signalling pathways in *Ustilago maydis*. Mol Gen Genet 260:193–198, 1998.
97. APF Bottoli, K Kertesz-Chaloupková, RP Boulianne, JD Granado, M Aebi, U Kües. Rapid isolation of genes from an indexed genomic library of *C. cinereus* in a novel *pab1*$^+$ cosmid. J Microbiol Methods 35:129–141, 1999.
98. I Uno, T Ishikawa. Biochemical and genetic studies on the initial events of fruitbody formation. In: K Wells, EK Wells, eds. Basidium and Basidiocarp. Evolution, Cy-

tology, Function and Development. New York: Springer-Verlag, 1982, pp 113–123.

99. S Swamy, I Uno, T Ishikawa. Regulation of the cyclic AMP metabolism by the incompatibility factors in *Coprinus cinereus*. J Gen Microbiol 131:3211–3217, 1985.
100. NB Raju, BC Lu. Meiosis in *Coprinus*. IV. Morphology and behaviour of spindle pole bodies. J Cell Sci 12:131–141, 1973.
101. D Moore, PJ Pukkila. *Coprinus cinereus*: an ideal organism for studies of genetics and developmental biology. J Biol Edu 19:31–40, 1985.
102. P Zambino, JV Groth, L Lukens, JR Garton, G May. Variation at the *b* mating type locus of *Ustilago maydis*. Phytopathology 87:1233–1239, 1997.
103. U Mittwoch. The genetics of *Coprinus lagopus*. I. The formation of monokaryons from dikaryons. J Genet 52:547–550, 1953.
104. JW Cowan. Recovery of monokaryons from veil cells of fruitbodies of *Coprinus lagopus* sensu Buller. Nature 204:1113–1114, 1964.
105. A Quintanilha. Contribution à l'étude génétique du phénomène de Buller. C R Acad Sci 13:425–486, 1933.
106. K Kimura. Diploidization in the Hymenomycetes. II. Nuclear behaviour in the Buller phenomenon. Biol J Okayama Univ 4:1–59, 1958.
107. KM Swiezynski. Exchange of nuclei between dikaryons in *Coprinus lagopus*. Acta Soc Bot Pol 30:535–552, 1961.
108. KM Swiezynski. Analysis of an incompatible di-mon mating in *Coprinus lagopus*. Acta Soc Bot Pol 31:169–184, 1962.
109. G May, JW Taylor. Patterns of mating and mitochondrial DNA inheritance in the agaric basidiomycete *Coprinus cinereus*. Genetics 118:213–220, 1988.
110. C Hui, H Yamamoto, T Ohta, K Takeo, Y Kitamoto. Nuclear selection in monokaryotic oidium formation from dikaryotic mycelia in a basidiomycete, *Pholiota nameko*. Mycoscience 40, 199–204, 1999.
111. A Butt. Perspective: sex, recombination, and the efficacy of selection—was Weismann right? Evolution 54:337–351, 2000.
112. MF Madelin. Studies on the nutrition of *Coprinus lagopus* Fr., especially as affecting fruiting. Ann Bot 20:307–330, 1956.
113. Y Koltin, AS Flexer. Alteration of nuclear distribution in *B*-mutant strains of *Schizophyllum commune*. J Cell Sci 4:739–749, 1969.
114. M Raudaskoski. The relationship between *B*-mating-type genes and nuclear migration in *Schizophyllum commune*. Fungal Genet Biol 24:207–227, 1998.
115. A Wild, R Hermann, T Leisinger. Isolation of an anaerobic bacterium which reductively dechlorinates tetrachloroethane and trichloroethene. Biodeg 7:507–511, 1997.

7

Blue Light Perception and Signal Transduction in *Neurospora crassa*

Hartmut Linden
University of Konstanz, Konstanz, Germany

1 INTRODUCTION

Living organisms are confronted with rapid environmental changes in the course of their development, such as changes in light, temperature, humidity, and nutrient supply. The sensing of these environmental changes is a vital capacity. An important environmental cue is light which is perceived by photoreceptors. Following perception, an internal signal is generated which is subsequently transported via a signal transduction chain and finally causes the observed response and the acclimatization of the organisms to environmental light conditions. The signal transduction chains usually comprise protein components (e.g, protein kinases, protein phosphatases, and G-proteins) and second-messenger molecules (e.g., Ca^{2+} and cyclic AMP). In higher plants light is essential for photosynthesis whereas in fungi normal growth and development are possible in the complete absence of light. Nevertheless, light is also an important signal for fungi and affects many aspects of fungal development and physiology.

One of the model organisms for the investigation of light regulation is the ascomycete *Neurospora crassa* [1–3]. This fungus reveals many advantageous features such as a rather small genome (47 megabases) [4], fast heterotrophic growth, straightforward genetics, and easy transformation with foreign DNA.

Genetic transformation usually leads to the random and stable integration of foreign DNA into the *Neurospora* genome. The high transformation rate together with the fact that integration of foreign DNA into the genome occurs randomly has been exploited for the cloning of various genes by insertional mutagenesis [5–7]. An abundance of mutants have been isolated and mapped on the seven *Neurospora* chromosomes which make the fast genetic mapping of new mutants possible [8].

A very important reason for the choice of *Neurospora crassa* as a model organism for blue light perception and signal transduction was the fact that this process seems to be, in comparison to higher plants, less complex in fungi. For instance, higher plants perceive and respond to a broad spectrum of light and contain distinct photoreceptors for the perception of UV-B, UV-A, blue, green, red, and far-red light, whereas *Neurospora* perceives light in the blue/UV light range only [9–11]. Moreover, the fact that only a limited number of putative light signal transduction mutants have been isolated in *Neurospora* today further indicates this minor complexity. The photobiological investigation of light perception, the analysis of the blue light responses, and the molecular characterization of several components of the signal transduction pathway provided a wealth of information about this process in *N. crassa*. In addition, more recent research revealed a strong interaction between the *Neurospora* circadian clock and the light perception and signal transduction pathway. This is due to the dependency of the circadian clock on light for resetting and entrainment of the circadian cycle. Furthermore, the two signal transduction pathways seem to share protein components and members of the blue light signal transduction pathway such as the white collar 2 protein (WC2) also seem to represent components of the circadian clock. The present chapter will focus on blue light regulation; the interplay between the blue light signal transduction and the circadian clock will be covered by another contribution in this volume (this volume, Chapter 8 by Bell-Pedersen).

The present chapter first of all gives an overview of the regulation of *Neurospora* development by light and then summarizes our present knowledge about light perception and signal transduction in *N. crassa*. It will furthermore review the induction of *Neurospora* carotenogenesis by light as an example of a biosynthetic pathway which seems to be entirely regulated by light. Additionally, it will concentrate on recent advances in the biochemical investigation of the blue light signaling pathway.

2 LIGHT AND DEVELOPMENT

During *Neurospora crassa* development three different sporulation pathways have been described [12]. The two vegetative sporulation pathways lead to the formation of macro- and microconidia whereas the sexual reproduction pathway results in the formation of ascospores (Fig. 1). Macroconidia are multinucleate

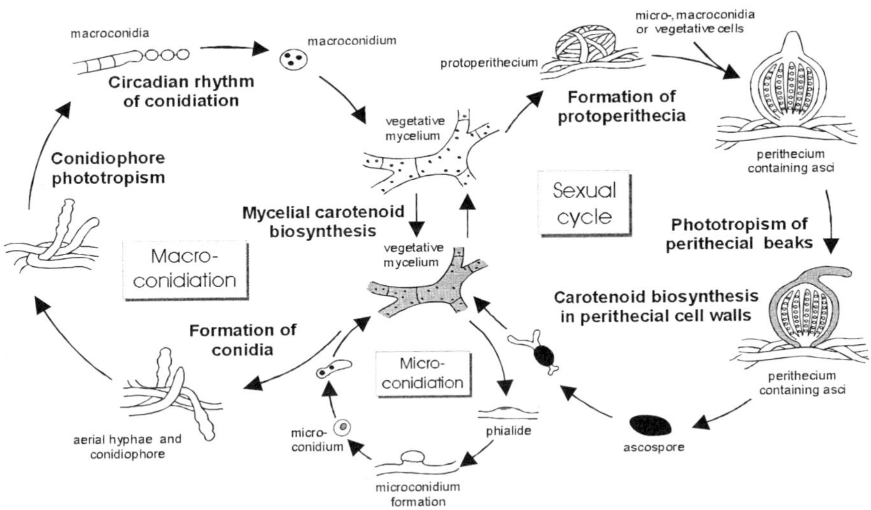

FIGURE 1 Blue light control of *Neurospora* development. *Neurospora* reveals two asexual sporulation pathways (macro- and microconidiation) as well as the sexual life cycle (modified according to Ref. 12). During these life cycles vegetative and sexual development is subject to blue light control as indicated in bold type.

and their formation is induced primarily by desiccation and carbon deprivation. Macroconidiation (normally called conidiation) occurs in the wild type also in the dark and in mutants that are "blind" toward light [13,14]. However, blue light was shown to be essential for maximal conidiation in the wild type. In the light, the formation of conidiophores that result from apical budding is restricted to the illuminated surface of mycelial pads whereas conidia are formed on both surfaces in the dark [15]. The *Neurospora* circadian clock, which is responsible for rhythmic conidiation, is entrained by blue light, and illumination results in the suppression and in phase shifts of rhythmic conidiation [10]. For the formation of the uninucleate microconidia an influence of light has not been reported [16]. It has been shown, however, that several conidiation-specific genes are also expressed in microconidia [17]. Some of the latter genes such as *con6* and *con10* are subject to light induction which suggests a possible role of light on microconidiation as well. Blue light also affects the sexual development of *N. crassa* by inducing the formation of the female sexual organs called protoperithecia [18]. Similar to the situation for macroconidiation, protoperithecia are constitutively produced in the dark. However, their formation is strongly induced by light and the increased production of protoperithecia was not observed in blind mutants

[19]. Another effect of blue light on sexual development is the phototropism of the perithecial beaks which play a role in the expulsion of ascospores [20]. The most prominent blue light response in *Neurospora* is the biosynthesis of the orange-colored carotenoids in mycelia [21,22]. In contrast, carotenogenesis is constitutive in macroconidia.

Most of the *N. crassa* blue light responses are brought about by transcriptional activation of genes. Thus, upregulation of carotenoid biosynthesis genes results in the induction of carotenogenesis (see Sec. 3). Several light-regulated conidiation genes (*con* genes) seem to participate in light induction of macroconidiation [23–25]. Another well-studied example is the light induction of the central circadian clock protein frequency (FRQ) which is responsible for entrainment and resetting of the circadian clock by light [26]. The blue light–regulated genes can be divided into early light–regulated genes and late light–regulated genes. The carotenoid biosynthesis genes *al1* (for albino), *al2*, and *al3*, the conidiation genes *con5* and *con10* as well as the *wc1* gene reveal a mRNA peak at ~20–30 min after onset of light and are early light–regulated genes [24,27–29]. In contrast, late light–regulated genes such as the clock-controlled genes *ccg1*, *ccg2*, and *ccg9* show a maximal expression after 90–120 min of light [30–32]. Some physiological and biochemical blue light responses have been identified which are probably independent of gene regulation. For instance, light was reported to induce the hyperpolarization of the cell membrane as well as the ADP ribosylation of proteins [33,34]. Light also leads to the phosphorylation of proteins in *N. crassa*. Not only the blue light regulatory proteins white collar 1 and 2 (WC1 and WC2, discussed in detail below) but also a *Neurospora* nucleoside diphosphate kinase (NDK1) is subject to light-dependent phosphorylation [35–38]. Light-induced protein phosphorylation occurs rapidly following the onset of light, thus excluding the involvement of transcriptional gene activation. Moreover, instead of representing simply a blue light response, the light-dependent protein phosphorylation was suggested in both cases to play a role in the internal transport of the blue light signal itself.

3 LIGHT REGULATION OF CAROTENOID BIOSYNTHESIS

The biosynthesis of carotenoids in *Neurospora* has been extensively studied during the last decades. The biochemical analysis of biosynthetic enzymes, the examination of structural mutants as well as the isolation and characterization of almost all biosynthetic genes resulted in the elucidation of the biosynthetic pathway, as depicted in Figure 2. Several albino mutants have been isolated which correspond to three different loci (*al1*, *al2*, and *al3*) [8]. In contrast to other carotenoid mutants, most of the albino mutants reveal a white phenotype and are defective in carotenoid biosynthesis not only in mycelia but also in macroconidia. The albino genes were cloned and shown to encode geranylgeranly diphosphate synthase (*al3*), phytoene synthase (*al2*), and phytoene desaturase (*al1*) [39–41]. The

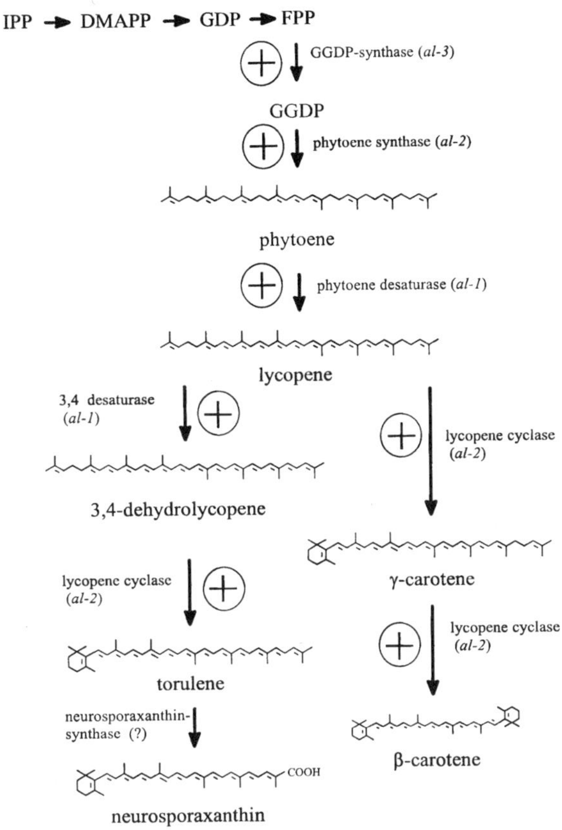

FIGURE 2 Light regulation of the carotenoid biosynthetic pathway in *Neurospora crassa*. The biosynthetic enzymes as well as the corresponding genes (in parentheses) are indicated. Light-regulated biosynthesis steps are marked by +. Several intermediates were omitted. IPP, isopentenyl pyrophosphate; DMAPP, dimethylallyl pyrophosphate; GDP, geranyl diphosphate; FPP, farnesyl diphosphate; GGDP, geranylgeranyl diphosphate.

GGDP synthase converts farnesyl diphosphate (FPP) into geranylgeranyl diphosphate (GGDP) whereas the genuine substrate for GGDP synthase seems to be a shorter prenyl diphosphate such as dimethylallyl pyrophosphate (DMAPP) [42]. Subsequently, GGDP is converted into phytoene by the gene product of the *al2* gene [22].

The *Neurospora* phytoene desaturase catalyzes the conversion of phytoene into lycopene; the same enzyme seems to mediate also the additional desaturation for the synthesis of 3,4-dehydrolycopene which was shown by complementation analysis of the *al1* gene in *Rhodobacter capsulatus* [43,44]. Interestingly, in spite of the high number of carotenoid biosynthesis mutants isolated from *Neurospora* to date, a lycopene-accumulating mutant has never been identified. This question was recently solved by complementation analysis using the phytoene synthase gene (*crtYB*) from the heterobasidiomycetous yeast *Xanthophyllomyces dendrorhous* [45]. The corresponding gene product not only converted GGDP into phytoene but was also capable of converting lycopene into β-carotene. The same results were obtained using the phytoene synthase gene (*carRP*) from *Mucor circinelloides* [46].

The *crtYB* gene, the *carRP* gene, and the *N. crassa al2* gene reveal an overall sequence similarity which suggested that *Neurospora* phytoene synthase also catalyzes the cyclization of lycopene and 3,4-dehydrolycopene. By complementation experiments using the *Neurospora* phytoene synthase, the latter presumption has recently been confirmed (G. Sandmann, personal communication, 2000). The acidic pigment neurosporaxanthin represents the final product of carotenoid biosynthesis in *N. crassa* which represents the major carotenoid under certain growth conditions [47]. A corresponding gene has not been isolated to date. Other genes, such as *ylo1* and *vvd*, have previously been implicated as putative structural genes of carotenoid biosynthesis in *N. crassa* owing to changes in carotenoid composition and quantities in the respective mutants [8]. However, the *ylo1* gene was recently shown to encode an aldehyde dehydrogenase whereas the *vvd* protein seems to represent a regulatory protein involved in photoadaptation (see Sec. 7) [48].

The kinetics of carotenoid accumulation in *Neurospora crassa* in response to light pulses and under continuous illumination was investigated by Schrott at the beginning of the 1980s [49,50]. It was shown that *Neurospora* reveals a biphasic fluence response curve for the blue light induction of carotenogenesis which was proposed to be due to the depletion of the photoreceptor and/or components of the signal transduction pathway. The subsequent replacement of the latter signaling components would then lead to the observed second phase of the fluence response curve. As indicated in Figure 2 almost the entire biosynthetic pathway is subject to a coordinated regulation by light, which seems to be mainly due to the transcriptional activation of the albino genes. All three albino genes are early light-regulated genes and reveal a transient induction pattern [27,39,41]. After light induction for 30 min, mRNA levels decrease and become undetectable after 100–120 min of light.

Dark-grown mycelia accumulate phytoene which is rapidly converted into coloured carotenoids following a light induction [47]. The accumulation of phytoene is the result of a low expression of *al3* and *al2* genes even in the dark [40,41]. In addition to the light-dependent expression of the albino genes in *Neurospora* mycelia, all three genes were found to be under developmental control as well [28,51]. The *al1*, *al2*, and *al3* genes revealed increased mRNA steady-state levels shortly after the induction of macroconidia formation, and high transcript levels were observed at later stages of conidiation. The high expression of the three carotenoid biosynthesis genes during conidiation in the dark is consistent with the constitutive accumulation of carotenoids in macroconidia. The constitutive carotenoid biosynthesis in conidia was shown to be important for the protection of conidia against UV light and against photooxidative damage [52,53]. Nevertheless, transcript levels of all three genes are still photoinducible even in macroconidia [54]. In addition to the regulation by light and development, at least the *al1* and *al3* genes are also under the control of the circadian clock and reveal a rhythmic expression pattern [51] (C. Schwerdtfeger, H. Linden, unpublished results).

4 LIGHT PERCEPTION

Despite the extensive investigation of the photoperception and light signal transduction process over the past 50 years, the *Neurospora crassa* blue light photoreceptor has not been identified. The action spectra which were recorded for the light induction of carotenoid biosynthesis and for the circadian rhythm of conidiation indicated a flavin- or carotene-type photoreceptor [10,11]. Subsequently, carotenoids were excluded as blue light receptors owing to the normal blue light responses observed in carotenoid biosynthesis mutants [55]. On the contrary, the role of flavins in blue light perception was emphasized by the findings that flavin-deficient *Neurospora* mutants revealed a reduced sensitivity for several blue light responses [56]. More recent reports show that the higher-plant blue light photoreceptors such as the cryptochrome family and phototropin are flavin-type photoreceptors and thus provide further support for this hypothesis [57]. The involvement of several putative candidates such as *Neurospora* DNA photolyase and nitrate reductase as photoreceptors was disproved [58–60].

A first fungal opsin gene, *nop1*, has recently been isolated from *N. crassa*, and the gene product was shown to bind retinal following the heterologous expression in yeast [61,62]. However, Δ*nop1* strains did not reveal deficiencies in any blue light-regulated process, and the function of this rhodopsin in *Neurospora crassa* remains elusive. In addition, the white collar proteins which have an essential function in the transduction of the blue light signal have been implicated in the perception of blue light (see Sec. 5).

5 BLUE LIGHT SIGNAL TRANSDUCTION PATHWAY

5.1 Genetic Dissection of the Light-Signaling Pathway

Several mutants defective in light perception and signal transduction have been isolated and characterized. Many of these display minor deficiencies in light perception and seem to affect only a limited number of blue light responses, but not all. For example, the light-insensitive mutants *lis1*, *lis2*, and *lis3* showed a decreased sensitivity toward light for the photosuppression of circadian conidiation, but other blue light–regulated processes such as carotenoid biosynthesis and entrainment of the circadian rhythm were not affected [63]. For the flavin-deficient mutants *rib1* and *rib2*, reduced light sensitivity was observed mainly for the photosuppression of the circadian rhythms whereas the lack of flavins seemed to have a minor effect on other blue light responses [56].

To isolate additional regulatory mutants which affect blue light induction of carotenogenesis in *Neurospora*, a selection system was applied [64,65]. In this selection system the light-regulated *al3* promoter was fused to the coding region of the *mtr* gene which encodes a transporter protein for the uptake of amino acids in *Neurospora*. Following the transformation of this construct into an mtr^-/trp^- strain, growth of the transformant in the presence of tryptophan was dependent on light owing to the light-dependent expression of the *mtr* gene. In contrast, upon addition of a poisonous amino acid analogue, growth was inhibited in the light. The selection system was subsequently applied not only for the isolation of mutants defective in blue light signaling but also for mutants which revealed a light-grown phenotype and constitutive carotenoid biosynthesis even in the dark (*ccb1* and *ccb2*) [64]. In comparison to the wild type, the *ccb1* mutant showed an increased expression of some light-regulated genes following a light induction. This indicated a putative role of the corresponding gene product as a transcriptional repressor of light-regulated genes. The *ccb2* gene product instead was proposed to act during the developmental process of spore formation. The selection system was further used for the isolation of mutants which revealed decreased sensitivity in the light. The so-called blue light regulatory mutants *blr1* and *blr2* had decreased mRNA levels for several light-regulated genes and revealed a pale-orange phenotype in the light [65]. The *blr* mutants were implicated in the light signal transduction pathway; the corresponding genes, however, have not been isolated.

Another screening approach was applied which aimed at the isolation of mutants with an elevated expression of the conidiation-specific gene *con10* [66]. The *con10* gene is not only expressed during the formation of conidia but is also strongly light regulated. One of the isolated mutants revealed an increased expression of the *con6* and *con10* genes as well as enhanced carotenogenesis. The mutant was subsequently shown to represent an allele of the previously identified vivid mutant (*vvd*) [67]. A further characterization of *vvd* led to the conclusion that the VVD protein is required for the photoadaptation process in *N. crassa*. The only mutants isolated so far which seemed to be completely blind for all *Neurospora* blue light responses are the white collar mutants [8,68]. The phenotype of the white collar mutants is indistinguishable from wild-type phenotype when grown in the dark. However, when growth is performed in the light, these mutants reveal pigmented conidia whereas the mycelia are white (this phenotype led to the name ''white collar''). The white collar phenotype was subsequently shown to be due to a specific defect in the light induction of carotenoid biosynthesis genes, which suggested that the white collar genes encode regulatory proteins of the blue light sensing system. Despite extensive screening for white collar mutants, all the mutants isolated today fall in two segregation groups [8,19,64]. It was therefore concluded that *wc1* and *wc2* genes represent the only two nonredundant loci of *Neurospora crassa* which participate in blue light signaling. Moreover, these findings indicated that the blue light signal transduction chain may be rather short and possibly consists of only two protein components, WC1 and WC2.

5.2 Molecular Analysis of *wc1* and *wc2* Genes

Both the *wc1* and the *wc2* genes have been isolated and characterized. The *wc1* gene was cloned by chromosome walking whereas the cloning of *wc2* gene was carried out using an insertional mutagenesis approach [7,29]. The *wc1* and *wc2* genes encode 128- and 57-kDa proteins consisting of 1167 and 530 amino acids, respectively (Fig. 3). The white collar proteins share several common features.

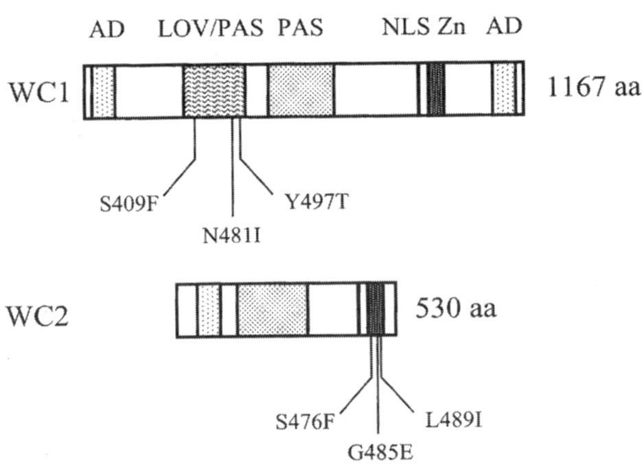

FIGURE 3 Schematic representation and domains of WC1 and WC2. The position of the following domains are indicated: Zn, zinc finger binding domains; NLS, putative nuclear localization signals (nuclear targeting domains); AD, putative activation domains; LOV and PAS domains. The locations of single amino acid substitutions identified in individual *wc1* and *wc2* mutants are depicted below the schematic representation of WC1 and WC2, respectively. The numbers show the position of the mutation, whereas the first and last letters indicate the wild-type and mutant amino acids, respectively.

Both proteins contain a single putative zinc finger DNA binding domain. In addition, putative transcriptional activation domains were characterized in WC1 and WC2 proteins. Nuclear targeting signals suggested a localization of WC1 and WC2 in the nucleus. In addition, other domains were identified in both proteins which revealed a similarity to a domain called PAS (for Per-ARNT-SIM). The PAS domains usually consist of two degenerate direct repeats of 50 amino acid, called PAS A and PAS B [69]. The WC2 PAS domain, however, differs from other PAS domains reported so far, including WC1, since it does not comprise the usual PAS A and PAS B repeats, but seems to consist of only one PAS repeat [7]. In addition to the PAS A and PAS B domains, a third PAS domain has recently been described for WC1, the so-called PAS C domain [70]. The first WC1 PAS domain has also been referred to as LOV domain (for light oxygen and voltage) because of the similarity to the LOV1 and LOV2 domains of the higher-plant photoreceptor phototropin [71].

PAS domains have been identified in proteins from mammals, insects, plants, fungi, and bacteria and have been shown to play a role in protein–protein interactions of regulatory proteins. In addition, PAS/LOV domains are important signaling modules which occur in proteins involved in the sensing of light, redox potential, small ligands, and oxygen [69]. It is intriguing that not only bacterial but also several higher-plant photoreceptors represent PAS proteins. Thus, the bacterial photoreceptor PYP (photoactive yellow protein, *Ectothiorhodospira halophila*) and the higher-plant photoreceptors for blue and red light (phototropin and the phytochromes) belong to the superfamily of PAS proteins [72,73]. Whereas the phytochrome PAS repeats seem to be involved in the activation of downstream signaling components, it has been shown for some of the photoreceptors that the PAS domain is also capable of binding a light-sensing chromophore [73]. The presence of the PAS motifs in both *Neurospora* blue light regulatory proteins led to the hypothesis that WC1 and/or WC2 may be directly involved in the sensing of blue light and may therefore represent the *Neurospora* blue light photoreceptors [3,74].

The perception of blue light by the white collar proteins and their function as transcription factors would result in a direct coupling of light perception with transcriptional activation. Such a direct targeting of light signals to a transcription factor has recently been described for the higher-plant photoreceptor phytochrome [75]. However, the binding of a chromophore to either WC1 or WC2 has not been shown.

Evidence for the important functions of WC1 and WC2 domains in blue light signaling came from DNA sequencing analysis of *wc1* and *wc2* mutant strains (Fig. 3). For WC1, three single amino acid substitutions were identified which reside in the WC1 PAS A domain and which resulted in a blind phenotype [76]. The analysis of several *wc2* mutants led to the characterization of three single amino acid substitutions in the putative zinc finger binding region, each leading to the expression of a WC2 protein defective in blue light signaling [7] (H. Linden, C. Schwerdtfeger, unpublished results).

The expression of *wc1* and *wc2* genes is regulated by light. The light-dependent increase in mRNA steady-state levels of *wc1* was dependent on functional WC1 and WC2 proteins and revealed similar kinetics when compared to other fast light–regulated genes [29]. The light induction of *wc2*, however, also occurred in a *wc1* and *wc2* mutant background, and a posttranscriptional regulatory mechanism independent of WC1 and WC2 proteins was proposed [7].

5.3 Biochemical Characterization of WC1 and WC2

The features of WC1 and WC1 pointed at a role of these proteins as blue light–regulated transcription factors. In support of this idea it was shown in bandshift experiments that both WC1 and WC2 are capable of binding the light-regulated promoter of the carotenoid biosynthesis gene *al3* [7,29]. In other experiments it was observed that WC1 and WC2 are able to form homo- and heterodimers in vitro and that the interaction was dependent on the presence of WC1 and WC2 PAS domains [76]. These results suggested that dimerization of the white collar proteins may be important for blue light signaling. In addition, the presence of protein–protein interaction domains in the white collar proteins may also make the interaction with other regulatory proteins possible (see Sec. 7).

Furthermore, the in vivo regulation of WC1 and WC2 proteins was investigated using specific antisera against WC1 and WC2, respectively [35]. The WC1 protein was detected as a 150-kDa protein in dark-grown *Neurospora* wild-type mycelia, whereas the detection of WC2 revealed an immunoreactive band of about 70 kDa. Upon light induction additional immunoreactive WC1 bands were observed with a lower mobility during SDS gel electrophoreses. In another publication, it was reported that WC2 also showed lower mobility forms and it was concluded that both proteins become modified in response to light [36]. The treatment with lambda phosphatase indicated that the posttranslational modification of WC1 and WC2 was due to protein phosphorylation. In the case of WC1, several phosphorylated forms with different mobility's were detected upon light induction, suggesting that WC1 is subject to hyperphosphorylation.

The light-dependent phosphorylation of WC1 and WC2 revealed different kinetics. Under constant light the phosphorylation of WC1 was transient whereas the phosphorylation of WC2 seemed to be stable. Interestingly, the transient appearance of the phosphorylated WC1 proteins paralleled the transient increase in transcript levels of early light–regulated genes. It is not known whether phosphorylation of WC1 is involved in the activation of WC1 or whether it represents a signal for WC1 degradation. Nevertheless, these results suggested a correlation between the light-dependent phosphorylation of WC1 and the signal flow through the blue light signaling pathway.

The stable phosphorylation of WC2 in the light led to the assumption that the phosphorylation of WC2 may be involved in other processes such as the control of dimerization or the light regulation of the circadian rhythm in *Neurospora*. When the protein biosynthesis inhibitor cycloheximide was added, a reduced stability of WC1 under constant illumination was observed, suggesting an increased protein turnover of WC1 in the light [35]. In contrast, WC2 seemed to be stable under these conditions. The presence of WC1/WC2 heterodimers was investigated by immunoprecipitation in dark-grown and in light-induced wild-type mycelia. In corroboration with the in vitro interaction studies, the white collar proteins were shown to assemble into a white collar complex also in vivo. The WC1/WC2 complexes were detected in dark-grown mycelia, and no changes were observed following the induction by light. However, the immunoprecipitation experiments did not allow the detection and quantification of WC1 and WC2 homodimers, respectively, and therefore a possible role of the homodimers in the signaling process could not be examined.

A nuclear localization of the white collar proteins was postulated owing to their putative function as blue light–regulated transcription factors. The latter assumption was supported by the presence of nuclear localization domains in both polypeptides (Fig. 3). A cellular fractionation technique was applied to determine the subcellular localization of WC1 and WC2 in *Neurospora* [36]. The WC1 protein was localized exclusively in the nucleus, whereas WC2 was detected in both the nuclear and cytoplasmic fractions. Nuclear-localized WC1 and WC2 polypeptides were phosphorylated in response to light, whereas cytoplasmic WC2 did not reveal the light-induced phosphorylation. Recently, a regulatory mechanism which includes the light-dependent nuclear translocation of the photoreceptors phytochrome has been proposed for the higher plant [77]. However, no major changes in the localization of WC1 and WC2 were observed upon illumination, which indicated that the nuclear localization of the white collar proteins is independent of light. Owing to these results, blue light signaling by a light-driven nuclear import of WC1 and/or WC2 was excluded. Several white collar mutants were subsequently applied in order to examine the phosphorylation and localization of the WC1 and WC2 proteins in various mutant backgrounds. The phosphorylation of WC1 was abolished in a *wc1* mutant which revealed a single amino acid substitution in the WC1 PAS A/LOV domain. Similarly, the phosphorylation of WC2 was no longer detected in a *wc2* mutant background. As a consequence, the phosphorylation of WC1 and WC2 was correlated with functional WC1 and WC2 proteins, respectively.

The light-specific phosphorylation of WC1 also occurred in a *wc2* mutant background, which suggested that a functional WC2 protein is not necessary for WC1 phosphorylation. On the other hand, a functional WC1 polypeptide was essential for the phosphorylation of WC2 in response to light. These results indicated an epistatic relationship between WC1 and WC2 with WC2 acting downstream of WC1 in the signal transduction pathway of blue light. The nuclear localization of WC1 and WC2 was independent of the presence of functional WC2 or WC1 proteins, respectively, which was shown by the analysis of the WC1/WC2 localization in various *wc1* and *wc2* mutant backgrounds. It was concluded that the nuclear transport of WC1 and WC2 takes place independently of the heterodimerization of WC1 and WC2.

5.4 The Function of WC1 and WC2 in Blue Light Signal Transduction: A Model

How do WC1 and WC2 proteins work in the signaling of blue light in *Neurospora crassa*? Figure 4 presents a preliminary model and summarizes the experimental data. Both WC1 and WC2 are localized in the nucleus in the dark. They form WC1/WC2 heterodimers which, however, do not support the transcription of light-regulated genes. Upon light induction, both proteins become phosphory-

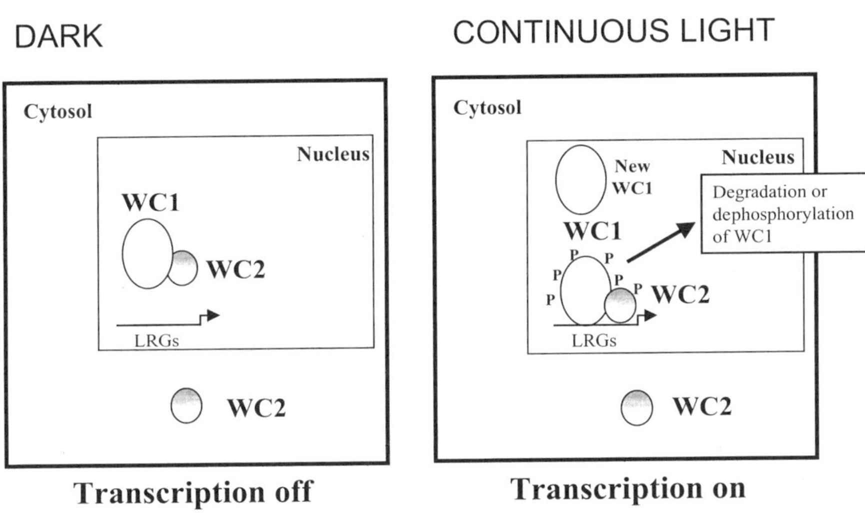

FIGURE 4 Function of WC1 and WC2 in the transcriptional activation of blue light–regulated genes: a hypothetical model. The heterodimeric WC1/WC2 complex is localized in the nucleus already in the dark. While WC1 is localized in the nucleus only, WC2 was also localized in the cytosol. In response to light, transcription of light-regulated genes (LRGs) starts and both WC proteins are phosphorylated. The phosphorylated WC1 proteins are subject to either dephosphorylation or degradation whereas WC2 is stable. The WC1 polypeptides synthesized as a consequence of the light induction are not phosphorylated under continuous light conditions.

lated, and transcription of light-regulated genes starts. After 20–30 min of light induction, phosphorylated WC1 proteins are either degraded or dephosphorylated, and transcription of light-regulated genes is turned off at the same time. Following induction by light, which also results in the transcriptional activation of the *wc1* gene itself, new WC1 proteins are biosynthesized. However, newly synthesized or dephosophorylated WC1 proteins no longer become phosphorylated. In contrast to WC1, the WC2 proteins are also localized in the cytoplasm. However, cytoplasmic-localized WC2 proteins are not subject to phosphorylation in response to light.

6 BLUE LIGHT–REGULATED PROMOTERS IN *NEUROSPORA*

The induction of the blue light signaling pathway results in the transcriptional activation of light-regulated genes which was confirmed for several genes by the application of the transcriptional inhibitor actinomycin D as well as by nuclear run-on experiments [27,39]. The nuclear localization of WC1 and WC2, their structural features, and the in vitro interaction of both proteins with the light-inducible *al3* promoter suggested that WC1 and WC2 represent blue light-regulated transcription factors. Consequently, an interaction of WC1 and/or WC2 with the light-regulated promoters seems to take place which results in the activation of transcription as indicated in Figure 4. Although this interaction has not yet been confirmed in vivo, several light-regulated promoters were examined and putative light-specific *cis* elements were characterized. For example, the *al3* promoter has been analyzed and a so-called *al3* proximal element (APE) was identified [78]. The APE motif was also present in other light-regulated promoters and deletion of this *cis* element abolished light induction. However, the APE element is not conserved in all light-regulated genes in *Neurospora* indicating the presence of other *cis* elements for light induction. The light-regulated promoter of the *con10* gene does not seem to contain a positive light element but only revealed two dark repression sites [79]. The authors put forward a hypothesis where light acts to relieve dark repression. Analysis of the clock-controlled gene *ccg2* identified several putative light-responsive elements in the promoter [80,81]. An interesting result from the promoter studies was the finding that these promoters contain distinct regulatory *cis* elements for light induction and for circadian clock and developmental regulation.

7 PHOTOADAPTATION AND DESENSITIZATION OF THE BLUE LIGHT SIGNALING PATHWAY

Most of the blue light–regulated genes in *Neurospora crassa* reveal a transient induction pattern following light induction. Thus, in spite of the continuous presence of light, the mRNA steady-state levels become undetectable after prolonged illumination. This phenomenon is called photoadaptation and is ubiquitous in prokaryotic and eukaryotic organisms such as bacteria, fungi, plants, and animals. Nevertheless, photoadaption has been thoroughly examined only in vertebrates, e.g., the photoadaptation of the visual systems [82]. In vertebrates, desensitization of the receptors and/or signaling pathway is achieved by various mechanisms. Common mechanisms are receptor phosphorylation, binding of inhibitors to the receptor (arrestin), receptor sequestration and degradation. Furthermore, the increased or decreased levels of second messenger compounds such as Ca^{2+} result in a feedback regulation attenuating the signal flow through the signaling pathway. Photoadaptation in *Neurospora* also seems to be an active process of down-regulation and desensitization of the photoreceptor and signaling pathway; it is not due simply to the destruction of the photoreceptor and to the depletion of signaling compounds. This suggestion is supported by the fact that photoadaptation can be overcome by a second illumination using higher light intensities [83,84]. A second induction with higher light intensities results again in the transient induction of light-regulated genes, indicating that *Neurospora crassa* is able to respond and photoadapt to different light intensities. However, desensitization of the photosensory system was not only observed under continuous illumination but also occurred as the consequence of a short light pulse [50,83,84]. Following the induction by a short light pulse of 1 min, *Neurospora* revealed a temporary insensitivity and was incapable of responding to light pulses of the same light intensity. However, light responsiveness was gradually restored and full responsiveness was recovered after dark incubation for about 2 h.

There is only limited information available concerning the molecular basis of photoadaptation and desensitization of light signaling in *Neurospora crassa*. Applying a pharmacological approach an involvement of protein kinase C in the desensitization process was proposed [83]. Inhibition of protein kinase C resulted in sustained transcript levels of the light-regulated *al3* gene under continuous light conditions. Moreover, the addition of inhibitors of protein kinase C to photoadapted *Neurospora* cultures abolished photoadaptation. The phosphorylation of GST-WC1 fusion proteins in vitro using *N. crassa* protein extracts suggested a possible function of protein kinase C in the light-dependent phosphorylation of WC1. The corresponding protein kinase gene has not been characterized today. A *Neurospora* mutant with a defect in photoadaptation of light-regulated conidiation genes has recently been isolated [66]. This *vvd* mutant (for vivid coloration) reveals a deep red phenotype in comparison to the wild type which is due to the increased accumulation of carotenoids in the light. In this mutant, the transiency of light-induced genes was found to be abolished, which resulted in the constitutive expression of various light-regulated genes under continuous illumination [66,84].

In addition, the *vvd* mutant revealed other defects in photoadaptation. For example, the *vvd* mutant did not show the temporary insensitivity of the blue light signaling pathway following a short light pulse. Furthermore, the *vvd* mutant was incapable of differentiating between and of adapting to low and high light intensities. That both photoadaptation of carotenoid biosynthesis genes and the adaptation of light-regulated genes involved in conidiation were abolished, indicated a more general function of the *vvd* gene product in photoadaptation. The *vvd* gene was recently cloned and reported to code for a small PAS protein of ~20 kDA [85]. The VVD PAS domain may allow the direct protein–protein interaction of VVD with components of the blue light signal transduction pathway and may result in the observed desensitization.

8 CONCLUDING REMARKS AND OUTLOOK

The data on the function and regulation of the white collar proteins certainly provide a profound insight into the blue light signal transduction pathway of *Neurospora crassa*. Nonetheless, the identification and characterization of additional components of the blue light signaling pathway are necessary before we can fully understand the underlying regulation. First of all, the lack of identification of the *Neurospora* blue light photoreceptor continues to represent a large gap in our knowledge. That there are only two white collar loci in *Neurospora crassa* may be indicative of a model where WC1 and WC2 represent the only protein components of the signaling pathway. A thorough biochemical analysis of WC1 and WC2, the identification of putative chromophores, and a site-directed mutagenesis approach may further clarify this question.

Alternatively, perception of light may be carried out by a family of photoreceptors with overlapping functions, and this redundancy would prevent the genetic identification by mutagenesis. In this case, new screenings for leaky light signal transduction mutants as well as other approaches such as the yeast two-hybrid technique are needed to dissect the initial steps of blue light perception. An example of another component of this signaling pathway is the protein kinase involved in light-induced WC1 and WC2 phosphorylation. The identification of these components will be facilitated by the recent advancements that have been made in the sequencing of the *Neurospora* genome. Several *Neurospora* sequencing projects are under way such as the *Neurospora* Genome Project at the University of New Mexico, the *Neurospora crassa* cDNA Project at the University of Oklahoma, and the German *Neurospora crassa* Genome project. The wealth of DNA sequence data will help in the search for new signal transduction candidates. Furthermore, the application of the microarray technique will lead to the identification of new light-regulated genes and will result in a much better understanding of the regulation of *Neurospora* development by light.

ACKNOWLEDGMENTS

I am grateful to the members of my laboratory for helpful suggestions on the manuscript and to E. O'Halloran for her help in the preparation of the manuscript.

In addition, I thank Dr. G. Sandmann for communicating unpublished data. Research in the author's laboratory is supported by a grant from Deutsche Forschungsgemeinschaft (LI 819/2-1).

REFERENCES

1. FR Lauter. Molecular genetics of fungal photobiology. J Genet 75:375–386, 1996.
2. H Linden, P Ballario, G Macino. Blue light regulation in *Neurospora crassa*. Fungal Genet Biol 22:141–150, 1997.
3. H Linden, P Ballario, G Arpaia, G Macino. Seeing the light: news in *Neurospora* blue light signal transduction. Adv Genet 41:35–54, 1999.
4. MJ Orbach, D Vollrath, RW Davis, C Yanofsky. An electrophoretic karyotype of *Neurospora crassa*. Mol Cell Biol 8:1469–1473, 1988.
5. S Kang, RL Metzenberg. Insertional mutagenesis in *Neurospora crassa*: cloning and molecular analysis of the $preg^{+}$ gene controlling the activity of the transcriptional activator NUC-1. Genetics 133:193–202, 1993.
6. C Cogoni, G Macino. Gene silencing in *Neurospora crassa* requires a protein homologous to RNA-dependent RNA polymerase. Nature 399:166–169, 1999.
7. H Linden, G Macino. White collar 2, a partner in blue-light signal transduction, controlling expression of light-regulated genes in *Neurospora crassa*. EMBO J 16: 98–109, 1997.
8. DD Perkins, A Radford, D Newmeyer, M Bjorkmann. Chromosomal loci of *Neurospora crassa*. Microbiol Rev 46:426–570, 1982.
9. C Fankhauser, J Chory. Light control of plant development. Annu Rev Cell Dev Biol 13:203–229, 1997.
10. ML Sargent, WR Briggs. The effects of light on a circadian rhythm of conidiation in *Neurospora*. Plant Physiol 42:1504–1510, 1967.
11. EC DeFabo, RW Harding, W Shropshire. Action spectrum between 260 and 800 nanometers for the photoinduction of carotenoid biosyntheses in *Neurospora crassa*. Plant Physiol 57:440–445, 1976.
12. ML Springer. Genetic control of fungal differentiation: the three sporulation pathways of *Neurospora crassa*. Bioessays 15:365–374, 1993.
13. FR Lauter, CT Yamashiro, C Yanofsky. Light stimulation of conidiation in *Neurospora crassa*: studies with the wild-type strain and mutants *wc-1*, *wc-2* and *acon-2*. J Photochem Photobiol B 37:203–211, 1997.
14. H Ninnemann. Photostimulation of conidiation in mutants of *Neurospora crassa*. J Photochem Photobiol 9:189–199, 1991.
15. RW Siegel, SS Matsuyama, JC Urey. Induced macroconidia formation in *Neurospora crassa*. Experientia 24:1179–1181, 1968.
16. R Maheshwari. Microconidia of *Neurospora crassa*. Fungal Genet Biol 26:1–18, 1999.
17. ML Springer, C Yanofsky. Expression of *con* genes along the three sporulation pathways of *Neurospora crassa*. Genes Dev 6:1052–1057, 1992.
18. F Degli-Innocenti, U Pohl, VEA Russo. Photoinduction of protoperithecia in *Neurospora crassa* by blue light. Photochem Photobiol 37:49–51, 1983.
19. F Degli-Innocenti, VEA Russo. Isolation of new white collar mutants of *Neurospora*

crassa and studies on their behavior in the blue light-induced formation of protoperithecia. J Bacteriol 159:757–761, 1984.
20. RW Harding, S Melles. Genetic analysis of phototropism of *Neurospora crassa* perithecial beaks using white collar and albino mutants. Plant Physiol 72:996–1000, 1983.
21. M Zalokar. Studies on biosynthesis of carotenoids in *Neurospora crassa*. Arch Biochem Biophys 50:71–80, 1954.
22. RW Harding, RV Turner. Photoregulation of the carotenoid biosynthetic pathway in albino and white collar mutants of *Neurospora crassa*. Plant Physiol 68:745–749, 1981.
23. FR Lauter, VE Russo, C Yanofsky. Developmental and light regulation of *eas*, the structural gene for the rodlet protein of *Neurospora*. Genes Dev 6:2373–2381, 1992.
24. FR Lauter, VE Russo. Blue light induction of conidiation-specific genes in *Neurospora crassa*. Nucleic Acids Res 19:6883–6886, 1991.
25. FR Lauter, C Yanofsky. Day/night and circadian rhythm control of *con* gene expression in *Neurospora*. Proc Natl Acad Sci USA 90:8249–8253, 1993.
26. SK Crosthwaite, JJ Loros, JC Dunlap. Light-induced resetting of a circadian clock is mediated by a rapid increase in frequency transcript. Cell 81:1003–1012, 1995.
27. S Baima, G Macino, G Morelli. Photoregulation of the *albino-3* gene in *Neurospora crassa*. J Photochem Photobiol B 11:107–115, 1991.
28. C Li, TJ Schmidhauser. Developmental and photoregulation of *al-1* and *al-2*, structural genes for two enzymes essential for carotenoid biosynthesis in *Neurospora*. Dev Biol 169:90–95, 1995.
29. P Ballario, P Vittorioso, A Magrelli, C Talora, A Cabibbo, G Macino. White collar-1, a central regulator of blue light responses in *Neurospora*, is a zinc finger protein. EMBO J 15:1650–1657, 1996.
30. G Arpaia, JJ Loros, JC Dunlap, G Morelli, G Macino. The interplay of light and the circadian clock. Independent dual regulation of clock-controlled gene ccg-2(eas). Plant Physiol 102:1299–1305, 1993.
31. G Arpaia, JJ Loros, JC Dunlap, G Morelli, G Macino. Light induction of the clock-controlled gene *ccg-1* is not transduced through the circadian clock in *Neurospora crassa*. Mol Gen Genet 247:157–163, 1995.
32. D Bell-Pedersen, ML Shinohara, JJ Loros, JC Dunlap. Circadian clock–controlled genes isolated from *Neurospora crassa* are late night- to early morning-specific. Proc Natl Acad Sci USA 93:13096–13101, 1996.
33. K Mizoguchi, K Oda, K Hasunuma. Blue light induced ADP ribosylation of 38 and 56 kDa proteins in the soluble fraction of mycelia of *Neurospora crassa*. J Photochem Photobiol B 35:197–201, 1996.
34. TV Potapova, NN Levina, TA Belozerskaya, MS Kritsky, LM Chailakhian. Investigation of electrophysiological responses of *Neurospora crassa* to blue light. Arch Microbiol 137:262–265, 1984.
35. C Talora, L Franchi, H Linden, P Ballario, G Macino. Role of a white collar–1–white collar–2 complex in blue-light signal transduction. EMBO J 18:4961–4968, 1999.
36. C Schwerdtfeger, H Linden. Localization and light-dependent phosphorylation of

white collar 1 and 2, the two central components of blue light signaling in *Neurospora crassa*. Eur J Biochem 267:414–422, 2000.
37. K Oda, K Hasunuma. Light signals are transduced to the phosphorylation of 15 kDa proteins in *Neurospora crassa*. FEBS Lett 345:162–166, 1994.
38. Y Ogura, Y Yoshida, K Ichimura, C Aoyagi, N Yabe, K Hasunuma. Isolation and characterization of *Neurospora crassa* nucleoside diphosphate kinase NDK-1. Eur J Biochem 266:709–714, 1999.
39. TJ Schmidhauser, FR Lauter, VE Russo, C Yanofsky. Cloning, sequence, and photoregulation of *al-1*, a carotenoid biosynthetic gene of *Neurospora crassa*. Mol Cell Biol 10:5064–5070, 1990.
40. MA Nelson, G Morelli, A Carattoli, N Romano, G Macino. Molecular cloning of a *Neurospora crassa* carotenoid biosynthetic gene (albino-3) regulated by blue light and the products of the white collar genes. Mol Cell Biol 9:1271–1276, 1989.
41. TJ Schmidhauser, FR Lauter, M Schumacher, W Zhou, VE Russo, C Yanofsky. Characterization of *al-2*, the phytoene synthase gene of *Neurospora crassa*. Cloning, sequence analysis, and photoregulation. J Biol Chem 269:12060–12066, 1994.
42. G Sandmann, N Misawa, M Wiedemann, P Vittorioso, A Carattoli, G Morelli, G Macino. Functional identification of *al-3* from *Neurospora crassa* as the gene for geranylgeranyl pyrophosphate synthase by complementation with *crt* genes, in vitro characterization of the gene product and mutant analysis. J Photochem Photobiol B 18:245–251, 1993.
43. GE Bartley, TJ Schmidhauser, C Yanofsky, PA Scolnik. Carotenoid desaturases from *Rhodobacter capsulatus* and *Neurospora crassa* are structurally and functionally conserved and contain domains homologous to flavoprotein disulfide oxidoreductases. J Biol Chem 265:16020–16024, 1990.
44. AH Goldie, RE Subden. The neutral carotenoids of wild-type and mutant strains of *Neurospora crassa*. Biochem Genet 10:275–284, 1973.
45. JC Verdoes, P Krubasik, G Sandmann, AJJ van Ooyen. Isolation and functional characterisation of a novel type of carotenoid biosynthetic gene from *Xanthophyllomyces dendrorhous*. Mol Gen Genet 262:453–461, 1999.
46. A Velayos, AP Eslava, EA Iturriaga. A bifunctional enzyme with lycopene cyclase and phytoene synthase activities is encoded by the *carRP* gene of *Mucor circinelloides*. Eur J Biochem 267:5509–5519, 2000.
47. RW Harding, PC Huang, HK Mitchel. Photochemical studies of the carotenoid biosynthetic pathway in *Neurospora crassa*. Arch Biochem 129:696–707, 1969.
48. Y Wan, H Liu, C Li, TJ Schmidhauser. Genome analysis on linkage group VI of *Neurospora crassa*. Fungal Genet Biol 21:329–336, 1997.
49. EL Schrott. Fluence response relationship of carotenogenesis in *Neurospora crassa*. Planta 150:174–179, 1980.
50. E Schrott. The biphasic fluence response of carotenogenesis in *Neurospora crassa*: temporal insensitivity of the photoreceptor system. Planta 151:371–374, 1981.
51. G Arpaia, A Carattoli, G Macino. Light and development regulate the expression of the *albino-3* gene in *Neurospora crassa*. Dev Biol 170:626–635, 1995.
52. SAC Morris, RE Subden. Effect of ultraviolet radiation on carotenoid containing and albino strains of *Neurospora crassa*. Mutat Res 22:105–109, 1974.

53. SA Thomas, ML Sargent, RW Tuveson. Inactivation of normal and mutant *Neurospora crassa* conidia by visible light and near-UV: role of $^1O_2$, carotenoid composition and sensitizer location. Photochem Photobiol 33:349–354, 1981.
54. C Li, MS Sachs, TJ Schmidhauser. Developmental and photoregulation of three *Neurospora crassa* carotenogenic genes during conidiation induced by desiccation. Fungal Genet Biol 21:101–108, 1997.
55. VEA Russo. Are carotenoids the blue-light photoreceptor in the photoinduction of protoperithecia in *Neurospora crassa*. Planta 168:56–65, 1986.
56. J Paietta, ML Sargent. Photoreception in *Neurospora crassa*: correlation of reduced light sensitivity with flavin deficiency. Proc Natl Acad Sci USA 78:5573–5577, 1981.
57. M Ahmad. Seeing the world in red and blue: insight into plant vision and photoreceptors. Curr Opin Plant Biol 2:230–235, 1999.
58. M Shimura, Y Ito, C Ishii, H Yajima, H Linden, T Harashima, A Yasui, H Inoue. Characterization of a *Neurospora crassa* photolyase-deficient mutant generated by repeat induced point mutation of the *phr* Gene. Fungal Genet Biol 28:12–20, 1999.
59. TA Belozerskaya, SS Burikhanov, EK Chernyshova, MS Kritsky, NP Lvov. Does nitrate reductase play a key role in the photoinduction of carotenoid synthesis in *Neurospora crassa*. Fungal Genet Newslett 29:14–15, 1982.
60. J Paietta, ML Sargent. Blue light responses in nitrate reductase mutants of *Neurospora crassa*. Photochem Photobiol 35:853–855, 1982.
61. JA Bieszke, EN Spudich, KL Scott, KA Borkovich, JL Spudich. A eukaryotic protein, NOP-1, binds retinal to form an archaeal rhodopsin-like photochemically reactive pigment. Biochem 38:14138–14145, 1999.
62. JA Bieszke, EL Braun, LE Bean, S Kang, DO Natvig, KA Borkovich. The *nop-1* gene of *Neurospora crassa* encodes a seven transmembrane helix retinal-binding protein homologous to archaeal rhodopsins. Proc Natl Acad Sci USA 96:8034–8039, 1999.
63. J Paietta, ML Sargent. Isolation and characterization of light-insensitive mutants of *Neurospora crassa*. Genetics 164:11–21, 1983.
64. H Linden, M Rodriguez-Franco, G Macino. Mutants of *Neurospora crassa* defective in regulation of blue light perception. Mol Gen Genet 254:111–118, 1997.
65. A Carattoli, E Kato, M Rodriguez-Franco, WD Stuart, G Macino. A chimeric light-regulated amino acid transport system allows the isolation of blue light regulator (*blr*) mutants of *Neurospora crassa*. Proc Natl Acad Sci USA 92:6612–6616, 1995.
66. LB Shrode, LD White, DJ Ebbole. The *vvd* gene is required for light-adaptation of the *N. crassa* conidiation-induced genes, *con-6* and *con-10*. 20th Fungal Gentic Conference, Asilomar, California, 1999.
67. DD Perkins, BS Margolin, EU Selker, SD Haedo. Occurence of repeat induced point mutation in long segmental duplications of *Neurospora*. Genetics 147:125–136, 1997.
68. RW Harding, W Shropshire. Photocontrol of carotenoid biosynthesis. Annu Rev Plant Physiol 31:217–238, 1980.
69. BL Taylor, IB Zhulin. PAS domains: internal sensors of oxygen, redox potential, and light. Microbiol Mol Biol Rev 63:479–506, 1999.
70. K Lee, JJ Loros, JC Dunlap. Interconnected feedback loops in the *Neurospora* circadian system. Science 289:107–110, 2000.

71. E Huala, PW Oeller, E Liscum, IS Han, E Larsen, WR Briggs. *Arabidopsis* NPH1: a protein kinase with a putative redox-sensing domain. Science 278:2120–2123, 1997.
72. R Kort, WD Hoff, M Van West, AR Kroon, SM Hoffer, KH Vlieg, W Crielaand, JJ Van Beeumen, KJ Hellingwerf. The xanthopsins: a new family of eubacterial blue-light photoreceptors. EMBO J 15:3209–3218, 1996.
73. C Fankhauser, J Chory. Light receptor kinases in plants! Curr Biol 9:R123–126, 1999.
74. P Ballario, G Macino. White collar proteins: PASsing the light signal in *Neurospora crassa*. Trends Microbiol 5:458–462, 1997.
75. JF Martinez-Garcia, E Huq, PH Quail. Direct targeting of light signals to a promoter element-bound transcription factor. Science 288:859–863, 2000.
76. P Ballario, C Talora, D Galli, H Linden, G Macino. Roles in dimerization and blue light photoresponse of the PAS and LOV domains of *Neurospora crassa* white collar proteins. Mol Microbiol 29:719–729, 1998.
77. MM Neff, C Fankhauser, J Chory. Light: an indicator of time and place. Genes Dev 14:257–271, 2000.
78. A Carattoli, C Cogoni, G Morelli, G Macino. Molecular characterization of upstream regulatory sequences controlling the photoinduced expression of the *albino-3* gene of *Neurospora crassa*. Mol Microbiol 13:787–795, 1994.
79. LM Corrochano, FR Lauter, DJ Ebbole, C Yanofsky. Light and developmental regulation of the gene *con-10* of *Neurospora crassa*. Dev Biol 167:190–200, 1995.
80. R Kaldenhoff, VEA Russo. Promoter analysis of the *bli-7/eas* gene. Curr Genet 24: 394–399, 1993.
81. D Bell-Pedersen, JC Dunlap, JJ Loros. Distinct *cis*-acting elements mediate clock, light, and developmental regulation of the *Neurospora crassa eas (ccg-2)* gene. Mol Cell Biol 16:513–521, 1996.
82. EN Pugh, S Nikonov, TD Lamb. Molecular mechanisms of vertebrate photoreceptor light adaptation. Curr Opin Neurobiol 9:410–418, 1999.
83. G Arpaia, F Cerri, S Baima, G Macino. Involvement of protein kinase C in the response of *Neurospora crassa* to blue light. Mol Gen Genet 262:314–322, 1999.
84. C Schwerdtfeger, H Linden. Blue light adaptation and desensitization of light signal transduction in *Neurospora crassa*. Mol Microbiol 2001 (in press).
85. C Heintzen. *vvd* encodes a novel PAS protein involved in light perception for the *Neurospora* circadian clock. Neurospora 2000 Conference, Asilomar, California, 2000.

8

Circadian Rhythms in *Neurospora crassa*

Deborah Bell-Pedersen
Texas A&M University, College Station, Texas

1 INTRODUCTION

Circadian (daily) rhythms are biological rhythms that are observed in virtually all eukaryotes [reviewed in 1] and in some prokaryotes [reviewed in 2, 3]. The circadian rhythm we are all most familiar with is our daily sleep–wake cycle. Daily rhythms in biochemical, cellular, and behavioral activities are produced and controlled by a rhythm generator composed of one or more oscillators, herein referred to as the ''clock.'' The clock generates daily rhythmicity in a wide variety of processes, ranging from the control of development in fungi, cell division in the marine protist *Gonyaulax*, and photosynthesis in plants, to cognitive functions in people [reviewed in 1].

Circadian rhythms, by virtue of their ubiquity and importance in human mental and physical well-being, have been the subject of extensive research. Today, hundreds of laboratories worldwide use a variety of methods and organisms to study the circadian system. Despite this diversity, the field is unified by the fact that circadian rhythms in all organisms share the same defining properties, which in turn likely reflects similarities among clock mechanisms and ancestry. These properties include:

1. Persistence of circadian rhythms under constant conditions with a free-running period length of ~24 h.

2. The ability of the clock to be reset (entrained) in a time-dependent manner by environmental stimuli.

3. Compensation of period length for changes in an organisms natural environment. For example, when an organism is placed in varying temperatures within its physiological range, the period of the rhythm does not change. Here, the period is said to be ''temperature compensated.''

Together, these properties are key for a biological timing mechanism that responds rapidly to multiple environmental cues to maintain an appropriate phase relationship with environmental cycles. These circadian properties may be intrinsic to a single oscillator or, more likely, generated by interactions between multiple oscillators.

The circadian clock not only measures the passage of time (like an hourglass), but also endows organisms with the ability to anticipate dependable cyclic changes that occur in the environment, such as recurrent changes in light intensity, temperature, and humidity. A prime example here is a plant. To save energy, a plant needs only to produce the enzymes responsible for photosynthesis when the sun is up. The clock provides a way for the plant to anticipate the sun's arrival so that it can gear up production of photosynthetic enzymes just before dawn. Experiments have also demonstrated that a circadian clock, whose intrinsic period closely matches that of environmental cycles, improves the fitness of cells [4]. Thus, the capacity to predict and prepare for environmental changes has likely provided an adaptive advantage for organisms and accordingly has probably led to the ubiquity of clocks within the biological world.

For the circadian clock system to provide an internal measure of external time and allow anticipation, the endogenous free-running period needs to be reset each day to precisely 24 h. This is accomplished by sensing environmental time cues (termed *zeitgebers*; from the German for time-giver) and shifting the phase of the rhythm appropriately. The two most pervasive zeitgebers are considered to be the daily light/dark and temperature cycles. The clock responds differently to the zeitgebers when applied at particular times within the circadian cycle. Moreover, the intensity and duration of the entraining signal influence the magnitude of a phase shift. For example, in *Neurospora crassa*, light signals perceived in the early subjective night are interpreted as dusk, and the clock delays to dusk. Light signals given during the late subjective night are interpreted as dawn, and the clock advances to dawn. In many organisms, although apparently not in *Neurospora* [5], there is an extended time during the subjective day in which the organisms' clock is insensitive to an entraining light signal [6].

Temperature compensation is one of the characteristic features of a circadian clock [7]. To prevent the clock from responding inappropriately when temperatures vary, it makes sense that an accurate clock requires a mechanism to maintain its rate at different ambient temperatures. For example, our sleep–wake cycles do not change with the seasons. Yet, because a circadian clock can be entrained by temperature transitions (pulses or cycles), the system is not entirely temperature independent.

From this introduction to circadian clocks, it should be apparent to the reader that the circadian system is quite complex. To resolve such a complex circadian system into its individual components, considerable effort has centered on characterizing the genes and proteins responsible for rhythmicity. Organisms in which mutants can be readily isolated and the nature of the mutations determined provide essential experimental models for uncovering the mechanisms behind circadian rhythmicity. The fungus *N. crassa* easily fulfills this need, and studies in this organism have provided major contributions to our knowledge of circadian biology. Importantly, the basic properties and mechanisms of the circadian system appear to be conserved across species [reviewed in 8], a circumstance that provides justification for the use of simple model systems for clock studies.

To formulate clear experimental questions aimed at understanding the complex circadian system, the mechanism can be conceptually divided into thee basic parts: input pathways; the central clock composed of one or more oscillators; and output pathways (Fig. 1). Based on these divisions, three fundamental questions pertaining to the genetic basis of circadian rhythmicity are typically raised:

1. What are the components of the oscillators and how do they function to keep accurate time (the clock)?
2. What are the signaling pathways though which the cellular clock is synchronized to the external world (input)?
3. What genes are regulated by the clock and how is control achieved (output)?

Complicating this simple view, however, are examples of feedback from the clock to the input pathways, as well as feedback from output genes to the

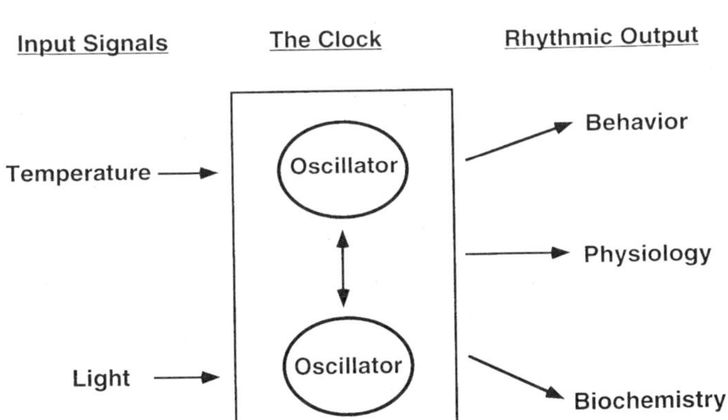

FIGURE 1 Simple view of a circadian clock system. See text for details.

clock [9]. Therefore, while these questions have provided the basic framework for early genetic and molecular studies of clocks, it is clear that the circadian system involves multiple levels of feedback control that likely contribute to the robustness of the system. This level of complexity is even becoming evident in the relatively simple eukaryote *N. crassa* [10].

General principles have emerged from physiological, genetic, and molecular studies of clocks from a wide variety of model organisms, and although these will be described in more detail below, they can be summarized as follows: (1) At the center of the clock system lies one or more circadian oscillators comprised of transcription/translation-based feedback loops containing both positive and negative elements; (2) environmental input signals to the clock rapidly alter the level or activity of an essential oscillator component, thereby changing the dynamics of the feedback loop; and (3) rhythmic output can be manifested though transcription factors that activate or repress the expression of genes in a time-of-day–specific fashion. However, despite the extraordinary progress being made in deciphering the circadian system, there are still many mysteries that remain to be solved.

2 *N. CRASSA* AS A MODEL SYSTEM FOR STUDYING CLOCKS

Neurospora is particularly suited for clock analyses because it displays a circadian rhythm that can be easily assayed in the laboratory—a rhythm in asexual spore development (conidiation) [11–13]. Clock regulation of development likely provides a fitness advantage to the organism, possibly by producing spores at the time of day when they are more readily dispersed or are less affected by harmful UV light. As vegetative hyphae grow across the surface of an agar medium, once each day the clock signals the production of aerial hyphae which grow away from the surface and, after ~12 h of development, bud to give rise to the readily observable fluffy orange conidiospores.

The circadian conidiation rhythm is typically assayed on 30 to 40-cm-long cylindrical glass tubes (called race tubes) that are bent upward at both ends to contain an agar growth medium (Fig. 2) [12]. Conidia are inoculated at one end of the race tube and the cultures are germinated in constant light at 25°C for ~1 day. The growth front is then marked and the race tube is transferred to constant dark (25°C), which synchonizes the cells and sets the clock to dusk (Circadian Time 12). (The concept of circadian time [CT] was developed to allow comparison of the properties of circadian rhythms in organisms having different endogenous periods, whereby the period is divided into 24 equal parts, with each part defined as one circadian hour. By convention, CT0 represents subjective dawn and CT12 represents subjective dusk.) Every 24 h, the growth front is marked

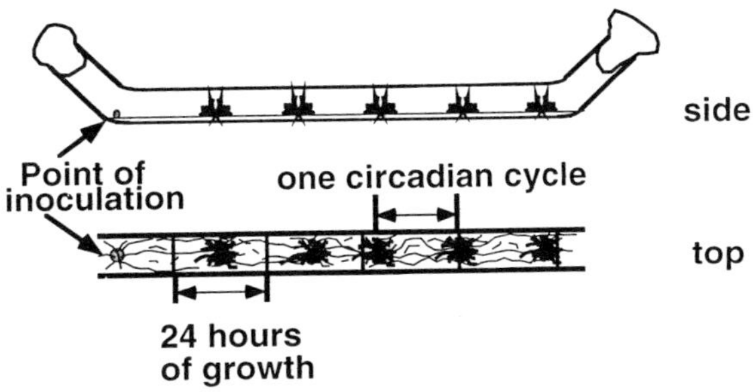

FIGURE 2 *Neurospora* circadian rhythm of conidiation viewed on a race tube. To monitor the circadian rhythm of conidiation, conidia are inoculated at one end of a race tube. After a day of growth in constant light, the position of the growth front is marked (shown as a solid black line) and the culture is placed in constant dark at 25°C. Following transfer to darkness, the growth front is marked every 24 h with the aid of a red light. The growth rate is constant, and the positions of the orange conidial bands (separated by undifferentiated surface hyphae) relative to the marked growth fronts allow determination of period and phase of the rhythm. (From Ref. 86.)

under a red safe light, which is known to have no entraining effect on the clock [12]. During vegetative growth on the agar surface, some time in the late evening the clock initiates macroconidiation, beginning with the production of aerial hyphae that eventually bud to give rise to the conidiospores. The cells that are not determined to differentiate continue to grow down the tube as undifferentiated vegetative hyphae and the cycle renews (see http://www.mrs.umn.edu/~goochv/Circadian/neur.mov for a video of the *Neurospora* circadian rhythm). At the conclusion of an experiment, the center of each conidiation zone (called a band) is marked. The pattern of the conidiation bands can be analyzed later at leisure because they act as a "fossil record" of the state of the clock at the time the conidia were produced. Because vegetative growth in *Neurospora* occurs at a constant rate, the period of the rhythm can be calculated from the distance between consecutive bands, and the phase of the rhythm can be determined from the position of the bands relative to the growth fronts.

The conidiation rhythm exhibits all of the key characteristics of a circadian oscillation. The cycle occurs once every 21.5 h in constant dark at 24°C [11]; the period of the rhythm is the same between 18°C and 30°C and is thus temperature compensated [14]; and the conidiation rhythm can be entrained by various light–dark cycles [5,12,14] and reset by temperature pulses [15,16]. In practice, the conidiation rhythm is monitored in strains carrying the *band* (*bd*) mutation, which allows a clearer visualization of the rhythm as compared to wild-type strains [12]. The *bd* mutation does not appear to affect the clock mechanism itself, but only the overt expression of the underlying oscillator(s) by rendering cells insensitive to CO_2 buildup in the culture tubes.

3 ISOLATION OF CIRCADIAN RHYTHM MUTANTS IN *NEUROSPORA*

To begin to describe components of the *Neurospora* circadian system, cultures were mutagenized and assayed on race tubes to identify mutations that altered the conidiation rhythm [17,18]. In addition, several strains defective in known biochemical processes were examined for effects on the conidiation rhythm [13]. From these initial studies, >20 loci were found to affect the canonical features of the conidiation rhythm, including period and temperature compensation (Table 1). The results from mutant analyses of *Neurospora* rhythms suggested that many genes and gene products are capable of affecting the operation of the circadian clock.

Much of the initial efforts in characterizing clock components concentrated on the *frequency* (*frq*) gene because it was represented by multiple alleles, and because mutations in *frq* resulted in altered periods. Some *frq* alleles have periods ranging from 16 to 29 h, null alleles are arrhythmic, and some of the alleles have lost temperature compensation [17–20] (Table 1). In addition, no other phenotypes are readily observed in the *frq* mutant strains. Together, these data supported the idea early on that *frq* encodes a central circadian clock component.

The question is still open for some of the other genes identified by mutation as to whether they encode clock components, or if the effects on period in the mutants are indirect, possibly reflecting defects in output pathways from the clock. Thus, the future characterization of these additional clock-affecting loci will likely provide important information on the mechanisms of the *Neurospora* circadian clock. In addition, because a number of the mutant alleles affect temperature compensation of the clock, these mutant loci may provide clues to the process of temperature compensation—the "black box" of circadian biology. Answers to these questions will likely await cloning of the genes and biochemical analyses of their encoded products. Currently the *frq*, *wc1*, *wc2*, *prd2*, *prd4*, and *prd6* genes have been cloned, although studies on the *prd* genes are still in their infancy [21].

Furthermore, additional oscillator components are also likely to be involved in rhythm generation. The initial genetic screens did not reach saturation, and lethal mutations would have been missed in the screens. New screens have now been initiated for temperature-sensitive mutants that affect rhythmicity [22], arrhythmic mutants, and suppressors of existing mutants (D. Bell-Pedersen, unpublished data). These screens will likely uncover additional clock-affecting genes. Currently, our understanding of the molecular aspects of the *Neurospora* clock comes primarily from extensive studies of the *frq* and *wc* genes. However, as new genes are discovered and known loci are cloned and analyzed, it is likely that additional features of the circadian system and its relationship with the environment and cellular metabolism will be revealed.

4 MOLECULAR ANALYSIS OF THE *FRQ* AND *WC* GENES

The *frq* gene was cloned several years ago [23] and was found to encode multiple transcripts [20] (Fig. 3). Two sense transcripts of 4 and 4.5 kb possess the potential to encode the FRQ protein (see below). The function of the other transcript, an antisense transcript of 4.5 kb, remains a mystery as no significant ORFs are present and no clock-specific activity has yet been associated with it [18]. Both of the sense transcripts have long 5′ untranslated regions containing upstream ORFs and encode two forms of the FRQ protein. These include a long form of 989 amino acids that initiates at the first ATG (ATG1) and a shorter form of 890 amino acids that initiates at a third ATG (ATG3) [24,25]. While both forms are necessary for conidial rhythms within the physiological temperature range of *Neurospora* [26], no distinct activities have been assigned to the different forms. In addition, no known motifs are evident in the first 100 amino acids that might suggest functional differences. Therefore, unless otherwise indicated, FRQ is used here collectively to represent both the long and short forms.

The biochemical function of FRQ is unknown, although it contains several signature motifs that are consistent with its being involved in gene regulation. These motifs include a nuclear localization signal (NLS), a helix-turn-helix motif, and conserved acidic and basic regions [20,27,28] (Fig. 3). Additional motifs include a TG/SG repeated amino acid sequence that is also found in the *Drosophila* PER protein [23], a central component of the fly clock; however, the importance of this motif in FRQ function has not been examined. The sequence motifs are also found to be essentially conserved in FRQ homologs isolated from distantly related fungal species [27,29,30].

Accumulation of *frq* mRNA occurs in a rhythmic fashion, and while it is generally assumed that regulation occurs at the level of transcription initiation, this has not been confirmed experimentally. Both *frq* mRNA and FRQ protein levels cycle with a 22-h period in wild-type strains grown in constant darkness, and the period of the oscillation is appropriately altered in both short and long period mutant strains [31]. Interestingly, the levels of *frq* mRNA in the long period frq^7 mutant strain are significantly higher than in wild-type strains [31], suggesting less efficient turnover of *frq* mRNA in the mutant. The *frq* gene is also light inducible. Light pulses that are effective in phase-shifting the conidiation rhythm rapidly induce *frq* transcription, resulting in high levels of transcript accumulation [32]. Circadian rhythmicity is lost in strains constantly expressing *frq* from the inducible *qa2* promoter at an ectopic locus, indicating that the abundance of *frq* needs to oscillate for the clock to function [31]. In addition, manipu-

TABLE 1 Rhythm Mutants in *Neurospora crassa*

Gene	Allele name	Period (h) at 25°C	Temperature compensation	Dominance	Growth rate (GR) mm/day[a] Comments	References
wild-type		21.5	+		GR 39.2	
frequency	frq^{1}	16.5	+	semidominant	GR 36.1	88
	frq^{2}	19.3	+	semidominant	GR 37.2	88, 89
	frq^{3}	24.0	−	semidominant	GR 36.7	88, 89
	frq^{7}	29.0	−	semidominant	GR 36.9	88–90
	frq^{9}	variable	−	recessive		19, 91
	frq^{10}	variable	−	recessive	gene disruption	20
	frq^{11}	arrhythmic at 30°C	−		temperature sensitive >30°C	24
period-1	*prd-1*	25.8	−	recessive	GR 24.7	92, 93
period-2	*prd-2*	25.5	+	recessive	GR 33.3	90, 93
period-3	*prd-3*	25.1	−	recessive	GR 31.2	90, 93
period-4	*prd-4*	18.0	−	semidominant	GR 38.3	90, 93
period-6	*prd-6*	18.0 at 22°C	+	recessive	GR 27.7 temperature sensitive >21°C	94
chrono	*chr*	23.5	+	semidominant	GR 37.7	90, 93
white collar-1	*wc1 ER53*	arrhythmic		recessive	light insensitive	33, 34
white collar-2	*wc2 ER33*	arrhythmic		recessive	light insensitive	33, 34
	wc2 ER24	29.7 at 25°C	−	temperature sensitive		35; M. Collett and J. Dunlap, unpublished
arginine-13	*arg13*	19[b]	+	recessive	amino acid requirer	13
chain elongation	*cel*	variable[c]	−	recessive	fatty acid synthase deficient	95

choline-1	*chol1*	variable[c]	–		phosphatidylcholine deficient	82
cytochrome a-5	*cya5*	19	+		cytochrome aa3 deficient	96
cytochrome b-2	*cyb2*	18	+		cytochrome b deficient	96
cytochrome b-3	*cyb3*	20	+		cytochrome b deficient	13
cytochrome-4	*cyt4*	20	+		cytochrome aa3 and b deficient	13
cysteine-4	*cys4*	19[b]	+	recessive	amino acid requirer	93, 97
cysteine-9	*cys9*	variable[b]	–		thioredoxin reductase	98
cysteine-12	*cys12*	19[b]	+	recessive	amino acid requirer	93, 97
female fertility-1	*ff1 (glp3)*	19				96
maternally inherited	*mi2, mi3, mi5*	18–19	+		cytochrome oxidase subunit 1 deficient	96
oligomycin resistant	*oli*	18–19	+	semidominant	mitochondrial ATPase subunit 9	99
phenyl-alanine-1	*phe1*	19[b]	+		ergosterol synthesis deficient	13
	rhy1	arrhythmic at 30°C	+		temperature sensitive >30°C	22
unknown-18	*un18*	24.5 at 22°C			temperature sensitive >22°C, RNA polymerase subunit	100

[a] The growth rate was measured at 25°C on standard race tube media containing 1 × Vogel's salts, 0.3% glucose, 0.5% arginine.
[b] Period length is reduced by increasing starvation for the required supplement.
[c] The period length of these strains can be altered by changing the supplementation of the medium.
Source: Ref. 87.

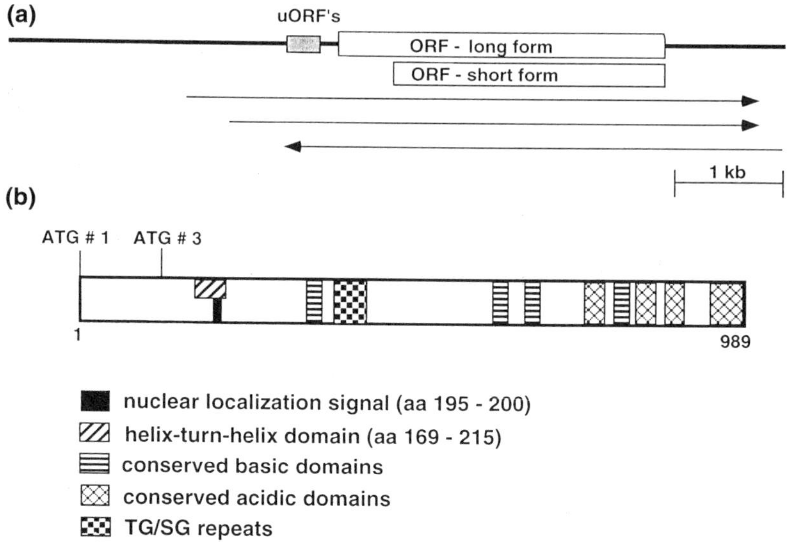

FIGURE 3 Schematic view of the *frq* gene and FRQ protein. (a) A 7.7-kb DNA fragment containing the *frq* gene. This is the minimal fragment that is capable of rescuing rhythmicity of FRQ null mutations. Both the long and the short ORFs and the upstream ORFs (uORFs) are indicated. The *frq* transcripts are shown below the fragment. The exact locations of the start and stop sites for each of the transcripts are not yet known. (b) Structural domains of FRQ are indicated, along with the alternative initiation codons used to produce the long (989 amino acids) and short (890 amino acids) forms of FRQ. (From Refs. 18 and 87.)

lation of *frq* levels in the cell from high to low, independent of time of day, causes the phase of the oscillator to be reset to dusk (the initial low point in the *frq* mRNA cycle). Thus, the level of *frq* defines the phase (state) of the clock.

Constant high levels of *frq* transcripts are observed at all times of the day in strains that have a deletion of the FRQ coding region, and repression of the native *frq* gene is observed in strains bearing an ectopic overexpressing version of *frq*. These results provided the first demonstration that FRQ protein is part of a negative-feedback loop that regulates the timing of its own synthesis [31]. Consistent with this notion, nuclear localization of FRQ is required for *frq* molecular rhythms and overt circadian rhythmicity [28].

Activation of *frq* transcription requires the products of the *white collar 1* (*wc1*) and *white collar 2* (*wc2*) genes [33]. These genes are involved in all known light responses in *Neurospora* [34] and are required for *frq* photoinduction and overt circadian rhythms [33]. Lesions in either gene were also found to prevent accumulation of *frq* transcripts in the dark, thereby preventing sustained *frq* mRNA and protein cycling. Recently, a temperature-sensitive allele of *wc2*, ER24 [35], was found to have both period and temperature compensation defects at 25°C (M. Collett and J. Dunlap, personal communication, 2000). In addition, the mutation results in reduced *frq* expression and long period *frq* oscillations. Together the data indicate that WC1/WC2 act each day at subjective dawn within the feedback cycle to activate *frq* transcription, at the time when FRQ protein has fallen below a critical level needed for repression.

In contrast to *frq* mRNA, the transcripts from *wc1* and *wc2* do not accumulate rhythmically. Despite this, WC1 protein is produced rhythmically with a peak time of accumulation around CT18, 180° out of phase with FRQ [36]. Recent experiments indicate that the WC1 rhythm results from FRQ acting at a posttranscriptional level to promote accumulation of WC1 at specific times of day. Thus, FRQ appears to have dual roles in the clock system—one to indirectly repress its own synthesis though WC1 and WC2, the other to positively affect WC1 accumulation. These multiple levels of control may strengthen the amplitude of the rhythm by increasing the resolution between the different phases of the oscillation. WC1 is present almost exclusively in the nucleus, whereas WC2 can be observed in both the nucleus and cytoplasm [37–39]. Both proteins are phosphorylated in response to light, and it is suggested that phosphorylation alters their activity, since the kinetics of WC1 phosphorylation correlates with light-induced gene expression [38].

One current model for the *frq* feedback loop is shown in Figure 4. At dawn, both *frq* mRNA and protein levels are low; however, the amount of *frq* transcript is increasing [25]. WC1 and WC2, which bind to each other though their PAS domains [37] and form a complex in vivo [39], activate *frq* transcription [33]. About 4–5 h later, *frq* mRNA reaches peak accumulation (just before noon) and the two forms of FRQ accumulate. A 4- to 6-h delay in maximal FRQ protein levels relative to the peak in *frq* mRNA is observed, wherein *frq* message levels begin to fall prior to FRQ protein reaching maximal accumulation. Soon after FRQ protein is synthesized, it enters the nucleus [28] and rapidly acts (within 3 h) [40] to keep *frq* mRNA levels low. This likely occurs by interfering with WC1/WC2, as experiments demonstrate an interaction between WC2 and FRQ (D. Denault, J.J. Loros, J.C. Dunlap, personal communication, 2000) [41]. For the rest of the day, and into the early evening, FRQ remains at sufficient levels in the nucleus to keep *frq* turned off and to act at a posttranscriptional level to increase the accumulation of WC1 in the evening [36]. Turnover of FRQ involves progressive phosphorylation over the day [25,42]. Once FRQ is fully phosphorylated, the protein is degraded. When the levels of FRQ fall below a critical level, *frq* can no longer be efficiently repressed and can in turn be reactivated via WC1 and WC2 [33] to complete the cycle.

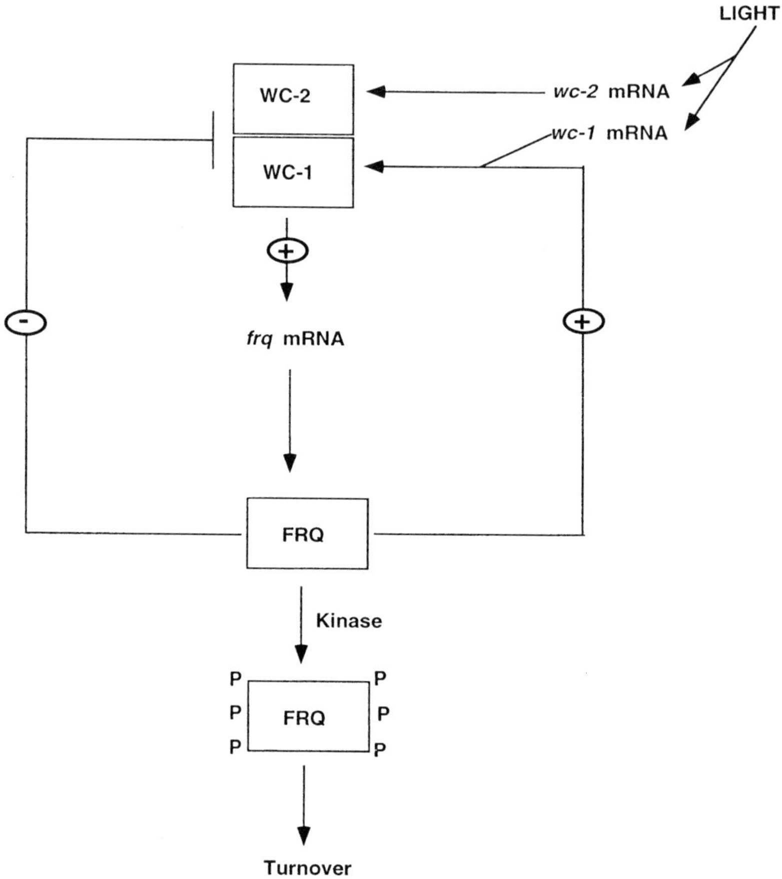

Figure 4 A model for the *Neurospora* FRQ-based oscillator. The specifics of the feedback loop are described in the text. Not included in this model are interactions between the FRQ-based oscillator and other putative oscillators in the cell (see also Fig. 1). (From Ref. 36.)

Time delays imposed within the molecular feedback loop are necessary to achieve stable circadian rhythms of gene expression. For *frq*, posttranscriptional regulation likely contributes to the time lags. Currently, we only have clues to the underpinnings of these temporal features. First, *frq* mRNA contains a rather long 5′ untranslated region (>1 kb) with six short upstream open reading frames (uORFs). Although deletion of the uORFs does not appear to eliminate overt rhythmicity or *frq* cycling (N. Garceau and J. Dunlap, personal communication, 1997), it is possible that uORFs participate in FRQ translational regulation under specific growth conditions [25]. Once FRQ protein is made, repression of *frq* transcription occurs rapidly and is completed within 3–6 h [40], whereas FRQ phosphorylation and decay takes up to 14–18 h. So, for most of the day, *frq* transcript levels are low and FRQ is present. A delay in nuclear entry of FRQ may only play a minor role in the time lags, since FRQ enters the nucleus within a few hours after synthesis [28]. A role for phosphorylation in FRQ turnover has been established [42] and points to an involvement of protein kinases in clock function. Investigations are under way to determine the nature of the kinases; however, obvious parallels between clocks in other organisms point to casein kinase I as a good candidate [43].

What we have unmasked regarding the molecular nature of the *Neurospora* clock is found to be, at least in overall outline, very similar to how oscillators are put together in other organisms. A transcription/translation-based feedback loop consisting of both positive and negative interacting loops has been shown to be present and required for clock activity in flies and mice, and similar clock components have even now been identified in humans [44].

5 SETTING THE TEMPERATURE LIMITS OF THE *NEUROSPORA* CONIDIATION RHYTHM

Experiments in *Neurospora* have suggested that temperature effects on the clock may be related to translational regulation of FRQ, wherein the growth temperature of the cultures determines how much of the long and short forms of FRQ are produced [26]. Either form can suffice for clock activity at some, but not all, temperatures, and elimination of either form reduces the temperature range permissive for rhythmicity. Specifically, at high temperatures (approaching 30°C) the total level of FRQ rises and translational initiation at ATG1 is favored, whereas at lower temperatures (~18°C) translational initiation at ATG3 is favored. The increased overall levels of FRQ observed at high temperature indicate that quantities sufficient for clock activity at low temperature are not adequate at higher temperatures. In addition, at low temperature, equivalent amounts of the long form of FRQ protein will not suffice for rhythmicity and at high temperature the short form will not suffice, implying that the two forms do in fact differ qualitatively as well as quantitatively. Together these data indicate that the temperature limits permissive for rhythmicity (between 18°C and 30°C) are influenced by overall FRQ levels (too little or too much FRQ stops the clock), and that activity at temperature extremes is determined by the different forms of FRQ.

6 RESETTING THE CLOCK BY ENVIRONMENTAL INPUT SIGNALS

The *frq* gene has been an indispensable tool for investigating the clock mechanism, and also for understanding the molecular effects of entraining stimuli on the clock [26,31–33,45]. In *Neurospora*, a flavin-mediated response involving a blue light photoreceptor (which has not been identified) is observed in all known light-regulated events [46,47], including light resetting of the circadian clock [5,14,48]. While various light input pathways exist in different organisms, all models for rhythmic entrainment propose that light acts to rapidly alter the activity of a central clock component to cause phase resetting [49]. Consistent with this prediction, the levels of *frq* mRNA increase within 5 min after a brief light pulse, and light induction can occur at any time of the day [32]. Normal light induction of *frq* requires the products of the *wc* genes [33]. Both WC1 and WC2 are present at some level in the dark, and it has been suggested that light causes a rapid modification of the preexisting proteins, possibly phosphorylation, to promote their dimerization [37]. The WC1/WC2 heterodimers are then able to activate transcription of light-regulated gene promoters, including the *frq* promoter.

A direct correlation was found between the light-induced levels of *frq* transcript and the magnitude of phase shifts in the conidiation rhythm resulting from the same light treatment [32]. These data provide an important link between the clock and the light input pathway. In addition, these data help to explain how a single light pulse can produce either a phase advance or a phase delay of a rhythm. Specifically, a light pulse given in the late night to early morning (when *frq* mRNA levels are either low or are rising) rapidly causes mRNA levels to reach their typical midday levels, resulting in an advance of the cycle (Fig. 5). Alternatively, a light pulse administered in the late day to early evening (when *frq* levels are falling) slows transcript decline and delays the next cycle. A light pulse during midday (when *frq* is already at peak levels) does not cause an appreciable change in the phase of the conidiation rhythm.

A light-to-dark transition sets the clock to dusk. Based on the feedback model (Fig. 4), one would predict that this transition causes a rapid decline in the levels of *frq* mRNA. This is exactly what is observed [32]. Similarly, the ability of the clock to be entrained to 24 h by a 12-h light–dark cycle implies that some aspect of the endogenous feedback loop is lengthened under these conditions. Based on the data showing light induction of *frq* mRNA, it seems likely that induction of *frq* via WC1 and WC2 is the portion of the loop that is prolonged.

WC1 and WC2 bear sequence similarity to the GATA family of transcription factors found in fungi and vertebrates, and both have been shown to bind to consensus GATA elements within the promoters of blue light–regulated genes in *Neurospora* [50,51]. As indicated above, both WC1 and WC2 contain PAS dimerization domains involved in homodimerization and heterodimerization of the proteins [37,52]. The PAS domain was first identified as a common motif among the *Drosophila* clock protein PER, mammalian ARNT (a dimerization partner of the dioxin receptor), and *S*IM (the product of the single-minded gene) [53]. This domain has since been shown to be involved in dimerization of PER

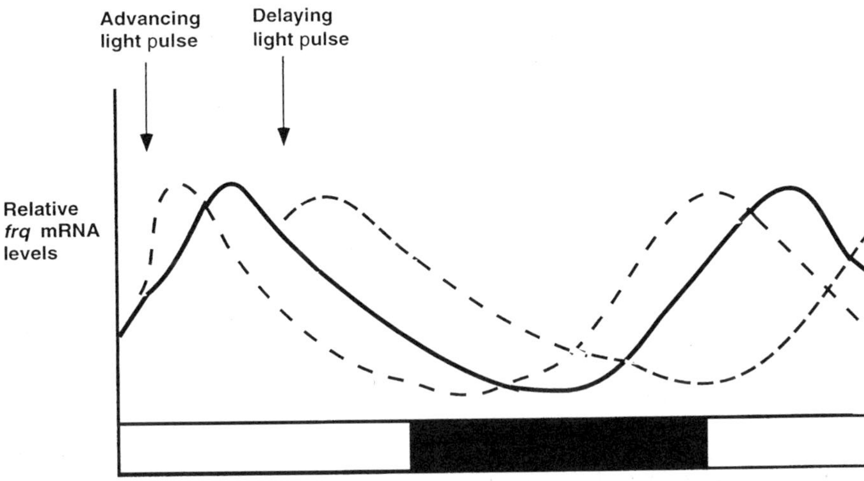

Figure 5 Diagram representing how light resets the FRQ-based feedback loop. Light causes a rapid increase in *frq* mRNA. A light pulse given in the early morning when *frq* levels are rising (advancing light pulse) causes the *frq* message to increase earlier than in cultures that have not seen light, and the clock advances to the midday phase. A light pulse given in the late day, when *frq* levels are declining (delaying light pulse), causes the a delay in the trough of *frq* message, and the clock delays back to the midday phase. The white bars below indicate subjective daytime and the black bars indicate subjective nighttime. The rhythm of *frq* mRNA in constant dark conditions is shown as a heavy solid black line, and the changes observed after a light pulse are shown as dotted lines. (From Ref. 32.)

to a second clock component, TIM [54]. The PAS motif, which generally consists of paired repeat sequences and is often coupled to a helix-loop-helix domain, is found in proteins involved in both the circadian clock [44,55] and in photoreception in bacteria and plants [56–58]. This suggests a possible evolutionary link between the clock's ability to anticipate daily light dark cycles and light-responsive proteins in prokaryotes.

Temperature changes also reset the *Neurospora* clock. Temperatures tend to increase at sunrise and decrease at dusk. So not surprisingly, a temperature step up resets the clock to dawn, and a temperature step down resets the clock to dusk [15,16,59]. However, unlike light treatments which result in increased *frq* message accumulation, temperature effects appear to be primarily mediated though translational control of FRQ. Temperature steps alter both the form and overall amount of FRQ in the cell [60]. Following a temperature shift, the change in phase of the oscillator is suggested to result from a rapid change in the amount

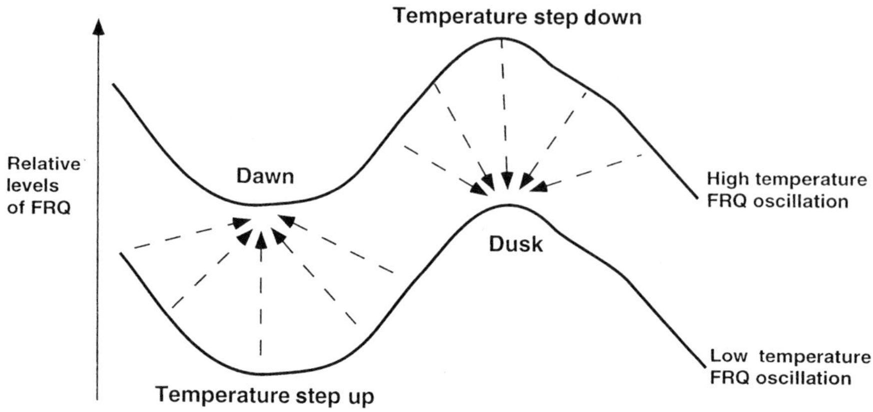

FIGURE 6 Diagram representing how temperature resets the FRQ-based oscillator. FRQ protein cycles at lower levels at low temperature (bottom curve) and at higher levels at high temperature (top curve). When the cultures are raised from low to high temperature, the clock is reset to the time corresponding to the low point in the new cycle (arrows pointing up), near dawn. When the temperature is changed from high to low, the clock is reset to the time corresponding to the high point in the new cycle (arrows pointing down), near dusk. (From Refs. 44 and 87.)

of FRQ (Fig. 6). When cells are shifted from 21°C to 28°C, the overall levels at which FRQ cycles are raised, resulting in the lowest level of FRQ in the cycle at 28°C being higher than any level at 21°C. Because dawn corresponds to the lowest point in the FRQ cycle, a step up would always initially be seen and interpreted as the lowest point of the FRQ cycle, and the clock would therefore be reset to dawn. In other words, the clock is reset to the time corresponding to the actual amount of FRQ in the cell as perceived in the context of the new temperature. Based on the feedback model (Fig. 4), one possible interpretation of these data is that at higher temperature, FRQ is less efficient in its ability to repress WC1 and WC2. This would allow *frq* mRNA levels to increase (mimicking dawn) despite the high levels of FRQ protein. Another implication from these experiments is that immediately after the shift, no synthesis or turnover of clock components is required [60]. Thus, unlike light changes, temperature changes reset the circadian oscillator immediately and from within the loop.

Although in nature light and temperature likely act synergistically to entrain the clock, light is typically considered to be the dominant entraining stimulus. Therefore, it was quite unexpected to find that under conditions in which light and temperature were able to compete, temperature steps were found to be more influential in entraining the *Neurospora* clock than light [45,60]. For example, when cultures were grown in constant saturating bright light (LL) at 25°C or 30°C and then transferred to constant dark (DD) at 30°C, the LL-to-DD transfer set the clock to dusk as predicted from the light-to-dark transition [60]. However, in cultures grown in LL at temperatures of 18°C or below, transfer to 30°C DD set the clock to dawn, as would be expected if temperature were the prevailing signal. At dawn, both *frq* mRNA and FRQ protein are low, whereas at dusk *frq* mRNA is low and FRQ protein levels are high. Therefore, at the molecular level, these data imply that the level or activity of FRQ protein is the overriding signal that sets the clock.

7 OUTPUT FROM THE CLOCK

Organisms have a clock to temporally control a vast array of cellular activities, yet little is known about how this regulation takes place, or of the clock proteins responsible for signaling time information to the rest of the cell. To describe circadian output pathways in *Neurospora*, genes that are rhythmically expressed (i.e., controlled by the clock) but that do not affect oscillator function when inactivated, were first targeted for isolation. The term "clock-controlled genes" (*ccgs*) was coined to describe them [61]. To date, eight *ccgs* have been identified as part of the output pathways by directed approaches [61,62], and expression of several additional genes has been shown to be rhythmic with circadian periods [63,64]. Aside from gene expression, a number of other clock outputs have been described in *Neurospora*, including oscillations in small molecules [13,65]. Verification of clock regulation for most of the genes was achieved by demonstrating that the period of the *ccg* mRNA abundance rhythm equals the period of the strain examined. Specifically, in the long period frq^7 background, which has an endogenous period of 29 h, the period of the peak in levels of *ccg* mRNAs approaches 29 h and eventually cycles 180° out of phase with the wild-type strain [61,62,66]. In all cases examined, the clock was shown to function normally in strains containing inactivated copies of the *ccgs*, demonstrating that they are part of an output pathway and are not involved in oscillator function [66–69] (Table 2).

In the initial screens for rhythmically expressed genes in *Neurospora*, only a few times of day were compared and the screens were not saturating; thus, the *ccgs* likely represent a small sampling of clock-regulated genes in *Neurospora*. Experiments using transcriptional profiling with DNA microarrays are currently being initiated, and this analysis will provide a means to determine the full extent of clock regulation of gene expression in *Neurospora*. In addition, changes in rhythmicity of the *ccgs* can be catalogued by comparing wild-type versus clock mutant strains to help distinguish among the different output signalling pathways.

Most of the known *Neurospora ccgs* peak in transcript accumulation in the late night to early morning, but they differ in overall expression levels and in

TABLE 2 Summary of *Neurospora* Clock-Controlled Genes

Gene	Average peak[a]	Identity[b]	Regulation Devel.[c]	Regulation Light	Ref.
ccg1	CT3	unknown	+	+	61
eas (ccg2)	CT22	hydrophobin	+	+	61, 66, 72
ccg4	CT5	pheromone	+	+	62
ccg6	CT19	unknown	+	+	62
ccg7	CT21	GAPDH	—	−	62, 69
ccg8	CT20	unknown	−	−	62
ccg9	CT19	trehelose synthase	+	+	62 68
cmt (ccg12)	CT18	CuMT	−	−	62, 101
al-3[d]	CT10	GGPPS	+	+	63
con6	ZT20	unknown	+	+	64
con10	ZT20	unknown	+	+	64

[a] The peak in message accumulation varies slightly in different experiments [62] and with the exception of *al3, con6,* and *con10* was determined from Northern blots of the same rhythmic RNA probed with the indicated *ccg*. The peak in *al3* message accumulation was estimated by us from Northern blots presented in Arpaia et al. [63], whereas *con6* and *con10* were shown to peak ~20 h after a light pulse representing zeitgeber time (ZT) 20 [64].
[b] Abbreviations are as follows: GAPDH, glyceraldehyde 3-phosphate dehydrogenase; CuMT, copper metallothionein; GGPPS, geranylgeranyl pyrophosphate synthase.
[c] Developmental and light regulation of the ccgs: + indicates increased transcription following developmental induction and light treatment;—indicates no effect.
[d] Only the longer *al3c* transcript has been demonstrated to be rhythmic [63].
Source: Ref. 87.

amplitude of the rhythm [62]. The approximate time of peak accumulation of the *ccgs* coincides with the time of initiation of conidiation, suggesting a role for some or all of these genes in the developmental pathway. In fact, many of the *ccgs* are induced during development and *eas(ccgs2)* is known to encode a component of the conidiospore (see below). This finding is interesting because the *ccg* were identified using mycelia grown in liquid shaking cultures in which the clock functions normally but development is curtailed [70]. This suggests that the initial steps of conidiation progress in liquid culture leading to some level of expression of the developmental genes. This could be by either positive or negative regulation of gene expression at specific times of day by the clock. For instance, some of the conidiation genes might normally be, by default, transcriptionally active, but repressed by the clock most times of the day. In this example, environmental signals that lead to robust conidiation at any time of day (e.g., light or C and N starvation) would therefore be suggested to override repression of conidiation genes by the clock. Isolating mutant strains that produce abundant conidia in liquid shake cultures could identify such a repressor. In fact, it appears that the clock exerts both positive and negative regulation of output genes. Examination of the *ccgs* for mRNA accumulation in a FRQ^+ versus a FRQ^- strain demonstrates that some *ccgs* accumulate significantly higher levels of message when FRQ is present in cells as compared to when FRQ is absent. Other *ccgs* show the opposite pattern of mRNA accumulation, indicating both positive and negative regulation by the clock (D. Bell-Pedersen, unpublished).

Not all of the *ccgs* (including *ccg7, -8*, and *-12*) are induced by light or developmental signals, indicating that clock-regulated output pathways distinct from conidiation exist in *Neurospora*. Indeed it was found that *ccg7* encodes glyceraldehyde 3-phosphate dehydrogenase (GAPDH), a key enzyme in glycolysis and gluconeogenesis [69], *ccg*12 encodes copper metallothionein, involved in metal storage and detoxification [62], and *ccg9* encodes trehalose synthase, important for stress protection [68]. Furthermore, *ccg4* has sequence similarity to the mating-type-specific ''alpha-like'' pheromone precursor gene *Mfl/l* from the Chestnut Blight fungus, *Cryphonectria parasitica* [71] (D. Bell-Pedersen, D. Ebbole, N. Van Alfen, unpublished data). These data suggest that *ccg4* encodes a pheromone involved in mating and supports a role for the *Neurospora* circadian clock in some aspects of the sexual cycle. Thus, even in this relatively simple eukaryote, the output pathways appear diverse.

The most highly characterized *Neurospora ccg* at both the biochemical and molecular levels is the *eas(ccg2)* gene. The *eas(ccg2)* gene encodes a small hydrophic protein (called a hydrophobin) that covers the outer surfaces of spores rendering them hydrophic and easily dispersed in air [66,72]. Nuclear run-on experiments demonstrated that *eas(ccg2)* is transcriptionally regulated by the circadian clock [73], implicating the involvement of *cis*-acting regulatory elements mediating temporal control. Subsequent dissection of the *eas(ccg2)* promoter localized a positive-*a*ctivating *c*lock *e*lement (ACE) to within a 45-bp fragment, found to be distinct from other light and developmental elements regulating its expression [74]. Using an unregulated promoter/reporter system, it was shown that the ACE element is sufficient to confer high amplitude rhythmicity on the reporter gene. The ACE sequences are being used to biochemically identify upstream regulatory factors responsible for cycling in attempts to trace the output pathway back to the oscillator (D. Bell-Pedersen, Z. Lewis, J. J. Loros, J. C. Dunlap, unpublished data). Using a labeled 68-bp *eas(ccg-2)* probe containing the ACE, factors present in nuclear extracts from light-grown (LL) *Neurospora* were found to interact specifically with these sequences. Examination of the binding factors at different times in the circadian day in either frq^+ (22-h period) or frq^7 (29-h period) strains revealed that the amount of binding and the mobility of the complexes changes with time. These data suggest that the amount or activity of the factors, modification of the factors, or the addition of accessory factors is rhythmic and is consistent with these proteins having a role in clock control of the *eas(ccg2)* gene.

Circadian regulation of *eas(ccg2)* appears to be through positive activation by the clock. Deletion of ACE results in constant low-level transcript accumulation over the course of the day, and the maximal level of factor binding to the ACE element occurs in the morning, at the time of day when *eas(ccg2)* mRNA is at its peak [74].

One mechanism by which some *ccgs* are predicted to be rhythmically controlled is directly though transcription factors that are known to be components of the oscillator. In fact, this was recently demonstrated in mice. The positive PAS-containing CLOCK/BMAL heterodimers were found to activate transcription of the rhythmically expressed arginine vasopressin gene [75]. In addition, CLOCK was shown to directly regulate circadian expression of the transcription factor DBP [76]. It is not known if the positive elements (WC1 and WC2) and/or the negative element (FRQ) of the *Neurospora* FRQ-based oscillator directly regulates rhythmicity of any of the output genes.

In several systems it has been demonstrated that output pathways feed back on the central oscillator [9,77,78]. Mutations in known *Neurospora ccgs*, however, have not been shown to affect the period of the rhythm. Even mutations that abolish conidiation at early stages do not abolish aerial hyphae formation (A. Correa and D. Bell-Pedersen, unpublished data) [79], although there are no mutations in genes that are known to specifically abolish aerial hypha formation. However, Ramsdale and Lakin-Thomas [65] recently provided the first suggestion of feedback from an output to an oscillator in *Neurospora*. They demonstrated circadian rhythms in diacylglycerol (DAG) levels and showed that DAG levels are high in a *chol1* mutant strain that has a long, noncircadian period of 60 h on minimal media lacking choline, suggesting that a correlation might exist between DAG levels and period. The addition of membrane-permeable DAG and inhibitors of DAG kinase further lengthened the period in this strain, hinting that DAG may feedback on the time-keeping mechanism to lengthen the period.

8 COMPLEXITY OF THE *NEUROSPORA* CIRCADIAN SYSTEM

Under most growth conditions, sustained conidiation rhythms are lost in the absence of the FRQ protein. However, under certain media and temperature conditions, FRQ-deficient strains display a conidiation rhythm that ranges between 12 and 30 h [19,20]. To explain this residual rhythmicity, the presence of additional oscillators in the *Neurospora* cell has been invoked [8,45]; however, the nature of the putative additional oscillator(s) has not been established.

One hypothesis was that if the residual rhythmicity in *frq*-less strains results from low-amplitude, uncompensated, or damped oscillations, perhaps an entraining cycle could bestow an amplifying effect on the rhythm. Indeed, null mutants of *frq* were found to entrained by temperature cycles [45]. These data suggested that the entrainment has allowed a cryptic, temperature-entrainable oscillator to be uncovered in the absence of the *frq*-based feedback loop [45,80]. Further support for multiple oscillators comes from double-mutant studies of *chol1* or *cel* and *frq* or *wc* nulls. The double-mutant strains are arrhythmic with full supplementation, but display a long period rhythm on media where the period lengthening effects of the *cel* or *chol1* mutation are observed [81–83]. With appropriate supplementation, the *cel* and *chol1* mutations can cause a robust long-period conidiation rhythm (albeit outside of the circadian range) in *frq*-null (or *wc*-null) strains with the same period as the *cel* and *chol1* single mutants. These data provide additional evidence for the existence of a second oscillator, and further suggest a linkage of this oscillator to cellular metabolism [83]. The two oscillators are likely to be coupled, since the period of the system is affected by the *frq* allele. For example, the short period *frq*^{1} allele shortens the long period observed in the *chol1* or *cel* backgrounds [82]. However, when FRQ is absent, the rhythms lose some circadian characteristics, including light entrainment and compensation for changes in temperature and metabolic state.

Spatial complexity is another consideration in the *Neurospora* circadian system. Previously, Dharmananda and Feldman [84] demonstrated light-sensitive circadian oscillators in different parts of the fungal mycelium. Furthermore, Lakin-Thomas et al. [83,85] reported rhythms in both the determination and differentiation stages of conidiation. Interestingly, these two rhythms do not appear to be tightly coupled to each other.

In summary, these data indicate that the circadian system comprises a population of oscillatory systems. However, while a multiple oscillator model can be imposed on the physiological and genetic data, the lack of molecular data still holds the connection between the FRQ oscillator and the rest of the cell a mystery. In particular, all of what we know about the independent role of the other oscillator(s) is derived from the ability, though mutation and genetically engineered strains, to manipulate or to eliminate altogether the FRQ feedback loop. Until the other oscillators can be similarly manipulated, we are constrained to modeling and phenomenology. Thus, one goal now is to identify components of the other oscillator(s), and we may already have some clues. Genetic data indicate a possible role of the *prd6* gene in coupling of the FRQ-based oscillator to a temperature-dependent metabolic oscillator [21]. Mutations in *prd6* have an increased range of temperature compensation, suppress the temperature compensation defects of other mutations, and are resistant to some media conditions previously shown to affect period.

9 SUMMARY AND FUTURE PROSPECTS

Solving the mechanisms of the circadian clock has become an important goal, mainly because of the ubiquity of clocks and their role in many organisms' lives, including humans. The past few years have seen significant advances in our understanding of the mechanisms of circadian rhythmicity, with the molecular genetic analysis of clocks in *Neurospora* continuing to provide major contributions to the story. Clock genes have been identified though mutation, and are currently being studied to determine their potential roles in generating rhythms. Eventually, these studies will allow more specific comparisons among clock systems in diverse organisms.

These studies have also reaffirmed that the circadian system is complex, likely comprising more than one oscillator that drives diverse output rhythms. We know little about the signaling pathways from the environment that reset the oscillator(s) or about signaling to the output genes. Genetic and biochemical screens, as well as functional genomic studies, are under way to identify critical components residing in these pathways. In addition, the biochemical function of FRQ is still unknown, and determination of this will likely provide important information into the workings of the clock.

Some additional questions include the following. What is the role of the two forms of FRQ and are these involved in the temperature compensation mechanism of the oscillator? What are the kinases that phosphorylate FRQ? How many genes are regulated by the clock and at what times of day? So, despite the extraordinary progress being made, there is still plenty to keep us busy for a long, long time.

ACKNOWLEDGMENTS

I am grateful to Dr. Richard Gomer for his helpful comments on the chapter, and members of my laboratory for discussions. I also thank members of the *Neurospora* clock community for sharing ideas and unpublished data. Studies in the author's laboratory are supported by NIH (RO1 GM58529-01, P01 NS39546) and a Texas A&M University Interdisciplinary Grant.

REFERENCES

1. LN Edmunds. Cellular and Molecular Bases of Biological Clocks. New York: Springer-Verlag, 1988.
2. CH Johnson, SS Golden. Circadian programs in cyanobacteria: adaptiveness and mechanism. Annu Rev Microbiol 53:389–409, 1999.
3. T Kondo, M Ishiura. The circadian clock of cyanobacteria. Bioessays 22:10–15, 2000.
4. Y Ouyang, CR Andersson, T Kondo, SS Golden, CH Johnson. Resonating circadian

clocks enhance fitness in cyanobacteria. Proc Natl Acad Sci USA 95:4475–4480, 1998.

5. S Dharmananda. Studies of the Circadian Clock of *Neurospora crassa*: Light Induced Phase Shifting. Santa Cruz: University of California, 1980.
6. AT Winfree. Integrated view of resetting a circadian clock. J Theor Biol 28:327–374, 1970.
7. JW Hastings, BM Sweeney. On the mechanism of temperature independence in a biological clock. Proc Natl Acad Sci USA 43:804–811, 1957.
8. J Dunlap. Circadian rhythms. An end in the beginning. Science 280:1548–1549, 1998.
9. ED Herzog, GD Block. Keeping an eye on retinal clocks. Chronobiol Int 16:229–247, 1999.
10. D Bell-Pedersen, SK Crosthwaite, PL Lakin-Thomas, MW Merrow, M Vinsjevik. The *Neurospora* circadian clock—simple or complex? Phil Trans R Soc. In press.
11. CS Pittendrigh, BG Bruce, NS Rosensweig, ML Rubin. Growth patterns in *Neurospora*. Nature 184:169–170, 1959.
12. ML Sargent, WR Briggs, DO Woodward. Circadian nature of a rhythm expressed by an invertaseless strain of *Neurospora crassa*. Plant Physiol 41:1343–1349, 1966.
13. PL Lakin-Thomas, GG Cote, S Brody. Circadian rhythms in *Neurospora crassa*: biochemistry and genetics. Crit Rev Microbiol 17:365–416, 1990.
14. H Nakashima, JF Feldman. Temperature-sensitivity of light-induced phase shifting of the circadian clock of *Neurospora*. Photochem Photobiol 32:247–251, 1980.
15. VD Gooch, RA Wehseler, CG Gross. Temperature effects on the resetting of the phase of the *Neurospora* circadian rhythm. J Biol Rhythms 9:83–94, 1994.
16. L Rensing, A Kallies, G Gebauer, S Mohsenzadeh. The effects of temperature change on the circadian clock of *Neurospora*. Ciba Found Symp 183:26–41, 1995.
17. JF Feldman. Genetic approaches to circadian clocks. Annu Rev Plant Physiol 33: 583–608, 1982.
18. JC Dunlap. Genetics and molecular analysis of circadian rhythms. Annu Rev Genet 30:579–601, 1996.
19. JJ Loros, JF Feldman. Loss of temperature compensation of circadian period length in the *frq-9* mutant of *Neurospora crassa*. J Biol Rhythms 1:187–198, 1986.
20. BD Aronson, KA Johnson, JC Dunlap. Circadian clock locus frequency: protein encoded by a single open reading frame defines period length and temperature compensation. Proc Natl Acad Sci USA 91:7683–7, 1994.
21. LW Morgan, JF Feldman, D Bell-Pedersen. Genetic interactions between clock mutations in *Neurospora crassa*: can they help us to understand complexity? Phil Trans R Soc. In press.
22. B Chang, H Nakashima. Isolation of temperature sensitive rhythm mutants in *Neurospora crassa*. Genes Genet Syst 73:71–73, 1998.
23. CR McClung, BA Fox, JC Dunlap. The *Neurospora* clock gene *frequency* shares a sequence element with the *Drosophila* clock gene period. Nature 339:558–562, 1989.
24. H Nakashima, K Onai. The circadian conidiation rhythm in *Neurospora crassa*. Semin Cell Dev Biol 7:765–774, 1996.

25. NY Garceau, Y Liu, JJ Loros, JC Dunlap. Alternative initiation of translation and time-specific phosphorylation yield multiple forms of the essential clock protein FREQUENCY. Cell 89:469–476, 1997.
26. Y Liu, NY Garceau, JJ Loros, JC Dunlap. Thermally regulated translational control of FRQ mediates aspects of temperature responses in the *Neurospora* circadian clock. Cell 89:477–486, 1997.
27. MT Lewis, JF Feldman. The putative FRQ clock protein of *Neurospora crassa* contains sequence elements that suggest a nuclear transcriptional regulatory role. Protein Sequence Data Anal 5:315–323, 1993.
28. C Luo, JJ Loros, JC Dunlap. Nuclear localization is required for function of the essential clock protein FRQ. EMBO J 17:1228–1235, 1998.
29. MW Merrow, JC Dunlap. Intergeneric complementation of a circadian rhythmicity defect: phylogenetic conservation of structure and function of the clock gene *frequency*. EMBO J 13:2257–2266, 1994.
30. MT Lewis, LW Morgan, JF Feldman. Analysis of *frequency (frq)* clock gene homologs: evidence for a helix-turn-helix transcription factor. Mol Gen Genet 253:401–414, 1997.
31. BD Aronson, KA Johnson, JJ Loros, JC Dunlap. Negative feedback defining a circadian clock: autoregulation of the clock gene *frequency*. Science 263:1578–1584, 1994.
32. SK Crosthwaite, JJ Loros, JC Dunlap. Light-induced resetting of a circadian clock is mediated by a rapid increase in *frequency* transcript. Cell 81:1003–1012, 1995.
33. SK Crosthwaite, JC Dunlap, JJ Loros. *Neurospora wc-1* and *wc-2*: transcription, photoresponses, and the origins of circadian rhythmicity. Science 276:763–769, 1997.
34. RW Harding, SW Shopshire Jr. Photocontrol of carotenoid biosynthesis. Annu Rev Plant Physiol 31:217–238, 1980.
35. F Degli-Innocenti, VE Russo. Isolation of new *white collar* mutants of *Neurospora crassa* and studies on their behavior in the blue light–induced formation of protoperithecia. J Bacteriol 159:757–761, 1984.
36. K Lee, JJ Loros, JC Dunlap. Interconnected feedback loops in the *Neurospora* circadian system. Science 289:107–110, 2000.
37. P Ballario, C Talora, D Galli, H Linden, G Macino. Roles in dimerization and blue light photoresponse of the PAS and LOV domains of *Neurospora crassa* WHITE COLLAR proteins. Mol Microbiol 29:719–729, 1998.
38. C Schwerdtfeger, H Linden. Localization and light-dependent phosphorylation of *WHITE COLLAR-1 and -2*, the two central components of blue light signaling in *Neurospora crassa*. Eur J Biochem 267:414–422, 2000.
39. C Talora, L Franchi, H Linden, P Ballario, G Macino. Role of a WHITE COLLAR-1–WHITE COLLAR-2 complex in blue-light signal transduction. EMBO J 18: 4961–4968, 1999.
40. MW Merrow, NY Garceau, JC Dunlap. Dissection of a circadian oscillation into discrete domains. Proc Natl Acad Sci USA 94:3877–3882, 1997.
41. M Merrow, L Franchi, Z Dragovic, M Gorl, J Johnson, M Brunner, G Macino, T Roenneberg. Circadian regulation of the light input pathway in *Neurospora crassa*. EMBO J 20:307–315, 2001.
42. Y Liu, J Loros, JC Dunlap. Phosphorylation of the *Neurospora* clock protein FRE-

QUENCY determines its degradation rate and strongly influences the period length of the circadian clock. Proc Natl Acad Sci USA 97:234–239, 2000.
43. B Kloss, JL Price, L Saez, J Blau, A Rothenfluh, CS Wesley, MW Young. The *Drosophila* clock gene double-time encodes a protein closely related to human casein kinase I epsilon. Cell 94:97–107, 1998.
44. JC Dunlap. Molecular bases for circadian clocks. Cell 96:271–290, 1999.
45. M Merrow, M Brunner, T Roenneberg. Assignment of circadian function for the *Neurospora* clock gene *frequency*. Nature 399:584–586, 1999.
46. E Klemm, H Ninnemann. Photoreceptors for circadian rhythms. Photochem Photobiol Rev 4:207–265, 1979.
47. J Paietta, ML Sargent. Photoreception in *Neurospora crassa*: correlation of reduced light sensitivity with flavin deficiency. Proc Natl Acad Sci USA 78:5573–5577, 1981.
48. BJ Fritz, S Kasai, K Matsui. Blue light photoreception in *Neurospora* circadian rhythm: evidence for involvement of the flavin triplet state. Photochem Photobiol 51:607–610, 1990.
49. CS Pittendrigh. Circadian rhythms and the circadian organization of living things. Cold Spring Harbor Symp Quant Biol 25:159–184, 1960.
50. P Ballario, P Vittorioso, A Magrelli, C Talora, A Cabibbo, G Macino. WHITE COLLAR-1, a central regulator of blue light responses in *Neurospora*, is a zinc finger protein. EMBO J 15:1650–1657, 1996.
51. H Linden, P Ballario, G Macino. Blue light regulation in *Neurospora crassa*. Fungal Genet Biol 22:141–150, 1997.
52. P Ballario, G Macino. WHITE COLLAR proteins: PASsing the light signal in *Neurospora crassa*. Trends Microbiol 5:458–462, 1997.
53. ZJ Huang, I Edery, M Rosbash. PAS is a dimerization domain common to *Drosophila* period and several transcription factors. Nature 364:259–262, 1993.
54. N Gekakis, L Saez, AM Delahaye-Brown, MP Myers, A Sehgal, MW Young, CJ Weitz. Isolation of timeless by PER protein interaction: defective interaction between timeless protein and long-period mutant PERL. Science 270:811–815, 1995.
55. AJ Millar. Circadian rhythms: PASsing time. Curr Biol 7:R474–R476, 1997.
56. GE Borgstahl, DR Williams, ED Getzoff. A structure of photoactive yellow protein, a cytosolic photoreceptor: unusual fold, active site, and chromophore. Biochemistry 34:6278–6287, 1995.
57. M Koh, G Van Driessche, B Samyn, WD Hoff, TE Meyer, MA Cusanovich, JJ Van Beeumen. Sequence evidence for strong conservation of the photoactive yellow proteins from the halophilic phototrophic bacteria *Chromatium salexigens* and *Rhodospirillum salexigens*. Biochemistry 35:2526–2534, 1996.
58. BL Taylor, IB Zhulin. PAS domains: internal sensors of oxygen, redox potential, and light. Microbiol Mol Biol Rev 63:479–506, 1999.
59. CD Francis, ML Sargent. Effects of temperature perturbations on circadian conidiation in *Neurospora*. Plant Physiol 64:1000–1004, 1979.
60. Y Liu, M Merrow, JJ Loros, JC Dunlap. How temperature changes reset a circadian oscillator. Science 281:825–829, 1998.
61. JJ Loros, SA Denome, JC Dunlap. Molecular cloning of genes under control of the circadian clock in *Neurospora*. Science 243:385–388, 1989.

62. D Bell-Pedersen, ML Shinohara, JJ Loros, JC Dunlap. Circadian clock–controlled genes isolated from *Neurospora crassa* are late night- to early morning-specific. Proc Natl Acad Sci USA 93:13096–13101, 1996.
63. G Arpaia, A Carattoli, G Macino. Light and development regulate the expression of the *albino-3* gene in *Neurospora crassa*. Dev Biol 170:626–635, 1995.
64. FR Lauter, C Yanofsky. Day/night and circadian rhythm control of con gene expression in *Neurospora*. Proc Natl Acad Sci USA 90:8249–8253, 1993.
65. M Ramsdale, PL Lakin-Thomas. sn-1,2-diacylglycerol levels in the fungus *Neurospora crassa* display circadian rhythmicity. J Biol Chem 275:27541–27550, 2000.
66. D Bell-Pedersen, JC Dunlap, JJ Loros. The *Neurospora* circadian clock–controlled gene, *ccg-2*, is allelic to *eas* and encodes a fungal hydrophobin required for formation of the conidial rodlet layer. Genes Dev 6:2382–2394, 1992.
67. KM Lindgren. Characterization of *ccg-1*, a Clock-Controlled Gene of *Neurospora crassa*. HANOVER, NH: Dartmouth Medical School, 1994.
68. ML Shinohara, A Correa, D Bell-Pedersen, JC Dunlap, JJ Loros. *Neurospora clock-controlled gene-9 (ccg-9)* encodes trehelose synthase: circadian regulation of stress responses and development. Euk Cell. In press.
69. ML Shinohara, JJ Loros, JC Dunlap. Glyceraldehyde-3-phosphate dehydrogenase is regulated on a daily basis by the circadian clock. J Biol Chem 273:446–452, 1998.
70. J Perlman, H Nakashima, JF Feldman. Assay and characteristics of circadian rhythmicity in liquid cultures of *Neurospora crassa*. Plant Physiol 67:404–407, 1981.
71. L Zhang, RA Baasiri, NK Van Alfen. Viral repression of fungal pheromone precursor gene expression. Mol Cell Biol 18:953–959, 1998.
72. FR Lauter, VE Russo, C Yanofsky. Developmental and light regulation of *eas*, the structural gene for the rodlet protein of *Neurospora*. Genes Dev 6:2373–2381, 1992.
73. JJ Loros, JC Dunlap. *Neurospora crassa* clock-controlled genes are regulated at the level of transcription. Mol Cell Biol 11:558–563, 1991.
74. D Bell-Pedersen, JC Dunlap, JJ Loros. Distinct *cis*-acting elements mediate clock, light, and developmental regulation of the *Neurospora crassa eas (ccg-2)* gene. Mol Cell Biol 16:513–521, 1996.
75. X Jin, LP Shearman, DR Weaver, MJ Zylka, GJ de Vries, SM Reppert. A molecular mechanism regulating rhythmic output from the suprachiasmatic circadian clock. Cell 96:57–68, 1999.
76. JA Ripperger, LP Shearman, SM Reppert, U Schibler. CLOCK, an essential pacemaker component, controls expression of the circadian transcription factor DBP. Genes Dev 14:679–689, 2000.
77. VM Cassone, WS Warren, DS Brooks, J Lu. Melatonin, the pineal gland, and circadian rhythms. J Biol Rhythms 8:S73–S81, 1993.
78. E Gwinner, M Hau, S Heigl. Melatonin: generation and modulation of avian circadian rhythms. Brain Res Bull 44:439–444, 1997.
79. CL Martens, ML Sargent. Circadian rhythms of nucleic acid metabolism in *Neurospora crassa*. J Bacteriol 117:1210–1215, 1974.
80. H Iwasaki, JC Dunlap. Microbial circadian oscillatory systems in *Neurospora* and *Synechococcus*: models for cellular clocks. Curr Opin Microbiol 3:189–196, 2000.

81. PL Lakin-Thomas, S Brody. Circadian rhythms in *Neurospora crassa*: interactions between clock mutations. Genetics 109:49–66, 1985.
82. PL Lakin-Thomas. Choline depletion, *frq* mutations, and temperature compensation of the circadian rhythm in *Neurospora crassa*. J Biol Rhythms 13:268–277, 1998.
83. PL Lakin-Thomas, S Brody. Circadian rhythms in *Neurospora crassa*: lipid deficiencies restore robust rhythmicity to null *frequency* and *white-collar* mutants. Proc Natl Acad Sci USA 97:256–261, 2000.
84. S Dharmananda, JF Feldman. Spatial distribution of circadian clock phase in aging cultures of *Neurospora crassa*. Plant Physiol 63:1049–1054, 1979.
85. PL Lakin-Thomas. Circadian rhythms: new functions for old clock genes. Trends Genet 16:135–142, 2000.
86. JF Feldman. Genetics of circadian clocks. Bioscience 33:426–431, 1983.
87. D Bell-Pedersen. Understanding circadian rhythmicity in *Neurospora crassa*: from behavior to genes and back again. Fungal Genet Biol 29:1–18, 2000.
88. JF Feldman, MN Hoyle. Isolation of circadian clock mutants of *Neurospora crassa*. Genetics 75:605–613, 1973.
89. GF Gardner, JF Feldman. The *frq* locus in *Neurospora crassa*: a key element in circadian clock organization. Genetics 96:877–886, 1980.
90. GF Gardner, JF Feldman. Temperature compensation of circadian period length mutants of *Neurospora crassa*. Plant Physiol 68:1244–1248, 1981.
91. JJ Loros, A Richman, JF Feldman. A recessive circadian clock mutation at the *frq* locus of *Neurospora crassa*. Genetics 114:1095–1110, 1986.
92. JF Feldman, CA Atkinson. Genetic and physiological characteristics of a slow-growing circadian clock mutant of *Neurospora crassa*. Genetics 88:255–265, 1978.
93. JF Feldman, GF Gardner. Genetic analysis of the circadian clock of *Neurospora*. In: M Suda ed. Biological Rhythms and Their Central Mechanism. Amsterdam: Elsevier, 1979, pp 57–66.
94. LW Morgan, JF Feldman. Isolation and characterization of a temperature-sensitive circadian clock mutant of *Neurospora crassa*. Genetics 146:525–530, 1997.
95. D Mattern, S Brody. Circadian rhythms in *Neurospora crassa*: effects of saturated fatty acids. J Bacteriol 139:977–983, 1979.
96. S Brody, L MacKensie, L Chuman. Circadian rhythms in *Neurospora crassa*: the effects of mitochondrial mutations and inhibitors. Genetics 116:S30, 1987.
97. JF Feldman, R Widelitz. Manipulation of circadian periodicity in cysteine auxotrophs of *Neurospora crassa*. Am Soc Microbiol 1977, Abstract 158.
98. K Onai, H Nakashima. Mutation of the *cys-9* gene, which encodes thioredoxin reductase, affects the circadian conidiation rhythm in *Neurospora crassa*. Genetics 146:101–110, 1997.
99. C Diekmann, S Brody. Circadian rhythms in *Neurospora crassa*: oligomycin-resistant mutations affect periodicity. Science 207:896–898, 1980.
100. K Onai, S Katagiri, M Akiyama, H Nakashima. Mutation of the gene for the second-largest subunit of RNA polymerase I prolongs the period length of the circadian conidiation rhythm in *Neurospora crassa*. Mol Gen Genet 259:264–271, 1998.
101. K Munger, UA Germann, K Lerch. The *Neurospora crassa* metallothionein gene. Regulation of expression and chromosomal location. J Biol Chem 262:7363–7367, 1987.

9

Sexual Development in Ascomycetes
Fruit Body Formation of *Aspergillus nidulans*

Gerhard H. Braus, Sven Krappmann, and Sabine E. Eckert*
Georg-August-Universität, Göttingen, Germany

1 INTRODUCTION

Sexual development is the classical parameter for the taxonomical classification of animals, plants, and fungi. This is reflected in the nomenclature of numerous taxons of the fungal kingdom. The typical feature of many ascomycetes as a result of their sexual reproductive cycle is a saclike bag, which is called the ascus (Gr. *askos* = sac, goat skin). The ascus is filled with ascospores (Gr. *askos*; *spora* = seed, spore), which are formed in a process originally termed "free-cell formation" [1]. These ascospores are meiospores and are the final products of a complex series of events. The number of ascospores within an ascus varies between one and >1000 depending on the species. In many ascomycetes, ascosporogenesis results in either four or eight ascospores. Many ascomycetes do not seem to reproduce sexually, which is a principal problem for taxonomists. Since ascomycetes—and many other fungi—were originally classified primarily on the basis of their sexual reproduction, an artificial classification group was

* *Current affiliation*: Imperial College of Science, Technology and Medicine, London, England

created for organisms lacking a sexual life phase [2]. This classification as Deuteromycota or *Fungi imperfecti* denotes their status as second (deutero)-class members in comparison to sexually reproducing organisms which were considered as ''perfect.'' For some fungi, this taxon was only a transient classification until proper conditions for sexual reproduction were discovered. When morphological, biochemical, and genetic criteria are applied, many members of this group are found to be closely related to ascomycetes. Thus, species without recorded sexual cycle are now usually grouped in their respective related ''teleomorphic'' families and designated as ''anamorphic'' species. For more than a century it was assumed that many deuteromycetes, including many *Aspergillus* spp., have completely lost their sexual cycle [2,3]. For some imperfect fungi this view might still hold true; for others, e.g., the well-studied human pathogen *Candida albicans*, however, recent genomics studies have challenged this view owing to the discovery of mating type–specific genes. It was also shown that *C. albicans* could be forced to mate, suggesting that necessary elements of a sexual cycle are retained in its genome. Since the genome sequencing project of *C. albicans* also uncovered homologs of genes required for meiosis, a full sexual cycle could—if only rarely—happen in nature [4,5]. Presumably, the analyses of additional fungal genomes will soon shed more light on the presence or absence of sexual life phases of numerous anamorphic fungi.

The sexual life phase represents a specific reproduction cycle which is one of the most important biological processes due to the potential of rearranging genetic information. Furthermore, many ascomycetes exhibit a parasexual cycle by which recombination can be accomplished without sexual reproduction [6]. Meiotic events generally include the rearrangement of chromosomes as well as recombination events within chromosomes or the repair of specific genes. Sexual reproduction includes a series of distinguishable events:

1. A first prerequisite of this process is to move two compatible nuclei in close proximity. Therefore, specialized mating structures, which are the result of *sexual differentiation*, are often formed.
2. For the rearrangement of genetic material, haploid nuclei of the sexual partners have to be brought together in one cell. The process is called *plasmogamy*. The resulting dikaryotic cell is termed heterokaryotic when different nuclei are fused and heteroplasmic when both partners provide cytoplasm.
3. The third step of sexual reproduction is the formation of a diploid by fusion of the two nuclei in a process called *karyogamy*.
4. *Meiosis* reduces the diploid genome to haploidy in a complex series of events which includes DNA replication, recombination, and regrouping of chromosomes.
5. Finally, during *ascosporogenesis*, the haploid nuclei have to be packaged into ascospores as a starting point for a new individual. Since ascospores often function as dormant spores, they can play an important role for survival

until growth can be resumed under more favorable conditions. Some fungi lacking a regular sexual cycle produce alternative survival structures—e.g., sclerotia.

The ascomycetes have developed different modes to produce asci. Hemiascomycetes, which primarily live as yeasts, normally form naked asci. Euascomycetes predominantly form hyphae, and there are only a few exceptional filamentous ascomycetes that do not develop fruit bodies. The fruit body or *ascocarp* (Gr. *askos* = sac; *karpos* = fruit) surrounds the asci in a characteristic manner. Three major different morphological structures of fruiting bodies can be distinguished (Fig. 1): *Cleistothecia* (Gr. *kleistos* = closed; *theke* = case) are completely closed structures enveloping the asci. *Perithecia* (Gr. *peri* = around; *theke* = case) are more or less closed, but at maturity a pore or ostiole is provided through which the ascospores can be released. *Apothecia* (Gr. *apotheke* = storehouse) are open ascocarps. In some species, asci are formed in a cavity within a cushion of somatic interwoven hyphae, resulting in a pseudothecium.

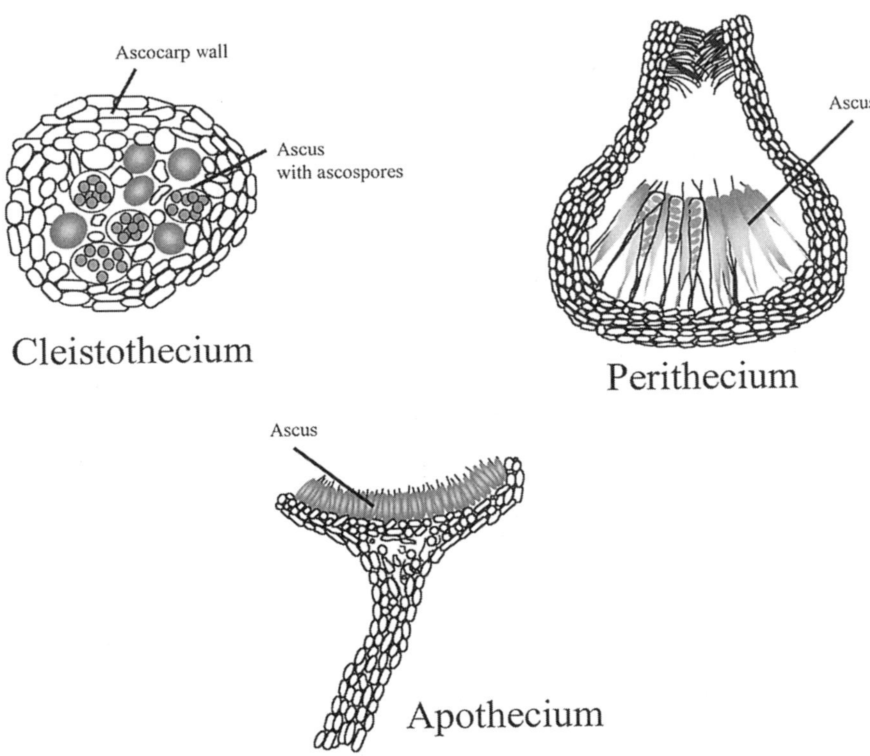

FIGURE 1 Morphological structures of fruit bodies from ascomycetes.

An important characteristic of the sexual life of an ascomycete is the potential choice of mating partners. Some fungi require an individual different from themselves for the sexual cycle. These fungi are *heterothallic*. Often, there are two different *mating types* at a single gene locus which regulate compatibility [7]. When two alternative DNA sequences located at the same chromosomal locus determine whether mating can take place, the different mating types are called *idiomorphs*. In contrast, other fungi are self-fertile and therefore able to complete the sexual life cycle in the absence of a mating partner. These fungi are *homothallic*. Some yeasts, including *Saccharomyces cerevisiae* and *Schizosaccharomyces pombe* and a few filamentous ascomycota as Chromocrea, Sclerotinea, and Glomorella [8], are homothallic despite possessing different mating types. They undergo regular mating-type switching using two silent copies of the mating-type regulatory genes as template. Switching of the mating type can function in a transpositionlike event including the excision of the active gene from the mating-type locus and its replacement by a new synthesized copy of the silent locus determining the opposite mating type [9,10]. Other species, e.g., *Neurospora tetrasperma*, are secondary homothallics or pseudohomothallics. They behave as homothallic forms since the binucleate ascospores contain one nucleus of each mating type. These different mechanisms ensure the presence of a mixture of the two mating types within a population.

Sexual reproduction and consequently the formation of ascospores usually follow a determined time course, although external conditions dictate whether the sexual reproductive pathway can be initiated. Crucial environmental factors include physical parameters such as temperature and light, the nutritional status (e.g., the availability of nitrogen and carbon sources), and the presence of an appropriate compatible sexual partner perceived by the interplay between sexual pheromones from one partner and suitable receptors of the other partner. All environmental parameters have to interact with the genetic determinants of the individual fungus. This crosstalk decides whether the sexual program will be initiated or alternative programs such as hyphal growth, dimorphic changes between hyphal and yeast growth, the asexual sporulation program, or the parasexual cycle are to be launched.

The ascomycetes are presumably the largest group among the fungi. For most members of this group our knowledge concerning the sexual life is rather scarce and rarely advances to the molecular level. In this chapter, we will primarily focus on sexual differentiation, fruit body formation, and ascosporogenesis, which are processes specific for filamentous fungi. We will not discuss plasmogamy, karyogamy, or meiosis, which can be found and studied in other, similar model systems. Besides a growing number of more recently established systems, there are two major model organisms of basic research among the filamentous fungi which have been studied for decades.

Neurospora crassa and its relatives of the genera Podospora and Sordaria from the family of the Sordariaceae serve as one of these two model systems.

They represent a presumably monophyletic group of the Pyrenomycetes class. Sordariaceae preferentially develop perithecia, whereas cleistothecia are formed rarely. The natural habitat of *N. crassa* is decaying or burnt vegetation, and species of the genus *Neurospora* are also known as red or orange bread mold, since in the past they caused considerable contamination in bakeries. *Neurospora* may also invade laboratories and literally lift the lids of Petri dishes. It contaminates by rapid growth and production of enormous numbers of orange, air-dispersed conidiospores. *Neurospora* research had its roots in 1840, when Louis Pasteur was called to advise the French army regarding the infestation of Parisian bakeries by this mold. A plethora of interesting discoveries emerged from work with this organism and its relatives and made it one of the most important model systems among fungi [11].

The analysis of *N. crassa* mutant strains carrying defects in amino acid biosynthetic genes by Beadle and Tatum lead to the one-gene/one-enzyme hypothesis, which is only one of many highlights of *Neurospora* research [12]. In particular, a large number of developmental studies have relied on the easily cultivable sordariaceous species. *N. crassa* is a heterothallic outcrossing fungus, whereas *Podospora anserina*, another intensively studied fungus of this group, is described as pseudohomothallic. However, for the genetic analysis of fruit body formation, it can be advantageous to use a homothallic fungus which allows the generation of developmental mutants without the requirement of the cooperative interaction of two strains of opposite mating types. *Sordaria macrospora*, a common fungus on dung, is an example of a homothallic fungus from this group [13]. The development of optimized molecular tools has recently resulted in an intensive analysis of fruit body formation in *S. macrospora* [14–17].

A second model of basic research is *Aspergillus nidulans*. This homothallic organism of the order of the Eurotiales represents the more heterogeneous group of the cleistothecia formers or Plectomycetes. In this chapter, we will focus on *A. nidulans*, a filamentous fungus with a well-studied program for asexual sporulation. Sexual reproduction of this ascomyceteous model organism for simple developmental processes has recently become a focus of attention and research. Thus, molecular information on some aspects of the sexual fruit body formation is available, which we will summarize and compare with data obtained from *N. crassa* and its relatives.

2 SEXUAL DEVELOPMENT AND FRUIT BODY FORMATION—A SURVEY

2.1 *Aspergillus nidulans* and Its Relatives

Aspergillus spp. are ubiquitous, with habitats covering a broad range of climatic zones. All species can exist more or less saprophytically in the soil, and a number of them can infect different hosts as opportunistic parasites. Many *Aspergillus* spp. are classified as Deuteromycetes or anamorphic Trichocomaceae because sexual processes have not been observed for them. Among these, several are of economic and medical importance: *A. flavus* is a producer of the secondary metabolite aflatoxin [18] and causes spoilage of plant seed–based foodstuffs and animal fodder, making the organism an important target for research on secondary metabolites. *A. fumigatus* accounts for a large proportion of fungal infections in humans, threatening immunocompromised patients, which can hardly be treated [19]. *A. oryzae*, *A. sojae*, and *A. awamori* are employed for food refining by proteolytic activities, e.g., for the production of soy sauce and sake [reviewed in 20]. *A. niger* is a common soil fungus employed for its ability to secrete large amounts of substances into the surrounding medium [21]. Citric acid production by *A. niger* is the main industrial source of this important chemical compound. The secretion of other substances, especially homologous and heterologous proteins [22,23], is also a major subject of *Aspergillus* research.

The scientific representative of the gender is *Aspergillus* (= *Emericella*) *nidulans*. The synonym *Emericella* refers to its ability to propagate sexually. This homothallic fungus is able to form fruit bodies in the absence of a partner in a process called selfing, which is the development of cleistothecia in homokaryons with two identical parent nuclei fusing and subsequently undergoing meiosis. This process results in meiospores with genotypes identical to the (single) parent nucleus. In contrast to asexual Aspergilli, the sexual life cycle permits the application of classical genetic techniques by marker combination via crossing, and determination of gene location via mapping of mutations [24]. *A. nidulans* is a saprophytic soil organism which can act as opportunistic pathogen. It causes systemic fungal infections with a significant lower incidence than *A. fumigatus*. *A. nidulans* was established as a genetic model organism in the 1950s [25]. It has a relatively small, haploid genome of 30 Mbp, spread over eight chromosomes, and forms uninucleate conidiospores. The *A. nidulans* mycelium can exist as a homo- as well as as a heterokaryon, the latter containing two genetically different sorts of nuclei after fusion of vegetative hyphae, and as diploids which are spontaneously generated at low frequency [26,27]. Heterokaryon formation in the parasexual phase of this fungus is an alternative, nonmeiotic mechanism for the recombination of genetic information. Alternatively, the combination of heterokaryon and diploid formation enables the fungus to form haploid nuclei with novel genetic variations by mitotic crossover in the spontaneously generated diploids and subsequent rehaploidization [28]. Easy cultivation and a short generation cycle as well as established molecular methods such as transformation make this organism a favorite subject for research. In conclusion, *A. nidulans* genetics, including gene regulation, biochemistry, the synthesis of secondary metabolites as antibiotics, cell biology, and development, has been extensively studied for >50 years.

2.2 Sexual Phase in the Life of a Filamentous Ascomycete

Many aspects in the life cycle of a filamentous ascomycete follow a similar pattern (Fig. 2), although individual species may deviate considerably, e.g., in the shape of the fruiting bodies (see above and Fig. 1). Ascosporogenesis generates mature ascospores within the ascus of the ascocarp which are eventually released and spread. Dry spores exhibit relative stability and preserve well over long periods of time. Under favorable conditions they germinate, depending on the nutrient status of the matrix. Germination of spores begins with the uptake of water and nutrients leading to isotropic swelling of the spores [29]. The first mitotic divisions of the nucleus are followed by unipolar formation of a primary germ tube [30]. Cell division and polar growth by tip extension lead to the formation of the multinucleate vegetative mycelium.

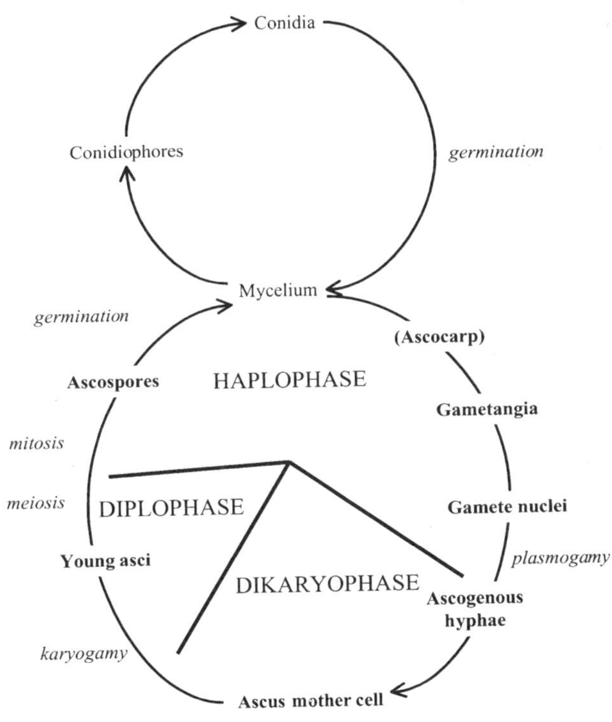

FIGURE 2 Schematic representation of an idealized life cycle of ascomycetes. The vegetative cycle (top half) results in the formation of asexual spores (conidia), whereas sexual development (bottom half) yields fruit bodies in which ascospores are formed.

The mycelium bears conidiogenous cells which produce large numbers of the asexual, mitotic spores called *conidiospores* or *conidia*. The formation of conidia and their carrier structures, the *conidiophores*, has been intensively studied in *A. nidulans* as a model process for differentiation of filamentous fungi [31–33]. The environmental conditions that are prerequisites of this development include a variety of factors such as light or aeration [34,35]. A medium/air interface is also important since the differentiation process requires a targeted three-dimensional growth instead of random branching of the vegetative mycelium for nutrient acquisition. The conidia are the main dispersal form of this fungus, rapidly forming new colonies. Depending on its internal developmental schedule as well as environmental conditions, the mycelium typically forms functional sex organs called *gametangia*. In many filamentous ascomycetes, including *N. crassa* and its relatives, the gametangia of one of the sexual partners are degenerated. During the sexual process, the ascospores within the ascocarp are formed. In nature—depending on the fungus, the substrate, and environmental parameters—the filamentous ascomycete passes the winter preferably in the mycelium or ascospore stage and less frequently as conidia.

2.3 Fungal Fruit Bodies—The Products of Sexual Development

Most filamentous ascomycetes produce meiotically derived ascospores associated with the development of a sexual fruit body. The structures of fruiting bodies containing the final products of sexual development differ within the ascomycota. The vase-shaped perithecia of *N. crassa* and *S. macrospora* contain linear asci, in which ascospores are ordered according to their genesis. This is a major advantage in genetics for directly studying the results of a single meiotic event. Each perithecium contains hundreds of asci. Mature ascospores are released by bursting of the ascus wall or can be expelled actively from the perithecium through the ostiole aperture. The meiotically reproducing *Aspergillus* spp. form closed, spherical fruit bodies with defined walls (envelopes) [24]. These cleistothecia are surrounded by a special cell type, the thick-walled, globose *Hülle cells* [36]. Eight ascospores are enclosed by the ascus, which resembles a spherical container. Under laboratory conditions, cleistothecia and ascospores reach maturity ~100 h after the initial spore germination.

While the production of the conidiospores is morphologically similar in all *Aspergillus* spp., the range of differentiation programs varies from species completely devoid of sexual development like *A. niger*, acleistothecial Hülle cell producers (e.g., *A. raperi*), cleistothecial species devoid of Hülle cells (e.g., *A. ornatus*), to *A. nidulans* with its complete developmental program. One heterothallic species with mating types, *A. heterothallicus*, is described for the genus [37]. Acleistothecial species were proposed to be derived from cleistothecial ancestors which have lost these reproductive functions [3]. It is assumed that meiospore production is an evolutionary old trait in *Aspergillus* spp., dispensable for propagation but retained by a few species. It is not obvious why *A. nidulans* can propagate via both spore types and even produce meiospores without obligate mating. The advantages of homothallism might lie in an extended possibility to develop different types of spores, which could be adapted to different survival conditions. In some fungi, e.g., *Blumeria graminis*, the sexual spores have a better capacity of survival than asexual spores, but in *A. nidulans* both spore types seem to show a similar capability of withstanding drought, radiation, and high temperatures. The preference of mating over selfing suggests that genetic variation is advantageous [38]. A conclusive answer on why *A. nidulans* undergoes the resource-consuming process of ascospore formation can not be given at present.

2.4 Fertilization and Fruit Body Development

Sexual development typically starts after conidiophore differentiation, when conidia production is well under way. In fungi with two different mating types, an *antheridium* cell (''male'') fuses with an *ascogonium* (''female'') to give a dikaryotic hypha. In the homothallic fungus *A. nidulans*, wild-type sexual development is initiated either by mating of two strains or by selfing. Mating types are not known for this fungus, and no antheridium or ascogonium structures can be observed. In analogy to other filamentous ascomycetes, it is assumed that an *A. nidulans* cell functionally equivalent to an ascogonium fuses to a second cell equivalent to an antheridium [24]. Random spore analyses originally suggested that a single fertilization event within each protocleistothecium results in a mature cleistothecium [25]. Recent octad analysis of the asci of single cleistothecia demonstrated, however, that cleistothecia of *A. nidulans* are not necessarily the result of a single fertilization event but can be the consequence of two or more fertilizations [39]. In the onset of this event within the homokaryon, ~50 h after spore germination, the first morphological changes indicating the development of the fruit body are visible. The fused hyphae are surrounded by growing, unordered mycelium, which forms an increasingly packed ''*nest*'' and differentiates to the multinucleate Hülle cells which support development of the cleistothecia. From these nestlike structures, the name of the species (Lat. *nidulans* = nest former) is derived [36].

The *A. nidulans* development shows significant differences when compared to the Neurospora/Sordaria group. A prerequisite for the sexual cycle of the heterothallic *N. crassa* or the homothallic *S. macrospora* is the formation of protoperithecia, which are primordia of the fruit body representing the female structure. Protoperithecia are spherical and ~50 μm in diameter. Several hyphae protrude from the surface. One of these hyphae, the trichogyne, is connected to the ascogonial cell as the female gametangium. Fertilization requires the donation of a ''male'' nucleus, which can be taken up by the ''female'' cell. In the heterothallic organisms, fertilization is initiated by the contact of the trichogyne with conidia or mycelium of the opposite sex. While a single colony of *N. crassa* is unable to complete the sexual cycle, an ascospore of *S. macrospora* can form perithecia from protoperithecia by selfing, similar to *A. nidulans*. In the heterothallic species, a recognition mechanism, which identifies the two nuclei of opposite sex and presumably employs the products of mating-type genes, has to operate before the migration of the nuclear pairs into the ascogenous hyphae. While no mating-type genes are known to date in *A. nidulans*, the homothallic *S. macrophora* expresses mating-type genes which are highly similar to the mating loci of the heterothallic *N. crassa* [40]. These genes are able to complement mating-type defects in their heterothallic counterparts. In species like *N. crassa* and its relatives, deletion studies have shown that the products of the mating-type genes are not the factors directly responsible for the morphogenesis of sexual structures like the protoperithecium [41,42].

Further on in development, the trichogyne grows toward the cell of the opposite mating type, responding to attractants which are not yet defined. The nucleus of the mating partner is handed over to the trichogyne by wall fusion and is then transported to the ascogonium. In *A. nidulans* nests, dikaryotic hyphae are formed by the fertilization events and subsequently undergo an extended series of coordinated cellular and nuclear divisions. The surrounding mycelium which formed the nest is subject to another step of differentiation, the formation of the cleistothecial envelope. Spaces between the hyphae are filled with cleistin, a substance not characterized to date [43]. As demonstrated by electron microscopy studies, the hyphae forming the envelope are morphologically modified as cells flatten and fuse to form a dense, interwoven layer [24]. This cleistothecial envelope is capable of expansion to suit the spatial needs of the growing dikaryotic mycelium. The developing cleistothecium grows out of the surrounding nest hyphae and Hülle cells, while the dikaryotic mycelium undergoes a switch from the coordinated nuclear and cellular division of the ascogeneous hyphae to the formation of the so-called *croziers* (Fig. 3).

In a large number of ascomycetes, one of the binucleate cells of the ascogenous hypha elongates and bends over to form this hook-shaped structure. Crozier formation ensures that two nuclei of opposite mating types are positioned in the top cell, which will later develop the ascus. Every crozier can form a zygote by karyogamy. Crozier formation is similar in *A. nidulans* and Neurospora/Sordaria. Two nuclei are trapped in the topmost crozier cell by a series of divisions which require exact nuclear positioning and cell wall insertion. In every single crozier, a nuclear fusion event (karyogamy) forms a diploid nucleus ~70–80 h after spore germination [25]. This short zygote stage is immediately followed by meiosis, which results in four nuclei. After meiosis, one round of mitosis produces eight nuclei which are separated from each other by membranes. Another

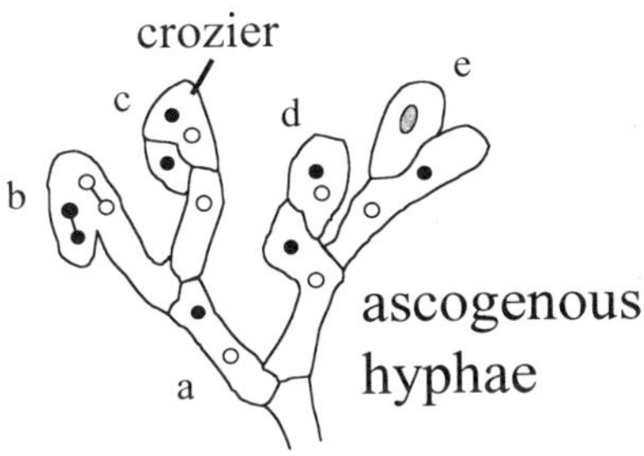

FIGURE 3 Illustration of crozier formation. In the ascogeneous, heterokaryotic hyphae (a) containing two haploid nuclei of opposite mating type, a hook-shaped structure is formed in which nuclei divide synchronously (b). The penultimate, dikaryotic cell of the ascogonium forms the top crozier cell (c) in which, after fusion of the end cell and basal cell (d), karyogamy (e) and further ascus development take place.

round of mitosis yields the eight binucleate ascospores organized in an octad of an *A. nidulans* ascus. Mature cleistothecia of wild-type strains can reach a size of ~200 μm and usually contain ~80,000 viable ascospores. The ascospores are red owing to the accumulation of a characteristic red pigment called asperthecin. During the end of cleistothecia maturation, the nest hyphae degenerate and detach from the cleistothecial envelope.

Crossing two *A. nidulans* strains requires a heterokaryon. Two strains can form anastomoses, cytoplasmic bridges between hyphae for the exchange of nuclei, when growing sufficiently close to each other [25]. For strain-crossing experiments, heterokaryons, especially of closely related strains, are usually transferred to selective medium to maintain stability of the heterokaryon during the prolonged process of fruit body formation. Heterokaryons tend to form segments of homokaryon within the colony if not subjected to selection. In the heterokaryon, a recognition mechanism probably exists to identify nuclei derived from different parent strains. Octad analysis revealed that unlike haploid nuclei preferentially fuse to the prezygotic diploid nucleus. When heterokaryons are formed between nuclei of different genetic background, recombinant asci derived from opposite nuclei are formed exclusively. In contrast, *A. nidulans* strains that differ at only a single genetic marker randomly fuse nuclei for zygote formation [39]. This nuclear recognition mechanism is yet uninvestigated in *A. nidulans*, and it will be interesting to see whether genes with similarities to mating-type loci are involved.

Although *A. nidulans* strains can be crossed without regard for mating types, not all strains are capable of mating with any other strain. Heterokaryon incompatibility limits mating to strains with compatible *het* loci. Compatible strains forming an *h-c* group carry the same set of alleles at all of these loci and are usually quite similar, if not identical, in their genetic background. *A. nidulans* wild isolates of the Birmingham strain collection were classified into 19 *h-c* groups. The Fungal Genetics Stock Center A4 derivatives used in most laboratories are known as Glasgow strains and form the 20th group [44,45]. Formation of a viable heterokaryon between members of different *h-c* groups usually requires strong selection or protoplast fusion [46]. Nevertheless, such fused incompatible strains are unaffected in asexual and sexual spore formation [47].

2.5 Determinants Influencing Fruit Body Formation in *A. nidulans*

2.5.1 Environmental Factors Affecting Sexual Development

Every organism grows and reproduces within a distinct microenvironment which may change its properties rapidly in time and space. Depending on these environmental circumstances, the microorganism has to decide whether to initiate a specific differentiation program. Regarding the fruit body formation of ascomycetes, a variety of physical parameters have to be checked and a number of preconditions fulfilled before the sexual cycle of a filamentous fungus can be initiated and finally completed (Fig. 4).

The acquisition of developmental competence is *time* dependent. In the laboratory, cleistothecia production through selfing usually starts 50 h after spore germination in *A. nidulans* strains. Mature fruit bodies can be observed after ~100 h in the wild type as earliest possible time point [24]. This implies that intrinsic factors responsible for this strict timing play a major role in development.

Usually, *light* influences sexual development. Incubation of *A. nidulans* wild type in the dark for the first 24 h after inoculation leads to denser fruit bodies and less conidiation than incubation in the light. In the dark, cleistothecia formation can also be initiated at an earlier time point than in light-grown colonies [48]. Conidiation is promoted by red light pulse; fruit body formation, by far-red light. The effect of a red light pulse can be reversed by a far-red light pulse and vice versa, depending on the velvet factor (see below). In *N. crassa*, blue light is the predominant signal to induce several developmental processes [reviewed in 49] including protoperithecia formation, biosynthesis of carotenoids in perithecial walls, and the phototropism of perithecial beaks [50–52].

Fruit body formation of *A. nidulans* is limited by *air exchange*. During development, this is a critical factor determining the cleistothecial density [35,53]. *A. nidulans* wild-type strains produce ~2000 fruit bodies per cm^2 on Petri dishes while developing approximately double the amount of cleistothecia and fewer conidiospores when air exchange is limited by wrapping the plates with tape. This phenomenon is attributed to a reduction of the CO_2 content of

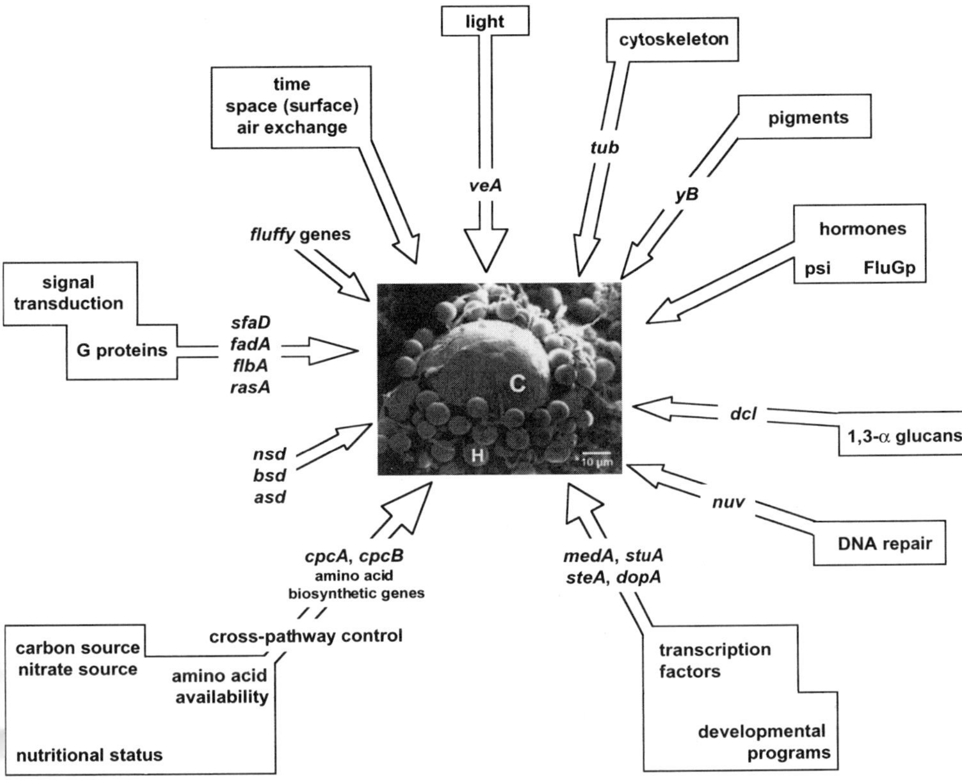

FIGURE 4 Determinants influencing fruit body formation in *A. nidulans*. Environmental factors are shown in boxes, whereas relevant genetic determinants are presented in the inner circle (see text for details). The electron micrograph showing a mature *A. nidulans* cleistothecium (C) surrounded by Hülle cells (H) was taken by K. Adler, Gatersleben, Germany.

the air inhibiting or stalling fruit body development, presumably because CO_2 is required for carbon metabolism of the developing colonies [48].

Fruit body formation generally depends on a *surface*. Usually, a mycelium does not differentiate in submerged culture but can be induced by transfer to solid medium. Sexual development requires a constant surface on which to develop. This medium/air interface can be on solid or liquid medium. Whereas conidiation in *A. nidulans* can be forced in liquid culture by offering a special media composition low in nitrogen [54], a surface is indispensable for cleistothecia formation [53].

The sexual developmental program requires specific precursors provided by specialized metabolic pathways. Although the *nutritional status* is an important factor for the differentiation programs of *A. nidulans* wild-type strains, few observations have been recorded to date [55]. Trace elements such as manganese seem to affect the sexual program [56]. Major nutritional compounds play a crucial role. On medium with reduced carbon source (0.8% glucose), fruit body development was reduced or blocked compared to growth on 3% glucose [57]. Limiting the carbon source presumably inhibits fruit body formation through a lack of α-1,3-glucan components necessary for the formation of the cleistothecial cell walls or as energy storage compounds (see below). According to Han *et al.* [58], the carbon source of the medium also affects the number of ascospores produced in wild-type strains. Low nitrogen levels inhibit cleistothecia development [56]. This is in contrast to other fungi, e.g., the fission yeast *S. pombe*, where nitrogen starvation is a prerequisite for mating [59].

In addition, the availability of *amino acids*, the translational precursors of proteins, influences developmental programs of *A. nidulans*. While an *argB* mutant strain was reported to be deficient in cleistothecia formation, excess arginine was found to inhibit the ascospore production of the wild type [55]. Auxotrophic strains defective in the tryptophan biosynthesis pathway are dependent on an external supply and show developmental blocks in conidiation as well as cleistothecia formation [60,61]. A systematic analysis of strains carrying mutations in four different tryptophan biosynthetic genes showed that fruit body formation could be restored by high concentrations of tryptophan and was promoted by supplementation with indole or auxin. However, fertility of ascospores of the tryptophan auxotrophic strains could be only partially restored. Decreasing tryptophan supplementation resulted in a stepwise loss of the potential to differentiate. The lowest amount of tryptophan is required for mycelial growth, higher amounts for conidiation, and the highest amount for the formation of fruit bodies [62]. The targeted deletion of the tryptophan synthase-encoding gene *trpB* showed that these effects are not the result of specific alleles carrying single point mutations but depend on the activity of the gene product [63]. Similar defects of sexual development were also found for histidine auxotrophic mutant strains [64].

In *A. nidulans*, a *hormone* was reported to influence sexual development, the so-called *psi* factor (= *p*recocious *s*exual *i*nducer), an endogenous mixture of hydroxylinoleic acid moieties [65,66]. The application of the psi factor extracted from growth medium of *A. nidulans* to a confluent plate culture strongly inhibits asexual sporulation and induces premature sexual sporulation. In addition, a yellow pigment is released to the medium. The different interconvertible compounds of psi—psiA, psiB, and psiC (Fig. 5)—might have different functions supporting either asexual or sexual spore formation [24]. The most active compound is psiC (5,8-dihydroxylinoleic acid). psiB, 8-hydroxylinoleic acid, might be an intermediate of the biosynthesis. It can lyse the hyphae of certain

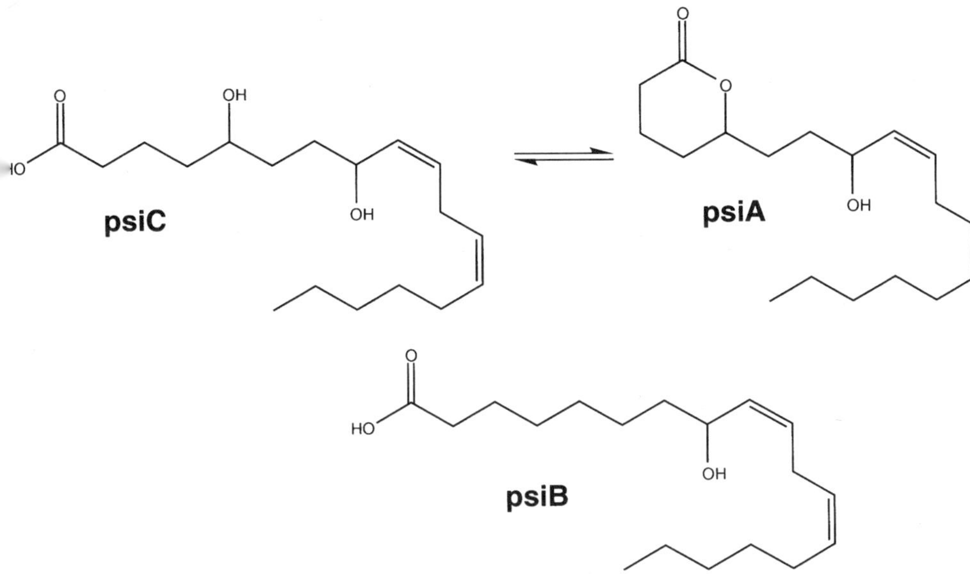

FIGURE 5 Chemical structures of psi compounds. psiC (5,8-dihydroxylinoleic acid) and its cyclic δ-lactone, psiA, are shown. psiB (8-hydroxylinoleic acid) is presumed to be an intermediate of psi factor biosynthesis.

oomycetes, suggesting that the membrane is the target. psiA is the cyclic δ-lactone of psiC and is less active. The exact mode of influence of psi on development is still unknown, but it has been suggested that psi may change membrane properties to promote the special hyphal fusions that give rise to the dikaryon or that a specific receptor for the hormone exists which might be coupled to a signal transduction pathway. Additionally, other linoleic compounds were found to have similar sporogenic effects on *A. nidulans* [67].

2.5.2 Genetic Determinants Regulating Fruit Body Development

Perception of the environmental status as well as signal transduction relies on specific systems encoded by genetic determinants. From the ~8000 genes encoded by the *A. nidulans* genome, an estimated 6000 are required for "housekeeping" biochemical functions. A large proportion of the remainder of the genes is expected to be required for developmental and differentiation processes, e.g., the detection of environmental signals, signal transduction processes within developmental programs, or altered gene expression for coordinating the action of general and specialized biosynthetic enzymes.

Genes Involved in the Perception of Environmental Signals. The influence of light [37] was found to be dependent on a protein encoded by the gene *veA* (=*ve*lvet) [34]. The VeA protein is thought to be the regulator of gene expression at the switch point determining whether to emphasize conidiation or sexual development. The regulatory genes *brlA* and *abaA* encode transcription factors induced and required for asexual spore development, and both genes were shown to be transcribed only upon irradiation [34]. Additionally, characterization of *fluG* mutations suggests that VeAp interacts with FluGp, which is assumed to be part of a signal transduction pathway influencing the input signal and regulating the initiation of conidiation [68,69] (see foregoing). The *veA* gene product itself is not well characterized so far and seems to include a phytochromelike substance [34]. Depending on exposure to light, the protein VeAp has two states, corresponding to either a red light–induced state promoting conidiation or a far-red light–triggered state promoting fruit body formation. Switching the wavelength of light can change the regulatory properties of VeAp. The dependency on light for conidiation is abolished in *veA1* mutant strains, which additionally show increased transcription of the *brlA* regulatory gene of asexual differentiation, resulting in profuse conidiation in a light-independent manner [70]. *veA1* mutant strains also show increased conidia density, while their sexual development time is doubled from 100 h to 200 h and fruit body density is drastically reduced from $>2000/cm^2$ to $\sim 200/cm^2$. However, the fruit bodies formed are wild type size and contain normal amounts of fertile ascospores. Furthermore, *veA1* strains display a conditional temperature phenotype with respect to cleistothecia formation as they are acleistothecial at temperatures >42°C [71].

In many microorganisms sporulation is induced by nutrient limitation. In *A. nidulans*, poor nutritional conditions inhibit cleistothecia formation (discussed below). However, a secondary, intracellular starvation signal was found to correlate with the onset of asexual developmental processes. Self-induced starvation was demonstrated for the conidiation pathway by temporal misexpression of the regulatory *brlA* and *abaA* gene products which leads to a general, drastic reduction of protein and RNA levels of numerous metabolic genes [72]. Amino acids are precursors of translation and are required to synthesize new proteins, e.g., enzymes for the metabolic rebuilding of the fungus. Additionally, their levels could influence other biosynthetic pathways or constitute the building blocks for secondary metabolites. As mentioned above, amino acid auxotrophy resulting from biosynthesis gene defects can lead to impaired or stalled fruit body formation [e.g., 55,61,62].

Auxotrophic strains unable to form fruit bodies show a derepressed cross-pathway control system that is generally induced by amino acid starvation [62]. In numerous ascomycetes, imbalances (both shortage/starvation and excess) in the pool of amino acids will lead to induction of the cross-pathway system [73–75]. This is illustrated by the finding that overexpression of a single gene for histidine biosynthesis in *A. nidulans* blocks development and leads to an acleistothecial phenotype [76]. The cross-pathway control is a complex transcriptional network coordinating the biosynthesis of protein precursors like amino acids or charged tRNAs [77,78]. Starvation for a single amino acid induces the network. In the yeast *S. cerevisiae*, it has been shown that uncharged tRNAs induce a signal transduction pathway which ultimately results in the synthesis of Gcn4p, a transcriptional activator protein [77]. Homologs of the *GCN4* gene have been described from filamentous fungi including *Aspergillus*, and are functionally exchangeable [79,80]. In *A. nidulans*, amino acid limitation results in impaired sexual fruit body formation.

Growth under amino acid starvation conditions permits the initiation of the sexual development program but blocks fruit body formation before completion of meiosis. The arrest results in *microcleistothecia* which are filled with hyphae. They are considerably smaller than mature cleistothecia and completely covered with nest material. At the microcleistothecia state, hyphae are short and swollen, a phenotype which is described for the dikaryotic hyphae. Supplementation of amino acids results in release of the block and completion of development to mature ascospores. Overexpression of the *cpcA* gene encoding the *Aspergillus* *c*ross-*p*athway *c*ontrol transcriptional activator, the homolog of Gcn4p, results in a similar block even in the absence of amino acid limitation, suggesting that intrinsic signals affect the developmental program [81]. Even a partially activated cross-pathway control caused by deleting the *cpcB* gene, which acts in repressing the network in the presence of amino acids, results in the formation of microcleistotecia as final products of sexual development [81,82]. The biological advantage of such a block in fruit body formation upon amino acid starvation seems to be an economical consequence of an anticipated lack of building material.

fluG is the only gene which has been identified to be responsible for the production of an extracellular, psi-independent developmental signal that seems to be primarily essential for the initiation of conidiation. The gene encodes an enzyme with similarities to prokaryotic glutamine synthase I enzymes, but does not seem to be involved in glutamine biosynthesis. The extracellular signal might therefore resemble an amino acid. This signal depends on the activity of *fluG*, seems to initiate the developmental program, and is thought to be received by an as-yet-unknown receptor [83].

Signal Transduction. Although numerous signal transduction pathways are described and conserved among eukaryotes, the signaling compounds necessary for sexual development in *A. nidulans* are hardly known. This might be partially due to the fact that there is a spatiotemporal order of the two programs of asexual and sexual development subsequent to hyphal growth. However, both reproduction pathways seem to share signal transduction compounds. In addition, conidiation starts significantly earlier than sexual development, so intrinsic signals play an important role for the decision of the fungus to start cleistothecia formation. Accordingly, there has to be crosstalk between regulatory proteins specific for the asexual cycle and the sexual cycle of development. Some environmental parameters, which have to be perceived and translated into internal signals, are essential for both differentiation pathways. Numerous mutant strains altered in genes essential for both spore-producing developmental programs which have been described as defective in sexual sporulation, had originally been isolated for other characteristic phenotypes.

These genes include e.g. some of the *flu* (= *flu*ffy) genes or their suppressors [e.g., 84], which influence the development of the vegetative mycelium as well as sporulation programs [85]. *flu* mutations generate colonies with profuse aerial hyphae, giving them the appearance of cotton wool [86]. The aerial hyphae are stalks which lack the clearly defined length and further developmental program of wild-type conidiophores. Genetic analysis of *flu* genes and their suppressors revealed several elements of signaling pathways. They seem to be involved in the transmission of the extracellular signals including the one dependent on FluGp (see above), which initiate developmental programs. Although initially described as defective in asexual sporulation, the *flu* phenotype is typically correlated with the inability to perform the sexual cycle, indicating that the gene products exert a connecting role between the two developmental programs.

In addition, elements of heterotrimeric G-proteins have been identified which are involved in development. The *sfaD* gene encodes the β-subunit [84], *fadA* the α-subunit of a heterotrimeric G-protein [87]. The *flbA* gene, a homolog of yeast *SST2*, encodes an RGS protein (= *r*egulator of *G*-protein *s*ignaling) and seems to antagonize the action of the heterotrimeric G-protein [88]. The major role of this G-protein might be to decide between growth as vegetative mycelium and the initiation of a developmental program like sporulation. It is unclear in *A. nidulans* whether the isolated heterotrimeric G-protein is connected to the cAMP-dependent PKA (*p*rotein *k*inase *A*) pathway. This connection exists in budding yeast between response to the nutritional situation in the environment and initiation of a development program, the filamentous pseudohyphal growth [89,90]. The gene product of *flbE* is another protein which is presumably involved in signal transduction and required for developmental processes, but its exact molecular function is unknown [33].

Besides the FadAp/SfaDp heterotrimeric G-protein, another GTP/GDP-binding molecular switch has been described in *A. nidulans*. A small G-protein encoded by a homolog of the *ras* genes, *A-ras* or *rasA*, has been shown to be essential for regulating an ordered developmental program. The active GTP-bound and the inactive GDP-bound form of the protein have been mimicked by the construction of dominant alleles with the appropriate mutations. Large amounts of active RasAp protein inhibit development at the early stage of germ tube formation, although nuclear division proceeds. In the absence of a carbon source, asexual spores overexpressing the dominant active *rasA* allele displayed swelling, adhesion, and nuclear reorganization [91]. More complex levels of differentiation can be reached by decreasing RasAp activity. The lowest activity seems to correlate with sexual development [92]. Even the overexpression of constitutively inactive *rasA* alleles resulted in an acleistothecial phenotype, suggesting that this molecular switch has to be predominantly inactivated before spore production and fruit body formation in particular can be performed [81].

In the yeasts *S. cerevisiae* and *S. pombe*, the MAPK (*m*itogen-*a*ctivated *p*rotein *k*inase) cascades are highly conserved pathways, responding to different stimuli by launching different cellular programs including mating and differentiation and, in *S. pombe*, meiosis [93]. There is also crosstalk between the MAPK pathway and the PKA pathway in the budding yeast in the control of mating/filamentous growth and stress response [94]. Surprisingly, MAPK modules and MAP kinases have not been identified in *A. nidulans*. However, the recent finding of a transcription factor which presumably is the target of MAP kinase signal transduction pathways and that is required for cleistothecia production suggests that MAPK modules are also involved in *A. nidulans* development (see below) [95].

Transcription Factors Involved in Sexual Development. Several transcription factors have been identified which seem to be involved in the choice between hyphal growth and the two spore-forming differentiation programs. Some of these transcription factors are thought to be targets of the signaling cascades (see above). These factors might also be a prerequisite for sexual development, which normally occurs only when conidiation is already highly progressed. Many of these genes have been identified through mutant strains which are unable to perform any sporulation program. The *flbD* gene encodes a DNA-binding protein with similarities to the proto-oncogene c-Myb. *flbB* encodes a DNA-binding protein with a bZIP dimerization and DNA binding domain. *flbC* encodes a protein containing two C_2/H_2 zinc finger domains, which suggests that it also binds DNA [33,96].

Several key regulators of development have been identified which seem to be specific for the asexual pathway. The most prominent regulatory genes are *brlA* and *abaA*. *brlA* encodes another C_2/H_2 zinc finger protein and overexpression of either *flbC* or *flbD* in submerged hyphae activates its expression [33,96]. *abaA* encodes a protein with an ATTS/TEA domain which is conserved among other members of the family (ATTS = *A*baAp, *T*ec1p from budding yeast, *T*EF-1 from simian virus 40 enhancer factor *s*equence; TEA = *TE*F1, *Te*c1p, *A*baAp). The *abaA* gene product acts downstream of BrlAp. These two key regulatory proteins seem to be specific for condition without obvious influence on cleistothecia formation. In contrast to this, *medA* and *stuA* are two modifier genes of development where mutant alleles exist which exhibit clear effects on asexual as well as on sexual differentiation with the *medA* gene product having a more general role influencing both sporulation pathways. The corresponding wild-type protein is responsible for the correct temporal expression of both transcripts of the asexual regulator *brlA* and also functions as coactivator required for normal levels of *abaA* expression [97]. Mutations in the *medA* (= *med*usa) gene result in aberrant conidiophores with branching chains of metulae, delayed conidial differentiation, and frequent reinitiation of secondary conidiophores [98].

With respect to sexual development, *medA* mutant strains produce only Hülle cells during the sexual cycle, but it is unknown how the defect in *medA* expression stalls cleistothecia formation at this stage. *stuA* mutant strains are completely acleistothecial and exhibit spatially abnormal conidiophores with spore production from the vesicles [99]. StuAp, a transcriptional regulator, probably acts by influencing expression of developmentally regulated genes of asexual as well as sexual development [100,101]. *stuA* gene function is required for early events of asexual reproduction until the completion of conidiophore differentiation. It is also necessary for the spatial distribution of other gene products like BrlAp or AbaAp. In comparison, StuAp seems to be required for the correct spatial, and MedAp for the correct temporal, expression of the *brlA* gene. StuAp shows similarities to the *S. cerevisiae* proteins Swi4p and Mbp1p in the DNA-binding region. These transcription factors have important roles in cell cycle progression of yeast [102]. Furthermore, *stuA* is homologous to the *PHD1* gene of *S. cerevisiae*, a factor described to affect pseudohyphal growth [103].

Another transcription factor which is involved in asexual spore formation, the sexual cycle, and which in addition also seems to affect hyphal growth, is encoded by the *dopA* locus [104]. A temperature-sensitive allele of this gene has been described earlier as $aco586^{ts}$ [53,71], and homologs of *dopA* have been found in the genomes from yeast to man. DopAp includes three leucine zipperlike domains and a carboxyterminal domain similar to the C/EBP transcription factors. Whereas deletion of the yeast homolog is lethal, deletion of *dopA* in *A. nidulans* results in several morphologically distinguishable defects: vegetative hyphae show an abnormal morphology, conidiophores display aberrant morphogenesis, and the sexual cycle is abolished, suggesting a very early block in sexual development. The analysis of mutant strains implies that there is a genetic interaction between DopAp and the G-protein encoded by *rasA* described above. In the asexual pathway, DopAp primarily affects the different transcripts of the *brlA* locus [104].

Two additional transcription factors have been identified which seem to be primarily important for the sexual cycle, although their exact role is yet unclear. One is the gene product of *nsdD*, identified via complementation of a UV-induced acleistothecial mutant strain [105]. Deletion of *nsdD* prevents fruit body development or the formation of Hülle cells. Overexpression of *nsdD* by increasing the copy number represses the asexual sporulation program. The gene encodes a deduced protein of 460 amino acids, which shows similarities in its C-terminal part to the GATA-binding transcription factor family and the light regulators encoded by *wc1* and *wc2* of *N. crassa* [106,107]. The N-terminal part has no significant similarity with any other protein currently found in the databases.

The *A. nidulans steA* gene encodes a homeodomain C_2/H_2 zinc finger transcription factor required for sexual reproduction [95]. A mutant strain carrying a *steA* deletion is sterile and differentiates neither ascogenous tissue nor fruiting bodies. Like *medA* mutant strains, the *steA* deletion strain is able to develop Hülle cells. *STE12* of the budding yeast *S. cerevisiae* is the homolog of *steA* and encodes a transcription factor which is regulated by the MAPK signal transduction pathway. This suggests that such a regulatory cascade might also be involved in *A. nidulans* development. Yeast Ste12p regulates cell identity, karyogamy, and morphogenesis. In addition, it interacts with the product of *TEC1*, which is the yeast homolog of *abaA*, the second major key regulator of asexual development in *A. nidulans*. Ste12p/Tec1p promote filamentous growth in yeast [94]. Although SteAp function seems to be restricted to the sexual cycle, experimental evidence suggests some crossregulation to other morphological events of the fungus. For example, the deletion strain shows derepressed levels of the *medA* transcript, the regulatory gene involved in both sporulation programs of the fungus [95].

In other fungi, few transcription factors influencing sexual development are known. The *S. macrospora pro1*$^+$ (= *pro*toperithecia) gene product, which contains a DNA-binding domain found in fungal C_6 zinc finger transcription factors, is involved in early events of fruiting body production. The mutant strain is able to form protoperithecia, but no ascus primordia are detectable. *pro1*$^+$ homologs were found to be present in other sexually propagating filamentous ascomycetes related to *Sordaria*. Southern experiments using the *pro1*$^+$ gene as probe showed heterologous hybridization, suggesting that there might be a yet uncharacterized gene with some similarity in *A. nidulans* [17].

Specific Gene Expression. The three-dimensional *A. nidulans* cleistothecium is the most complex structure this fungus is able to form. The process is highly energy and material consuming [48]; parts of hyphae have to be dissolved and locally rearranged. The sexual cycle consists of several developmental programs which are interconnected: the formation of ascogenous hyphae, asci, and ascospores; the formation of Hülle cells; and the formation of the fruit body envelope surrounding the asci. Since several *A. nidulans* mutant strains exhibit only Hülle cells, the formation of this specific cell type can be uncoupled from the other processes. Furthermore, sexual and asexual development seem to be interconnected as the two programs presumably share regulatory elements necessary for both sporulation programs. Accordingly, cleistothecia formation depends on the regulation of a large number of genes.

A number of mutant alleles of genes which might play a role in cleistothecia formation have not been analyzed in molecular detail. These include a number of *acl* (*acl*eistothecial) mutations, which have been described by Clutterbuck [98]. Other mutations were obtained by UV irradiation mutagenesis in a systematic study of *A. nidulans* mutant strains defective in sexual development by Han and coworkers: 20 *nsd* (= *n*ever in *s*exual *d*evelopment) mutants, which include a strain with the *nsdD* mutation mentioned above, have been isolated and represent four different complementation groups. These mutant strains produce neither cleistothecia nor Hülle cells. More than 100 *bsd* (= *b*locked in *s*exual *d*evelopment) mutants are arrested at different steps of fruit body formation, and 100 *asd* (= *a*bnormal *s*exual *d*evelopment) mutant strains show differences in morphology or the timing of development [58,108,109].

For the rearrangement of genetic information among different nuclei within the ascogenous hyphae, the formation of a heterokaryon is important. The *A. nidulans* strain *aclA1* is unable to form cleistothecia. It is assumed that this is due to its inability to form a heterokaryon, but the exact mechanism has to be elucidated [57]. *nuv* mutant strains are affected in recombination-specific genes. Some of these strains have developmental blocks or are acleistothecial [110], suggesting that an intact recombination apparatus is required for the sexual program.

The formation of fruit bodies is dependent on the integrity of elements of the cytoskeleton. Accordingly, *tubB* encoding a tubulin is specifically required for sexual development and is assumed to be essential for meiosis since deletion of *tubB* completely prevents ascosporogenesis by selfing [111]. The proteins SamAp and SamBp are described to be involved in microtubule-dependent processes and also seem to play a role in fruit body formation [112]. The copper-dependent laccase II is found only in the Hülle cells. *yB* mutant strains defective in this enzyme, encoding a phenol oxidase, are acleistothecial, emphasizing the importance of this Hülle cell–specific enzyme for development. Activity of this enzyme is induced during cleistothecia formation, presumably owing to changes in gene expression during cleistothecia production [113,114]. The molecular details of this regulation and the specific transcription factors remain to be elucidated.

Many auxotrophic mutants are unable to enter the sexual pathway. This presumably reflects the fact that the formation of cleistothecia is a major metabolic effort for a filamentous fungus. Therefore, the metabolic pattern and the activity of metabolic enzymes have to be reprogrammed in comparison to other life phases and are critical for the differentiation program. The correlation between levels of amino acids dependent on the expression of their respective biosynthetic genes and fruit body formation has been discussed above.

Another example of gene expression directly influencing cleistothecia production involves the carbohydrate metabolism. It is assumed that the requirement of α-1,3-glucan for the formation of cell walls of the cleistothecium depends on increased α-1,3-glucanase activity. α-1,3-glucanase activity correlates with the density of cleistothecia. An *A. nidulans* mutant strain exhibiting an increased density of cleistothecia has been isolated which showed increased α-1,3-glucanase activity [57]. However, a regulation of gene expression has not yet been shown. Transcription of another gene involved in carbohydrate metabolism, the phosphoglucomutase-encoding *pgmB* of *A. nidulans*, was found to be low during hyphal growth and in the sexual phase of development, but was significantly increased during the asexual stage of the *A. nidulans* life cycle [115], suggesting that there is no specific role of this gene in cleistothecia formation.

In other fungi, few factors influencing sexual development are known. In *S. macrospora*, an ATP citrate lyase encoded by *acl1* was described to be essential for fruiting body maturation. *acl1* is specifically induced at the beginning of the sexual cycle and produces acetyl-CoA, which probably is a prerequisite for fruiting body formation during later stages of sexual development [16].

With the entire genome sequence of *A. nidulans* on its way, the establishment of transcriptional profiles will be the most prominent primary approach to determine the genes which are changed in their expression pattern during the course of development. This has to be verified by constructing *A. nidulans* strains where such genes are specifically mutated. The broad range of available classical and molecular tools make *A. nidulans* here to an especially suitable model organism.

Development stage–specific hybridization will give a much clearer picture of the changes in expression of the numerous genes expected to take part in the differentiation processes. The analysis of genes and their expression has to be complemented by the analysis of altered protein patterns to determine the role of posttranscriptional events. This might identify additional regulatory factors not characterized to date. In addition, more detailed information on the regulation of primary biosynthetic genes and the expressed proteins can be expected. The crosstalk between metabolism and development is of considerable interest, which is demonstrated by the association of differentiation processes and drastic changes in expression, self- induced internal starvation, and the careful balance of carbohydrate metabolism and amino acid biosynthesis gene expression.

3 OUTLOOK

Among filamentous ascomycetes, *Aspergillus nidulans* is an important model system for developmental processes. Various tools and methods required for molecular biology have been developed. In addition, numerous methods of classical genetics can be applied. Whereas in the past many results on the molecular level have primarily focused on conidiation, the field of the analysis of cleistothecia formation, representing the most complex three-dimensional structure this fungus is able to build, is broadening. Our present knowledge consists mainly of many different pieces of a jigsaw puzzle, which wait to be correctly rearranged. Information concerning the potential of different developmental programs in filamentous fungi will increase within the next years. In analogy to other fields of fungal research, it is predictable that this will also have a great impact to understand more complicated and complex processes in development of higher eukaryotes.

ACKNOWLEDGMENTS

We thank Silke Busch and Oliver Draht for carefully proofreading the manuscript. This work was supported by the Deutsche Forschungsgemeinschaft, the Fonds der Chemischen Industrie, and the Volkswagen-Stiftung.

REFERENCES

1. RA Harper. Kerntheilung und Freie Zellbildung im Ascus. Jahrb Wiss Bot 30:249–284, 1897.
2. HA de Bary. Comparative Morphology and Biology of the Fungi, Mycetozoa and Bacteria. Oxford: Oxford University Press, 1887.
3. DM Geiser, WE Timberlake, ML Arnold. Loss of meiosis in *Aspergillus*. Mol Biol Evol 13:809–817, 1996.
4. CM Hull, RM Raisner, AD Johnson. Evidence for mating of the 'asexual' yeast *Candida albicans* in a mammalian host. Science 289:307–310, 2000.
5. BB Magee, PT Magee. Induction of mating in *Candida albicans* by construction of MTLa and MTLα strains. Science 289:310–313, 2000.
6. G Pontecorvo. The parasexual cycle in fungi. Annu Rev Microbiol 128:162–171, 1956.
7. E Coppin, R Debuchy, S Arnaise, M Picard. Mating types and sexual development in filamentous fungi. Microbiol Mol Biol Rev 61:411–428, 1997.
8. DD Perkins. Mating-type switching in filamenouts ascomycetes. Genetics 115: 215–216, 1987.
9. JE Haber. Mating-type gene switching in *Saccharomyces cerevisiae*. Annu Rev Genet 32:561–569, 1998.
10. JZ Dalgaard, AJS Klar. Orientation of DNA replication establishes mating type switching pattern in *S. pombe*. Nature 400:181–184, 1999.
11. JW Taylor, B Bowman, ML Berbee, TJ White. Fungal model organisms: phylogenetics of *Saccharomyces*, *Aspergillus* and *Neurospora*. Syst Biol 42:440–457, 1993.
12. DD Perkins. *Neurospora*: the organism behind the molecular revolution. Genetics 130:687–701, 1992.
13. K Esser, J Straub. Genetische Untersuchungen an *Sordaria macrospora* Auersw.: Kompensation und Induktion bei genbedingten Entwicklungsdefekten. Z Entwicklungsl 89:729–746, 1958.
14. M Walz, U Kück. Transformation of *Sordaria macrospora* to hygromycin B resistance: characterization of transformants by electrophoretic karyotyping and tetrad analysis. Curr Genet 28:88–95, 1995.

15. S Pöggeler. Sequence characteristics within nuclear genes from *Sordaria macrospora*. Fungal Genet Newsl 44:41–44, 1997.
16. M Nowrousian, S Masloff, S Pöggeler, U Kück. Cell differentiation during sexual development of the fungus *Sordaria macrospora* requires ATP citrate lyase activity. Mol Cell Biol 19:450–460, 1999.
17. S Masloff, S Pöggeler, U Kück. The *pro1*(+) gene from *Sordaria macrospora* encodes a C_6 zinc finger transcription factor required for fruiting body development. Genetics 152:191–199, 1999.
18. MF Dutton. Enzymes and aflatoxin biosynthesis. Microbiol Rev 52:274–295, 1988.
19. AA Brakhage, B Jahn, A Schmidt. *Aspergillus fumigatus*. Basel: Karger, 1999.
20. K Kitamoto, K Gomi, K Goto, S Hara. Genetic transfer applied to traditional sake brewing. Biotechnol Genet Eng Rev 9:89–125, 1991.
21. EA Bodie, B Bower, RM Berka, NS Dunn-Coleman. Economically important organic acid and enzyme products. In: SD Martinelli, JR Klinghorn, eds. Prog Ind Microbiol. Amsterdam: Elsevier, 1994, Vol 29, pp 561–602.
22. DB Archer, DJ Jeenes, DA Mackenzie. Strategies for improving heterologous protein production from filamentous fungi. Antonie Van Leeuwenhoek 65:245–250, 1994.
23. PJ Punt, G Veldhuisen, CA van den Hondel. Protein targeting and secretion in filamentous fungi. A progress report. Antonie Van Leeuwenhoek 65:211–216, 1994.
24. SP Champe, DL Nagle, LN Yager. Sexual sporulation. In: SD Martinelli, JR Klinghorn, eds. Prog Ind Microbiol. Amsterdam: Elsevier, 1994, Vol 29, pp 429–454.
25. G Pontecorvo, JA Roper, LM Hemmons, KD MacDonald, AWJ Bufton. The genetics of *Aspergillus nidulans*. Adv Genet 5:141–239, 1953.
26. E Käfer. Meiotic and mitotic recombination in *Aspergillus* and its chromosomal aberrations. Adv Genet 19:33–131, 1977.
27. A Upshall. Naturally occurring diploid isolates of *Aspergillus nidulans*. J Gen Microbiol 122:7–10, 1981.
28. AJ Clutterbuck. Sexual and parasexual genetics of *Aspergillus* species. Biotechnology 23:3–18, 1992.
29. C d'Enfert, BM Bonini, PD Zapella, T Fontaine, AM da Silva, HF Terenzi. Neutral trehalases catalyse intracellular trehalose breakdown in the filamentous fungi *Aspergillus nidulans* and *Neurospora crassa*. Mol Microbiol 32:471–483, 1999.
30. SD Harris, AF Hofmann, HW Tedford, MP Lee. Identification and characterization of genes required for hyphal morphogenesis in the filamentous fungus *Aspergillus nidulans*. Genetics 151:1015–1025, 1999.
31. BL Miller. Brushing up on bristles: complex genes and morphogenesis in molds. Trends Genet 9:293–295, 1993.
32. WE Timberlake, AJ Clutterbuck. Genetic regulation of conidiation. In: SD Martinelli, JR Klinghorn, eds. Prog Ind Microbiol. Amsterdam: Elsevier, 1994, Vol 29, pp 383–427.
33. TH Adams, JK Wieser, JH Yu. Asexual sporulation in *Aspergillus nidulans*. Microbiol Mol Biol Rev 62:35–54, 1998.
34. JL Mooney, LN Yager. Light is required for conidiation in *Aspergillus nidulans*. Genes Dev 4:1473–1482, 1990.

35. AJ Clutterbuck. *Aspergillus nidulans*. In: RC King ed. Handbook of Genetics. New York: Plenum, 1974, pp 447–510.
36. E Eidam. Zur Kenntniss der Entwicklung bei den Ascomyceten III. Cohn Beitr Biol Pflanzen 3:392–411, pl 20–22, 1883.
37. KB Raper, DI Fennell. The Genus *Aspergillus*. Baltimore: Williams and Wilkins, 1965.
38. DM Geiser, ML Arnold, WE Timberlake. Sexual origins of British *Aspergillus nidulans* isolates. Proc Natl Acad Sci USA 91:2349–2352, 1994.
39. B Hoffmann, SE Eckert, S Krappmann, GH Braus. Sexual diploids of *Aspergillus nidulans* do not form by random fusion of nuclei in the heterokaryon. Genetics 157:141–147, 2001.
40. S Pöggeler, S Risch, U Kück, HD Osiewacz. Mating-type genes from the homothallic fungus *Sordaria macrospora* are functionally expressed in a heterothallic ascomycete. Genetics 147:567–580, 1997.
41. GN Bistis. Chemotrophic interactions between trichogynes and conidia of opposite mating type in *Neurospora crassa*. Mycologia 73:959–975, 1981.
42. E Coppin, S Arnaise, V Cotamine, M Picard. Deletion of the mating-type sequences in *Podospora anserina* abolishes mating without affecting vegetative functions and sexual differentiation. Mol Gen Genet 241:667–673, 1993.
43. SP Champe, LD Simon. Cellular differentiation and tissue formation in the fungus *Aspergillus nidulans*. In: EF Rossomando, S Alexander, eds. Morphogenesis: An Analysis of the Development of Biological Structure. New York: Marcel Dekker, 1992, pp 255–276.
44. RB Dales, JH Croft. Investigation of the *het* genes that control heterokaryon incompatibility between members of heterokaryon-compatibility (*h-c*) groups A and G1 of *Aspergillus nidulans*. J Gen Microbiol 136:1717–1724, 1990.
45. MM Anwar, JH Croft, RB Dales. Analysis of heterokaryon incompatibility between heterokaryon-compatibility (*h-c*) groups R and GL provides evidence that at least eight *het* loci control somatic incompatibility in *Aspergillus nidulans*. J Gen Microbiol 139:1599–1603, 1993.
46. JF Peberdy. Fungal protoplasts: isolation, reversion and fusion. Annu Rev Microbiol 33:21–39, 1979.
47. AC Butcher. The relationship between sexual outcrossing and heterokaryon incompatibility in *Aspergillus nidulans*. Heredity (Lond) 24:621–631, 1969.
48. BJM Zonneveld. Biochemistry and ultrastructure of sexual development in *Aspergillus*. In: JE Smith, JA Pateman eds. Genetics and Physiology of *Aspergillus*. London: Academic Press, 1977, pp 59–80.
49. H Linden, P Ballario, G Macino. Blue light regulation in *Neurospora crassa*. Fungal Genet Biol 22:141–150, 1997.
50. F Degli-Innocenti, JAA Chambers, VEA Russo. Photoinduction of protoperithecia in *Neurospora crassa* by blue light. Photochem Photobiol 37:49–51, 1983.
51. DD Perkins. Photoinduced carotenoid synthesis in perithecial wall tissue of *Neurospora crassa*. Fungal Genet Newsl 35:38–39, 1988.
52. RW Harding, S Melles. Genetic analysis of the phototrophism of *Neurospora crassa* perithecial beaks using white collar and albino mutants. Plant Physiol 72: 996–1000, 1983.

53. DE Axelrod, M Gealt, M Pastushok. Gene control of developmental competence in *Aspergillus nidulans*. Dev Biol 34:9–15, 1973.
54. J Galbriath, JE Smith. Induction of asexual sporulation in *Aspergillus niger* in submerged liquid culture. J Gen Microbiol 53(suppl), 1968.
55. Serlupi-Crescenzi, MB Kurtz, SP Champe. Developmental defects resulting from arginine auxotrophy in *Aspergillus nidulans*. J Gen Microbiol 129:3535–3544, 1983.
56. BJ Zonneveld. Sexual differentiation in *Aspergillus nidulans*: the requirement for manganese and its effect on alpha-1,3 glucan synthesis and degradation. Arch Microbiol 105:101–104, 1975.
57. BJ Zonneveld. Alpha-1,3 glucan synthesis correlated with alpha-1,3 glucanase synthesis, conidiation and fructification in morphogenetic mutants of *Aspergillus nidulans*. J Gen Microbiol 81:445–451, 1974.
58. DM Han, YJ Han, KS Chae, KY Jahng, YH Lee. Effects of various carbon sources on the development of *Aspergillus nidulans* with $velA^+$ or *velA1* allele. Korean J Mycol 22:332–337, 1994.
59. H Inoue, C Shimoda. Induction of trehalase activity on a nitrogen-free medium: a sporulation-specific event in the fission yeast, *Schizosaccharomyces pombe*. Mol Gen Genet 183:32–36, 1981.
60. MM Yelton, JE Hamer, ER de Souza, EJ Mullaney, WE Timberlake. Developmental regulation of the *Aspergillus nidulans trpC* gene. Proc Natl Acad Sci USA 80:7576–7580, 1983.
61. E Käfer. The anthranilate synthetase enzyme complex and the trifunctional *trpC* gene of *Aspergillus*. Can J Genet Cytol 19:723–738, 1977.
62. SE Eckert, B Hoffmann, C Wanke, GH Braus. Sexual development of *Aspergillus nidulans* in tryptophan auxotrophic strains. Arch Microbiol 172:157–166, 1999.
63. SE Eckert, E Kübler, B Hoffmann, GH Braus. The tryptophan synthase encoding gene *trpB* of *Aspergillus nidulans* is regulated by the cross-pathway control system. Mol Gen Genet 263:867–876, 2000.
64. S Busch, B Hoffmann, O Valerius, K Starke, K Düvel, GH Braus. Regulation of the *Aspergillus nidulans hisB* gene by histidine starvation. Curr Genet 38:314–322, 2001.
65. SP Champe, P Rao, A Chang. An endogenous inducer of sexual development in *Aspergillus nidulans*. J Gen Microbiol 133:1383–1387, 1987.
66. SP Champe, AA El-Zayat. Isolation of a sexual sporulation hormone from *Aspergillus nidulans*. J Bacteriol 171:3982–3988, 1989.
67. AM Calvo, LL Hinze, HW Gardner, NP Keller. Sporogenic effect of polyunsaturated fatty acids on development of *Aspergillus* spp. Appl Environ Microbiol 65: 3668–3673, 1999.
68. BN Lee, TH Adams. *fluG* and *flbA* function interdependently to initiate conidiophore development in *Aspergillus nidulans* through *brlA* beta activation. EMBO J 15:299–309, 1996.
69. LN Yager, HO Lee, DL Nagle, JE Zimmerman. Analysis of *fluG* mutations that affect light-dependent conidiation in *Aspergillus nidulans*. Genetics 149:1777–1786, 1998.

70. JL Mooney, DE Hassett, LN Yager. Genetic analysis of suppressors of the *veA1* mutation in *Aspergillus nidulans*. Genetics 126:869–874, 1990.
71. LN Yager, MB Kurtz, SP Champe. Temperature-shift analysis of conidial development in *Aspergillus nidulans*. Dev Biol 93:92–103, 1982.
72. TH Adams, WE Timberlake. Developmental repression of growth and gene expression in *Aspergillus*. Proc Natl Acad Sci USA 87:5405–5409, 1990.
73. M Carsiotis, RF Jones. Cross-pathway regulation: tryptophan-mediated control of histidine and arginine biosynthetic enzymes in *Neurospora crassa*. J Bacteriol 119: 889–892, 1974.
74. M Piotrowska. Cross-pathway regulation of ornithine carbamoyltransferase synthesis in *Aspergillus nidulans*. J Gen Microbiol 116:335–339, 1980.
75. MS Sachs, C Yanofsky. Developmental expression of genes involved in conidiation and amino acid biosynthesis in *Neurospora crassa*. Dev Biol 148:117–128, 1991.
76. O Valerius, O Draht, E Kübler, K Adler, B Hoffmann, GH Braus. Regulation of *hisHF* transcription of *Aspergillus nidulans* by adenine and amino acid limitation. Fungal Genet Biol 32:21–31, 2001.
77. MS Sachs. General and cross-pathway controls of amino acid biosynthesis. In: R Brambl, GA Marzluf, eds. The Mycota III. Biochemistry and Molecular Biology. Berlin: Springer-Verlag, 1996, pp 315–345.
78. AG Hinnebusch. General and pathway-specific regulatory mechanisms controlling the synthesis of amino acid biosynthetic enzymes in *Saccharomyces cerevisiae*. In: EW Jones, JR Pringle, JR Broach, eds. The Molecular and Cellular Biology of the Yeast *Saccharomyces*. Vol 2: Gene Expression. Cold Spring Harbor, NY: Cold Spring Harbor Laboratory Press, 1992, pp 319–414.
79. JL Paluh, MJ Orbach, TL Legerton, C Yanofsky. The cross-pathway control gene of *Neurospora crassa*, *cpc-1*, encodes a protein similar to GCN4 of yeast and the DNA-binding domain of the oncogene *v-jun*–encoded protein. Proc Natl Acad Sci USA 85:3728–3732, 1988.
80. C Wanke, S Eckert, G Albrecht, W van Haringsveldt, PJ Punt, CAMJJ van den Hondel, GH Braus. The *Aspergillus niger GCN4* homologue *cpcA* is transcriptionally regulated and encodes an unusual leucine zipper. Mol Microbiol 23:23–33, 1997.
81. B Hoffmann, C Wanke, SK Kirchner, GH Braus. *c-Jun* and *RACK1* homologs regulate a control point for sexual development in *Aspergillus nidulans*. Mol Microbiol 37:28–41, 2000.
82. B Hoffmann, HU Mösch, E Sattlegger, IB Barthelmess, A Hinnebusch, GH Braus. The WD protein Cpc2p is required for repression of Gcn4 protein activity in yeast in the absence of amino acid starvation. Mol Microbiol 31:807–822, 1999.
83. BN Lee, TH Adams. The *Aspergillus nidulans fluG* gene is required for production of an extracellular developmental signal and is related to prokaryotic glutamine synthase I. Genes Dev 8:641–651, 1994.
84. S Rosen, JH Yu, TH Adams. The *Aspergillus nidulans sfaD* gene encodes a G protein beta subunit that is required for normal growth and repression of sporulation. EMBO J 18:5592–5600, 1999.
85. J Wieser, JH Yu, TH Adams. Dominant mutations affecting both sporulation and

sterigmatocystin biosynthesis in *Aspergillus nidulans*. Curr Genet 32:218–224, 1997.
86. J Wieser, BN Lee, J Fondon 3rd, TH Adams. Genetic requirements for initiating asexual development in *Aspergillus nidulans*. Curr Genet 27:62–69, 1994.
87. TH Adams, WA Hide, LN Yager, BN Lee. Isolation of a gene required for programmed initiation of development by *Aspergillus nidulans*. Mol Cell Biol 12: 3827–3833, 1992.
88. JH Yu, J Wieser, TH Adams. The *Aspergillus* FlbA RGS domain protein antagonizes G protein signaling to block proliferation and allow development. EMBO J 15:5184–5190, 1996.
89. H-U Mösch, GR Fink. Dissection of filamentous growth by transposon mutagenesis in *Saccharomyces cerevisiae*. Genetics 145:671–684, 1997.
90. E Kübler, H-U Mösch, S Rupp, MP Lisanti. Gpa2p, a G-protein subunit, regulates growth and pseudohyphal development in *Saccharomyces cerevisiae* via a cAMP-dependent mechanism. J Biol Chem 272:20321–20323, 1997.
91. N Osherov, G May. Conidial germination in *Aspergillus nidulans* requires RAS signaling and protein synthesis. Genetics 155:647–656, 2000.
92. T Som, VSR Kolaparthi. Developmental decisions in *Aspergillus nidulans* are modulated by Ras activity. Mol Cell Biol 14:5333–5348, 1994.
93. F Banuett. Signalling in the yeasts: an informational cascade with links to the filamentous fungi. Microbiol Mol Biol Rev 62:249–274, 1998.
94. H-U Mösch, RL Roberts, GR Fink. Ras2 signals via the Cdc42/Ste20/mitogen-activated protein kinase module to induce filamentous growth in *Saccharomyces cerevisiae*. Proc Natl Acad Sci USA 93:5352–5356, 1996.
95. MA Vallim, KY Miller, BL Miller. *Aspergillus* SteA (Sterile 12-like) is a homeodomain-C_2/H_2-Zn^{+2} finger transcription factor required for sexual reproduction. Mol Microbiol 36:290–301, 2000.
96. J Wieser, TH Adams. *flbD* encodes a Myb-like DNA-binding protein that coordinates initiation of *Aspergillus nidulans* conidiophore development. Genes Dev 9: 491–502, 1995.
97. TM Busby, KY Miller, BL Miller. Suppression and enhancement of the *Aspergillus nidulans medusa* mutation by altered dosage of the *bristle* and *stunted* genes. Genetics 143:155–163, 1996.
98. AJ Clutterbuck. A mutational analysis of conidial development in *Aspergillus nidulans*. Genetics 63:317–327, 1969.
99. KY Miller, TM Toennis, TH Adams, BL Miller. Isolation and transcriptional characterization of a morphological modifier: the *Aspergillus nidulans* stunted (*stuA*) gene. Mol Gen Genet 227:285–292, 1991.
100. JR Dutton, S Johns, BL Miller. StuAp is a sequence-specific transcription factor that regulates developmental complexity in *Aspergillus nidulans*. EMBO J 16: 5710–5721, 1997.
101. J Wu, BL Miller. *Aspergillus* asexual reproduction and sexual reproduction are differentially affected by transcriptional and translational mechanisms regulating stunted gene expression. Mol Cell Biol 17:6191–6201, 1997.
102. KY Miller, J Wu, BL Miller. *stuA* is required for cell pattern formation in *Aspergillus*. Genes Dev 6:1770–1782, 1992.

103. CJ Gimeno, GR Fink. Induction of pseudohyphal growth by overexpression of *PHD1*, a *Saccharomyces cerevisiae* gene related to transcriptional regulators of fungal development. Mol Cell Biol 14:2100–2112, 1994.
104. RC Pascon, BL Miller. Morphogenesis in *Aspergillus nidulans* requires Dopey (DopA), a member of a novel family of leucine zipper-like proteins conserved from yeast to humans. Mol Microbiol 36:1250–1264, 2000.
105. KS Chae, JH Kim, Y Choi, DM Han, KY Jahng. Isolation and characterisation of a genomic DNA fragment complementing an *nsdD* mutation of *Aspergillus nidulans*. Mol Cells 5:146–150, 1995.
106. P Ballario, P Vittorioso, A Magrelli, C Talora, A Cabibbo, G Macino. White collar-1, a central regulator of blue light responses in *Neurospora*, is a zinc finger protein. EMBO J 15:1650–1657, 1996.
107. H Linden, G Macino. White collar-2, a partner in blue-light signal transduction, controlling expression of light-regulated genes in *Neurospora crassa*. EMBO J 16: 98–109, 1997.
108. DM Han, YJ Han, YH Lee, KY Jahng, SH Jahng, KS Chae. Inhibitory conditions of asexual development and their application for the screening of mutants defective in sexual development. Korean J Mycol 18:225–232, 1990.
109. DM Han, YJ Han, JH Kim, KY Jahng, YS Chung, JH Chung, KS Chae. Isolation and characterisation of *nsd* mutants in *Aspergillus nidulans*. Korean J Mycol 22: 1–7, 1994.
110. F Osman, B Tomsett, P Strike. The isolation of mutagen-sensitive *nuv* mutants of *Aspergillus nidulans* and their effects on mitotic recombination. Genetics 134:445–454, 1993.
111. KE Kirk, NR Morris. The *tubB* alpha-tubulin gene is essential for sexual development in *Aspergillus nidulans*. Genes Dev 5:2014–2023, 1991.
112. M Krüger, R Fischer. Integrity of a Zn finger-like domain in SamB is crucial for morphogenesis in ascomycetous fungi. EMBO J 17:204–214, 1998.
113. TE Hermann, MB Kurtz, SP Champe. Laccase localized in Hulle cells and cleistothecial primordia of *Aspergillus nidulans*. J Bacteriol 154:955–964, 1983.
114. M Scherer, R Fischer. Purification and characterization of laccase II of *Aspergillus nidulans*. Arch Microbiol 170:78–84, 1998.
115. B Hoffmann, SK LaPaglia, E Kübler, M Andermann, SE Eckert, GH Braus. Developmental and metabolic regulation of the phosphoglucomutase-encoding gene *pgmB* of *Aspergillus nidulans*. Mol Gen Genet 262:1001–1011, 2000.

10

Sexual Development in Basidiomycetes

Erika Kothe
Friedrich-Schiller-Universität, Jena, Germany

1 INTRODUCTION

1.1 Sexual Development and Mushroom Formation

Among the basidiomycetes, the mushroom-forming fungi are of commercial value and thus the production of fruiting bodies has been studied in detail. Mushroom growers are interested in the basic mechanisms underlying fruit body formation in order to improve the yield and to find stable conditions under which high crop yields can be obtained on a regular basis. The induction of fruit body development under defined and sterile conditions, which is prerequisite to study the molecular clues for mushroom formation, is dependent on the growth conditions and therefore easiest with saprophytic fungi. A wide range of basidiomycetes, however, are able to fructificate only in association with a host plant because they live in close association with the plant either in a mutualistic symbiosis or as phytopathogens. These fungi include many edible fungi like the ectomycorrhizal boletes, the wood-rotting, parasitic honey agaric *Armillaria*, and other valuable edible species. The fungi living in close contact to a host plant, e.g., ectomycorrhizal species [this volume, Chapter 12 by Duplessis et al.], are not able to form fruit bodies without the signals from their host tree. Only in very rare instances have we been able to identify such host signals.

This is also true for most phytopathogenic fungi, including the heterobasidiomycetous rust and smut fungi where host signals are needed to allow advance in development, e.g., for formation of infection structures, but also for the induction of the diploid stage that is needed to produce teliospores, e.g., in *Ustilago* [this volume, Chapter 14 by Banuett]. Therefore, for research purposes mushroom-forming saprophytes have been used in many instances. In this chapter the prerequisites of fruit body formation and development of fruit bodies in homobasidiomycetes with special emphasis on mushroom-forming fungi will be discussed in detail.

1.2 What Makes a Mushroom?

The developmental processes necessary for the production of mushrooms and also within the fruit bodies for meiosis and sexual reproduction are dependent on the formation of a diploid stage. Usually, the diploid stage is short in basidiomycetes and immediately precedes meiosis. From the encounter of a possible mate to meiosis, a prolonged dikaryotic stage in which both nuclei from the parental strains are found closely connected but unfused generally is observed in basidiomycetes.

Establishment of a dikaryon competent for fruit body formation depends on the mating of two haploid strains. In the easiest case, from the basidiospores such haploid, monokaryotic mycelia carrying one nucleus per cell are established (Fig. 1). Since all nuclei are identical, this mycelium is also called a homokaryon. When two mycelia of different mating types come into contact, anastomoses between the two mycelial types allow the exchange of nuclei, and a dikaryon can be established in which the two nuclei of both parental strains are found in every cell, typically associated but unfused. The dikaryon is able to develop fruit bodies with the hymenium in which karyogamy takes place. Karyogamy results in a diploid cell which usually immediately enters meiosis resulting in four haploid basidiospores.

The described life cycle is typical for many homobasidiomycetes, with the two mushroom-forming, saprophytic fungi *Schizophyllum commune*, a white rot fungus, and the ink cap *Coprinus cinereus* being widely used as model organisms. Besides these heterothallic fungi with different mating types, there are homothallic fungi that can form fruit bodies and sexual spores without the need for mating. A reason may be found in genes usually activated only upon mating between two different mating types. If those genes are constitutively expressed in a monokaryotic strain, it may lead to fruit body formation without the need of mating. Such mutants are termed homokaryotic fruiters and can be found rarely but regularly in heterothallic species. And there are secondary homothallic species like the commercially grown white button mushroom, or champignon, *Agaricus bisporus*, in which only two basidiospores are formed, which may contain two

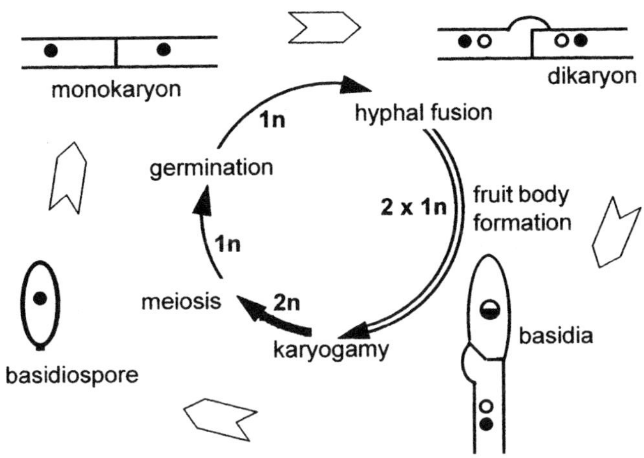

FIGURE 1 Typical homobasidiomycete life cycle. Besides the capability of some basidiomycetes to form asexual reproductive structures [this volume, Chapter 6 by Kües et al.], homobasidiomycetes generally propagate by sexual reproduction. Two haploid monokaryotic mycelia of different mating types fuse to allow the formation of a dikaryon in which every cell contains two different nuclei, one of each mate. The dikaryon typically forms clamp connections to maintain this distribution of nuclei. On a dikaryotic mycelium fruit body formation can be induced and only in the hymenium of the developing basidia karyogamy occurs resulting in a diploid cell which immediately enters meiosis leading to the formation of four haploid basidiospores. These can germinate to yield the monokaryotic mycelia.

nuclei of different mating type. The germinating basidiospores in this case can form a dikaryon directly. However, such cases can be seen as variations on the general theme of heterothallic mating systems found in most basidiomycetes.

In the following sections the making of a mushroom will be discussed following the developmental stages. The mating system governing sexual reproduction in basidiomycete fungi is the first prerequisite and includes the recognition of mating partners by the mating-type genes (sections 2–8). The signal obtained by encountering a compatible mating partner is then transmitted to allow differential gene expression (sections 9–11). The genes involved in sexual reproduction can be found as differentially expressed genes involved in the formation or maintenance of a stable dikaryon, often associated with the formation of clamp connections. However, mating type–specific, differentially expressed genes can themselves be regulators or may have a function in integrating the intracellular signal with environmental factors (section 12). These environmental signals, e.g., light or, in case of pathogenic fungi, a host signal, are necessary for fruit body formation, meiosis, and spore formation to occur (sections 13–16).

All these stages in formation of mushrooms have been analyzed over the past decade on the level of molecular genetics. Many genes involved in the different steps of fruit body development have been analyzed from different basidiomycetes in order to understand the general and unifying processes that make a mushroom. This chapter will try to define such steps in development on a molecular basis calling on different model fungi. In these model organisms separate steps have been investigated because the unique organization of the fungus made it especially well suited to examine single sequences in the development of fruit bodies and sexual reproduction. From these pieces a general picture has to be built, painting the backdrop for mushroom development.

2 MATING SYSTEM

2.1 Tetrapolar Mating Types in the Basidiomycetes

Most heterothallic homobasidiomycetes such as *Schizophyllum commune* or *Coprinus cinereus* are tetrapolar. The term has been coined for a system of multiple mating types—in *S. commune* there are >20,000 different mating types observed in nature, for *C. cinereus* a total of 12.800 mating types has been calculated—in which among crosses between sibling haploid strains all coming from spores of one fruit body, four different mating interactions can be observed. The four different interactions can be explained by two independent, genetically unlinked mating factors called *A* and *B*: the fully compatible mating with $A\neq B\neq$, the incompatible mating with $A= B=$, and the two semicompatible interactions $A\neq B=$ and $A= B\neq$. This feature of semicompatible interactions allowed the identification of processes governed by the *A* mating-type genes as contrasted to those that required the action of *B* genes [1]. The *A* genes are necessary for the formation of clamp cells and conjugate nuclear division, while *B* genes are responsible for the exchange of nuclei (Fig. 2). The final step of fusion in clamp formation and the further development of fruit bodies need the activity of both loci.

Successful mating is therefore dependent on the two sets of mating genes in the *A* and *B* loci. The two compatible strains undergoing a productive mating must contain different mating type loci in both *A* and *B*. And for both the *A* and the *B* loci multiple allelic specificities exist in nature. In a first attempt, recombination analyses have shown that more than one locus is responsible for *A*-dependent development, and for *B* there are also at least two independent but functionally redundant loci. The recombination analyses for *S. commune* have revealed two genetically linked but independent loci for *A* and *B* each, which have been called $A\alpha$, $A\beta$, $B\alpha$, and $B\beta$. Each of these four loci is multiallelic, and from isolation

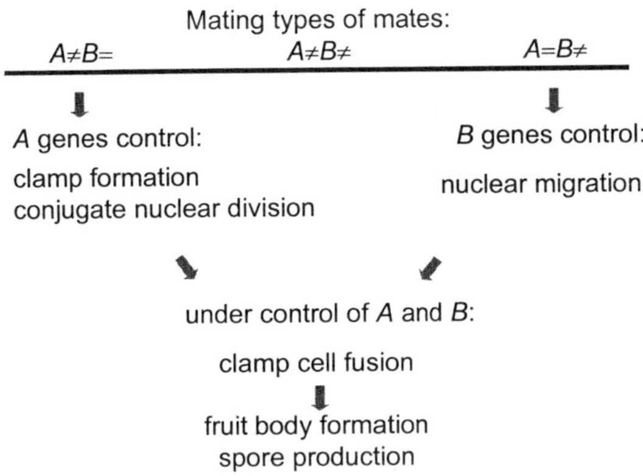

FIGURE 2 Two different sets of genes regulate basidiomycete sexual development. While genes of one locus, called *A* in *Schizophyllum commune* and *Coprinus cinereus*, regulate the formation of clamp cells, the *B* loci control the exchange of nuclei needed to establish the dikaryotic mycelium. Only if both pathways of development are turned on in a mating interaction between strains differing in both *A* and *B* loci, can a fertile dikaryon be established.

of strains from nature a variability of $A\alpha$ with nine specificities, $A\beta$ with 32, $B\alpha$ with nine, and $B\beta$ again with nine allelic specificities has been calculated [2]. This makes for 23,328 possible independent mating types in wild-type strains in the population of *S. commune*.

In the ink cap mushroom *C. cinereus* a high number of independent mating-type alleles of the two factors *A* and *B* can also be observed, with ~160 different *A* factors and 80 different *B* specificities [1]. In *C. cinereus* the partial loci making up the *A* factor, as well as the *B* loci, are linked more closely than has been observed in *S. commune*. Detailed molecular analyses have shown that four *A* loci and three *B* loci can be designated [3,4].

The reason for such an extensive redundancy can be found in the enhancing of outbreeding versus inbreeding. Among sibling spores, only one of four interactions is fully compatible and can distribute by producing sexual spores. In a mating of two unrelated strains, almost all encounters will allow the production of sexual spores as 98% of the encountered putative mating partners will have compatible mating types. In order to evolve, such a system must represent a strong evolutionary bias.

2.1.1 The *A* Loci

In both model homobasidiomycetes in which mating-type genes have been analyzed, in *Schizophyllum commune* and *Coprinus cinereus*, two and (based on sequence data in *C. cinereus*) up to four separate sets of divergently transcribed genes are found that can provide functional *A* mating type genes [for review see 3,4].

The specificity of any *A* gene can be tested in transformation experiments. If a gene encoding a subunit from a different *A* mating type is transformed into a monokaryotic cell, this cell will allow the transcription of genes for clamp cell formation, and unfused clamp cells can be observed in the phenotype of the transformed cell. This phenotype has been used to clone the *A* genes.

The first mating-type genes cloned and sequenced were the genes *z* and *y* found in the *Aα* locus of *S. commune* [5]. These genes, as well as the *Aβ* genes of *S. commune* and *C. cinereus* homologs, code for homeodomain-containing transcription factors which can enter the nucleus and regulate transcription of target genes (Fig. 3). To do so, heterodimers between the two different gene

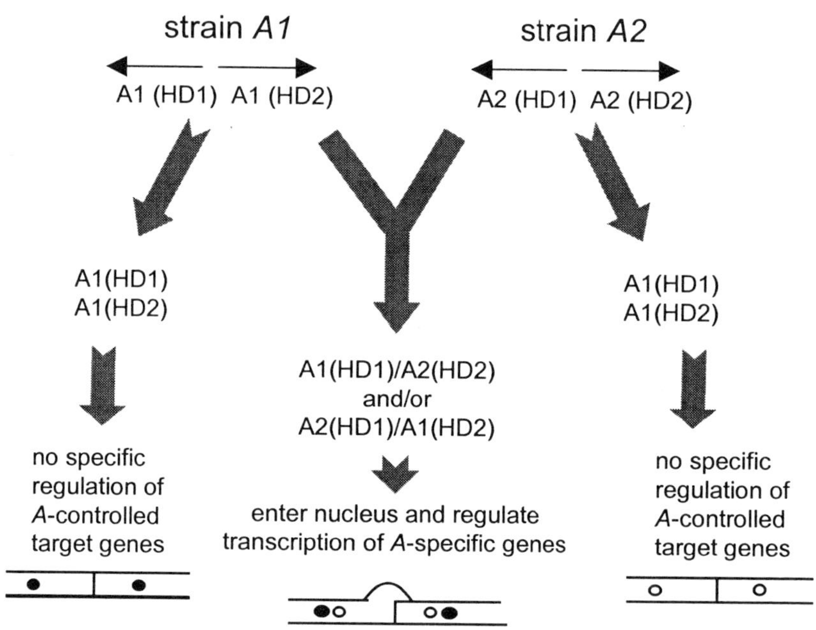

FIGURE 3 A typical *A* locus contains two divergently transcribed genes encoding homeodomain transcription factors of HD1 and HD2 types. A heterodimer HD1/HD2 of the two gene products is only active in inducing *A*-regulated development (resulting in dikaryon formation in matings with different *B* genes between the two strains) if the two proteins are from different allelic specificity, e.g., from strain *A1* and strain *A2*, respectively.

products of one set must be formed and enter the nucleus. The target genes are only responsive to a heterodimer that consists of subunits from two different specificities. This allows the regulation to be dependent on the fusion of two cells of different mating type, while in monokaryotic cells a heterodimer, even if it is formed, is inactive in regulating target genes and thereby allowing haploid-specific gene expression to continue [6].

In accordance with the role of the *A* genes as master regulators of development, these genes are transcribed constitutively at a basal level. They show sequence similarity to genes of *Saccharomyces cerevisiae* that are part of the mating-type system there. The genes *a1* and *α2* of the yeast are transcribed only in a and α cells, respectively. Only in a mated diploid are both proteins found in a common cytoplasm. They form a heterodimer and this is able to regulate transcription, leading to derepression of diploid-specific genes. The *α2* homodimer, instead, is able to repress a-specific functions in haploid α and diploid a/α cells.

As with the yeast, the basidiomycete genes encode homeodomain transcription factors with the three helices typical for this class of proteins [6]. However, there is a difference between the two different gene products encoded in one sublocus: while one shows the completely conserved motif of 60 amino acids forming the three helices with the consensus sequence WF X N X R, the other one shows a different spacing pattern between the first and second helix and the consensus sequence in the third helix involved in DNA recognition is divergent with WF X D X R. Accordingly, the two gene products of a sublocus can be placed into two groups—HD1 and HD2 proteins. While HD2 proteins contain the fully conserved homeodomain sequence, HD1-class proteins share the divergent motif.

This structure of the mating-type genes implies that one HD1 and one HD2 protein of allelic versions of the same sublocus must form a heterodimer that is the regulator of *A*-specific development [for review see 6]. Thus, for example the HD1 protein Z4 from the locus *Aα4* in *S. commune* must associate with the HD2 protein Y2 of *Aα2*. In *C. cinereus* the situation is even more complex, given that up to four subloci are encoded in the *A* factor with the a-pair, b-pair, c-pair (for which still only one of the functional proteins has been identified), and d-pair, each of which (possibly with exception of the c-pair) encodes both an HD1 and an HD2 protein [3,6]. Multiple allelic specificities within each of the pairs can explain easily the multitude of different *A* specificities observed in nature.

The function of the two proteins seems to be a different one [for review: 6]. While the HD1 protein is thought to contain the functional signal for nuclear localization, HD2 is the protein that binds to the specific DNA target sites in regulated genes [7]. In accordance with this, the DNA-binding domain in HD1 can be deleted, and a fusion protein providing HD1 nuclear localization and HD2

DNA-binding capacity has been found to be the cause for a mutation rendering a mutant *C. cinereus* strain constitutively activated in *A*-regulated development because the dimerization is no longer necessary in the fusion protein [8].

However, in the wild-type situation dimerization has to occur before the complex enters the nucleus and regulates target genes. The dimerization domains have been localized in the N-terminus of the proteins, and a whole array of interaction sites are contributing to the overall dimerization or aversion of the two proteins. In the C-terminal part of the proteins, sites are located that may also allow binding [9]; however, these sites do not contribute to specificity of this interaction. In a yeast two hybrid system therefore with the C-termini interaction may be encountered that is not depending on the allelic difference of the two subunits. The full-length proteins, however, are dependent on the different allelic origin for heterodimer formation.

As with *S. cerevisiae*, it is expected that the HD1 proteins can also form homodimers which may have a function in the regulation of oidia formation in *C. cinereus*. In accordance with this function, the HD1 protein contains a C-terminal activation domain that may regulate transcription. Since the HD1 protein is also the one that carries the nuclear localization signal, the regulation of a different set of genes by the homodimer would give an elegant explanation to the question of how such an elaborate system has evolved in the first place. The force for this extensive diversification of mating-type genes is expected to be the enhanced outbreeding rate, as discussed above.

2.1.2 The *B* Genes

From ascomycetes, pheromones had been known to be involved in mating-partner attraction. For example, the formation of mating protractions called shmooh cells in *Saccharomyces cerevisiae* is induced by pheromone of the opposite mating type, and the direction into which the protrusion is formed is a response to the pheromone gradient secreted by the potential mate. Also, in phragmobasidiomycetes (including *Tremella*), pheromones had been shown to exist. This was corroborated for the heterobasidiomycete *Ustilago maydis*—which shows a tetrapolar mating type—where pheromones have also been found [for review: 10].

In the homobasidiomycetes, fusion can be observed among all mating types. Indeed, anastomoses are formed even within a monokaryon, and it has been argued that these connections are necessary to allow nutrient distribution throughout an extensive mycelium [1]. It therefore came as a surprise when in *Schizophyllum commune* and *Coprinus cinereus* pheromone receptors and pheromones were found to be encoded in the mating-type loci of the *B* factors [11–13]. The function of these homobasidiomycetous pheromones, however, seems not to be primarily the attraction of mates. Rather, in the mushrooms nuclear migration, which is governed by the *B* factor, is under control of the pheromone system [4].

A function of pheromones is nuclear migration can be observed in mating interactions where two mates are opposed and nuclear migration is scored in each mate. This is easily achieved by scoring the formation of dikaryotic cells with clamp connections, which can only be formed in *Schizophyllum* when two different nuclei are present in one cell. If both mates are normal, wild-type monokaryons, both will be able to transport the mate's nuclei throughout the entire preformed mycelium, and at the new growth zones on both sides dikaryotic hyphae are formed.

If one of the two mates lacks pheromone genes, the induction of nuclear migration in the mate is not induced and hence in the mate no nuclear migration and no formation of dikaryotic clamped hyphae is seen (Fig. 4). This behavior is termed a unilateral nuclear acceptor, as the pheromoneless mutant cell is able to take up and transport the mate's nuclei but does not donate nuclei to the wild-type mate [1,4]. If the mutant contains pheromone but no receptor genes, it will behave as a unilateral nuclear donor in that the mate is induced and accepts the nuclei of the mutant strain, but in the mutant no nuclear migration is possible and therefore no clamped hyphae will be seen on the mutants side of the mating. These observations have been reproduced in numerous variations, using not only mutant strains but also transformants, and in each case the behavior was as proposed using this model of nuclear migration being induced by pheromone and receptor interactions.

The *B* loci have been cloned and sequence analyses revealed at the two loci $B\alpha$ and $B\beta$ of *S. commune* and at three loci of *C. cinereus* pheromone receptor systems each of which is multiallelic [11–14] (Fig. 5). Thus, one receptor gene in each locus (called, e.g., *bar1* for the $B\alpha1$ receptor, and *bbr2* for the $B\beta2$ receptor in *S. commune*) is able to accept pheromones of every other specificity, while the two to six self pheromones encoded in the same locus (called e.g., *bap2(1)* for the first pheromone of the $B\beta2$ locus) are excluded from inducing the downstream signal transduction cascade. The different loci within one cell are functionally separate. Thus, for example in *S. commune*, $B\alpha$ pheromones are able to induce $B\alpha$ receptors and $B\beta$ pheromones are able to induce $B\beta$ receptors.

In contrast to the *A* genes, the transcription of *B* genes is induced upon mating [11,15]. Some time after the initial rise in mRNA levels following the contact between mates, transcript levels are decreasing and basal transcript accumulation is seen again.

Pheromone Perception. The presence of a compatible pheromone is perceived via the pheromone receptor. This signal is then transduced within the cell and leads to a change in expression of target genes. To allow signal transduction, the pheromone must be recognized by a seven-transmembrane domain receptor (Fig. 6). This class of proteins is involved in recognition of extracellular signals in different organisms from halobacteria—where bacteriorhodopsin is a light

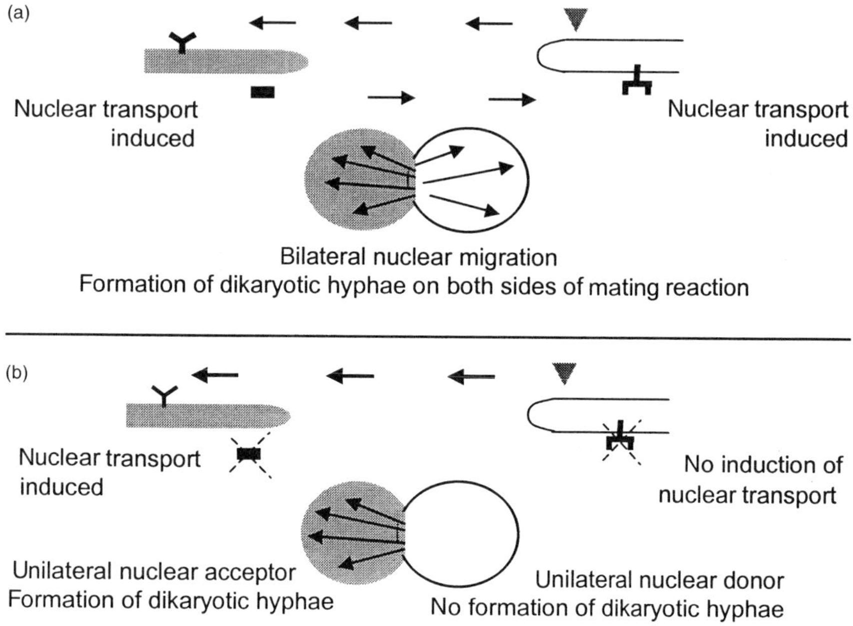

Figure 4 Nuclear migration is easily scored in matings by the occurrence of clamped dikaryotic mycelium on either side of a mating reaction. Reciprocal pheromone stimulation will result in formation of dikaryotic hyphae on both performed mycelia (a). Mutations in the pheromones in the strain depicted on the left (or loss of receptor in the strain on the right) will result in unilateral nuclear migration and formation of dikaryotic hyphae only on the left side of the mating shown (b). Only at this side will activated receptor molecules be found that allow nuclear transport, making the strain a unilateral nuclear acceptor in this interaction. The pheromone response will then lead to *B*-regulated development.

driven ion pump—to mammalian receptors for somatostatin, bradykinin, bitter taste, or odor perception. Most of these systems are able to react upon binding of one very specific molecule. With odor perception, however, it was shown that receptor molecules can react to multiple small molecules, which can include, for one and the same receptor, alcohols and carboxy acids with short carbohydrate backbones [16]. Other components, which may vary only in chain lengths from those recognized, will not produce a signal in such a receptor. In a similar fashion, the basidiomycete pheromone receptors are able to distinguish between rather similar molecules, answering to one lipopeptide pheromone with signal transduction and altered expression profile while a second lipopeptide pheromone is not

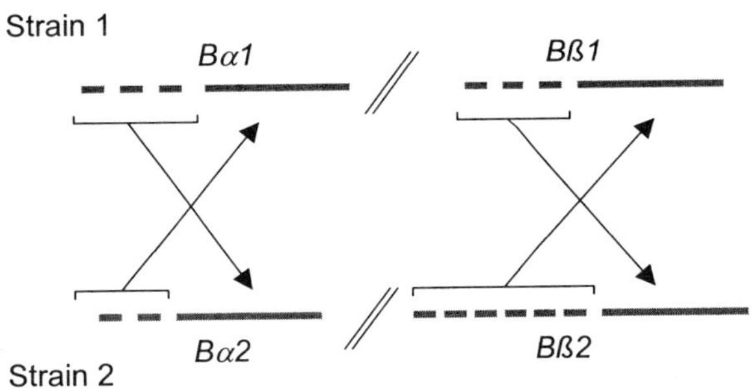

Figure 5 The *Schizophyllum commune* mating type loci $B\alpha$ and $B\beta$ are located on the same chromosome at ~50 kb distance [17]. Each locus encodes a pheromone receptor (large bar) and multiple pheromones (short bars). At least one of the pheromones (brackets) in each locus is able to induce the receptor of a different specificity (arrow). Self-receptors are not induced and no crossreactivity between the α and β loci is seen. Thus, pheromones of strain 1 will induce pheromone response in strain 2 but not in strain 1 itself.

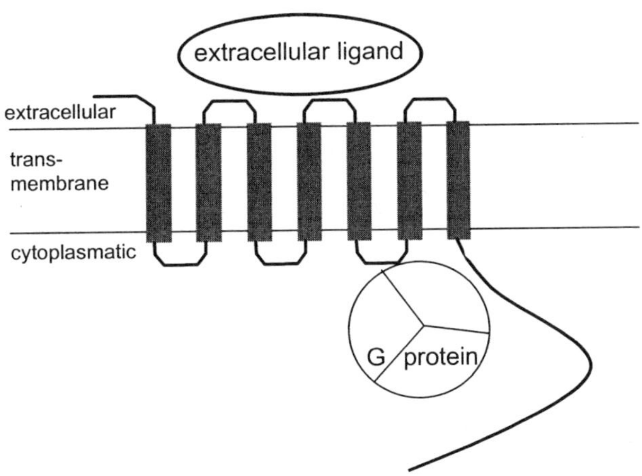

Figure 6 The pheromone receptors of basidiomycetes belong to the class of G-protein linked, seven-transmembrane domain receptors which are known to transmit an extracellular signal such as a hormone or pheromone binding at the extracellular side of the membrane into the cell. Upon ligand interaction, a heterotrimeric G-protein is released to relate the signal to an intracellular signal transduction pathway.

able to elicit the signal. Thus, the system is different from the ascomycete pheromone perception, where only one pheromone has to be recognized by a single receptor and only two alternative sets of pheromone and receptor are governing mate attraction.

In the homobasidiomycetes *Schizophyllum commune* and *Coprinus cinereus*, every locus codes for only one pheromone receptor while several pheromone genes are present. The activation spectra for the *S. commune* genes were assayed in transformation experiments in which a strain was used as the recipient that carries no pheromone or receptor genes itself [17]. The resulting transformant that now carried one specific *B* gene only was then tested in mating reactions, and the resulting phenotype, e.g. of unilateral mating, was analyzed for productive interaction of pheromones and receptors. Prior to the identification of such a B_{null} transformation recipient strain, transformation experiments were only possible with strains carrying native *B* loci. Then, a cell transformed with, e.g., a receptor gene would show induction of *B*-regulated development if the receptor expressed from the introduced gene was stimulated by one of the pheromones produced by the recipient cell itself. In this way, no test matings were necessary but activation of a receptor gene could be tested by transformation into strains of all different specificities.

It could be shown that each receptor can be induced by pheromones of all different allelic mating specificities, but not with any of the self pheromones encoded in the same specificity locus [for *S. commune* see 12,13,18,19]. In every *B* locus of a given specificity, multiple pheromone genes are found and each pheromone gene has its own activation spectrum. All pheromones together should be able to cover activation of all other receptors, but a single pheromone activates between one and six other specificities only (cf. Fig. 5).

The question of specificity domains in both receptor and pheromone thus arises. In case of the receptors, it has been shown that two allelic version of *Bα1* specificity are very similar, differing only in the intracellular C-terminus of the protein [20]. Among three different specificities, however, a larger number of different amino acids have been found. Chimeric pheromone receptor genes of the *Bα* locus of *S. commune* have been investigated to identify such specificity regions. Receptors of *Bα1* and *Bα2* specificity encoded by the *bar1* and *bar2* genes, respectively, were combined in domain swapping experiments and it could be shown that *Bα1* specificity resides in the third extracellular domain of the receptor and adjacent parts within the transmembrane domains 6 or 7 [19]. For the *Bα2* specificity, however, specificity is encoded in noncontiguous parts of the receptor.

In addition, with the chimeric receptors new phenotypes were observed. One of these phenotypes was constitutive, thus no longer dependent on the binding of pheromone. Such a constitutive phenotype was also observed for a specific amino acid substitution in *C. cinereus* [8]. The mutation Q229P in transmembrane domain 6 leads to constitutive activation of the receptor. This is interesting since in the analogous yeast receptor mutant the replacement of the native proline to leucine leads to constitutive activation, while in *C. cinereus* the introduction of proline leads to constitutive activation.

Another phenotype observed in the *S. commune* chimeric receptors was a nonselective receptor which no longer discriminated among pheromones of different specificities but still needed activation by binding of any of the pheromones [20].

Use of a Yeast Reporter System to Define Specificity Regions in Homobasidiomycete Pheromone Receptors. The similarity of the general pheromone recognition in yeast and the homobasidiomycetes led to the idea that the multiallelic pheromone recognition of homobasidiomycetes could be studied in yeast, linking the yeast signal transduction cascade and the easy scoring using yeast reporter genes with the activation of signal transduction by pheromones binding to basidiomycetous pheromone receptor molecules.

In yeast the pheromone signal is transduced via a heterotrimeric G-protein that binds to intracellular regions of the receptor including the third intracellular domain between transmembrane domains 5 and 6. To use the yeast downstream cascade of signal transduction, linking of the homobasidiomycete receptor and the yeast G-protein must be achieved. Indeed, yeast G-protein can bind to the basidiomycete receptor and signal transduction is seen depending on the respective basidiomycete pheromone being present. The pheromone was either applied externally using culture filtrates of *Schizophyllum* and expression of a pheromone pathway reporter gene was scored [21] or a pheromone gene was expressed from a different yeast transformant and yeast mating events were scored [18]. However, the G-protein binding is rather ineffective, resulting in a high background and low coupling rates. A better result could be achieved by exchanging the third intracellular domain of the basidiomycete receptor for the sequence found in the yeast pheromone receptor [21]. Then, coupling is better and supernatant of *S. commune* cultures could be used to induce the yeast signal transduction cascade when the *S. commune* receptor was expressed in yeast. The specificity of the receptor was retained as far as tested, possibly allowing mutagenesis approaches in the yeast background. These experiments will allow the deduction of specificity domains more clearly with single amino acids being pinpointed for specific interactions between one pheromone and a receptor. The best coupling between pheromone receptor of a basidiomycete and the yeast signal transduction cascade was achieved in a third set of experiments. Here, a recombinant G-protein was cloned into yeast together with the *C. cinereus* receptor [8]. Obviously, the heterologous G-protein is able to fill the role of the yeast

G-protein and at the same time allows a better binding to the *Coprinus* pheromone receptor.

Basidiomycete Pheromones. The pheromones of basidiomycetes are of the lipopeptide type as seen with the yeast **a**-factor. A short peptide of seven to 14 amino acids is modified with a isoprenoid lipophilic tail that renders the product hydrophobic. A farnesyl moiety is added to a cystein which is present four positions from the stop codon in a farnesylation motif at the C-terminus of the peptide chain. This C-a-a-X farnesylation motif (C, cystein; a, aliphatic amino acid; X, unspecified) is best studied in yeast, where the last amino acid, X = alanine, serine, glutamine, or methionine, is used to signify for the farnesyl tail [22]. In *S. commune* farnesylation and carboxymethylation could be shown to be necessary for full activity of pheromones [18] (Fig. 7).

In *Ustilago maydis* this lipophilic addition has been investigated in more detail. Synthetic peptides have been used that were linked to different lipophilic tails [23]. While the natural component is a farnesyl moiety, an alkyl tail was also found to be active. Indeed, a C14 tail was even more active than the natural component by a factor of 10. An even longer carbohydrate (C16) tail had a weaker effect in the bioassay, which argues for a limited length in the range of the natural

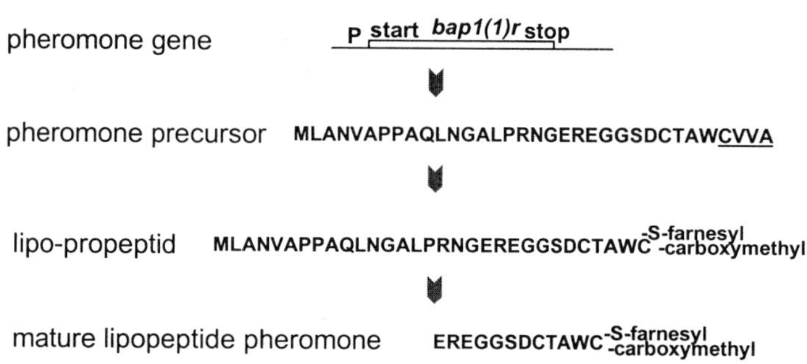

FIGURE 7 Homobasidiomycete pheromone precursors carry a C-terminal signature (underlined) which resembles the C-a-a-X motif that signifies for isoprenylation in yeast. Since farnesylated and carboxymethylated pheromones have been found to be more active in a heterologous assay using yeast to express the basidiomycete pheromone receptors [18], it is assumed that the native basidiomycete pheromones are also modified by farnesylation and carboxymethylation. N-terminal processing is expected to yield a mature pheromone of 7–14 amino acids length. Here, the first of the *S. commune* *Bα1* pheromones is shown as an example for pheromone processing. The N-terminal processing site is predicted from alignments with other homobasidiomycete pheromones.

component. The isoprenoid chains are thus not necessary for pheromone function, while addition of a lipophilic component is essential. In addition, the carbohydrates have not been added via a sulfur bridge to cystein as seen with the natural farnesyl tail, but rather as α-aminohexadecanoic acid derivatives. Since additions of the type described are easier to obtain, this finding now allows the use of synthetic pheromones for the identification of specificity sites in pheromones.

The amino acid sequence found in the mature pheromone is obtained form a larger precursor that is formed from a gene encoding ~50 amino acids. An N-terminal processing is necessary to allow formation of the short chain that is farnesylated to form the mature lipopeptide. While in yeast a Kex-dependent processing is observed, the respective protease in basidiomycetes has not been identified. However, from alignments of the amino acid sequences deduced form the pheromone genes of *S. commune* and *C. cinereus*, a putative splice site could be designated which allows prediction of the native lengths of pheromone amino acid chains. The predicted length was then tested using synthetic pheromones of *C. cinereus*, and indeed that predicted peptide showed highest activities in the induction of the yeast downstream reporter genes in the interspecific test system [8] (see Fig. 7).

The lipopeptides have to be exported to the outside of the cell in order to be able to induce separate cells. In yeast, the **a**-factor is transported by a specific ABC-transporter, encoded by the *STE6* gene. Whether or not a similar transport system is necessary for pheromone excretion in basidiomycetes is not clear. In *S. commune* a peptide transporter has been identified as a gene specifically induced for transcription upon mating [24]. This would be a very good candidate for a pheromone transporter. However, in deletion mutants induction of mates for nuclear migration was as fast as in matings of the wild type. Thus, this gene is most likely not involved in pheromone transport. In *C. cinereus* a gene has been detected associated with the *B* locus that belongs to another class of multidrug transporters. Whether or not this gene is involved in pheromone transport awaits gene disruption experiments [14]. Further evidence can only be drawn from the fact that the heterologously expressed *S. commune* pheromone is excreted from yeast cells. Yeast cells expressing a receptor activated by the pheromone expressed in a different yeast strain form a mating protrusion. Thus, pheromone has to be excreted from the pheromone expressing transformant. The use of Ste6 as the transporter for excretion of the *Schizophyllum commune* pheromone is a possible explanation for this phenomenon. However, *ste6* deletion mutants were also able to induce formation of shmooh cells in mating tests. Thus, a Ste6-independent mechanism of pheromone transport must be assumed in yeast [18].

2.1.3 Analysis of Mutant Mating-Type Genes

For both *A* and *B* genes the function has been sought to be identified by mutagenesis approaches during the 1970s. In a screening for altered specificities for both

A and *B* factors of homobasidiomycetes mutant alleles were identified that were induced for the respective activity. The A_{con} (or for *Coprinus cinereus* A_{mut}) mutants showed formation of unfused clamp cells while B_{con} strains were displaying the phenotype known for $B_{``on"}$ matings which is, for *Schizophyllum commune*, flat in appearance, lacking aerial mycelium, and with continuous nuclear migration and hyphal distortions [for review see 25]. Combination of A_{con} and B_{con} mutations in one strain results in a clamped mycelium able to fruit independent of mating in both *S. commune* and *C. cinereus*.

Molecular analyses of the mutations have revealed for the A_{mut} mutant of *Coprinus* an interesting fusion of genes encoding the two subunits of the heterodimeric homeodomain transcription factor. By gene fusion a protein has been constructed in this mutant that combines the active parts of the two subunits in one peptide [26]. This peptide is able to enter the nucleus and is an active transcriptional regulator thus resulting in *A*-specific gene expression.

For *B*, B_{con} mutants have been analyzed in *S. commune* and it could be shown that one of the pheromones genes is altered in such a mutant such that the former self-specificity now is induced by the mutant pheromone [17]. The specific amino acid substitution encoded in the mutant pheromone leads to exchange of the wildtype valine in the pheromone Bbp2(1) to alanine in the mutant pheromone Bbp2(1−1) (T.J. Fowler, personal communication, 2000). Thus, a single amino acid substitution is sufficient to destroy the self-recognition by the Bbr2 receptor.

In this original constitutive mutant, a complete set of different secondary mutants had been constructed that is reverted from the ''flat'' phenotype to a normal, monokaryonlike phenotype. In these revertants, all possible explanations for the loss of activation of the receptor encoded within the same locus have been realized: the mutant pheromone, the receptor or both, can be deleted or rendered nonfunctional, or the receptor is changed in that it now is no longer able to be activated by the mutant pheromone. Indeed, this most interesting variation on the theme has been identified, and a short, in-frame deletion within the receptor is sufficient for exclusion of the mutant pheromone [17]. This finding again shows that only very limited alterations within a receptor molecule are sufficient to alter the specificity profile of the receptor.

3 FROM SIGNAL TO DIFFERENTIAL GENE EXPRESSION

3.1 Homo- Versus Heterobasidiomycetes

The existence of two mating factors, *A* and *B*, with different function and multiple sets of genes, seems to hold true with new mating-type genes identified from other homobasidiomycetes (e.g., *Pleurotus ostreatus*).

As with the tetrapolar mating system described above, the heterobasidiomycetes contain heterodimer homeodomain transcription factors encoded at one locus called *b* in *Ustilago maydis*, and pheromone receptor and pheromone genes encoded at a second locus termed *a* (this volume, Chapter 14 by Banuett). In contrast to the former systems, only one locus for each function is found. Also in contrast to the homobasidiomycetes, the heterobasidiomycetes show multiallelic mating types only at one of the two loci, the *b* locus encoding the homeodomain transcription factors, while the *a* locus encodes a biallelic pheromone receptor system [see also 27]. The *a* genes encode a pheromone receptors (*pra1* and *pra2* for *a1* and *a2* specificity, respectively) with sequence similarity to the *STE3* gene of *S. cerevisiae*. The respective pheromones *mfa1* and *mfa2* are, like the yeast **a**-factor, farnesylated lipopeptides. As with yeast, the pheromones are necessary to allow fusion of two mates.

The mating-type loci of *Ustilago maydis* were used to investigate the mating-type genes of other smut fungi. It is interesting to note that hybridization of both loci occurred, even if *U. hordei* is not tetrapolar but a bipolar fungus. It could subsequently be shown that the bipolar nature of *U. hordei* is due to linkage of the two loci rather than loss of one locus [28]. Both loci, *b* and *a*, could be analyzed and are functionally similar to those in *U. maydis*.

Again using *U. maydis*, a target gene for the heterodimer transcription factor could be identified and the homeodomain factor was shown to function as a positive regulator [29]. A negative regulatory role could also be shown to be involved in the signaling pathway (this volume, Chapter 14 by Banuett).

3.2 Signal Transduction Cascades

The signal obtained by binding of a pheromone to the compatible pheromone receptor is conveyed by a signal cascade which is presumed to be analogous to that found in *Saccharomyces cerevisiae*. In *Ustilago maydis* this picture begins to be filled with substance. The binding of a ligand supposedly induces the GTPase function of a bound heterotrimeric G-protein which is released and dissociates into the $\beta\gamma$ and the α subunits. Genes for Gα subunits and for members of the MAP/MAPK kinase families have been identified and characterized in *U. maydis* as well as genes linked to cAMP signaling. As with yeast, signal transduction with basidiomycetes is not a single pathway; rather, an integration of different information is needed to allow fruit body formation and sexual reproduction [for review see 27]. The genes identified seem to indicate three independent signal transduction pathways in *U. maydis*: a pheromone response MAP kinase pathway, an environmental response MAP kinase pathway, and a cAMP-dependent pathway. Crosstalk and integration of the pathways establishes the expression pattern for pheromone response [this volume, Chapter 14 by Banuett]. One of the transcription factors mediating these signals is the factor Prf1 which binds to sequences called pheromone response elements, or *PRE.* The transcription factor itself seems likely to be activated by phosphorylation, thus closing the circle to the signal transduction pathways. In *Coprinus cinereus* a gene, *pcc1*, has been analyzed that leads to clamp cell formation. This gene shows sequence similarity to Prf1 of *U. maydis* [30]. Genes for adenylate cyclases have been isolated from mushroom-forming fungi including *C. cinereus* and *Agaricus bisporus*.

3.3 Differentially Regulated Genes

Having established that mating-type genes via the signal transduction pathways directly or indirectly lead to differential expression of target genes, these target genes are sought. From *Ustilago maydis* it is known that the mating-type genes themselves are such target genes since the pheromone, receptor, and transcription factor genes all have pheromone response elements which establish control via the pheromone response pathway by Prf1 binding to the sequences called pheromone response elements. In various approaches differentially expressed genes were identified.

Among the genes regulated differentially in response to the mating-type genes, there are genes involved in pathogenicity (*U. maydis*), asexual reproduction (*Coprinus cinereus*), and a whole array of genes which seem to be somehow integrated into the intracellular system of redistributing and reorganizing the cells to allow for the growth of fruit bodies. Among such differentially transcribed genes in *Schizophyllum commune* are genes that encode putative peptide transporters and putative translational control elements, as well as general intracellular regulatory units that modulate the activity of GTPases such as ras- and rholike proteins.

In *Agaricus bisporus* nine differentially regulated cDNAs have been identified. Three of these could be characterized by sequence similarity to other fungal genes. The three genes putatively code for a mitochondrial ATP synthase, a cell divisional control protein, and a cytochrome P450. In the shiitake mushroom, *Lentinula edodes*, 13 fragments could be identified and placed for putative function. Besides MAP kinase and signal transduction proteins, they encoded proteins for membrane transport, sugar metabolism, mitochondrial origin, intracellular trafficking, protein degradation, and cell cycle control [31]. This seems to give a good apprehension of what to expect in the changes of cell metabolism observed with fruit body formation.

This comprehensive view is supported by samples from other fungi where protein or gene regulation pattern have been observed during fruit body formation. The cell metabolism changes such that sugar metabolism is altered (*S. com-*

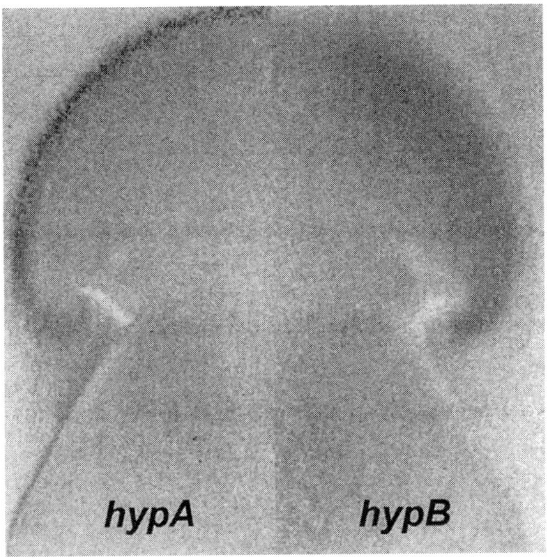

FIGURE 8 Differential expression of two different *Agaricus* hydrophobin genes analyzed by in situ Northern hybridization. Thin sections of the paraffin-embedded young mushroom were hybridized with the antisense probes for the *hypA* and *hypB* gene, as indicated. A control using a sense strand did not show hybridization. (From Ref. 33.)

mune [32]). In addition, mitochondria are produced more proficiently and they seem to have enhanced activities, which seem necessary to supply the energy needed for fruit body development.

One general feature is that especially the cell walls are altered. Many differentially regulated genes seem to have a function in production of certain cell wall proteins or cell wall lytic enzymes. A lot of attention has focused on hydrophobins, small amphipathic proteins that aggregate to form a protective layer on the cell wall at hydrophilic/hydrophobic interphases. Hydrophobins expressed differentially in primordia or developing fruit bodies can be distinguished from those of vegetative aerial mycelium [see 33,34]. Also, hydrophobins specific to the peel tissue could be detected (Fig. 8). Another type of protein involved in making contact with surfaces is lectins. Such lectins have been identified as specifically expressed proteins in different basidiomycetes [cf. 6]. The turnover of proteins that accompanies the restructuring is linked to ubiqitin and differentially regulated protease activity (e.g., *S. commune* [35]).

4 ENVIRONMENTAL FACTORS

The influence of plant signals has already been addressed. Appressorium formation, for example, is stimulated by a thigmotropic signal. Using *Uromyces* rust fungi, it could be shown that physical stimuli on oil–collodion membranes are necessary and sufficient for the induction of genes specifically transcribed during infection. Investigation of protein profiles altered upon induction of infection structures exhibited that superoxide dismutase is one of the earliest enzymes appearing upon induction [36]. The pathway may include cAMP signaling as the cAMP-dependent protein kinase catalytic subunit is required for appressorium formation in the rice blast fungus *Magnaporthe grisea* [37]. However, the interrelations between the thigmotropic signal and the specific activation of enzymes are unknown.

Plant signals are not only involved in interactions between pathogenic fungi and their hosts. Rather, symbiotic relationships between ectomycorrhizal basidiomycetes and trees are also obligatory, dependent on the exchange of signals that also are under investigation (this volume, Chapter 12 by Duplessis et al.). It turned out that there, too, hydrophobins have a central role in the interaction of different cells.

Chemical substances like phenols or simply wounding are able to induce fruit body formation [for review: 38]. Another inductor is UV (320–400 nm) and blue light (400–520 nm), which in *Coprinus cinereus* in conjunction with the pathways regulated by one of the mating factors is able to induce sexual development [39]. The wavelengths active in inducing fruit body development have been analyzed, e.g., in *Schizophyllum commune*, but a blue light receptor is still unknown for basidiomycetes. In many other fungi light is needed for fruit body induction, and sometimes a light/dark scheme is needed to prevent abnormal fruit body development. The light signal possibly works through a cAMP-dependent pathway since increase of intracellular cAMP levels is the fastest reaction to blue light in *S. commune* and *C. cinereus*. Another factor needed for fruiting is low CO_2. High CO_2 can not only prevent fruiting but also suppresses expression of fruiting-specific genes in *S. commune* [38].

5 CELLULAR FUNCTIONS

5.1 Nuclear Exchange

One of the most interesting features governed by the mating-type loci is the exchange of nuclei between the two mating partners in a reciprocal nuclear migration. This process is under the control of the *B* loci, and in *Schizophyllum commune* it is very easily observed since the semicompatible $A= B\neq$ matings show a constant nuclear migration that leads to a phenotype called ‘‘flat’’ since it does not produce aerial mycelium and is easily distinguished on a microscopic level by distorted cell walls [1]. It is thought that the production of cell wall lytic enzymes in this mating reaction, which is needed to dissolve the septa to allow passage of the nuclei, is turned on and overexpression of the cell wall lytic enzymes leads to partial weakening of the cell wall, allowing the turgor to form cytoplasmic protrusions along the lateral cell wall. Therefore, the glucanases and chitinases are expected to be upregulated in flat matings. It has been shown in different fungi that the cell wall composition varies between monokaryotic and dikaryotic strains. This might be explained by the fact that septa once formed in the dikaryon are much more stable against the attack of R-glucanase and chitinase in order to prevent the phenotype that is seen in the flat matings (*S. commune* [see 38]).

Investigation of flat matings or their genetic counterparts obtained by mutation or by transforming a strain with a compatible *B* gene have not yet yielded molecular evidence for specifically regulated nuclear migration genes. The exchange of the two nuclei coming from the different mates is dependent on microtubuli [22]. It is a very fast process that is on the upper limit of microtubular transport known so far, regardless of the fact that at least some nuclear divisions have to be performed during the nuclear passage. However, tubulin genes seem not to be upregulated in the corresponding cellular background [40]. The nuclear exchange is not accompanied by an exchange of mitochondria, which pleads for a specific recognition of nuclei for transport [cf. this volume, Chapter 9 by Braus et al.].

5.2 Dikaryon and Clamp Connections

The two nuclei in a dikaryon are typically held together by microtubuli and using microtubule-destructive agents the two nuclei can be separated. It has been shown that in *Schizophyllum commune*, growth stimuli also may lead to separation of the two nuclei as in aerial hyphae; the mean distance between the two nuclei is larger than in liquid growth medium. At the same time, a ''homokaryonlike'' expression pattern can be found which suggests that nuclear distance is a key regulator for developmental genes [41].

In fully compatible matings the nuclear exchange is stopped when the hyphal tips of the mating partner have been inoculated with the new nuclei. These multikaryotic hyphal tips then lose nuclei by inducing unfused pseudoclamp formation and by branching until only the two different nuclei are left [1]. After that, the binucleate state is maintained by producing fused clamps at every cell division. In *Coprinus cinereus* mutations in clamp cell formation have been identified. These mutations affecting clamp formation but localized outside of the *A* mating-type loci are reminiscent of mutations described as modifier mutations that modify the general master regulatory pathways of the *A* and *B* mating-type gene in *S. commune* [1]. Also, it has been shown for *C. cinereus* that the two different nuclei change position after each conjugate division in that the nucleus entering the clamp makes the other nucleus pass, thereby changing places [42]. However, not all true dikaryotic basidiomycetes show clamp connections. How the nuclear distribution in these cases is maintained is not understood.

5.3 Fruit Body Formation

The formation of a fruit body is dependent on the interaction of the mating-type genes and the ensuing signal transduction processes discussed above (Fig. 9). However, it has been noted that haploid monokaryotic strains may also be able to form fruit bodies. This ability is obtained by mutation and segregates like a single-locus gene. This observation seems to hint at a downstream target of the mating types that, if turned on by mutation, is in itself able to induce the development of fruit bodies. Such a gene, *frt1*, could be identified from *Schizophyllum commune*, and it was shown that the gene product contains a nucleotide-binding

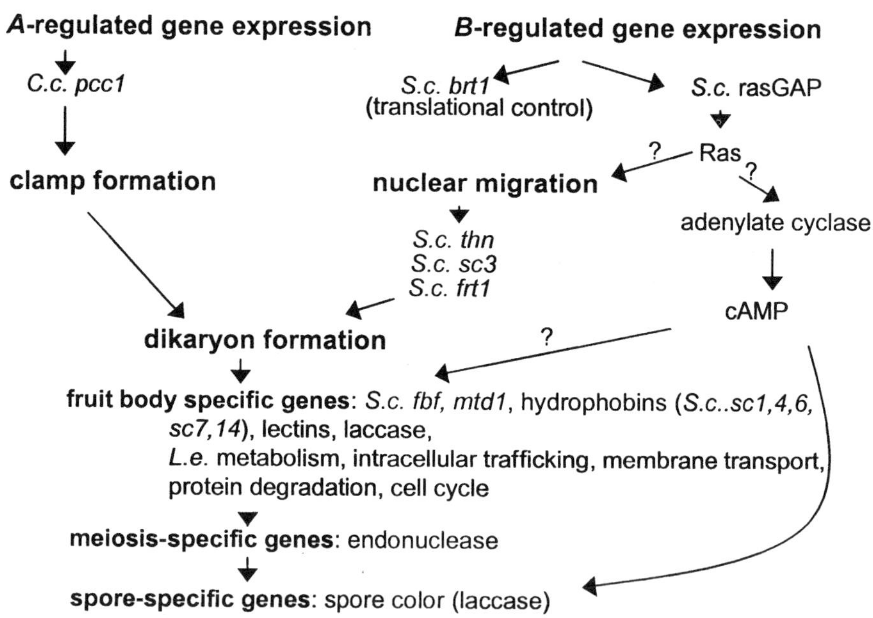

FIGURE 9 The minimal number of pathways that can be postulated in homobasidiomycetes for the signal transduction in the formation of fruit bodies and production of spores. Genes or proteins that have been found to be regulated at the different levels are indicated (*S.c.*, *Schizophyllum commune*; *C.c.*, *Coprinus cinereus*; *L.e.*, *Lentinus edodes*). For further explanations, see text.

site [43]. This makes the gene a likely regulator of other gene's expression. Indeed, the expression of dikaryon-specific genes was turned on in a strain into which the fruiting-inducing gene *frt1* had been transformed. Using *frt1* as bait in a yeast two-hybrid screen, other genes related to fruit body production in *S. commune* are sought. Another putative downstream target gene that seems to be a regulator is a mutation observed regularly. The mutation *fbf* in *S. commune* leads to loss of the capability to from fruit bodies [see 38].

Another gene linked to the pathway of fruit body formation is the thin (*thn*) gene of *S. commune*. Here, the phenotype of strains carrying the trait shows no aerial mycelium formation, which is attributed to the failure to express one single gene involved in coating the hyphae with hydrophobin. Indeed, knockout mutations in the vegetative *Sc3* hydrophobin gene show no aerial mycelium formation [44]. This is necessary for the hyphae to extend from the medium into the air, and hence aerial mycelium formation is missing. In addition, mutant strains are defective in fruit body formation. The *thn* gene therefore is considered to be a regulator involved in control of *sc3* expression. The investigation of a new transposable element in *S. commune* allowed the cloning of the *thn1* gene [45]. The gene codes for a putative regulator of G-protein signaling and shows homology to the yeast *sst2* gene, which is known to be involved in regulating the pheromone receptor response yielding a hypersensitive phenotype in the mutant yeast strains.

A role for cAMP-dependent signaling in fruiting can be postulated from the fact that cAMP can induce fruiting in low concentrations working through cAMP-dependent protein kinase in *C. cinereus* monokaryon and dikaryon. An active role for cAMP in fruiting was also established for other basidiomycetes. As expected for a morphogenetic substance, high levels of cAMP led to abnormal fruit body development in *S. commune* [32].

The differential expression of hydrophobins linked to fruit body formation has already been discussed. As with other basidiomycetes, in *Agaricus bisporus* multiple hydrophobin genes have been identified and differential expression could be shown very elegantly in the mushrooms using in situ hybridization techniques [33] (see Fig. 8). The hydrophobins specifically expressed during fruit body formation seem to play a role in the formation of pseudoparenchymatic tissue that is glued together by a matrix encoating the hyphae that is excreted from the cells. Also lectins, discussed above, presumably play a role in this process of tissue formation.

A role for laccases in oxidative crosslinking of hyphae in fruit body tissue has also been discussed based on the finding that higher laccase activities seem to be connected to fruiting in *S. commune* and *L. edodes* or polypores [see 38].

The morphogenetic processes involved in fruit body formation have been analyzed in *C. cinereus*. After initial stages of fruit body primordia have been built, in mushrooms like the ink cap the stipe must expand in a short time to its multiple original length. It could be shown that the cells elongate by water uptake. In the cap, then, the peel tissue has to be separated from the cap tissue. A development including programmed cell death has been shown to form a layer of lysing cells between the two tissues that leads to separation of the peel tissue from the original cap and between the gills [see 6]. For the pileus, the tissue that bears the hymenium in which spore production occurs, it could be shown that a recessive gene, *ich1*, is necessary for development in *C. cinereus* [46]. In *Agaricus bisporus*, morphogenetic cell death is also involved in mushroom formation [47]. The cap color is investigated in *A. bisporus* where a commercial interest is based on the sale of ''native'' brown-capped versus the white-capped variety. It has been shown that one recessive locus, *ppc1*, is responsible for the brownish cap color.

5.4 Meiosis and Spore Formation

In the hymenium developing within the fruit bodies, karyogamy and meiosis occur. This is dependent on the possibility to study synchronous cells, which is hard to accomplish with fungi since the basidia all represent different developmental stages within the hymenium. However, one natural system, the fruit body development in *Coprinus cinereus*, has the advantage of being highly synchronous in meiosis and hence has been studied in detail [48].

The progression through meiosis has been visualized using fluorescence in situ hybridization in *C. cinereus*. It could be shown that pairing of homologous chromosomes occurs rapidly after karyogamy. By 4 h after karyogamy all chromosomes were at least partially paired. After pairing, condensation of chromosomes further increased. After 6 h postkaryogamy, essentially all meiotic nuclei were in pachytene with still stable compaction.

For DNA synthesis prior to meiosis, DNA polymerase activity is needed. A polymerase activity specific to meiotic prophase has been reported, while an inactive form can be detected in somatic cells by immunoblotting. Thus, the role of this specific enzyme in meiotic recombination, repair, or synthesis can be assumed.

Another enzymatic activity needed during meiosis is the ligase. A pachytene ligase active in the recombination process has been purified and characterized. This enzyme was indistinguishable from one purified from zygotene and presumably involved in chromosome pairing, or from ligase purified from mitotic cells [49].

In recombination, endonuclease activity is needed. An endonuclease has been purified from fruiting caps of *C. cinereus*. The gene is differentially expressed with high expression in fruit body primordial prior to premeiotic S-phase through early pachytene. In late pachytene the level of mRNA dropped drastically and was virtually undetectable in the stage of sterigmata formation [50]. Two more nucleases, which are single-strand specific and believed to be preparing for recombination, have been identified from meiotic tissues of *C. cinereus*.

Related to recombination are the processes of DNA repair. Both repair and recombination need double-strand breaks that must be mended. Again, the natural synchrony of *Coprinus cinereus* proves helpful for the investigation of DNA repair during meiosis. And it may prove a system more closely reflecting the processes in complex eukaryotes and therefore can be investigated in addition to the experiment performed on the yeast *Saccharomyces cerevisiae* [51]. The different functional *recA* homologs and meiosis-specific recombinases and the phenotypes of the respective mutations have been analyzed in *C. cinereus*. Most of the knockout strains are UV or radiation sensitive, which is in good accordance with the genes needed for recombination and repair. Systems to investigate recombination through rDNA gene conversion or by PCR have been devised. Another system that is easily tractable is the heterobasidiomycetes *Ustilago maydis*. Here, recombinases have also been studied.

After meiotic divisions are performed the nuclei enter the spores and spore maturation takes place. In *Schizophyllum commune* this will take ~40–45 h. The sporulation is inhibited by ammonium or glutamine in *C. cinereus* and during sporogenesis ubiquitin is differentially expressed [52]. Spore pigmentation, which may include activity of laccases, seems to be under control of a cAMP-dependent signal transduction cascade.

6 CONCLUSIONS

Even though the molecular processes that underlie mushroom formation in basidiomycetes are only poorly understood, the stage is now set for a more thorough understanding. Key players have been identified and the interactions of these players may now be described in sufficient detail by using the molecular tools now available for the investigation of developmental processes in basidiomycetes. Especially fruitful seems an approach in which different pathways known to be involved in regulation of development in other model fungi such as *Saccharomyces cerevisiae*, *Neurospora crassa*, or *Aspergillus nidulans* are extrapolated for their function in the mushroom fungi. Use of data from the genome projects of the mentioned organisms will greatly enhance such studies. As this research is conducted, the differences between the ascomycetes and basidiomycetes will be focused and in these differences there might be the key to understanding the processes actually involved in mushroom formation.

Mushrooms have been used as human food from ancient times. Growing mushrooms for human consumption started in Asia. In Europe the beginnings of mushroom cultivation were as early as the 17th century, when *Agaricus bisporus* was grown in France on a commercial basis. While other mushrooms are also commercially produced today, including the well-known shiitake, matsutake, or oyster mushrooms, the white button mushroom, in the United States, also called pizza mushroom, *A. bisporus* still tops the list. From 300,000 tons of crop in 1970 the fresh weight produced in 1997 increased to 2 million tons. These numbers alone show why interest in the formation of mushrooms will stay high. And this will apply not only to consumers and producers. Also, we still have too little knowledge on the basic processes involved, which can only be investigated by true basic research.

ACKNOWLEDGMENTS

The author wishes to thank Dr. K. K. Klein and Dr. U. Kües for critical reading of the manuscript. Dr. T. J. Fowler supplied material prior to publication, and Dr. J. Baars kindly helped by providing Figure 8, for which De Groot, Roeven, Van Griensven, and Visser are gratefully acknowledged.

REFERENCES

1. JR Raper. Genetics of Sexuality in Higher Fungi. New York: Ronald Press, 1966.
2. Y Koltin, J Stamberg. Suppression of a mutation disruptive to nuclear migration in *Schizophyllum* by a gene linked to the *B* incompatibility factor. J Bacteriol 109:594–598, 1972.
3. U Kües, LA Casselton. The origin of multiple mating-types in mushrooms. J Cell Sci 104:227–230, 1993.
4. E Kothe. Tetrapolar fungal mating types: sexes by the thousands. FEMS Microbiol Rev 18:65–87, 1996.
5. L Giasson, CA Specht, C Milgrim, CP Novotny, RC Ullrich. Cloning and comparison of $A\alpha$ mating-type alleles of the basidiomycete *Schizophyllum commune*. Mol Gen Genet 218:72–77, 1989.
6. U Kües. Life history and developmental processes in the basidiomycete *Coprinus cinereus*. Microbiol Mol Biol Rev 64:316–353, 2000.
7. A Spit, RH Hyland, EJC Mellor, LA Casselton. A role for heterodimerization in nuclear localization of a homeodomain protein. Proc Natl Acad Sci USA 95:6228–6233, 1998.
8. NS Olesnicky, AJ Brown, SJ Dowell, LA Casselton. A constitutively active G-protein-coupled receptor causes mating self-compatibility in the mushroom *Coprinus*. EMBO J 18:2756–2763, 1999.
9. Y Asada, C Yue, J Wu, GP Shen, CP Novotny, RC Ullrich. *Schizophyllum commune* $A\alpha$ mating-type proteins, Y and Z, form complexes in all combinations in vitro. Genetics 147:117–123, 1997.
10. M Bölker, R Kahmann. Sexual pheromones and mating responses in fungi. Plant Cell 5:1461–1469, 1993.
11. SF O'Shea, PT Chaure, JR Halsall, NS Olesnicky, A Leibbrandt, IF Connerton, LA Casselton. A large pheromone and receptor gene complex determines multiple *B* mating type specificities in *Coprinus cinereus*. Genetics 148:1081–1090, 1998.

12. LJ Vaillancourt, M Raudaskoski, CA Specht, CA Raper. Multiple genes encoding pheromones and a pheromone receptor define the *Bβ1* mating-type specificity in *Schizophyllum commune*. Genetics 146:541–551, 1997.
13. J Wendland, LJ Vaillancourt, J Hegner, KB Lengeler, KJ Laddison, CA Specht, CA Raper, E Kothe. The mating-type locus *Bα1* of *Schizophyllum commune* contains a pheromone receptor gene and putative pheromone genes. EMBO J 14:5271–5278, 1995.
14. JR Halsall, MJ Milner, LA Casselton. Three subfamilies of pheromone and receptor genes generate multiple *B* mating specificities in the mushroom *Coprinus cinereus*. Genetics 154:1115–1123, 2000.
15. M Raudaskoski, M Färdig, M Uuskallio. The structure of pheromone and receptor gene transcripts in *Bα1* and *Bβ1* mating-type loci of *Schizophyllum commune*. In: LJLD van Griensven, J Visser, eds. Proceedings of the Fourth Conference on the Genetics and Cellular Biology of Basidiomycetes. Horst, Netherlands: Mushroom Experimental Station, 1998, pp 119–124.
16. B Malnic, J Hirono, T Sato, LB Buck. Combinatorial receptor codes for odors. Cell 96:713–23, 1999.
17. TJ Fowler, MF Mitton, CA Raper. Gene mutations affecting specificity of pheromone/receptor mating interactions in *Schizophyllum commune*. In: LJLD van Griensven, J Visser, eds. Proceedings of the Fourth Conference on the Genetics and Cellular Biology of Basidiomycetes. Horst, Netherlands: Mushroom Experimental Station, 1998, pp 130–134.
18. TJ Fowler, SM DeSimone, MF Mitton, J Kurjan, CA Raper. Multiple sex pheromones and receptors of a mushroom-producing fungus elicit mating in yeast. Mol Biol Cell 10:2559–1572, 1999.
19. S Gola, J Hegner, E Kothe. Chimeric pheromone receptors in the basidiomycete *Schizophyllum commune*. Fungal Genet Biol 30:191–196, 2000.
20. J Wendland, E Kothe. Allelic divergence at *Bα1* pheromone receptor genes of *Schizophyllum commune*. FEMS Microbiol Lett 145:451–455, 1996.
21. J Hegner, C Siebert-Bartholmei, E Kothe. Ligand recognition in multi-allelic pheromone receptors from the basidiomycete *Schizophyllum commune* studied in yeast. Fungal Genet Biol 26:190–197, 1999.
22. M Raudaskoski. The relationship between *B*-mating-type genes and nuclear migration in *Schizophyllum commune*. Fungal Genet Biol 24:207–227, 1998.
23. M Koppitz, T Spellig, R Kahmann, H Kessler. Lipoconjugates: structure–activity studies for pheromone analogues of *Ustilago maydis* with varied lipophilicity. Int J Pept Protein Res 48:377–390, 1996.
24. KB Lengeler, E Kothe. Mated: a putative peptide transporter of *Schizophyllum commune* expressed in dikaryons. Curr Genet 36:159–164, 1999.
25. CA Raper. Controls for development and differentiation in the dikaryon of basidiomycetes. In: J Bennett, A Ciegler, eds. Secondary Metabolism and Differentiation in Fungi. New York: Marcel Dekker, 1983, pp 195–238.
26. U Kües, B Göttgens, R Stratmann, WV Richardson, SF O'Shea, LA Casselton. A chimeric homeodomain protein causes self-compatibility and constitutive sexual development in the mushroom *Coprinus cinereus*. EMBO J 13:4054–4059, 1994.

27. R Kahmann, C Basse, M Feldbrügge. Fungal-plant signalling in the *Ustilago maydis*–maize pathosystem. Curr Opin Microbiol 2:647–650, 1999.
28. G Bakkeren, JW Kronstad. Conservation of the *b* mating-type gene complex among bipolar and tetrapolar smut fungi. Plant Cell 5:123–136, 1993.
29. T Romeis, A Brachmann, R Kahmann, J Kämper. Identification of a target gene for the bE-bW homeodomain protein complex in *Ustilago maydis*. Mol Microbiol 37: 1–14, 2000.
30. Y Murata, M Fujii, ME Zolan, T Kamada. Molecular analysis of *pcc1*, a gene that leads to *A*-regulated sexual morphogenesis in *Coprinus cinereus*. Genetics 149: 1753–1761, 1998.
31. GS Leung, M Zhang, WJ Xie, HS Kwan. 2000. Identification by RNA fingerprinting of genes differentially expressed during the development of the basidiomycete *Lentinula edodes*. Mol Gen Genet 262:977–990.
32. MN Schwalb. Effect of adenosine 3′,5′-cyclic monophosphate on the morphogenesis of fruit bodies in *Schizophyllum commune*. Arch Mikrobiol 96:17–20, 1974.
33. PWJ de Groot, RT Roeven, LJLD van Griensven, J Visser, PJ Schaap. Different temporal and spatial expression of two hydrophobin-encoding genes of the edible mushroom *Agaricus bisporus*. Microbiology 145:1105–1113, 1999.
34. FHJ Schuren, JGH Wessels. Two genes specifically expressed in fruiting dikaryons of *Schizophyllum commune*: homologies with a gene not regulated by mating-type genes. Gene 90:199–205, 1990.
35. MN Schwalb. Developmentally regulated proteases from the basidiomycete *Schizophyllum commune*. J Biol Chem 252:8435–8439, 1977.
36. JS Lamboy, RC Staples, HC Hoch. Superoxide dismutase: a differentiation protein expressed in *Uromyces* germlings during early appressorium development. Exp Mycol 19:284–296, 1995.
37. TK Mitchell, RA Dean. The cAMP-dependent protein kinase catalytic subunit is required for appressorium formation and pathogenesis by the rice blast fungus *Magnaporthe grisea*. Plant Cell 7:1869–1878, 1995.
38. JGH Wessels. Fruiting in the higher fungi. Adv Microbiol Physiol 34:147–202, 1993.
39. U Kües, JD Granado, R Hermann, RP Boulianne, K Kertesz-Chaloupkova, M Aebi. The *A* mating type and blue light regulate all known differentiation processes in the basidiomycete *Coprinus cinereus*. Mol Gen Genet 260:81–91, 1998.
40. P Russo, JT Juuti, M Raudaskoski. Cloning, sequence and expression of a beta-tubulin-encoding gene in the homobasidiomycete *Schizophyllum commune*. Gene 119:175–182, 1992.
41. TA Schuurs, HJP Dalstra, JMJ Scheer, JGH Wessels. Positioning of nuclei in the secondary mycelium of *Schizophyllum commune* in relation to differential gene expression. Fungal Genet Biol 23:150–161, 1998.
42. M Iwasa, S Tanabe, T Kamada. The two nuclei in the dikaryon of the homobasidiomycete *Coprinus cinereus* change position after each conjugate division. Fungal Genet Biol 23:110–116, 1998.
43. JS Horton, GE Palmer, WJ Smith. Regulation of dikaryon-expressed genes by *FRT1* in the basidiomycete *Schizophyllum commune*. Fungal Genet Biol 26:33–47, 1999.
44. M-A van Wetter, HAB Wösten, JGH Wessels. Sc3 and Sc4 hydrophobins have dis-

tinct roles in formation of aerial structures in dikaryons of *Schizophyllum commune*. Mol Microbiol 36:201–210, 2000.

45. TJ Fowler, MF Mitton. *Scooter*, a new active transposon in *Schizophyllum commune*, has disrupted two genes regulating signal transduction. Genetics 156:1585–1594, 2000.
46. H Muraguchi, T Kamada. The *ich1* gene of the mushroom *Coprinus cinereus* is essential for pileus formation in fruiting. Development 125:3133–3141, 1998.
47. MH Umar, LJLD van Grienven. Morphogenetic cell death in developing primordia of *Agaricus bisporus*. Mycologia 89:274–277, 1997.
48. PJ Pukkila, BM Yashar, DM Binninger. Analysis of meiotic development in *Coprinus cinereus*. Symp Soc Exp Biol 38:177–194, 1984.
49. S Matsuda, K Sakaguchi, K Tsukada, H Teraoka. Characterization of DNA ligase from the fungus *Coprinus cinereus*. Eur J Biochem 237:691–697, 1996.
50. S Charlton, R Boulianne, YC Chow, BC Lu. Cloning and differential expression during the sexual cycle of a meiotic endonuclease-encoding gene from the basidiomycete *Coprinus cinereus*. Gene 122:163–169, 1992.
51. WJ Cummings, ME Zolan. Functions of DNA repair genes during meiosis. Curr Top Dev Biol 37:117–140, 1998.
52. T Kanda, N Tanaka, T Takemaru. Ubiquitin immunoreactivity shows several proteins varying with development and sporulation in the basidiomycete *Coprinus cinereus*. Biochem Cell Biol 68:1019–1025, 1990.

11

Spore Killers
Meiotic Drive Elements That Distort Genetic Ratios

Namboori B. Raju
Stanford University, Stanford, California

1 INTRODUCTION

Spore killers (*Sk*) in fungi are chromosomal elements that distort allele ratios of *Sk* and *Sk*-linked genes. They are expressed postmeiotically, causing the death of ascospores that do not receive the killer element. Ascospore death occurs when one parent carries the killer element (Sk^K) and the other parent carries the sensitive counterpart (Sk^S). The best-studied examples are found in the filamentous ascomycetes *Neurospora* and *Podospora*. Turner and Perkins [1] first showed that ascospore death in certain crosses of *Neurospora sitophila* and *N. intermedia* resulted from the action of spore killers. In crosses of killer × sensitive, each ascus produces four large, black, viable ascospores (Sk^K) and four small, hyaline, inviable, ascospores (Sk^S). Earlier, Padieu and Bernet [2] had described ascospore death in crosses between *Podospora* strains and had attributed it to two independently segregating ascospore abortion factors. In fact, the *Podospora* abortion factors showed all the characteristics of spore killers [3]. Additional spore killers have recently been found in wild *P. anserina* populations [4]. They were also found in *Gibberella fujikuroi* (*Fusarium moniliforme*) [5] and *Cochliobolus heterostrophus* [6].

Spore killers show a striking resemblance to the previously described segregation-distorting meiotic drive systems in *Drosophila*, mouse, tomato, wheat, and several other animals and plants [1,7]. Ascomycete spore killers provide the most direct demonstration of meiotic drive because all four products of individual meioses are held together in intact asci and are often conspicuously displayed in the form of black and white ascospores (Fig. 1). Furthermore, in fungi, neither fertilization of gametes nor the diploid phase intervenes between meiosis and manifestation of a spore killer.

Criteria have been described for the detection and analysis of spore killers and for distinguishing them from other causes of spore abortion [8]. The genetic basis of spore killer death is quite different from that of other causes of ascospore abortion, such as autonomous ascospore maturation defects and deficiencies resulting from the segregation of chromosome rearrangements [9–11].

Ascospore death that superficially resembles that of spore killers has been found in *Coniochaeta tetraspora* [12]. In this homothallic ascomycete, four of the eight incipient uninucleate ascospores die and degenerate in every ascus. All surviving progeny cultures are self-fertile and again produce four viable and four aborted ascospores, generation after generation. Ascospore death is developmen-

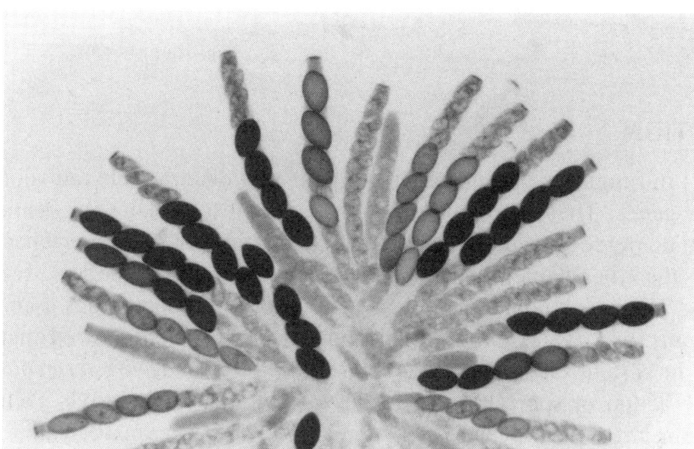

FIGURE 1 *N. crassa*. A rosette of maturing asci from $Sk2^K \times Sk2^S$ (wild type). Mature asci show four large, black ascospores (viable) and four small, white ascospores (aborted); the four dead ascospores contain the wild-type, sensitive chromosome. The asci that do not show the 4B:4W pattern are still immature. All mature asci show 4B:4W first-division segregation pattern because of a recombination block in the centromere-proximal region on linkage group III, where spore killer is located. (From Ref. 25.)

tally programmed, resulting from a genetic or epigenetic modification in one of the two nuclei that make up the diploid zygote nucleus of each ascus [13].

The chromosomal spore killers described here bear no relation to the killer systems of *Saccharomyces* or *Ustilago*, which involve RNA plasmids responsible for secreting a toxic protein [14].

I will first summarize normal ascus development in the eight-spored prototypic species *N. crassa* and in the four-spored species *N. tetrasperma* and *P. anserina* (Fig. 2). Second, I will briefly recount the discovery of spore killers in *N. sitophila* and *N. intermedia*, followed by a description of spore killer characteristics. Interaction of killer and sensitive nuclei in the heterokaryotic ascospores in the four-spored asci of *N. tetrasperma* and *P. anserina* will then be described. Finally, the mode of action of spore killers will be considered, and consequences of spore killers to the species at the population level will be discussed.

2 ASCUS DEVELOPMENT IN *NEUROSPORA* AND *PODOSPORA*

N. crassa is an eight-spored heterothallic species; mating occurs only between strains of opposite mating type (*mat A* and *mat a*). When a killer (K) is crossed to a sensitive (S) strain, the nuclei proliferate in the premeiotic ascogenous hyphae, which give rise to asci within the developing perithecium. Two haploid nuclei of opposite mating type fuse in the ascus initial, and the zygote nucleus immediately undergoes meiosis and a postmeiotic mitosis in the common cytoplasm of the developing ascus. The resulting eight nuclei are sequestered into separate ascospores—four killer and four sensitive (Fig. 2). A second mitosis occurs in the young ascospores, and additional mitoses occur after the spores become pigmented. All eight ascospores are held in linear order in the narrow ascus, and the segregation of killer and sensitive alleles at the first division of meiosis is reflected in the order of ascospores (4K:4S). See Davis [15] and Raju [16] for background information and photographs.

In *N. tetrasperma* and *P. anserina* the asci are four-spored, and each ascospore encloses two nuclei of opposite mating type. Thus, single-ascospore cultures are self-fertile, and such species are said to be pseudohomothallic or secondarily homothallic. In *N. tetrasperma*, different alleles of centromere-linked genes (e.g., mating type or spore killer) that segregate at the first division of meiosis become enclosed in each ascospore because of overlapping and pairwise alignment of spindles at the second and third divisions (Fig. 2) [17,18]. In *P. anserina*, ascus development is programmed differently so that alleles that segregate at the second division of meiosis (e.g., mating type) become enclosed together in each of the four ascospores. In this species the second division spindles do not overlap, as in *N. tetrasperma*, but they are aligned in tandem as in *N. crassa* [19]. Contrary to what happens in *N. tetrasperma*, the ascospores of *P. anserina* will be homo-

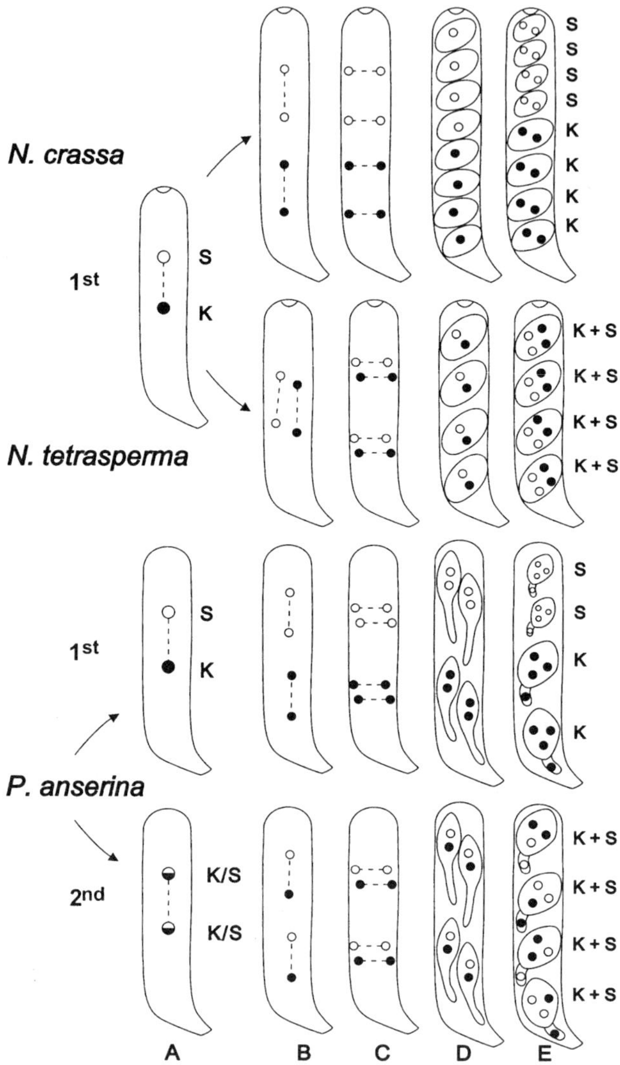

FIGURE 2 A schematic diagram of ascus development in *Neurospora crassa*, *N. tetrasperma*, and *Podospora anserina*. *N. crassa* is eight-spored and heterothallic. The latter two species are four-spored and pseudohomothallic. Killer (K) and sensitive (S) "alleles" are shown segregating at the first division of meiosis in *N. crassa* and *N. tetrasperma*, and at both first and second divisions in *P. anserina* (A). Spindles at the second division (B) are aligned in tandem in *N. crassa* and *P. anserina*, and their alignment is parallel and pairwise in *N. tetrasperma*. The four spindles at the postmeiotic mitosis (C) are

karyotic for alleles of centromere-linked genes that show first-division segregation, and heterokaryotic for genes far from centromere that show second-division segregation (Fig. 2) [20].

3 SPORE KILLERS IN *NEUROSPORA*

3.1 Discovery of Spore Killers in *N. sitophila* and *N. intermedia*

N. sitophila and *N. intermedia*, like *N. crassa*, are heterothallic and eight-spored [21,22]. In the process of developing species testers for *N. sitophila* in the early 1970s, my colleague Barbara Turner crossed strains from Arlington (Virginia) with strains from Nigeria. To her surprise, all mature asci showed a 4B:4W ascospore pattern (see Fig. 1). The black ascospores (B) are full-size and viable, and the white ascospores (W) are small and inviable. The 4B:4W asci were produced in reciprocal crosses between the two parents, regardless of which was used as female. Moreover, all asci were 4B:4W again when viable f_1 progeny from 4B:4W asci were backcrossed to the Nigeria parent, but 8B:0W asci resulted when the same f_1 progeny were backcrossed to the Arlington parent. From these genetic results, Turner and Perkins [1] inferred that the Arlington and Nigeria strains differed from one another in specificity of ascospore killing: the Arlington strains behave as killers (Sk^K) and the Nigeria strains are sensitive to killing (Sk^S). Apparently, the Nigeria genotype fails to carry out some ascospore maturation function(s) whenever it is heterozygous with the Arlington genotype. The spore killer in the Arlington strains was named $Sk1^K$; the sensitive counterpart in the Nigeria strains was called $Sk1^S$. When additional strains of *N. sitophila* were tested, some resembled the Arlington strains and others resembled the Nigeria strains in killing specificity.

---

aligned across the ascus, equidistant in *N. crassa* and pairwise in *N. tetrasperma* and *P. anserina*. Subsequently, eight uninucleate ascospores are delimited in *N. crassa* and four binucleate ascospores in *N. tetraperma* and *P. anserina* (D). A second postmeiotic mitosis occurs in the young ascospores, which now become binucleate in *N. crassa* and four-nucleate in *N. tetrasperma* and *P. anserina* (E). In *N. crassa*, four of the eight ascospores that are homokaryotic for the killer allele survive, and the four that carry the sensitive allele abort and degenerate. In *N. tetrasperma*, all four ascospores are heterokaryotic for killer and sensitive alleles and are thus viable. In *P. anserina*, first-division segregation results in two ascospores that are homokaryotic for the killer (viable) and two that are homokaryotic for the sensitive (inviable). Second-division segregation of killer and sensitive alleles results in all four heterokaryotic (K + S), viable ascospores. (From Ref. 19.)

Ascospore abortion in the 4B:4W asci from Arlington × Nigeria could not be due to an autonomous ascospore maturation defect in one of the parents, because all ascospores are black and viable both in crosses of Arlington × Arlington and in crosses of Nigeria × Nigeria. Also, the 4B:4W asci cannot be attributed to a chromosome rearrangement because crosses heterozygous for a rearrangement are expected to generate asci with varying numbers of defective ascospores (0, 2, 4, 6, or 8). No chromosome rearrangement type is known that is capable of producing exclusively 4B:4W asci [9,23,24].

Spore killer factors called $Sk2^K$ and $Sk3^K$ were subsequently found in another species, *N. intermedia*. These resemble $Sk1^K$ in all properties, but they are much less frequent than is $Sk1^K$. $Sk2^K$ was originally discovered in *N. crassa* strains into which a *nit4* mutation had been introgressed from *N. intermedia*, and $Sk2^K$ was then traced back to a single *N. intermedia* ancestor, a strain isolated from forest soil in Brunei (Borneo) by J.H. Warcup [1]. A different spore killer ($Sk3^K$) was found in a *N. intermedia* strain from Papua New Guinea. Although $Sk3^K$ resembles $Sk2^K$ in chromosomal location and killing behavior, the two differ in killing specificity and resistance, and each is sensitive to killing by the other [1]. $Sk2^K$ and $Sk3^K$ are extremely rare. Of 2500 *N. intermedia* strains from around the world that have been tested, killers were found at only five sites—in Borneo, Java, and Papua New Guinea.

No spore killers have been found in 467 strains from wild populations of *N. crassa*. However, $Sk2^K$ and $Sk3^K$ have been introgressed from *N. intermedia* into *N. crassa* for purposes of detailed genetic analyses. $Sk1^K$ could not be introgressed from *N. sitophila* into *N. crassa* or *N. intermedia* because interspecific crosses were sterile. Thus, it is not known whether $Sk1^K$ is homologous to $Sk2^K$ or $Sk3^K$.

3.2 Expression of Spore Killer in the Ascus

All three *Neurospora* spore killers are very efficient. Killing of Sk^S ascospores occurs in 100% of asci; the survivors are all killers (Sk^K). (Exceptions due to genes conferring resistance to killing will be considered later.) Spore killer death is identical in reciprocal crosses regardless of whether Sk^K is used as male or female. *Spore killer-2* (in *N. crassa* background) is the best studied of the killers and it provides a model for the other spore killers. The characteristics of spore killer expression in the maturing asci, summarized here for *Sk2*, apply also to *Sk3* and *Sk1*. Killing occurs only when a killer is heterozygous in crosses of $Sk2^K \times Sk2^S$. As with other segregation distorting systems, spore killer death is expressed postmeiotically. Early ascus development, meiosis, and a postmeiotic mitosis are completely normal, and eight uninucleate ascospores are delimited [16,25]. Following another normal mitosis in all eight ascospores, the four ascospores that contain $Sk2^S$ stop developing and abort; they remain small, hyaline,

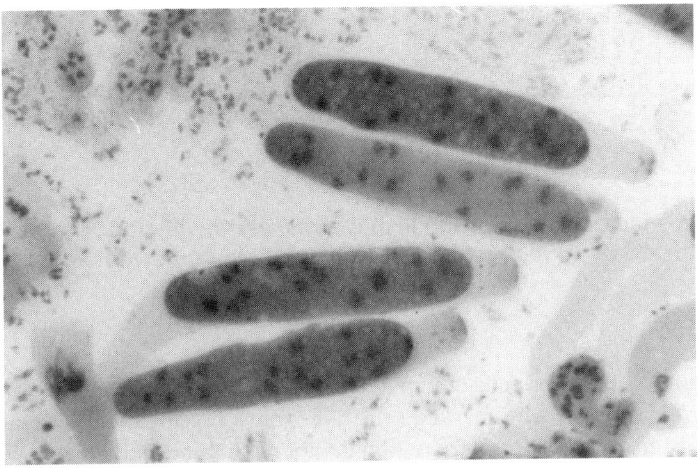

FIGURE 3 *N. crassa*. The developmental mutant *Banana* (*Ban*) produces giant ascospores, which enclose all four products of meiosis and their mitotic derivatives. In the giant ascospores of $Sk2^K$ Ban^+ × $Sk2^S$ *Ban*, the Sk^S nuclei are sheltered by Sk^K during ascospore maturation; their survival has been demonstrated by analysis of progeny nuclei. (From Ref. 26.)

and inviable. Only the four $Sk2^K$ ascospores develop and mature normally (Figs. 1, 2).

Markers unlinked to *Sk2* show normal segregation, and alleles from the sensitive parent are recovered in the viable Sk^K progeny, indicating that meiosis and allele segregation are normal. Spore killer does not kill itself: homozygous crosses of $Sk2^K \times Sk2^K$ resemble the normal crosses of $Sk2^S \times Sk2^S$, where all eight ascospores are black and viable.

An $Sk2^S$ nucleus that would otherwise die is rescued if an $Sk2^K$ nucleus is also included in the same ascospore. This was first shown in the developmental mutant *Banana (Ban)*, which produces giant ascospores containing all four products of meiosis and their mitotic derivatives (Fig. 3) [25,26]. Similar rescue of sensitive nuclei has been observed in the normally heterokaryotic ($Sk^K + Sk^S$) ascospores of the pseudohomothallic species *N. tetrasperma* [27].

3.3 Chromosomal Basis of Spore Killers

When *Sk2* is heterozygous in killer × sensitive crosses of *N. crassa* or *N. intermedia*, 100% of asci show first-division segregation, with four viable Sk^K ascospores at one end and four dead Sk^S ascospores at the other end (Fig. 1). $Sk2^K$ is linked to centromere markers in linkage group III. The spore killer element was initially

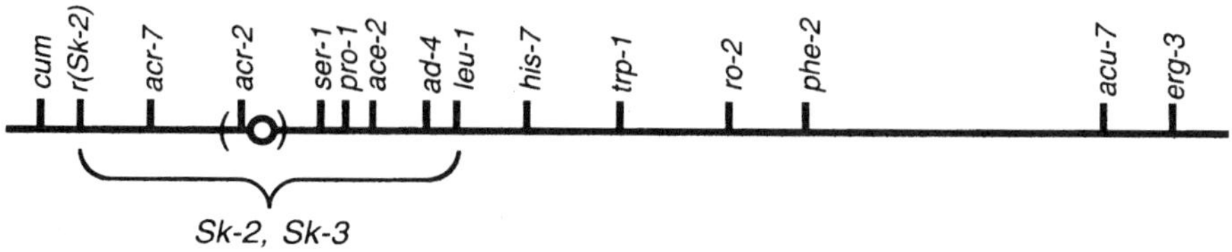

FIGURE 4 Genetic map of linkage group III in *N. crassa*, showing the recombination block from *r(Sk2)* to *leu1* when $Sk2^K$ or $Sk3^K$ is heterozygous. (From Ref. 3.)

thought of as a single gene at a locus completely linked to the centromere. However, crosses of $Sk2^K$ or $Sk3^K$ with marked sensitive strains revealed that several markers on linkage group III centromere region do not recombine and are always inherited together. The spore killer element is therefore considered a complex or haplotype rather than a single gene [1]. The recombination block in linkage group III extends over 30 map units, from about *cum (cumulus)* in the left arm to *leu1* in the right arm (Fig. 4) [28]. Recombination in the interval is reduced to $<10^{-5}$, although it is completely normal outside the block and in other linkage groups. $Sk2^K$ and $Sk3^K$ affect recombination in the same interval.

Despite the recombination block, three genetic markers have been inserted into the *Sk2* killer complex by selective plating of large numbers of ascospores. When marked $Sk2^K$ strains are crossed with other $Sk2^K$ strains, the marker sequences and crossing-over frequencies are normal [28]. Furthermore, chromosome pairing at pachytene appears to be normal both in heterozygous killer × sensitive and homozygous killer × killer crosses [29; N.B. Raju, unpublished]. Thus, the observed recombination block in heterozygous crosses cannot be due to gross chromosome inversions, although small inversions in the intervals between markers cannot be ruled out.

Unlike *Sk2* and *Sk3* in *N. crassa* and *N. intermedia*, *Sk1* in *N. sitophila* does not completely block recombination with the centromere in killer × sensitive crosses. Up to 5% of asci show second-division segregation patterns resulting from a crossing over in the *Sk1*–centromere interval [25].

3.4 *Spore Killer-3* and Its Interaction with *Sk2* in *Neurospora*

The above description of *Sk2* applies equally well to *Sk3* [1]. Like $Sk2^K$, $Sk3^K$ does not kill itself, and $Sk3^K$ blocks recombination across the same intervals as does $Sk2^K$. The main difference between $Sk2^K$ and $Sk3^K$, and the primary basis for identifying them as different killers, is their specificity of killing and resistance. Unlike crosses homozygous for $Sk2^K$ or for $Sk3^K$, where there is no killing, all eight ascospores are usually killed in crosses between $Sk2^K$ and $Sk3^K$, because each killer is mutually sensitive to killing by the other. Thus, the genotype of the two nonallelic killers may be written fully as $Sk2^K$ $Sk3^S$ and $Sk2^S$ $Sk3^K$. When $Sk2^K$ and $Sk3^K$ nuclei are enclosed in a heterokaryotic ascospore, both nuclei survive because of mutual rescue. This was first observed in crosses of $Sk2^K$ *Ban* × $Sk3^K$ Ban^+ where each ascus produces a single giant ascospore in which all four products of meiosis are enclosed in the common cytoplasm. In a subsequent study using *N. tetrasperma*, progeny analysis of heterokaryotic ($Sk2^K$ + $Sk3^K$) ascospores showed that both $Sk2^K$ and $Sk3^K$ nuclei survive and are capable of killing when crossed to sensitive testers [27].

3.5 Spore Killer Behavior After Transfer into *N. tetrasperma*

No spore killers have been identified in natural populations of *N. tetrasperma*, though they may be present but difficult to detect. Nevertheless, $Sk2^K$ and $Sk3^K$ from *N. intermedia* have been introgressed into *N. tetrasperma* by way of *N. crassa*. This was done mainly to examine the interaction between killer and sensitive nuclei and between $Sk2^K$ and $Sk3^K$ in the normally heterokaryotic ascospores [27].

Although the asci of *N. tetrasperma* are mostly four-spored (Fig. 5), a small proportion of asci contain five or six spores. In these asci one or two large binucleate ascospores are replaced by a pair (or pairs) of small uninucleate ascospores. These exceptional, small, homokaryotic ascospores provide a basis for testing whether *N. tetrasperma* is sensitive to $Sk2^K$ and $Sk3^K$ from *N. intermedia*.

The frequency of asci with homokaryotic small ascospores is greatly increased in *N. tetrasperma* crosses heterozygous for the dominant gene *E (eight spore)* [18,30]. *E* is unlinked to *Sk*. Behavior of *Sk2* and *Sk3* was examined in crosses with or without *E*. Crosses were compared that were homozygous for each of the killers ($Sk2^K$ × $Sk2^K$ or $Sk3^K$ × $Sk3^K$), heterozygous for each killer ($Sk2^K$ × $Sk2^S$ or $Sk3^K$ × $Sk3^S$), and heterozygous for both killers in the intercross $Sk2^K$ × $Sk3^K$. The results were as predicted from behavior of *Sk2* and *Sk3* in *N. crassa*.

Because the spore killers are centromere linked, Sk^K and Sk^S segregate at the first division of meiosis, just as do the mating-type idiomorphs *mat a* and *mat A*, which are centromere linked in another chromosome. Consequently, each large ascospore is heterokaryotic (Sk^K + Sk^S) and there is no killing because presence of a homologous killer nucleus in the same ascospore protects the sensitive nucleus from being killed. However, in asci where one or more large ascospores are replaced by a pair of small ascospores, one small ascospore of each pair contains an unprotected Sk^S nucleus, and this aborts. The only small ascospores to survive are those that contain the Sk^K nucleus (Fig. 5) [27].

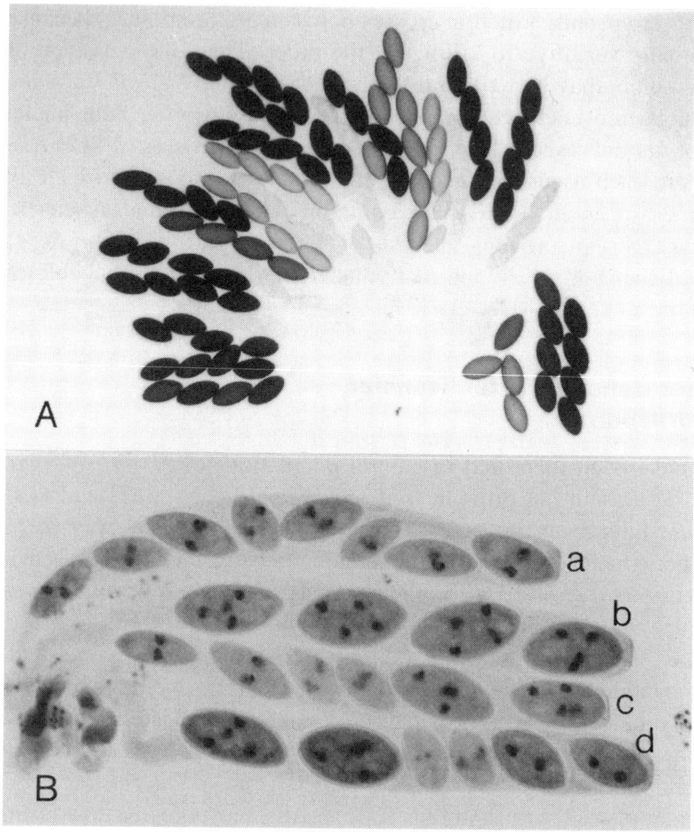

Figure 5 *N. tetrasperma*. (A) Four-spored asci from wild type. All four ascospores are heterokaryotic for centromere-linked genes like mating type (or *Sk*). Asci with all nonblack spores are still immature. (B) Four- to eight-spored asci from $Sk2^K \times Sk2^S$ *E*. The developmental gene *E* causes many asci to produce more than four spores. Because of a mitosis in the young ascospores, the small ascospores contain two nuclei and the large ascospores contain four nuclei. The heterokaryotic large ascospores carrying both killer and sensitive nuclei grow to full size and mature normally. The homokaryotic small ascospores carrying killer nuclei also enlarge and mature. Only the homokaryotic small ascospores carrying sensitive nuclei abort and shrink. The genotypes of immature ascospores can be inferred based on previous observations of developing asci and progeny testing. The inferred genotypes for the individual asci shown here, from left to right, are: (a) S, S, K, S, K, S, K, K; (b) (K + S), (K + S), (K + S), (K + S); (c) K, K, S, S, (K + S), (K + S); (d) (K + S), (K + S), S, S, K, K. (From Ref. 27.)

When $Sk2^K$ and $Sk3^K$ are intercrossed, each large ascospore encloses two nuclei, one that is $Sk2^K$ ($Sk3^S$) and one that is ($Sk2^S$) $Sk3^K$. There is no killing, because each nucleus is rescued by being enclosed in the same ascospore with a nucleus that contains the homologous killer element. The resulting cultures are self-fertile heterokaryons of constitution ($Sk2^K$ $Sk3^S$ + $Sk2^S$ $Sk3^K$). In contrast, all small ascospores from the same cross abort because they carry either $Sk2^K$ $Sk3^S$ or $Sk2^S$ $Sk3^K$ in homokaryotic condition, and each unsheltered killer is vulnerable to killing by the other—$Sk2^K$ $Sk3^S$ by $Sk2^S$ $Sk3^K$ and $Sk2^S$ $Sk3^K$ by $Sk2^K$ $Sk3^S$—just as when the two are intercrossed in *N. crassa* [27].

3.6 Nonkiller Strains That Are Insensitive to Killing

In addition to killers and sensitives, resistant strains were found that neither kill nor are killed. The resistance [*r(Sk)*] was shown to be due to genes that were linked to the killer elements in linkage group III. In *N. intermedia*, three types of nonkiller *r(Sk)* strains were found—*r(Sk2)* only, *r(Sk3)* only, and doubly resistant *r(Sk2) r(Sk3)*. The doubly resistant strains were found wherever *r(Sk2)* and *r(Sk3)* occur in the same population. Strains resistant to $Sk2^K$, to $Sk3^K$, or to both are frequent in parts of the world where killers were found, and few or no resistant strains were found in areas where there were no killers [31]. In *N. crassa*, where no killers were found, strains from widely scattered regions nevertheless show resistance to $Sk2^K$. Strains resistant to $Sk3^K$ have not been found. When *r(Sk2)* from *N. crassa* was crossed to *r(Sk3)* from *N. intermedia*, no recombinants were obtained among 300 progeny. In *N. sitophila*, by contrast, killers are fairly common but resistant strains are very rare [31].

3.7 Spore Killers in Natural Populations of *Neurospora*

Spore killers have been found only among natural isolates, not as laboratory variants. The discovery of *Neurospora* spore killers was a byproduct of studies involving natural isolates of the heterothallic species *N. sitophila*, *N. intermedia*, and *N. crassa*. Since the discovery of *Neurospora* spore killers in the 1970s, a worldwide collection of *Neurospora* isolates have been screened for spore killer polymorphisms [31,32]. Spore killer strains were found to be fairly common in *N. sitophila*. Among 469 strains collected on outdoor burnt substrates from many countries, 77 were killers ($Sk1^K$); all others were sensitive to killing ($Sk1^S$). The killers and sensitives were not uniformly distributed in various geographical regions, however. In wild populations of *N. sitophila*, killers were present in Pacific islands, Australia, and eastern Asia, but with the exception of one site, were absent from collections in the Americas, Africa, India, and Malaya.

In contrast to the situation in *N. sitophila* (~16% killers), killer strains are extremely rare in *N. intermedia*. Among a total of 2500 *N. intermedia* isolates,

$Sk2^K$ was found in only four strains, two from Borneo, one from Java, and another from Papua New Guinea. $Sk3^K$ was found only once, in Papua New Guinea [32,33]. No spore killers have been found in *N. crassa*, despite screening 467 wild-collected isolates. At least 100 wild isolates of *N. tetrasperma* have been screened for the presence of spore killers using Dodge's *E mat A* and *E mat a* spore killer–sensitive testers. All crosses produced many five- to eight-spored asci because of *E*, but there was no indication of death of homokaryotic small ascospores due to spore killers [N.B. Raju, unpublished]. It is conceivable that spore killers may indeed be present in cryptic condition, but tests adequate to detect their presence have not been made. Drive elements that showed <100% killing would probably be difficult to recognize.

4 SPORE KILLERS IN *PODOSPORA*

4.1 Discovery of Spore Killers in *P. anserina*

The discovery of spore killers in *P. anserina* actually predates that of other fungal spore killers. Padieu and Bernet [2] analyzed a cross between two wild-collected *Podospora* strains, one of which contained two unlinked genes, *a* and *b*, that resulted in ascospore death. The cross produced up to 90% of asci with fewer than four viable ascospores. Although their results were not interpreted in terms of spore killers, Turner and Perkins [3] recast the *Podospora* results in terms of *Neurospora*-like spore killers, with $a1 = Sk1^S$, $a2 = Sk1^K$, $b1 = Sk2^K$, and $b2 = Sk2^S$. The original *Podospora* cross was apparently doubly heterozygous for spore killers at two unlinked loci. Ascospores survive only if they contain at least one $Sk1^K$ and one $Sk2^K$ factor. Any ascospore that does not contain a killer element at each locus is killed.

In the last 10 years, efforts have been made to find spore killer polymorphisms among natural populations of *P. anserina* [34]. When Van der Gaag et al. [4] sampled a *P. anserina* population from Wageningen, the Netherlands, they found 23 of 99 isolates to be killers. Six different spore killers (symbolized Psk^K) were recognized, based on the frequency of asci (45–95%) with two viable and two dead spores, and on the interaction between different killers.

Psk1 and *Psk2* produced >70% two-spored asci, whereas *Psk4*, *Psk6*, and *Psk7* showed 45–54% two-spored asci. The frequency of two-spored asci in *Psk3* was highly variable, however. In addition to the 23 wild isolates from the Netherlands, three of the six *Podospora* strains (T, Y, and Z) that were isolated in 1937 in Picardy, France, proved to be killers. Two of these were of types found also in the Wageningen population. Strain T from France was one of the strains previously characterized as a spore killer [2–4]. This isolate was recently renamed *P. comata* on the basis of morphological and molecular data [35].

4.2 Expression of Spore Killers in the Ascus

Detection of spore killers is much easier in *P. anserina* than it is in *N. tetrasperma*, because of differences in programming of ascus development in the two pseudohomothallic species [19,27]. The spore killers, detected by Van der Gaag et al. [4], are not tightly centromere linked. When asci are heterozygous for a spore killer (*Psk*^{K}/*Psk*^{S}), second-division segregation frequencies (signaled by asci with four viable spores) range from 5% to 55% for various spore killers. The frequency of asci with four viable spores thus provides a measure of the distance of a killer element from the centromere. In asci where there is no crossover proximal to *Psk*, the spore killer alleles segregate at the first division of meiosis and will result in all four homokaryotic ascospores: two *Psk*^{K} and two *Psk*^{S}. In these asci, the two homokaryotic *Psk*^{K} ascospores survive but the two homokaryotic *Psk*^{S} ascospores are killed (Figs. 2, 6). Thus, the percentage of two-spored asci has been used as a diagnostic feature for the detection and analysis of spore killers in *P. anserina*.

Selfings and backcrosses of progeny from two-spored and four-spored asci of killer × sensitive crosses confirmed that the observed ascospore death in *Podospora* is due to *Neurospora*-like spore killers. The results of Van der Gaag et al. [4] for *Psk7*^{K} × *Psk7*^{S} are summarized here: The parental cross produced both two-spored asci (54%) and four-spored asci (46%). Selfing of two-spored ascus progeny always produced normal four-spored asci (no killing), because all four ascospores are homokaryotic for the killer (*Psk7*^{K}). Selfing of the four-spored ascus progeny from the parental cross showed spore killing and produced >50% two-spored asci, indicating that the original progeny were heterokaryotic for the killer (*Psk*^{K} + *Psk*^{S}). Backcrosses of two-spored ascus progeny to the sensitive parent (killer × sensitive) showed spore killing, whereas backcrosses to the killer parent (killer × killer) gave normal four-spored asci. Backcrosses of the four-spored ascus progeny to both parents showed that each of the four ascospores is heterokaryotic both for spore killer and for mating-type idiomorphs. The general characteristics of *Psk7* are equally valid for other *Podospora* spore killer types. These results fit the rules of behavior that have been established for *Sk2*^{K} and *Sk3*^{K} in *Neurospora*.

4.3 Chromosomal Basis

The spore killers discovered by Padieu and Bernet [2] were located on different chromosomes and showed independent segregation. However, all but one of the six killer types studied by Van der Gaag et al. [4] appear to be linked in the same chromosome. Killers of the different types showed marked differences in second-division segregation frequencies. Recombination is clearly not blocked in the interval between killer factor and centromere, in contrast to the situation with

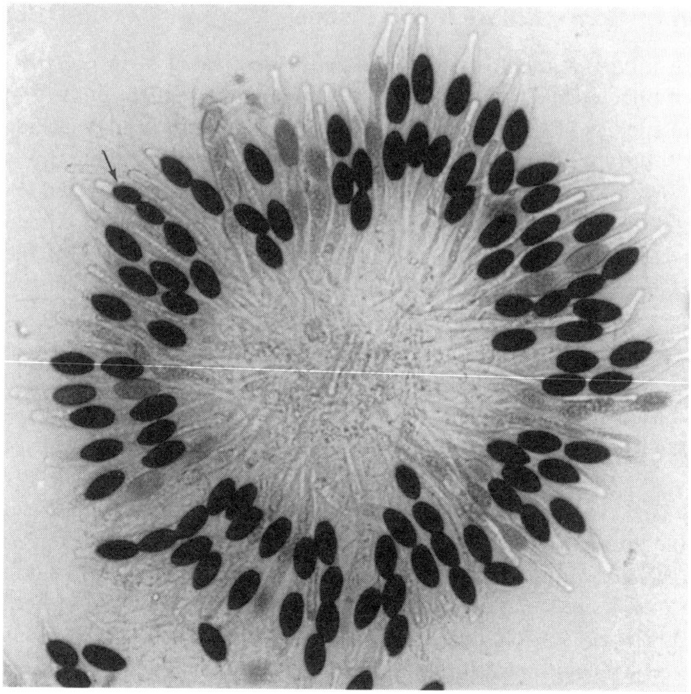

FIGURE 6 Spore killer expression in *Podospora anserina*. A cross of $Psk2^K \times Psk2^S$ shows ~70% two-spored asci resulting from first-division segregation of *Psk*. The asci were initially four-spored, but only the two homokaryotic Psk^K ascospores matured and the two homokaryotic Psk^S ascospores have already aborted and disintegrated. The minority of asci in which all four spores matured have resulted from the second-division segregation at *Psk*. A rare ascus (arrow) shows two large heterokaryotic black ascospores (K + S) and two small homokaryotic black ascospores (K). Two other small homokaryotic sensitive ascospores must have aborted and disintegrated in this ascus. (From Ref. 4.)

Sk2 and *Sk3* in *Neurospora*. The centromere linkage differences between the linked killers in *Podospora* suggest that they are at different, nonhomologous loci. However, if crossing over proximal to the locus were suppressed to different degrees in the different types, second-division segregation frequencies might differ even though the killer factors were at the same locus.

4.4 Interaction Between Different Spore Killer Types

The *Podospora* wild isolates from the Netherlands were initially classified on the basis of the frequency of asci with two aborted spores in crosses to standard sensitive testers (Fig. 6). The spore killer strains were subsequently grouped into six different killer types based on the interaction between various killer strains. A seventh spore killer type (*Psk5*) was found in the French strain Y. Van der Gaag et al. [4] have shown that various strains belonging to the same killer type give a reproducible frequency of two-spored asci when crossed to the same sensitive tester, and various killer strains of the same spore killer type show no killing when intercrossed (e.g., $Psk1^K \times Psk1^K$). However, killer strains of different spore killer types show killing when intercrossed [see Tables 1 and 3 in Ref. 4]. This result is similar, though not identical, to the killing behavior of $Sk2^K \times Sk3^K$ in *Neurospora* [3,27].

Analysis of intercrosses between various *Podospora* spore killer types goes beyond that of a single *Neurospora* intercross ($Sk2^K \times Sk3^K$), however. *Neurospora* spore killers $Sk2^K$ and $Sk3^K$ kill each other when intercrossed—i.e., all eight ascospores are killed in eight-spored asci of *N. crassa* and *N. intermedia*. Similarly, in crosses of $Sk2^K\ E \times Sk3^K$ in *N. tetrasperma*, all homokaryotic, small ascospores ($Sk2^K$ or $Sk3^K$) are killed because of mutual killing (or sensitivity), and all heterokaryotic, large ascospores ($Sk2^K + Sk3^K$) survive because of mutual rescue. In contrast, no mutually sensitive killer strains were found in the Wageningen populations of *P. anserina*. Spore killer types in *Podospora* show either dominant epistasis or mutual resistance [4]. For example, *Psk1* and *Psk7* are mutually resistant and show dominant epistasis to other killer types, whereas *Psk4* and *Psk6* are also mutually resistant but they are sensitive to killing by *Psk1*, *Psk2*, and *Psk7*. *Psk2* is intermediate in its killing hierarchy: It kills both *Psk4* and *Psk6* but is sensitive to killing by *Psk1* and *Psk7*.

4.5 Population Aspects

Spore killer strains are relatively frequent in the natural populations of *P. anserina* from the Netherlands. A killer was present in 23 of the 99 strains examined. This is similar to the frequency of killer strains in *N. sitophila* (16%) and *G. fujikuroi* (~50%). The 23 killers fall into six different spore killer types. *Psk1* was found in nine strains. *Psk2*, *Psk3*, *Psk4*, *Psk6*, and *Psk7* were found, respectively, in five strains, four, one, three, and one [4]. *Psk5* was represented by one of the French strains of Padieu and Bernet [2]. All other *Podospora* strains were sensitive to killing. Unlike *Neurospora*, no neutral or resistant strains that are not themselves killers have been found in *Podospora*. However, spore killer strains of one killer type may be resistant to killing by a different spore killer type, because of dominant epistatic or mutually resistant interactions [4].

5 HETEROKARYOTIC RESCUE IN PSEUDOHOMOTHALLIC SPECIES

In *N. tetrasperma*, up to 100% of sensitive meiotic products are sheltered from killing in heterokaryotic ($Sk^K + Sk^S$) ascospores because of tight centromere linkage of both of the spore killers, *Sk2* and *Sk3*. Once sheltered in the self-fertile heterokaryotic progeny, the sensitive nuclei (Sk^S) are assured of future sheltering at each sexual cycle. Only the exceptional sensitive nuclei that are sequestered into homokaryotic small ascospores are condemned to death. In contrast, attaining the same degree of sheltering in *Podospora*, where ascus programming is different, would require a single obligate crossover in 100% of asci. While theoretically possible (as is the case with mating types), the highest frequency of second-division segregation (% four-spored asci) for *Podospora* spore killers was 55%; usually, the frequency was between 5% and 25%. Thus sheltering of sensitive nuclei is far more efficient in *N. tetrasperma* than in *P. anserina*. The sheltering of sensitive nuclei in heterokaryotic ascospores led Raju and Perkins [27] to suggest that four-spored pseudohomothallic species may have evolved from their eight-spored progenitors to counter deleterious effects of invading spore killer elements.

6 SPORE KILLERS IN *GIBBERELLA FUJIKUROI* AND *COCHLIOBOLUS HETEROSTROPHUS*

Spore killers resembling those of *Neurospora* and *Podospora* have been found in two other heterothallic ascomycetes: *Gibberella fujikuroi* (*Fusarium moniliforme*) [5,36,37] and *Cochliobolus heterostrophus* [6,8]. The species are eight-spored, and four of the eight spores are killed in crosses of killer × sensitive, just as in the eight-spored asci of *Neurospora* described above. Early ascus development and nuclear divisions resemble those of *N. crassa*, except that the ascospores are not linearly ordered in the ascus. *Sk* in *Gibberella* maps to linkage group 5 [38]. Whether there is a recombination block in the interval between spore killer and the centromere, resulting in the segregation of killer and sensitive alleles at the first-division of meiosis, has not been determined in either fungus. For a summary of ascus development, spore killer description, and ascus photographs for *G. fujikuroi* and *C. heterostrophus*, see Raju [8].

Kathariou and Spieth [5] examined 225 wild isolates of *G. fujikuroi* for spore killer polymorphisms, using strains from southern Europe, North America, and Central America. Over 80% of the isolates from Europe and the Americas were killers. Sensitives (15%) were more frequent in Europe than in the Americas. The remaining 5% of isolates, when crossed with killers or sensitives, produced a mixture of asci, some with eight and some with four mature ascospores; these were designated as Sk^{Mx}. In another study, >50% of field isolates of *G. fujikuroi* from the midwestern United States were sensitive to killing [36,37].

7 PUTATIVE SPORE KILLERS IN OTHER FUNGI

Many naturally occurring spore anomalies have been described that have not been tested to determine whether asci or basidia with half of the spores aborted reflect a segregating Mendelian gene, a chromosome rearrangement, or a spore killer–like drive element. For example, half of the nuclei abort following meiosis and a postmeiotic mitosis in the basidiomycete *Pleurotus* [39]. Two of the four meiotic products abort in crosses between *Ustilago* spp. [40]. Four of the eight ascospores shrink and degenerate in the apomictic Pyrenomycete *Podospora arizonensis* [41], and in the Discomycetes *Octosporus alpestris* and *O. phagospora* [42].

8 OUTSTANDING PROBLEMS

8.1 Chromosomal Basis and Evolutionary Significance

Fungal spore killers and other animal and plant meiotic drive elements have been found only among natural populations, not as laboratory variants. Also, our attempts to revert the killer to nonkiller by mutagenesis have not been successful [N.B. Raju, unpublished]. Conceivably, *Sk* is not a single gene but is a complex chromosomal entity evolved over an extended period of time and as such does not readily mutate or recombine. Similarly, *Drosophila* segregation distorter chromosome and mouse *t* haplotypes have been known to be riddled with duplications or inversions, and suppress recombination when heterozygous [7].

Once a spore killer element is fixed in a population, it would no longer be detected unless outcrossing to a sensitive population occurred. Killers would be detected either while a new element is on its way to fixation or in populations where a stable polymorphism—consisting of killers, sensitives, and resistant strains—is established. The evolutionary significance of spore killers in *P. anserina* and in *N. tetrasperma* has been discussed by Nauta and Hoekstra [34] and Van der Gaag et al. [4]. Apparently, these pseudohomothallic species could use selfing as the first line of defense against newly arrived spore killers from spreading through sensitive populations. Even when a sensitive strain occasionally outcrosses with a killer, the sensitive nuclei are not likely to be harmed because of heterokaryotic rescue. The heterokaryotic rescue was shown to be very efficient (100%) for the introgressed *Sk2* and *Sk3* in *N. tetrasperma* because of a recombination block in the *Sk* centromere region [27], but the rescue is far less efficient for the naturally occurring spore killers in *P. anserina*, where first-division segregation of *Psk* alleles leads to death of two of the four homokaryotic ascospores [4].

8.2 Mode of Action

Spore killer–induced death of Sk^S ascospores is expressed only when a killer is crossed with a sensitive strain, and only after the killer and sensitive nuclei are sequestered into separate ascospores. Killer factors are undetectable in the vegetative phase or in homozygous $Sk^K \times Sk^K$ crosses. Death of Sk^S ascospores becomes apparent only during spore development. Although the exact time of primary killer function is not known, it must already be in place prior to spore delimitation. Developmental events are complex during stages when it might occur, from fertilization through formation of the ascospore wall [11].

Many *Neurospora* gene mutations are known that have defects in ascospore differentiation, pigmentation, or viability [10,11,26,43,44]. Deficiencies resulting from segregation of chromosome rearrangements also produce defective, inviable ascospores that resemble those carrying Sk^S nuclei [9,23]. Thus, normal ascospore maturation requires many gene functions which, if defective, would render the ascospores inviable. Any one of these could perhaps be a target for inactivation by spore killer.

In a heterozygous cross, the killer and sensitive elements coexist in a common cytoplasm from the time of fertilization in the ascogonium until ascospore walls are formed in the developing asci. In the ascogonium and ascogenous hyphae, killer and sensitive chromosomes are located in different nuclei. Then, at karyogamy in the young ascus, they are brought together briefly into a single diploid zygote nucleus. Their presence in the same nucleus is transient, lasting only until they segregate into separate nuclei at the first or the second meiotic division. The haploid meiotic products, two killer and two sensitive, then undergo an apparently normal postmeiotic mitosis within the common ascus cytoplasm before the daughter nuclei (and the surrounding cytoplasm) are sequestered into eight homokaryotic ascospores (4K:4S) in heterothallic species, or into four heterokaryotic ascospores (K + S) in *N. tetrasperma*. In *P. anserina*, ascospores are homokaryotic in some asci (2K:2S) and heterokaryotic (K + S) in others. In asci from crosses of $Sk^K \times Sk^S$, killer and sensitive ascospores receive the same common cytoplasm. It is only the nuclei that differ. Only after enclosure by the ascospore wall do the ascospores carrying sensitive nuclei stop developing and die. Development and maturation are normal in ascospores that are homokaryotic or heterokaryotic for killer nuclei.

Apparently, Sk^S ascospores are somehow rendered unable to perform one or more functions essential for ascospore maturation. Loss of function could be either genetic or epigenetic. The fact that a sensitive nucleus survives and remains functional when enclosed in the same ascospore with a killer nucleus suggests that the inactivating change is epigenetic rather than genetic.

It should be noted that rescue of Sk^S nuclei in heterokaryotic ascospores occurs neither because the sheltered Sk^S nuclei become immune to killing nor because Sk^K is somehow rendered harmless to Sk^S. Progeny analysis showed that the sheltered Sk^S nuclei from heterokaryotic ($Sk^K + Sk^S$) ascospores are unchanged and remain sensitive to killing in tests with a killer strain. Likewise, the killer nuclei that rescued the Sk^S nuclei in the heterokaryotic spores are also unchanged and are capable of causing the death of homokaryotic Sk^S ascospores [25,27]. Thus, if modification of the sheltered Sk^S nuclei occurs, it is temporary and readily reversible.

The question remains, why the Sk^K nucleus is itself immune to inactivation. A likely model is suggested by the two-component meiotic drive systems known for *Drosophila* and the mouse. In both these organisms, distorter and activator elements are present at different linked loci within the killer complex [45,46]. In such a model, spore killer chromosomes would be distorter positive and activator insensitive, and sensitive chromosomes would be distorter negative and activator sensitive. To explain observations with *Neurospora* and *Podospora*, the protective function of the activator-insensitive element would necessarily extend to a sensitive nucleus that was located in the same ascospore, preventing it from being killed.

Although the specific functions targeted by meiotic drive are very different in fungi than in flies or mammals, the drive systems in all three probably share certain general characteristics. Thus, existence of a chromosome segment within which recombination is blocked is thought to reflect a long evolutionary history during which an ''arms race'' between distorting element and suppressor has resulted in accumulation of multiple modifiers [45,47,48]. After >50 years of intense study, basic genetic elements of the drive systems in *Drosophila* and in the mouse have now been identified, cloned molecularly, and sequenced, and their mode of action is at least partially understood [45,46].

Meiotic drive in fungi has been known for a much shorter time and knowledge is much less advanced. Conventional genetic analysis in *Neurospora* has been hampered by the recombination block. Availability of the *Neurospora* genome sequence, soon to be completed, may be expected to speed molecular analysis of the spore killer phenomenon. Molecular findings already made in the better-known organisms may be useful as a guide. Similarities in pattern may emerge, but profound differences in the particular examples that underlie that pattern are to be expected.

ACKNOWLEDGMENTS

I thank David Perkins for suggestions on the manuscript, David Jacobson for help with Figures 4 and 6, and Fons Debets for making Figure 6 available from Ref. 4. The work in our laboratory is supported by MCB-9728675 from the National Science Foundation.

REFERENCES

1. BC Turner, DD Perkins. Spore killer, a chromosomal factor in *Neurospora* that kills meiotic products not containing it. Genetics 93:587–606, 1979.

2. E Padieu, J Bernet. Mode d'action des gènes responsibles de l'avortment de certaine produits de la méiose chez l'ascomycète *Podospora anserina*. C.R. Hebd Seances Acad Sci, Sér D, 264:2300–2303, 1967. (English translation in Turner and Perkins, 1991 [3].)
3. BC Turner, DD Perkins. Meiotic drive in *Neurospora* and other fungi. Am Nat 137: 416–429, 1991.
4. M van der Gaag, AJM Debets, J Oosterhof, M Slakhorst, JAGM Thijssen, RF Hoekstra. Spore killing meiotic drive factors in a natural population of the fungus *Podospora anserina*. Genetics 156:593–605, 2000.
5. S Kathariou, PT Spieth. Spore killer polymorphism in *Fusarium moniliforme*. Genetics 102:19–24, 1982.
6. CR Bronson, M Taga, OC Yoder. Genetic control and distorted segregation of T-toxin production in field isolates of *Cochliobolus heterostrophus*. Phytopathology 80:819–823, 1990.
7. TW Lyttle. Segregation distorters. Annu Rev Genet 25:511–557, 1991.
8. NB Raju. Ascomycete spore killers: chromosomal elements that distort genetic ratios among the products of meiosis. Mycologia 86:461–473, 1994.
9. DD Perkins. Chromosome rearrangements in *Neurospora* and other filamentous fungi. Adv Genet 36:239–398, 1997.
10. DD Perkins, A Radford, MS Sachs. The *Neurospora* Compendium: Chromosomal Loci. San Diego: Academic Press, 2000.
11. NB Raju. Genetic control of the sexual cycle in *Neurospora*. Mycol Res 96:241–262, 1992.
12. RF Cain. Studies of soil fungi. III. New species of *Coniochaeta*, *Chaetomidium*, and *Thielavia*. Can J Bot 39:1231–1239, 1961.
13. NB Raju, DD Perkins. Programmed ascospore death in the homothallic ascomycete *Coniochaeta tetraspora*. Fungal Genet Biol 30:213–222, 2000.
14. RB Wickner. Killer systems in *Saccharomyces cerevisiae*. In: JN Strathern, EW Jones, JR Broach, eds. The Molecular Biology of the Yeast *Saccharomyces*: Life Cycle and Inheritance. Cold Spring Harbor, NY: Cold Spring Harbor Laboratory, 1981, pp 415–444.
15. RW Davis. *Neurospora*: Contributions of a Model Organism. Oxford: Oxford University Press, 2000.
16. NB Raju. Meiosis and ascospore genesis in *Neurospora*. Eur J Cell Biol 23:208–223, 1980.
17. BO Dodge. Nuclear phenomena associated with heterothallism and homothallism in the ascomycete *Neurospora*. J Agr Res 35:289–305, 1927.
18. NB Raju. Functional heterothallism resulting from homokaryotic conidia and ascospores in *Neurospora tetrasperma*. Mycol Res 96:103–116, 1992.
19. NB Raju, DD Perkins. Diverse programs of ascus development in pseudohomothallic species of *Neurospora*, *Gelasinospora* and *Podospora*. Dev Genet 15:104–118, 1994.
20. G Rizet, C Engelmann. Contribution à l'étude génétique d'un ascomycète tétrasporé: *Podospora anserina* (Ces.) Rehm. Rev Cytol Biol Veg 11:201–304, 1949.
21. CL Shear, BO Dodge. Life histories and heterothallism of the red bread-molds of the *Monilia sitophila* group. J Agr Res 34:1019–1042, 1927.

22. FL Tai. Two new species of *Neurospora*. Mycologia 27:328–330, 1935.
23. DD Perkins, EG Barry. The cytogenetics of *Neurospora*. Adv Genet 19:133–285, 1977.
24. DD Perkins, NB Raju. Three-to-one segregation from reciprocal translocation quadrivalents in *Neurospora* and its bearing on the interpretation of spore-abortion patterns in unordered asci. Genome 38:661–672, 1995.
25. NB Raju. Cytogenetic behavior of spore killer genes in *Neurospora*. Genetics 93: 607–623, 1979.
26. NB Raju, D Newmeyer. Giant ascospores and abnormal croziers in a mutant of *Neurospora crassa*. Exp Mycol 1:152–165, 1977.
27. NB Raju, DD Perkins. Expression of meiotic drive elements spore killer-2 and spore killer-3 in asci of *Neurospora tetrasperma*. Genetics 129:25–37, 1991.
28. JL Campbell, BC Turner. Recombination block in the spore killer region of *Neurospora*. Genome 29:129–135, 1987.
29. M Bojko. Presence of abnormal synaptonemal complexes in heterothallic species of *Neurospora*. Genome 30:697–709, 1988.
30. BO Dodge. A new dominant lethal in *Neurospora*: the *E* locus in *Neurospora tetrasperma*. J Hered 20:467–474, 1939.
31. BC Turner. Geographic distribution of spore killer strains and strains resistant to killing in *Neurospora*. Fungal Genet Biol 32:93–104, 2001.
32. BC Turner, DD Perkins, A Fairfield. *Neurospora* from natural populations: A global study. Fungal Genet Biol 32:67–92, 2001.
33. DD Perkins, BC Turner. *Neurospora* from natural populations: toward the population biology of a haploid eukaryote. Exp Mycol 12:91–131, 1988.
34. MJ Nauta, RF Hoekstra. Evolutionary dynamics of spore killers. Genetics 135:923–930, 1993.
35. L Belcour, M Rossignol, F Koll, CH Sellem, C Oldani. Plasticity of the mitochondrial genome in *Podospora*. Polymorphism for 15 optional sequences: group-I, group-II, intronic ORFs and an intergenic region. Curr Genet 31:308–317, 1997.
36. GS Sidhu. Genetics of *Gibberella fujikuroi*. J Hered 75:237–238, 1984.
37. GS Sidhu. *Gibberella* spp., pathogens of many crop species. In: GS Sidhu, ed. Genetics of Plant Pathogenic fungi. Advances in Plant Pathology, Vol 6. New York: Academic Press, 1988, pp 159–167.
38. JR Xu, JF Leslie. A genetic map of *Gibberella fujikuroi* mating population A (*Fusarium moniliforme*). Genetics 143:175–189, 1996.
39. AM Slézec. Les étapes de la méiose chez les Pleurotes des ombellifères. Cryptogr Mycol 7:235–265, 1986.
40. J Nielsen. Experiments on vegetative dissociation of the dikaryon and on lysis of hybrid sporidia of the cross *Ustilago avenae* × *U. kolleri*. Can J Bot 46:487–496, 1968.
41. HR Mainwaring, IM Wilson. The life cycle and cytology of an apomictic *Podospora*. Trans Br Mycol Soc 51:663–677, 1968.
42. RWG Dennis. British Ascomycetes. Varduz, Liechtenstein: J Cramer, 1978.
43. AM Srb, M Basl, M Bobst, JV Leary. Mutations in *Neurospora crassa* affecting ascus and ascospore development. J Hered 64:242–246, 1973.

44. NB Raju. A mutant of *Neurospora crassa* with abnormal croziers, giant ascospores, and asci having multiple apical pores. Mycologia 79:696–706, 1987.
45. B Ganetzky. Tracking down a cheating gene. Am Sci 88:128–135, 2000.
46. J Schimenti. Segregation distortion of mouse *t* haplotypes. Trends Genet 16:240–243, 2000.
47. LD Hurst, A Atlan, BO Bengtsson. Genetic conflicts. Q Rev Biol 71:317–364, 1996.
48. LM Silver. The peculiar journey of a selfish chromosome: mouse *t* haplotypes and meiotic drive. Trends Genet 9:250–254, 1993.

12

Living Together Underground
A Molecular Glimpse of the Ectomycorrhizal Symbiosis

Sébastien Duplessis, Denis Tagu, and Francis Martin
Centre INRA de Nancy, Champenoux, France

1 INTRODUCTION

Within the rhizosphere, which hosts a large and diverse community of prokaryotic and eukaryotic microbes that compete and interact with each other and with plant roots, mycorrhizal fungi are almost ubiquitous. The ectomycorrhizal hyphae and the root tips form a novel composite organ, so-called mycorrhiza, which is the site of nutrient and carbon transfer between the two symbionts. This association allows terrestrial plants to grow efficiently in suboptimal environments [1]. Among the various types of mycorrhizal symbioses, the endomycorrhizal, ectomycorrhizal, and ericoid associations are found on most annual and perennial plants (probably >90%). About two-thirds of these plants are symbiotic with arbuscular mycorrhizal glomalean fungi [2]. Ericoid mycorrhizas are ecologically important, but mainly restricted to heathlands [3]. While a relatively small number of plants, ~8000, form ectomycorrhiza, their global importance is amplified by their wide occupancy of terrestrial ecosystems. Trees of Betulaceae, Cistaceae, Dipterocarpaceae, Fagaceae, Pinaceae, Myrtaceae, Salicaceae, and several tribes

FIGURE 1 The ectomycorrhizal symbiosis. (A) A seedling of Douglas fir (*Pseudotsuga menziesii*) colonized by the ectomycorrhizal basidiomycete *Laccaria bicolor*. The fungal mycelium has developed ectomycorrhizas on the root system and has produced a fruiting body above ground. (Photograph courtesy of P. Frey-Klett.) (B) Transverse section of a *Eucalyptus/Pisolithus* ectomycorrhiza showing the extramatrical hyphae (EM), the mantle (M); the fungal hyphae have begun to penetrate between the epidermal cells (E) of the root cortex (C) to form the Hartig net (HN). Epidermal cells (E) are radially enlarged. CC, central cylinder. (Photograph courtesy of B. Dell.)

in Fabaceae are ectomycorrhizal plants (Fig. 1A), dominating boreal, temperate, Mediterranean, and some subtropical forest ecosystems [1]. Within days after their emergence in the upper 10 cm of the soil profiles (e.g., organic humus and mor layer), most of the short roots of these ectomycorrhizal shrubs and trees are colonized by ectomycorrhizal fungi, and in most cases symbiotic colonization is close to 100% [4]. The fungal mycelium and the root tips form a novel composite organ, so-called ectomycorrhiza, which is the site of nutrient and carbon transfer between the two symbionts [5,6].

Ectomycorrhiza is structurally characterized by (1) the presence of an extensive extramatrical mycelial web prospecting the soil and gathering nutrients, (2) a mantle of fungal hyphae ensheating the root and mainly acting as a storage compartment, and (3) a network of hyphae growing in the apoplastic space of the rhizodermis (in angiosperms) and cortex (in conifers) root cells (Fig. 1B). The fungus gains access to sugars from the plant while plant nutrient, and water uptake is mediated via the fungus (Fig. 2). In addition, the establishment of the symbiosis is requested for the completion of the fungal life cycle (i.e., formation of fruiting bodies) (Fig. 1A). The formation of the symbiosis requires several days and induces major morphological changes including a novel spatial tissue organization, changes in cell shape, and the generation of different cell types [7,8].

Ectomycorrhiza formation therefore involves a series of complex and overlapping ontogenic processes in the mycobiont and the host plant: increased rhizogenesis, enhanced hyphae branching, aggregation of the proliferating hyphae, arrest of meristematic activity in roots surrounded by the fungal mantle, and radial elongation of epidermal cells. These morphological changes are accompanied by the onset of novel protein patterns [9] and metabolic organizations [10,11] in fungal and plant cells, leading to the functioning symbiosis. What could be the molecular basis of such a progressive, highly organized ontogenic process? What is the role of cell-to-cell signaling in symbiosis development? How many genes control ectomycorrhiza development-as distinct from providing the housekeeping functions of the fungal and plant cells? After a brief overview of the evolution, biology and anatomy of ectomycorrhiza, these are some of the most important questions that will be tackled in the present chapter.

2 MYCORRHIZAS ARE ANCESTRAL SYMBIOTIC INTERACTIONS

The first mycorrhizal associations must have been derived from earlier types of plant–fungus interactions, such as the fungus *Geosiphon pyriforme* forming endocytobiosis with *Nostoc* (Cyanobacteria) [12] and endophytic fungi found in the bryophite-like precursors of vascular plants [13]. Structures similar to arbuscular mycorrhiza have been observed in plant fossils from the Early Devonian [14], whereas fossil ectomycorrhiza have been found in the Middle Eocene [15]. Based on phylogenetic analysis of the rRNA gene, it has been suggested that ectomycorrhizal basidiomycetes evolved convergently from saprophytic ancestors [16]. The switch between saprophytic and mycorrhizal life styles likely happened many times during evolution of fungal lineages, as revealed by recent molecular phylogenetic analyses [17]. This may have facilitated evolution of ectomycorrhizal lineages with a broad range of physiological and ecological functions reflecting partly the activities of their disparate saprotrophic ancestors. These symbioses have had major consequences for the diversification of both the mycobionts and their hosts [18]. It remains to be determined whether the development of different lateral root structures (actinorhiza, mycorrhiza, mycorrhizal nodules) are gov-

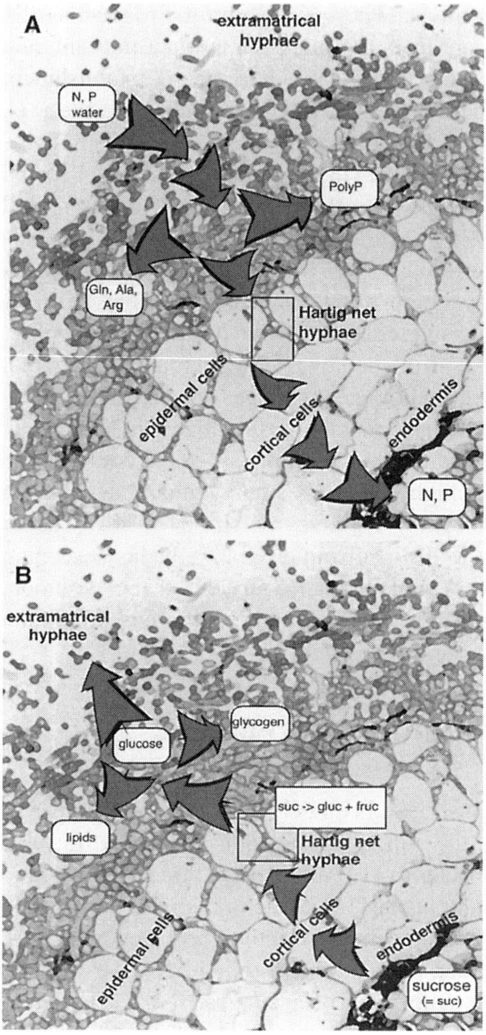

FIGURE 2 Ectomycorrhiza: a mutualistic symbiosis. (A) Extramatrical hyphae prospect the soil, gather nutrients (N, P, H_2O) and translocate them to the mantle. Mineral and organic N are assimilated and the synthesized amino acids (Gln, Ala, and Arg), together with inorganic P polymers (PolyP, polyphosphates) are accumulated. Amino acids and Pi are then transferred through the intraradicular hyphal net (the Hartig net) to the root cells and the other host tissues. (B) On the other hand, sucrose (suc) downloaded to the root cells is degraded by apoplastic invertases in glucose (gluc) and fructose (fruc). Glucose is then translocated to the fungal compartment through the Hartig net hyphae. The carbohydrate is then stored in the mantle as glycogen and lipids, or transported to the extramatrical hyphae.

erned by the same set of genes [19]. Current ectomycorrhizal fungal species (~6000) mainly belong to homobasidiomycetes (agarics, bolets), although many species are found within the ascomycetes (truffles, terfez) and zygomycetes.

3 A BENEFICIAL SYMBIOSIS

Ectomycorrhizal communities are taxonomically diverse [20,21] and are likely able to maintain a large degree of functional diversity [22]. Although a few tree/fungus combinations are unique, a great many different fungi can combine with a great many different trees. A single host tree could simultaneously interact with dozens of fungal species [4], and this high symbiont diversity likely allows ectomycorrhizal associations to use most N and P forms present in forest soils [23]. The symbiosis between trees and soilborne ectomycorrhizal fungi results in an intimate relationship between the plant and its symbiotic partner (Fig. 1B). It provides several benefits to both the host plant and its fungal associate(s). The prospecting and absorbing extraradical hyphal web (1000 m of hyphae/m of root) captures soil minerals (phosphate, nitrogen, water, micronutrients) [1] and organic nitrogen [24,25] and assimilates and translocates a large proportion of them to the growing plant [1,24] (Fig. 2A). Ectomycorrhizal fungi affect not only mineral and water uptake, but also adaptation to adverse soil chemical conditions [5] and susceptibility to diseases [1], and contribute substantially to plant productivity [6]. On the other hand, the fungus within the root is protected from competition with other soil microbes and therefore is a preferential user of the plant carbon (~20% of the host photoassimilates) (Fig. 2B). Mycorrhizal fungi represent an interface in the soil–plant system and have the ability to regulate plant metabolism. In addition, they constitute links in the chain of transfers by which carbon and nitrogen move between plant and soil compartments [26,27] and can thus influence carbon and nitrogen cycling rates in host plants and forest ecosystems [23,28,29].

4 ECTOMYCORRHIZA ONTOGENESIS: THE DANCE IS THE SAME, THE COUPLES ARE DIFFERENT

Morphological and anatomical changes that accompany ectomycorrhiza development have been studied and described in great detail in various associations (e.g., *Picea abies/Amanita muscaria* [7,30]; *P. abies/Hebeloma crustuliniforme* [31]; *Eucalyptus/Pisolithus* [8,32–34]; *Alnus rubra/Alpova diplophloeus* [35]; and *Betula pendula/Paxillus involutus* [36]). The mature organization of ectomycorrhiza varies with the host and fungal species [37]. In addition, a survey of almost any natural fungal population will reveal a considerable range in phenotypes [38]. However, although some of the details vary, early stages of ectomycorrhiza development have well-characterized, similar morphological transitions. In an effort

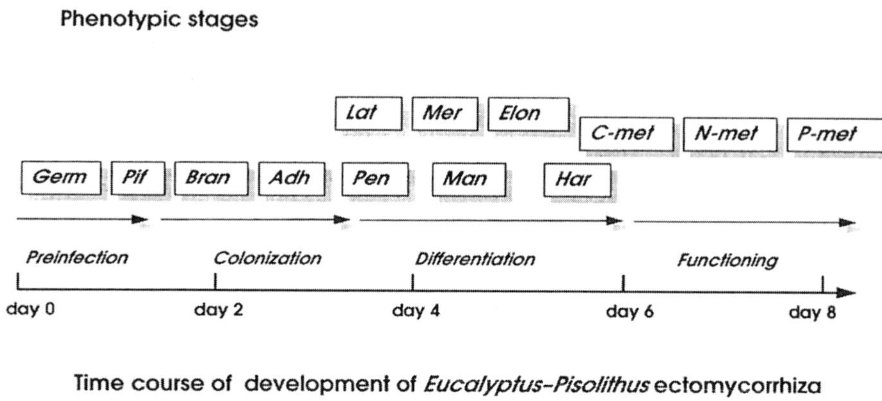

FIGURE 3 Main phenotypic stages of ectomycorrhiza development. Abbreviations: *Germ*, spore germination; *Pif*, preinfection growth of hyphae; *Bran*, hyphae branching on to root surface; *Adh*, attachment of hyphae on root surface and formation of the adhesion pads; *Pen*, penetration between epidermal cells; *Man*, hyphae aggregation to form mantle; *Har*, differentiation of the Hartig net; *Lat*, increased formation of lateral roots; *Mer*, changes in meristematic activity; *Elon*, radial elongation of epidermal cells; *C-met*, *N-met*, and *P-met*—changes in carbon, nitrogen, and phosphate metabolisms, including transfer between symbionts.

FIGURE 4 The different interactions between the host root and the ectomycorrhizal fungus and the main morphogenetic stages observed during ectomycorrhiza development. (A) The preinfection stage: (A1) host root releases chemicals (e.g., flavonoids, cytokinines) in the rhizosphere able to alter the morphology of the compatible ectomycorrhizal fungus (e.g., enhanced hyphal branching); (A2) conversely, the hyphae releases various compounds (auxins, alkaloids) eliciting changes in the root morphology (e.g., increased rhizogenesis, decay of root hairs). (B) The colonization stage: running hyphae attach to the root surface and then experience drastic morphological changes, such as tip swelling, leading to a fingerlike structure on the root epidermal cell. Hyphae initiate their aggregation between host cells to form hyphal webs. (C) Morphogenesis per se: massive and rapid aggregation of hyphae around the root lead to the formation of a pseudoparenchyma, the mantle; penetration between epidermal cells and cortical cells; and formation of the Hartig net with concomitant coordinated alteration in the root structure. Original drawings, Armoise Conseil.

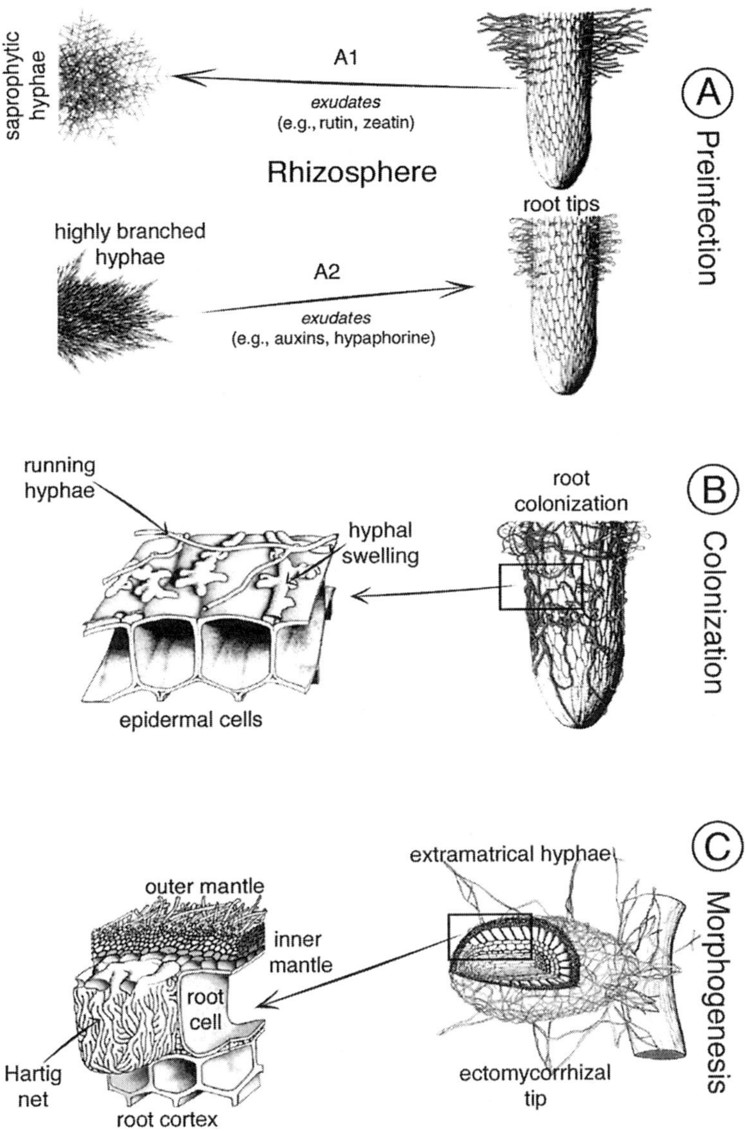
saprophytic hyphae
A1
exudates
(e.g., rutin, zeatin)
Rhizosphere
root tips
highly branched hyphae
A2
exudates
(e.g., auxins, hypaphorine)
A Preinfection
running hyphae
hyphal swelling
root colonization
epidermal cells
B Colonization
extramatrical hyphae
outer mantle
inner mantle
root cell
Hartig net
root cortex
ectomycorrhizal tip
C Morphogenesis

to provide a useful framework in which to categorize existing data on gene expression and accommodate future efforts to categorize existing natural fungal variants and future experimental mutants, we have subdivided ectomycorrhiza formation in discrete stages from the preinfection (rhizospheric) phase to the morphogenesis per se (Fig. 3) [39].

Spore germination and saprophytic growth of the hyphae are initiated in the rhizosphere (Fig. 4A). In most natural situations, however, hyphae are invading a newly emerging root from previously established ectomycorrhiza. Extensive preinfection branching of hyphae requires the presence of host plant roots (Fig. 4A). After contact, hyphal growth on the root surface initiates swelling of the hyphal tips (pads) and formation of dense, fingerlike structures (Fig. 4B). Hyphae aggregate initially to form wefts and then ensheath the lateral roots (Fig. 4B). After root penetration, intraradical hyphae proliferate and form a coenocytic structure in the root apoplastic space (i.e., the Hartig net) (Fig. 4C). This intraradical fungal web is active in nutrient transfer, and an active traffic of carbohydrates promotes extensive growth of external hyphal web that gathers nutrient in soil. These morphological changes are concomitant with accelerated nuclear division, cytoskeletal rearrangements, and synthesis of differentiation-related genes and proteins (see below). Growth and differentiation of the plant root and fungal hyphae must be tightly coordinated. This multistep development therefore implies the existence of a developmental strategy for building up an ectomycorrhiza that early on imposes a basic scheme, on top of which subsequent species-specific customizations occur. Behind every aspect of ectomycorrhiza development there must likewise be genetic control from the earliest proliferation of hyphae to the buildup of the complicated symbiotic structure.

Since fungal mutants affected in their ability to form ectomycorrhiza are not available, one approach to identify the genetic processes that trigger and regulate ectomycorrhiza development is to look for natural variation in symbiosis structure. It has been shown that natural populations of sib monokaryotic and dikaryotic strains of *Laccaria bicolor* [40], *H. cylindrosporum* [41], and *Pisolithus* [42] vary greatly in their ability to form mycorrhizas. Some *L. bicolor* variants undergo morphological changes that signal the onset of mycorrhiza formation but fail to complete the development process and do not move on to the next stage [43]. They have been classified into different basic categories: intraradical hyphal network not formed; hyphal network formed but not developed further; Hartig net development normal but failure of mantle to form. This suggests that the morphogenetic programs for the differentiation of the mantle and the Hartig net are partly independent and they likely involve different sets of genes. This has been recently confirmed by the differential effect of auxin transport inhibitors on the formation of the mantle and the Hartig net [44]. Variation in mycorrhizal structures appears to be genetically determined, which should make it possible to identify the loci that contribute to this variation.

5 SIGNALING CHEMICALS IN THE RHIZOSPHERE AND IN SYMBIOTIC TISSUES

In almost all plant–microbe interactions exchange of signals between the partners is the earliest step in a series of interaction events leading to contact at the host surface and subsequent development of the microbial structures in the host plant tissues [45] (Fig. 4). Signaling processes must exist to bring the mycobiont into the vicinity of susceptible host roots. This mechanism has been poorly investigated in ectomycorrhizal symbioses, and only a limited set of chemical signals produced by either the host or symbiont have been identified [46–49]. Only the broad outlines of the signaling processes have been defined, but the little that is known suggests that some intriguing similarities exist between ectomycorrhizal associations and other plant–microbe symbioses [45,50,51].

No specific chemicals able to attract ectomycorrhizal fungi toward the root surface have been identified, although their occurrence has been suggested [52]. However, there is evidence that host root exudates contain more than one kind of metabolite that can stimulate hyphal growth and/or morphological features of the colonizing ectomycorrhizal hyphae, and several factors likely help the partners match each other. Host plants secrete continuously a spectrum of chemicals able to attract rhizospheric microbes. Within these compounds, C20 diterpene abietic acid is able to stimulate spore germination of ectomycorrhizal bolets, such as *Suillus* spp. [46]. Among the secreted compounds are phenolic substances, especially flavonoids. The ectomycorrhizal *Pisolithus* spp. respond to traces of eucalypt flavonoids (e.g., the flavonol rutin quercetin-3-rutinoside [49]) by enhanced growth (Fig. 4A). Conjugates of flavonoids, such as rutin, are more soluble in water than their aglycones; thus, they diffuse readily and can be hydrolyzed to more active metabolites [53]. The presence of such compounds probably increases the possibility of the interaction. Interestingly, increased branching of the endomycorrhizal *Gigaspora rosea* in the presence of root exudates is enhanced by the rutin aglycone, quercetin [54]. Cytokinin, such as zeatin, presents in the rhizosphere can alter hyphal branching of *Pisolithus* and mimics some of the earliest steps of the ectomycorrhizal interaction (Fig. 4A) [49]. The branching is more numerous and compact in the presence of the phytohormone, and this bushy type of hyphal branching pattern likely increases the chance for the hyphae to enter in contact with the root surface. In addition to altering fungal morphology, zeatin interacts with the metabolism of alkaloid in the hyphae. The presence of zeatin results in the increased accumulation and secretion of hypaphorine, a tryptophan betaine [55] able to trigger morphological changes in eucalyptus roots (e.g., arrest of root hair growth) [47,48].

Although the colonization of emerging root tips by ectomycorrhizal hyphae is often initiated from older mycorrhizal parts of the root system, ectomycorrhizal fungi may be widely dispersed in the different soil horizons. It is tempting to speculate that extensive gradients of mixed chemicals in the soil provide the mechanisms by which the mycobionts are initially recruited to the root from the general rhizosphere populations of fungi. Then, after the hyphae accumulate in the mucigel that is adjacent to the root surface, the symbiosis can further proceed through the hyphae attachment and colonization (Fig. 4B). It is not known how these rhizospheric chemicals may play a role in setting up the ectomycorrhizal associations. In addition to attracting and stimulating ectomycorrhizal mycelium (and other symbiotic microbes), these plant metabolites, such as flavonoids, have numerous other activities (e.g., antimicrobial activities, modification of plant growth) [56], confirming a lack of specificity.

The presence of multiple nonspecific signals is ecologically consistent with the lack of specificity of the ectomycorrhizal symbiosis. It is highly improbable that the wide range of ectomycorrhizal trees species secrete a single universal signaling chemical to which all ectomycorrhizal fungi respond. Individual fungal species may sense one signal or a set of specific signals within a complex cocktail of plant chemicals and would respond according to what is secreted by any given host plant. The composition and concentration of the signaling compounds mixture that is secreted in the host tree rhizosphere are probably crucial. Further investigation is under way to identify additional root chemicals involved in the alteration of hyphal morphology to fully understand signalling and recognition processes in ectomycorrhizas.

On the other hand, ectomycorrhizal fungi have the potential to morphologically alter the host root through refined intervention in the developmental programme of the host plant. Hyphae enter the root preferentially at the elongation or differentiation zone and then migrate intercellularly toward the exodermis (in most angiosperms) (Fig. 1B) and to the endodermis in conifers. Intraradicular proliferation of the Hartig net hyphae implies host cell wall openings away from the hyphal tips [57]. Cell wall loosening and breakdown are likely involved in this apoplastic progression of the syncitial mycobiont. Multiple evidences have been provided that auxins, such as indole-3-acetic acid (IAA), play a role in this process and in additional early stage processes of ectomycorrhiza development [50,58,59]. The hypothesis that ectomycorrhizal fungi disturb root tissues by secreting IAA (and, in its wake, ethylene) is relatively old [60,61]. IAA released by ectomycorrhizal fungi elicits similar root responses as those induced by ectomycorrhiza formation including an enhanced rhizogenesis and dichotomous branching of pine roots [62,63]. It has been suggested that the intraspecific variations in symbiotic structures of *L. bicolor–Pinus banksiana* mycorrhizas are related to the differences in IAA-synthesizing activity among the various fungal isolates [43]. Tryptophan released in root exudates could be sufficient to trigger the increased biosynthesis of IAA or a homolog of this phytohormone in ectomycorrhizal fungi [64]. Pine inoculated with mutant of *H. cylindrosporum* strains overproducing IAA produced an increased number of ectomycorrhizal roots [65], which presented a strikingly altered morphology (i.e., hyphae proliferation leading to a multiseriate Hartig net) [57]. It has been suggested that this Hartig net hypertrophy was the result of an increased cell wall loosening resulting from local increase in IAA concentration [57]. The presence of increased concentration of IAA as a result of fungal colonization of root tissues has not been experimentally demonstrated.

The local increase in the auxin concentration could also be realized by stimulating the influx or inhibiting the efflux of auxins in the colonized root zone [66]. To investigate the role of polar auxin transport in ectomycorrhiza development, Douglas fir (*Pseudotsuga mensiesii*) seedlings were exposed to the phytotropin triiodobenzoic acid (TIBA) [44] (Rincon and Le Tacon, unpublished results). Subsequently roots were inoculated with the ectomycorrhizal basidiomycete *L. bicolor*. In lateral roots treated with TIBA, cross sections revealed that TIBA inhibited the formation of the fungal mantle [44]. Alternatively, the auxin transport inhibitor N-(1-naphthyl)phtalamic acic (NPA) induced similar alteration of the mycorrhiza development [67]. The failure of *L. bicolor* to aggregate to form the mantle in the phytotropin-treated seedlings points to a prominent role of polar auxin transport in early stages of mycorrhiza development. TIBA and NPA are known to block the basipetal transport of IAA in the outer cortex and epidermis through the inactivation of the auxin efflux carrier PIN (pin-formed) complexes [66]. The pleiotropic effects of auxin secreted by ectomycorrhizal fungi, the negative impact of phytotropins on the formation of symbiosis tissues, and the data obtained with *Hebeloma* mutants overproducing IAA are strong indications for a crucial role of auxin (and ethylene) in ectomycorrhiza morphogenesis.

The most parsimonious explanation of this set of data obtained through different approaches is that hyphae proliferation to form the mantle and the growth of hyphae through plant walls is accompanied by a fungus-induced, local increase of the auxin concentration. This enhanced auxin concentration appears to be reached through both hyphae secretion and alteration of plant-synthesized auxin transport. A local accumulation of auxin during the early stages of ectomycorrhiza development is consistent with the expression of the auxin downregulated transcripts *adr-6* and upregulation of the auxin-induced glutathion-*S*-transferase, *EgHypar*, in ectomycorrhizal tissues [68,69]. Whether tryptophan and/or other components of the plant exudates induce an IAA amplification loop in the rhizospheric mycelium, the IAA synthesis must be tightly controlled or compensated by other factors since above a certain level, exogenously supplied IAA inhibits root development. The latter may explain the observed arrest of root meristematic activity in mature mycorrhiza [33].

The fungal alkaloid hypaphorine, a betaine of tryptophan, is the major indolic compound isolated from the ectomycorrhizal fungus *Pisolithus* [55]. It is produced in larger amounts by this fungus during mycorrhiza development [47] and upon triggering by root exudates and zeatin [47] (Lagrange and Lapeyrie, unpublished results). Hypaphorine acts as an IAA antagonist [48] and it affects root hairs of *Eucalyptus* seedlings by reducing their elongation rate (Fig. 4A), while it has no activity on root elongation and development [70]. This indolic alkaloid induces drastic changes in the tubulin and actin cytoskeletons. For example, actin microfilaments, which extend as long cables in untreated eucalypt root hairs, are markedly induced to form thicker bundles (Ditengou et al., unpublished results) following the application of hypaphorine.

It thus seems that auxins and their derivatives and antagonists were master keys in ectomycorrhiza development. Although the above summarizes the scarce current knowledge on signaling processes in plant–ectomycorrhizal fungi associations, it does not explain why a particular tree establishes a symbiosis with a certain type of mycobiont or why most host plants can interact with hundreds of ectomycorrhizal fungi. Most probably the solutions to these puzzles lie in the nature of signals and receptors themselves. Plants and fungi excrete a wide range of more or less attractive compounds (e.g., flavanoids, alkaloids). Both partners possess one too many types of signal receptors/sensors that may bind with a number of rhizospheric excreted signals. In turn, signals/sensors complexes activate/repress expression of downstream genes including components of the signaling pathways, such as the ras GTPase and serine/threonine kinases (Duplessis and Martin, unpublished results). Once the fungal hyphae are within the root, other trophic and developmental inputs, from both symbionts, are likely necessary for successful symbiosis. In entering their novel niche, the colonizing hyphae need to adjust to their new environment. One essential modification is the alteration of the cell surface leading to the insulation of the mycobiont and/or changes in the permeability of the cell surface allowing the symbiotic traffic.

6 CHANGING THE NATURE OF THE FUNGAL AND PLANT SURFACE

After chemotropism and exchange of rhizospheric signals, the earliest stages of ectomycorrhiza formation is characterized by the colonization of the root cap (Fig. 4B). At this stage, the hyphae are likely at a saprophytic stage [32–34]. The symbiotic fungal infection is initiated in a discrete zone behind the growing root apex and in advance of the region where the primary cortex begins to deteriorate as the root matures [32,33]. The colonizing hyphae secrete various types of extracellular material, much of which is composed of chitosans, β-1,3-glucans, and proteins [71–74]. Although the precise mechanisms that govern this range of cell–cell interactions have not been fully defined, a number of specific cell surface molecules have been identified as critical elements in the interaction. The colonization of host surfaces by microorganisms often requires specific polymer interactions between microbial ligands (so-called adhesins) and host receptors

[75]. Adhesins are often found on a dense network of radially projecting fibrils, so-called fimbriae, which are bridging the partners.

In ectomycorrhiza, fungal attachment to the epidermal cells involves a polysaccharide mucigel and the secretion of oriented fibrillar materials, containing polysaccharides and glycoproteins, in which the whole of the sheath eventually becomes embedded [76]. Cytological observations showed that these orientated fimbriae, containing ConA-recognized glycoproteins, are likely involved in the adhesion of the hyphae on the root surface [77]. A layer of extracellular fibrillar polymers is present in the extracellular matrix of the free-living mycelium of *L. bicolor* [77] and *Pisolithus* [76,78] even before the interaction with the root. However, at the contact sites between hyphae and root surface, an increased secretion of these extracellular fibrillar polymers takes place in compatible ectomycorrhizal associations. Reorganization of the extracellular fibrillar polymers occurs, observed on microscopic sections as an accumulation and orientation of the extracellular polymeric fimbriae toward the host cell. In contrast, isolates of *P. tinctorius* with delayed symbiosis development do not secrete this fibrillar material [76,78]. This fibrillar material can bring about better contact or adhesion and lead to a better colonization.

A lectin purified from *Lactarius deterrimus* fruiting bodies preferentially bound to root hairs and tips of lateral roots of *Picea abies* [79], suggesting that lectin–polysaccharide recognition plays a role in the fungal adhesion. However, Lapeyrie and Mendgen [80] showed a low binding of fluorescein isothiocyanate–labeled lectins to the surface of free-living mycelium of *Pisolithus*. In addition, no change in fluorescein isothiocyanate–lectin binding during the interaction of *Pisolithus* with *Eucalyptus* roots was observed, indicating that lectins play a minor, if any, role in this symbiosis.

Major changes in cell wall structure during the colonization process have not been observed in ectomycorrhizal associations [74]. However, several of the symbiont responses to ectomycorrhiza development appear to be correlated with alterations in gene expression of cell wall proteins [81]. Both preferential synthesis and downregulation of polypeptide biosynthesis have been observed. Within these symbiosis-regulated (SR) proteins, two families have been characterized in detail: the hydrophobins and the 32-kDa symbiosis-regulated acidic polypeptides ($SRAP_{32}$).

6.1 Hydrophobins: Proteins That Function at the Symbiotic Interface?

Twenty-two (~3%) of the expressed sequence tags (EST) of *Eucalyptus/Pisolithus* ectomycorrhiza characterized by subtractive suppression hybridization (SSH) and cDNA array analyses shared a significant similarity with the cysteine-rich hydrophobins [69]. Hydrophobins are small, secreted, moderately hydrophobic proteins with a conserved spacing of eight cysteine residues [82–84]. They are either excreted in the medium or trapped in the cell wall. In the latter case, hydrophobin assemblages confer an increased hydrophobicity to the hyphal surface which allows the adhesion to host surfaces or between hyphae [85]. Hydrophobins have escaped detection until recently because they are tightly bound to cell wall polymers and could only be released by using concentrated formic acid or trifluoroacetic acid [86,87]. Hydrophobins have been widely found in various species of ascomycetes and basidiomycetes species [83]. They have been involved in emergence of aerial hyphae [82,85], fruiting body and conidia formation, desiccation tolerance, pathogenesis, and symbiosis [83,84]. It is clear that hydrophobins display a wide range of functions in different fungal species and within a single species. They have been recruited by several biotrophic fungi for surface interactions associated with the infection of their plant or animal host and they may play such a role in ectomycorrhiza [88]. The expression of the transcripts *hydPt1* and *hydPt2*, coding for *Pisolithus* hydrophobins, is strongly upregulated during the symbiosis formation [89].

Additional hydrophobin genes, *hydPt3* to *hydPt8*, which shows 47–52% homology with other *Pisolithus* hydrophobins [69], have been identified in *Eucalyptus/Pisolithus* mycorrhiza. The expression of their transcripts is increased six- to eightfold in the symbiotic tissues. Nonwettable and water-repellent mycorrhizas of *Eucalyptus* are often found in air pockets in soil [90]. The most likely explanation for this lies in the observed deposition of hydrophobins in fungal cell walls [87]. However, *Pisolithus* hyphae simultaneously express several types of hydrophobins in their walls during the formation of the symbiosis. Whether these different hydrophobins are involved in the adhesion of hyphae on the root surface, the mechanical penetration of hyphae between root cells, or the aggregation of hyphae to form the mantle [81,88] remains to be determined by selective gene inactivation.

6.2 Symbiosis-Regulated Polypeptides with Adhesin-Type Motif

Many SRAPs, observed in soluble protein extracts of *Eucalyptus/Pisolithus* ectomycorrhizas [9,91], are also abundantly accumulated in *Pisolithus* cell walls. A family of cell wall acidic polypeptides (so-called $SRAP_{32}$) composed of at least six isoforms with different charge and/or molecular mass has been isolated [92]. These polypeptides are encoded by a multigene family and a dozen of slightly different sequences have been identified (Sorin, Voiblet, and Tagu, unpublished results). The $SRAP_{32}$ proteins showed no significant homology with known proteins, but the central part of the deduced protein contains an Arg-Gly-Asp (RGD) motif. The RGD motif was first discovered in fibronectin as a cell attachment site [93] and was subsequently found to be the recognition sequence for a number of integrin receptors [94]. The presence of the RGD motif suggests that $SRAP_{32}$ are coding for adhesion proteins [92]. Comparing the upregulation of $SRAP_{32}$ transcript levels and the increased concentration of $SRAP_{32}$ polypeptides in ectomycorrhiza [9,91,92] showed that there was a good correlation between changes in protein synthesis and transcript levels [69,92]. Immunogold labeling confirmed that $SRAP_{32}$ and immunocross-reacting $SRAP_{31}$ are localized in cell walls of the free-living and symbiotic hyphae. These proteins could be found mainly associated with the flocculent material covering the hyphal surface, but they are never observed in the thin, delicate filaments or fimbriae-bridging hyphae or the fungal cells to the surface of the root.

Cell wall proteins are known to form crosslinked networks with other proteins and polysaccharides in fungal walls [75], but the structural properties and the functional significance of such networks are not known. It is tempting to speculate that high levels of $SRAP_{32}$, degradation of mannoproteins [72], and increased levels of hydrophobins [89] take place simultaneously to modify the molecular architecture of protein networks in a manner that allows new developmental fates for both fungal cell adhesion and root colonization by the fungus [75,81]. Further investigation of the structure and regulation of SR wall proteins will provide a more complete picture of their role in developing ectomycorrhizal tissues.

7 ESTs AND cDNA ARRAYS FOR GENE EXPRESSION ANALYSIS

It is increasingly clear that developmental pathways leading to the ectomycorrhizal symbiosis can be considered as modular (Fig. 3), and that developmental transitions are accompanied by global changes in the expression of specific complements of genes under the control of rhizospheric and intracellular signals (see above) [39,50]. To date, it is not possible to predict the number of symbiosis-specific fungal and plant genes. However, owing to the fact that ectomycorrhizas are widespread, a significant number of mycorrhiza-specific genes must exist. A goal of primary importance is to achieve a comprehensive description of the mechanisms induced in both symbionts at each stage of the symbiosis development. This molecular sketching of the ectomycorrhiza development should be carried out simultaneously on different associations to identify common ''molecular signatures'' typical of this symbiosis. Over the last decade, changes in gene expression have mainly been studied by using two-dimensional gel electrophoresis. Up- and downregulated proteins have been found in *Pisolithus/Eucalyptus* [9,91], *Amanita muscaria/Picea abies* [95], *Paxillus involutus/Betula pendula* [96], and *Suillus bovinus/Pinus sylvestris* [97] associations. These investigations confirmed that ectomycorrhizal development leads to an alteration of gene expression in both symbionts and to the synthesis of ectomycorrhiza-specific proteins (ectomycorrhizins) [9,91]. However, this approach was limited because only a restricted set of proteins can be visualized on 2D gels.

While such studies have been fruitful in the past, their potential use in developing a complete and accurate understanding of the symbiotic factors during the course of the mycorrhiza development is limited. Although modulation of the symbiont interactions clearly implies posttranslational modifications, regulation of transport mechanisms, and protein degradation, key mechanisms in fungal gene regulation take place at the transcriptional level. Consequently, detection of even subtle gene expression modulations will provide a comprehensive framework for studying events which affect cellular differentiation and metabolism, and regulation on a genomic scale. The fact that ectomycorrhizal fungi are not yet amenable to gene inactivation has prevented the application of forward genetics to decipher ectomycorrhiza development. Therefore, alternative molecular techniques for the identification of SR genes have been developed. Subtractive cDNA hybridization and differential mRNA display were used to identify plant and fungal genes that are induced upon symbiosis development in ectomycorrhizal associations involving *Pisolithus/Eucalyptus* [68,69], *L. bicolor/Pinus resinosa* [98,99], and *A. muscaria/P. abies* [100].

These investigations confirmed that ectomycorrhiza development is accompanied by striking changes in gene expression at the transcriptional level and allowed the identification of a dozen SR genes (Fig. 5). For example, hexose transporters of the symbionts in the *A. muscaria/P. abies* [101] and *P. involutus/B. pendula* [102] symbioses are regulated. The fungal gene coding for a hexose transporter, *AmMst1*, was upregulated [101,104], whereas the *Picea* hexose transporter was slightly downregulated [104]. Similarly, the expression of *B. pendula* hexose and sucrose transporters BpSUC1, BpHEX1, and BpHEX2 had been downregulated in mycorrhizal roots [102]. As stressed by the authors, the downregulation of expression of these transporters is not compatible with the increased carbon fluxes taking place in the roots as a result of the carbon drain imposed by the mycobiont. Other transporters are likely involved in the symbiotic traffic. The *A. muscaria* phenylalanine ammonium lyase gene *AmPAL* is likely regulated in ectomycorrhiza through changes in nitrogen and sugar levels [103]. Whether gene expression of these metabolic genes is controlled by sugar-dependent regulation or by symbiosis-related developmental signals is not known. Nehls et al. [104] have suggested that the expression of hexose transporter gene *AmMst1* is only regulated by the hexose concentration of the symbiotic apoplastic space of *Amanita/Populus* mycorrhiza. These findings illustrated the drastic molecular changes experienced by the partners during the mycorrhiza development and functioning [105–107].

To identify cellular functions expressed in the symbiosis on a wider scale, EST programs have been developed on several ectomycorrhizal fungi (*A. muscaria, H. cylindrosporum, P. tinctorius, Tuber borchii*) [69,108,109] (Bonfante,

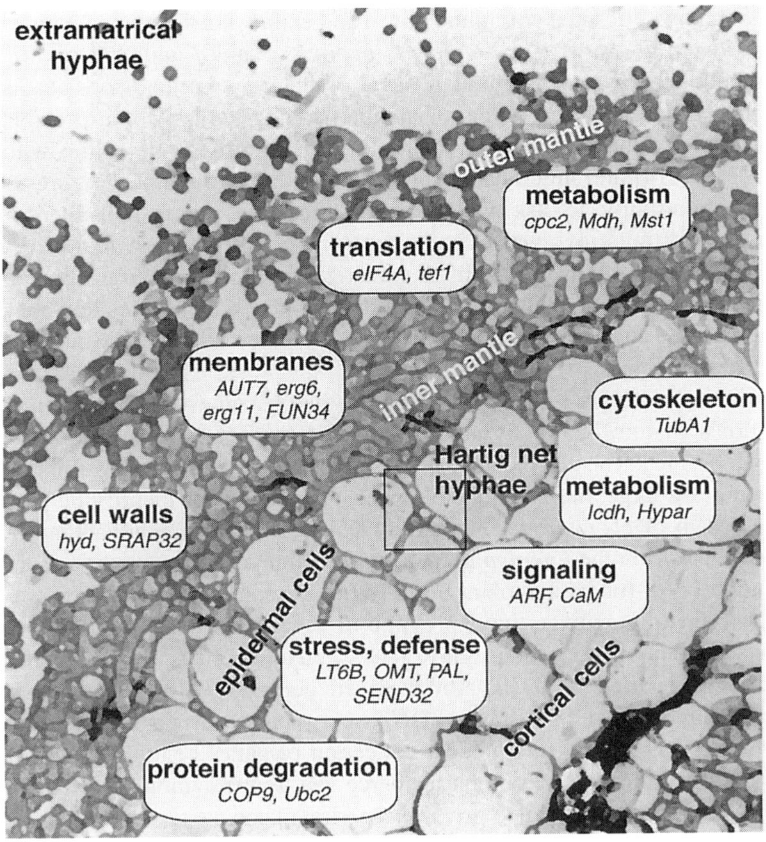

FIGURE 5 Regulation of gene expression in the ectomycorrhizal symbiosis. This figure compiles the known upregulated genes in various types of ectomycorrhizas (*Pisolithus/E. globulus* [68,69,87,89,92,119,120]; *Paxillus involutus/Betula pendula* [121]; *L. bicolor/Pinus resinosa* [98,99]; and *A. muscaria/P. abies* [100]). *ARF*, ADP-ribosylation factor; *AUT7*, vesicular transport and autophagocytosis; *CaM*, calmodulin; *COP9*, constitutive photomorphogenic subunit (related to proteasome); *cpc2*, cross-pathway control WD-repeat protein; *eIF4A*, elongation initiation factor 4A (dead-box helicase); *erg6*, δ-(24)-sterol c-methyltransferase; *erg11*, sterol-14-alpha-demethylase; FUN34, transmembrane protein; hyd, hydrophobins; *Hypar*, hypaphorine- and auxin-regulated glutathion-*S*-transferase; *Icdh*, NADP-isocitrate dehydrogenase; *LT6B*, salt-stress–induced LT6B protein; *Mdh*, mitochondrial malate dehydrogenase; *Mst1*, monosaccharide transporter; *OMT*, O-methyltransferase; *PAL*, phenylammonia lyase; *SEND32*, senescence down-regulated protein; *SRAP32*, 32-kDa symbiosis-regulated acidic polypeptides; *tef1*, translation elongation factor 1α; *TubA1*, α-tubulin; *Ubc2*, ubiquitin-conjugating enzyme E2.

Nehls, and Sentenac, personal communication) and ectomycorrhizal associations (*Eucalyptus/Pisolithus* [69]; *B. pendula/P. involutus*, Tunlid and Söderström, personal communication). As the number of ESTs increases, comparisons across genera, species, ecotypes, and strains of symbiotic fungi will become possible through ''digital Northern'' [110]. With multiple EST programs dealing with pathogenic [111–114] and mutualistic fungi, we will have in the near future an unparalleled opportunity to ask which genetic features are responsible for common/divergent traits involved in pathogenesis and symbiosis. A few of the many possible breakthroughs will be in characterization of common transduction networks, identification of novel surface proteins that play critical roles in plant–fungus interactions, and new insights into unique metabolic routes critical for mycorrhiza functioning.

Quantitative analysis of the transcriptome has become possible through ''hybridization signature'' methods which allow large-scale measurement of gene expression [115–117]. The cDNA array analyses are currently providing efficient means of acquiring large amounts of biological information for identifying processes involved in plant–microbe interactions [114,118]. To take advantage of the available ESTs from the *Eucalyptus/Pisolithus* ectomycorrhiza, we have constructed miniarrays of fungal and plant ESTs [69]. These miniarray analyses provided a tool to broadly analyze the expression of several hundred genes during the symbiosis development, to identify SR genes, and to identify candidate genes for further, more detailed, analysis. About 80 SR genes (17%) were identified by differential screening of 480 arrayed cDNAs between free-living partners and symbiotic tissues [69]. Even this modest collection of genes begins to provide an indication of symbiosis environment as perceived by the symbionts (Fig. 5). Within the cellular functions which are strikingly regulated by symbiosis development, we have identified cell wall and membrane synthesis, stress and defense responses, protein degradation (in plant cells), and protein synthesis (in hyphae) (Fig. 5). EST/cDNA array analyses confirmed that most members of the hydrophobin and $SRAP_{32}$ gene families are dramatically upregulated (up to eightfold) during fungal mantle formation (see Sec. 6.2). *Egubc2*, which encodes a ubiquitin-conjugating (E2) enzyme, and *EgCop9a*, coding for a subunit of the proteasome-related complex, are highly upregulated in ectomycorrhizal tips, confirming that symbiosis development induces drastic plant protein degradation [9]. Protein degradation may be a result of stress conditions experienced by the roots colonized by massive amount of hyphae.

These data suggest a highly dynamic environment in which symbionts are sending and receiving signals, exposed to high levels of stress conditions and remodeling tissues. A striking result of these studies is the fact that all genes investigated are common to the nonsymbiotic and symbiotic stages. At the developmental stage studied, symbiosis development does not induced the expression of ectomycorrhiza-specific genes, but a marked change in the gene expression in the partners.

8 CONCLUSIONS

As outlined in this review, ectomycorrhiza development influences both plant and fungus gene expression in a pleiotropic manner. A range of fungal tissues differentiates that can be distinguished by a combination of anatomical, cytological, and molecular features. On the other hand, root tips proliferate and root cells experience major alteration in their shape and gene expression. Advances of recent years have provided insights on the molecular basis of ectomycorrhiza morphogenesis. With the identification of several developmentally regulated proteins and genes and a description of their expression and activities, the ground is now set for recasting earlier models of symbiosis development in molecular terms.

It is apparent from this brief review, however, that there is a vast complexity of genetic programs with overlapping expression patterns. This includes induction of plant defense/stress reactions, the downexpression of plant protein biosynthesis, the initiation of lateral roots by fungal auxins, the morphogenetic switches of the fungal hyphae, and the establishment of novel cell walls and extracellular matrices. Among the many remaining challenges is the elucidation of mechanisms and inducer molecules that integrate the actions of these multiple programs of gene expression in generating a mature symbiotic organ. Studies in areas such as the identification of chemicals and genes/proteins involved in cell–cell interactions and control of cell expression at the level of signal transduction will be the source for many answers. A comparative study of gene expression in different types of ectomycorrhizas using the molecular approaches including genomics and gene inactivation might reveal to what extent similarities and differences in the various types of ectomycorrhizas are the result of variation in the basic mechanisms underlying the respective developmental programs and the effects of the different trophic and environmental cues.

ACKNOWLEDGMENTS

S.D. was supported by a Doctoral Scholarship from the Ministère de l'Education Nationale de la Recherche et de la Technologie. We also appreciated partial support from the Groupement de Recherches et d'Etude des Génomes, the INRA Collaborative Research Programme in Microbiology, and the INRA GenoPop research grant. We thank Drs. Frank Ditengou, Hubert Lagrange, Frédéric Lapeyrie, and Catherine Voiblet for stimulating discussions.

REFERENCES

1. SE Smith, DJ Read. Mycorrhizal Symbiosis. 2nd ed. London: Academic Press, 1997.
2. AH Fitter, B Moyersoen. Evolutionary trends in root–microbe symbioses. Phil Trans R Soc Lond B 351:1367–1375, 1996.

3. DJ Read. Mycorrhizas in ecosystems. Experientia 47:376–390, 1991.
4. AFS Taylor, F Martin, DJ Read. Fungal diversity in ectomycorrhizal communities of Norway spruce [*Picea abies* (L.) Karst.] and beech (*Fagus sylvatica* L.) along North–South transects in Europe. In: ED Schulze, ed. Carbon and Nitrogen Cycling in European Forest Ecosystems, Ecological Studies 142. Berlin: Springer-Verlag, 2000, pp 343–365.
5. AA Meharg, JWG Cairney. Co-evolution of mycorrhizal symbionts and their hosts to metal-contaminated environments. Adv Ecol Res 30:69–112, 2000.
6. TS Grove, F Le Tacon. Mycorrhiza in plantation forestry. Adv Plant Pathol 9:191–227, 1993.
7. I Köttke, F Oberwinkler. The cellular structure of the Hartig net: coenocytic and transfer cell-like organization. Nord J Bot 7:85–95, 1987.
8. HB Massicotte, RL Peterson, CA Ackerley, AE Ashford. Ontogeny of *Eucalyptus pilularis–Pisolithus tinctorius* ectomycorrhizae. II. Transmission electron microscopy. Can J Bot 65:1940–1947, 1987.
9. JL Hilbert, G Costa, F Martin. Ectomycorrhizin synthesis and polypeptide changes during the early stage of eucalypt mycorrhiza development. Plant Physiol 97:977–984, 1991.
10. B Botton, M Chalot. Nitrogen assimilation: enzymology in ectomycorrhizas. In: AK Varma, B Hock, eds. Mycorrhiza: Structure, Molecular Biology and Function. 2nd ed. New York: Springer-Verlag, 1999, pp 333–372.
11. R Hampp, C Schaeffer. Mycorrhiza—carbohydrate and energy metabolism. In: AK Varma, B Hock, eds. Mycorrhiza: Structure, Molecular Biology and Function. 2nd ed. New York: Springer-Verlag, 1999, pp 273–303.
12. H Gehrig, A Schüβler, M Kluge. *Geosiphon pyriforme*, a fungus forming endocytobiosis with *Nostoc* (Cyanobacteria), is an ancestral member of the Glomales: evidence by SSU rRNA analysis. J Mol Evol 43:71–81, 1996.
13. M-A Selosse, F Le Tacon. The land flora: a phototroph–fungus partnership? Trends Ecol Evol 13:15–20, 1998.
14. TN Taylor, W Remy, H Hass, H Kerp. Fossil arbuscular mycorrhizae from the Early Devonian. Mycologia 87:560–573, 1995.
15. BA LePage, RS Currah, RA Stockey, GW Rothwell. Fossil ectomycorrhizae from the Middle Eocene. Am J Bot 84:410–412, 1997.
16. TD Bruns, TM Szaro, M Gardes, KW Cullings, JJ Pan, DL Taylor, TR Horton, A Kretzer, M Garbelotto, Y Li. A sequence database for the identification of ectomycorrhizal basidiomycetes by phylogenetic analysis. Mol Ecol 7:257–272, 1998.
17. DS Hibbett, LB Gilbert, MJ Donoghue. Evolutionary instability of ectomycorrhizal symbioses in basidiomycetes. Nature 407:506–508, 2000.
18. TR Horton, TD Bruns. The molecular revolution in ectomycorrhizal ecology: peeking into the black-box. Mol Ecol 10:1855–1871, 2001.
19. E Duhoux, G Rinaudo, HG Diem, F Auguy, D Fernandez, D Bogusz, C Franche, Y Dommergues, B Huguenin. Angiosperm *Gymnostoma* trees produce root nodules colonized by arbuscular mycorrhizal fungi related to Glomus. New Phytol 149: 115–125, 2001.
20. M Gardes, TD Bruns. Community structure of ectomycorrhizal fungi in a *Pinus*

muricata forest: above- and below-ground views. Can J Bot 74:1572–1583, 1996.
21. L Jonsson, A Dahlberg, MC Nilsson, O Zackrisson, O Karen. Ectomycorrhizal fungal communities in late-successional Swedish boreal forests, and their composition following wildfire. Mol Ecol 8:205–215, 1999.
22. JWG Cairney. Intraspecific physiological variation: implications for understanding functional diversity in ectomycorrhizal fungi. Mycorrhiza 9:125–135, 1999.
23. T Wallenda, C Strober, H Högbom, H Schinkel, E George, P Högberg, DJ Read. Nitrogen uptake processes in roots and mycorrhizas. In: ED Schulze, ed. Carbon and Nitrogen Cycling in European Forest Ecosystems, Ecological Studies 142. Berlin: Springer-Verlag, 2000, pp 122–143.
24. M Chalot, A Brun. Physiology of organic nitrogen acquisition by ectomycorrhizal fungi and ectomycorrhizas. FEMS Microbiol Rev 22:21–44, 1998.
25. F Martin, S Lorillou. Nitrogen acquisition and assimilation in ectomycorrhizal systems. In: H Rennenberg, W Eschrich, H Ziegler, eds. Trees—Contributions to Modern Tree Physiology. Leiden, Netherlands: Backhuys, 1997, pp 423–429.
26. MW Simard, DA Perry, MD Jones, DD Myrold, DM Duralland, R Molina. Net transfer of carbon between ectomycorrhizal tree species in the field. Nature 388: 579–582, 1997.
27. B Wu, K Nara, H Taizo. Can ^{14}C-labeled photosynthetic products move between *Pinus densiflora* seedlings linked by ectomycorrhizal mycelia? New Phytol 149: 137–146, 2001.
28. P Vivin, F Martin, J-M Guehl. Acquisition and within-plant allocation of ^{13}C and ^{15}N in CO_2-enriched *Quercus robur* plants. Physiol Plant 98:89–96, 1996.
29. AH Fitter, A Heinemeyer, PL Staddon. The impact of elevated CO_2 and global climate change on arbuscular mycorrhizas: a mycocentric approach. New Phytol 147:179–187, 2000.
30. I Köttke, F Oberwinkler. Root–fungus interactions observed on initial stages of mantle formation and Hartig net establishment in mycorrhizas of *Amanita muscaria* on *Picea abies* in pure culture. Can J Bot 64:2348–2354, 1986.
31. I Brunner, C Scheidegger. Effects of high nitrogen concentrations on ectomycorrhizal structure and growth of seedlings of *Picea abies* (L.) Karst. New Phytol 129: 83–95, 1995.
32. GA Chilvers. Low power electron microscopy of the root cap region of eucalypt mycorrhizas. New Phytol 67:663–668, 1968.
33. DP Horan, GA Chilvers, FF Lapeyrie. Time sequence of the infection process in eucalypt ectomycorrhizas. New Phytol 109:451–458, 1988.
34. J Dexheimer, J Gerard, P Genet. Study of transformations of the root system of *Eucalyptus globulus* associated with *Pisolithus tinctorius*. I. Aptitude to mycorrhization of different kinds of roots. Phytomorphology 44:235–245, 1994.
35. HB Massicotte, CA Ackerley, RL Peterson. Ontogeny of *Alnus rubra–Alpova diplophloeus* ectomycorrhizae. II. Transmission electron microscopy. Can J Bot 67:201–210, 1989.
36. A Brun, M Chalot, RD Finlay, B Söderström. Structure and function of the ectomycorrhizal association between *Paxillus involutus* (Batsch) Fr. and *Betula pendula* (Roth.). I. Dynamics of mycorrhiza formation. New Phytol 129:487–493, 1995.

37. R Agerer. Colour Atlas of Ectomycorrhizae. 7th ed. Schwäbisch Bmünd, Germany: Einhorn-Verlag, 1993–1998.
38. G Gay, R Marmeisse, P Fouillet, M Bouletreau, JC Debaud. Genotype/nutrition interactions in the ectomycorrhizal fungus *Hebeloma cylindrosporum* Romagnesi. New Phytol 123:335–343, 1993.
39. F Martin, D Tagu. Developmental biology of a plant-fungus symbiosis: the ectomycorrhiza. In: AK Varma, B Hock, eds. Mycorrhiza: Structure, Molecular Biology and Function. 2nd ed. Berlin: Springer-Verlag, 1999, pp 51–73.
40. BR Kropp, BJ McAfee, JA Fortin. Variable loss of ectomycorrhizal ability in monokaryotic and dikaryotic cultures of *Laccaria bicolor.* Can J Bot 65:500–504, 1987.
41. JC Debaud, G Gay, A Prevost, J Lei, J Dexheimer. Ectomycorrhizal ability of genetically different homokaryotic and dikaryotic mycelia of *Hebeloma cylindrosporum.* New Phytol 108:323–328, 1988.
42. MS Lamhamedi, AJ Fortin, HH Kope, BR Kropp. Genetic variation in ectomycorrhiza formation by *Pisolithus arhizus* on *Pinus pinaster* and *Pinus banksiana.* New Phytol 115:689–697, 1990.
43. KK Wong, Y Piché, D Montpetit, BR Kropp. Differences in the colonization of *Pinus banksiana* roots by sib-monokaryotic and dikaryotic strains of ectomycorrhizal *Laccaria bicolor.* Can J Bot 67:1717–1726, 1989.
44. C Karabaghli-Degron, B Sotta, M Bonnet, G Gay, F Le Tacon. The auxin transport inhibitor 2,3,5-triiodobenzoic acid (TIBA) inhibits the stimulation of in vitro lateral root formation and the colonization of the tap-root cortex of Norway spruce (*Picea abies*) seedlings by the ectomycorrhizal fungus *Laccaria bicolor.* New Phytol 140: 723–733, 1998.
45. AM Hirsch, MJ McFall-Ngai. Fundamental concepts in symbiotic interactions: light and dark, day and night, squid and legume. J Plant Growth Regul 19:113–130, 2000.
46. N Fries, K Serck-Hanssen, L Häll Dimberg, O Theander. Abietic acid, an activator of basidiospore germination in ectomycorrhizal species of the genus *Suillus* (Boletaceae). Exp Mycol 11:360–363, 1987.
47. T Béguiristain, F Lapeyrie. Host plant stimulates hypaphorine accumulation in *Pisolithus tinctorius* hyphae during ectomycorrhizal infection while excreted fungal hypaphorine controls root hair development. New Phytol 136:525–532, 1997.
48. FA Ditengou, F Lapeyrie. Hypaphorine from the ectomycorrhizal fungus *Pisolithus tinctorius* counteracts activities of indole-3-acetic acid and ethylene but not synthetic auxins in eucalypt seedlings. Mol Plant–Microbe Interact 13:151–158, 2000.
49. H Lagrange, C Jay-Allemand, F Lapeyrie. Rutin, the phenolglycoside from eucalyptus root exudates, stimulates *Pisolithus* hyphal growth at picomolar concentrations. New Phytol 149:349–355, 2001.
50. F Martin, F Lapeyrie, D Tagu. Altered gene expression during ectomycorrhizal development. In: G Caroll, P Tudzynski, eds. The Mycota, Vol V, Part A, Plant Relationships. Berlin: Springer-Verlag, 1997, pp 223–242.
51. SJ Barker, D Tagu. The roles of auxins and cytokinins in mycorrhizal symbioses. J Plant Growth Regul 19:144–154, 2000.
52. DP Horan, GA Chilvers. Chemotropism: the key to ectomycorrhizal formation? New Phytol 116:297–301, 1990.

53. HA Hartwig, DA Phillips. Release and modification of nod-gene-inducing flavonoids from alfalfa seeds. Plant Physiol 95:804–807, 1991.
54. G Nagahashi, DD Douds. Partial separation of root exudate components and their effects upon the growth of germinated spores of AM fungi. Mycol Res 104:1453–1464, 2000.
55. T Béguiristain, R Côté, P Rubini, C Jay-Allemand, F Lapeyrie. Hypaphorine accumulation in hyphae of the ectomycorrhizal fungus, *Pisolithus tinctorius*. Phytochemistry 40:1089–1091, 1995.
56. NL Paiva. An introduction to the biosynthesis of chemicals used in plant–microbe communication. J Plant Growth Regul 19:131–143, 2000.
57. L Gea, L Normand, B Vian, G Gay. Structural aspects of ectomycorrhiza of *Pinus pinaster* (Ait.) Sol. formed by an IAA-overproducer mutant of *Hebeloma cylindrosporum* Romagnési. New Phytol 128:659–670, 1994.
58. N Gogala. Regulation of mycorrhizal infection by hormonal factors produced by hosts and fungi. Experientia 47:331–340, 1991.
59. H Beyrle. The role of phytohormones in the function and biology of mycorrhizas. In: AK Varma, B Hock, eds. Mycorrhiza: Structure, Molecular Biology and Function. Springer Berlin Heidelberg New York, 1995, pp 365–390.
60. V Slankis. Effect of napthaleneacetic acid on dichotomous branching of isolated roots of *Pinus sylvestris*. Physiol Plant 3:40–43, 1950.
61. JM Ulrich. Auxin production by mycorrhizal fungi. Physiol Plant 13:429–443, 1960.
62. LA Rupp, KW Mudge. Ethephon and auxin induce mycorrhiza-like changes in the morphology of root organ cultures of Mugo pine. Physiol Plant 64:316–322, 1985.
63. LA Rupp, KW Mudge, FB Negm. Involvement of ethylene in ectomycorrhiza formation and dichotomous branching of roots of Mugo pine seedlings. Can J Bot 67: 477–482, 1989.
64. WT Frankenberger, M Poth. Biosynthesis of indole-3-acetic acid by the pine ectomycorrhizal fungus *Pisolithus tinctorius*. Appl Environ Microbiol 53:2908–2913, 1987.
65. G Gay, L Normand, R Marmeisse, B Sotta, JC Debaud. Auxin overproducer mutants of *Hebeloma cylindrosporum* Romagnési have increased mycorrhizal activity. New Phytol 128:645–657, 1994.
66. K Palme, L Gälweiler. PIN-pointing the molecular basis of auxin transport. Curr Opin Plant Biol 2:375–381, 1999.
67. H Tranvan, Y Habricot, E Jeannette, G Gay, B Sotta. Dynamics of symbiotic establishment between an IAA-overproducing mutant of the ectomycorrhizal fungus *Hebeloma cylindrosporum* and *Pinus pinaster*. Tree Physiol 20:123–129, 2000.
68. U Nehls, T Béguiristain, F Ditengou, F Lapeyrie, F Martin. The expression of a symbiosis-regulated gene in eucalypt roots is regulated by auxins and hypaphorine, the tryptophan betaine of the ectomycorrhizal basidiomycete *Pisolithus tinctorius*. Planta 207:296–302, 1998.
69. C Voiblet, S Duplessis, N Encelot, F Martin. Identification of symbiosis-regulated genes in *Eucalyptus globulus–Pisolithus tinctorius* ectomycorrhiza by differential hybridization of arrayed cDNAs. Plant J 25:181–191, 2001.
70. FA Ditengou, T Béguiristain, F Lapeyrie. Root hair elongation is inhibited by hy-

paphorine, the indole alkaloid from the ectomycorrhizal fungus *Pisolithus tinctorius*, and restored by IAA. Planta 211:722–728, 2000.
71. J Dexheimer, JC Pargney. Comparative anatomy of the host–fungus interface in mycorrhizas. Experientia 47:312–320, 1991.
72. D Tagu, F Martin. Molecular analysis of cell wall proteins expressed during the early steps of ectomycorrhiza development. New Phytol 133:73–85, 1996.
73. R Balestrini, MG Hahn, P Bonfante. Location of cell-wall components in ectomycorrhizae of *Corylus avellana* and *Tuber magnatum*. Protoplasma 191:55–69, 1996.
74. P Bonfante. At the interface between mycorrhizal fungi and plants: the structural organization of cell wall, plasma membrane and cytoskeleton. In: B Hock, ed. Fungal Associations Mycota IX. Berlin: Springer-Verlag, 2001.
75. WL Chaffin, JL Lopez-Ribot, M Casanova, D Gozalbo, JP Martinez. Cell wall and secreted proteins of *Candida albicans*: identification, function, and expression. Microbiol Mol Biol Rev 62:130–180, 1998.
76. F Lapeyrie, J Lei, N Malajczuk, J Dexheimer. Ultrastructural and biochemical changes at the pre-infection stage of mycorrhizal formation by two isolates of *Pisolithus tinctorius*. Ann Sci For 46s:754s–757s, 1989.
77. J Lei, KKY Wong, Y Piché. Extracellular concanavalin A–binding sites during early interactions between *Pinus banksiana* and two closely related genotypes of the ectomycorrhizal basidiomycete *Laccaria bicolor*. Mycol Res 95:357–363, 1991.
78. J Lei, F Lapeyrie, N Malajczuk, J Dexheimer. Infectivity of pine and eucalypt isolates of *Pisolithus tinctorius* (Pers.) Coker & Couch on roots of *Eucalyptus urophylla* S.T. Blake in vitro. II. Ultrastructural and biochemical changes at the early stage of mycorrhiza formation. New Phytol 116:115–122, 1990.
79. M Giollant, J Guillot, M Damez, M Dusser, P Didier, E Didier. Characterization of a lectin from *Lactarius deterrimus*. Research on the possible involvement of the fungal lectin in recognition between mushroom and spruce during the early stages of mycorrhizae formation. Plant Physiol 101:513–522, 1993.
80. F Lapeyrie, K Mendgen. Quantitative estimation of surface carbohydrates of ectomycorrhizal fungi in pure culture and during *Eucalyptus* root infection. Mycol Res 97:603–609, 1993.
81. F Martin, P Laurent, D De Carvalho, C Voiblet, R Balestrini, P Bonfante, D Tagu. Cell wall proteins of the ectomycorrhizal basidiomycete *Pisolithus tinctorius*: identification, function, and expression in symbiosis. Fungal Genet Biol 27:161–174, 1999.
82. JGH Wessels. Wall growth, protein excretion and morphogenesis in fungi. New Phytol 123:397–413, 1993.
83. MJ Kershaw, NJ Talbot. Hydrophobins and repellents: proteins with fundamental roles in fungal morphogenesis. Fungal Genet Biol 23:18–33, 1998.
84. HAB Wösten, M Richter, JM Willey. Structural proteins involved in emergence of microbial aerial hyphae. Fungal Genet Biol 27:153–160, 1999.
85. HAB Wösten, JM Willey. Surface-active proteins enable microbial aerial hyphae to grow into the air. Microbiology 146:767–773, 2000.
86. OMH De Vries, MP Fekkes, HAB Wösten, JGH Wessels. Insoluble hydrophobin

complexes in the walls of *Schizophyllum commune* and other filamentous fungi. Arch Microbiol 159:330–335, 1993.

87. D Tagu, R De Bellis, R Balestrini, OMH De Vries, G Piccoli, V Stocchi, P Bonfante, F Martin. Immuno-localization of the hydrophobin HYDPt-1 from the ectomycorrhizal basidiomycete *Pisolithus tinctorius* during colonization of *Eucalyptus globulus* roots. New Phytol 149:127–135, 2001.
88. D Tagu, I Köttke, F Martin. Hydrophobins in ectomycorrhizal symbiosis: hypothesis. Symbiosis 25:5–18, 1998.
89. D Tagu, B Nasse, F Martin. Cloning and characterization of hydrophobins—encoding cDNAs from the ectomycorrhizal basidiomycete *Pisolithus tinctorius.* Gene 168:93–97, 1996.
90. PA Vesk, AE Ashford, A-L Markovina, WG Allaway. Apoplasmic barriers and their significance in the exodermis and sheath of *Eucalyptus pilularis–Pisolithus tinctorius* ectomycorrhizas. New Phytol 145:333–346, 2000.
91. T Burgess, P Laurent, B Dell, N Malajczuk, F Martin. Effect of the fungal isolate aggressivity on the biosynthesis of symbiosis-related polypeptides in differentiating eucalypt ectomycorrhiza. Planta 195:408–417, 1995.
92. P Laurent, C Voiblet, D Tagu, D De Carvalho, U Nehls, R De Bellis, R Balestrini, G Bauw, P Bonfante, F Martin. A novel class of ectomycorrhiza-regulated cell wall polypeptides in *Pisolithus tinctorius.* Mol Plant–Microbe Interact 12:862–871, 1999.
93. MD Pierschbacher, E Ruoslahti. Cell attachment activity of fibronectin can be duplicated by small synthetic fragments of the molecule. Nature 309:30–33, 1984.
94. E Ruoslahti. RGD and other recognition sequences for integrins. Annu Rev Cell Dev Biol 12:697–715, 1996.
95. M Guttenberger, R Hampp. Ectomycorrhizins—symbiosis-specific or artifactual polypeptides from ectomycorrhizas? Planta 188:129–136, 1992.
96. P Simoneau, JD Viemont, JC Moreau, DG Strullu. Symbiosis-related polypeptides associated with the early stages of ectomycorrhiza organogenesis in birch (*Betula pendula* Roth). New Phytol 124:495–504, 1993.
97. M Tarkka, SS Niini, M Raudaskoski. Developmentally regulated proteins during differentiation of root system and ectomycorrhiza in Scots pine (*Pinus sylvestris*) with *Suillus bovinus.* Physiol Plant 104:449–455, 1998.
98. SJ Kim, J Zheng, ST Hiremath, GK Podila. Cloning and characterization of a symbiosis-related gene from an ectomycorrhizal fungus *Laccaria bicolor.* Gene 222: 203–212, 1998.
99. SJ Kim, D Bernreuther, M Thumm, GK Podila. LB-AUT7, a novel symbiosis-regulated gene from an ectomycorrhizal fungus, *Laccaria bicolor*, is functionally related to vesicular transport and autophagocytosis. J Bacteriol 181:1963–1967, 1999.
100. U Nehls, S Mikolajewski, M Ecke, R Hampp. Identification and expression-analysis of two fungal cDNAs regulated by ectomycorrhiza and fruit body formation. New Phytol 144:195–202, 1999.
101. U Nehls, J Wiese, M Guttenberger, R Hampp. Carbon allocation in ectomycorrhizas: identification and expression analysis of an *Amanita muscaria* monosaccharide transporter. Mol Plant–Microbe Interact 11:167–176, 1998.

102. DP Wright, JD Scholes, DJ Read, SA Rolfe. Changes in carbon allocation and expression of carbon transporter genes in *Betula pendula* Roth. colonized by the ectomycorrhizal fungus *Paxillus involutus* (Batsch) Fr. Plant Cell Environ 23:39–49, 2000.
103. U Nehls, M Ecke, R Hampp. Sugar- and nitrogen-dependent regulation of an *Amanita muscaria* phenylalanine ammonium lyase gene. J Bacteriol 181:1931–1933, 1999.
104. U Nehls, A Bock, M Ecke, R Hampp. Differential expression of the hexose-regulated fungal genes *Am PAL* and *Am Mst1* within *Amanita/Populus* ectomycorrhizas. New Phytol 150:583–590, 2001.
105. D Blaudez, M Chalot, P Dizengremel, B Botton. Structure and function of the ectomycorrhizal association between *Paxillus involutus* and *Betula pendula*. II. Metabolic changes during mycorrhiza formation. New Phytol 138:543–552, 1998.
106. JWG Cairney, RM Burke. Physiological heterogeneity within fungal mycelia: an important concept for a functional understanding of the ectomycorrhizal symbiosis. New Phytol 134:685–695, 1996.
107. R Hampp, C Schaeffer, T Wallenda, C Stülten, P Johann, W Einig. Changes in carbon partitioning or allocation due to ectomycorrhiza formation: biochemical evidence. Can J Bot 73:s548–s556, 1995.
108. D Tagu, F Martin. Expressed sequence tags of randomly selected cDNA clones from *Eucalyptus globulus–Pisolithus tinctorius* ectomycorrhiza. Mol Plant–Microbe Interact 8:781–783, 1995.
109. C Voiblet, F Martin. Identifying symbiosis-regulated genes in *Eucalyptus globulus–Pisolithus tinctorius* ectomycorrhiza using suppression subtractive hybridization and cDNA arrays. In: Molecular Plant–Microbe Interaction, 9th International Congress, Amsterdam, July 25–30, 1999, pp 208–213.
110. RM Ewing, A Ben Kahla, O Poirot, F Lopez, S Audic, JM Claverie. Large-scale statistical analyses of rice ESTs reveal correlated patterns of gene expression. Genome Res 9:950–959, 1999.
111. M Hahn, K Mendgen. Characterization of in planta induced rust genes isolated from a haustorium-specific cDNA library. Mol Plant–Microbe Interact 10:427–437, 1997.
112. S Kamoun, P Hraber, B Sobral, D Nuss, F Govers. Initial assessment of gene diversity for the oomycete pathogen *Phytophthora infestans* based on expressed sequences. Fungal Genet Biol 28:94–106, 1999.
113. D Qutob, PT Hraber, BWS Sobral, M Gijzen. Comparative analysis of expressed sequences in *Phytophthora sojae*. Plant Physiol 123:243–253, 2000.
114. PM Schenk, K Kazan, I Wilson, JP Anderson, T Richmond, SC Somerville, JM Manners. Coordinated plant defense responses in *Arabidopsis* revealed by microarray analysis. Proc Natl Acad Sci USA 97:11655–11660, 2000.
115. L Wodicka, HL Dong, M Mittmann, MH Ho, DJ Lockhart. Genome-wide expression monitoring in *Saccharomyces cerevisiae*. Nat Biotechnol 15:1359–1367, 1997.
116. T Desprez, J Amselem, M Caboche, H Höfte. Differential gene expression in *Arabidopsis* monitored using cDNA arrays. Plant J 14:643–652, 1998.
117. T Girke, J Todd, S Ruuska, J White, C Benning, J Ohlrogge. Microarray analysis of developing *Arabidopsis* seeds. Plant Physiol 124:1570–1581, 2000.

118. J Györgyey, D Vaubert, JI Jiménez-Zurdo, C Charon, L Troussard, A Kondorosi, E Kondorosi. Analysis of *Medicago truncatula* nodule expressed sequence tags. Mol Plant–Microbe Interact 13:62–71, 2000.
119. E Carnero Diaz, F Martin, D Tagu. Eucalypt α-tubulin: cDNA cloning and increased level of transcripts in ectomycorrhizal root system. Plant Mol Biol 31:905–910, 1996.
120. V Boiffin, M Hodges, S Galvez, R Balestrini, P Bonfante, P Gadal, F Martin. Eucalypt NADP-dependent isocitrate dehydrogenase. cDNA cloning and expression in ectomycorrhizae. Plant Physiol 117:939–948, 1998.
121. L Feugey, DG Strullu, P Poupard, P Simoneau. Induced defence responses limit Hartig net formation in ectomycorrhizal birch roots. New Phytol 144:541–547, 1999.

13

Development and Molecular Biology of Arbuscular Mycorrhizal Fungi

Philipp Franken
Max-Planck-Institut für Terrestrische Mikrobiologie, Marburg/Lahn, Germany

Gerrit Kuhn
University of Lausanne, Lausanne, Switzerland

Vivienne Gianinazzi-Pearson
UMR INRA/Université de Bourgogne, Dijon, France

1 INTRODUCTION

Mycorrhiza, which literally means "fungus root," was introduced into the literature by Frank [1] and describes the phenomenon of an intimate symbiotic association of certain groups of soilborne fungi with absorbing organs (roots and rhizoids) of higher plants. Different types of mycorrhiza are distinguished depending on the plants and the fungi that are involved [2]. Ectomycorrhiza are formed by the roots of woody plants and a great variety of fungi, predominantly belonging to the Basidiomycota. Arbutoid and monotropoid ectendomycorrhiza are restricted to a few plant species. Among the endomycorrhiza, the ericoid and orchidoid mycorrhiza can only be found in the ericaceae and orchidaceae. Arbuscular mycorrhizas (AM) contrast with all these in that they represent a very widespread type of endomycorrhiza and some 80% of vascular land plant families as hosts for the fungal endosymbionts [3]. In contrast, only ~130 species of AM fungi are described. Six genera are distributed in four families and grouped in the order Glomales, Zygomycota [4]. *Gigaspora* and *Scutellospora* (Gigasporaceae) belong to the suborder Gigasporineae, while *Glomus*, *Sclerocystis* (Glomaceae), and *Acaulospora* and *Entrophospora* (Acaulosporaceae) are clustered in the Glomineae. The Glomales are very ancient microorganisms compared to other true fungi. Fossil data and molecular phylogenetic analyses indicate that their origin dates back to the Ordovicion-Devonion era some 460 to 400 Myr ago [5–7]. This period coincides with the colonization of the land by the plants, and AM fungi might have been essential for this process [8]. They became an integral part of most terrestrial ecosystems [9], and it was recently shown in microcosms how their biodiversity can influence aboveground biomass and plant species distribution [10]. Three basic mechanisms are discussed to be functional in influencing plant growth and development. AM fungi are able as ''biofertilizers'' to take up nutrients like phosphate, nitrate, or trace elements from the soil, to transport them into the roots, and to exchange these nutrients against carbohydrates, which are delivered by the plant [11]. As ''bioprotectors,'' they can induce plant resistance against root pathogens [12] or raise the plant's tolerance to abiotic stress like drought [13] or heavy metal contamination of the soil [14]. In addition, AM fungi influence plant hormone levels in the plant acting as ''bioregulators'' [15]. These activities result in most cases in a positive growth response of the plant. This phenomenon, the so-called mycorrhiza effect, has led to the use of AM fungi in plant production systems, in particular in horticulture [16] and in acclimatization of plantlets produced by micropropagation [17].

The application of these micro-organisms in plant production is developing in several countries, but because they are obligate symbionts and cannot be grown in pure culture [18], inoculum production on a large scale is difficult and its quality requires rigorous control. To understand the reason for their inability to grow without a host is useful in facilitating their application in plant production. AM fungi possess a fascinating life cycle, which is worth per se investigating. Furthermore, we are just beginning to get some insight into the complexity of the genomes of the Glomales [19] and there is a strong need for learning more about the biology of these obligate symbionts. In this chapter, we discuss the different stages of their development and highlight first results from research on the molecular biology of the Glomales.

2 PRESYMBIOTIC DEVELOPMENT

Although AM fungi are obligate biotrophs, they are able to germinate and to show a limited hyphal growth without the host plant. These stages are summarized under the term *presymbiotic development* (Fig. 1).

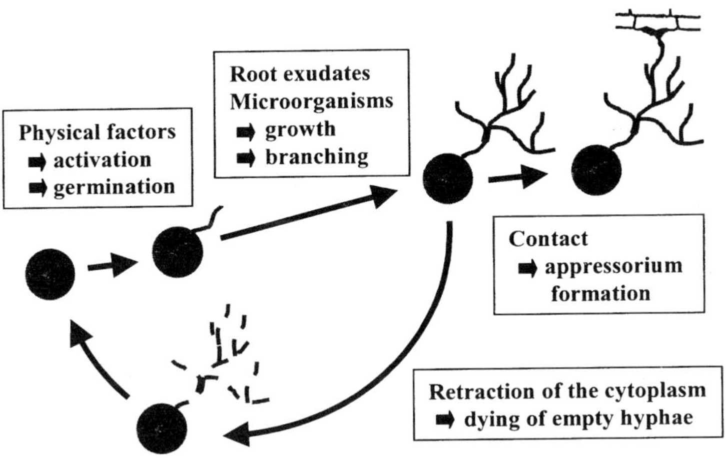

FIGURE 1 Schematic drawing of the presymbiotic phase of AM fungal development. Dormant spores are activated and start to germinate under the influence of physical factors. Further development can be influenced by root exudates or certain soil microorganisms. When hyphae come into contact with a plant root, they form an appressorium and enter the symbiotic phase. If the fungus does not find a host, the protoplasm is retracted, empty hyphae are septated off, and the fungus becomes dormant again.

2.1 Dormancy

Asexually formed chlamydospores are considered to be the dominant propagule of AM fungi. The production of zygospores has been reported once in *Gigaspora decipiens* [20], but it has never been repeated or observed in other species. The dormant chlamydospores are able to germinate and to start the symbiotic interaction with a host plant after staying in the soil for years [21]. They possess a characteristic morphology and cell wall structure, on which the taxonomy of the Glomales is based [4,22]. The chlamydospores are large single cells (50–600 μm diameter), and it has been shown for *Gig. margarita* that the cytoplasm can be divided into two areas [23]. One probably functions as a storage compartment with lipids, protein bodies, and glycogen, while the other contains many nuclei, which are blocked in the G0/G1 phase [24]. The high number of nuclei [25] is peculiar to the single cell spores of AM fungi, and calculation of the nuclear DNA content in spores [24,26] has resulted in the assumption of a relatively large genome size (10^8–10^9 nt). This characteristic feature has made chlamydospores a useful source of genomic DNA for phylogeny studies [27], for the construction of genomic libraries [8,29,30] or for the amplification of protein-encoding genes [31–33]. Interestingly, spores of the Glomales also contain amounts of RNA, which can be used for molecular analyses such as PCR cloning [34,35], RNA accumulation studies [36], and cDNA library construction [37].

Another particular feature of a number of AM fungi is the presence of bacterialike organisms (BLOs) inside the cytoplasm of the spores [38]. These bacteria were initially characterized as belonging to the genus *Burkholderia* [39]. The genus was thereafter identified by PCR in different species of the Gigasporineae, while other species harbored bacteria not belonging to that genus or no bacteria at all [40,41]. Interestingly, there is no correlation between the phylogeny of the Glomales and the occurrence of BLOs [40,41]. A genomic library established from DNA of *Gig. margarita* turned out to contain a high number of clones of prokaryotic origin [30]. This library was used to screen for certain genes of BLO origin, and two interesting operons were cloned and characterized. One operon contains an ORF for a putative phosphate transporter [42]; the other encodes a *vacB* gene described as being functional in colonization of eukaryotic cells [43]. Another clone found in this library interestingly contains a sequence with homologies to *Nif* genes, which are involved in nitrogen fixation in other bacteria [44]. All these findings raise questions concerning the life cycle and the metabolism of these BLOs with regard to the existence of a free-living stage, or how the fungus and the prokaryote interact in phosphate and nitrogen nutrition. Unfortunately, it is not possible to culture the bacteria without their host.

In addition to these BLOs, which seem to be an integral part of at least some AM fungal species, dormant spores are also subject to colonization by parasites probably owing to their high content in nutrients. Lee and Koske [45] found 44 different fungal species and eight actinomycetes in *Gig. gigantea* spores isolated from a sand dune. More recently, in a different study, where new AM fungal strains were isolated, one parasite, *Piriformospora indica*, turned out to be itself a plant root colonizer [46], and further experiments showed that this basidiomycete promotes plant growth [47]. Although it is not known whether these additional parasites colonize only senescent or also living AM fungal spores, the latter might be a source of microorganisms worth studying in more detail.

2.2 Germination

Germination of glomalean spores under axenic conditions was first reported by Mosse [48]. In some species, this developmental step seems to be induced by abiotic factors like temperature and humidity, while the dormant spores of others are not able to germinate for a certain period. Thereafter they change into a new physiological stage, called quiescence [21]. RNA accumulation studies have indicated the existence of another step between quiescence and germination [34].

Spores of *Gig. rosea* do not contain detectable amounts of RNA immediately after isolation from the soil. When they are stored for ~1 week in water at 4°C, however, they become activated, and RNA accumulates to rather large amounts (2 μg/500 spores) without germination. In contrast, spores or sporocarps of *Glomus mosseae* show no changes in RNA amounts from time point zero until 3 months after isolation from soil (Bütehorn and Franken, unpublished). Experiments with different inhibitors suggest that RNA stored during dormancy or synthesized at spore activation is used during germination for the translation of different proteins [49,50]. This has also been shown in other fungi like the zygomycete *Mucor* [51]. In addition to translation, the cell cycle is also activated [24,25]. However, in contrast to the symbiotic stage, only a small number of nuclei are in mitosis, while most of them are arrested in the G_0–G_1 phase [24].

2.3 Presymbiotic Mycelium

Most AM fungi are able to develop a more or less extended mycelium after spore germination without physical contact with a host plant. This can be observed on water agar and is not dependent on the addition of nutrients [49]. Neither carbohydrates nor mineral nutrients permit independent development, although they may prolong the limited hyphal growth [49]. Hyphal growth stops after ~2–3 weeks, if the AM fungus does not meet a host root. Logi et al. [52] used video microscopy of *G. caledonium* to show that under such circumstances the protoplasm in the mycelium is retracted from the tip backward in the direction of the spore, and that the empty regions are separated from the rest of the hypha by a septum. This probably helps the fungi to survive in the absence of a host, because the remaining mycelium is still able to form infection structures even at ~6 months. A particular feature of the two genera *Gigaspora* and *Scutellospora* is the development of auxiliary cells, which form on presymbiotic mycelium independently of host roots. These structures have been suggested as possible infective propagules [53] or reminiscent of spores [4], but their true role in the life cycle of AM fungi is not clear.

The failure of independent growth has often been attributed to limitations in uptake or metabolism of carbon without the host. Bago et al. [54] have tested this hypothesis by feeding *G. intraradices* spores with different labeled carbon sources. It turned out that the asymbiotic hyphae could take up hexose and acetate for their metabolism. The labeling patterns indicated that they also use internal storage lipids for hexose synthesis. Another interesting result of this work was the observation that dark fixation of CO_2 takes place. This would explain why CO_2-enriched conditions lead to an enhancement of asymbiotic hyphal development [55]. The major difference between asymbiotic and symbiotic hyphae, which were analyzed by Pfeffer et al. [56], is the lack of lipid biosynthesis. Bago et al. [54] therefore suggest that the switch of catabolism to anabolism of storage lipids is the limiting step for AM fungi to fulfill their life cycle in the absence of a host. Cloning and expression analysis of the genes encoding the corresponding enzymes involved in fatty acid biosynthesis could provide proof to support this hypothesis.

RNA accumulation pattern analyses of presymbiotic hyphal development using the differential RNA display technique have indicated that the RNA that is synthesized during activation of *Gig. rosea* spores is sufficient for all following steps of germination and hyphal development, because no significant changes could be observed [57]. The same is true for *Scutellospora castanea* (Fig. 2) while differential occurrence of cDNA fragments has been observed in *G. mosseae*, indicating regulation on the transcriptional level for this fungus from the Glomineae [57]. EST libraries have therefore been constructed from activated spores of the two Gigasporineae species, because they should contain cDNA fragments of all genes necessary for presymbiotic development [37].

A similar library was constructed by Lanfranco et al. [58]. They incubated spores of *Gig. margarita* for 12 days in water and extracted the RNA for cDNA library construction from the developed presymbiotic mycelium. Requena et al. [59], in contrast, subtracted the cDNA of presymbiotic hyphae of *G. mosseae* by cDNA obtained from extraradical mycelium before cloning, in order to enrich for genes that are induced during the early stages of the fungal life cycle. Sequence analysis has in all cases revealed similarities to genes coding for proteins involved in basic functions like translation and protein processing, primary metabolism and transport processes, cell cycle, replication or chromatin structure, cell structure, or signal transduction.

The developmental program of the fungus totally changes close to a host root. Hyphal growth loses apical dominance and shows strongly enhanced branching [60]. This phenomenon can only be observed in the presence of roots of host plants, and not with those of nonhosts [61]. The hypothesis that a host root produces certain factors that are recognized by the fungus has been supported by experiments showing that root exudates influence hyphal development in a similar manner [62–65]. The nature of the inducive compounds is unknown, but there have been several reports indicating that they may belong to the flavonoid class of molecules, in comparison to the symbiosis between roots of legumes and nodule-forming bacteria [66–69]. However, maize mutants deficient in flavonoid synthesis develop mycorrhiza to the same extent as the corresponding wild type [70], and partial purification of an active fraction clearly showed that it is not a compound produced via the flavonoid pathway [71].

Another line of investigation has been based on the phenomenon that phosphate-deficient plants show higher mycorrhizal colonization than those that are fertilized with sufficient amounts of phosphate [62,72]. Analysis of root exudates showed the occurrence of a UV-fluorescing compound in P-starved plants, which disappears after fertilization or mycorrhization [73]. The compound extracted

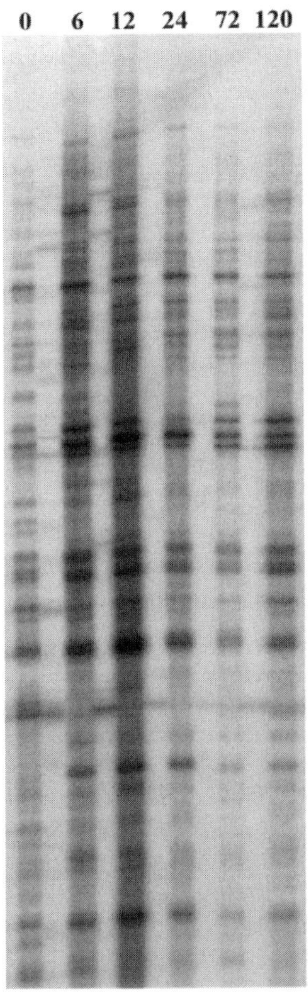

Figure 2 Differential display analysis of mRNA from spores of *Scutellospora castanea* kept for 1 week at 4°C and used directly or after incubation for 6h, 12, 24, 72, and 120 h on water agar.

from TLC plates clearly showed a positive effect on branching of *S. castanea* [74]. First analysis indicated that the chemical principle might be a salicylic acid derivative. Application of salicylic acid affected hyphal elongation and branching of the AM fungus *Gig. rosea* (Fig. 3a), and molecular studies of this phenomenon have been carried out by differential display analysis [75]. Addition of salicylic acid induced changes in the pattern of transcripts (Fig. 3b), while no changes in RNA accumulation were observed in untreated controls, as pointed out above.

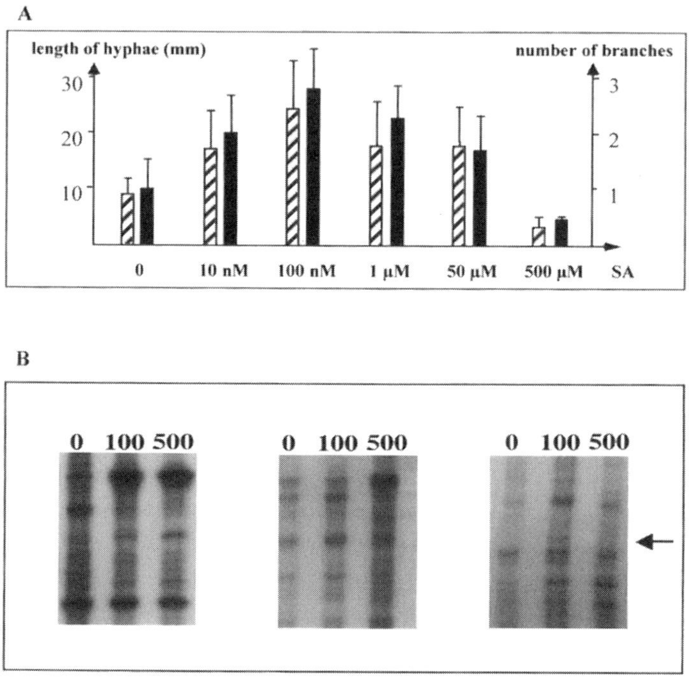

Figure 3 Effect of salicylic acid on spores of *Gigaspora rosea*. (A) 100 spores were incubated on water agar with or without the addition of 10 nM to 500 µM salicylic acid, and developing hyphal length and number of branches were measured after 6 days. (B) In parallel, 100 spores were harvested after three days of incubation in 0, 100 nM, or 500 µM salicylic acid, RNA extracted and differentially displayed. Three examples of polyacrylamide gels are shown, where differences in cDNA profiles can be observed. The fragment marked with an arrow was used for further analysis.

Differentially occurring cDNA fragments were cloned, and the expression pattern of the corresponding genes was verified. One of the cDNA fragments (Fig. 4) showed similarity to a gene with unknown function from *Schizosaccharomyces pombe*.

There is also a positive influence of microorganisms on presymbiotic development of AM fungi similar to that reported for root exudates [76]. Interaction between a *Bacillus subtilis* strain, isolated as a plant growth promoting rhizobacteria (PGPR), and *G. mosseae* has been used for molecular analysis [77]. One fungal gene that has been isolated, *GmFox2*, encodes a fatty acid oxidase. Analysis of the encoded protein showed that GMFOX2 harbors all possible domains,

TTCGGCGATAGTGATGATGATATTTCAAGCGATGAAGAAGAAGAGGAAGATGATACAGCA 60

F G D **S** D D D I **S** S D **E** E E E **E D** D **T** A

GCACTTTTATTAGAGTTGGAAAAAATTAAAAAAGAGCGTGCAGAAGAGAAAGAAAGAATG 120

A **L L** L **E L E** K **I K** K **E R** A E **E** K E R M

GAGCTAGAAAAGATGGAATCTGCCGAACGACAAAGACAAGAAGAAGCTATGAACGGAAAT 180

E L **E K** M E S A **E** R Q R Q **E** E A M N **G N**

CCCCTGCTTCGTAAGGATTTCAGTGTAAAAAGAAGATGGGACGATGACGTAATCTTCAAG 240

P **L L** R **K** D **F** S **V K R R W D** D **D V** I **F** K

AATCAAGCACGTGGAGTAGATGATAAACCTAAAAAAAGGTTTATTAATGACACTCTTCGC 300

N Q A R **G** V **D D** K **P** K K R **F** I **N D** T **L R**

AGCGATTTTCATCGAAAGTTTATGAACAAATACGTGAAATGATGTGTATAATACATGGAT 360

S D **F H** R **K F** M N K Y **V** K

GTCTAAATTATTTTGCCTTTGACGTAATTTAAATGGCTAAATTATTTTTAGCAGATTAAA 420

TCGTAAATAATTTGGGATGAATGAAATAAAGTTATTTGGTCATAGTAGTCGATCGCAAAA 480

Figure 4 The fragment marked with an arrow in Figure 3 was cut from the gel, reamplified, and cloned. Differential expression of the corresponding gene was verified by hybridization to labeled fungal cDNA, and the insert was sequenced. The figure shows the cDNA fragment and the deduced amino acid sequence. Amino acids that were identical to an ORF in *Schizosaccharomyces pombe* are indicated in bold and underlined.

in contrast to its homologs in humans and *Neurospora crassa*. This suggests that the common ancestor of this gene is possibly closer to the AM fungus than to the other organisms, and further indicates the ancient nature of the mycosymbiont. In humans, downregulation of the corresponding protein 17-HDS IV leads to activation of the mitogenic cycle and increases the amount of estrogens [78]. *GmFox2* is also downregulated, when growth of the AM fungus is enhanced by the bacterium. Future research will show whether similar molecules act as internal signals in the same manner in AM fungi.

That flavonoids are structurally related to estrogens may explain the apparently contradictory results concerning their implication in AM development. If flavonoids are not the primary signal, they would not be a necessary compound of the branch-inducing fraction of root exudates [71], and it would not matter, if they are not produced by the plant [70]. However, a flavonoid-related compound could act as a secondary messenger, and this would explain why an antiestrogen inhibits stimulation of presymbiotic development by certain flavonoids [79].

Another gene has been isolated in the study of the effect of *B. subtilis* on *G. mosseae* [80]. This gene encodes TOR2, a protein that has been shown in yeast to be involved in control of the cell cycle and the actin cytoskeleton. *GmTOR2* is, like its homolog in yeast, not regulated on the transcriptional level. Application of the drug rapamycin, which acts against the cell cycle controlling activity of TOR2, does not influence germination of spores, but stops further development of germ tubes [80]. This supports the hypothesis of Bianciotto and Bonfante [24] and Becard and Pfeffer [25] that nuclear replication is not a prerequisite for germination, but is necessary for presymbiotic hyphal growth.

3 SYMBIOTIC DEVELOPMENT

Symbiotic development starts with the physical contact of the fungus and the root. It involves appressorium formation, extra- and intracellular colonization, the formation of arbuscules, and the spread of the fungus into the soil where it completes its life cycle with the development of new spores (Fig. 5).

3.1 Appressorium Development

Development of appressoria is probably the key step in fungal recognition of potential hosts [81]. Since this has been extensively studied in the rice pathogen *Magnaporthe grisea* [82], one could pose the question: ''Why analyze it in such

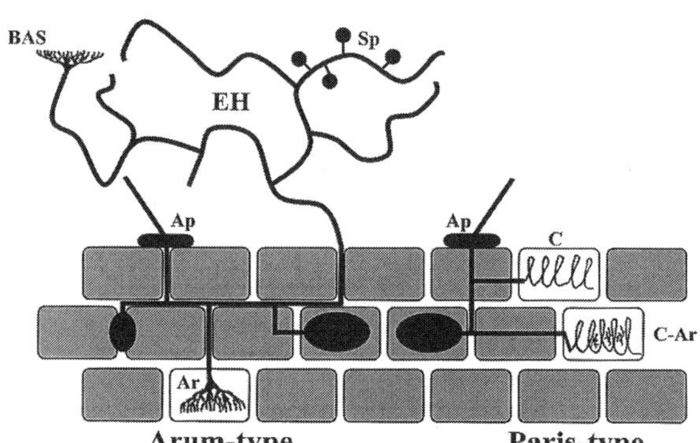

FIGURE 5 Scheme of AM fungal symbiotic development. Appressoria (Ap) are developing at the surface of the root from where an infective hypha grows. AM colonization of the Arum type is first extra- and intracellularly. In the inner cortex cells the arbuscules (Ar) are formed. In the Paris type, the AM fungus colonizes the root from cell to cell, where it forms large coils (C) which can harbor small intercalated arbuscules (C-Ar). AM fungi of the suborder Glomineae also develop inter- or intracellular vesicles (●). In the meantime, extraradical hyphae (EH) of the fungus spread into the soil. Branched absorbing structures (BAS) and new spores (Sp) can develop on this mycelium.

a complicated system as the AM symbiosis?'' However, in contrast to other organisms, AM fungal appressoria never develop on artificial surfaces, but need a plant cell. Interestingly, these infection structures do not form either on intact roots of nonhosts [83] just as arbuscules do not form in host roots grafted to a nonhost shoot [84]. Appressorium-like structures have, however, been observed on detached roots of a nonhost [83] and on isolated root cell walls [85]. These studies suggest the existence of an interaction between an obligate positive stimulus of root cell walls and a negative factor in the shoots of nonhost plants. Another interesting feature is the absence of a septum separating the appressorium from the rest of the hypha outside the root. It is therefore not possible for an AM fungus to build up high pressure and use this mechanical force to invade host tissues, as described for pathogenic fungi [86]. Certain plant mutants could provide experimental systems to study the molecular basis of mechanisms of appressorium development and function in AM fungi. Early myc$^-$ mutants of *Pisum sativum* [87] and *Medicago truncatula* [88] stop fungal development after appressorium formation. Transcript comparisons between extracts of the fungus on such mutants and extracts from presymbiotic hyphae could permit identification of appressorium-induced AM fungal genes.

3.2 Root Colonization

Plant host tissues are colonized by AM fungi in a specific mode. After entering the root from the appressorium, they first grow intercellularly or cross outer cells with linear or simple coiled hyphae [89]. When they reach the inner cortex, the fungal symbionts totally change their mode of colonization. They penetrate the plant cell wall and extensively ramify to form a highly branched haustorium, called arbuscule, which is the central and name-giving structure of the symbiosis. This Arum type of development is the one that is most studied, although the Paris type seems to occur more frequently [90]. In the Paris type, the intercellular phase is absent and hyphae grow from cell to cell, where they develop large coils with small intercalated arbuscules. The type of mycorrhiza that is formed seems to be determined by the genomes of both partners, and is not under environmental control [90,91]. To what extent these different structures influence the function of the symbiosis is not clear. Other typical structures of AM fungi are vesicles, which are formed only by members of the Glomineae and are considered to be storage organs [92].

How these symbiotic micro-organisms penetrate and colonize the roots is an open question. As pointed out above, mechanical forces probably play a minor role owing to the coenocytic nature of the AM fungi, and enzymatic activities should therefore be involved. Hydrolytic enzymes have been proposed to be involved in the process of root penetration and colonization [93], and the production of pectinases, cellulases, and xyloglucanases by glomalean fungi has been described [94–96]. The role of wall-degrading hydrolytic enzymes in root colonization is in most cases unclear. Only polygalacturonase has been specifically detected at the arbuscule interface [97], and endoxyloglucanase activity has been correlated with percentage root colonization [98]. One fungal endoglucanase [99] and a lipolytic enzyme [100] have been purified and could now be used for microsequencing and cloning of the corresponding genes.

Physiological studies on the metabolism of AM fungi during the symbiotic stage have revealed that hexoses like glucose and fructose are taken up by hyphae inside the root [101,102], but not by extraradical mycelium [56]. In contrast to the presymbiotic phase described above, these sugars are converted not only into carbohydrates but also into lipids. To study enzymatic activities of symbiotic fungal hyphae, Saito [103] digested colonized roots with a mixture of cellulase and pectinase. Comparisons of enzymatic activities of symbiotic and presymbiotic hyphae indicated that assimilation of sugars is much higher during the symbiosis. Further analysis using the extracted hyphae showed that when glucose was supplied to the symbiotic hyphae, P efflux was enhanced suggesting a coupling between phosphate and sugar exchange [104,105].

A few fungal genes expressed during root colonization have been identified by nontargeted approaches comparing RNA accumulation patterns in control roots and mycorrhiza. Interesting similarities have been found, e.g., to a cruciform DNA-binding protein [106] or to transcriptional regulators [107]. It is not clear, however, if these genes are differentially expressed, since expression patterns are compared in most cases to control roots, where the fungal endosymbiont is absent.

To identify genes that are expressed when arbuscules are differentiated within host cortical cells, Lapopin et al. [108] carried out differential RNA display analyses of wild-type *P. sativum* and a mutant which showed aborted arbuscule development, both inoculated with the AM fungus *G. mosseae*. Among the cDNA fragments corresponding to differentially expressed genes, one was of fungal origin, but showed no similarity to any known sequence. Further RNA accumulation studies by RT-PCR showed that this gene is highly expressed in *G. mosseae*–colonized wild-type *P. sativum* roots, but only weakly in the inoculated mutant, and only extremely low levels of transcript could be detected in extraradical symbiotic hyphae, dormant spores, or presymbiotic mycelium (Lapopin, Gianinazzi-Pearson, and Franken, unpublished). Current experiments are directed to cloning of the entire gene in order to obtain insight into its function. In contrast to this, a fungal gene identified in *M. truncatula*/*G. mosseae* mycorrhiza is expressed during the whole life cycle of the AM fungus (dormant spores, presymbiotic development, intra- and extraradical hyphae) [57]. In this case, the whole cDNA was cloned, but again no similarity to any known sequence has been found. A phosphoglycerate kinase (PGK) gene from *G. mosseae* has been identified and cloned [109]. Further studies have revealed a significantly higher accumulation of the encoded protein during the symbiotic stage compared to presymbiotic development [110]. This may be responsible for a higher glycolytic activity in the intraradical hyphal cells which have to deal with the carbon flux from root cells, and where it has been shown that hexoses like glucose and fructose are taken up by the AM fungus [101]. This agrees also with the finding that hexokinase activity was only detected inside the root, but not in germinating spores of *Gig. margarita* [103]. The PGK-encoding gene from *G. mosseae* is probably regulated by sugar metabolism as in other fungi. Cloning and analysis of the promoter have revealed two sequence motives with homology to carbon source–controlled, upstream-activating elements in *Saccharomyces cerevisiae* which might be involved in such a regulation [111]. Paterson and Harrier [112] have recently started to clone genes for glucose sensors from *G. mosseae* in order to further analyze the regulation of the *Pgk* gene on the molecular level.

Nutrient exchange between two AM partners probably involves transport of phosphate, nitrate, and trace elements from the fungus to the plant [11] across the periarbuscular membrane at the arbuscule interface [113]. Carbohydrates are transported in the opposite direction from plant to fungal cells. H^+-ATPase activity, which could drive the active uptake of these carbohydrates by the AM fungus, has been detected at the membrane of intraradical hyphae [114]. Ferrol et al. [33] have recently cloned five such ATPase-encoding genes from *G. mosseae* by PCR. Future analysis of the expression patterns of these genes might give a clue about the spatiotemporal functioning of carbohydrate transfer from plant to fungus. Kaldorf et al. [31] concentrated on nitrogen transport from the fungus to the plant and identified a nitrate reductase-encoding gene in the AM fungus *G. intraradices*. In situ hybridization showed that this gene is induced in arbuscules [115]. Interestingly, the fungal gene seems to substitute the nitrate reductase gene of the plant, which shows reduced RNA accumulation during the symbiosis. This is a first hint that plant and fungus complement each other at the molecular level during their symbiotic interaction.

Another important constituent of the symbiotic interface is the cell wall of the fungus, since this is the contact site with the plant cell. Chitin is present in all AM fungi, but β(1–3) glucans seem to be restricted to the Glomineae [116]. However, chitin is probably differentially distributed in hyphae, with a chitinase-sensitive amorphous form accumulating to a greater extent in the fine branches of the arbuscules [117,118]. Lanfranco et al. [119] used RT-PCR with degenerated primers to amplify fragments of chitin synthase genes from two different AM fungi. In further studies, they isolated a genomic clone from *G. versiforme* [32] and studied the expression of several gene copies in *Gig. margarita* [120]. This analysis revealed the induction of two of the chitin synthase genes during the symbiotic stage, which might be related to enhanced chitin production associated with the prolific growth of the fungus within host root tissues.

3.3 Extraradical Hyphae and Spore Production

As AM fungi colonize the root cortex, they also develop mycelium into the surrounding soil and explore the environment for nutrients and for roots of other host plants. It has been shown that mineral nutrients or carbohydrates can be exchanged via the AM hyphal network between plants of different species [121]. Read [122] formulated the term ''Plants on the Web'' to describe the fact that in a given ecosystem probably all mycotrophic plants are interconnected by a hyphal network in the soil. Extraradical symbiotic hyphae differ from the presymbiotic mycelium in that they produce new spores and form so-called branched absorbing structures (BAS). The latter are small groups of dichotomous hyphae, which cannot be observed in monoculture but are formed shortly after the establishment of the symbiosis [123]. Their branching pattern recalls arbuscules, and there are suggestions that they may be involved in nutrient uptake from the soil. Sometimes these structures subtend developing spores so that new spores are formed at the tip of the branches of the BAS.

Phosphate is the most studied, and seems to be the most important, nutrient supplied by the fungus to the plant in AM [113]. Screening a mycorrhiza cDNA library with a phosphate transporter gene from yeast, Van Buuren and Harrison [124] isolated one clone which was shown to belong to the fungus *G. versiforme*. Heterologous expression in a yeast mutant confirmed that it codes for a high-affinity phosphate transporter. The gene was therefore called *Gvgpt*. RT-PCR experiments revealed that the gene is only expressed in the extraradical mycelium and not in fungal structures inside the root. It might therefore be involved in phosphate uptake from the soil into fungal hyphae. In fact, a correlation has been found between expression levels of this phosphate transporter gene in different AM fungal isolates and the ability to supply the plant with phosphate, underlining the importance of this gene (S. Burleigh, personal communication). AM fungi have been reported to differ spatially in their ability to take up phosphate [125]. For example, *G. caledonium* is more efficient relatively far from the plant, whilst *S. calospora* preferentially obtains phosphate in the root compartment. The existence of a synergism between different AM fungi in exploiting the soil has an important impact for the ecological success of mycorrhiza-dependent plant species.

Attempts are being made to clone AM fungal genes from extraradical hyphae using nontargeted approaches. Recently, a number of EST sequences were detected in a cDNA library of mycelium from a two-compartment system of *G. intraradices* [126], and RNA accumulation patterns have been compared by differential display analysis in extraradical hyphae from pot cultures supplied with low or high phosphate concentrations [127]. In the latter study, an ATPase-encoding gene was identified from *G. intraradices* which was downregulated by phosphate.

Knowledge about spore development in AM fungi is relatively scarce and has been reviewed by Bianciotto and Bonfante [128]. Monoxenic or root organ cultures provide a useful system for future work on this issue. AM fungi were first cultured on excised roots by Mosse [48], and later established on prolific root organ cultures transformed with *Agrobacterium rhizogenes* [129]. Since then, a number of AM fungal isolates have been grown in such cultures where they fulfill their life cycle and develop all structures that can be found in the soil [130–133]. Root organ cultures are therefore a convenient system to produce masses of spores for molecular studies, under conditions free of contamination by other soil microorganisms. However, it is necessary to determine to what extent such artificial cultures are relevant to the physiology, biochemistry, and molecular biology of the natural symbiotic development of AM fungi. For example, when the mycoparasite *Trichoderma harzianum* interacts with the AM fungus *G. intraradices* in monoxenic culture, the mycoparasite infects and kills the symbiont [134]. In pot cultures, however, the AM fungus survives perfectly, and the hyphal mass of *T. harzianum* declines [135].

4 CONCLUSION

As can be seen from this review, research about the molecular biology of AM fungal development is still rather limited, but more information should become available in the near future especially through the application of nontargeted analyses. Several EST-sequencing projects are running which include cDNA libraries of mycorrhizal tissues (e.g., http://sequence.toulouse.inra/fr/M.truncatula.html). Screening of corresponding cDNA arrays with probes obtained from dormant and germinating spores, presymbiotic mycelium, or extraradical hyphae will provide information about genes associated with the different stages of fungal development. In parallel, targeting genes with physiological functions in the different steps of the symbiosis could give clues to the differences like hexose uptake or lipid biosynthesis between symbiotic and presymbiotic mycelium. Sequence information is being collected with the respective annotations in a genetic archive of fungal isolates registered by the Banque European de Glomales (http://wwwbio.ukc.ac.uk/beg). In our opinion, at least one fungus from the suborder Gigasporineae and one from the Glomineae should be analyzed in greater detail, since all data collected up to now point to major differences in their biology. A synergism between different AM fungi may lead to the phenomenon that fungal biodiversity enhances overall biomass and species richness of plant populations [10]. Consequently, further research into the molecular biology of AM fungi will also contribute to a better understanding of the ecological significance of these mycosymbionts in both natural and agricultural ecosystems.

ACKNOWLEDGMENTS

The results were obtained by projects supported by the DFG (SFB 395) and by the EU-funded BEG-net (contract No. B104-CT97-2225).

REFERENCES

1. AB Frank. Ueber die auf Wurzelsymbiose beruhende Ernährung gewisser Bäume durch unterirdische Pilze. Bericht der deutschen botanischen Gesellschaft 3:128–148, 1885.
2. R Molina, H Massicotte, JM Trappe. Specificity phenomena in mycorrhizal symbiosis: community-ecological consequences and practical implications. In: MF Allen, ed. Mycorrhizal Functioning: An Integrative Plant–Fungal Process. London: Chapman & Hall, 1992, pp 357–423.
3. JL Harley, EL Harley. A check list of mycorrhiza in the British flora. New Phytol 105:1–102, 1987.
4. JB Morton, GL Benny. Revised classification of arbuscular mycorrhizal fungi (Zygomycetes): a new order, Glomales, two new suborders, Glomineae and Gigasporineae, and two new families, Acaulosporaceae and Gigasporaceae, with an emendation of Glomaceae. Mycotaxon 37:471–491, 1990.
5. SB Stubblefield, TN Taylor, JM Trappe. Vesicular–arbuscular mycorrhizae from the triassic of Antarctica. Am J Bot 74:1904–1911, 1987.
6. L Simon, J Bousquet, RC Lévesque, M Lalonde. Origin and diversification of endomycorrhizal fungi and coincidence with vascular land plants. Nature 363:67–69, 1993.
7. D Redecker, R Kodner, LE Graham. Glomalean fungi from the Ordovician. Science 289:1920–1921, 2000.
8. KA Pirozynski, DW Malloch. The origin of land plants: a matter of mycotrophism. Biosystems 6:153–164, 1975.
9. DJ Read, DH Lewis, AH Fitter, IJ Alexander. Mycorrhizas in Ecosystems. Oxford: CAB International, 1992.
10. MGA van der Heijden, JN Klironomos, M Ursic, P Moutoglis, R Streitwolf-Engel, T Boller, A Wiemken, IR Sanders. Mycorrhizal fungal diversity determines plant biodiversity, ecosystem variability and productivity. Nature 396:69–72, 1998.
11. H Marschner, B Dell. Nutrient uptake in mycorrhizal symbiosis. Plant Soil 159: 89–102, 1994.
12. C Azcon-Aguilar, JM Barea. Arbuscular mycorrhizas and biological control of soil-borne plant pathogens—an overview of the mechanisms involved. Mycorrhiza 6: 457–464, 1996.
13. N Requena. Mycorrhizal symbiosis and management of the mycorrhizosphere as a tool for the restoration of degraded semi-arid ecosystems. Rec Res Dev Microbiol 1:267–276, 1997.
14. U Galli, H Schüepp, C Brunold. Heavy metal binding by mycorrhizal fungi. Physiol Plant 92:364–368, 1994.

15. N Gogala. Regulation of mycorrhizal infection by hormonal factors produced by hosts and fungi. Experientia 47:331–340, 1991.
16. C Azcon-Aguilar, JM Barea. Applying mycorrhiza biotechnology to horticulture—significance and potential. Sci Hort 68:1–24, 1997.
17. P Lovato, V Gianinazzi-Pearson, A Trouvelot, S Gianinazzi. The state of art of mycorrhiza and micropropagation. Adv Hort Sci 10:46–52, 1996.
18. B Mosse. Some studies relating to 'independent' growth of vesicular–arbuscular endophytes. Can J Bot 66:2533–2540, 1988.
19. V Gianinazzi-Pearson, D van Tuinen, E Dumas-Gaudot, H Dulieu. Exploring the genome of glomalean fungi. In: B Hock, ed. The Mycota IX. Berlin: Springer-Verlag, 2000, pp 3–17.
20. IC Tommerup, K Sivasithamparam. Zygospores and asexual spores of *Gigaspora-decipiens* an arbuscular mycorrhizal fungus. Mycol Res 94:897–900, 1990.
21. IC Tommerup. Spore dormancy in vesicular–arbuscular mycorrhizal fungi. Trans Br Mycol Soc 81:37–45, 1983.
22. C Walker, JM Trappe. Names and epithets in the Glomales and Endogonales. Mycol Res 97:339–344, 1993.
23. P Bonfante, R Balestrini, K Mendgen. Storage and secretion processes in the spore of *Gigaspora margarita* Becker & Hall as revealed by high-pressure freezing and freeze substitution. New Phytol 128:93–101, 1994.
24. V Bianciotto, P Bonfante. Evidence of DNA replication in an arbuscular mycorrhizal fungus in the absence of the host plant. Protoplasma 176:100–105, 1993.
25. G Becard, PE Pfeffer. Status of nuclear division in arbuscular mycorrhizal fungi during in vitro development. Protoplasma 194:62–68, 1993.
26. M Hosny, V Gianinazzi-Pearson, H Dulieu. Nuclear DNA content of 11 fungal species in Glomales. Genome 41:422–428, 1998.
27. P Franken. Trends in molecular studies of AM fungi. In: A Varma, B Hock, eds. Mycorrhiza. Heidelberg: Springer-Verlag, 1998, pp 37–49.
28. A Zézé, H Dulieu, V Gianinazzi-Pearson. DNA cloning and screening of a partial genomic library from an arbuscular mycorrhizal fungus, *Scutellospora castanea*. Mycorrhiza 4:251–254, 1994.
29. P Franken, V Gianinazzi-Pearson. Phage cloning of ribosomal RNA genes from the arbuscular mycorrhizal fungi *Glomus mosseae* and *Scutellospora castanea*. Mycorrhiza 6:167–173, 1996.
30. ML van Buuren, L Lanfranco, S Longato, D Minerdi, MJ Harrison, P Bonfante. Construction and characterization of genomic libraries of two endomycorrhizal fungi: *Glomus versiforme* and *Gigaspora margarita*. Mycol Res 103:955–960, 1999.
31. M Kaldorf, W Zimmer, H Bothe. Genetic evidence for the occurence of assimilatory nitrate reductase in arbuscular mycorrhizal and other fungi. Mycorrhiza 5:23–28, 1994.
32. L Lanfranco, L Garnero, P Bonfante. Chitin synthase genes in the arbuscular mycorrhizal fungus *Glomus versiforme*: full sequence of a gene encoding a class IV chitin synthase. FEMS Microbiol Let 170:59–67, 1999.
33. N Ferrol, JM Barea, C Azcon-Aguilar. The plasma membrane H^+-ATPase gene

family in the arbuscular mycorrhizal fungus *Glomus mosseae*. Curr Genet 37:112–118, 2000.

34. P Franken, L Lapopin, G Meyer-Gauen, V Gianinazzi-Pearson. RNA accumulation and genes expressed in spores of the arbuscular mycorrhizal fungus *Gigaspora rosea*. Mycologia 89:295–299, 1997.
35. P Franken, B Ressin, L Lapopin, V Gianinazzi-Pearson, S Gianinazzi. PCR cloning of genes from arbuscular mycorrhizal fungi. In: A Varma, B Hock, eds. The Mycorrhizal Manual. Heidelberg: Springer-Verlag, 1998, pp 401–412.
36. B Bütehorn, V Gianinazzi-Pearson, P Franken. Quantification of β-tubulin RNA expression during asymbiotic and symbiotic development of the arbuscular mycorrhizal *Glomus mosseae*. Mycol Res 103:360–364, 1999.
37. M Stommel, P Mann, P Franken. Construction and analysis of an EST library using RNA from dormant spores of the arbuscular mycorrhizal fungus *Gigaspora rosea*. Mycorrhiza 10:281–285, 2001.
38. B Mosse. Honey-coloured, sessile Endogone spores. II. Changes in fine structure during spore development. Arch Microbiol 74:129–145, 1970.
39. V Bianciotto, C Bandi, D Minerdi, M Sironi, HV Tichy, P Bonfante. An obligately endosymbiotic mycorrhizal fungus itself harbors obligately intracellular bacteria. Appl Environ Microbiol 62:3005–3010, 1996.
40. R Perotto, P Bonfante. Bacterial associations with mycorrhizal fungi: close and distant friends in the rhizosphere. Trend Microbiol 5:496–501, 1997.
41. M Hosny, D van Tuinen, F Jacquin, P Füller, B Zhao, V Gianinazzi-Pearson, P Franken. Arbuscular mycorrhizal fungi and bacteria: how to construct prokaryotic DNA-free genomic libraries from the Glomales. FEMS Lett 170:425–430,1999.
42. JM Ruiz-Lozano, P Bonfante. Identification of a putative P-transporter operon in the genome of a *Burkholderia* strain living inside the arbuscular mycorrhizal fungus *Gigaspora margarita*. J Bacteriol 181:4106–4109, 1999.
43. JM Ruiz-Lozano, P Bonfante. A *Burkholderia* strain living inside the arbuscular mycorrhizal fungus *Gigaspora margarita* possesses the *vacB* gene, which is involved in host cell colonization by bacteria. Microbiol Ecol 39:137–144, 2000.
44. D Minerdi, R Fani, R Gallo, A Boarino, P Bonfante. Nitrogen fixation genes in an endosymbiotic *Burkholderia*. Appl Environ Microbiol 67:725–732, 2001.
45. P-J Lee, RE Koske. *Gigaspora gigantea*: parasitism of spores by fungi and actinomycetes. Mycol Res 98:458–466, 1994.
46. S Verma, A Varma, K-H Rexer, A Hassel, G Kost, A Sarbhoy, P Bisen, B Bütehorn, P Franken. *Piriformospora indica*, gen. nov. sp. nov., a new root-colonizing fungus. Mycologia 90:896–903, 1998.
47. A Varma, S Verma, Sudah, N Sahay, B Bütehorn, P Franken. *Piriformospora indica*, a cultivable plant growth–promoting root endophyte. Appl Environ Microbiol 65:2741–2744, 1999.
48. B Mosse. The establishment of vesicular–arbuscular mycorrhiza under aseptic conditions. J Gen Microbiol 27:509–520, 1962.
49. CM Hepper. Germination and growth of *Glomus caledonius* spores: the effects of inhibitors and nutrients. Soil Biol Biochem 2:269–277, 1979.
50. JP Beilby. Effects of inhibitors on early protein, RNA, and lipid synthesis in germi-

nating vesicular–arbuscular mycorrhizal fungal spores of *Glomus caledonius*. Can J Microbiol 29:596–601, 1983.
51. JH Linz, M Orlowski. Stored mRNA in sporangiospores of the fungus *Mucor racemosus*. J Bacteriol 150:1136–1144, 1982.
52. C Logi, C Sbrana, M Giovannetti. Cellular events involved in survival of individual arbuscular mycorrhizal symbionts growing in the absence of the host. Appl Environ Microbiol 64:3473–3479, 1998.
53. F Pons, V Gianinazzi-Pearson. Observations on extramatrical vesicles in *Gigaspora margarita* in vitro. Trans Br Mycol Soc 84:168–170, 1985.
54. B Bago, PE Pfeffer, DD Douds, J Brouillette, G Becard, Y Shachar-Hill. Carbon metabolism in spores of the arbuscular mycorrhizal fungus *Glomus intraradices* as revealed by nuclear magnetic resonance spectroscopy. Plant Physiol 121:263–271, 1999.
55. G Becard, DD Doud, PE Pfeffer. Extensive in vitro hyphal growth of vesicular–arbuscular mycorrhizal fungi in the presence of CO_2 and flavonols. Appl Environ Microbiol 58:821–825, 1992.
56. PE Pfeffer, DD Douds, G Becard, Y Shachar-Hill. Carbon uptake and the metabolism and transport of lipids in an arbuscular mycorrhiza. Plant Physiol 120:587–598, 1999.
57. P Franken, N Requena, B Bütehorn, F Krajinski, G Kuhn, L Lapopin, P Mann, D Rhody, M Stommel. Molecular analysis of the arbuscular mycorrhiza symbiosis. Arch Agron Soil Sci 45:271–286, 2000.
58. L Lanfranco, S Gabella, P Bonfante. EST as a useful tool for studying gene expression in arbuscular maycorrhizal fungi. In: HC Weber, S Imhof, D Zeuske, eds. Programs, Abstracts and Papers of the Third International Congress on Symbiosis. Marburg, Germany: Philipps University, 2000, pp 108–115.
59. N Requena, B Achatz, P Mann, P Franken. Isolation of genes expressed during the early stages of arbuscular mycorrhizal fungus *Glomus mosseae* by suppressive subtractive hybridisation (SSH). In: HC Weber, S Imhof, D Zeuske, eds. Programs, Abstracts and Papers of the Third International Congress on Symbiosis. Marburg, Germany: Philipps University, 2000, p 176.
60. M Giovannetti, S Citernesi. Time-course of appressorium formation on host plants by arbuscular mycorrhizal fungi. Mycol Res 97:1140–1142, 1993.
61. M Giovannetti, C Sbrana, C Logi. Early processes involved in host recognition by arbuscular mycorrhizal fungi. New Phytol 127:703–709, 1994.
62. KS Elias, GR Safir. Hyphal elongation of *Glomus fasciculatus* in response to root exudates. Appl Environ Microbiol 53:1928–1933, 1987.
63. B Balaji, MJ Poulin, H Vierheilig, Y Piché. Responses of an arbuscular mycorrhizal fungus, *Gigaspora margarita*, to exudates and volatiles from the Ri T-DNA-transformed roots of nonmycorrhizal and mycorrhizal mutants of *Pisum sativum* L Sparkle. Exp Mycol 19:275–283, 1995.
64. K Tawaraya, S Watanabe, E Yoshida, T Wagatsuma. Effect of onion (*Allium cepa*) root exudates on the hyphal growth of *Gigaspora margarita*. Mycorrhiza 6:57–59, 1996.
65. H Vierheilig, M Althug, R Engelstreitwolf, P Mader, A Wiemken. Studies on the attractional effect of root exudates on hyphal growth of an arbuscular mycorrhizal

fungus in a soil compartment–membrane system. Plant Soil 203:137–144, 1998.

66. V Gianinazzi-Pearson, B Branzanti, S Gianinazzi. In vitro enhancement of spore germination and early hyphal growth of a vesicular–arbuscular mycorrhizal fungus by host root exudates and plant flavonoids. Symbiosis 7:243–255, 1989.
67. S Chabot, R Bel-Rhlid, R Chênevert, Y Piché. Hyphal growth promotion in vitro of the VA mycorrhizal fungus, *Gigaspora margarita* Becker & Hall, by the activity of structurally specific flavonoid compounds under CO_2-enriched conditions. New Phytol 122:461–467, 1992.
68. R Bel Rhild, S Chabot, Y Piché, R Chenevert. Isolation and identification of flavonoids from Ri T-DNA-transformed roots (*Daucus carota*) and their significance in vesicular–arbuscular mycorrhiza. Phytochemistry 33:1369–1371, 1993.
69. MJ Poulin, R Bel-Rhlid, Y Piché, R Chênevert. Flavonoids released by carrot (*Daucus carota*) seedlings stimulate hyphal development of vesicular–arbuscular mycorrhizal fungi in the presence of optimal CO_2 enrichment. J Chem Ecol 19:2317–2327, 1993.
70. G Bécard, LP Taylor, DD Douds, PE Pfeffer, LW Doner. Flavonoids are not necessary plant signal compounds in arbuscular mycorrhizal symbioses. Mol Plant–Microbe Interact 8:252–258, 1995.
71. M Buee, M Rossignol, A Jauneau, R Ranjeva, G Bécard. The pre-symbiotic growth of arbuscular mycorrhizal fungi is induced by a branching factor partially purified from plant root exudates. Mol Plant–Microbe Interact 13:693–698, 2000.
72. K Tawaraya, K Hashimoto, T Wagatsuma. Effect of root exudate fractions from P-deficient and P-sufficient onion plants on root colonisation by the arbuscular mycorrhizal fungus *Gigaspora margarita*. Mycorrhiza 8:67–70, 1998.
73. P Franken, F Gnädinger. Analysis of parsley arbuscular endomycorrhiza: infection development and mRNA levels of defense-related genes. Mol Plant–Microbe Interact 7:612–620, 1994.
74. P Franken, N Requena. Molecular approaches to arbuscular mycorrhiza functioning. In: B Hock, ed. The Mycota IX. Berlin: Springer-Verlag, 2000, pp 19–28.
75. P Franken, B Bütehorn, G Kuhn, L Lapopin, H Roussel, N Requena, V Gianinazzi-Pearson. RNA accumulation patterns in AM fungi. In: U Ahonen-Jonnarth, E Danell, P Fransson, O Karen, B Lindahl, I Rangel, R Finlay, eds. Abstracts of the Second International Conference on Mycorrhizae. Uppsala, Sweden: SLU Service/Repro, 1998.
76. N Requena. Mycorrhizal interactions in the rhizosphere. In: A Varma, ed. Microbes: For Health, Wealth and Sustainable Environment. New Delhi: Malhotra Publishing House, 1998, pp 725–738.
77. N Requena, P Füller, P Franken. Molecular characterisation of *GmFOX2*, an evolutionary highly conserved gene from the mycorrhizal fungus *Glomus mosseae*, down-regulated during interaction with rhizobacteria. Mol Plant–Microbe Interact 12:934–942, 1999.
78. JF Carstensen, JG Tesdorpf, M Kaufmann, MM Markus, B Husen, F Leenders, F Jakob, Y Delaunoit, J Adamski. Characterization of 17-β-hydroxysteroid dehydrogenase IV. J Endocrinol 150:S3–S12, 1996.

79. M-J Poulin, J Simard, J-G Catford, F Labrie, Y Piche. Responses of symbiotic endomycorrhizal fungi to estrogens and antiestrogens. Mol Plant–Microbe Interact 10:481–487, 1997.
80. N Requena, P Mann, P Franken. A homologue of the cell-cycle check-point TOR2 from *Saccharomyces cerevisiae* exists in the arbuscular mycorrhizal fungus *Glomus mosseae*. Protoplasma 212:89–98, 2000.
81. RC Staples, V Macko. Formation of infection structures as a recognition response in fungi. Exp Mycol 4:2–16, 1980.
82. JE Hamer, NJ Talbot. Infection-related development in the rice blast fungus *Magnaporthe grisea*. Curr Opin Microbiol 1:693–697, 1998.
83. M Giovannetti, L Avio, C Sbrana, AS Citernesi. Factors affecting appressorium development in the vesicular–arbuscular mycorrhizal fungus *Glomus mossese*. New Phytol 123:115–122, 1993.
84. V Gianinazzi-Pearson, S Ginaninazzi. Influence of intergeneric grafts between host and non-host legumes on formation of vesicular-arbuscular mycorrhiza. New Phytol 120:505–508, 1992.
85. G Nagahashi, DD Douds Jr. Appressorium formation by AM fungi on isolated cell walls of carrot roots. New Phytol 136:299–304, 1997.
86. NP Money, RJ Howard. Confirmation of a link between fungal pigmentation, turgor pressure, and pathogenicity using a new method of turgor measurement. Fungal Genet Biol 20:217–227, 1996.
87. G Duc, A Trouvelot, V Gianinazzi-Pearson, S Gianinazzi. First report of non-mycorrhizal mutants (myc-) obtained in pea (*Pisum sativum* L.) and fababean (*Vicia faba* L.). Plant Sci 60:215–222, 1989.
88. M Sagan, D Morandi, E Tarenghi, G Duc. Selection of nodulation and mycorrhizal mutants in the model plant *Medicago trunculata* (Gaertn) after gamma-ray mutagenesis. Plant Sci 111:63–71, 1995.
89. V Gianinazzi-Pearson, S Gianinazzi. Morphological interactions and functional compatibility between symbionts in vesicular arbuscular endomycorrhizal associations. In: S Scannerini, ed. Cell to Cell Signals in Plant, Animal and Microbial Symbiosis. Berlin: Springer-Verlag, 1988, pp 73–84.
90. FA Smith, SE Smith. Tansley review No. 96 Structural diversity in (vesicular)–arbuscular mycorrhizal symbioses. New Phytol 137:373–388, 1997.
91. J Jacquelinet-Jeanmougin, V Gianinazzi-Pearson, S Gianinazzi. Endomycorrhiza in the Gentianaceae I. The fungi associated with *Gentiana lutea* L. New Phytol 95: 663–666, 1987.
92. C Walker. AM or VAM: what's in a word? In: A Varma, B Hock, eds. Mycorrhiza: Structure, Function, Molecular Biology and Biotechnology. Berlin: Springer-Verlag, 1995, pp 25–26.
93. J Jacquelinet-Jeanmougin, V Gianinazzi-Pearson, S Gianinazzi. Endomycorrhiza in the Gentianaceae II. Ultrastructural aspects of symbionts relationship in *Gentiana putea* L. Symbiosis 3:269–286, 1987.
94. I Garcia-Romera, JM Garcia-Garrido, JA Ocampo. Pectolytic enzymes in the vesicular–arbuscular mycorrhizal fungus *Glomus mosseae*. FEMS Lett 78:343–346, 1991.
95. I Garcia-Romera, JM Garcia-Garrido, JA Ocampo. Cellulase production by the

vesicular–arbuscular mycorrhizal fungus *Glomus mosseae* (Nicol. & Gerd.) Gerd. and Trappe. New Phytol 121:221–226, 1992.
96. A Rejon-Palomares, JA Ocampo, I Garcia-Romera. Production of xyloglucanase enzyme by different *Glomus* strains. In: U Ahonen-Jonnarth, E Danell, P Fransson, O Karen, B Lindahl, I Rangel, R Finlay, eds. Abstracts of the Second International Conference on Mycorrhizae. Uppsala, Sweden: SLU Service/Repro, 1998.
97. R Perotto, V Bettini, F Favaron, P Alghisi, P Bonfante. Polygalacturonase activity and location in arbuscular mycorrhizal roots of *Allium porrum* L. Mycorrhiza 5: 157–163, 1995.
98. JM Garcia-Garrido, M Tribak, A Rejon-Palomares, JA Ocampo, I Garcia-Romera. Hydrolytic enzymes and ability of arbuscular mycorrhizal fungi to colonize roots. J Exp Bot 51:1443–1448, 2000.
99. JM Garcia-Garrido, I Garcia-Romero, MD Parra-Garcia, JA Ocampo. Purification of an arbuscular mycorrhizal endoglucanase from onion roots colonized by *Glomus mosseae*. Soil Biol Biochem 28:1443–1449, 1996.
100. ML Gaspar, R Pollero, M Cabello. Partial purification and characterization of a lipolytic enzyme from spores of the arbuscular mycorrhizal fungus *Glomus versiforme*. Mycologia 89:610–614, 1997.
101. MZ Solaiman, M Saito. Use of sugars by intraradical hyphae of arbuscular mycorrhizal fungi revealed by radiorespirometry. New Phytol 136:533–538, 1997.
102. M Saito. Regulation of arbuscular mycorrhiza symbiosis—hyphal growth in host roots and nutrient exchange. Jpn Agr Res Q 31:179–183, 1997.
103. M Saito. Enzyme activities of the internal hyphae and germinated spores of an arbuscular mycorrhizal fungus, *Gigaspora margarita*, Becker & Hall. New Phytol 129:425–431, 1995.
104. MZ Solaiman, T Ezawa, T Kojima, M Saito. Polyphosphates in intraradical and extraradical hyphae of an arbuscular mycorrhizal fungus, *Gigaspora margarita*. Appl Environ Microbiol 65:5604–5606, 1999.
105. M Saito, MZ Solaiman. Phosphate efflux from intraradical hyphae of arbuscular mycorrhiza in vitro and their implication on P translocation. In: HC Weber, S Imhof, D Zeuske, eds. Programs, Abstracts and Papers of the Third International Congress on Symbiosis. Marburg, Germany: Philipps University, 2000, p 181.
106. SH Burleigh, MJ Harrison. A cDNA from the arbuscular mycorrhizal fungus *Glomus versiforme* with homology to a cruciform DNA binding protein from *Ustilago maydis*. Mycorrhiza 7:301–306, 1998.
107. G Delp, SJ Barker, SE Smith. Isolation by differential display of three partial cDNAs potentially coding for proteins from the VA mycorrhizal fungus *G. intraradices*. Mycol Res 104:293–300, 2000.
108. L Lapopin, V Gianinazzi-Pearson, P Franken. Comparative differential display analysis of arbuscular mycorrhiza in *Pisum sativum* and a mutant defective in late stage development. Plant Mol Biol 41:669–677, 1999.
109. LA Harrier, F Wright, JE Hooker. Isolation of the 3-phosphoglycerate kinase gene of the arbuscular mycorrhizal fungus *Glomus mosseae* (Nicol. & Gerd.) Gerdemann & Trappe. Curr Genet 34:386–392, 1998.
110. L Harrier, J Sawczak. Detection of the 3-phosphoglycerate kinase protein of *Glomus mosseae*. Mycorrhiza 10:81–86, 2000.

111. LA Harrier. Isolation and sequence analysis of the arbuscular mycorrhizal fungus *Glomus mosseae* 3-phosphoglycerate kinase gene promoter region. DNA SEQ 11: 463–473, 2001.
112. L Paterson, LA Harrier. Isolation of carbon sensing gene homologues the arbuscular mycorrhizal fungus *Glomus mosseae*. In: HC Weber, S Imhof, D Zeuske, eds. Programs, Abstracts and Papers of the Third International Congress on Symbiosis. Marburg, Germany: Philipps University, 2000, p 169.
113. SE Smith, V Gianinazzi-Pearson. Phosphate uptake and vesicular–arbuscular activity in mycorrhizal *Allium cepa* L.: effect of photon irradiance and phosphate nutrition. Aust J Plant Physiol 17:177–188, 1990.
114. V Gianinazzi-Pearson, SE Smith, S Gianinazzi, FA Smith. Enzymatic studies on the metabolism of vesicular–arbuscular mycorrhizas. V. Is H^+-ATPase a component of ATP-hydrolysing enzyme activities in plant-fungus interfaces? New Phytol 117: 61–74, 1991.
115. M Kaldorf, E Schmelzer, H Bothe. Expression of maize and fungal nitrate reductase genes in arbuscular mycorrhiza. Mol Plant–Microbe Interact 11:439–448, 1998.
116. V Gianinazzi-Pearson, MC Lemoine, C Arnould, A Gollotte, JB Morton. Localization of β(1–3) glucans in spore and hyphal walls of fungi in the Glomales. Mycologia 86:478–485, 1994.
117. P Bonfante-Fasolo, A Faccio, S Perotto, A Schubert. Correlation between chitin distribution and cell wall morphology in the mycorrhizal fungus *Glomus versiforme*. Mycol Res 94:157–165, 1990.
118. MC Lemoine, A Gollotte, V Gianinazzi-Pearson. Localization of β(1–3) glucan in walls of the endomycorrhizal fungi *Glomus mosseae* (Nicol. and Gerd.) Gerd. and Trappe and *Acaulospora laevis* Gerd. and Trappe during colonization of host roots. New Phytol 129:97–105, 1995.
119. L Lanfranco, ML van Buuren, L Longato, L Garnero, MJ Harrison, P Bonfante. Chitin synthase genes in arbuscular mycorrhizal fungi (*Glomus versiforme* and *Gigaspora margerita*). In: TM Szaro, TD Bruns, eds. Abstracts of the First International Conference on Mycorrhizae. Berkeley: University of California, 1996, pp 74.
120. L Lanfranco, M Vallino, P Bonfante. Expression of chitin synthase genes in the arbuscular mycorrhizal fungus *Gigaspora margarita*. New Phytol 142:347–354, 1999.
121. GJ Bethlenfalvay, M Reyes-Solis, SB Camel, R Ferrera-Cerrato. Nutrient transfer between the root zones of soybean and maize plant connected by a common mycorrhizal mycelium. Physiol Plant 82:423–432, 1991.
122. DJ Read. Biodiversity—plants on the web. Nature 396:22–23, 1998.
123. B Bago, C Azconaguilar, A Goulet, Y Piche. Branched absorbing structures (BAS)—a feature of the extraradical mycelium of symbiotic arbuscular mycorrhizal fungi. New Phytol 139:375–388, 1998.
124. MJ Harrison, ML van Buuren. A phosphate transporter from the mycorrhizal fungus *Glomus versiforme*. Nature 378:626–629, 1995.
125. FA Smith, I Jakobsen, SE Smith. Spatial differences in acquisition of soil phosphate between two arbuscular mycorrhizal fungi in symbiosis with *Medicago truncatula*. New Phytol 147:357–366, 2000.
126. H Sawaki, M Saito. Expressed genes in extraradical hyphae of an arbuscular mycor-

rhizal fungus, *Glomus intraradices* in symbiotic phase. FEMS Microbiol Let 195: 109–113, 2001.

127. J Nielsen. Differential mRNA display and temporal expression analysis of extraradical AMF hyphae under different P-nutrition regimes. In: HC Weber, S Imhof, D Zeuske, eds. Programs, Abstracts and Papers of the Third International Congress on Symbiosis. Marburg, Germany: Philipps University, 2000, p 158.
128. V Bianciotto, P Bonfante. Presymbiotic versus symbiotic phase in arbuscular endomycorrhizal fungi. In: A Varma, B Hock, eds. Mycorrhiza. Heidelberg, Germany: Springer-Verlag, 1998, pp 229–251.
129. G Becard, JA Fortin. Early events of vesicular-arbuscular mycorrhiza formation on Ri T-DNA transformed roots. New Phytol 108:211–218, 1988.
130. S Declerck, D Strullu, C Plenchette, T Guillemette. Entrapment of in vitro produced spores of *Glomus versiforme* in alginate beads—in vitro inoculum potentials. J Biotechnol 48:51–57, 1996.
131. S Declerck, DG Strullu, C Plenchette. Monoxenic culture of the intraradical forms of *Glomus* sp. isolated from a tropical ecosystem—a proposed methodology for germplasm collection. Mycologia 90:579–585, 1998.
132. TE Pawlowska, DD Douds, I Charvat. In vitro propagation and life cycle of the arbuscular mycorrhizal fungus *Glomus etunicatum*. Mycol Res 103:1549–1556, 1999.
133. V Karandashov, I Kuzovkina, HJ Hawkins, E George. Growth and sporulation of the arbuscular mycorrhizal fungus *Glomus caledonium* in dual culture with transformed carrot roots. Mycorrhiza 10:23–28, 2000.
134. A Rousseau, N Benhamou, I Chet, Y Piché. Mycoparasitism of the extramatrical phase of *Glomus intraradices* by *Trichoderma harzianum*. Phytopathology 86:434–443, 1996.
135. H Green, J Larsen, PA Olsson, DF Jensen, I Jakobsen. Suppression of the biocontrol agent *Trichoderma harzianum* by mycelium of the arbuscular mycorrhizal fungus *Glomus intraradices* in root-free soil. Appl Environ Microbiol 65:1428–1434, 1999.

14

Pathogenic Development in *Ustilago maydis*

A Progression of Morphological Transitions That Results in Tumor Formation and Teliospore Production

Flora Banuett

University of California–San Francisco, San Francisco, California

1 INTRODUCTION

1.1 General Comments

Ustilago maydis (DeCandole) Corda is a basidiomycete fungus belonging to the class Ustilaginales, the smut fungi. This group consists of plant pathogenic fungi that attack >75 families of flowering plants, both dicots and monocots [1,2]. Smut diseases of monocots are the best known since they affect cereal grains (wheat, barley, sorghum, oats, maize, rice) and other monocots of economic importance such as sugar cane. *U. maydis* is the etiological agent of corn smut disease, or huitlacoche (as has been known since ancient times by the inhabitants of Mexico). There are only two known hosts of the fungus [3]: maize (*Zea mays* L.) and teosinte (*Zea mays* ssp. *parviglumis* and ssp. *mexicana*). The disease is characterized by tumors that occur in all aerial plant parts. *U. maydis* is related to the rusts, a group of plant pathogenic fungi of the class Uredinales, and also to *Cryptococus neoformans (Filobasidiella neoformans*), an opportunistic pathogenic fungus of immunocompromised patients (this volume, Chapter 19 by Lengeler and Heitman). It is more distantly related to the Tremellales and to other basidiomycetes such as *Schizophyllum commune* and *Coprinus cinereus*, two homobasidiomycetes which have been extensively studied [reviewed in 4].

U. maydis has been the subject of numerous studies since the late 1880s by Brefeld and others [for references see 3,5]. It was intensively studied from 1930 to 1950 at the University of Minnesota by pioneers such as Hanna, Kernkamp, Christensen, Rowell, and DeVay [see 3]. The last two demonstrated the genetic basis for mating-type specificity, a true genetics classic (see Sec. 3). Later, studies of Holliday [6,7], Puhalla [8,9], Day and Anagnostakis [10], and Day et al. [11] rekindled interest in this fungus. Development of the first *E. coli–U. maydis* shuttle vector and a transformation procedure for introduction of exogenous DNA [12] and demonstration of gene disruption by homologous recombination [13,14] ushered in the era of molecular genetics in *U. maydis*.

1.2 Useful Features

U. maydis is one of the most genetically tractable fungal pathogens, and some of the features that characterize it are listed below [see reviews by 7,15,16]:

1. *U. maydis* has a unicellular haploid phase with a fast generation time (120 min in rich medium) comparable to that of *S. cerevisiae* (90 min) and *S. pombe* (120 min), the best-studied microbial eukaryotes. *U. maydis* haploids, like *S. cerevisiae* and *S. pombe*, form compact colonies on different solid media, allowing the application of standard microbiological techniques. The unicellular form also grows well in liquid media at a range of temperatures, with a maximum of ~35°C (Banuett, unpublished).
2. *U. maydis* has a small genome size (~20 Mb), making it possible to sequence its entire genome in a short time and to apply genome wide strategies in its study.
3. The haploid phase can be mutagenized with chemical mutagens or UV irradiation. Mutants exist affecting biosynthetic pathways, DNA recombination, siderophore biosynthesis, signaling, growth in the plant, pathogenicity, teliospore germination, motor proteins, transcription factors, sensitivity to pesticides, chitin synthases, and morphology of the haploid form.
4. DNA-mediated transformation with two types of shuttle vectors, integrating and autonomously replicating, has made it possible to clone genes by functional complementation.
5. The existence of several dominant selectable markers makes it possible to introduce multiple knockouts into the same cell. These markers are phleomycin (*ble*), hygromycin (*hyg*), carboxin (*cbx*), and noursethricin (*nat*).
6. Homologous recombination allows the generation of null mutations or

other mutations at the wild-type locus. Gene replacement by homologous recombination can occur with a frequency of 20–65%. Thus, gene knockouts can be generated with relative ease.

7. The existence of meiosis allows segregation analysis and the generation of double and triple mutant strains by crosses.

8. Stable diploids can be easily constructed, allowing dominance/recessiveness tests to be carried out. Diploids also allow testing the role of different genes in postfusion events and generation and maintenance of deleterious mutations in a heterozygous condition.

9. The fungus completes its life cycle in any plant part, unlike other smut fungi, which require sexual maturity of their hosts to complete their life cycles. Under controlled environmental conditions, the life cycle can be completed in 2–3 weeks. [For some representative references, see 17–37.]

This chapter is not intended to be a comprehensive review of *U. maydis*. Because of space limitations I can only do justice to some aspects of the work on this fungus, in particular, to signaling and interaction with the host. Therefore, I apologize to *U. maydis* colleagues whose important contributions have not been included. Because this section of the book deals with pathogenic development, I will start with a description of the infectious process and then review our current knowledge of the mating-type loci. I next review signaling, a critical aspect of the interaction with the host and of interaction among *Ustilago* cells. Lastly, I describe the identification of genes for filamentous growth and for interaction with the plant and describe a recent development that allows formation of teliosporelike structures in a maize callus system. This in vitro system in combination with microarray technology will likely accelerate the discovery of genes required for interaction of the fungus with the plant.

2 LIFE CYCLE AND THE INFECTIOUS PATHWAY

2.1 Life Cycle in Brief

Three major cell types characterize the life cycle of *U. maydis* (Figs. 1 and 2) [reviewed in 3,15,38]: a haploid unicellular form that is nonpathogenic; a dikaryotic filamentous form that is pathogenic; and a round diploid cell, the teliospore, that undergoes meiosis to produce the haploid cell. These morphological transitions entail conjugation, karyogamy, and meiosis, respectively, and are accompanied by changes in growth habit, ploidy, and pathogenicity. Therefore, pathogenicity, morphology, and growth habit are intimately intertwined processes. Understanding the life cycle of *U. maydis* entails understanding the mechanisms that regulate these transitions.

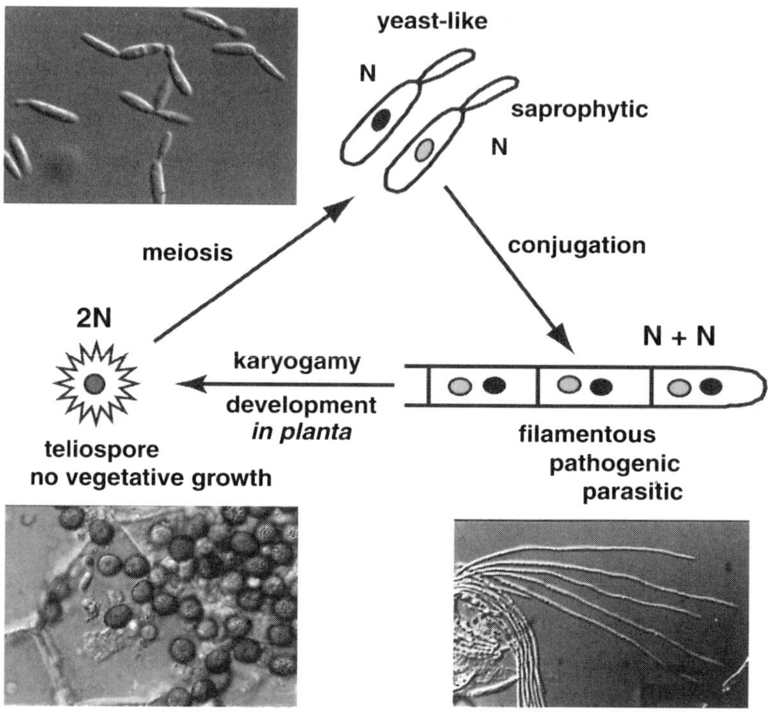

Figure 1 Morphological transitions in the life cycle of *U. maydis*. The life cycle of *U. maydis* is characterized by three basic forms (see text): (1) a unicellular haploid form that is nonpathogenic; (2) a filamentous dikaryotic form that is pathogenic; and (3) a round cell, the teliospore, that undergoes meiosis to produce the haploid form. These morphological transitions entail conjugation, karyogamy, and meiosis, and are regulated by the mating-type loci, by environment, by nutrients, and by plant signals (see text for details and references). The insets show photographs (clockwise beginning at top left corner) of the unicellular form budding, the filamentous form growing in vitro, and teliospores within plant cells. Hyphae do not branch in culture but branch profusely while growing in the plant (see Fig. 4).

2.2 The Infectious Pathway

The infectious process of *U. maydis* involves a progression of events that results in tumor induction and the production of teliospores (see Fig. 2). It is regulated by the mating-type loci, environmental conditions, nutrition, and the host plant.

The *U. maydis* infectious pathway can be divided into the following steps (Fig. 3) [5,39]:

1. Fusion of haploid cells, generation of the dikaryotic form, hyphal branching, and proliferation of the hyphae on the leaf surface.
2. Invasion of host cells by hyphae and proliferation of the fungus within host cells.
3. Alteration of growth control of host—induction of tumors.
4. Production of mucilaginous material, karyogamy, hyphal fragmentation, and cell rounding—reorganization of the machinery for polarized growth and secretion.
5. Deposition of a specialized cell wall and production of mature teliospores.
6. Teliospore germination, meiosis, and generation of the haploid form.

Different alleles at two unlinked mating-type loci (***a*** and ***b***) are necessary for completion of the life cycle (see Sec. 3) [reviewed in 7,15,16,38]. The existence of a saprophytic unicellular haploid form and the ability to construct diploids, which mimic the behavior of the dikaryon, have made it possible to study the early steps of the life cycle outside the plant and to determine the input of the mating-type loci in these events. These studies have shown that fusion of haploid cells is regulated by the ***a*** locus and that filamentous growth of the dikaryon is strictly dependent on the ***b*** locus. Filamentous growth in vitro requires also the ***a*** locus (see Sec. 3).

2.2.1 Fusion of Haploid Cells and Generation of the Filamentous Dikaryotic Form

In the laboratory, maize seedlings 5–10 days old are inoculated with a mixture of haploid strains that carry different ***a*** and ***b*** alleles using a hypodermic syringe. Cells form conjugation tubes [5] like those described by Snetselaar [40] and Snetselaar et al. [41] on water agar or by Banuett and Herskowitz [42] in low-nitrogen medium (see Sec. 3.1). Cell fusion occurs readily, as soon as 8 h after inoculation, and dikaryotic filaments arise from the clumps of yeastlike cells and proliferate on the leaf surface, attaining a substantial mass by 24 h. The hyphae branch as they grow (Fig. 3). Branching is observed as early as the hyphae emerge from the clusters of yeastlike cells [see 5 and references therein]. The fact that hyphae branch on the leaf surface is striking because hyphae in culture do not branch. These observations suggest that either the leaf surface or a plant hormone induces branching [5]. The leaf surface is known to exert an important control on development of several fungi [see, e.g., 43].

The first symptom of infection is chlorosis, a yellowing of the green tissue, which can be observed as early as 24 h after inoculation [3,5]. It has been reported that the fungus produces a toxin responsible for this symptom [44]. In maize varieties of the appropriate genotype, *U. maydis* induces production of anthocyanin pigmentation. Microscopic observation reveals pigmented cells but no fungal

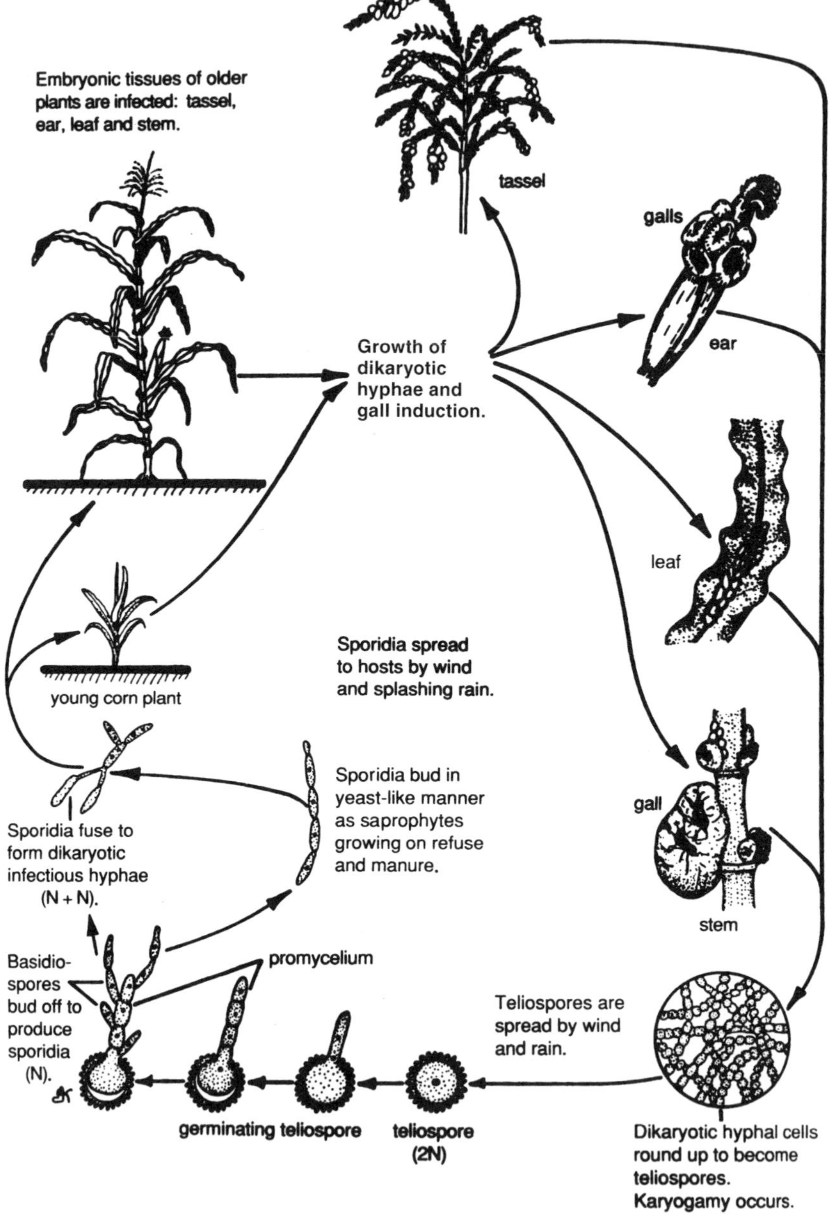

Embryonic tissues of older plants are infected: tassel, ear, leaf and stem.
tassel
galls
ear
Growth of dikaryotic hyphae and gall induction.
leaf
Sporidia spread to hosts by wind and splashing rain.
young corn plant
Sporidia bud in yeast-like manner as saprophytes growing on refuse and manure.
gall
stem
Sporidia fuse to form dikaryotic infectious hyphae (N + N).
Basidio-spores bud off to produce sporidia (N).
promycelium
Teliospores are spread by wind and rain.
germinating teliospore
teliospore (2N)
Dikaryotic hyphal cells round up to become teliospores. Karyogamy occurs.

hyphae, suggesting that the induction occurs at a distance [5,45]. The anthocyanin pigment may be part of a defense response of maize to fungal invasion.

2.2.2 Penetration of Host Cells by Hyphae

Penetration can occur by means of an appressoriumlike structure, directly through the plant cells, or in between plant cells [39,45]. Penetration can also occur through wounds [46] or stomata [5,42,47]. It is not known if the fungus produces enzymes that digest the cell wall at the point of penetration or if it uses mechanical force or both. Recent work by Cano-Canchola [48] indicates that there is differential induction of lytic enzymes associated with different phases of growth: pectate lyase and cellulase are induced by apical meristem in a pathogenic diploid strain (heterozygous for ***a*** and ***b***), whereas leaves induce xylanase and cellulase in a nonpathogenic haploid strain. An increase in pectate lyase accompanies chlorosis and teliospore production; increased expression of polygalacturonase accompanies anthocyanin production, tumors, and teliospore formation. The role of these enzymatic changes in symptom development remains to be elucidated.

2.2.3 Proliferation of the Fungus Within Host Cells

Alteration of Growth Control of Host and Formation of Tumors. Once inside plant cells, the hyphae continue branching and traverse from cell to cell (Fig. 4), increasing the fungal mass [5,39]. Growth can be intra- and intercellular [39,49]. Electron microscopic studies [39] show that hyphae do not break the host cell plasma membrane but rather appear to be in close apposition to it. Thus, exchange of nutrients and metabolites must occur across this boundary. The relationship of *U. maydis* with its hosts, maize and teosinte, is almost a commensal one. It is only at the time of teliospore production that the fungal mass replaces the plant tumor tissue [3].

FIGURE 2 Life cycle of *U. maydis*. The life cycle of *U. maydis* begins with fusion of haploid cells on the leaf surface and formation of the pathogenic dikaryon. Proliferation of this filamentous dikaryon leads to tumor formation on all aerial parts of the maize plant. Tumors on the vegetative parts cause stunting, and those on the ear reduce yield severely. Within the tumors, the hyphae undergo several morphological changes which result in formation of mature teliospores. Teliospores are dispersed by wind and water. Upon landing on a leaf surface, they germinate by formation of a short filament, the promycelium, where meiosis takes place. The teliospore can also start an infectious cycle without undergoing meiosis (see [3]). It is hypothesized that reciprocal signalling governs the interaction of *U. maydis* with its hosts, maize and teosinete. (From Ref. 1.)

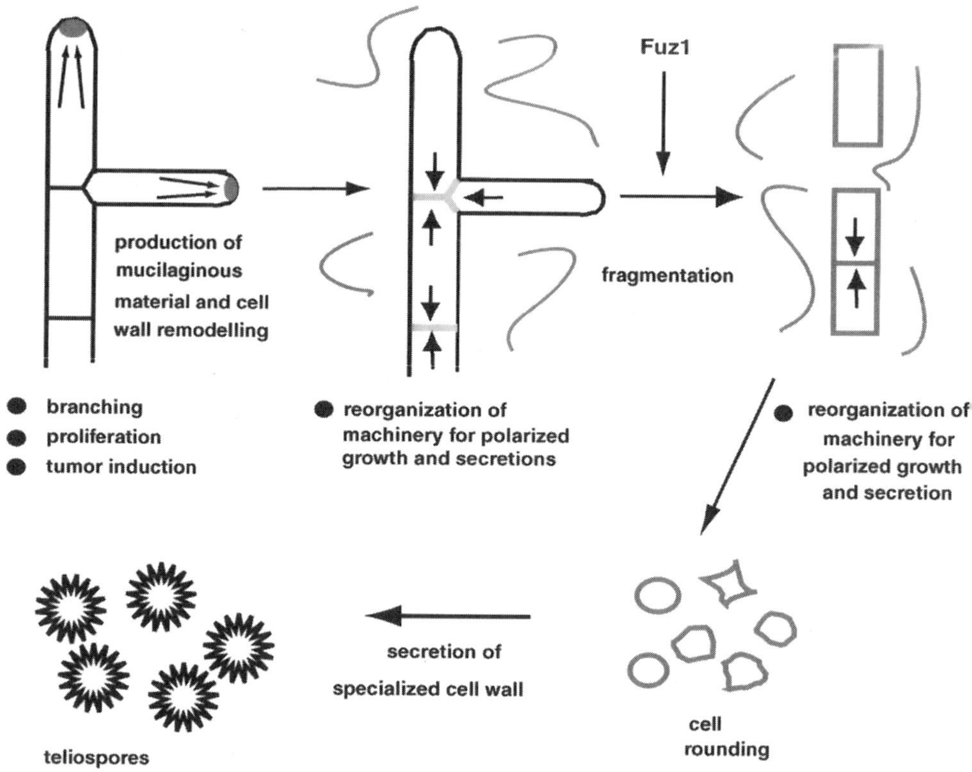

FIGURE 3 The infectious cycle—a developmental program regulated by plant signals. Several discrete steps characterize the developmental program that ensues upon fusion of two haploid cells to produce the pathogenic filamentous dikaryon: proliferation and branching of hyphae on the leaf surface followed by invasion of host cells and proliferation within the plant cells; tumor induction and production of mucilaginous material within which the hyphae undergo fragmentation. Cylindrical cells, produced by fragmentation, undergo cell rounding and a specialized cell wall is deposited, to produce the mature teliospore. The machinery for polarized growth and secretion may be reorganized during these morphological transitions under the influence of plant signals. Execution of this developmental program is strictly dependent on the plant and suggests that plant signals are crucial for fungal development. (See text for details.)

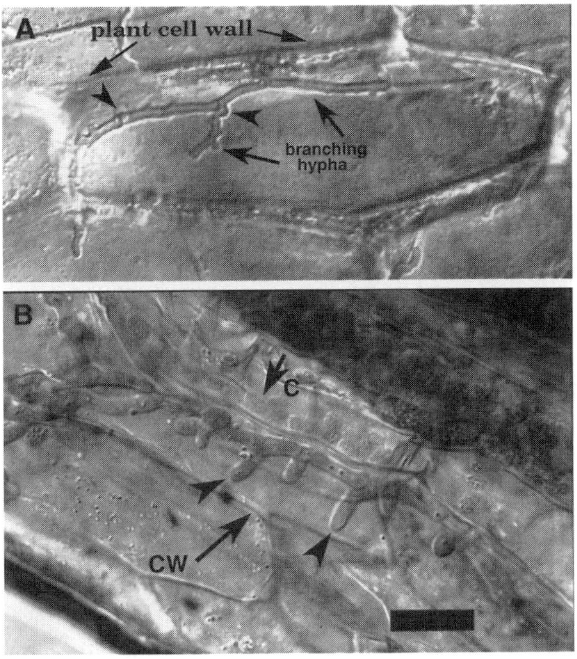

FIGURE 4 Growth of hyphae in the plant. *U. maydis* hyphae branch profusely when growing in the plant, in contrast with growth in culture, where no branching occurs, suggesting that plant signals may influence hyphal branching. **(A)** A single plant cell with a hypha showing several branch primordia (arrowheads). This hypha arose after inoculation of maize seedlings with wild-type strains. **(B)** Hyphae formed after inoculation with *fuz1*$^{-}$ strains. Their appearance is no different from that of wild-type hyphae. Arrowheads point to branches. CW, cell wall of host cell, C, chloroplasts. Magnification in **A** and **B** is different. (From Ref. 5.)

Proliferation within the host cells leads to tumor induction. The trigger that causes the fungus to induce tumors is not known but may involve activation of a MAPK cascade pathway. Tumor formation can be discerned as early as 3 days but normally at 5 days postinfection [5]. Tumors enlarge and continue to form on different plant parts: leaf sheath, leaf blade, stems, and floral parts. Tumor formation is not synchronous: some arise early, others late. Infection of floral parts can lead to an interesting phenomenon in which tassels also produce kernels, and ears also produce tassels [3]. Thus, the fungus may alter the normal hormonal control of flowering, just as it may alter the normal hormonal control of cell growth and division during vegetative growth. Tumors result from changes in cell enlargement and cell division of the host cells [5,49], suggesting a veritable transformed state. However, it has not been demonstrated that these cells are capable of hormone-independent and fungus-independent growth in culture.

The Mechanism by Which the Fungus Induces Tumors. The mechanism by which the fungus alters the growth control of the host is not known. Several plausible mechanisms have been proposed [see 15]: (1) The fungus alters host hormonal control by production of auxins and cytokinins and perhaps other plant hormones; (2) the fungus activates endogenous plant reserves of hormones, resulting in an increased level or imbalance of auxin and cytokinins; or (3) the fungus transfers a segment of DNA that codes for hormones or activators of inactive sources of plant hormones. Several reports indicate that *U. maydis* produces both auxin and cytokinin in culture [50–55]. However, production of these hormones in culture by *U. maydis* is not indicative of their role in tumor induction. Many microorganisms inhabiting the rhizosphere and the phyllosphere produce plant hormones in culture yet do not induce tumors.

The pathway by which *U. maydis* produces the auxin indole acetic acid (IAA) is not known. In plants, IAA is synthesized from tryptophan by several different pathways [56]. One pathway involves indole-3 acetaldehyde dehydrogenase, which catalyzes conversion of indole acetic aldehyde (IAAld) to IAA. Basse et al. [54] reported the identification of two potential indole-3 acetaldehyde dehydrogenases in *U. maydis*, Iad1 and Iad2. The purified *U. maydis* enzymes catalyze the same reaction as the plant enzymes. The gene for one of the *U. maydis* enzymes, *iad1*, was cloned. Deletion of this gene does not affect production of IAA in culture by *U. maydis*, suggesting redundancy in pathways for IAA production. Deletion of *iad1* does not affect filamentous growth or pathogenicity [54].

Recent work by Sosa-Morales et al. [55] reports the isolation of *U. maydis* mutants with diminished IAA synthetic capacity in culture. Disruption of the genes responsible for hormone biosynthesis will allow a critical evaluation of the role of hormone production in tumor induction by *U. maydis*.

2.2.4 Production of Mucilaginous Material, Karyogamy, Hyphal Fragmentation, and Cell Rounding—Reorganization of the Machinery for Polarized Growth and Secretion

As the tumors enlarge, a mucilaginous material is produced in the tumors. The fungal hyphae are embedded in this material and appear bloated and twisted, not straight as in previous stages. The hyphae branch profusely, some of their tips have a lobed appearance, and most remarkably, the hyphae fragment (Figs. 3 and 5A) [5]. Calcofluor staining clearly shows the presence of fragments composed of two, three, or four cylindrical cells. Fragmentation occurs within the tumorous cells [5], although some researchers [39,47] have reported this process to occur

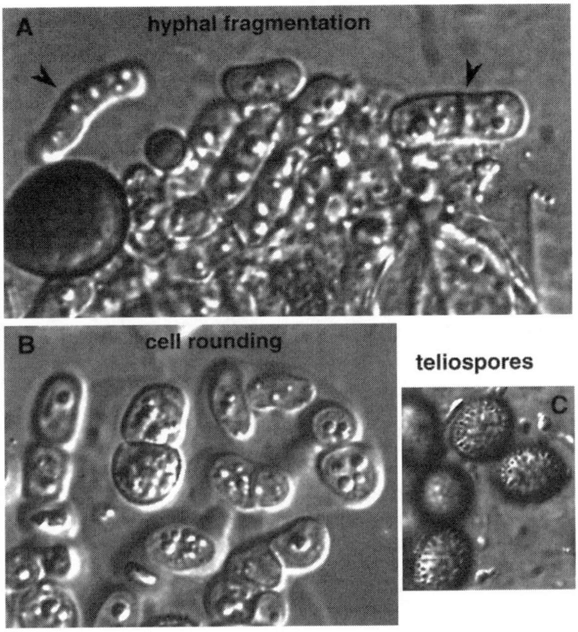

FIGURE 5 Hyphal fragmentation and cell rounding. Wild-type hyphae become embedded in a mucilaginous material within the tumors. They fragment and the cylindrical fragments undergo cell rounding and deposit a specialized cell wall resulting in mature teliospores. (For details, see text.) **(A)** Cylindrical hyphal fragments (arrowheads). The black round spot at the bottom left is a mature teliospore. **(B)** The individual cells produced by fragmentation are seen in the process of cell rounding. These cells are embedded in a mucilaginous material. **(C)** Mature teliospores. Note the echinulated cell wall. Magnification in the three panels is different.

in between cells. Karyogamy appears to occur just before fragmentation and marks the beginning of the diploid phase.

The origin of the mucilaginous material is not known. It may serve as an osmotic stabilizer during the morphological transitions that ensue, which are accompanied by extensive cell wall remodeling. A weak cell wall may render cells susceptible to lysis in the aqueous environment of the host cell.

The fragmentation process suggests a reorganization of the secretory machinery from the hyphal tip, where growth and secretion normally take place, to the septal region, where enzymes required for dissolution of the septa and for new cell wall assembly are needed during fragmentation (Fig. 3). Single cells produced by fragmentation become rounded (Figs. 3 and 5B) [5]. The transition from a cylindrical to a rounded cell entails a change to uniform secretion as the cell wall is loosened and remodeled.

I suggest that the different morphological changes that ensue from the time of hyphal fragmentation to production of the round teliospore are a consequence of reorganization of the machinery responsible for polarized growth and secretion and that this reorganization is regulated by plant signals.

2.2.5 Deposition of a Specialized Cell Wall and Production of Mature Teliospores

A yellow-brown cell wall becomes apparent as the cells become rounded. This wall becomes dark brown and echinulated (with spikes) as the process continues (Figs. 3 and 5C) [5,39]. The cell wall of the teliospore is very different from that of the hypha or the yeast cell–type [57,58]; it is thicker and apparently consists of three layers [58]. This specialized wall protects the teliospore against adverse environmental conditions. Teliospores can survive in the soil for many years and are a very important source of inoculum in nature [see 3]. They can germinate directly on the leaf surface and start the infectious process.

2.2.6 Teliospore Germination, Meiosis, and Generation of the Haploid Form

Germination of the teliospore entails breakdown of the cell wall and formation of a short filament, the promycelium (Fig. 2). The nucleus migrates to this filament and undergoes meiosis [59,60]. After meiosis takes place, the promycelium is divided into three or four cylindrical compartments, each being the primary meiotic product. Haploid yeastlike cells are produced from each of these compartments by mitotic divisions, thereby completing the life cycle (Fig. 2) [59,60; reviewed in 3].

Competence to undergo meiosis is acquired by passage through the plant. Little is known about the signals involved in triggering germination and conferring competence for meiosis. In ascomycetes and basidiomycetes, karyogamy occurs in a specialized cell and is, in most cases, followed immediately by meiosis [see 61]. Karyogamy may be the trigger for initiation of meiosis in *U. maydis*, but further steps of meiosis may be arrested by a plant signal that inactivates, for example, a cyclin-dependent kinase (Cdk) or a MAPK pathway necessary for completion of meiosis. Upon germination, signals such as nutrients or osmoticum or the leaf surface may relieve the block and allow completion of meiosis.

2.2.7 The *a* Locus and Pathogenicity

Different ***a*** and ***b*** alleles are necessary for completion of the life cycle (see Sec. 3) [reviewed in 7,15,16,38]. The role of these loci, independently of each other, in filamentous growth and pathogenicity was analyzed unequivocally with a set of isogenic diploid strains [62]. This work shows that an ***a***= ***b***≠ diploid strain is severely defective in formation of filaments in vitro (Table 1). Inoculation of

TABLE 1 Properties of Diploids and Haploids Used in the Analysis of Filamentous Growth and Pathogenicity

Strain Genotype/ designation	Filament formation		Tumor Induction	Ploidy[b]	Filament formation after pheromone treatment	Ref.
	In vitro	In planta				
***a*≠ *b*≠**[a] (FBD12)	+	+	+	2N	NA	62
***a*≠ *b*=**[a] (FBD12-3)	–	–	–	2N	–	62,74,76
***a*= *b*≠**[a] (FBD12-17)	–	+	+	2N	+	32,62,74,76
***a*= *b*=**[a] (FBD12-174)	–	–	–	2N	ND	42
a1 mfa2 bW1 bE2 (SG200)	+	ND	+	N	NA	32
a1 bW2 bE1 (CL13)	–	ND	+	N	+	32

NA, not applicable; ND, not determined.
[a] Several different isolates with the same genotype are available. These diploid strains are isogenic derivatives of FBD12 or FBD11.
[b] 2N = diploid strain; N = haploid strain.

plants with this strain leads to tumor induction that is no different from the response exhibited by inoculation with a diploid strain heterozygous at both loci (***a*≠ *b*≠**). Furthermore, the strain produces teliospores that undergo normal meiosis and produce the expected ratios of segregants.

Similar observations with respect to tumor induction have been made with genetically engineered haploids [see 23,32,63]. Development of ***a*= *b*≠** diploids in planta was compared with that of ***a*≠ *b*≠** diploids and with that of a mixture of two haploid strains with different ***a*** and ***b*** alleles. The ***a*= *b*≠** diploid strain follows the same course of development as the control strains [5]. These observations support the contention that different ***a*** alleles are not necessary for tumor induction. Although it is clear that heterozygosity at ***a*** is not necessary for pathogenicity, it has not been determined if either ***a*** allele is sufficient for pathogenicity—that is, whether a functional ***a*** allele is necessary for pathogenicity. An ***a*≠ *b*=** strain does not form filaments in vitro or in planta and is not pathogenic [5,6,8,9,11]. Table 1 summarizes the properties of different diploid and haploid strains used in assessment of filamentous growth and pathogenicity.

2.3 Genes Required for Progression of the Infectious Cycle

2.3.1 *fuz1*, a Gene Specifically Required for Hyphal Fragmentation

The goal of genetic analysis of the infectious process is to identify the functions required for the orderly progression of the events just described. It is expected that some genes when mutated will cause a specific arrest in this developmental pathway, similar to how mutations in yeast sporulation or in the cell cycle cause arrest at particular steps [reviewed in 64,65]. Indeed this is the case with $fuz1^-$ mutants [5]. Inoculation of plants with ***a1 b1*** $fuz1^-$ and ***a2 b2*** $fuz1^-$ strains leads to smaller and fewer tumors (a reduced tumor response) and the absence of teliospores [17]. A time course of infection with mutant strains was compared with that of wild-type strains. Even though $fuz1^-$ mutants do not form filaments on charcoal agar, they produce filaments in the plant (Fig. 4B), indicating that filament formation in vitro does not correlate with filament formation in planta. This observation is true for several other mutants unable to form filaments in vitro, $fuz2^-$, $fuz7^-$, $rtf1^-$ (Banuett, unpublished) [unpublished work cited in 23,32], and suggests that in *U. maydis*, as in other pathogenic fungi, for example, *Candida* [66,67] and *Cryptococcus* [68], there are alternative pathways for filamentous growth that are activated by different signals: nutrients, host environment, and other signals. The $fuz1^-$ filaments in the plant are no different from those formed by wild-type strains: they branch, invade, and proliferate in the same manner and even induce some tumor response, as indicated above. At the stage where wild-type hyphae fragment, the $fuz1^-$ hyphae do not (Fig. 3) [5]. Eventually the hyphae die out, and small tumors are produced that are completely devoid of fungal material. The block imposed by $fuz1^-$ is absolute: no further steps are executed.

fuz1 is the first gene identified that blocks progression of the infectious process. Interestingly, haploid *fuz1⁻* strains in culture exhibit a cell separation defect. Thus, *fuz1* appears to regulate events at the junction of cells: the mother-bud neck region in budding cells and the septum separating cylindrical cells of the hyphae. The *fuz1* gene codes for a Zn-finger protein with similarities to other fungal proteins involved in morphogenesis (Banuett, in preparation). Other *fuz⁻* mutants obtained in the same screen as *fuz1⁻* (see Sec. 5) may identify other genes required for progression of specific steps in the infectious pathway.

2.3.2 *rum1* and *ubc1* Are Also Required for Progression of the Infectious Cycle

Two other genes have been shown to block progression of the infectious cycle: *ubc1* and *rum1*. The *ubc1* gene codes for the regulatory subunit of cAMP-dependent protein kinase and is required for pathogenicity [69]. *ubc1* mutants do not induce tumors, but proliferate within the host. Whether the mutant accumulates less fungal mass than a threshold needed for tumor induction is not known. *ubc1* is described further in Section 4.

The *rum1* gene codes for a protein with similarity to the human retinoblastoma binding protein 2. It is involved in negative regulation of ***b***-regulated genes and is also required for progression of the infectious cycle [70]. Its block was placed just before teliospore cell wall deposition [70].

Several other genes have been identified which are required for pathogenicity, in the broadest sense of the word. Except for *fuz1*, *ubc1*, and *rum1*, it is not known if mutation in these genes blocks an early step (e.g., hyphal formation or proliferation on the leaf surface) or a later step (e.g., invasion or proliferation within host cells). Some of these genes are described in Sections 4 and 5. Understanding the complex interactions that occur during the infectious process of *U. maydis* will require the use of classical and novel approaches (see Sec. 5).

3 THE MATING TYPE LOCI: MASTER REGULATORS OF THE LIFE CYCLE

Two unlinked mating-type loci, ***a*** and ***b***, regulate various aspects of the *U. maydis* life cycle: cell fusion, filamentous growth, and pathogenicity. In this section I review our current knowledge of both loci.

3.1 The *a* Mating-Type Locus

3.1.1 Molecular Structure and Organization

The ***a*** locus has two alleles, ***a1*** and ***a2*** [71,72]. Cloning and sequencing demonstrated that ***a1*** and ***a2*** encode allele-specific pheromone precursors and receptors [73,74] (Fig. 6). ***a1*** contains *pra1* (receptor) and *mfa1* (pheromone precursor), whereas ***a2*** contains *pra2* and *mfa2* and additional genes not found in the ***a1***

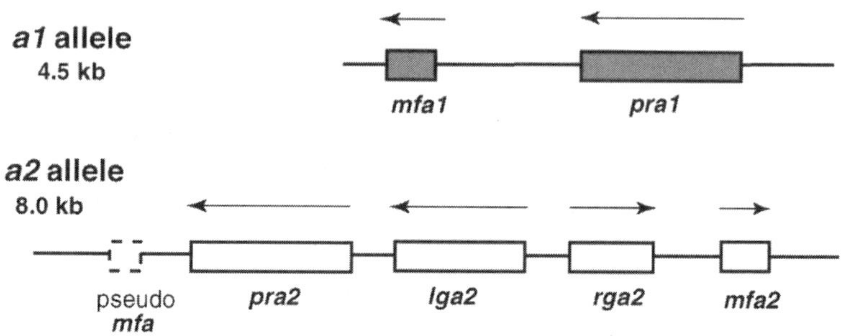

Figure 6 Molecular structure and organization of the ***a*** locus. The ***a*** locus has two alleles, ***a1*** and ***a2***, encompassing 4.5 and 8.0 kb, respectively. Each allele contains genes for allele-specific pheromone and receptor. *mfa1* and *mfa2* code for pheromone precursors (see Fig. 7). *pra1* and *pra2* code for putative seven-transmembrane G protein–coupled receptors. *rga2* and *lga2* code for polypeptides of 158 and 215 amino acids, respectively, of unknown function. Arrows indicate direction of transcription.

allele: *rga2*, *lga2*, and a pseudo-pheromone gene (Fig. 6). The functions of these additional genes are not known. Their deletion does not cause any discernible phenotype [75]. The receptor genes, *pra1* and *pra2*, code for presumptive G-protein-coupled receptors (GPCRs) with similarities to the pheromone receptors of the yeasts: *STE3* of *S. cerevisiae* and *map3* of *S. pombe*. GPCRs are a family of seven transmembrane receptors that bind diverse ligands, including pheromones. *mfa1* and *mfa2* code for pheromone precursors that are modified by prenylation and carboxymethylation at the CAAX box located at the carboxy terminus [76] (Fig. 7). A similar modification takes place in the precursor of **a** factor of *S. cerevisiae* and **M** factor of *S. pombe* [77a,78, and references therein]. The *U. maydis* precursors are processed further by proteolysis to produce the mature pheromones: **a1** is 13 amino acids and **a2** is 9 amino acids [76] (Fig. 7).

The finding that the ***a*** locus codes for components of a pheromone response pathway led to the hypothesis that activation of the receptor by pheromone activates a MAPK cascade signal transduction pathway similar to that involved in the pheromone response pathway of the yeasts [reviewed 80,81]. Several components of this signal transduction pathway have been identified and are discussed in Section 4.

3.1.2 The *a* Locus Governs Fusion of Haploid Cells

The ***a*** locus was proposed [71,72] to regulate cell fusion. However, unequivocal demonstration became possible only after development of assays for the pheromone response [41,42]. Two independent assays were developed which measure

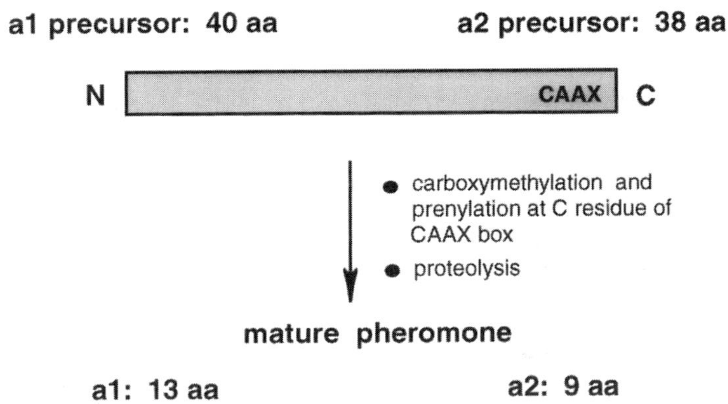

FIGURE 7 Processing of pheromone precursors to produce lipopeptides. The pheromones of *U. maydis* are synthesized as precursors of 40 and 38 amino acids for ***a1*** and ***a2***, respectively, which contain a CAAX box at their C terminus. The CAAX box is a substrate for modification by prenylation and carboxymethylation. These modifications are essential for activity of the pheromones [76]. This type of modification was first demonstrated in basidiomycete fungi (*Rhodosporidim* and *Tremella* [77 and references therein] and is found in many types of proteins, including mammalian Ras and the small GTP-binding protein Cdc42 [see 79 and references therein]. The modified pheromone precursors of *U. maydis* are processed further by proteolysis to yield lipopeptides of 13 and 9 amino acids for ***a1*** and ***a2***, respectively. All basidiomycete pheromones identified thus far are of the lipopeptide class. Secretion of these lipopeptides is likely to occur via a nonclassical secretory pathway involving an ABC-type transporter [77a].

the response of a strain to the presence of another strain: a water agar assay (confrontation assay) [40,41], and a liquid assay containing low nitrogen [42]. If strains with different ***a*** alleles, regardless of the ***b*** allele present, are placed in close proximity or cocultured, conjugation tubes (sinuous filaments distinct from the straight dikaryotic hyphae) are formed (Fig. 8). If the cells carry identical ***a*** alleles, no conjugation tubes form. The conjugation tubes direct growth toward conjugation tubes of cells of opposite mating type and fuse at their tips. The nuclei migrate through these tubes to establish the dikaryon, as inferred from still pictures (Fig. 8) [41,42]. Formation of conjugation tubes and fusion are ***a*** locus dependent but independent of the ***b*** locus. The fate of the fusion product is determined by the ***b*** locus: if different ***b*** alleles are present, the dikaryon grows as a straight hypha; if identical ***b*** alleles are present, the hyphae are contorted and multinucleate, and do not grow extensively [41,42]. The contorted hyphae ob-

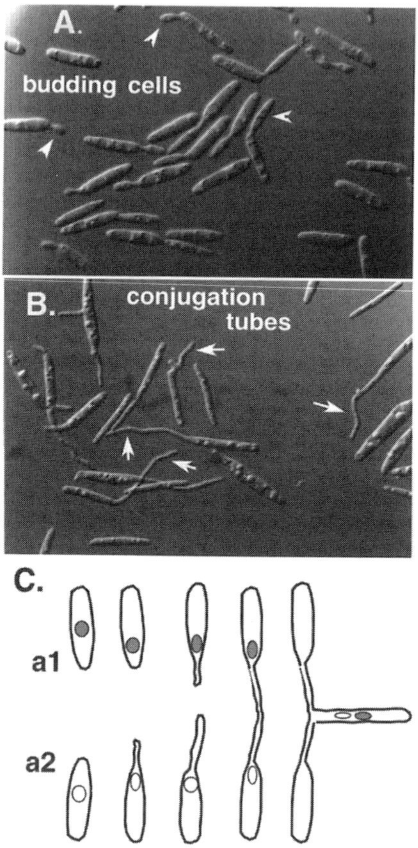

Figure 8 Conjugation tube formation. Conjugation tubes are the morphological response of *U. maydis* cells to the presence of pheromones. If **a1** cells are cocultured or in close proximity with **a2** cells, conjugation tubes form at the apex of the cells and grow toward cells of opposite mating type, regardless of the ***b*** allele present. Conjugation tubes fuse at their tips, and nuclei migrate through the conjugation tubes to establish the dikaryotic cell. **(A)** Cells grown in rich medium bud at an angle. In rich medium cells do not respond to pheromones. **(B)** Conjugation tubes formed after ***a1 b1*** and ***a2 b1*** cells are cocultured in low-nitrogen medium [42]. **(C)** Diagram depicting the steps leading to conjugation tube formation and cell fusion.

served when the strains carry identical ***b*** alleles are similar to those observed when ***a*≠ *b*=** diploids are cultured in low-nitrogen medium [42].

The water agar assay is simple and qualitative; the liquid assay is quantitative and synchronous. The response is rapid in both assays; quantitation using the liquid assay indicates that >90% of the cells respond by 3 h of coculture [42]. Using the water agar assay, it was shown that the response is asymmetrical: **a2** cells respond faster than **a1** cells [41] (Fig. 8). The reason for this difference is not known, but asymmetry in response to pheromones has also been observed in another dimorphic Basidiomycete, *Rhodosporidium toruloides*. In *R. toruloides*, when cells of the ***A*** mating type are in close proximity to cells of the ***a*** mating type, the ***a*** cells show precocious development of conjugation tubes, whereas the ***A*** cells show a delayed response [82]. It was proposed that ***A*** cells produce the pheromone constitutively, leading to a faster response in ***a*** cells [82], which produce their pheromone only upon stimulation by **A** cells.

3.1.3 Pheromone-Inducible Genes Contain Pheromone Response Elements (PREs)

Under noninducing conditions, for example, when ***a1*** and ***a2*** cells are grown separately on charcoal agar, a basal level of mRNA for the pheromone (*mfa1* or *mfa2*) and receptor genes (*pra1* or *pra2*) is detected by Northern analysis [75]. Upon pheromone stimulation, expression of these genes is induced; in addition, the expression of the ***b*** genes (***bW1 bE1*** or ***bW2 bE2***; see Sec. 3.2) is also induced [75]. The level of induction has not been quantitated but appears to be different for the different genes [75]. Several repeats of a 9-bp motif (ACAAAGGGA) are found in the ***a1*** and ***a2*** alleles, in the ***b*** genes, and upstream of *prf1* (see below and Sec. 4) [74,75,83]. This repeat was shown to be responsible for pheromone induction: a reporter gene (GUS) regulated by the wild-type arrangement of *mfa1* PREs or by tandem ACAAAGGGA repeats is activated when treated with the appropriate pheromone [83]. Therefore this element was designated PRE, for pheromone response element. A transcription factor of the HMG-box class, Prf1, binds specifically to this sequence [23] (see Sec. 4).

3.1.4 The *a* Locus Governs Filamentous Growth—An Autocrinelike Response

An unexpected result was that different ***a*** alleles are necessary for maintenance of filamentous growth [62]. Because the dikaryon contains both ***a*** alleles and thus has two different receptors and produces two different pheromones capable of self-stimulation, the requirement for different ***a*** alleles indicates that filamentous growth is controlled by an autocrinelike response [38,74,76]. This observation has been corroborated with genetically engineered haploids that contain a receptor and the pheromone that activates it (e.g., the *pra2* receptor and the *mfa1* pheromone) and nonallelic ***b*** genes (e.g., ***bW1 bE2***) [see 32] (Table 1).

It has also been shown that an ***a***= ***b***≠ diploid, which is not filamentous on charcoal agar [62], produces filaments upon stimulation with purified pheromone (Table 1) [76]. This result clearly demonstrates that the pheromones are involved in filamentous growth on charcoal agar. These observations provide the first evidence for the role of pheromones in filamentous growth in fungi. Because ***a***= ***b***≠ diploids form filaments in planta and are pathogenic, it is postulated that a plant signal activates filamentous growth [5]. One possibility is that this putative signal mimics the pheromone lipopeptide produced by the fungus.

3.2 The *b* Mating-Type Locus

3.2.1 Molecular Structure and Organization

The ***b*** locus is multiallelic, with 25 naturally occurring alleles. Different ***b*** alleles are necessary for filamentous growth and pathogenicity, and furthermore, any combination of different ***b*** alleles is active in promoting pathogenic development [6–11,71,72]. The molecular basis of this self-nonself recognition puzzle was shown to be the result of protein–protein interactions (see below). The requirement for heterozygosity at the ***b*** locus for filamentous growth is more stringent than that for ***a*** [62,76].

Cloning and sequencing of four different ***b*** alleles showed that each allele codes for a polypeptide containing a homeodomain motif. It was thus proposed that interaction of polypeptides coded by different ***b*** alleles generates an active heteromultimer that activates genes for filamentous growth and pathogenicity [84]. Analysis of additional ***b*** alleles confirmed these observations [63]. This analysis indicates that self-nonself recognition is due to interaction of polypeptides.

Deletion analysis of the ***b*** locus led to the finding that the ***b*** locus is complex (Fig. 9), each allele containing two divergently transcribed genes—***bW*** and ***bE***. These genes are separated by a short intergenic region of 200 bp where no recombination occurs and thus are inherited as a single genetic unit [85]. The ***b1*** allele contains ***bW1*** and ***bE1; b2*** contains ***bW2*** and ***bE2***, etc. Both genes code for homeodomain proteins that share similarity only in the homeodomain region. Interestingly, the two polypeptides share the same organization (Fig. 9): a variable region at the amino terminus, and a constant region for the remainder of the protein, where the homeodomain motif is located [63,84,85]. Incisive genetic analysis by Gillissen et al. [85] demonstrates that self-nonself recognition is due to interaction of nonallelic polypeptides in the dikaryon. For example, in a ***b1/ b2*** dikaryon, the **bW2** polypeptide interacts with the **bE1** polypeptide to form **bW2-bE1**, and **bW1** interacts with **bE2** to form **bW1-bE2**. Therefore, the dikaryon contains two active combinatorial activities, either of which is sufficient to trigger pathogenic development [85]. The number of possible active heterodimers—>600—is large, whereas that of inactive combinations is small—25 [see

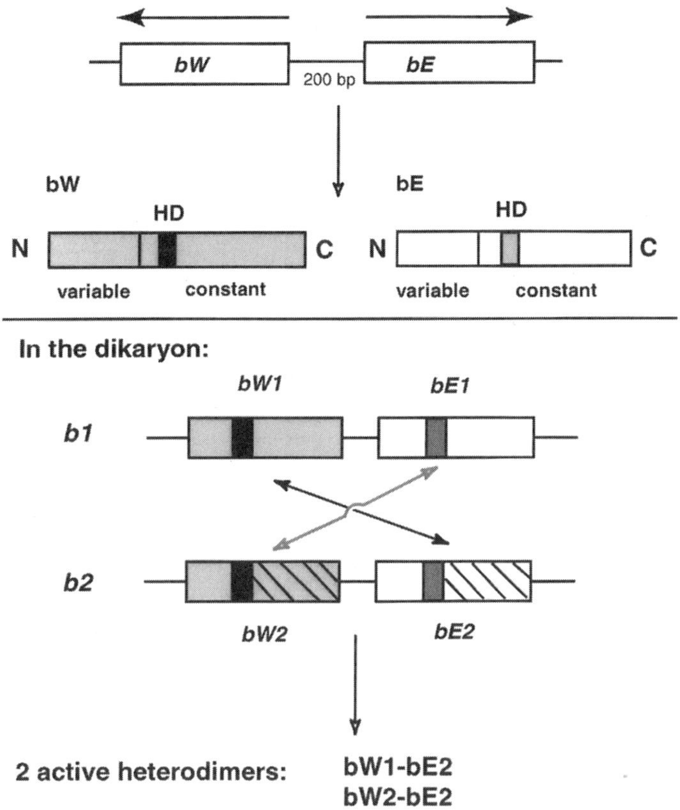

FIGURE 9 Molecular structure and organization of the ***b*** locus. (Top panel) The ***b*** locus has 25 naturally occurring alleles. Any combination of different ***b*** alleles triggers pathogenic development. Each allele consists of two divergently transcribed genes, ***bW*** and ***bE***, separated by a short intergenic region of 200 bp. Each gene codes for a polypeptide containing a homeodomain motif; the **bW** and **bE** polypeptides share no similarity except for the homeodomain motif. **bW** and **bE** polypeptides have the same structural organization: a variable region at the amino terminus and a constant region for the remainder of the protein. The **bW** variable region encompasses approximately the first 140 amino acids, whereas the **bE** variable region encompasses approximately the first 120 amino acids. The homeodomain is in the constant region, distal to the variable region (see text for details and references). (Bottom panel) In the dikaryon, nonallelic polypeptides interact via their variable regions to produce two different heterodimers, **bW1-bE2** and **bW2-bE1**, either one of which is sufficient for pathogenic development. The ***b*** locus is not essential for vegetative growth of haploids but is essential for pathogenic development and completion of the life cycle (see text for details and references).

38,85,86]. Deciphering the rules that generate this large number of active heterodimers remains a challenge in studies of protein–structure relationships.

Just as the **a1**-α2 heterodimeric homeodomain protein is the hallmark of the **a**/α diploid cell type in *S. cerevisiae* [87], the **bW-bE** combinatorial activity is the hallmark of the filamentous pathogenic dikaryon of *U. maydis*. The **a1**-α2 heterodimer consists of two different homeodomain polypeptides, encoded by the *MATa* and *MAT*α alleles, respectively, and regulates expression of haploid-specific genes (*hsg*) [reviewed in 87,88]. The **bW-bE** heterodimer is proposed to regulate genes for filamentous growth and pathogenicity. A binding site for the active ***b*** protein has been recently identified (see below). The molecular analysis of the *U. maydis* ***b*** locus provides the framework in which to understand and analyze the mating-type loci of other basidiomycetes, *S. commune* and *C. cinereus*, which also code for multiallelic combinatorial homeodomain proteins [89; reviewed in 4,15].

3.2.2 Specificity Determinants

The variable region of each polypeptide is proposed to be the specificity determinant. Analysis of chimeric constructs and mutations in the variable region support this contention [90–92]. In studies using the yeast two-hybrid system, it was shown that the variable region mediates interaction of nonallelic polypeptides, for example, **bW1** and **bE2**, but not of allelic polypeptides, for example, **bE1** and **bW1**, under the conditions used [86]. It is proposed that the interaction results in formation of an active heterodimer. The interactions of nonallelic polypeptides were also demonstrated biochemically using a His-tagged ***b*** protein. Attempts at purification of these polypeptides using *E. coli* expression vectors resulted in production of highly insoluble proteins (F. Banuett, unpublished; W. Lao, personal communication) [86]. A recent development has allowed the successful purification of an active ***b*** protein (see below).

Despite these advances, it remains to be deciphered how the variable regions discriminate self from nonself. Addressing this challenge may require determination of the crystal structure of the heterodimer, perhaps as a cocrystal with its DNA-binding site.

3.2.3 Targets of the *b* Locus

Because the ***b*** locus is the major pathogenicity determinant, the identification of its targets is key in understanding the molecular mechanisms by which it governs how the fungus induces disease. One possible candidate target for the ***b*** heterodimer is the *rtf1* gene, a putative inhibitor of tumor induction [17] (see Sec. 5). Other possible candidates are described below.

Because genes at the ***a*** locus exhibit ***b*** locus–dependent expression, it was proposed that these genes could be targets of the active ***b*** heterodimer [75,93]. The pheromone and receptor genes (*mfa1*, *mfa2*, *pra1*, and *pra2*) are downregulated in the dikaryon, whereas *lga2*, in the ***a2*** allele, is dramatically upregulated. Another putative candidate is the *prf1* gene, whose expression is also downregulated in the **b1-b2** dikaryon. In order to facilitate biochemical analysis, Romeis et al. [93] constructed a series of **bW-bE** fusion proteins that might be active and soluble, and thus circumvent the need to produce a heterodimer in vitro.

Precedent for an active protein resulting from fusion of two different homeodomain proteins is provided by analysis of a naturally occuring variant at the ***A*** locus in *C. cinereus* [reviewed in 4]. In *U. maydis*, **bE1** and **bW2** were fused head to tail with a linker region in between (Fig. 10), which yields a hybrid protein that is active in triggering pathogenic development. In other constructs, the hybrid protein contains a deletion of the variable region of **bE1** or a deletion of the variable region of **bW2**. Both hybrids are still active in vivo. A fusion protein lacking the variable regions of both **bE1** and **bW2** is also active in vivo [93] (Fig. 10), as is a smaller derivative of this protein used in the in vitro studies

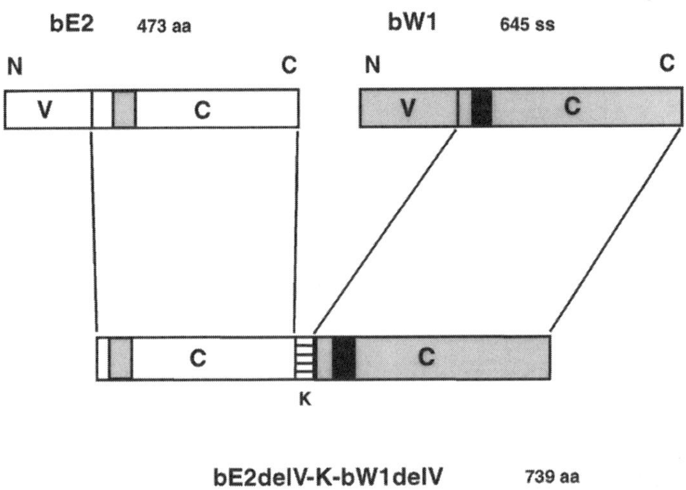

FIGURE 10 Fusion of two homeodomain polypeptides generates a biologically active b protein. The **bE1** and **bW2** polypeptides were fused head to tail with a small linker in between. This fusion protein and derivatives thereof, in which the variable region of **bE1** or **bW2** or of both is deleted, are active in triggering pathogenic development [93]. A derivative of the fusion protein shown here is active in vivo and in vitro binding assays with a fragment containing the upstream region of the *lga2* gene. This protein binds a site, *bbs1*, of ~29 bp, which contains direct repeats of the sequence motif AC/GTGTG and also sequence motifs found in the *hsg* operator recognized by **a1**-α2 in *S. cerevisiae.*

described below. These observations demonstrate that the variable regions are dispensable once nonallelic polypeptides have been fused. Strikingly, fusion of polypeptides from the same allele, for example, **bE2** and **bW2**, creates an active protein. Thus, the variable region appears to be required only for discrimination of self-nonself, as had been inferred from other analyses [90–92].

The smaller derivative of the **bW2-bE1** fusion protein [94] was used for in vitro binding studies with a fragment containing the *lga2* upstream region. This fusion protein binds to a region of ~29 bp located 150–178 bp upstream of the *lga2* ATG start codon [94]. This region, designated *bbs1* (for ***b*** protein–binding site), contains a direct repeat of the sequence motif AC/GTGTG separated by four nucleotides [94]. Mutation of the repeat motif abolishes in vitro binding. *bbs1* also contains sequence motifs with resemblance to the *hsg* operator site recognized by **a1**-α2 in *S. cerevisiae* [88]. Both contain the sequences GATG and ACA separated by 9 bp [88,94]. In the case of the *S. cerevisiae* site, the spacing of these sequences is critical for binding by **a1**-α2. The role for in vitro and in vivo activity of the different sequence elements in the *U. maydis bbs1* site is under investigation.

The presence of the two homeodomains is necessary for in vivo and in vitro activity of the various **bE1-bW2** fusion proteins [93,94] and also for the native ***b*** heterodimer [95]. In contrast, in *S. commune* and *S. cerevisiae*, only one of the two homeodomains of the heterodimer is necessary for activity [96,97].

These results provide the first description of the regulatory site recognized by the ***b*** protein. Future experiments should provide further insights as to the minimal *cis*-acting regulatory element necessary and sufficient for ***b***-regulated expression. It is also now possible to define the in vivo binding site for the ***b*** protein using chromatin immunoprecipitation and microarrays [97a].

4 THE PHEROMONE RESPONSE: MAPK CASCADE AND cAMP SIGNALING

4.1 The Pheromone Response Pathway

4.1.1 MAPK Cascade Components

The finding that the ***a*** locus codes for pheromones and receptors (Fig. 6) led to the hypothesis that a MAPK cascade signal transduction pathway is involved in the pheromone response of *U. maydis* by analogy with that in the yeasts [38,74, 76; reviewed in 80]. Because different ***a*** alleles are necessary for filamentous growth in vitro, this signal transduction pathway is postulated to be also necessary for filamentous growth [38,74,76].

The MAPK cascade is a highly conserved module that mediates transduction of signals generated at the cell surface to the nucleus (Fig. 11) [reviewed in 80,81]. This module consists of three serine/threonine protein kinases—

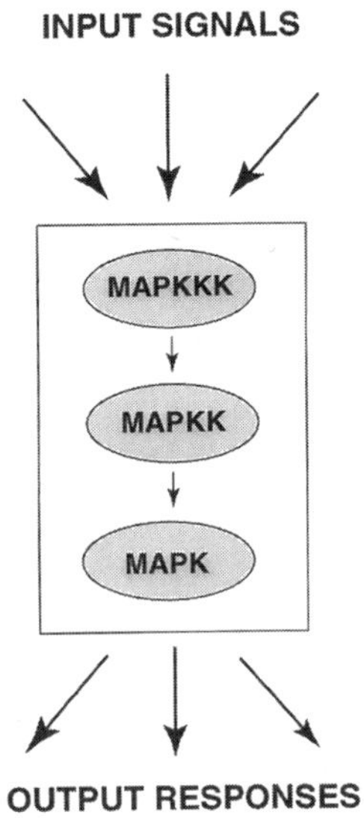

Figure 11 The MAPK cascade module is a highly conserved module in all eukaryotes and consists of three serine/threonine protein kinases: MAPKKK, MAPKK, MAPK. It mediates signal transduction from the surface of the cell to the nucleus and can be activated by a variety of signals, resulting in diverse output responses. Activation by sequential phosphorylation is key to the transduction of input signals. The MAPK is activated by the dual specificity serine/threonine tyrosine protein kinase MAPKK, which in turn is activated by the serine/threonine protein kinase MAPKKK, and it in turn can be activated by various proteins, including $p65^{PAK}$ protein kinase (not shown) [see Ref. 80].

MAPKKK or MEKK, MAPKK or MEK, and MAPK or ERK. Sequential activation by phosphorylation is key in transduction of a signal. The MAPK is activated by the dual specificity serine/threonine tyrosine kinase MAPKK, which in turn is activated by the serine/threonine kinase MAPKKK. The latter is activated in response to the input of a signal (Fig. 11). Eukaryotic cells contain multiple MAPK cascade modules, and a given MAPK cascade can be activated by multiple signals. Both *S. cerevisiae* and *S. pombe* contain several distinct MAPK cascades that are activated by different stimuli, including pheromones, stress, and nitrogen starvation [reviewed in 80,81]. The targets of the MAPK are diverse, but normally include a transcription activator that is responsible for expression of genes involved in the response [reviewed in 80,81].

The search for MAP kinase cascade components in *U. maydis* led to the identification of a MAPK kinase (MEK), designated Fuz7 [98]. Deletion of *fuz7* indicates that it is necessary for ***a*** locus–dependent processes [98]: conjugation tube formation, cell fusion, and filamentous growth. These observations support a role for *fuz7* in the pheromone response. The most surprising result is that *fuz7* is also required for pathogenicity, an ***a*** locus–independent process [98]. Because different ***a*** alleles are not required for pathogenicity [15,32,62], the results with *fuz7* led to the hypothesis that this MAPK kinase is activated by a plant signal and that this activation is necessary for induction of tumors [98]. This work with *fuz7* demonstrates, for the first time, that components of a MAPK cascade play a key role in pathogenicity of fungi. These results have been subsequently corroborated in a number of plant and animal pathogenic fungi: *Candida* [67,99], *Cryptococcus* [reviewed in 68], *Cochliobolus* [100], *Botrytis* [101], *Magnaporthe* [101a], and *Colletotrichum* [102].

The nature of the putative plant signals that activate the MAPK cascade and other fungal processes in the plant and the fungal sensors that perceive these signals is not known. A recent development, described in Section 5, is likely to accelerate identification of these signals and sensors.

Other MAPK cascade components have been identified in *U. maydis*: a MAPK (Kpp2 or Ubc3) [103,104] and a MAPKKK (Ubc4 and Kpp4) [105] (P. Müller and R. Kahmann, personal communication). *kpp2* and *kpp4* were identified by PCR with degenerate primers [104]. *ubc3* and *ubc4* were identified in a screen for mutants that suppress the filamentous phenotype of an adenylyl cyclase mutant (see Sec. 4.2) [103,105]. Kpp4 is necessary for mating and for pathogenicity (P. Müller and R. Kahmann, personal communication). No information was provided for Ubc4 [105]. Kpp2/Ubc3 is required for conjugation tube formation, filamentous growth, and basal and pheromone-induced expression of *mfa1*, indicating a role in the pheromone response [103,104]. In addition, Kpp2/Ubc3 is required for pathogenicity [104]. Thus, *fuz7*, *kpp2/ubc3*, and *kpp4* participate in ***a*** locus–dependent and –independent processes [98,103,104] (P. Müller and R. Kahmann, personal communication). Work to be described in Section 4.2 suggests that Ubc4, Fuz7, and Ubc3/Kpp2 are part of the same MAPK cascade.

4.1.2 Gα-Protein Subunits

G protein–coupled receptors are associated with heterotrimeric G proteins, which consist of α, β, and γ subunits. In the inactive state, the Gα subunit is bound to GDP. Upon binding of the ligand, for example, pheromone, the receptor is activated, which leads to exchange of GDP for GTP on Gα and dissociation of Gα-GTP from Gβγ. Either or both Gα-GTP and Gβγ can activate downstream effectors in different signal transduction pathways.

Four Gα protein subunits were identified in *U. maydis* [32]: *gpa1*, *gpa2*, *gpa3*, and *gpa4*. Deletion of *gpa1*, *gpa2*, or *gpa4* does not result in any discernible phenotype; double-mutant analysis was not reported. Given that fungi appear to have multiple Gα protein subunits, it would not be surprising if some of these Gα protein subunits have partially overlapping functions. In contrast, deletion of *gpa3* leads to several phenotypes, reflecting a role in ***a*** locus–dependent and –independent processes. *gpa3* is necessary for basal and pheromone-induced expression of *mfa1* and *mfa2* and also for filamentous growth, indicating that *gpa3* participates in ***a*** locus–dependent processes, and thus is part of the pheromone response [32]. This role is further supported by the observation that a strain containing an activated allele of *gpa3 (gpa3*)* leads to increased expression of *mfa1*, similar to that observed in pheromone-stimulated cells [32]. *gpa3* is also required for pathogenicity [32]. Thus, Gpa3, like Fuz7, Kpp4, and Ubc3/Kpp2, is also required for ***a*** locus–independent processes, and is likely to be activated in response to plant signals during the interaction of the fungus with its host [32].

In contrast to *gpa3*, *fuz7* does not affect the basal or pheromone-induced expression of *mfa1* and *mfa2*. Several possibilities can be invoked to explain these results: (1) *fuz7* is not part of the pheromone MAPK cascade but rather is part of a morphogenetic MAPK cascade, as proposed by Regenfelder et al. [32]; (2) there is redundancy at this level of the MAPK cascade, as observed in some of the MAPK pathways in *S. cerevisiae* and *S. pombe* [reviewed in 80,81]; or (3) the absence of Fuz7 allows a MAPKK from another pathway to usurp its place, as has been recently reported in *S. pombe* [106]. Although expression of *mfa1* and *mfa2* provides a quantitative measure of the response of these genes to pheromone stimulation, it may not accurately reflect all the events in the pheromone response, which include morphological changes leading to conjugation tube formation. Future studies using microarray analyses should provide a more definitive picture of the components of the pathway and would likely help identify a reporter gene.

4.1.3 A Transcription Activator That Regulates Pheromone Response Genes

A common target of MAPKs is a transcriptional activator. In *S. cerevisiae*, Ste12, a homeodomain-type protein, is the target of the MAPK Fus3. In *S. pombe*, Ste11, an HMG-box protein, may be the target of the MAPK Spk1. Fus3 and Spk1 are the MAPKs of the pheromone response in these yeasts [reviewed in 80,81].

In *U. maydis* pheromone stimulation results in increased expression of several genes—*mfa1*, *mfa2*, *pra1*, *pra2*, *rga2*, *lga2*, *prf1*, and the ***b*** genes (***bE*** and ***bW***) [23,75,83]. The transcription activator responsible for basal and pheromone-stimulated expression appears to be Prf1, a member of the HMG class of transcription activators (Fig. 12) [23]. Prf1 binds the pheromone response element, PRE (ACAAAGGGA), located in the ***a1*** and ***a2*** alleles and in the ***b*** genes (see Sec. 3) [22,83]. Prf1 binds specifically to all of the pheromone response elements in the ***a1*** and ***b2*** alleles that perfectly match the consensus sequence. Those with

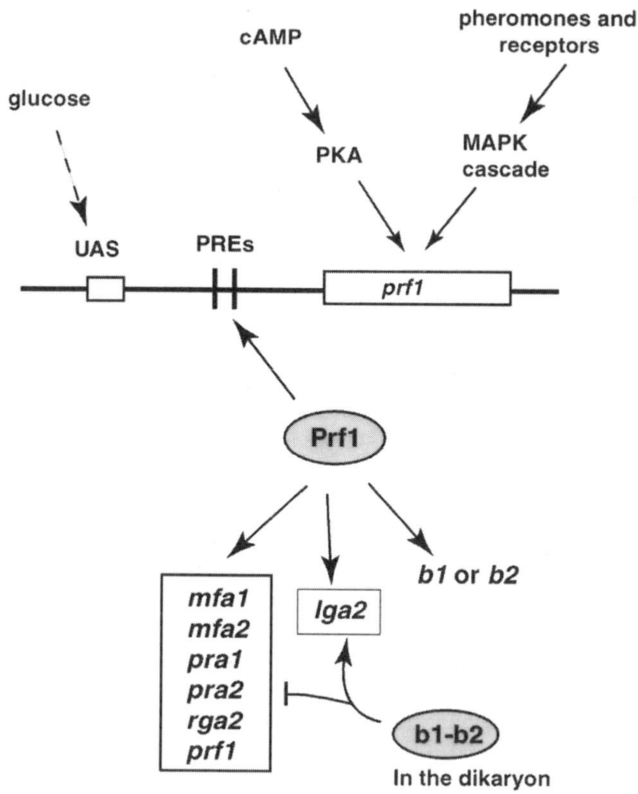

Figure 12 Prf1, the transcriptional activator of pheromone response genes, is activated by multiple signals. Prf1 is an HMG-class transcription activator and regulates expression of pheromone response genes by binding to PREs located in these genes [23]. Prf1 itself is upregulated upon pheromone stimulation; it binds PREs located in its 5′ region. Prf1 is regulated not only at the transcriptional level but also posttranscriptionally, most likely by phosphorylation. Two candidates for activation by phosphorylation of Prf1 are the MAP kinase Kpp2/Ubc3 and Ard1, the catalytic subunit of cAMP-dependent protein kinase. Consensus sites for both of these proteins exist in Prf1, but direct biochemical evidence for interaction of the above proteins does not exist. In addition, Prf1 may be regulated by nutritional conditions. It contains a UAS in its 5′ region that appears to be involved in sensing carbon source.

mismatches are bound less efficiently [23]. Therefore, it is highly likely that Prf1 is the transcription factor that regulates expression of pheromone response genes upon pheromone stimulation by directly binding to the PREs.

The basal level of *prf1* mRNA is induced ~20-fold upon pheromone stimulation. Two PREs that match perfectly the consensus are located in the 5′ regulatory region of *prf1*, suggesting that Prf1 stimulates its own synthesis upon pheromone treatment [23]. Supporting the role of Prf1 in the pheromone response is the observation that strains deleted for *prf1* do not produce or respond to pheromones, as assayed with tester strains [23]. In addition, Prf1 is required for filament formation on charcoal agar.

prf1 is also required for pathogenicity as assayed in a haploid solopathogenic strain (***a1 bW2 bE1*** vs. ***a1 bW2 bE1*** Δ*prf1*) [23]. Interestingly, the nonpathogenic phenotype is suppressed by constitutive expression of an active ***b*** protein. These results indicate that one role of Prf1, and of the pheromone response, is the activation of ***b*** locus expression in haploid cells. Upon cell fusion the dikaryon formed will be poised to have a high level of the active ***b*** protein. This high level of ***b*** is presumably necessary for regulation of ensuing processes.

Although the above observations indicate a critical role for Prf1 in the pheromone response, it remains to be elucidated how Prf1 is activated. One possibility is that Prf1 is the direct target of a MAPK. However, there is no direct biochemical evidence indicating that a MAPK phosphorylates Prf1. Inactivation of the consensus sites for phosphorylation by a MAPK and also of a presumed MAPK docking site reduced but did not block Prf1 activity [104]. Therefore, either Prf1 is not the direct target of a MAPK or, more likely, Prf1 requires activation by multiple mechanisms (Fig. 12). The existence of consensus sites for cAMP-dependent protein kinase in Prf1 suggests that PKA may be involved in activation of Prf1. Direct biochemical evidence should help elucidate the mechanisms of Prf1 activation.

In *S. pombe*, Stell is an HMG-box protein that is activated by the Sty MAPK cascade in response to several stimuli, including stress and nutritional starvation [reviewed in 78,80]. Under nitrogen starvation, the Stell protein activates expression of accessory transcription activators. These accessory proteins act with Stell in an initial burst of transcription of pheromone response genes. The pheromones thus produced then activate the pheromone response pathway, including regulators necessary for subsequent steps (entry into meiosis) [reviewed in 78,80]. Because nutritional conditions are necessary for the pheromone response in *U. maydis* [42,98,107], it is possible that the pheromone response in *U. maydis* is as complex as that in *S. pombe*, necessitating the input of accessory proteins for a full response. These accessory proteins could be activated by Prf1 itself or in response to other signals, for example, nitrogen starvation, carbon source, etc.

It remains to be determined how nutrition, cAMP, and pheromones influence each other during the pheromone response of *U. maydis*. Identification of a reporter gene for the pheromone response, using, for example, microarray technology, would greatly aid such studies.

4.2 Components of the cAMP Pathway

In *U. maydis* as in other eukaryotes, this pathway consists of adenylyl cyclase (designated Uac1 in *U. maydis*), which catalyzes conversion of ATP to cAMP. cAMP activates cAMP-dependent protein kinase (PKA), a tetrameric protein consisting of two regulatory subunits (designated Ubc1 in *U. maydis*) and two catalytic subunits (in *U. maydis* the major one is designated Ard1; a second minor one is called Uka1) [reviewed in 108]. In the absence of cAMP, the tetrameric complex is inactive; the regulatory subunit prevents the catalytic subunit from phosphorylating its substrates. Activation occurs when cAMP binds to the regulatory subunit, causing release of the catalytic subunit, which is then free to act on its many substrates. Phosphodiesterase (Pde) catalyzes the conversion of cAMP to AMP but has not been identified in *U. maydis*.

The cAMP pathway appears to be necessary for growth habit of haploid cells and for pathogenicity. In *U. maydis*, haploid strains deficient in adenylyl cyclase, *uac1*, exhibit a filamentous phenotype on charcoal agar [109] (see Sec. 5). This filamentous phenotype is independent of an active ***b*** heterodimer and of different ***a*** alleles. These results suggest that low cAMP promotes filamentous growth and high cAMP promotes yeastlike growth. The fact that *uac1*$^-$ haploid strains are filamentous but not pathogenic indicates that filamentous growth per se is not sufficient for pathogenicity. Second-site suppressor mutations that abolish the filamentous phenotype of *uac1*$^-$ strains led to the identification of the *ubc1* gene, which codes for the regulatory subunit of cAMP-dependent protein kinase [109]. Since *ubc1* mutants are expected to have high PKA activity, and since high levels of cAMP promote high PKA activity, it was reasoned that addition of cAMP to the medium should suppress the filamentation of *uac1*$^-$ strains. Indeed this is the case [109]. These results support the contention that cAMP is an important determinant of growth habit. Whether the filamentous phenotype of *uac1*$^-$ mutants reflects activation of a ***b***-dependent pathway or a ***b***-independent pathway remains to be determined.

ubc1$^-$ mutants exhibit a cell separation defect. Addition of cAMP to wild-type cells produces a *ubc1*$^-$ phenocopy [109,110]. These results suggest that high or unregulated PKA activity interferes with separation of cells during budding. Ubc1 is required for pathogenicity; *ubc1*$^-$ mutants do not induce tumors but form hyphae which grow within the plant [69,111]. These results suggest that high PKA activity or unregulated PKA activity is detrimental to tumor formation and perhaps other aspects of the infectious cycle. Indeed, recent work by Krüger et al. [111a] indicates that inappropiate activation of the cAMP pathway by either mutation in the Gα subunit (see Sec. 4.3) or in *ubc1* results in a drastic reduction of fungal mass within the plant and arrest during formation of teliospores. Thus, they conclude that cAMP is tightly regulated during the infectious process.

The above results led to the prediction that mutations inactivating the catalytic subunit of cAMP-dependent protein kinase (PKA) would result in filamentous growth. This was shown to be the case. Two genes encode catalytic subunits: *adr1* and *uka1* [111]. The *adr1* gene had been previously identified in a screen for mutants resistant to the fungicide vinclozolin [112]. Disruption of *adr1* leads to constitutive haploid filamentation, just as *uac1*$^-$ mutants, whereas Δ*uka1* strains exhibit yeastlike growth. The *adr1*$^-$ *uka1*$^-$ double mutant exhibits diminished filamentation compared to the *adr1*$^-$ mutant [111]. This indicates that the two catalytic subunits have different roles: *adr1* inhibits filamentous growth whereas *uka1* stimulates it, albeit not strongly (Fig. 13). In *S. cerevisiae*, three genes (*TPK1*, *TPK2*, and *TPK3*) code for the catalytic subunit of cAMP-dependent protein kinase. These genes exert antagonist effects on pseudohyphal growth; *TPK2* stimulates, whereas TPK3 inhibits [113,114], similarly to the situation observed in *U. maydis*. *adr1* appears to be the major contributor of PKA activity in *U. maydis* [111]. Measurement of PKA activity in *adr1*$^-$ and *uka1*$^-$ mutants showed that Δ*uka1* mutants have almost wild-type levels, whereas Δ*adr1* mutants have reduced levels. *adr1* is necessary for pathogenicity, whereas *uka1* is not [111].

The gene for phosphodiesterase has not been identified, but it is predicted that deletion of this gene in a wild-type strain may result in a cell separation defect similar to the phenotype of *ubc1*$^-$ strains.

4.3 Links Between the cAMP and the Pheromone Response Pathways

Several results point to a connection between the pheromone MAPK pathway and the cAMP pathway [reviewed in 108,115]. I discuss evidence that suggests that these pathways act in parallel and appear to converge on Prf1, the transcription activator that regulates expression of pheromone response genes. Some of the evidence is discussed below.

4.3.1 The cAMP Pathway Inhibits a MAPK Cascade

Some of the second-site suppressors of the filamentous phenotype of *uac1*$^-$ strains code for components of the pheromone response MAPK cascade described earlier. These are the MAPKKK Ubc4 (also identified independently as Kpp4 by P. Müller and R. Kahmann, personal communication), the MAPKK Ubc5 (previously identified as Fuz7 [98]), and the MAPK Ubc3 (also identified independently as Kpp2 [104]). The fact that mutations in *fuz7*, *ubc3*, and *ubc4* sup-

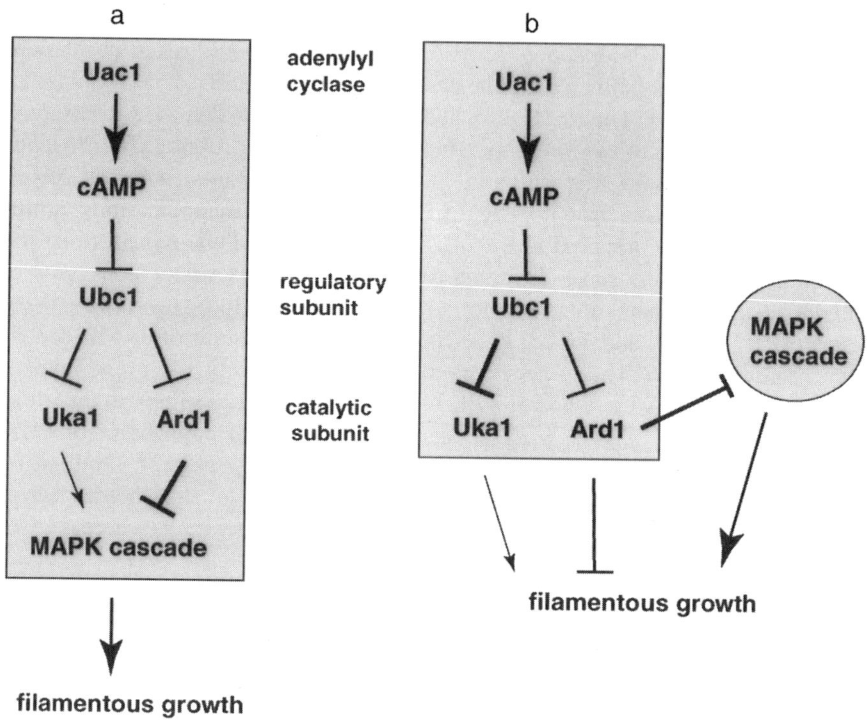

Figure 13 The cAMP pathway in *U. maydis* regulates filamentous growth and pathogenicity. This pathway consists of Uac1 (adenylyl cyclase), Ubc1 (the regulatory subunit of cAMP-dependent protein kinase), and Ard1 and Uka1 (two different catalytic subunits of cAMP-dependent protein kinase). Because mutations in *uac1* lead to a filamentous phenotype in haploid cells, it is proposed that *uac1* inhibits filamentous growth. Mutations in *ard1* and *ubc1* support this contention. The two isoforms of the catalytic subunit have antagonistic effects: Ard1 inhibits, whereas Uka1 stimulates filamentous growth, although the latter seems to play a minor role (see text for details). Second site suppressor mutations of the *uac1*⁻ filamentous phenotype led to the identification of the same MAPK module that mediates the pheromone response and that is required for pathogenicity. This observation suggests that uac1 inhibits this MAPK module. This MAPK cascade may lie downstream of the cAMP pathway (panel a) or in parallel to the cAMP pathway (panel b).

press a *uac1*$^-$ mutation suggests, first of all, that Fuz7, Ubc3, and Ubc4 are part of the same MAPK cascade module, and secondly, that the cAMP pathway inhibits this MAPK cascade module. In principle, the MAPK cascade could lie either downstream of the cAMP pathway (Fig. 13a) or parallel to the cAMP pathway (Fig. 13b). Work in *S. cerevisiae*, *S. pombe*, and *C. neoformans* provides precedent for these two pathways acting in parallel [113,114,116, reviewed in 78; this volume, Chapter 19 by Lengeler and Heitman]. Additional evidence supporting the notion that there is crosstalk between these pathways is discussed below.

4.3.2 cAMP Levels Determine Level of Expression of Pheromone Genes

As discussed above, Gpa3 regulates expression of pheromone genes. Several observations indicate that Gpa3 may regulate expression of pheromone genes by regulating adenylyl cyclase and cAMP levels. In addition to the various phenotypes already described, a Δ*gpa3* mutant also exhibits an elongated cell morphology, which can be suppressed by addition of cAMP [110]. These results suggest that Δ*gpa3* cells have low levels of cAMP and that *gpa3* controls cAMP production, most likely by regulating adenylyl cyclase. One possibility is that Gpa3 stimulates adenylyl cyclase activity, as shown for Gna1, a Gα protein subunit of *Neurospora crassa* [117]; alternatively, Gpa3 may regulate adenylyl cyclase protein levels, as has been demonstrated for Gna3, another Gα protein in *N. crassa* [118]. Analysis of a Δ*uac1 gpa3** (activated allele) double mutant indicates that *gpa3* acts upstream of *uac1*, supporting a role in regulation of adenylyl cyclase.

Additional evidence supports the existence of crosstalk between the cAMP pathway and the pheromone response pathway. cAMP affects pheromone mRNA levels: addition of 6 mM cAMP to wild-type cells results in a higher basal level of *mfa1* expression than in wild-type cells not exposed to cAMP [110]. Higher concentrations were inhibitory. Furthermore, *uac1* and *ubc1* mutants affect pheromone expression levels: in Δ*uac1* mutants the basal level is less than in wild-type strains, whereas in Δ*ubc1* mutants the level is similar to that observed in pheromone-stimulated cells [110]. Thus, cAMP levels are important determinants of pheromone expression.

Gpa3 could be activated in response to nutrients, and in turn it would activate Uac1, resulting in increased cAMP levels, and higher PKA activity. PKA may then activate, by phosphorylation, a transcription factor, perhaps Prf1 or another protein that acts together with Prf1. Activated Prf1 (or Prf1 complex) then increases expression of pheromone response genes. This increased level of pheromones is likely to activate the pheromone MAPK cascade, resulting in a final burst of induction of pheromone response genes. Thus, the pheromone response would occur in steps, as proposed for the pheromone response in *S. pombe* [reviewed in 78,80]. If the cAMP pathway inhibits the pheromone MAPK (Fig. 13), then it must be proposed that once there is an initial induction of phero-

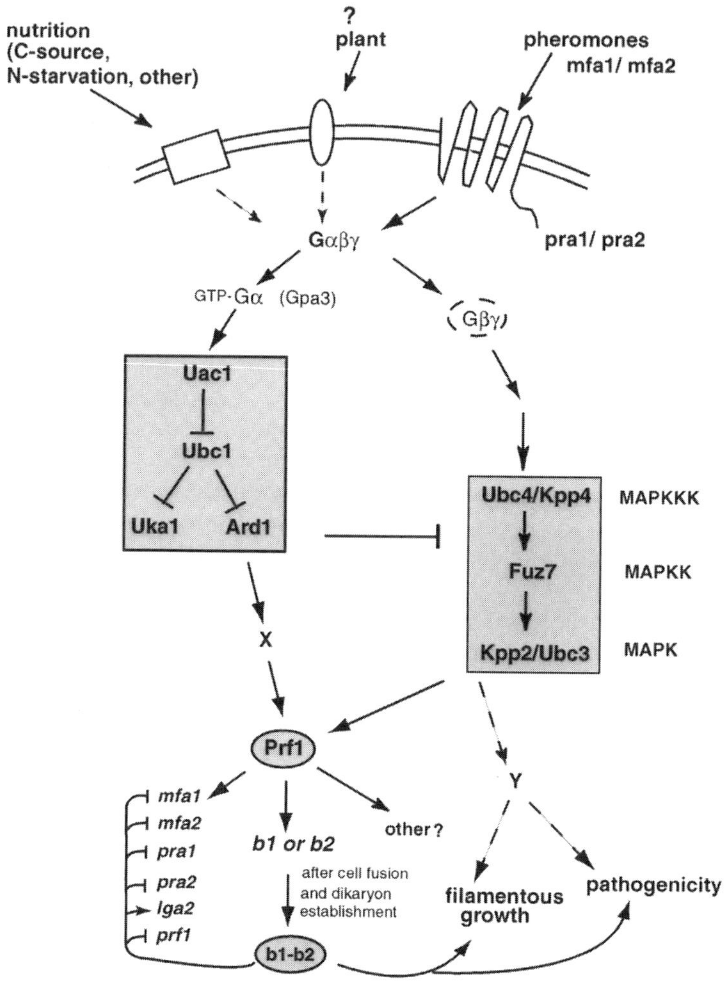

FIGURE 14 A MAPK cascade and the cAMP pathway regulate the pheromone response, filamentous growth, and pathogenicity in *U. maydis*. *U. maydis* responds to various signals: plant signals, nutritional cues, and pheromones. G protein–coupled receptors (GPCR), of the seven-transmembrane family, are activated by different signals. Upon binding of the ligand, for example, pheromone, the receptor is activated, which leads to exchange of GDP for GTP on Gα and dissociation of Gα-GTP from Gβγ. Either or both Gα-GTP and Gβγ can activate downstream effectors. Gβγ has not been identified. Gα is proposed to activate adenylyl cyclase, resulting in activation of Ard1, which in turn may activate Prf1 and inhibit a MAPK. Prf1 regulates expression of pheromone response genes. Because the pheromone response requires a

mone response genes, the cAMP pathway is downregulated. This speculative model is based on limited evidence. Much remains to be done to figure out how these pathways talk to each other.

It is also possible that multiple inputs regulate activity of Gpa3—nutritional signals (nitrogen starvation, carbon source), a pheromone signal, and plant signals—and that these different inputs may result in different levels of activation of adenylyl cyclase (Fig. 14). Critical evaluation of the effect of mutations awaits identification of a reporter gene for each of these pathways.

4.3.3 Nutritional Sensing and the Pheromone Response

In most fungi, for example, *N. crassa*, *M. grisea*, and *C. neoformans*, sexual development occurs only under low nitrogen conditions; *S. cerevisiae* is the exception. In *U. maydis*, low-nitrogen conditions are necessary for the pheromone response, which includes conjugation tube formation, cell fusion, and filamentous growth [42,98]. Low nitrogen could result in increased expression of pheromone response genes, as in *S. pombe*, by activation of a nitrogen starvation sensing signaling pathway (as noted earlier in this section). Alternatively, nitrogen starvation conditions may alter cAMP levels, which would then affect expression of pheromone genes. In *S. pombe*, low nitrogen level causes a reduction in cAMP levels [119]. It is not known if nitrogen concentration has any effect on cAMP levels in *U. maydis*.

Other nutritional inputs, for example, carbon source, appear to be mediated by Prf1, as indicated by the recent identification of a carbon source–regulated UAS in the upstream region of *prf1* (the transcription activator of pheromone

---

nutritional input and pheromones, it is possible that the nutritional input leads to activation of a transcription activator other than Prf1, which then interacts with Prf1 to achieve full pheromone induction (see text). The cAMP pathway may be downregulated, after an initial burst of expression of pheromone response genes, to allow activation of the MAPK cascade. Once cell fusion takes place and the dikaryon is established, the ***b*** heterodimer downregulates pheromone response genes and activates genes for filamentous growth and pathogenicity. The MAPK cascade may have a role in pathogenicity independently of Prf1 (shown by dashed lines). Because cAMP pathway components are necessary for pathogenicity, it is possible that ***b*** modulates activity of cAMP-dependent protein kinase during plant infection. Only the pheromone receptors have been identified. The other putative sensors may belong to the GPCR class, in which case they could interact with Gpa3, or to other classes of membrane proteins and interact with proteins other than the heterotrimeric G protein. Much remains to be elucidated in the pheromone response of *U. maydis*. This model is highly speculative.

response genes) and of a protein, Ncp1, which binds this UAS [107]. These observations support the contention that multiple signals converge on *prf1*, all of which may be necessary to achieve full activation of Prf1 (Figs. 12, 14) [23,107]. In *S. cerevisiae*, multiple signals appear to converge on the large promoter of the *FLO11* gene that codes for a flocculin necessary for pseudohyphal growth [116].

In summary, *prf1* seems to be regulated by [23,107] (1) pheromones, (2) the active ***b*** heterodimer, (3) carbon source and perhaps other nutritional conditions, and (4) cAMP. Future analyses should help elucidate how these different inputs regulate the pheromone response in *U. maydis*. A model for the pheromone and cAMP response is shown in Figures 12 and 14.

5 IDENTIFICATION OF GENES NECESSARY FOR FILAMENTOUS GROWTH AND PATHOGENICITY

In this section I describe some of the approaches to identify genes necessary for filamentous growth and interaction with the plant. I also describe a new development that allows formation of teliosporelike cells using a maize callus system and how this system can be exploited for multiple purposes: to identify reporter genes for different stages of filamentous growth and teliospore formation, in genetic screens to identify genes for filamentous growth and teliospore formation, and to identify the signals promoting these events.

Because much of the work in *U. maydis* relies on the behavior of strains on charcoal agar, I next describe this plate assay.

5.1 The Charcoal Plate Assay—An Assay for Filament Formation

Formation of dikaryotic filaments is a two-step process that requires, first, cell fusion to form a dikaryotic cell (establishment of the dikaryon), and second, filamentous growth of the dikaryon (maintenance). As described in Section 3, the mating-type loci govern filament formation [reviewed in 7,15,16]. Cell fusion requires different ***a*** alleles; filamentous growth requires different ***a*** and ***b*** alleles. Filament formation is assayed on charcoal agar, on which costreaking or cospotting of strains with different ***a*** and ***b*** alleles results in formation of a white fuzziness due to dikaryotic filaments (a Fuz^+ phenotype) (Fig. 15) [62]. If strains carry identical ***a*** alleles or identical ***b*** alleles, there is no filament formation (a Fuz^- phenotype) [62]. This assay is not a mating assay, as is often incorrectly described in the *Ustilago* literature, because it does not distinguish between the two steps of the process. For example, a mutation that interferes with the second step results in a Fuz^- phenotype just as a mutation which blocks the first step. Evaluation of filamentous growth (the second step) independently of cell fusion (the first step) requires the use of diploids, for example, heterozygous at both ***a*** and ***b*** or

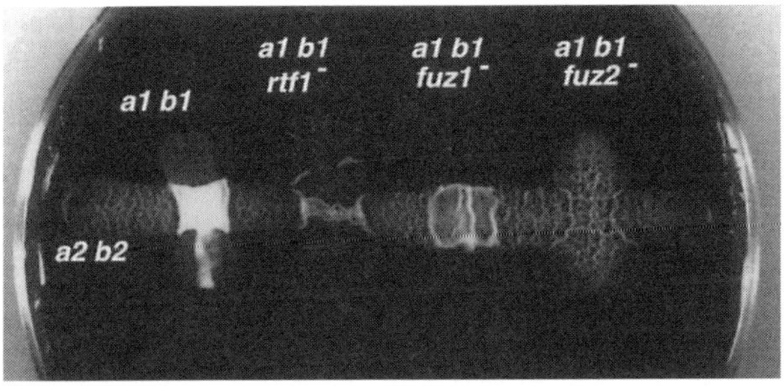

FIGURE 15 Charcoal agar assay. In this assay filament formation appears as a white fuzziness (a Fuz^+ phenotype) (leftmost cross-streak). If strains contain identical ***a*** or identical ***b*** alleles or mutations that affect cell fusion or filamentous growth, no filaments develop (a Fuz^- phenotype). Strain in horizontal line: wild-type ***a2 b2***. Strains in vertical lines, from left to right: wild-type ***a1 b1; a1 b1*** *fuz1*⁻; ***a1 b1*** *rtf1*⁻; and ***a1 b1*** *fuz2*⁻. (From Ref. 17.)

genetically engineered haploids containing different ***a*** and ***b*** alleles [see 32]. The properties of this ***a1/a2 b1/b2*** diploid or haploid carrying a mutation of interest are then determined [see, e.g., 23,32,62,98,104,111,120]. Cell fusion can be assayed as described in Section 3.

5.2 Genetic Approaches

5.2.1 Identification of *fuz* Genes

The plate assay was used in a screen to identify *fuz* genes necessary for filamentous growth (Fig. 15) [17]. Mutagenized haploid cells of one mating type are replica mated onto a lawn of a wild-type strain of opposite mating type and screeened for a Fuz^- phenotype. Mutants that do not form filaments were crossed to a wild-type strain to determine if the mutation was linked to the mating-type loci. This analysis depends on the formation of tumors and production of teliospores in crosses between wild-type and mutant strains. The mutants analyzed carry mutations unlinked to the mating-type loci and appear to be recessive with respect to tumor induction and teliospore formation. Four mutations analyzed in detail led to identification of three new genes: *fuz1*, *fuz2*, and *rtf1*. In addition to its requirement for filament formation on charcoal agar, *fuz2* is also required for teliospore germination. It is possible that formation of the promycelium (see Sec. 2) is impaired in *fuz2*⁻ mutants. *rtf1* is hypothesized to be an inhibitor of tumor formation: mutation in this gene bypasses the requirement for different ***b*** alleles for tumor formation. Thus, it is proposed that the active ***b*** protein inhibits *rtf1* in the dikaryon, thereby relieving the inhibitory effect of *rtf1* on tumor formation. In the haploid, where there is no active ***b*** protein, *rtf1* is expressed and tumor formation is inhibited [17]. *fuz1* was described in Section 2 [5,17]. Other Fuz$^-$ mutants are likely to identify functions that play roles in different aspects of the life cycle.

5.2.2 Candidate Gene Approach

In this strategy homologs of genes known to play key roles, for example, in signal transduction and pseudohyphal growth in *S. cerevisiae* and *S. pombe*, and also in hyphal morphogenesis in other fungi, are PCR amplified using degenerate primers based on sequence alignments. The identified genes are deleted, and their phenotype with respect to filamentous growth and pathogenicity is determined. This approach has led to the identification of several genes that when mutated confer a Fuz$^-$ phenotype in the charcoal plate assay: *fuz7* (MAPKK [98]), *gpa3* (Gα [32]), *prf1* (HMG box transcription activator [23]), *kpp2* (MAPK [104]), *kpp4* (MAPKKK; P. Müller and R. Kahmann, personal communication), *ukc1* [21], and *kin2* (kinesin [120]). All except the last two genes were described in Sections 3 and 4 and, as indicated, are also required for pathogenicity.

ukc1 codes for a serine/threonine protein kinase [21] with similarity to fungal protein kinases involved in morphogenesis: *cot1* of *Neurospora crassa*, *orb6* of *S. pombe*, and *TB3* of *Colletotrichum trifolii* [121–123]. It is required for wild-type morphology, filament formation, and pathogenicity: *ukc1*$^-$ cells are darkly pigmented, form clusters of round cells with extensions, and fail to form filaments. *ukc1*$^-$/*ukc1*$^-$ diploids are nonfilamentous and nonpathogenic [21].

kin2 codes for a motor protein of the kinesin family and is required for normal morphology of dikaryotic hyphae in vitro and for pathogenicity [120].

As can be seen, this approach has proven of great utility for identifying genes required for filamentous growth and pathogenicity. The next two methods have been less productive.

5.2.3 Restriction Enzyme–Mediated Integration (REMI) Mutagenesis

Transposon tagging is an ideal method to obtain mutants because it allows rapid identification of the mutated (tagged) gene. REMI can be viewed as a transposon-like strategy. In this method, a plasmid containing a selectable marker and lacking sequences homologous to *U. maydis* is introduced by transformation, in the presence of a restriction enzyme [124]. Under these conditions, the plasmid integrates randomly in the genome at sites cleaved by the restriction enzyme. Single integration events occur in 90% of the REMI events [124].

The advantage of this method over classical mutagenesis procedures is that sequence information of the disrupted gene is readily generated after rescue of the insert with the flanking regions. One drawback is that the procedure has to be repeated with different enzymes, in each case optimizing conditions, in order to generate a random collection of mutants. Several genes were reported to have been identified using this method, but additional information was not provided [125]. REMI has been successfully used in a number of fungi to identify a variety of genes [reviewed in 125].

A REMI-likc strategy was used to isolate mutants unable to form filaments on charcoal agar [126]. Analysis of one mutant led to identification of *myp1*, which codes for a novel protein of 1150 amino acids. Deletion of the gene confers a similar phenotype to that of the original mutant.

5.3 Molecular Biological Approaches—Subtractive Hybridization and Differential Display

Subtractive hybridization, with RNAs of diploid strains that exhibit nonfilamentous growth on charcoal agar versus diploids that are filamentous, was used to identify filamentous growth–specific genes. The *egl1* gene codes for a cellulase specifically expressed during the filamentous growth phase [127]. *egl1* is not required for filamentous growth or pathogenicity.

The *mig1* gene was identified using differential display with mRNAs isolated from infected and uninfected plants [128]. *mig1* is specifically induced during growth within the plant, from penetration of host cells to teliospore formation. It is not expressed on the leaf surface. The *mig1* gene codes for a novel hydrophilic protein of 185 amino acids. Sequences in the upstream region of this gene were shown to be necessary for induction in the plant. *mig1* is not necessary for pathogenicity. Both *egl1* and *mig1* may prove to be useful reporter genes for filamentous growth and growth in the plant, respectively.

5.4 High-Throughput Methods—Microarrays

As the sequence of fungal genomes becomes available, genomewide strategies for the analysis of hyphal growth, conidiation, development of sexual structures, growth under different nutritional conditions, and pathogenicity are likely to be the methods of choice. DNA microarrays make it possible to study expression of all of the genes of a genome in parallel. Analysis of a process in a global manner, for example, meiosis and sporulation or cell cycle in *S. cerevisiae*, became possible with completion of the genomic sequence and with improved and cheaper robotics and reagents [129–131]. Fortunately, knowledge of the genome sequence, particularly in a microorganism, is not absolutely required to take advantage of this powerful new strategy. Such analysis is being carried out with *Plasmodium falciparum*, the etiological agent of malaria, even before its genome sequence has been determined [132]. For an organism such as *U. maydis* with few and small introns (50–100 nucleotides), DNA microarrays can be constructed by PCR amplification of a small insert (1–2 kb) library constructed in a standard vector (i.e., pUC18). The PCR products are arrayed on glass slides or nylon membranes. These microarrays are then hybridized with cDNAs obtained from two different sets of conditions, each cDNA set synthesized with a different fluor. Up- or downregulated genes can be readily identified by determining the ratio of one fluor to another on a given spot on the microarray. The clone of interest can then be sequenced and analyzed further.

Microarray analysis should prove very powerful in combination with the in vitro system described next for analysis of steps during filament formation and teliospore formation and germination.

5.5 In Vitro Assay for Teliospore Formation

Growth of the filamentous dikaryon in culture is short-lived, and the morphological transitions that result in teliospore formation have never been observed to occur in culture (see Sec. 2). A recent exciting advance has been described [133], which will make fungal development more accessible for genetic dissection and for analysis using microarray technology. These studies show that *U. maydis* haploid cells carrying different ***a*** and ***b*** alleles form filaments that differentiate into teliosporelike structures when placed on top of a small-pore membrane in contact with embryonic maize callus (Fig. 16) [133].

If the cells carry identical ***a*** and ***b*** alleles, only yeastlike growth is observed [133]. Because the cells are separated from the maize callus by a small-pore membrane, the results indicate that the maize callus provides a diffusible substance that triggers differentiation. The putative teliospores that are produced appear to undergo meiosis, although the ratios of segregants obtained are skewed [133]. One possibility is that the teliosporelike structures are contaminated with parental material. Alternatively, there could be a growth disadvantage of some segregants. An interesting observation is that the maize callus cells undergo morphological changes (Fig. 16) [133]. These changes are not observed when *U. maydis* cells of only one mating type are present. Thus, it appears that not only plant signals but also fungal signals diffuse through the membrane.

This system makes it possible to identify diffusible signaling molecules of plant and fungal origin and to identify genes expressed at different times during this developmental program. RNA can be isolated at different times and hybridized to DNA microarrays to identify upregulated genes that could serve as reporter genes for different stages of fungal development. Likewise, microarrays of maize could lead to identification of genes that are important for growth control of the host. In combination with genetic screens, the in vitro system should prove very powerful in the dissection of filamentous growth and teliospore formation.

A combination of different strategies is likely to provide new insights on *U. maydis* development and its interaction with its host maize. *U. maydis* would be an excellent model plant pathogen for such studies.

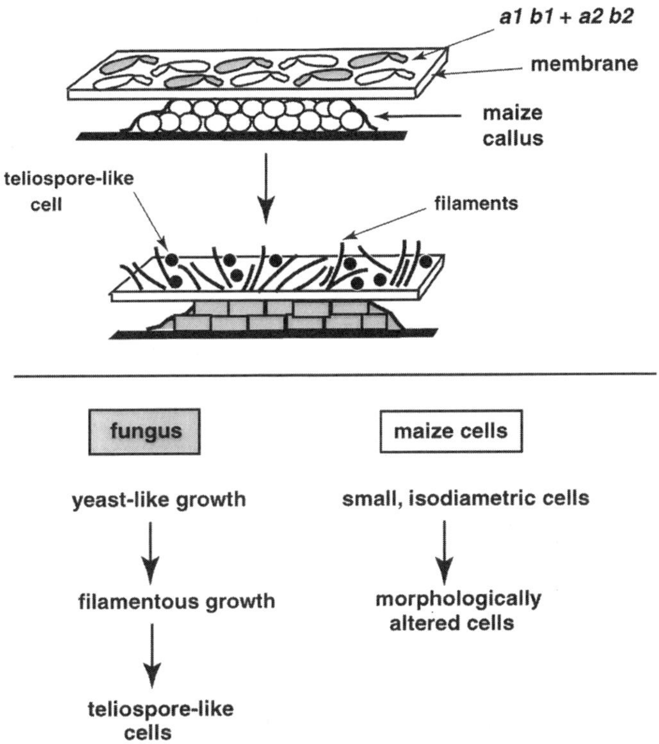

FIGURE 16 In vitro development of teliospores. The process by which *U. maydis* produces teliospores occurs only in the plant and thus is not easily accessible to manipulation [3,5,39]. Recently, formation of teliosporelike structures was accomplished using embryonic maize callus culture [133]. When *U. maydis* cells carrying different ***a*** and ***b*** alleles are placed on a small-pore membrane in contact with the callus tissue, the *U. maydis* cells fuse and form filaments which differentiate into teliosporelike cells. These teliosporelike cells appear to undergo meiosis. If the *U. maydis* cells carry identical ***a*** and ***b*** alleles, no teliospores form. Remarkably, the normally small isodiametric callus cells undergo morphological changes in the presence of *U. maydis* cells only if these cells carry different ***a*** and ***b*** alleles. These results indicate that signals from the maize callus as well as from the fungus are mediating the observed responses. (See text for details.)

6 CONCLUDING REMARKS

The last 10 years have witnessed remarkable progress in studies of *U. maydis*. The mating-type loci of this fungus were one of the first ones cloned and sequenced, after those of the yeasts, and provided a framework in which to analyze the mating-type loci of other basidiomycetes. The analysis of the mating-type loci also demonstrates, for the first time, that genes for pheromones and receptors reside at a major regulatory locus. Work in *U. maydis* has also demonstrated for the first time that components of the highly conserved MAPK cascade module are required for pathogenicity. These findings have now been corroborated in many pathogenic fungi, both animal and plant pathogens. In addition to the MAPK cascade, regulation of cAMP levels appears to be critical for pathogenicity. Thus a common theme is emerging in pathogenic fungi that both MAPK cascades and cAMP are key determinants of pathogenicity.

The challenge in *U. maydis* remains the elucidation of the mechanisms by which the fungus induces tumors in its host, and how the environment of the plant tumor influences the developmental steps that ensue, resulting in teliospore formation. Genes that are necessary for progression of specific steps in this developmental program hold the key to understanding how *U. maydis* interacts with maize and teosinte.

ACKNOWLEDGMENTS

I would like to thank Ira Herskowitz for comments on this manuscript, and Philip Müller and Regine Kahmann for communicating results prior to publication.

REFERENCES

1. CB Kenaga, EB Williams, RJ Green. Plant Disease Syllabus. Lafayette, IN: Balt, 1971.
2. GN Agrios. Plant Pathology. 4th ed. New York: Academic, 1994.
3. JJ Christenssen. Corn smut caused by *Ustilago maydis*. Am Phytopatholol Soc Monogr No. 2, 1963.
4. LA Casselton, NS Olesnicky. Molecular genetics of mating recognition in basidiomycete fungi. Microbiol Mol Biol Rev 62:55–70, 1998.
5. F Banuett, I Herskowitz. Discrete developmental stages during teliospore formation in the corn smut fungus, *Ustilago maydis*. Development 122:2965–2976, 1996.
6. R Holliday. The genetics of *Ustilago maydis*. Genet Res 2:204–230, 1961.
7. R Holliday. *Ustilago maydis*. In: RC King, ed. Handbook of Genetics, Vol 1. New York: Plenum, 1974, pp 575–595.
8. JE Puhalla. Compatibility reactions on solid medium and interstrain inhibition in *Ustilago maydis*. Genetics 60:461–474, 1968.
9. JE Puhalla. Genetic studies of the ***b*** incompatibility locus of *Ustilago maydis*. Genet Res 16:229–232, 1970.

10. PR Day, SL Anagnostakis. Corn smut dikaryon in culture. Nature 231:19–20, 1971.
11. PR Day, SL Anagnostakis, JE Puhalla. Pathogenicity resulting from mutation at the ***b*** locus of *Ustilago maydis*. Proc Natl Acad Sci USA 60:533–535, 1971.
12. J Wang, DW Holden, SA Leong. Gene transfer system for the phytopathogenic fungus *Ustilago maydis*. Proc Natl Acad Sci USA 85:865–869, 1988.
13. S Fotheringham, WK Holloman. Cloning and disruption of *Ustilago maydis* genes. Mol Cell Biol 9:4052–4055, 1989.
14. JW Kronstad, J Wang, SF Covert, DW Holden, GS McKnight, SA Leong. Isolation of metabolic genes and demonstration of gene disruption in the phytopathogenic fungus *Ustilago maydis*. Gene 79:97–106, 1989.
15. F Banuett. Genetics of *Ustilago maydis*, a fungal pathogen that induces tumors in maize. Annu Rev Genet 29:179–208, 1995.
16. R Kahmann, T Romeis, M Bölker, J Kämper. Control of mating and development in *Ustilago maydis*. Curr Opin Genet Dev 5:559–564, 1995.
17. F Banuett. Genes required for filamentous growth and tumor induction by the corn smut fungus *Ustilago maydis*. Proc Natl Acad Sci USA 88:3922–3926, 1991.
18. CW Basse, S Stumpferl, R Kahmann. Characterization of a *Ustilago maydis* gene specifically induced during the biotrophic phase: evidence for negative as well as positive regulation. Mol Cell Biol 20:329–339, 2000.
19. A Bottin, J Kämper, R Kahmann. Isolation of a carbon source–regulated gene from *Ustilago maydis*. Mol Gen Genet 253:342–352, 1996.
20. J Chavez-Ontiveros, AD Martínez-Espinoza, J Ruiz-Herrera. Double chitin synthetase mutants from the corn smut fungus *Ustilago maydis*. New Phytol 146:335–341, 2000.
21. F Dürrenberger, J Kronstad. The *ukc1* gene encodes a protein kinase involved in morphogenesis, pathogenicity, and pigment formation in *Ustilago maydis*. Mol Gen Genet 261:281–289, 1999.
22. SE Gold, G Bakkeren, JE Davies, JW Kronstad. Three selectable markers for transformation of *Ustilago maydis*. Gene 142:225–230, 1994.
23. HA Hartmann, R Kahmann, M Bölker. The pheromone response factor coordinates filamentous growth and pathogenicity in *Ustilago maydis*. EMBO J 15:1632–1641, 1996.
24. CS James, RS Burden, RST Loeffler, JA Hargreaves. Isolation and characterization of polyene-resistant mutants from the maize smut pathogen, *Ustilago maydis*, defective in ergosterol biosynthesis. J Gen Microbiol 138:1437–1443, 1992.
25. JPR Keon, GA White, JA Hargreaves. Isolation and characterization and sequence of a gene conferring resistance to the systemic fungicide carboxin from the maize smut pathogen *Ustilago maydis*. Curr Genet 19:475–481, 1991.
26. JPR Keon, S Jewitt, JA Hargreaves. A gene encoding gamma-adaptin is required for apical extension growth in *Ustilago maydis*.Gene 162:141–145, 1995.
27. TG Kinscherf, SA Leong. Molecular analysis of the karyotype of *Ustilago maydis*. Chromosoma 96:427–433, 1988.
28. M Kojic, WK Holloman. Shuttle vectors for genetic manipulations in *Ustilago maydis*. Can J Microbiol 46:333–338, 1999.
29. AD Martínez-Espinoza, C León, G Elizarraraz, J Ruiz-Herrera. Monomorphic nonpathogenic mutants of *Ustilago maydis*. Phytopathology 87:259–265, 1997.

30. BG Mei, AD Budde, SA Leong. *sid1*, a gene initiating siderophore biosynthesis in *Ustilago maydis*—molecular characterization, regulation by iron, and role in phytopathogenicity. Proc Natl Acad Sci USA 90:903–907, 1993.
31. K Onel, WK Holloman. Molecular cloning of a gene required for DNA replication in *Ustilago maydis*. Mol Gen Genet 253:463–468, 1997.
32. E Regenfelder, T Spellig, A Hartmann, S Lauenstein, M Bölker, R Kahmann. G proteins in *Ustilago maydis*: transmission of multiple signals? EMBO J 16:1934–1942, 1997.
33. BP Rubin, DS Li, WK Holloman. The *leu1* gene of *Ustilago maydis*. Gene 140: 131–135, 1994.
34. G Steinberg, M Schliwa, C Lehmler, M Bölker, R Kahmann, JR Mcintosh. Kinesin from the plant pathogenic fungus *Ustilago maydis* is involved in vacuole formation and cytoplasmic migration. J Cell Sci 111:2235–2246, 1998.
35. T Tsukuda, S Carleton, S Fotheringham, WK Holloman. Isolation and characterizatin of an autonomously replicating sequence from *Ustilago maydis*. Mol Cell Biol 8:3703–3709, 1988.
36. J Wang, AD Budde, SA Leong. Analysis of ferrichrome biosynthesis in the phytopathogenic fungus *Ustilago maydis*: cloning of an ornithine-N^5-oxygenase gene. J Bacteriol 171:2811–2818, 1989.
37. B Xoconostle-Cazares, CA Specht, PW Robbins, Y Liu, C León, J Ruiz-Herrera. *Umchs5*, a gene coding for a class IV chitin synthase in *Ustilago maydis*. Fungal Genet Biol 22:199–208, 1997.
38. F Banuett. *Ustilago maydis*, the delightful blight. Trends Genet 8:174–180, 1992.
39. KM Snetselaar, CW Mims. Light and electron microscopy of *Ustilago maydis* hyphae in maize. Mycol Res 98:347–355, 1994.
40. KM Snetselaar. Microscopic observation of *Ustilago maydis* mating interactions. Exp Mycol 17:345–355, 1993.
41. KM Snetsealaar, M Bölker, R Kahmann. *Ustilago maydis* mating hyphae orient their growth toward pheromone sources. Fungal Genet Biol 20:299–312, 1996.
42. F Banuett, I Herskowitz. Morphological transitions in the life cycle of *Ustilago maydis* and their genetic control by the ***a*** and ***b*** loci. Exp Mycol 18:247–266, 1994.
43. HC Hoch, RC Staples. Structural and chemical changes among the rust fungi during appressorium development. Annu Rev Phytopathol 25:231–247, 1987.
44. AM Abou-Zeid. Effect of *Ustilago maydis* (DC) Corda and its toxin on some maize cultivars. J Phytopathol 143:577–580, 1995.
45. WF Hanna. Studies in the physiology and cytology of *Ustilago zeae* and *Sorosporium reilianum*. Phytopathology 19:415–441, 1929.
46. JM Walter. The mode of entrance of *Ustilago zeae* into corn. Phytopathology 24: 1012–1020, 1934.
47. LJ Mills, JM Kotzé. Scanning electron microscopy of the germination, growth and infection of *Ustilago maydis* on maize. Phytopathol Z 102:21–27, 1981.
48. C Cano-Canchola, L Acevedo, P Ponce-Noyola, A Flores-Martínez, A Flores-Carreón, CA Leal-Morales. Induction of lytic enzymes by the interaction of *Ustilago maydis* with *Zea mays* tissues. Fungal Genet Biol 29:145–151, 2000.
49. JA Callow, IT Ling. Histology of neoplasms and chlorotic lesions in maize seed-

lings following the infection of sporidia of *Ustilago maydis* (DC) Corda. Physiol Plant Pathol 3:489–494, 1973.
50. JE Moulton. Extraction of auxin from maize, from smut tumors of maize and from *Ustilago zeae*. Bot Gaz 103:725–739, 1942.
51. FT Wolf. The production of indoleacetic acid by *Ustilago zeae*, and its possible significance in tumor formation. Proc Natl Acad Sci USA 38:106–111, 1952.
52. LJ Mills, J Van Staden. Extraction of cytokinins from maize, smut tumors of maize and *Ustilago maydis* cultures. Physiol Plant Pathol 13:73–80, 1978.
53. G Turian, RH Hamilton. Chemical detection of 3-indolyl-acetic acid in *Ustilago zeae* tumors. Biochim Biophys Acta 41:148–150, 1960.
54. CW Basse, F Lottspecih, W Steglich, R Kahmann. Two potential indole-3-acetaldehyde dehydrogenases in the phytopathogenic fungus *Ustilago maydis*. Eur J Biochem 242:648–656, 1996.
55. ME Sosa-Morales, F Guevara-Lara, VM Martínez-Juárez, O Paredes-López. Production of indole-3-acetic acid by mutant strains of *Ustilago maydis* (maize smut/huitlacoche). Appl Microbiol Biotechnol 48:726–729, 1997.
56. J Normanly, B Bartel. Redundancy as a way of life—IAA metabolism. Curr Opin Plant Biol 2:207–213, 1999.
57. J Ruiz-Herrera, CG León, A Carabez-Trejo, E Reyes-Salinas. Structure and chemical composition of the cell walls from the haploid yeast and mycelial forms of *Ustilago maydis*. Fungal Genet Biol 20:133–142, 1996.
58. JE Ramberg, DJ McLaughlin. Ultrastructural study of promycelial development and basidioposre initiation in *Ustilago maydis*. Can J Bot 58:1548–1561, 1980.
59. KL O'Donnell, KJ McLaughlin. Ultrastructure of meiosis in *Ustilago maydis*. Mycologia 76:468–485, 1984.
60. KL O'Donnell, KJ McLaughlin. Postmeiotic mitosis, basidiospore development and septation in *Ustilago maydis*. Mycologia 76:486–502, 1984.
61. CJ Alexopoulos, CW Mims. Introductory Mycology. 3rd ed. New York: Wiley, 1992.
62. F Banuett, I Herskowitz. Different ***a*** alleles of *Ustilago maydis* are necessary for maintenance of filamentous growth but not for meiosis. Proc Natl Acad Sci USA 86:5878–5882, 1989.
63. JW Kronstad, SA Leong. The ***b*** mating-type locus of *Ustilago maydis* contains variable and constant regions. Genes Dev 4:1383–1395, 1990.
64. DJ Lew, T Winert, JR Pringle. Cell cycle control in *Saccharomyces cerevisiae*. In: JR Pringle, JR Broach, EW Jones, eds. The Molecular and Cellular Biology of the Yeast *Saccharomyces*, Vol 3. Cold Spring Harbor, NY: Cold Spring Harbor Laboratory Press, 1997, pp 607–695.
65. M Kupiec, B Byers, ME Esposito, AP Mtichell. Meiosis and sporulation in *Saccharomyces cerevisiae*. In: JR Pringle, JR Broach, EW Jones, eds. The Molecular and Cellular Biology of the Yeast *Saccharomyces*, Vol 3. Cold Spring Harbor, NY: Cold Spring Harbor Laboratory Press, 1997, pp 889–1036.
66. BR Braun, AD Johnson. *TUP1*, *CPH1*, and *EFG1* make independent contributions to filamentation in *Candida albicans*. Genetics 155:57–67, 2000.
67. H Liu, J Kohler, GR Fink. Suppression of hyphal formation in *Candida albicans* by mutation of a STE12 homolog. Science 266:1723–1726, 1994.

68. P Wang, J Heitman. Signal transduction cascades regulating mating, filamentation, and virulence in *Cryptococcus neoformans*. Curr Opin Microbiol 2:358–362, 1999.
69. SE Gold, SM Brogdon, ME Mayorga, JW Kronstad. The *Ustilago maydis* regulatory subunit of a cAMP-dependent protein kinase is required for gall formation in maize. Plant Cell 9:1585–1594, 1997.
70. C Quadbeck-Seeger, G Wanner, S Huber, R Kahmann, J Kämper. A protein with similarity to the human retinoblastoma binding protein 2 acts specifically as a repressor for genes regulated by the ***b*** mating type locus in *Ustilago maydis*. Mol Microbiol 38:154–166, 2000.
71. JB Rowell, JE DeVay. Genetics of *Ustilago zeae* in relation to basic problems of pathogenicity. Phytopathology 44:356–362, 1954.
72. JB Rowell. Functional role of compatibility factors: an in vitro test for sexual compatibility with haploid lines of *Ustilago zeae*. Phytopathology 45:370–374, 1955.
73. EH Froeliger, SA Leong. The ***a*** mating-type alleles of *Ustilago maydis* are idiomorphs. Gene 100:113–122, 1991.
74. M Bölker, M Urban, R Kahmann. The ***a*** mating type locus of *U. maydis* specifies cell signaling components. Cell 68:441–450, 1992.
75. M Urban, R Kahmann, M Bölker. The biallelic ***a*** mating type locus of *Ustilago maydis*: remnants of an additional pheromone gene indicate evolution from a multiallelic ancestor. Mol Gen Genet 250:414–420, 1996.
76. T Spellig, M Bölker, F Lottspeich, RW Frank, R Kahmann. Pheromones trigger filamentous growth in *Ustilago maydis*. EMBO J 13:1620–1627, 1994.
77. R Akada, K Minomi, J Kai, I Yamashita, T Miyakawa, S Fukui. Multiple genes coding for precursors of Rhodotorucine A, a farnesyl peptide mating pheromone of the basidiomycetous yeast *Rhodosporidium toruloides*. Mol Cell Biol 9:3491–3498, 1989.
77a. GA Caldwell, F Naider, JM Becker. Fungal lipopeptide mating pheromones: a model system for the study of protein prenylation. Microb Rev 59:406–422, 1995.
78. M Yamamoto, Y Imai, Y Watanabe. Mating and sporualtion in *Schizosaccharomyces pombe*. In: JR Pringle, JR Broach, EW Jones, eds. The Molecular and Cellular Biology of the Yeast *Saccharomyces*, Vol 3. Cold Spring Harbor, NY: Cold Spring Harbor Laboratory Press, 1997, pp 1037–1106.
79. FL Zhang, PJ Casey. Protein prenylation: molecular mechanisms and functional consequences. Annu Rev Biochem 65:241–269, 1996.
80. F Banuett. Signalling in the yeasts: an informational cascade with links to the filamentous fungi. Microbiol Mol Biol Rev 62:249–274, 1998.
81. MC Gustin, J Albertyn, M Alexander, K Davenport. MAP kinase pathways in the yeast *Saccharomyces cerevisiae*. Microbiol Mol Biol Rev 62:1264–1300, 1998.
82. K Abe, I Kusaka, S Fukui. Morphological change in the early stages of the mating process of *Rhodosporidium toruloides*. J Bacteriol 122:710–718, 1975.
83. M Urban, R Kahmann, M Bölker. Identification of the pheromone response element in *Ustilago maydis*. Mol Gen Genet 251:31–57, 1996.
84. B Schulz, F Banuett, M Dahl, R Schlesinger, W Schäfer, I Herskowitz, R Kahmann. The ***b*** alleles of *U. maydis*, whose combinations program pathogenic development, code for polypeptides containing a homeodomain-related motif. Cell 60:295–306, 1990.

85. B Gillissen, J Bergemann, C Sandmann, B Schroeer, M Bölker, R Kahmann. A two-component regulatory system for self/non-self recognition in *Ustilago maydis*. Cell 68:647–657, 1992.
86. J Kämper, M Reichmann, T Romeis, M Bölker, R Kahmann. Multiallelic recognition: nonself-dependent dimerization of the bE and bW homeodomain proteins in *Ustilago maydis*. Cell 81:73–83, 1995.
87. I Herskowitz. Life cycle of the budding yeast *Saccharomyces cervisiae*. Microbiol Rev 52:536–552, 1988.
88. C Goutte, AD Johnson. Recognition of a DNA operator by a dimer composed of two different homeodomain proteins. EMBO J 13:1434–1442, 1994.
89. Y Asada, C Yue, J Wu, GP Shen, CP Novony, RC Ullrich. *Schizophyllum commune* *A*α mating-type proteins, Y and Z, form complexes in all combinations in vitro. Genetics 147:117–123, 1997.
90. M Dahl, M Bölker, B Gillissen, F Schauwecker, B Schroeer, R Kahmann. The ***b*** locus of *Ustilago maydis*: molecular analysis of allele specificity. In: H Hennecke, PDS Verma, eds. Advances in Molecular Genetics of Plant–Microbe Interactions, Vol 1. Amsterdam: Kluwer, 1991, pp 264–271.
91. AR Yee, JW Kronstad. Construction of chimeric alleles with altered specificity at the ***b*** incompatibility locus of *Ustilago maydis*. Proc Natl Acad Sci USA 90:664–668, 1993.
92. AR Yee, JW Kronstad. Dual sets of chimeric alleles identify specificity sequences for the *bE* and *bW* mating and pathogenicity genes of *Ustilago maydis*. Mol Cell Biol 18:221–232, 1998.
93. T Romeis, J Kämper, R Kahmann. Single-chain fusions of two unrelated homeodomain proteins trigger pathogenicity in *Ustilago maydis*. Proc Natl Acad Sci USA 94:1230–1234, 1997.
94. T Romeis, A Brachmann, R Kahmann, J Kämper. Identification of a target gene for the bE-bW homeodomain protein complex in *Ustilago maydis*. Mol Microbiol 37:54–66, 2000.
95. R Schlesinger, R Kahmann, J Kämper. The homeodomains of the heterodimeric bE and bW proteins of *Ustilago maydis* are both critical for function. Mol Gen Genet 254:514–519, 1997.
96. CI Robertson, KA Bartholomew, CP Novotny, RC Ullrich. Deletion of the *Schizophyllum commune* **A**α locus: the roles of **A**α *Y* and *Z* mating-type genes. Genetics 144:1437–1444, 1996.
97. AK Vershon, YS Jin, AD Johnson. A homeodomain protein lacking specific side chains of helix 3 can still bind DNA and direct transcriptional repression. Genes Dev 9:182–192, 1995.
97a. VR Iyer, CE Horak, CS Scafe, D Botstein, M Snyder, PO Brown. Genomic binding sites of the yeast cell-cycle transcription factors SBF and MBF. Nature 409:533–538, 2001.
98. F Banuett, I Herskowitz. Identification of Fuz7, a *U. maydis* MEK/MAPKK homologue required for ***a***-dependent and independent processes in the fungal life cycle. Genes Dev 8:1367–1378, 1994.
99. C Csank, K Schroppel, E Leberer, D Harcus, O Mohamed, S Meloche, D Thomas, M Whiteway. Roles of the *Candida albicans* mitogen-activated protein kinase ho-

molog, Cek1, in hyphal development and systemic candidiasis. Infect Immunol 66: 2713–2721, 1998.
100. S Lev, A Sharon, R Hadar, H Ma, RB Horwitz. A mitogen-activated protein kinase of the corn leaf pathogen *Cochliobolus heterostrophus* is involved in conidation, appressorium formation, and pathogenicity: diverse roles for mitogen-activated protein kinase homologs in foliar pathogens. Proc Natl Acad Sci USA 96:13542–13547, 1999.
101. A Zheng, M Campbell, J Murphy, S Lam, J-R Xu. The *BMP1* gene is essential for pathogenicity in the gray mold fungus *Botrytis cinerea*. Mol Plant Microbe Interact 13:724–732, 2000.
101a. JR Xu, JE Hamer. MAP kinase and cAMP signaling regulate infection structure fromation and pathogenic growth in the rice blast fungus *Magnoporthe grisea*. Genes Dev 10:2696–2706, 1996.
102. Y-K Kim, T Kawano, D Li, PE Kolattukudy. A mitogen-activated protein kinase kinase required for induction of cytokinesis and appressorium formation by host signals in the conidia of *Colletotrichum gloeosporioides*. Plant Cell 12:1331–1343, 2000.
103. ME Mayorga, SE Gold. A MAP kinase encoded by the *ubc3* gene of *Ustilago maydis* is required for filamentous growth and full virulence. Mol Microbiol 34: 485–497, 1999.
104. P Müller, C Aichinger, M Feldbrügge, R Kahmann. The MAP kinase Kpp2 regulates mating and pathogenic development in *Ustilago maydis*. Mol Microbiol 34: 1007–1017, 1999.
105. DL Andrews, JD Egan, ME Mayorga, SE Gold. The *Ustilago maydis ubc4* and *ubc5* genes encode members of a MAP kinase cascade required for filamentous growth. Mol Plant Microbe Interact 13:781–786, 2000.
106. R Sugiura, T Toda, S Dhut, H Shuntoh, T Kuno. The MAPK kinase Pek1 acts as a phosphorylation-dependent molecular switch. Nature 399:479–483, 1999.
107. HA Hartmann, J Krüger, F Lottspeich, R Kahmann. Environmental signals controlling sexual development of the corn smut fungus *Ustilago maydis* through the transcriptional regulator Prf1. Plant Cell 11:1293–1305, 1999.
108. JW Kronstad, A De Maria, D Funnell, RD Laidlaw, N Lee, M Moniz de Sá, M Ramesh. Signaling via cAMP in fungi: interconnections with mitogen-activated protein kinase pathways. Arch Microbiol 170:395–404.
109. SE Gold, G Duncan, K Barrett, J Kronstad. cAMP regulates morphogenesis in the fungal pathogen *Ustilago maydis*. Genes Dev 8:2805–2816, 1994.
110. J Krüger, G Loubradou, E Regenfelder, A Hartmann, R Kahmann. Crosstalk between cAMP and pheromone signalling pathways in *Ustilago maydis*. Mol Gen Genet 260:193–198, 1998.
111. F Dürrenberger, K Wong, JW Kronstad. Identification of a cAMP-dependent protein kinase catalytic subunit required for virulence and morphogenesis in *Ustilago maydis*. Proc Natl Acad Sci USA 95:5684–5689, 1998.
111a. J Krüger, G Loubradou, G Wanner, E Regenfelder, M Feldbrügge, R Kahmann. Activation of the cAMP pathway in *Ustilago maydis* reduces fungal proliferation and teliospore formation in plant tumors. Mol Plant–Microbe Interact 13:1034–1040, 2000.

112. AB Orth, M Rzhetskaya, EJ Pell, M Tien. A serine (threonine) protein kinase confers fungicide resistance in the phytopathogenic fungus *Ustilago maydis*. Appl Environ Microbiol 61:2341–2345, 1995.
113. X Pan, J Heitman. Cyclic AMP-dependent protein kinase regulates pseudohyphal differentiation in *Saccharomyces cerevisiae*. Mol Cell Biol 19:4874–4887, 1999.
114. LS Robertson, GR Fink. The three A kinases have specific signaling functions in pseudohyphal growth. Proc Natl Acad Sci USA 95:13783–13787, 1998.
115. R Kahmann, C Basse, M Felbrügge. Fungal–plant signalling in the *Ustilago maydis*–maize pathosystem. Curr Opin Microbiol 2:647–650, 1999.
116. S Rupp, E Summers, H Lo, H Madhani, G Fink. MAP kinase and cAMP filamentation signaling pathways converge on the unusually large promoter of the yeast *FLO11* gene. EMBO J 18:1257–1269.
117. FD Ivey, Q Yang, KA Borkovich. Positive regulation of adenylyl cyclase activity by a Gα_1 homolog in *Neurospora crassa*. Fungal Genet Biol 26:48–61, 1999.
118. AM Kays, PS Rowley, RA Baasiri, KA Borkovich. Regulation of conidiation and adenylyl cyclase levels by the Gα protein GNA-3 in *Neurospora crassa*. Mol Cell Biol 20:7693–7705, 2000.
119. T Isshiki, N Mochizuki, T Maeda, M Yamamoto. Characterization of a fission yeast gene, *gpa2*, that encodes a Gα subunit involved in the monitoring of nutrition. Genes Dev 6:2455–2462, 1992.
120. C Lehmler, G Steinberg, KM Snetselaar, M Schliwa, R Kahmann, M Bölker. Identification of a motor protein required for filamentous growth in *Ustilago maydis*. EMBO J 12:3464–3473, 1997.
121. O Yarden, M Plamann, DJ Ebbole, C Yanofsky. *cot-1*, a gene required for hyphal elongation in *Neurospora crassa*, encodes a protein kinase. EMBO J 11:2159–2166, 1992.
122. F Verde, DJ Wiley, P Nurse. Fission yeast *orb6*, a ser/thr protein kinase related to mammalian rho kinase and myotonic dystrophy kinase, is required for maintenance of cell polarity and coordinates cell morphogenesis with cell cycle. Proc Natl Acad Sci USA 95:7526–7531, 1998.
123. TL Buhr, S Oved, GM Truesdell, C Huang, O Yarden, MB Dickman. A kinase encoding a gene from *Colletotricum trifolii* complements a colonial growth mutant of *Neurospora crassa*. Mol Gen Genet 19:565–572, 1996.
124. M Bölker, HU Böhnert, KH Braun, J Görl, R Kahmann. Tagging pathogenicity genes in *Ustilago maydis* by restriction enzyme–mediated integration (REMI). Mol Gen Genet 248:547–552, 1995.
125. R Kahmann, C Basse. REMI (restriction enzyme mediated integration) and is impact on the isolation of pathogencity genes in fungi attacking plants. Eur J Plant Pathol 105:221–229, 1999.
126. L Giasson, JW Kronstad. Mutations in the *myp1* gene of *Ustilago maydis* attenuate mycelial growth and virulence. Genetics 141:491–501, 1995.
127. F Schauwecker, G Wanner, R Kahmann. Filament-specific expression of a cellulase gene in the dimorphic fungus *Ustilago maydis*. Biol Chem 376:617–625, 1995.
128. CW Basse, S Stumpfer, R Kahmann. Characterization of a *Ustilago maydis* gene specifically induced during the biotrophic phase: evidence for negative as well as positive regulation. Mol Cell Biol 20:329–339, 2000.

129. DA Lashkari, JL DeRisi, JH McCusker, AF Namath. Yeast microarrays for genome wide parallel genetic and gene expression analysis. Proc Natl Acad Sci USA 94: 13057–13062, 1997.
130. TL Ferea, PO Brown. Observing the living genome. Curr Opin Genet Dev 9:715–722, 1999.
131. S Chu, J DeRisi, M Eisen, J Mulholland. The transcriptional program of sporulation in budding yeast. Science 282:699–705, 1998.
132. RE Hayward, JL DeRisi, S Alfadhli, DC Kaslow, PO Brown, PK Rathod. Shotgun DNA microarrays and stage-specific gene expression in *Plasmodium falciparum* malaria. Mol Microbiol 35:6–14, 2000.
133. J Ruiz-Herrera, C León-Ramírez, JL Cabrera-Ponce, AD Martínez-Espinoza, L Herrera-Estrella. Completion of the sexual cycle and demonstration of genetic recombination in *Ustilago maydis* in vitro. Mol Gen Genet 262:468–472, 1999.

15

Pathogenic Development in *Magnaporthe grisea*

Fernando A. Tenjo
Duke University Medical Center, Durham, North Carolina

John E. Hamer
Paradigm Genetics, Inc., Research Triangle Park, North Carolina

1 INTRODUCTION

Rice is produced at worldwide levels of approximately half a billion tons annually. Humans consume >90% of the rice production, and half of the world's human population derives their major caloric intake from rice consumption [1]. Given the ongoing increases in human population, it is reasonable to predict that rice consumption and demand will increase for the next several decades [1]. Pathogens account for 15% of the rice losses ($33 billion between 1975 and 1990 [2]. The most damaging disease of the rice crop is the rice blast disease which is caused by the heterothallic fungus *Magnaporthe grisea* (Herbert) Barr (anamorph, *Pyricularia grisea*). This disease is responsible for the loss of 157 million tons of rice worldwide between 1975 and 1990 [3]. Despite the efforts and understanding of the biology of both the fungus and the plant, the disease is still a potential threat for rice production [3].

The advances in understanding the biology of both rice and *M. grisea* make the rice blast disease a tractable system for the study of plant–pathogen interactions [4–7]. Rice is a member of the monocotyledoneous grass family Poaceae that also includes maize, barley, and wheat. Despite their economic importance, molecular studies of monocots have been hindered by the large genome size of these plants [1]. Rice possesses a small genome size, which results in higher gene density relative to the other cereals [1,8]. In addition, a well-saturated genetic map has been developed which makes the use of positional cloning an alternative for the isolation and characterization of agronomically important genes. Efficient rice transformation and regeneration technologies are available, making the characterization of genes and their function possible [8,9]. Thus, rice is considered a model system for cereal genomics. Also, isogenic rice cultivars carrying blast disease resistance genes are accessible [10].

M. grisea parasitizes >50 grasses, including economically important crops like barley, wheat, rice, and millet, although individual field isolates are limited to infect one or a few host species [5]. Rice pathogens of *M. grisea* infect aerial parts of the rice plant, causing large, ellipsoid lesions on the leaves. The lesions on the neck can result in complete loss of the crop [4]. *M. grisea* is a haploid member of the Ascomycetes family (class Pyrenomycetes), is not an obligate parasite, and can be grown in a variety of culture media. A sexual cycle can be induced under laboratory conditions thanks to the availability of fertile strains, allowing the use of genetic analysis to study various aspects of the biology of the fungus and the analysis of mutants [11]. In addition, various genomic and cDNA libraries are available and genetic maps have been developed [12–16]. Transformation protocols have been developed and various auxotrophic and drug resistance markers are available [11,17–19]. The formation of infection-related structures (appressoria, germ tubes, infection hyphae) can be followed in vitro using artificial surfaces and plant surfaces [6,20]. A large international strain collection and various probes for population genetics studies are available [21,22]. Infections can be carried out in a variety of hosts [23], and the plants can be infected by spraying the spores on the plants, by injecting the spores, or by depositing mycelia on abraded leaf surfaces [4]. The infection process can be followed biochemically or cytologically, allowing the analysis of infection deficient mutant phenotypes [4,6,20,24].

2 INFECTION CYCLE

The conidia, or asexual spores, of *M. grisea* initiate the infection once they land on a rice leaf (Fig. 1). A spore tip mucilage contained in the compartment of the conidial apex is released when the conidium is hydrated and serves to bind the conidium to the rice leaf surface [25]. Rice leaves are coated with a waxy cuticle, making them extremely hydrophobic. The mucilage mediates the attachment to this repellent surface, and a partial characterization demonstrates that it contains protein, carbohydrate, and lipid components [6].

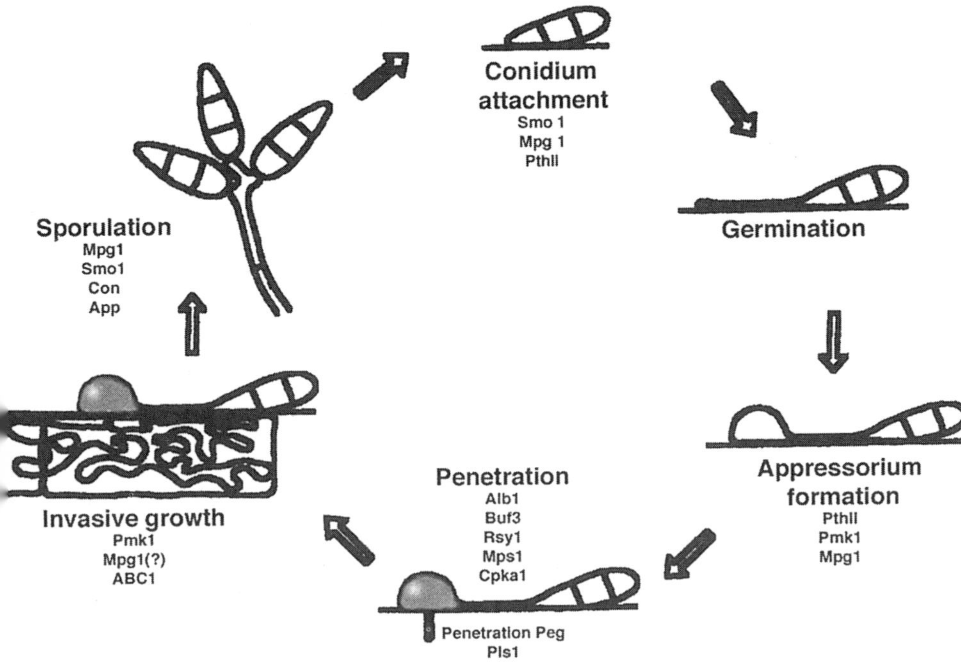

FIGURE 1 Infection cycle and key infection-related genes of *Magnaporthe grisea*. The diagram shows the different stages during the infection cycle of the rice blast fungus, and it also includes the genes involved in pathogenic development. The infection starts when a conidium lands on a rice leaf and attaches. Under high humidity the spore germinates and appressorium formation is initiated. After 8–12 h the appressorium melanizes and turgor pressure is generated within the appressorium. A high turgor pressure is required to drive the penetration of the rice leaf by a specialized hypha called the penetration peg. An infection hypha invades the rice leaf and symptoms can be observed 72 h after the infection starts. Sporulation occurs at the lesions, and conidia are spread to neighboring plants.

Once attached, the conidium germinates and produces a germ tube within 2 h. The germ tube normally emerges from one of the terminal cells of the conidium and grows apically. Within 4 h, growth ceases and the germ tube tip hooks and swells to form the appressorium [6,24]. During maturation, the appressorium becomes melanized except at a well-defined pore between the appressorium and the rice leaf [6]. The melanin layer is freely permeable to water but not to solute, and it allows the appressorium to establish and maintain a high internal turgor pressure [6,26]. Once melanization is complete, a narrow hypha called the *pene-*

tration peg enlarges through the pore, and the appressorium uses turgor pressure to drive the penetration peg through the host surface [6,26,27].

The penetration peg forms an infection hypha that invades and grows quickly throughout the rice leaf, and 72 h after the original inoculation, up to 10% of the biomass of an infected leaf is fungal material [28]. At this stage, the symptoms become evident and small oval lesions begin to appear and are accompanied by local chlorosis. The lesions grow and become necrotic before coalescing and covering most of the leaf surface. Conidiation at the large coalescent lesions occurs only when humidity exceeds 95%; 2000–6000 conidia can be produced each night for about 2 weeks. The conidia are carried by moist air to neighboring plants, thus spreading the disease [4].

3 GENERAL ASPECTS OF APPRESSORIUM FORMATION IN *M. GRISEA*

Appressorium formation in *M. grisea* has been extensively studied thanks to the experimental tractability of the fungus. The process can be followed in vitro using artificial and plant surfaces, and various studies have described in detail the morphologic changes that take place during appressorium development [6,20]. Thus, appressorium-deficient mutants can be characterized and their defect can be assigned to a particular stage of the formation or function of the appressorium [7].

Attachment of the conidium to a surface is an important factor in the development of an appressorium [25,29]. Germination of conidia only requires free water, and typically a single germ tube emerges from the apical and/or basal cell of the three-celled conidium [24]. After germination, the subsequent growth and differentiation of germ tubes may depend on the surface over which it occurs. Hardness of the substrate is an important factor for differentiation, as *M. grisea* conidia only form appressoria on solid surfaces, not on the surfaces of liquid or soft substrates [29].

It is generally accepted that hydrophobic surfaces trigger appressorium formation [25,30–32]. Lee and Dean [33] found that the frequency of appressorium development correlates with the degree of surface hydrophobicity. Hydrophilic substrates, such as glass surfaces, are considered noninductive substrates for appressorium formation [25,30]. However, an absolute correlation between hydrophobicity of the substrate and appressorium development cannot be made, since appressorium formation has been reported on noninductive surfaces [29,34]. This discrepancy is explained by differences in the experimental conditions such as different strains and kind of glass surfaces used. Despite these differences, it is believed that the hydrophobic polymers (Teflon, Polysterene) that trigger appressorium development resemble the waxy, hydrophobic surface of rice leaves

[25,30,32]. No special topographical characteristics of the surface are necessary, as has been observed with other appressorium-forming fungi [35].

Components of the rice leaf surface are able to elicit appressorium formation. Appressorial development is induced in conidia germinating on glass slides covered with rice leaf wax extracts [36]. The rice leaf cuticle also plays a role in the development of an appressorium, as cuticle monomers induce appressorium formation [32].

Once the spore germinates, the germ tube grows and a window of competence of appressorium formation has been observed. If the spore germinates on noninductive surfaces and it is not elicited to make an appressorium within 6–8 h, the germ tube will not commit to make appressorium [31]. Thus, the formation of an appressorium is the result of a series of events that result in a regulated developmental pathway.

Initial signs for appressorium formation are the arrest of growth of the germ tube, initial swelling at the apex, and hook formation [6]. These events occur at 4–6 h after germination. Then, the appressorium continues to swell, and the contents of the cytoplasm of the spore and germ tube are transported to the incipient appressorium. These contents include the accumulation of lipid droplets which appear to be deposited in a central vacuole of the developing appressorium [37,38]. A septum is formed between the appressorium and the germ tube. As maturation progresses, a melanin layer is deposited on the cell wall of the appressorium that covers most of the cell perimeter, except across the region in contact with the substratum. After melanization, this area of the cell surface appears to lack any cell wall, and is called the appressorium pore [6].

The nucleus of the cell that gives rise to the germ tube undergoes a mitotic division a short time after germination. One daughter nucleus moves through the germ tube and enters the forming appressorium where it is sealed in place by formation of the septum. The other nucleus remains in the conidium. Under in vitro conditions an appressorium completes expansion 4–8 h after conidia germinate [6].

Melanization of the appressorium is required for the penetration of the rice leaf. The use of chemical such as tricyclazole, which inhibits melanization, results in the production of unpigmented appressoria that are nonfunctional, and penetration fails to occur [6]. Melanin-deficient mutants are still able to make appressoria but they fail to penetrate the rice leaf [39]. The mutants are fully pathogenic if they are injected into leaf tissue. In other fungi, such as *Colletotrichum* spp., melanization is essential for the generation of fully functional appressoria as well [35,40].

The penetration of the rice leaf is accomplished by the ability of the appressorium to generate an enormous turgor pressure [27]. Turgor pressure is generated by the appressorium over a period of 24–48 h [27]. Penetration of artificial surfaces and rice leaves is inhibited by small decreases of turgor pressure [41]. Measurements of the turgor pressure estimate a value of at least 8.0 Mpa. This is the equivalent to 40 times the pressure in a car tire [42]. Appressorial melanin limits wall permeability, facilitating osmolyte accumulation and turgor generation within the cell [26]. Glycerol is the solute that provides the osmotic potential to generate the turgor pressure as it accumulates in excess of 3 M during generation of turgor by the appressorium [41]. Appressoria of melanin-deficient mutants are unable to accumulate glycerol, and a similar result is observed when wild-type appressoria are treated with tricyclazole, a melanin biosynthesis inhibitor [41].

The plant cell is then invaded by a specialized hypha that develops beneath the appressorium [6,20]. The role of extracellular enzymes in the penetration process is still under debate. Mutants obtained by gene deletion experiments of cutinase and xylanase are still pathogenic [35,43]. A residual enzymatic activity is still detected in both cases, which can be interpreted as the presence of isoenzymes that are regulated and expressed during penetration.

In conclusion, appressorium formation in *M. grisea* is a differentiation process that involves a series of morphological and physiological changes as a response to an appropriate infection site. The availability of an in vitro system to study appressorium development has been used to study the role of different genes during the various steps of appressorium development.

4 GENES INVOLVED IN APPRESSORIUM DIFFERENTIATION

Genes controlling the elaborate developmental pathway that the blast fungus undergoes during infection have been isolated by different strategies. Some of the genes affect appressorium formation while others are important for appressorium function (penetration).

4.1 The Melanin Layer

The importance of the melanin layer for appressorial function has been demonstrated genetically. Three genes involved in the melanin biosynthesis have been identified: *ALB1*, *RSY1*, and *BUF1* [39]. Strains carrying mutations of any of these genes are unable to synthesize melanin, make nonfunctional appresoria, and are nonpathogenic. These mutants are fully pathogenic when inoculated directly into the leaf tissue [39]. *ALB1* encodes a polyketide synthase, which is involved in the first step in the formation of melanin. The subsequent steps are a dehydration and reduction that are mediated by *RSY1* and *BUF1*, respectively [6]. *RSY1* encodes scytalone dehydratase, an extensively studied enzyme that has a known three-dimensional structure [44,45]. *BUF1* encodes a trihydroxynapthalene reductase, an NADPH-dependent dehydrogenase. Although mutants in

BUF1 are nonpathogenic they still produce scytalone, and a second tetrahydroxynaphthalene reductase gene has recently been cloned and shown to function in the production of scytalone. Binding studies demonstrate that this second enzyme is the primary physiological target for the fungicide tricyclazole [46].

4.2 Conidiation Genes

In some cases, genes that affect conidiation or conidial morphology also affect appressorium formation and function. The screening of mutants with a reduced or null pathogenicity phenotype has identified most of these genes.

4.2.1 SMO1

A mutation affecting conidial shape allowed the identification of a locus called *SMO1* (spore morphology) [47]. *smo*$^-$ mutants form spores, appressoria, and asci with an aberrant morphology. *smo*$^-$ mutants make appressorium on noninductive surfaces. The pathogenicity toward rice is diminished, and the number and size of the lesions are also reduced [48]. Unfortunately, efforts to clone *SMO1* by positional cloning or complementation have failed owing to a rearrangement tightly associated with the locus.

4.2.2 CON Genes

Six genes, *CON1*, *CON2*, *CON4*, *CON5*, *CON6*, and *CON7*, that control various steps in the sporulation pathway were identified using chemical and insertional mutagenesis [49]. Each of these mutations results in a unique profile of abnormal characteristics in sporulation and spore morphogenesis. *con5*$^-$ mutants do not form conidiophores, while *con6*$^-$ mutants produce abundant conidiophores but not conidia. Mutations at the *CON1*, *CON2*, *CON4*, and *CON7* loci result in the production of conidia with distinct cell shapes. *con2*$^-$ and *con4*$^-$ conidia produce fewer appressoria than wild type, and have a reduced pathogenicity on rice plants. Conidia of *con1*$^-$ and *con7*$^-$ display abnormal morphologies and do not form appressoria. Consequently, they are nonpathogenic even when plants are infected by wound inoculation [49,50]. Genetic analysis suggests that *CON2* is epistatic to *CON1* and *CON7*, *CON5* is epistatic to *CON6*, and *CON6* is epistatic to *CON7* [50]. The presence of a core pathway determining conidial development is indicated by these studies. *CON4* may affect spore development independently of this pathway [50].

4.2.3 *APP* Loci

UV mutagenesis experiments led to the identification of three genetic loci, *APP1*, *APP2*, and *APP3*, that affect appressorium formation [51]. *app1*$^-$ and *app2*$^-$ conidia have a reduced ability to make appressoria on hydrophobic surfaces. The addition of cAMP or cutin monomers can restore the ability of *app2*$^-$ conidia to make appresoria, while the addition of these inducers does not have any effect on *app1*$^-$ conidia. Allelism tests confirmed that they are separate loci, with *APP1* and *APP3* closely linked. *APP1* was mapped to the central region of the chromosome 2 of *M. grisea* by RFLP analysis, and various markers have been identified that will assist in the isolation of this gene by positional cloning [51].

The study of a spontaneous mutant defective in appressorium formation led to the definition of the gene *APP5* [52]. Crosses with mutants *app1*$^-$ and *app3*$^-$ revealed that *APP5* is a novel genetic locus. *app5*$^-$ mutants do not form appressoria on hydrophobic surfaces or on rice leaves, even in the presence of cAMP or cutin monomers. They are not pathogenic either by inoculating conidial suspensions by spraying or by injecting them into the rice leaf sheath. Experiments are under way to map *APP5* [52].

4.2.4 *APF1*

A spontaneous, nonpathogenic mutant resulted from a cross between two rice pathogenic *M. grisea* strains. The study of this mutant led to the identification of *APF1*, a gene with pleiotropic effects on spore morphology and on the ability to make appressorium [53]. Spores of *apf1*$^-$ are unable to make appressoria on inductive surfaces, and the addition of cAMP does not restore this capability. Thus, *APF1* may be acting downstream of the cAMP-signaling pathway, or it can be working in a cAMP-independent pathway for appressorium differentiation. In addition, *apf1*$^-$ mutants cannot infect rice plants by injection through the leaf sheath, and this suggests that the gene may be needed for invasive growth [53].

4.3 Sensing the Host Surface: Genes Involved in Surface Recognition and Signal Transduction Pathways

Appressorium formation is favored in *M. grisea* by contact with a hard, hydrophobic surface [30,32]. On noninductive surfaces, germinating conidia fail to proceed beyond the swelling and hooking of the germ tube (considered an early recognition phase), and the germ tube develops into a typical vegetative mycelium.

Appressorium formation can be induced on noninductive surfaces by the addition of cutin monomers and waxes 1,16-hexadecanediol being the most effective [32]. Furthermore, the addition of cAMP to germinating conidia growing on noninductive surfaces triggers appressorial development [33]. Thus, it is proposed that germinating conidia of *M. grisea* are able to respond to environmental cues and discriminate among surfaces to commit into appressorium formation [31,33]. The recognition of a hydrophobic surface results in signaling events that may mediate the differentiation and function of appressoria. Genes that play a role in this process have been cloned and their roles have been tested by genetic and biochemical approaches.

4.3.1 Genes Sensing the Surface

PTH11. Mutants with a reduced or suppressed pathogenicity were obtained by restriction enzyme insertional mutagenesis (REMI) [54]. *pth11⁻* mutants were identified by their defects in appressorium formation and pathogenicity. *pth11⁻* mutants have a reduced appressoria formation, and as a result their pathogenicity is also impaired. The formation of appressoria does not go beyond the swelling and hooking of the germ tube apex, resembling the growth of wild-type conidia on hydrophilic surfaces. Addition of cutin monomers to *pth11⁻* conidia growing on a hydrophobic surface does not initiate appressorium formation, suggesting that *pth11⁻* mutants are unable to recognize appressorium-inductive cues [55].

PTH11 encodes a novel transmembrane protein with multiple spanning regions. The protein not only localizes to the cell membrane but also is associated with intracellular compartments, especially vacuoles. This localization resembles the localization of the mating pheromone receptors and membrane transporters in *Saccharomyces cerevisiae*. Further characterization of *PTH11* in different *M. grisea* strains shows that *PTH11* can be acting as an inducer of appressorium formation on hydrophobic surfaces and as a repressor on noninductive surfaces [55].

MPG1. Another protein that may play a role in the perception of hydrophobic surfaces is encoded by the gene *MPG1*. *MPG1* was isolated in a screen for genes highly expressed in infected rice plants [56]. The expression of the gene is higher in early stages of appressorium formation and in the late stages of the infection process, a period that coincides with the onset of symptom development and conidiation from the lesions [56]. *mpg1⁻* mutants were obtained by gene replacement and they display a reduced conidiation, an impaired appressorium formation, and a reduced pathogenicity [56].

MPG1 encodes a hydrophobin [56], a new class of hydrophobic proteins that have been described in fungi, which are involved in conferring the hydrophobic character to various fungal structures and are also involved in various developmental processes including pathogenesis [57–59]. *mpg1⁻* mutants are "easily wettable," a characteristic associated with a reduced surface hydrophobicity, which also can be observed in other hydrophobin mutant fungi [56]. Electron microscopic studies of their conidial surface showed the absence of a rodlet layer that covers the spore in wild-type conidia. The presence of the rodlet layer is restored after transforming the *mpg1⁻* mutant with *MPG1*, suggesting that *MPG1* directs the formation of the rodlet layer on the conidial surface. Consistently, an epitope-tagged MPG1 protein can be detected on the spore surface and on the tip of conidiophores (Tenjo and Hamer, manuscript in preparation). Rodlet layers are present on the surface of fungal structures, making them hydrophobic [60].

Hydrophobins are the major components of the rodlet layer, and their formation is associated with the property of hydrophobins to self-assemble in high-molecular-weight complexes at interfaces between air and water or between any hydrophilic and hydrophobic interface [59].

The self-assembly of hydrophobins is also involved in the attachment of fungal structures to hydrophobic surfaces [61]. Consistently, *mpg1*⁻ mutants have a reduced attachment to hydrophobic surfaces [60]. Despite the low homology observed among hydrophobins, *mpg1*⁻ mutants can be complemented by various hydrophobins that are supposed to undergo self-assembly when expressed under the *MPG1* promoter [62]. The rodlet layer is present on the surface of the complemented transformants as well as the ability to make appressoria, suggesting that hydrophobin self-assembly is an important factor in the formation of appressoria [62]. Moreover, *mpg1*⁻ mutants seem to have a defect recognizing hydrophobic surfaces because their ability to make appressorium on various hydrophobic materials is impaired [63]. Since cAMP can restore the ability of *mpg1*⁻ mutants to make appressoria, it is suggested that self-assembly of MPG1 may be involved in the recognition and discrimination of hydrophobic surfaces that results in the signaling that triggers appressorium formation [60,63]. This process involves more components, because *mpg1*⁻ mutants are still able to develop appressorium [56,60].

4.3.2 Signal Transduction Pathways

Cyclic AMP (cAMP) Pathway. Addition of cAMP or analogs to conidia of *M. grisea* germinating on a noninductive surface induces appressorial development. Cellular levels of cAMP are regulated by adenylate cyclase and phosphodiesterase, enzymes that are responsible for its synthesis and degradation, respectively. Addition of phosphodiesterase inhibitors to germinating conidia also induces appressorial formation. Thus, it is possible that appressorium formation involves an increase in the cellular levels of cAMP [33].

Activation of adenylate cyclase is regulated by the α-subunit of trimeric GTP-binding (G) proteins. Membrane-bound G-proteins coupled to surface receptors are known to transduce surface signals to intracellular effectors, so they play a role in cell growth and differentiation in many eukaryotes. In filamentous fungi, G-proteins are also involved in pathogenicity [64].

Three genes that encode G-protein α-subunits have been cloned in *M. grisea*: *MAGA*, *MAGB*, and *MAGC* [65]. Deletion of either *MAGA* or *MAGC* does not affect pathogenicity, but *magC*⁻ mutants have a reduced conidiation. These genes appear to function in ascospore maturation and/or germination as perithecia are formed by *magA*⁻ or *magC*⁻ mutants but ascospores fail to germinate. On the other hand, deletion of *MAGB* reduces vegetative growth, conidiation, and appressorium formation. In addition, pathogenicity is impaired in *magB*⁻ mutants and mating is also affected [65]. Dominant negative mutants of *magB* also have severe phenotypes with defects in growth, sporulation, and pathogenicity [66]. In mammalian systems, adenylate cyclase activity is inhibited by G-protein α-subunits, so the levels of intracellular cAMP decrease. If an increased level of intracellular cAMP is required for appressorium formation, then the deletion of α-subunit genes may result in the removal of an inhibitor of adenylate cyclase, and consequently in an increase of intracellular cAMP. Surprisingly, appressorium formation is impaired in *magB*$^-$ mutants and this suggests that the mechanism of action of these proteins is different from their mammalian homologs. Furthermore, two Gα-subunit genes have been identified in *Neurospora crassa*: *GNA1* AND *GNA2* [67]. Deletion of *GNA1* affects various cellular processes, while *gna2* mutants display no obvious phenotypes [68]. *GNA1* is most identical to $G\alpha_I$ subfamily members that negatively regulate adenyl cyclase activity [67]. Surprisingly, adenyl cyclase activity is lost in *gna1*$^-$ mutants [68]. Thus, it is possible that structural differences of fungal adenyl cyclase account for these differences in regulation.

cAMP exerts its effect principally through the activation of cAMP-dependent protein kinase (PKA). Binding of cAMP to the regulatory subunit of an inactive enzyme leads to the release of the catalytic subunit of PKA (CPKA). The active catalytic subunit may activate phosphorylation cascades or migrate to the nucleus to phosphorylate target proteins [69]. As mentioned above, cAMP levels during appressorium formation may rise, which can result in PKA activation [33]. In fact, PKA activity can be detected in germinating conidia and increases during appressorium formation. PKA activity is higher in germinating conidia forming appressoria on a hydrophobic surface than in conidia growing vegetatively on a hydrophilic surface or in liquid culture [70].

The gene encoding *CPKA* has been cloned and mutants were obtained by gene disruption [71,72]. *cpka*$^-$ mutants are delayed in appressorium formation and they are unable to penetrate into the rice plant. Their ability to grow inside the rice plant is not affected, and other factors prior to infectious hyphae formation are normal such as nuclear migration into the appressorium, septa formation, melanization of the appressorium, and tight attachment to the rice leaf [72]. It is possible that *CPKA* is playing a role in the accumulation of glycerol necessary for the generation of turgor pressure within the melanized appressoria. Surprisingly, the formation of appressoria is induced in *cpka*$^-$ mutants growing on hydrophilic surfaces by the addition of cAMP, suggesting the presence of additional *CPKAs* involved in surface sensing [72]. Multiple *CPKA* genes exist in other fungi such as *Saccharomyces cerevisiae* [73] and *Ustilago maydis* [74].

The gene encoding adenylate cyclase (*MAC1*) was cloned and gene disruptants were analyzed to further characterize the role of this enzyme in appressorium formation [75,76]. *MAC1* is able to rescue a *Neurospora crassa* adenylate cyclase mutant, demonstrating that *MAC1* is a functional homolog to other fungal adenylate cyclases [75]. *mac1*$^-$ mutants have reduced conidiation and conidial germination, defects in vegetative growth, and an impaired mating ability [75,76]. Conidia of *mac1*$^-$ mutants are unable to make appressoria on inductive surfaces and are non-pathogenic. However, in the presence of cAMP, the formation of appressoria is restored [76].

Mutations were found that suppress some of the pleiotropic defects associated with the *mac1*$^-$ mutation [75]. *sum* (bypass suppressors of *mac1*$^-$) mutants form appressoria not only on hydrophobic surfaces but also on hydrophilic surfaces. However, pathogenicity is not restored in *sum* mutants. Among the mutations, one (*sum1-99*) is the result of a change of a conserved leucine by arginine in the cAMP-binding domain A of PKA regulatory subunit. A constitutive-active PKA is detected by PKA activity assays in *sum1-99* germlings, suggesting that the mutation has altered the proper regulation of the enzyme activity [75]. Thus, it is proposed that adenylate cyclase regulates both vegetative growth and pathogenesis but that different targets of cAMP mediate downstream effects specific for either cell morphogenesis or pathogenesis. The divergence may occur at the level of PKA regulation [75]. Three PKA catalytic subunits have been described in *S. cerevisiae* encoded by the genes *TPK1*, *TPK2*, and *TPK3* [73]. Various studies suggest that the three catalytic subunits are redundant for function. However, the PKA catalytic subunits have different roles in regulating filamentous growth: *TPK1* and *TPK3* primarily inhibit filamentous growth, whereas *TPK2* activates filamentous growth [77]. Attempts to detect additional genes encoding additional PKA catalytic subunits in *M. grisea* are ongoing.

MAP Kinase Pathways. One of the most-characterized signal transduction pathways in eukaryotes is the one involving MAP (mitogen-activated protein) kinases. As a response to external signals, MAP kinase (MAPK) is activated (phosphorylated) by a MAP kinase kinase (MEK), which in turn is activated by a MAP kinase kinase kinase (MEKK). Active MAPK then can be translocated to the nucleus where it interacts with various molecules, transcription factors generally, to direct new gene expression. *S. cerevisiae* has at least six identified pathways involving MAPKs that regulate various processes, such as mating, spore formation, and response of cells to osmolarity changes [78].

In filamentous fungi, various components of MAPK pathways have been cloned, and their role in various aspects of fungal development and pathogenesis has been established [74,79–82]. Three MAPK genes, *PMK1*, *MSP1*, and *OSM1*, have been cloned in *M. grisea* that play a role in appressorium formation and function [83–85]. The gene *PMK1* is highly homologous to the yeast MAPK genes *FUS3* and *KSS1* that are involved in the pheromone response pathway in yeast [83]. *MSP1* is similar to *SLT2* that is activated in response to membrane stress [84], and *OSM1* is most similar to *HOG1*, a yeast MAPK involved in the signal pathway that controls the response to high osmotic stress [85]. The role of *M. grisea* MAPKs in pathogenesis has been assessed by knockout of the corresponding gene [86].

PMK1 can complement the mating defect of a yeast *fus3/kss1* mutant, which demonstrates that it might work in a similar pathway in *M. grisea.* Surprisingly, *pmk1*$^-$ mutants do not have any defect in mating. *pmk1*$^-$ mutants do not display any defect in vegetative growth or conidiation; however, they arc nonpathogenic to rice plants infected by spraying or by inoculation on abraded leaf surfaces [83]. *pmk1*$^-$ mutants are unable to make appressoria, and when germlings growing on glass surfaces are stimulated with cAMP they can start appressorium differentiation, but this does not proceed beyond the swelling and hooking of the germ tube apex. Thus, *Pmk1* is required for appressorium formation and invasive growth but is dispensable for other aspects of growth and development. *Pmk1* may act downstream of the cAMP dependent pathway, opening the possibility of cooperation between cAMP and MAPK pathways [83].

Nuclear translocation has been observed for mammalian and yeast MAPKs once they are activated as a response to the appropriate stimulus. This migration has been correlated with the function of activated MAPKs as inducers of gene expression and cellular changes that occur as a response to the stimulus [87]. It is possible that active PMK1 migrate to the nucleus during appressorium formation and then interact with a transcription factor and/or downstream target genes required for appressorium development. To test this idea, a PMK1-GFP (green fluorescent protein) was constructed and used to transform a *pmk1*$^-$ strain to check the localization of PMK1 during appressorium development (Tenjo et al., manuscript in preparation). The resultant transformants were fully pathogenic and formed a mature appressorium. At initial stages of appressorium development PMK1-GFP was homogeneously distributed in the spore and in immature appressoria. PMK1-GFP appeared to be translocated to the nucleus at late stages of appressorium differentiation. Thus, PMK1 may be interacting with a transcription factor similar to STE12 (the final target of the MAPK mating and filamentous pathway in yeast) during the later stages of appressorium differentiation. This interaction may result in the activation of genes required for appressorium maturation and function (Tenjo et al., manuscript in preparation).

MPS1 is functionally related to *SLT2*, a yeast MAPK that is part of a signaling pathway in yeast that is responsible for cell wall remodeling in response to low osmolarity conditions and heat stress. *mps1*$^-$ mutants have a reduced conidiation and an impaired aerial hypha formation. In addition, msp1$^-$ fungal colonies growing on solid medium undergo progressive autolysis and also, the sensitivity to cell wall digesting enzymes is increased in *mps1*$^-$ mutants. Both defects can be suppressed by the addition of 1 M sorbitol. Female fertility is also affected in msp1$^-$ mutants. Appressorium formation is not affected in *mps1*$^-$ mutants, but they are nonpathogenic because of their inability to penetrate the rice leaf. Spraying conidial suspension on abraded leaf surfaces or injection into the leaf sheath results in lesions. These results suggest that MPS1 is required for the maintenance of cell wall integrity during various stages of fungal development and is required for the appropriate function of the appressorium [84].

The highly conserved MAPK HOG1 signal pathway controls cellular turgor in eukaryotes. Activation of the HOG1 pathway in yeast leads to the accumulation of glycerol which maintains the cellular turgor even after exposure to severe hyperosmotic stresses. *OSM1* is the functional homolog of *HOG1* in *M. grisea*, as *hog1*$^-$ mutants are complemented by *OSM1*. Deletion of *OSM1* does not result in apparent defects in vegetative growth, but a reduced conidiation can be detected. The response of hypha to hyperosmotic stress is impaired in *osm1*$^-$ mutants and this defect is related to a failure in the accumulation of arabitol in the mycelium. Also, *osm1*$^-$ mutants show reduced growth and morphological defects compared to wild type after exposure to hyperosmotic stress. Surprisingly, appressorium formation and pathogenicity are not affected in *osm1*$^-$ mutants. The accumulation of glycerol and turgor pressure are not affected in *osm1*$^-$ appressoria [85]. Glycerol is the major osmolyte that accumulates in the appressorium for the generation of the enormous turgor pressure that *M. grisea* uses to penetrate the rice leaf. Thus, it is possible that the *M. grisea* OSM1-mediated MAPK pathway acts independently of the signal transduction pathway that leads to appressorium turgor [85].

In summary, *M. grisea* possesses elements of signal transduction pathways that have been adapted to serve functions in pathogenicity such as appressorium formation, penetration, and invasive growth. The observations described above suggest that two of the pathways, PMK1 and MSP1, may be acting in concert to generate a fully functionally appressorium [84]. Even though the MAPKs isolated in *M. grisea* are functionally related to their yeast counterparts, the function of the enzymes is different in developmental processes and responses to environmental signals in the rice blast fungus. Thus, a conserved pathway may be used for other tasks that are the result of the life style of filamentous fungi.

Role of Other Signaling Pathways. Punchless, a nonpathogenic mutant, was obtained by plasmid-mediated insertional mutagenesis [88]. Punchless mutants show normal levels of conidiation, appressorium formation, and melanization, and their vegetative growth is not impaired. Appressoria developed by conidia from Punchless mutants are unable to penetrate rice or barley leaves as well as artificial substrates. This defect is explained by the inability to develop penetration pegs by Punchless appressoria. Furthermore, inoculation of rice leaves with Punchless conidia does not result in infection, suggesting that the mutants are blocked in an early step of the invasion process. The Punchless phenotype is the result of the inactivation of the gene *PLS1*. *PLS1* encodes a putative membrane protein that has all the characteristics of the tetraspanin family [88].

In animal cells the members of the tetraspanin family are involved in development, cell migration, and adhesion. Since in animal cells tetraspanin are associated to other proteins forming signaling complexes that mediate actin reorganization, it is possible that *PLS1* is required for the reorganization of the actin cytoskeleton at the base of the appressorium during the penetration process [6,88].

5 ROLES FOR SIGNALING PATHWAYS

Although both cAMP and MAPK pathways have now been identified as important for pathogenicity in *M. grisea*, downstream effectors of these pathways remain to be identified. Recently both the CPKA and PMK1 signaling pathways have been shown to be necessary for carbohydrate mobilization during appressorium formation, and thus perhaps enzymes in these pathways may be could candidates for genes regulated directly or indirectly by these pathways [37].

Transcription factors also appear to be likely targets for these kinase pathways, and gene sequences with similarities to candidate transcription factors (inferred from studies of their homologs in yeast) should be straightforward to identify. Finally, EST and genome sequencing projects should provide the necessary tools to allow more global surveys of gene expression during appressorium formation and infectious growth. These studies can employ appropriate signaling mutants to confirm the roles of specific genes in infection-related signaling pathways.

REFERENCES

1. SA Goff. Rice as a model for cereal genomics. Curr Opin Plant Biol 2:86–89, 1999.
2. B Baker, P Zambryski, B Staskawicz, SP Dinesh-Kumar. Signaling in plant–microbe interactions. Science 276:726–733, 1997.
3. RS Zeigler, SA Leong, PS Teng. Rice blast disease. Wallingford, Oxon, UK: CAB International in association with the International Rice Research Institute, 1994.
4. NJ Talbot. Having a blast: exploring the pathogenicity of *Magnaporthe grisea.* Trends Microbiol 3:9–16, 1995.
5. B Valent. Rice blast as a model system for plant pathology. Phytopathology 80:33–36, 1990.
6. RJ Howard, B Valent. Breaking and entering: host penetration by the fungal rice blast pathogen *Magnaporthe grisea.* Annu Rev Microbiol 50:491–512, 1996.
7. JE Hamer, NJ Talbot. Infection-related development in the rice blast fungus *Magnaporthe grisea.* Curr Opin Microbiol 1:693–697, 1998.
8. K Shimamoto. The molecular biology of rice. Science 270:1772–1773, 1995.
9. J Kyozuka, K Shimamoto. Rice transformation: methods and applications. Plant Micro Biotech Res Ser 3:53–64, 1995.
10. DH Chen, M de la Vina, T Inukai, DJ Mackill, PC Ronald, RJ Nelson. Molecular

mapping of the blast resistance gene, Pi44(t), in a line derived from a durably resistant rice cultivar. Theor Appl Genet 98:1046–1053, 1999.

11. B Valent, FG Chumley. Molecular genetic analysis of the rice blast fungus, *Magnaporthe grisea*. Annu Rev Phytopathol 29:443–467, 1991.
12. N Nitta, ML Farman, SA Leong. Genome organization of *Magnaporthe grisea*: integration of genetic maps, clustering of transposable elements and identification of genome duplications and rearrangements. Theor Appl Genet 95:20–32, 1997.
13. M Nishimura, S Nakamura, N Hayashi, S Asakawa, N Shimizu, H Kaku, A Hasebe, S Kawasaki. Construction of a BAC library of the rice blast fungus *Magnaporthe grisea* and finding specific genome regions in which its transposons tend to cluster. Biosci Biotechnol Biochem 62:1515–1521, 1998.
14. H Zhu, S Choi, AK Johnston, RA Wing, RA Dean. A large-insert (130 kbp) bacterial artificial chromosome library of the rice blast fungus *Magnaporthe grisea*: genome analysis, contig assembly, and gene cloning. Fungal Genet Biol 21:337–347, 1997.
15. ML Farman, SA Leong. Chromosome walking to the AVR1-CO39 avirulence gene of *Magnaporthe grisea*: discrepancy between the physical and genetic maps. Genetics 150:1049–1058, 1998.
16. H Zhu, BP Blackmon, M Sasinowski, RA Dean. Physical map and organization of chromosome 7 in the rice blast fungus, *Magnaporthe grisea*. Genome Res 9:739–750, 1999.
17. KA Parsons, FG Chumley, B Valent. Genetic transformation of the fungal pathogen responsible for rice blast disease. Proc Natl Acad Sci USA 84:4161–4165, 1987.
18. P Kachroo, A Potnis, BB Chattoo. Transformation of the rice blast fungus *Magnaporthe grisea* to benomyl resistance. World J Microbiol Biotechnol 13:185–187, 1997.
19. MJ Kershaw, G Wakley, NJ Talbot. Complementation of the Mpg1 mutant phenotype in *Magnaporthe grisea* reveals functional relationships between fungal hydrophobins. EMBO J 17:3838–3849, 1998.
20. MC Heath, RJ Howard, B Valent, FG Chumley. Ultrastructural interactions of one strain of *Magnaporthe grisea* with goosegrass and weeping lovegrass. Can J Bot 70:779–787, 1992.
21. L Babujee, SS Gnanamanickam. Molecular tools for characterization of rice blast pathogen (*Magnaporthe grisea*) population and molecular marker–assisted breeding for disease resistance. Curr Sci 78:248–257, 2000.
22. JE Hamer, L Farrall, MJ Orbach, B Valent, FG Chumley. Host species-specific conservation of a family of repeated DNA sequences in the genome of a fungal plant pathogen. Proc Natl Acad Sci USA 86:9981–9985, 1989.
23. G Lau, JE Hamer. Regulatory genes controlling MPG1 expression and pathogenicity in the rice blast fungus *Magnaporthe grisea*. Plant Cell 8:771–781, 1996.
24. TM Bourett, RJ Howard. In vitro development of penetration structures in the rice blast fungus *Magnaporthe grisea*. Can J Bot 68:329–342, 1990.
25. JE Hamer, RJ Howard, FG Chumley, B Valent. A mechanism for surface attachment in spores of a plant pathogenic fungus. Science 15:288–290, 1988.
26. NP Money, RJ Howard. Confirmation of a link between fungal pigmentation, turgor pressure, and pathogenicity using a new method of turgor measurement. Fungal Genet Biol 20:217–227, 1996.

27. RJ Howard, MA Ferrari, DH Roach, NP Money. Penetration of hard substrates by a fungus employing enormous turgor pressures. Proc Natl Acad Sci USA 15:11281–11284, 1991.
28. NJ Talbot, DJ Ebbole, JE Hamer. MPG1 encodes a fungal hydrophobin involved in surface interactions during infection-related development of *Magnaporthe grisea*. Plant Cell 8:985–999, 1996.
29. JZ Xiao, T Watanabe, T Kamakura, A Ohshima, I Yamaguchi. Studies on cellular differentiation of *Magnaporthe grisea*. Physicochemical aspects of substratum surfaces in relation to appressorium formation. Physiol Mol Plant Pathol 44:227–236, 1994.
30. YH Lee, RA Dean. Hydrophobicity of contact surface induces appressorium formation in *Magnaporthe grisea*. FEMS Microbiol Lett 115:71–75, 1994.
31. JL Beckerman, DJ Ebbole. MPG1, a gene encoding a fungal hydrophobin of *Magnaporthe grisea*, is involved in surface recognition. Mol Plant Microbe Interact 9:450–456, 1996.
32. RD Gilbert, AM Johnson, RA Dean. Chemical signals responsible for appressorium formation in the rice blast fungus *Magnaporthe grisea*. Physiol Mol Plant Pathol 48:335–346, 1996.
33. YH Lee, RA Dean. cAMP Regulates infection structure formation in the plant pathogenic fungus *Magnaporthe grisea*. Plant Cell 5:693–700, 1993.
34. TC Jelitto, HA Page, ND Read. Role of external signals in regulating the pre-penetration phase of infection by the rice blast fungus, *Magnaporthe grisea*. Planta 194: 471–477, 1994.
35. RA Dean. Signal pathways and appressorium morphogenesis. Annu Rev Phytopathol 35:211–234, 1997.
36. T Uchiyama, K Okuyama. Participation of *Oryza sativa* leaf wax in appressorium formation by *Pyricularia oryzae*. Phytochemistry 29:91–92, 1990.
37. E Thines, RWS Weber, NJ Talbot. MAP kinase and protein kinase A–dependent mobilization of triacylglycerol and glycogen during appressorium turgor generation by *Magnaporthe grisea*. Plant Cell 12:1703–1718, 2000.
38. RWS Weber, GE Wakley, E Thines, NJ Talbot. The vacuole as central element of the lytic system and sink for lipid droplets in maturing appressoria of *Magnaporthe grisea*. Protoplasma 216:101–112, 2001.
39. FG Chumley, B Valent. Genetic analysis of melanin-deficient, nonpathogenic mutants of *Magnaporthe grisea*. Mol Plant Microbe Interact 3:135–143, 1990.
40. JM Henson, MJ Butler, AW Day. The dark side of the mycelium: melanins of phytopathogenic fungi (review). Annu Rev Phytopathol 37:447–471, 1999.
41. JC Dejong, BJ McCormack, N Smirnoff, NJ Talbot. Glycerol generates turgor in rice blast. Nature 389:244–245, 1997.
42. RJ Howard, B Valent. Breaking and entering: host penetration by the fungal rice blast pathogen *Magnaporthe grisea*. Annu Rev Microbiol 50:491–512, 1996.
43. JA Sweigard, FG Chumley, B Valent. Disruption of a *Magnaporthe grisea* cutinase gene. Mol Gen Genet 232:183–190, 1992.
44. Z Wawrzak, T Sandalova, JJ Steffens, GS Basarab, T Lundqvist, Y Lindqvist, DB Jordan. High-resolution structures of scytalone dehydratase-inhibitor complexes crystallized at physiological pH. Proteins 35:425–439, 1999.

45. DB Jordan, GS Basarab, JJ Steffens, T Lundqvist, BR Pfrogner, RS Schwartz, Z Wawrzak. Catalytic mechanism of scytalone dehydratase from *Magnaporthe grisea*. Pest Sci 55:277–280, 1999.
46. JE Thompson, S Fahnestock, L Farrall, DI Liao, B Valent, DB Jordan. The second naphthol reductase of fungal melanin biosynthesis in *Magnaporthe grisea*—tetrahydroxynaphthalene reductase. J Biol Chem 275:34867–34872, 2000.
47. JE Hamer, B Valent, FG Chumley. Mutations at the SMO genetic locus affect the shape of diverse cell types in the rice blast fungus. Genetics 122:351–361, 1989.
48. JE Hamer, S Givan. Genetic mapping with dispersed repeated sequences in the rice blast fungus: mapping the SMO locus. Mol Gen Genet 223:487–495, 1990.
49. Z Shi, H Leung. Genetic analysis of sporulation in *Magnaporthe grisea* by chemical and insertional mutagenesis. Mol Plant Microbe Interact 8:949–959, 1995.
50. Z Shi, D Christian, H Leung. Interactions between spore morphogenetic mutations affect cell types, sporulation, and pathogenesis in *Magnaporthe grisea*. Mol Plant Microbe Interact 11:199–207, 1998.
51. H Zhu, DS Whitehead, YH Lee, RA Dean. Genetic analysis of developmental mutants and rapid chromosome mapping of APP1, a gene required for appressorium formation in *Magnaporthe grisea*. Mol Plant Microbe Interact 9:767–774, 1996.
52. SJ Chun, YH Lee. Genetic analysis of a mutation on appressorium formation in *Magnaporthe grisea*. FEMS Microbiol Lett 173:133–137, 1999.
53. D Silue, D Tharreau, NJ Talbot, PH Clergeot, JL Notteghem, MH Lebrun. Identification and characterization of apf1(−) in a non-pathogenic mutant of the rice blast fungus *Magnaporthe grisea* which is unable to differentiate appressoria. Physiol Mol Plant Pathol 53:239–251, 1998.
54. JA Sweigard, AM Carroll, L Farrall, FG Chumley, B Valent. *Magnaporthe grisea* pathogenicity genes obtained through insertional mutagenesis. Mol Plant Microbe Interact 11:404–412, 1998.
55. TM DeZwaan, AM Carroll, B Valent, JA Sweigard. *Magnaporthe grisea* Pth11p is a novel plasma membrane protein that mediates appressorium differentiation in response to inductive substrate cues. Plant Cell 11:2013–2030, 1999.
56. NJ Talbot, DJ Ebbole, JE Hamer. Identification and characterization of MPG1, a gene involved in pathogenicity from the rice blast fungus *Magnaporthe grisea*. Plant Cell 5:1575–1590, 1993.
57. MJ Kershaw, NJ Talbot. Hydrophobins and repellents—proteins with fundamental roles in fungal morphogenesis. Fungal Genet Biol 23:18–33, 1998.
58. DJ Ebbole. Hydrophobins and fungal infection of plants and animals. Trends Microbiol 5:405–408, 1997.
59. JG Wessels. Hydrophobins: proteins that change the nature of the fungal surface. Adv Micro Physiol 38:1–45, 1997.
60. NJ Talbot, MJ Kershaw, GE Wakley, VOMH De, JGH Wessels, JE Hamer. MPG1 encodes a fungal hydrophobin involved in surface interactions during infection-related development of *Magnaporthe grisea*. Plant Cell 8:985–999, 1996.
61. HA Wosten, FH Schuren, JG Wessels. Interfacial self-assembly of a hydrophobin into an amphipathic protein membrane mediates fungal attachment to hydrophobic surfaces. EMBO J 13:5848–5854, 1994.
62. MJ Kershaw, G Wakley, NJ Talbot. Complementation of the mpg1 mutant pheno-

type in *Magnaporthe grisea* reveals functional relationships between fungal hydrophobins. EMBO J 17:3838–3849, 1998.
63. JL Beckerman, DJ Ebbole. MPG1, a gene encoding a fungal hydrophobin of *Magnaporthe grisea*, is involved in surface recognition. Mol Plant-Microbe Interact 9:450–456, 1996.
64. M Bolker. Sex and crime: heterotrimeric G proteins in fungal mating and pathogenesis. Fungal Genet Biol 25:143–156, 1998.
65. S Liu, RA Dean. G protein alpha subunit genes control growth, development and pathogenicity of *Magnaporthe grisea*. Mol Plant–Microbe Interact 10:1075–1086, 1997.
66. EGC Fang, RA Dean. Site-directed mutagenesis of the magB gene affects growth and development in *Magnaporthe grisea*. Mol Plant–Microbe Interact 13:1214–1227, 2000.
67. GE Turner, KA Borkovich. Identification of a G-protein alpha-subunit from *Neurospora crassa* that is a member of the G(I) family. J Biol Chem 268:14805–14811, 1980.
68. FD Ivey, Q Yang, KA Borkovich. Positive regulation of adenylyl cyclase activity by a G alpha(i) homolog in *Neurospora crassa*. Fungal Genet Biol 26:48–61, 1999.
69. D Mochly-Rosen. Localization of protein kinases by anchoring proteins—a theme in signal transduction. Science 268:247–251, 1995.
70. SH Kang, CH Khang, YH Lee. Regulation of cAMP-dependent protein kinase during appressorium formation in *Magnaporthe grisea*. FEMS Microbiol Lett 170:419–423, 1999.
71. TK Mitchell, RA Dean. The cAMP-dependent protein kinase catalytic subunit is required for appressorium formation and pathogenesis by the rice blast pathogen *Magnaporthe grisea*. Plant Cell 7:1868–1878, 1995.
72. JR Xu, M Urban, JA Sweigard, JE Hamer. The CPKA gene of *Magnaporthe grisea* is essential for appressorial penetration. Mol Plant–Microbe Interact 10:187–194, 1997.
73. T Toda, S Cameron, P Sass, M Zoller, M Wigler. Three different genes in *Saccharomyces cerevisiae* encode the catalytic subunits of the cyclic AMP–dependent protein kinase. Cell 50:277–288, 1987.
74. J Kronstad, A de Maria, D Funnell, RD Laidlaw, N Lee, M Moniz de Sa, M Ramesh. Signaling via cAMP in fungi: interconnections with mitogen-activated protein kinase pathways. Arch Microbiol 170:395–404, 1998.
75. K Adachi, JE Hamer. Divergent cAMP signalling pathways regulate growth and pathogenesis in the rice blast fungus *Magnaporthe grisea*. Plant Cell 10:1361–1373, 1998.
76. WB Choi, RA Dean. The adenylate cyclase gene Mac1 of *Magnaporthe grisea* controls appressorium formation and other aspects of growth and development. Plant Cell 9:1973–1983, 1997.
77. X Pan, J Heitman. Cyclic AMP–dependent protein kinase regulates pseudohyphal differentiation in *Saccharomyces cerevisiae*. Mol Cell Biol 19:4874–4887, 1999.
78. MC Gustin, J Albertyn, M Alexander, K Davenport. MAP kinase pathways in the yeast *Saccharomyces cerevisiae*. Microbiol Mol Biol Rev 62:1264–1300, 1998.

79. MB Dickman, O Yarden. Serine/threonine protein kinases and phosphatases in filamentious fungi. Fungal Genet Biol 26:99–117, 1999.
80. HD Madhani, GR Fink. The control of filamentous differentiation and virulence in fungi. Trends Cell Biol 8:348–353, 1998.
81. JA Alspaugh, JR Perfect, J Heitman. Signal transduction pathways regulating differentiation and pathogenicity of *Cryptococcus neoformans*. Fungal Genet Biol 25:1–14, 1998.
82. KB Lengeler, RC Davidson, C D'Souza, T Harashima, WC Shen, P Wang, X Pan, M Waugh, J Heitman. Signal transduction cascades regulating fungal development and virulence. Microbiol Mol Biol Rev 64:746–785, 2000.
83. JR Xu, JE Hamer. MAP kinase and cAMP signaling regulate infection structure formation and pathogenic growth in the rice blast fungus *Magnaporthe grisea*. Genes Dev 10:2696–2706, 1996.
84. JR Xu, CJ Staiger, JE Hamer. Inactivation of the mitogen-activated protein kinase Mps1 from the rice blast fungus prevents penetration of host cells but allows activation of plant defense responses. Proc Natl Acad Sci USA 95:12713–12718, 1998.
85. KP Dixon, JR Xu, N Smirnoff, NJ Talbot. Independent signaling pathways regulate cellular turgor during hyperosmotic stress and appressorium-mediated plant infection by *Magnaporthe grisea*. Plant Cell 11:2045–2058, 1999.
86. JR Xu. MAP kinases in fungal pathogens. Fungal Genet Biol 31:137–152, 2000.
87. MH Cobb. MAP kinase pathways. Prog Biophys Mol Biol 71:479–500, 1999.
88. PH Clergeot, M Gourgues, J Cots, F Laurans, MP Latorse, R Pepin, D Tharreau, JL Notteghem, MH Lebrun. PLS1, a gene encoding a tetraspanin-like protein, is required for penetration of rice leaf by the fungal pathogen *Magnaporthe grisea*. Proc Natl Acad Sci USA 98:6963–6968, 2001.

16

Pathogenic Development of *Claviceps purpurea*

Birgitt Oeser, Klaus B. Tenberge, Sabine M. Moore, Martina Mihlan, Patrick M. Heidrich, and Paul Tudzynski
Westfälische Wilhelms-Universität, Münster, Germany

1 INTRODUCTION

Claviceps purpurea causes the ubiquitous ergot disease. This phytopathogenic ascomycete parasitizes ~400 species of grasses throughout the world [1], mainly rye, wheat, and barley as well as numerous forage and roadside grasses [2,3]. Such wide host range is unique in the genus, pointing to immense pathogenic potential. The fungus causes harvest losses due to replacement of host ovaries with the parasite's resting structures, the ergot-called sclerotia. The main problem, however, is not a severe loss in seed quantity but arises from complete ruin of seed quality due to the alkaloid content of the sclerotia. These secondary metabolites induce highly dangerous or even deadly ergotism in animals and man. A contamination of crops with ergots >0.3% by weight spoils the grain even for feeding [4]. Certainly, ergot alkaloids are produced worldwide on a large scale because of their high pharmacological value [5]. For these reasons, *C. purpurea* has been of interest ever since, and its importance, which peaked at the unfortunate notoriety of several severe epidemics of ergotism, called St. Anthony's fire, in the Middle Ages [6], will persist as long as cereals as the main nutritional basis to man and herbivorous livestock are affected. Around the turn of the new millennium the ergot disease even quickly increases worldwide owing to the use of highly susceptible male-sterile plants [7,8]. Reduction in grain yield and quality causes the permanent necessity for an expensive cleaning of attacked cereals to maintain a minimum of purity standard. Specific measures for reliable control of ergot are limited and closely depend on an overall understanding of host and pathogen biology.

In contrast to the limited knowledge of fundamental biology in many plant–parasite systems, numerous investigations add to a considerable body of research on *Claviceps*, recently reviewed by Tudzynski et al. [9], Křen and Cvak [5], Tenberge [10], Tudzynski [11], and, focused on cell wall degrading enzymes (CWDE), by ten Have et al. [12].

This article emphasizes recent advances in histopathology and molecular biology in *C. purpurea*. Although much further work is needed in this field of research, substantial knowledge and modern molecular methods open the opportunity to address unsolved hypotheses in this specific ergot–rye relationship and therewith contribute to general understanding of molecular mechanism in the interaction of hosts and pathogens.

2 GENERAL PATHOLOGICAL ASPECTS OF THE *CLAVICEPS*–RYE INTERACTION SYSTEM

2.1 Life Cycle

In nature, the parasitic life of *C. purpurea* starts with windborne ascospores landing on susceptible hosts in spring (Fig. 1). All arising stages of their life cycle can develop from one single spore; therefore, the ergot fungus is homothallic [13]. Typically, spores attach and germinate on the pistil surfaces of blooming host florets. Hyphae invade and colonize the ovary, grow down to the tip of the ovary axis, the rachilla, and establish a specific and persisting host–parasite frontier. The fungus never invades any part further down in the host but proliferates above this site. A sphacelial stroma grows profusely in the ovary, producing masses of anamorphic spores which are exuded into a syrupy fluid (Fig. 1). With this honeydew, the conidiospores are transferred to other blooming florets by rain splash, insect vectors, physical head-to-head contact between ears, and dripping down onto florets inserted below in the same ear [14–18]. Thereby *C. purpurea* spreads spatially in the field having used the plant gynoecia for its own proliferation. Next, honeydew production and condition usually cease when the formation of sclerotia starts. Sclerotia mature in ~5 weeks (Fig. 1). Finally, during autumn, instead of a caryopsis, a ripe sclerotium leaves the spike, making ergot a replacement tissue disease [19]. It serves for sexual reproduction and as a resting struc-

FIGURE 1 Life cycle of *C. purpurea*. The different stages depicted are: (1) germinating ascospore; (2) a rye floret at anthesis, exposing the stigma between the opened glumes; (3) an infected rye ovary during the colonization phase with withered stigma and style, long ovary cap hairs, and the rachilla (arrow); (4) a sphacelium; (5) a rye ear with honeydew (arrow) flowing out of infected florets; (6) a sphacelial stroma with phialidic conidiophores producing many anamorphous spores; (7) germinating conidiospore on the host ovary cap with subcuticular hyphal growth towards the cellular junction; (8) a mature rye ear with several sclerotia; (9) germinating sclerotium with stromata that differentiate perithecia (arrow) in the head periphery containing asci with ascospores.

ture for overwintering after having fallen to the ground or having been harvested together with the seed.

Sclerotia germinate in spring after a period of low temperature, which favors germination [16]. Germination results in 1–60 stromata, formed of stalks with spherical capitula (Fig. 1) that grow positively phototrophic [20] to reach the air. Ascogonia and antheridia develop in the periphery of the capitula and fuse to form dikaryotic ascogenous hyphae. The hyphae surrounding the fertilized ascogonia build flask-shaped perithecia within which karyogamy and meiosis occur, producing asci with thin, needlelike, hyaline, nonseptate ascospores. Under suitable moist conditions, eight ascospores are forcibly ejected through the apical pores of asci, which emerge through the ostiole of perithecia [21]. About 4 weeks after sclerotia germination, these ascospores represent the airborne primary inoculum and give rise to new infection foci.

2.2 Organ Specificity

Together with all natural ergot species, *C. purpurea* shares two conspicuous features, which obviously reflects special adaptations: all hosts are anemophilous monocotyledons and the fungal objectives are host gynoecia solely. This strict organ specificity is poorly understood and a matter of speculations, because the biological function of the targeted host organ is intended for plant sexual reproduction and shows distinctive adaptations.

Although the inoculum reaches every host surface area, in nature, successful ergot infection is strictly specific to florets. However, one can artificially induce sclerotia development by wounding and inoculating young tissue, e.g., on stalks [22], on the shoot apex of rye seedlings [23], or on nodes and internodes of rye [24]. These observations point to unique features of the pistil which appear to be indispensable for the establishment of infection. Since the ovary mainly serves as a host for the male gametophyte to favor and to guide pollen tube growth, pollen adhesion, pollen tube penetration, and growth processes that have been described in grasses [25–28] exhibit some striking similarities to fungal colonization. Furthermore, the two invaders interfere with each other when landing on the stigma at the same time. In particular, simultaneous pollination favors penetration of *C. purpurea* [29]; however, pollen tubes grow much quicker reaching the ovule in wheat in ~30 min [30]. Admittedly, the ovary appears to be dispensable for fungal development subsequent to primary infection, because sclerotia are formed even after artificial inoculation of florets, the ovaries of which have been removed previously [31] or after advanced kernel growth [32].

Another reason for organ specificity might be the exceptional molecular architecture of the monocotyledonous cell wall [33] certainly with some additional unique cell wall modifications in different pistil tissues [34,35] (see below), e.g., pistil epidermis with stigma hairs, ovary mesophyll, transmitting tissue, and integuments. Fungal growth mechanisms are likely adapted to these host cell walls.

2.3 Ergot Virulence and Host Susceptibility

Host floret biology is fundamentally important in every respect. Since the hosts are anemophilous, floret morphology and the time course of anthesis determine pollination and likewise access of fungal spores, especially of airborne primary inoculum. Florets can prevent infection completely by denying access of the inoculum to their pistils due to tightly closed bracts. Distinctive adaptations to pollination determine gaping and therewith affect susceptibility to ergot drastically. Pollination induces closing of florets and rapid withering of stigma and style. Susceptibility is highest at anthesis and declines afterwards but remains possible at a lower rate even after fertilization in rye, wheat or barley [16,36,37]. Single rye florets gape for several hours, then close tightly, sometimes leaving the stigmas exposed between the glumes. Regarding complete ears, florets open, starting near the top of the ear and progressing in a series basipetally over several days. Hence, rye fields are in bloom for ~2 weeks. While host male sterility or protogyny enhances gaping duration and consequently susceptibility [38], cleistogamy is most effective in avoiding ergot [39]. In conclusion, the timing of honeydew exudation together with secondary conidiation in relation to host blooming determines the ecological virulence of the pathogen in the field.

Active resistance is reported in only a few cases, such as spring and durum wheat [40]. In the experiments with four different *C. purpurea* strains, however, obscure ovary necrosis occurred only in some of the florets; others produced sclerotia in these hosts. Necrotic ovary response has also been found after double inoculation with two different *C. purpurea* strains [18]. The limited reports on resistance could point to special adaptations of the pathogen to the monocotyledonous host ovaries. Assuming that the fungus indeed mimics pollen tube growth with the genius of using components of the specific signal exchange of the pollen–stigma interaction, hyphae possibly grow unrecognized in the ovary and completely avoid host defense reaction of any type. This would suggest that the ergot fungi have coevolved, as assumed for anthracnose fungi's use of their host-ripening hormone [41]. One can speculate that successful defense reactions producing host resistance might interfere with the basic function of the ovary and therefore do not evolve, since no progenies arise. While being biotrophic parasites, *Claviceps* spp. are symbiotrophic at the same time. In exchange for a habitat and nutrition, their alkaloid content protects their hosts and thereby themselves against grazing animals [42], which might interfere with development of active resistance.

3 DEVELOPMENT OF FUNGAL STRUCTURES DURING HOST INFECTION

3.1 Infection Site and Route

Infection of a single host plant is naturally induced by spores landing on the pistil of open florets for which fewer than 10 conidia are sufficient [37]. Although the stigmas are exposed for catching windborne propagules, attachment and germination of *C. purpurea* are not restricted to this area. *C. purpurea* is able to penetrate the pistil epidermis anywhere. Employing scanning microscopy, spore adhesion, spore germination, and host cuticle penetration were found to occur on either part of the pistil surface (Figs. 1, 2) [9,15,16,19,43–47].

Following penetration in the outer epidermal wall of the pistil, the hyphae keep on growing toward the rachilla (Fig. 2). They grow either down the style in the transmitting tissue following the pollen tube path outside the ovule and leaving this way at the micropylar region in direction of the rachilla, or (Fig. 2a), after lateral entrance into the ovary, in the carpel mesophyll to the ovary basis [9,19,44]. The ovary wall gets completely colonized after ~6 days postinoculation (dpi). However, the ovule first remains uninvaded [16] owing to the integuments which appear to form a temporary barrier to the fungus [43,48]. However, ovules are additionally invaded through the chalazal region. Integuments typically collapse during development of caryopses. However, possibly owing to chitinases or chitin-binding lectins found in rye seeds [49,50], growing kernels remain noncolonized after late infection.

Fungal cells colonize the entire ovary wall, but in the ovarian axis the hyphae stop to spread in the plant tissue. No hyphae emerge beyond the rachilla tip. Thus, a narrow frontier between the fungal stroma and the noncolonized host tissue develops, which is finished ~6 days after infection and persists throughout the remaining life span [9,19,44,51]. Stoppage of fungal growth in the rachilla has been shown to be most likely caused by host phenolics that accumulate during infection at this site [44,52,53] and might inhibit fungal pectin-degrading enzymes [54].

The route that is usually used, however, depends on floral biology. Short floret gaping followed by stigma exposure between tightly closed glumes causes the pollen tube path to be of importance in nature. However, in male-sterile lines, other routes might be used as frequent. Additional routes in rye are obvious from successful infection with *C. purpurea* after previous ovary colonization with the bunt fungus *Tilletia* caries [55] or after fertilization with advanced kernel development [32], since in both cases the pollen tube path is blocked. This shows that, regardless of the infection route, the growth of the ergot fungus is strongly directed at the vascular tissue of the ovary, which itself represents the entrance but is dispensable [19,55] although incidentally consumed for overall nutrition.

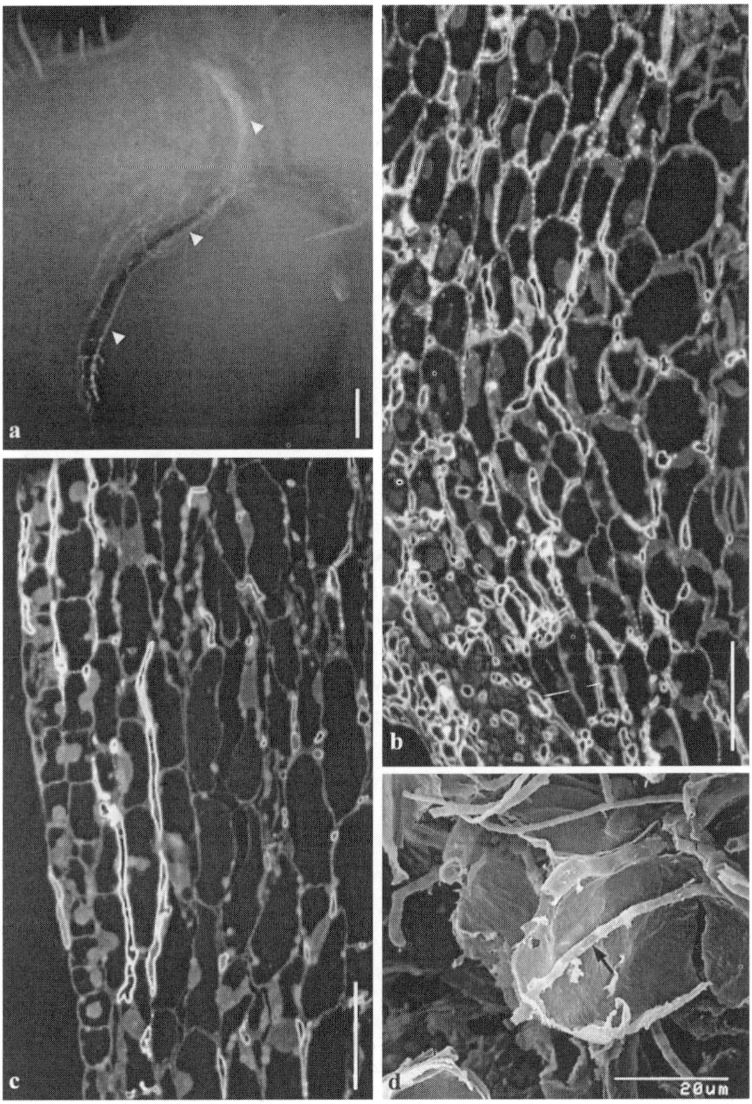

FIGURE 2 Hyphae of *C. purpurea* during the colonization of rye. Hyphal growth in the ovary is directed (a, b), mainly following the potential path of pollen tubes (a, arrowheads). In the carpel mesophyll, hyphal growth predominantly is intercellular (b–c) as well as intracellular (d, arrow). (a) Longitudinal hand section of fresh material, 3 dpi. (b, c) Longitudinal semithin sections from LR-White embedded samples, 5 dpi. (a–c) KOH-aniline blue staining. Fungal cell walls and callose exhibit bright fluorescence. (d) Scanning electron micrograph of the apical carpel mesophyll, 8 dpi. Scale bars: (a) 100 μm; (b, c) 50 μm.

3.2 Spore Adhesion, Primary Infection, and Colonization

The attachment process of *C. purpurea* spores is not investigated precisely. On stigma or style, the stigmatic fluid might offer hydrophilic conditions or, in case of honeydew-mediated transmittance, the syrupy fluid may support adhesion to the host surface. After attachment, conidia germination starts with the formation of one to several germ tubes, supported by dew periods, and is accomplished very quickly.

To study the penetration and colonization mechanisms of ergot fungi, early events of the infection process of *C. purpurea* were first documented and then cytochemically analyzed in detail in our laboratory. Spores attached everywhere on the pistil epidermis. Its outer epidermal wall comprises a faint cuticular membrane, measuring ~15 nm in thickness, and forms a continuous outer barrier of the host ovary. Germination results in one or two germ tubes, e.g., on the ovary cap (Fig. 1). Sometimes a limited external mycelium is formed. Next, the faint plant cuticle of the outer epidermal wall is directly penetrated. An indirect entry via natural openings is unimportant because the pistil is free of stomata and natural wounding was never observed. Infection hyphae originate either directly from the germ tube or from the external mycelium (Fig. 1). Since no changes in hyphal shape were apparent, it appears that *C. purpurea* penetrates without specialized infection structures (K.B. Tenberge, unpublished). At suitable sites, infection hyphae pass through the outer epidermal cell layer growing intercellularly into the anticlinal epidermal walls. However, hyphae may as well pass through the outer epidermal cell wall away from cellular junctions [9,44,47]. The intimate contact and abundant wrapping of superficial ergot hyphae with the host cuticle as seen in transmission electron microscopy (TEM) [9] (K.B. Tenberge, unpublished) might cause a mechanical disruption of the thin cuticle itself.

So far investigated, all ergot fungi penetrate directly into the anticlinal walls between epidermal host cells; however, the hyphae of some species develop special morphological structures prior to penetration. *C. gigantea* produces appressoria [56] and *C. fusiformis* bulbous infection structures at the tips of germ tubes [55]. Recent findings revealed new insight in signal events, e.g., the cloning of *pmk* genes (see below), raising the question on genetic control of the morphogenesis of these infection structures in *C. purpurea* and related species in relation to host range and virulence.

The mechanism of cuticle penetration still needs to be elucidated, although in most ergot fungi, the direct push of a infection hypha into the epidermal wall obviously matches the simpler type of penetration as classified by Mims [57]. The bulbous structures of *C. fusiformis* suggest the utilization of turgor pressure; however, it is an open question whether it is sufficient for penetration. According to Mendgen et al. [54], directly penetrating fungi that do not form appressoria clearly need cell wall–degrading enzymes for penetration. During penetration of grass stigma cuticles, tip growth of pollen tubes is mediated by cutin-degrading enzymes [26,58]; however, in case of ergot fungi, the secretions of such enzymes have not been demonstrated so far.

After penetration, ergot fungi live inside the ovary, i.e., endophytically, during the colonization phase (Fig. 2). Subcuticular hyphae, in fact, are located within the outer epidermal cell wall and then the fungi usually grow between epidermal cells into the host apoplast. Before tapping the vascular traces, fungal growth during the colonization phase has been reported to be exclusively intercellular, i.e., ectotrophic, in all ergot fungi investigated [19,20,44,55,59]. However, a limited intracellular growth has been documented electron microscopically in *C. purpurea* (Fig. 2d) [9,60]. Therefore, the mycelium of *C. purpurea* is ectotrophic but with limited endotrophism. The vegetative hyphae exhibit ultrastructural features typical for ascomycetous fungi. The thin fungal cell wall and the host cell wall build up an intimate zone of contact while both host and pathogen appear to be healthy [9]. This is particularly valid for intracellular hyphae and the penetrated host cells, and points to haustorial function. For this purpose, the endotrophic mycelium is well positioned in the chalazal region that serves for the host ovule nutrition. The interface of the intracellular hyphae, which are completely encapsulated by the host plasma membrane, has developed special adaptations [34,61]; however, nutrient uptake into hyphal cells has not been investigated.

3.3 Sphacelial Stromata for Secondary Propagation

The completion of the persistent frontier includes the active tapping of the xylem as well as the phloem and coincides with the exudation of honeydew, 6–9 dpi, depending on the specific infection conditions (Fig. 1). This indicates the end of the infection stage I (1–5 dpi) and the beginning of the sphacelial phase, stage II, and presents the first macroscopic evidence for infection. The sphacelial stroma (Fig. 1) is evident after 6 dpi and sporulation ceases approximately 11 dpi [19]. The sphacelial plectenchym is formed intercalarly [15] and accumulates lipids later in this phase. These filamentous hyphae are of the type normally found in axenic culture. At the base of the ovary, proliferation of the fungal cell starts [16]; hyphae accumulate beneath the host cortical layers and break through the epidermis toward the ovarian outer surface [19]. Finally, fungal cells cause ovary replacement, which is not necessary but a consequence of acropetal development [48]. Phialidic conidiophores emerge from the sphacelial stroma possibly favored by ergot hydrophobins (see below). Numerous oblong conidia are produced (Fig. 1). These conidia do not germinate in the honeydew, which is excreted simultaneously, owing to high osmotic pressure [1,16]. Some isolates of *C. purpurea* do not produce normal exudates of honeydew, and only a few spores were detected later. However, these strains produced normal sclerotia [62], indicating that abundant formation of honeydew is important for secondary infection but appears not to be a necessary prerequisite for sclerotial formation.

3.4 Ergot Sclerotia

After the secondary conidiation has ceased, the sclerotia start growing, and stage III starts. Within ~5 weeks postinoculation in *C. purpurea* [16], the maturity of sclerotia is achieved. They are oblong and measure 2–50 mm in length and a few millimetres in diameter. Sclerotia clearly grow epiphytic on top of the ovary stalk and they mostly emerge out of the florets, denoted stage IV, 20 dpi. Since the sclerotia are no longer enclosed between the glumes, energy must be provided to protect them from desiccation, UV radiation, and mycoparasitism [42]. The hard compact ergot consists of a plectenchymatous, whitish medulla consisting of special storage cells and the outer cortex that gets naturally pigmented dark purple.

First, at several places within the sphacelium, sphacelial hyphae differentiate into sclerotial hyphae; later, at the sclerotial base, sclerotial hyphae are formed directly [63]. The sclerotial plectenchym develops intercalarly above the stromatic fungal foot by a generative zone as previously did the sphacelium [43]. Remnants of sphacelial stroma with conidia as well as the ovary cap may persist on top of the growing sclerotia, but no internal conidiogenous locules were found. Hence the sclerotial hyphae are newly formed.

Purple pigmentation is the first sign of sclerotial development in *C. purpurea* ~12 dpi [63], but the trigger for the change from sphacelial into sclerotial growth is unknown [42]. It has been speculated that nutrition is a major factor, supported by the transition effect of certain amino acids found in axenic culture [64]. Changes in cytology of the cells coincide with increasing levels of lipids which are predominantly triglycerides with the fatty acid ricinoleate [65]. This increase in total lipid content from 10% to 30% of the dry weight is the first metabolic indicator of the morphogenesis of sclerotial cells [66]. The youngest fungal cells of the sclerotial stroma are longitudinally organized, distinct, frequently septate hyphae forming the prosenchymatous region at the proximal end of the sclerotium. Lipid content is evident in these storage cells; thus, they are packed with osmiophilic globules in contrast to the differentiating hyphae of the generative zone in the lower ergot region. In the medulla, distal from the prosenchymatous region, fungal cells from a region of a compact plectenchyma which is built out of bulbous storage cells interspersed with narrower hyphae [63]. The absorbing hyphae, which connect the sclerotium to the ovary stalk and form the stable host–parasite frontier, however, are neither of the sclerotial type nor typically sphacelial cells. They exhibit no parallel orientation, lack lipids, and contain large vacuoles [63]. A special function of the fungal foot is also suggested by xylanase protein which has been localized in the sclerotial phase at this host–parasite frontier exclusively, using tissue printing experiments [51].

Sclerotia are the only ergot structure containing alkaloids [67], and the pigmentation of the sclerotial cortex might protect these light sensitive alkaloids [1]. In axenic culture, differentiation of the sphaceliallike hyphae into sclerotiallike cells occurs [16]. They accumulate up to 40% of triglyceride/ricinoleate, but cells show no pigmentation and, as in the parasitic state, do not always produce alkaloids [66].

3.5 Assessment of Developmental Strategy

Throughout the ergot-infected pistil, host cells die subsequent to fungal exploitation of living tissue and possibly some due to induced senescence. *C. purpurea* never kills host cells in advance of colonizing with the intention to draw nutrition from the killed cells. In addition, owing to the unique pathogenesis pattern, the ergot fungi proliferate intercalarly in the ovary basis that inevitably results into an separation of the host ovary cap. Nevertheless, the separated and colonized tissue stays alive for a while, possibly with nutritional support from the sugary honeydew. Thus, ergot–grass interactions are classified as belonging to a pathosystem free of necrosis in fully susceptible hosts; cell death is not intended but inevitably induced after a while, similar to early host senescence by other biotrophs [42].

Like other ergot fungi, *C. purpurea* is a true holobiotroph which is an ecologically obligate parasite and in nature obtains nutrients only from living host tissue while managing to maintain host cell viability for extended periods, and serves as sink for plant metabolites. *C. purpurea*, in contrast to many other biotrophs, is able to survive outside living host tissue while growing saprophytically in axenic culture. *Claviceps* spp. likely take advantage of the most common source–sink system for synthate in a host by directly tapping the host's nutrition supply network [47] and exploiting the plant resources in a working sink (see below). However, the fungus likely utilizes principles to maintain the phloem synthate flow. In addition, the parasite might enhance the sink, which is supported by the suppression of seed development in noninfected florets of the same ear. Likewise, the inverse correlation of sclerotia size and number indicates a competition of sinks in one ear.

4 MOLECULAR ASPECTS OF HOST–PATHOGEN INTERACTION

The molecular genetics of the development of *C. purpurea* in planta and the interaction with its major host plant *S. cereale* has been studied in detail in recent years, using two different approaches: (1) focusing on specific aspects of the interaction (e.g., cell wall–degrading enzymes, cell wall components, AOS detoxification, etc.), by characterization and functional analysis of genes probably involved in these specific aspects, trying to find factors essential for pathogenicity and virulence, and (2) a more general approach, based on an EST library of *C. purpurea*–infected rye ovaries, i.e., an unbiased study of the fungal genes expressed in planta, with the chance to identify new, so far unconsidered factors.

4.1 Genes Coding for Potential Pathogenicity Factors: In Planta Expression Studies and Functional Analyses

4.1.1 Cell Wall–Degrading Enzymes (CWDE)

An important characteristic of *C. purpurea* is the inter- and intracellular growth in the cell wall apoplast of the histologically heterogeneous ovary. Splitting the middle lamella zones and direct breaching of host cell walls in numerous different grasses reveal that this pathogen is well adapted to the monocotyledonous cell wall habitat. Both features point to the pathogen's use of secretable CWDE that appears to be well controlled. Since grasses have developed a special cell wall type containing low amounts of pectins and considerably high amounts of glucurono-arabino-xylans (GAX) in addition to the major polysaccharide portion of cellulose [33], xylanases and cellulases, but also pectinases, are expected to be necessary for breaking down the major cell wall components during infection. The breakdown of cell walls is probably not only necessary for growth within the tissue; the cell wall material is also thought to be important for nutrition during colonization of the ovary supported by their complete use and also because cell wall extracts of ears stimulated growth in culture [24].

The molecular architecture of the host–pathogen interface has been studied in detail in this system, with emphasis on interaction-specific reactions, e.g., polymer alterations and protein secretion, at the electron microscopical level. This molecular cytological study is intensely coordinated with a molecular genetic approach in order to elucidate the role of CWDE in ergot pathogenicity. In the following a short update is presented of the information available for the major CWDE classes.

Pectinases. Host cell wall loosening during subcuticular and intercellular growth (Fig. 2) [9] indicated that actions of pectolytic enzymes, which Shaw and Mantle [44] have proved to be active in culture, in honeydew and in parasitic tissue extracts play a role in parasitism. In contrast to the model-based expectations, the two major pectin types, non-methyl-esterified and methyl-esterified galacturonan, have been found to be simultaneously present in considerable amounts in the cell walls along the usual infection path in healthy carpels, as visualized by immunogold labeling with the monoclonal antibodies JIM5 and

JIM7 [34]. During infection of rye, a local molecular pectin modification and degradation have been demonstrated for the host cell wall and the middle lamella zone at the interface of subcuticularly and intercellularly growing hyphae in situ. Chemical demethylation and immunogold labeling showed a high local content of galacturonan that, in late infection phases, was completely absent, providing evidence for the secretion of pectinolytic enzymes in planta. From on-section saponification studies, the local reactions were concluded to comprise an enzymatic demethylation mediated by pectin-methylesterases, converting pectin into the appropriate substrate of endopolygalacturonases for final degradation. The cellular junctions consist of high amounts of unesterified pectin. Therefore, polygalacturonase activity seems to be a fitting means to enable epidermis penetration and an entry into the middle lamella from the intercellular spaces which is not continuous along the infection route towards the rachilla.

Using a heterologous gene from *Aspergillus niger* as probe, two putative endo-PG genes were cloned and characterized [34]; they are closely linked in a head-to-tail arrangement and show 95% identity, pointing to a recent gene duplication event. By RT-PCR it could be shown that both genes are expressed throughout the first 3 weeks of infection, i.e., during the colonization phase and the early sclerotium development. The special head-to-tail arrangement allowed a one-step gene inactivation of both genes by a replacement approach, using a linear DNA fragment containing a phleomycin resistance cassette, flanked by the 5′ part of the first gene and the 3′ part of the second gene, respectively (B. Oeser, U. Müller, unpublished). The analysis of two independent double-mutant strains showed a drastic effect: these strains are significantly impaired in virulence, indicating an essential role of pectinolysis for pathogenicity. Restoration of the wild-type locus by an inverse replacement reversed the mutant's phenotype [110].

Cellulases. The ergot fungus actively and directly penetrates plant cell walls [60]. At the interface of intracellular hypha, the host cell wall is obviously lacking [9]. With the use of a specific enzyme-gold probe, a lack of β-1,4-glucan in host cell walls has been found at this site and also at host–pathogen interfaces of intercellular hyphae, pointing to the enzymatic action of cellulases in ergot infection [61]. However, cellulolytic activity could never be detected in liquid culture (only on solid medium using substrate staining) [68], indicating a strict regulation of cellulase activity. So far one gene has been cloned which probably is involved in cellulose degradation [61]: *cpcel1* probably encodes a cellobiohydrolase (lacking the substrate binding domain). The gene was shown by RT-PCR to be induced during the first days of infection of rye. Therefore, this cellobiohydrolase may be involved in the penetration and degradation of host cell walls. However, deletion of the gene by transformation with a replacement vector showed no effect on pathogenicity (U. Müller, P. Tudzynski, unpublished), indicating the presence of (an) additional cellulase gene(s) in *C. purpurea*.

Xylanases. β-1,4-Xylan, i.e., the substrate of a fungal β-1,4-xylanase used in the enzyme-gold technique, has been localized in rye ovary cell walls throughout the infection route [51], confirming that this major cell wall component of grass leaves is in fact a structural compound in ovary cell walls. The β-1,4-xylan is expected to represent only the backbone of GAX. Arabinofuranosyl epitopes, one of the possible side chains in GAX, were localized in ovary cell walls [51]. Absence of xylan in late infection stages and xylan alteration early in infection were visualized in TEM and after silver enhancement in light microscopy (LM) [69], strongly suggesting the secretion of xylanolytic activity by the fungus. In fact, xylanase activity could be detected in axenic culture, and—using three different heterologous antibodies in tissue printing experiments—the secretion of ergot xylanases during infection of rye has been localized in situ [51].

So far two putative endo-β-1,4-xylanase genes have been cloned and characterized in *C. purpurea: cpxyl1* and *cpxyl2*, probably coding for family G and family F enzymes, respectively [51]. By RT-PCR it could be shown that both genes are expressed in planta throughout the whole infection period. Using a gene replacement approach, single mutants (for both genes) and double mutants were obtained; they showed a significant reduction in total xylanase activity in axenic culture. Using IEF/activity staining, loss of corresponding xylanase activity bands was shown, confirming that *cpxyl1* and *cpxyl2* indeed encode xylanases. The effect of the deletions on pathogenicity was not nearly as pronounced as with the PG mutants, but at least the double mutants seem to show a slightly retarded development in planta [51] (J. Scheffer, A. Fleissner, P. M. Heidrich, B. Oeser, unpublished).

β-1,3-Glucanase. Plant synthates directed to the ovary are the main nutrition source to the fungus [52]. To exploit this natural sink, several enzymes are secreted and the persisting host–pathogen frontier is developed structurally for attaching and absorbing [19]. In sharp contrast to uninfected ovaries, common phloem callose was not found in infected ovaries at all or was distinctly reduced, as outlined by Tudzynski et al. [9]. This unblocking of sieve elements may be the reason for honeydew exudation due to increased flow of assimilates to the infected floret. The current opinion of the mechanisms is that ergot fungi enzymatically degrade the phloem callose by secreting β-1,3-glucanases, which have been purified from axenic cultures of *C. purpurea* [70]. This callase has been immunogold localized throughout the colonization phase and in the fungal secretion pathway proving the fungal origin of the β-1,3-glucanase activity found in infected ovaries and honeydew [47]. Immunogold electron microscopy documented that the secreted enzyme is diffusing into the host apoplast. The gold labeling over host periplasmic spaces showed that the enzyme reached the typical deposition sites of callose, pointing to an enzymatic suppression of putative plant defense reactions. The host phloem was colonized inter- and intracellularly. Hyphae penetrated into the pectic middle lamella of sieve plates and intense immunolabeling for β-1,3-glucanase in this area supports a phloem-unblocking hypothesis. In a mutagenesis approach mutants were detected that show a marked delay (>7 days) in honeydew exudation. Microscopical analysis revealed that in those mutants physical contact was established in normal time, indicating that physical contact alone is not sufficient for honeydew production.

Recently a putative mixed-link (β-1,3/1,4)-glucanase gene was identified within a EST library of in planta expressed genes of *C. purpurea* (see below), perhaps opening the way for a functional analysis also of this enzyme system.

Taken together, these data indicate that CWDEs play an important role in the successful colonization of rye ovarian tissue by *C. purpurea*.

4.1.2 Enzymes Involved in the Generation and Scavenging of Active Oxygen Species

One of the earliest defense reactions of plants against pathogens is the transient formation of active oxygen species (AOS: O_2^-, H_2O_2,· OH), in analogy to mammalian systems, this reaction is termed oxidative burst [71]. AOS play a direct role in early defense against pathogens by leading to a rapid reinforcement of the host cell wall (by crosslinking of cell wall proteins). In addition, AOS are probably involved in the induction of the so-called hypersensitive response (HR), triggering rapid necrosis at the infection site, and they may activate late defense reactive genes in the surrounding tissue.

The oxidative burst–derived AOS may also have direct impact on the pathogen, causing, e.g., membrane damage and inactivation of proteins. All living cells have developed protective systems against oxidative damage, including nonenzymic AOS scavenging mechanisms like the accumulation of ascorbic acid, mannitol, GSH, etc., and enzymes that rapidly detoxify AOS. These last include superoxide dismutases, which dismutate O_2^- to H_2O_2; catalases, which decompose H_2O_2 into water and O_2; and peroxidases, which can use a variety of substrates for the reduction of H_2O_2 to H_2O.

It has been shown that bacterial pathogens of mammals have highly efficient enzymatic AOS-scavenging systems. In *Staphylococcus aureus* secreted catalase activity seems to be correlated with its survival of phagocytosis as well as the mortalility levels of infected mice [72]; periplasmic Cu,Zn-superoxide dismutases obviously are involved in survival strategy of *Salmonella typhimurium* [73]. As of yet, there does not seem to be very much known about the importance of Cu,Zn-SODs in fungal–pathogen interactions. Although they have been implicated in the pathogenesis of *Aspergillus fumigatus* and *Cryptococcus neoformans* [74,75], it has yet to be determined if these proteins are significant virulence factors in their systems.

In pathogenic fungi/plant interactions, comparable analyses are rather limited. We have recently initiated a detailed study on the role of AOS in the interaction of the ergot fungus *Claviceps purpurea* and its main host crop plant, *Secale cereale*. Though in this biotrophic system no strictly incompatible reactions (and hence no HR) are known, the involvement of AOS in the interaction is implicated by several observations [see 10,76].

Especially, the presence of H_2O_2 in the rye tissue colonized by *C. purpurea* could be proved. Therefore we first focused on H_2O_2-decomposing enzyme systems. Detailed cytological and biochemical analyses by Garre et al. [76] showed that *C. purpurea*—as in other nonpathogenic ascomycetes—produces several catalase isozymes; in a combined IEF/native gel electrophoresis altogether four different catalases were identified in axenic culture, CAT A–D. Whereas CAT A is an intracellular form, CAT B seems to be mainly cell wall associated, and CAT C/D are secreted forms. The different catalase isoforms could be also detected in planta, CAT D even in honeydew of infected plants. This represents (to our knowledge) the first example of a secreted catalase in a phytopathogenic fungus. CAT C/D were shown to be encoded by one gene, *cpcat1*, as disruption of this gene resulted in loss of both isoforms, in axenic culture and in planta [77]. The gene was shown (by RT-PCR) to be expressed in all stages of infection (stages I–IV). The deletion mutants were not impaired in pathogenicity, demonstrating that this major extracellular catalase is not essential, at least not on rye and under the (near natural but still artificial!) inoculation conditions tested. Since with CAT B there is still a probable cell wall–associated catalase present, which could take over the protection against H_2O_2 in the mutants (though no increased activity of CAT B in the mutants could be detected), a second putative catalase gene, *cpcat2*, localized in the ergot alkaloid biosynthesis cluster [78] was cloned and disrupted (S. Moore, unpublished). Until a double mutant which lacks all secreted catalase activity has been generated and characterized, little can be concluded about the importance of these enzymes for the *Claviceps*/rye interaction.

An extracellular/cell wall associated SOD was recently detected by IEF/activity staining and preliminary immunolocalization experiments (S. Moore, B. Drauschke, M. von den Driesch, unpublished). Using a PCR approach based on degenerated primers derived from SOD sequences from various organisms, a gene could be cloned, *cpsod1*, showing significant homology to Cu-Zn-SODs. This gene was found also in an EST library (see Table 1). It was shown to be expressed throughout all phases of infection. Deletion of *cpsod1* led to disappearance of the major secreted SOD form, although the gene contains no apparent signal peptide. The deletion mutant showed some limited reduction of parasitic properties: the appearance of honeydew in some infection tests was retarded for 1 day and the amount of honeydew is reduced (111); it is questionable if this minor defect has any practical implications in nature. It shows, however, that also SOD1 activity is not really essential for pathogenicity in the laboratory test system.

TABLE 1 Parasitic *C. purpurea* EST Clones with Homology to a Protein with (I) or Without (II) Database-Specified Function

I. Homology to a protein with database-specified function

EST	Description of match with highest significance	HSP	P
Amino acid metabolism			
38n	α-keto-β-hydroxylreductoisomerase >gi|322996 (EC 1.1.1.86)–*Neurospora crassa*	345	1.6e-30
68n	>gi|223208|prf|0608196A aminotransferase, Asp (*Gallus gallus*)	238	9.4e-19
Energy metabolism			
10nB	>gi|6325281 mt member of the CDC48/PAS1/SEC18 family of ATPases *Saccharomyces cerevisiae*	109	3.6e-07
37n	>gi|1730118 fumarate hydratase (*Rhizopus oryzae*)	988	1.2e-98
54n	>gi|452311 glyceraldehyde-3-phosphat dehydrogenase (*Claviceps purpurea*) (2×)	338	1.1e-45
65n	cytochrome c oxidase polypeptide vibk, *Schizosaccharomyces pombe*	111	1.0e-05
76n	vacuolar ATPase sub D (*N. crassa*)	516	2.3e-81
102n	CG7834 gene product *Drosophila melanogaster*, electron transfer flavoprotein sub (2×)	695	1.3e-67
"284n"	ADP/ATP carrier protein, *S. cerevisiae*	217	4.8e-16
286n	gamma sub of mitochondrial ATP synthase, *S. cerevisiae*	673	2.9e-65
305n	vacuolar ATP synthase sub H, *Bos bovis*	102	4.8e-05
312n	acyl-CoA-binding protein type II, *S. cerevisiae*	121	9.1e-07
314n	6-phosphogluconate dehydrogenase, *S. cerevisiae*	293	1.5e-24
352nB	cytochrome c549 (*Fusarium oxysporum*)	308	1.4e-26
Lipid metabolism			
33a	putative phosphatidic acid phosphatase, *Arabidopsis thaliana*	191	4.4e-14
10nA	3-oxoacyl-[acyl-carrier-protein] reductase (*Brassica napus*) (2×)	103	1.5e-07
164n	acetoacetyl-CoA synthetase, *Rattus norvegicus*	327	9.1e-28
293n	SCS3 involved in inositol biosynthesis, *S. cerevisiae*	101	8.5e-12
342n	CDP-diacylglycerol synthetase homolog, *A. thaliana*	145	2.6e-16

Table 1 Continued

I. Homology to a protein with database-specified function

EST	Description of match with highest significance	HSP	P
Protein turnover/trafficking			
22a	vesicle coat component, *S. cerevisiae*	486	1.9e-45
24a	probable proteasome component precursor, *Sch. pombe* (2×)	527	8.6e-50
27n	peroxisomal membrane protein, *Penicillium citrinum* (3×)	454	4.7e-42
70n	vacuolar protease A, *N. crassa* (2×)	643	3.4e-71
125n	mannose-6-phosphate isomerase, *Emericella nidulans*	554	2.1e-74
163n	putative COPII coat component, *Sch. pombe*	155	1.4e-08
169n	polyubiquitin, *Manduca sexta*	289	3.9e-37
214n	YptA (GTP-binding protein), *Aspergillus niger* var. *awamorii*	796	2.7e-78
291n	vacuolar protein-sorting protein VPS3, *S. cerevisiae*	532	1.6e-49
318n	carboxylic acid transporter protein, *S. cerevisiae*	138	1.5e-07
Transcription factors/regulators/transcription			
21a	HAPB protein, *E. nidulans*	163	1.2e-10
104n	la costa protein, *Drosophila melanogaster*	136	9.5e-14
111n	swi6 protein, repression of silent mating type loci, *Sch. pombe*	165	2.8e-19
185n	cAMP response element-binding protein ATF2, *Rattus norwegicus*	168	5.9e-10
307n	transcription factor atf1, *Sch. pombe*	297	9.9e-25
359n	small nuclear ribonucleoprotein Sm D3, *Drosophila melanogaster*	117	2.4e-06
Cell division			
1a	histone H3, *E. nidulans* (3×)	516	9.1e-60
57n	cell cycle regulator p21 protein, *Sch. pombe*	277	4.9e-30
135n	homologous to a human proliferation-associated nucleolar protein, *S. cerevisiae*	1157	1.5e-116
"284n"	cell division control protein cdc 15, *Sch. pombe*	206	4.0e-13

Cell wall			
5a	mixed-linked glucanase precursor, *Cochliobolus carbonum*	91	1.8e-05
55nA	chitinase, *E. nidulans*	403	1.2e-36
61n	scytalone dehydratase, *Colletotrichum lagenarium*	304	2.4e-47
106n	beta-glucosidase, *Candida molischiana*	773	7.3e-76
112n	17beta-hydroxysteroid dehydrogenase (ketoreductase), *Cochliobolus lunatus*	273	7.1e-23
150n	cryparin (hydrophobin), *Cryphonectria parasitica* (2×)	169	5.4e-15
Cytoskeleton			
60n	probable Arp2–3 complex sub, *Sch. pombe*	737	4.8e-72
228n	actin-capping protein alpha-2 chain, *Gallus gallus*	396	6.5e-36
303n	cofilin related protein, *N. crassa*	473	4.5e-44
Protein synthesis			
"35a-rev"	40S ribosomal protein S2, *H. sapiens*	182	4.3e-13
55nB	ribosomal protein S25A, *S. cerevisiae*	277	2.6e-23
117n	ribosomal protein L12, *Rattus rattus*	583	1.0e-55
137n	ribosomal protein YmL17 precursor, mitochondrial, *S. cerevisiae*	157	2.8e-10
160n	elongation factor 3, *Candida albicans*	500	4.6e-65
207n	60S ribosomal protein L2, *Blumeria graminis* f. sp. *hordei*	703	1.9e-68
260n	60S ribosomal protein L3, *E. nidulans*	563	1.3e-53
285nA	eukaryotic translation initiation factor 3, *H. sapiens*	141	3.3e-08
"290n"	40S ribosomal protein S26E, *N. crassa*	529	5.3e-50
298n	50S sub ribosomal protein, *Kluyveromyces lactis*	151	1.1e-07
322n	cytoplasmic ribosomal protein S7, *Podospora anserina*	495	2.1e-46
341n	ribosomal protein L34.e.B, cytosolic, *S. cerevisiae*	422	1.2e-38
371n	eukaryotic peptide chain release factor GTP-binding sub, *Pichia pinus*	129	2.2e-06
377n	phenylalanyl-tRNA synthetase, beta sub, *S. cerevisiae*	805	3.0e-79

Table 1 Continued

I. Homology to a protein with database-specified function

EST	Description of match with highest significance	HSP	P
Secondary metabolism			
7n	CG2397 gene product, *D. melanogaster*, cytochrome P450	181	5.8e-12
259n	dimethyl-allyl-tryptophan-synthase, *Claviceps purpurea*[a]		
274n	probable multifunctional folic acid synthesis protein, *N. crassa*	569	1.3e-68
278nB	D-amino-acid oxidase, *Fusarium solani*	500	1.7e-89
323n	cytochrome P450 (*Gibberella fujikuroi*)	217	4.2e-16
334n	DNase 1 protein, *Metarhizium anisopliae* var. *anisopliae*	599	2.0e-57
368n	oxygenase (*Streptomyces argillaceus*)	305	2.4e-31
AOS			
186nB	peptide methionine sulfoxide reductase, *Methanobacterium thermoautotrophicum*	415	6.3e-38
"255n"	Cu-Zn superoxide dismutase, *Glomerella cingulata*	153	8.1e-13
Miscellaneous			
"35a-uni"	symbiosis-related protein, *Laccaria bicolor*	362	2.4e-51
48n	transposase-like protein, Ac-type transposon Tfo1, *Fusarium oxysporum*	170	9.5e-11
53n	RIC1 protein (*Phytophthora infestans*)	194	1.7e-14
232n	probable RNA-directed RNA polymerase, *Sch. pombe*	162	2.9e-16
248n	probable amiB protein, *Mycobacterium tuberculosis*, hydrolase	322	2.7e-49

II. Homology to a protein without database-specified function

EST	Description of match with highest significance	HSP	P
28a	hypothetical 107.9 KD protein in POL4-SRD1 intergenic region, *S. cerevisiae*	1015	1.7e-101
34a	glucose-repressible gene protein, *N. crassa*[b]	209	4.3e-16
74n	small glutamine-rich tetratricopeptide repeat containing protein, *S. cerevisiae*	383	1.6e-34
87n	CG12491 gene product, *D. melanogaster*	112	3.4e-05
91n	probable membrane protein YOL092w, *S. cerevisiae*	115	0.00074
105nB	CG16788 gene product, *D. melanogaster*	131	4.6e-05
145n	hypothetical protein, *H. sapiens*	162	7.6e-08
146nB	hypothetical protein, *Sch. pombe*	303	4.7e-26
159n	very hypothetical protein SPAC1687.07, *Sch. pombe*	175	4.8e-12
171n	glucose-repressible gene protein, *N. crassa* (2×)	200	3.9e-15
189n	hypothetical protein SPAC15E1.02c, *Sch. pombe*	239	2.9e-19
"193n"	CG13902 gene product, *D. melanogaster*	113	4.6e-05
256n	ESTs 0068 and 0162, *Leishmania major*	108	2.2e-05
285nB	hypothetical protein sll1188, *Synechocystis* sp.	122	9.6e-14
296n	blue light–induced protein BLI-3, *N. crassa*	346	9.6e-37
311n	hypothetical protein SPBC365.12c, *Sch. pombe*	130	1.3e-06
320n	heat shock protein 30, *N. crassa*	437	3.0e-40
330n	hypothetical protein, *Sch. pombe*	114	1.7e-09
"340n"	CGI-83 protein, *H. sapiens*	281	1.4e-33
347n	heat shock protein 70, *Ajellomyces capsulatus*	224	1.3e-16
352nA	cold shock protein CAPA, *Pseudomonas fragi*	243	1.1e-19
365n	hypothetical protein, *Sch. pombe*	132	6.5e-08

Abbreviations: HSP = high-scoring segment pair produced by BLAST search (sum of matches and mismatches); mt = mitochondrial; P = probability value produced by BLAST search (probability that a found HSP is found by chance); sub = subunit, "..." = clone probably containing more than one cDNA insert; A or B in clone name = clone with different coligated cDNA inserts. Underlined clones were found more than once.

[a] Identified by comparison to the alkaloid cluster of *Claviceps purpurea* strain P1.

[b] The amino acid sequence encoded by 34a has 64% homology to the one encoded by 171n.

To test if the combined loss of both major extracellular AOS scavenging activities has impact on pathogenicity, *cpcat1/cpsod1* double mutants were created (S. Joshi, unpublished). In these mutants the minor effect observed for Δ*cpsod1* was slightly more pronounced; i.e., the delay in honeydew production was clearly 1 day, and the amount of honeydew was significantly reduced. However, in spite of this retardation, normal sclerotia are formed; i.e., the infection is successful.

Since the defense system against AOS is highly complex in eukaryotic systems, we initiated a more general approach to identify genes induced under oxidative stress. Growth on medium with a high copper content was chosen for the induction of oxidative stress. Copper acts as a fenton catalyst in the generation of AOS [79] and induces transcription of Cu,Zn-SOD and catalase in yeast [80–82]. A cDNA library of RNA from copper-induced mycelium of *C. purpurea* strain 20.1 was established and differentially screened with copper-induced and copper-starved cDNA. In a first round of experiments 40 differentially hybridizing c-DNA clones were identified; Northern analysis showed that some of them were also induced by H_2O_2 in axenic culture (S. Joshi, unpublished), confirming that the copper stress approach was successful for the identification of genes induced by oxidative stress. Though the screening can by no means be considered as saturated, the list of clones obtained (see Table 2) is very interesting from various aspects:

1. There are clones which—according to the homology found in the BLAST analysis—could play a role in oxidative stress, e.g., clone X51, with

TABLE 2 Copper-Induced cDNAs That Show Homology to Other (Putative) Proteins

No.	Description of match with highest significance	HSP	P
X1	hypothetical protein B1D4.110 (imported), *Neurospora crassa*	120	2e-26
X12a	trihydrophobin, *Claviceps fusiformis*	72	2e-12
X23a	probable membrane transporter, *Schizosaccharomyces pombe*	70	3e-11
X34	conserved hypothetical protein SPCC1450.14c, *Sch. pombe*	58	le-07
X46	sorbitol dehydrogenase, *Bacillus subtilis*	136	5e-31
X51	multicopper oxidase fet3p, *Candida albicans*	174	6e-43
X62	probable glutamate 5-kinase, *Sch. pombe*	275	3e-73
X65	proteasome component Y7, *Saccharomyces cerevisiae*	113	4e-44

homology to multicopper oxidases, and clone X46, a potential sorbitol-dehydrogenase; these clones are also induced by H_2O_2.

2. Twenty of the differential cDNA clones stem from the same gene: *cph1*, coding for a class II hydrophobin, which has been twice detected in an EST-library (see Table 1). During infection it seems to be expressed in stages II–IV (Fig. 3). This gene is not induced by H_2O_2, but by copper; this is a new finding for a hydrophobin gene, and it will be highly interesting to analyze the regulation of this gene in more detail.

3. A large group of clones show no significant homology to any gene with a well-defined function; most of them are induced by H_2O_2, and two of them were also found in the in planta EST library (see below). This group could contain some surprises, and it will be interesting to compare them, e.g., with clones of unknown function from other pathogenic fungi. An interesting candidate here is the gene which is represented by altogether six cDNA clones in this sample (e.g., X9), which encodes a very small, cystein-rich, probably secreted protein, similar to the small secreted polypeptides in the *Cladosporium fulvum* system [83].

A few of these clones have already been analyzed in more detail:

1. X51a ("multicopper oxidase"): the corresponding gene (*cmpco1*) has been cloned and sequenced; it shows significant homology to the copper-induced iron transporter multicopper oxidase from *S. cerevisiae* (*Fet3*) [84]. Interestingly, closely linked (in inverse orientation), a putative iron transferase gene (*cpit1*) was found, showing high homology to the *fip1* gene of *Sch. pombe* which also there is closely linked to the *Fet3* homolog *fio1* [85]. Thus, in *C. purpurea* a high-affinity iron uptake system is present that is comparable to that in yeasts. Both genes are expressed in planta in all stages of infection, as shown by RT-PCR (S. Joshi, unpublished). As in yeast, the system is induced in axenic culture by low iron concentrations. Its potential role in oxidative stress response is confirmed by two observations: *cpmco1* is induced by H_2O_2, and both genes are upregulated in a *cpsod1* deletion mutant even in the presence of iron (S. Joshi, unpublished), as has been described for yeast [86]. A targeted deletion of *cpmco1* will clarify its importance for pathogenicity.

2. X46 ("dehydrogenase"): the corresponding gene has been cloned and sequenced. It shows high homology to Zn-binding dehydrogenases, in particular to sorbitol dehydrogenases. Since polyol metabolism has been implicated in both, AOS defense and pathogenicity (a deletion mutant of mannitoldehydrogenase in *C. fulvum* shows reduced resistance to oxidative stress in axenic culture and reduced pathogenicity) [87], and this gene is induced by H_2O_2, a more detailed analysis in the *C. purpurea* system seems justified.

During the EST analysis (see Sec. 4.2), two clones were found (Table 1) that were also detected among the copper stress clones (356nB = X9 and 132n = X2). One of them (132n) seems to be expressed only in stage II during infection

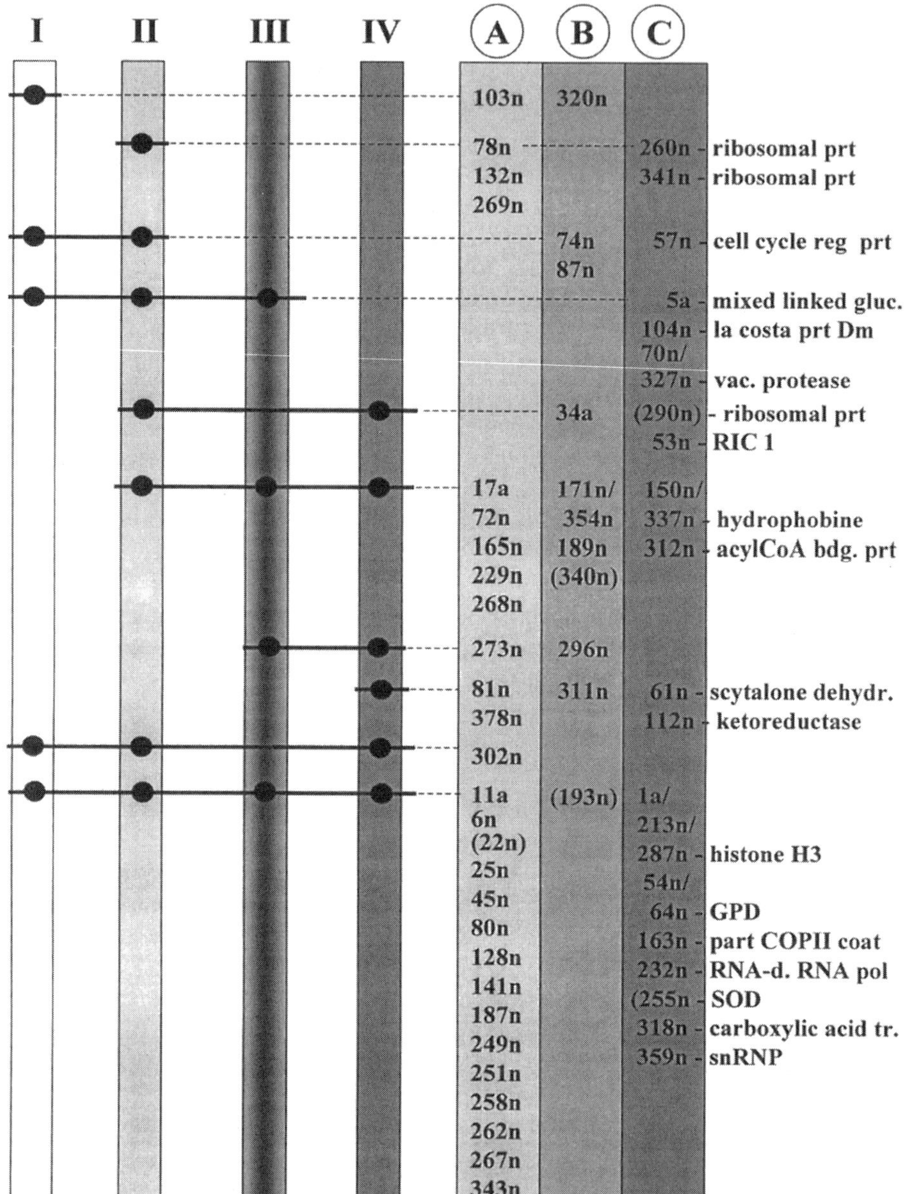
I
II
III
IV
A
B
C
103n
320n
78n
260n - ribosomal prt
132n
341n - ribosomal prt
269n
74n
57n - cell cycle reg prt
87n
5a - mixed linked gluc.
104n - la costa prt Dm
70n/
327n - vac. protease
34a
(290n) - ribosomal prt
53n - RIC 1
17a
171n/
150n/
72n
354n
337n - hydrophobine
165n
189n
312n - acylCoA bdg. prt
229n
(340n)
268n
273n
296n
81n
311n
61n - scytalone dehydr.
378n
112n - ketoreductase
302n
11a
(193n)
1a/
6n
213n/
(22n)
287n - histone H3
25n
54n/
45n
64n - GPD
80n
163n - part COPII coat
128n
232n - RNA-d. RNA pol
141n
(255n - SOD
187n
318n - carboxylic acid tr.
249n
359n - snRNP
251n
258n
262n
267n
343n

(see Fig. 3). Another EST clone (186nB, see Table 1) shows homology to peptide methioine sulfocide reductases, which mediate the reduction of protein sulfoxide methionyl residues back to methionine and therefore could be involved in AOS.

Moreover, another new and fascinating aspect of AOS-related gene expression in *C. purpurea* came from the EST data analysis: one of the potential transcription factors identified in the in planta expression analysis shows homology to an oxidative stress-related factor in mammalian cells (*atf*) (see Table 1). This gene (*cpatf1*) was shown to be induced in axenic culture of *C. purpurea* by H_2O_2. The corresponding gene has been cloned and characterized (S. Joshi unpublished); it is expressed in all stages of infection. A targeted inactivation of *cpatf1* will show which genes are controlled by this H_2O_2-induced transcription factor, and if its inactivation has an impact on pathogenicity. This could allow a more general approach to the understanding of the impact of AOS stress defense in the development of *C. purpurea* in planta.

4.1.3 Hydrophobins

Hydrophobins, a class of small (~120 amino acids) hydrophobic proteins that have been detected in several filamentous fungi [88], are characterized by their ability to form amphipathic layers at hydrophobic/hydrophilic interphases. They are abundant in aerial structures of fungi, like aerial hyphae, conidia, and fruiting bodies. They have also been implicated to play a role in the interaction of phytopathogenic fungi and their hosts, e.g., in *Magnaporthe grisea* and *Cryphonectria parasitica* [for review see 89]. During a differential cDNA screening aiming at identification of ergot–alkaloid biosynthesis-correlated genes in a submerse-producing (nonpathogenic) strain of *Claviceps fusiformis*, we detected a gene (*cfth1*) obviously coding for a new type of hydrophobin; it contains three hydrophobin domains seperated by asparagin/glycin repeats [90]. Detailed biochemical analysis resulted in the identification of the gene product, a full-size protein (i.e., being not processed down to single hydrophobin units) showing properties of a typical class II hydrophobin [91].

FIGURE 3 Expression analysis of the EST clones: All clones were probed separately with cDNA probes derived from infection stage I, II, III, or IV. Their hybridization was marked by big, black dots in the columns representing stages I–IV; e.g., in stage I and II expression of the clones 74n, 87n, and 57n was observed. 74n and 87 are located in column B (clones with homology to proteins without a database-specified function), 57n in column C (clones with homology to proteins without a database-specified function; see Table 1 for details). No expression of clones without any homology (column A) was observed for stages I and II.

Recently, a comparable gene was identified in *C. purpurea*, *cpph1*, which codes for a protein with an even higher complexity, a pentahydrophobin (T. Correia, B. Oeser, unpublished). It consists of five units showing significant homology to class II hydrophobins, interrupted by GN-repeat regions, which could form amphipathic α-helices; the amino terminus contains a glycin-rich region which could be involved in attaching the protein to the cell wall. The structure of *cpph1* is comparable to that of *cfth1* of *C. fusiformis*; the presence of long direct repeats within *cpph1* and the high homology of the internal three moduls suggest a recent generation of this gene from a tripartite precursor. By in situ hybridization it could be shown that *cpph1* is expressed in planta, especially in two areas: in external and penetrating hyphae, and in conidiogenic hyphae [92], indicating that these special proteins could be involved in establishment of fungus–host contact and coating of spores.

Using a gene replacement approach, two deletion mutants of *cpph1* could be obtained (G. Vautard-Mey, unpublished); preliminary experiments indicate that these mutants are not heavily impaired in their pathogenicity. Thus also this uniquely structured tissue-specifically expressed gene does not seem to be essential for the development of disease symptoms. As mentioned earlier, within an EST library of *C. purpurea* (see Table 1) a gene encoding a ''classical'' class II hydrophobin (*cph1*) has also been detected; the gene is expressed in most stages of infection and is induced by copper (see above), but a specific role has not yet been assigned to it.

4.1.4 Signal Chain Components

Pathogenic fungi, like all other organisms, respond to signals from the environment. These signals are transduced from the cell surface to the target genes, resulting in altered gene expression in response to changes in the environment. Recent studies have demonstrated that phytopathogenic fungi require specific signal transduction pathways for pathogenesis. Several genes coding for components of signal chains have been characterized from phytopathogens and functionally analyzed, e.g., for subunits of heterotrimeric GTP-binding proteins, various proteinkinases, and other compounds (e.g., of the cAMP transduction pathway) [for review see 93]. Since compounds of the MAP-kinase cascade were especially shown to have an impact on pathogenicity in various fungi (*Cochliobolus heterostrophus* [94]; *Magnaporthe grisea* [95]; *Botrytis cinerea* [96]), we initiated an analysis of this signaling cascade in *C. purpurea*. Using a degenerated primer approach, two putative MAP kinase genes were cloned and characterized ([112], G. Vutard-Mey, B. Oeser, unpublished): *cpmk1* and *cpmk2*, showing significant homology to the MAP kinase genes *mpk1* and *mps1* of *M. grisea*, respectively [95,97]. Both genes are expressed in planta throughout the whole colonization phase; since the corresponding *M. grisea* genes have both been shown to be essential for pathogenicity, though with different specific effects, we chose two approaches to study the function of these homologs in *C. purpurea*: (1) Knockout mutants were obtained by a gene replacement approach (the deletion mutants are under investigation); (2) complementation of the corresponding *M. grisea* deletion mutants, to see if the signal transduction pathway in these divergent fungi are comparable.

The fascinating aspect of signaling mutants which are affected in their pathogenicity is that they present invaluable tools to study the very first parts of the respective signal chain (i.e., the receptors) and to identify target genes, which together obviously make up the pathogenicity phenotype. This raises the analyses from a ''hit-and-run'' strategy to a more global understanding of the interaction system. In *C. purpurea* it will be especially interesting to learn more about the strict organ specificity and about the mechanisms which guide the infection hyphae along the path of the pollen tube, down to the vascular system.

4.2 Random Approach: In Planta–Expressed Genes (EST)

In order to obtain a deeper insight into the genes/factors involved in development of *C. purpurea* on rye, we initiated an unbiased analysis of in planta–expressed genes of the fungus. Rye plants were inoculated by pipetting conidia suspension into the floral cavities at the point of anthesis, an infection test which is very near to natural conditions. At 5, 10, 15, and 20 dpi (corresponding to stages I–IV; see Sec. 3), infected ovaries were harvested and immediately lyophilized. RNA was prepared and used for cDNA synthesis. Mixed RNA aliquots of all four stages were used for the establishment of a cDNA library (in λZAPII).

After in vivo excision of the pBluescript SK(−) plasmids, they were checked (by Southern analysis) for the presence of rye sequences. All clones seemed to be bona fide *C. purpurea* clones and were sequenced from both sides—i.e., analysis of ESTs (expressed sequence tags) was initiated.

After sequencing all ESTs were organized with the help of DNATools by S. Rasmussen (*http://www.crc.dk/phys/dnatools.htm*). DNA Tools is a program package for routine handling and analysis of DNA and protein sequences. We used it to find the multiple occurrence of ESTs, to compare the ESTs to other *C. purpurea* sequences in our laboratory (e.g., sequences found in a differential screening of a copper-induced cDNA bank; see Sec. 4.1.2), and to link information about ESTs to the sequence of the ESTs themselves. All ESTs were analyzed with the programs Blastx/Blastn (as provided by NCBI—*http://www.ncbi.nlm.nih.gov/BLAST/*—via DNATools) and WU-Blastx (as provided by BCM *http://dot.imgen.bcm.tmc.edu:9331/*). Matches with probability (*P*) values $<10^{-5}$ and a high-scoring segment pair (HSP) >100 were considered significant. There were three categories of matches: (1) homology to proteins *with* a database-specified function (33%) (Table 1 part I); (2) homology to proteins *without* such a function (e.g., putative ORFs from genome sequencing projects, proteins without a clearly defined cellular role) (13%) (Table 1 part II); and (3) no homology found (54%). The percentages of these three categories are similar to the values reported for other EST banks (e.g., *N. crassa*: 98; *Mycosphaerella graminicola*: 99). About a fourth of the ESTs in each category had a match to the dbEST databank as provided by NCBI.

To link the cDNA clones to a specific stage, all cDNA clones were dot-blotted and separately hybridized with cDNA probes derived from infection stage I (~5 dpi; few visible symptoms); II (~10 dpi; honeydew and conidia production; sphacelium); III (~15 dpi; sclerotial development starting; sphacelium > sclerotium); and IV (~20 dpi; progressed sclerotial development) (see Sec. 3). Expression of only 58 clones was observed (Fig. 3) despite the fact that of course all clones are derived from active genes since they are cDNA clones. Lack of hybridization might be due to the limits of the dot-blot procedure.

The EST analysis might help to identify genes involved in the two most interesting stages of *C. purpurea* during its parasitic development. These stages of particular importance for the fungus are (1) its colonization of the rye ovary to firmly establish its presence, and (2) its metabolic switch from the non-alkaloid-producing sphacelium to the alkaloid-producing sclerotium, which serves as resting structure and place for sexual reproduction.

Genes participating in the colonization would be expected to be expressed early (in stage I, II, and maybe III, as III contains still more sphacelial than sclerotial tissue). Among those clones (see Fig. 3) are four with no corresponding homology found so far, three with homology to proteins without a database-specified function as well as clones 5a and 104n. As interesting as the future assignment of function for those seven clones might be, one can already speculate on the function of 5a and 104n, accepting that a found homology tells something about the role of a clone. 5a shows homology to the so far only sequenced mixed-linked glucanase (MLG1 from the maize pathogen *Cochliobolus carbonum* [100] which probably can degrade cereal β-glucans. In *C. carbonum* a *mlg1* knockout had no effect on the pathogenicity of the fungus on maize. 104n might be generally involved in correct development, as mutations in its *Drosophila melanogaster* homolog ''la costa'' lead to defects in larval segment polarity [101].

EST clones triggering the switch to sclerotial growth should be expressed from stage III on. Only two of these kind were found (see Fig. 3): 273n and 296n. No function is assigned to these clones. Only based on the homology of 296n to a blue light–induced protein of *Neurospora crassa* could one speculate on the regulation of sclerotia formation by blue light.

Among the genes whose expression could not be linked to expected infection patterns, one has to try identify potential genes of interest, e.g., for the pathogenic development. Four clones will be mentioned here: 106n, 334n, 35a-uni, and 53n. They all occur in interactions between a fungal partner and a plant (or an insect)—ranging from symbiosis to parasitism.

106n shows homology to an extracellular β-glucosidase gene from *Candida molischiana*, which functions at low pH and whose function in *Candida* is unknown [102]. Both clone 106n and the *Candida* gene show further high homology to glucosidases, which are involved in the degradation of saponins. Saponins are thought to protect plants, which contain these substances, from microbial attack. Potential pathogens might need special glucosidases to overcome this protection [103–105].

334n shows homology as well to DNaseI from the insect pathogen *Metarhizium anisopliae* var. *anisopliae*, found also as EST clone, as well to DNaseI from the bean pathogen *Fusarium solani* f. sp. *phaesoli*. In the latter case it obviously acts as elicitor, turning the plant against the fungal invader [106]. Yet, in the more attenuated interaction of ergot and rye it might very well act in favor of the fungus.

35a-uni is homologous to *LB-AUT7*, a symbiosis-regulated and -related gene from the ectomycorrhizal fungus *Laccaria bicolor* [107]. It is not expressed in the free-living *Laccaria*. It is probably involved in an ectomycorrhizal-typical increase in vesicle production and turnover during symbiotic interaction. The authors assume that it may play an important role in preparing *Laccaria* for entry into the plant host root. It can be pointed out that also *Claviceps* has close contact to its host plant (see Sec. 3.2). Whether 35a-uni is similiarily not expressed in axenic culture needs to be determined.

53n probably encodes a short peptide of ~56 amino acids, which shows significant homology to the *Ric1* product of *Phytophthora infestans*. It was isolated from a late blight potato interaction cDNA library [108]. 53n and Ric1 show further homology to stress-induced plant proteins. The *Ric1* product is induced under high-salt or high-pH conditions. As pH and osmotic changes will be encountered by the invading fungus, the authors hypothesize that RIC1 might be important for the interaction. Interestingly, *Ric1* was found in 25% of 22 investigated cDNAs, whereas no expression of 53n was observed (see Fig. 3). On the other hand, *Phytophthora infestans* is a much more aggressive pathogen than *Claviceps*.

5 PERSPECTIVES

The cytological and general pathological aspects summarized in the first chapters of this overview show that the interaction of *C. purpurea* and *S. cereale* represents an interesting model system:

1. It is a well-balanced, subtle interaction; *C. purpurea* as a biotrophic pathogen causes only limited damage and is able to reach its goals (tapping of vascular tissue, production of spores, and survival structures) without causing efficient defense reactions.

2. It is a very specific interaction: only the ovarian tissue of grasses is attacked, and the fungus follows a strict routine of infection, including a directed, straightforward growth in the plant tissue, mostly following the path of the pollen tube.

The biochemical and molecular investigations have substantiated the first aspect: in contrast to a necrotrophic fungus like *Botrytis cinerea* [11], *C. purpurea* produces only very limited amounts of cell wall–degrading enzymes in axenic culture, confirming a strict regulation of the synthesis of hydrolytic enzymes in this system. This balanced, subtle interaction could explain why inactivation of CWDE-encoding genes has an effect on pathogenicity in this system, though in other model systems, like *Cochliobolus carbonum*, even multiple knockouts of CWDE genes never yielded a phenotype in planta [reviewed in 93,109]. However, it should be emphasized that so far only the double PG knockout has such a dramatic effect that the fungus would probably hardly survive in nature—indicating a prominent role of pectin degradation for the successful colonization. In all other cases (including the various AOS enzymes), the effects of knockouts observed were only minor (though significant), indicating a great flexibility of the system. Obviously, *C. purpurea* also operates with several safety systems. Therefore single surgical operations will only in rare cases have a dramatic effect.

The molecular analyses outlined in this review have opened at least two approaches to study the complexity of the system at a higher level: the analysis of signal chains involved in the interaction, and the characterization of putative transcription factors like *cpatf1*, which could control whole sets of genes. In addition, the EST analysis, especially the stage-specific expression patterns have yielded a wealth of material that needs further scrutiny.

Major problems to be addressed based on the presently available information are (1) the importance of oxidative stress for the interaction; (2) the nature and role of plant signals triggering the colonization process and the host and organ specificity; (3) the strategy of the fungus to gain plant-derived nutrition supplied by the vascular system; and (4) the strategy of the fungus to avoid major plant defense reactions.

The *C. purpurea*/rye system combines the fascinating characteristics of a highly envolved interaction system resulting from a long coevolutionary process with a comparably easy experimental access and thus represents an interesting and promising model system for the study of host-pathogen interaction.

ACKNOWLEDGMENTS

We thank the Deutsche Forschungsgemeinschaft and the European Community (TMR network ''Cerepath'') for funding, and Frau A. Kammerahl for typing the manuscript.

REFERENCES

1. WA Taber. Biology of *Claviceps*. In: AL Demain, NA Solomon, eds. Biology of Industrial Microorganisms. London: Benjamin Cummings, 1985, pp 449–486.
2. WP Campbell. Studies on ergot infection in gramineous hosts. Can J Bot 35:315–320, 1957.
3. AR Loveless. Conidial evidence for host restriction in *Claviceps purpurea*. Trans Br Mycol Soc 56:419–434, 1971.
4. GN Agrios. Plant Pathology. 3rd ed. San Diego: Academic Press, 1988.
5. V Křen, L Cvak. Ergot—The Genus *Claviceps*. In: Medicinal and Aromatic Plants—Industrial Profiles, Vol 6. Amsterdam: Harwood Academic Publishers, 1999.
6. H Mielke. Studien über den Pilz Claviceps purpurea (Fries) Tulasne unter Berücksichtigung der Anfälligkeit verschiedener Roggensorten und der Bekämpfungsmöglichkeit des Erregers. Mitteilungen aus der Biologischen Bundesanstalt für Land- und Forstwirtschaft, Berlin-Dahlem, H. 375. Berlin: Parey Buchverlag, 2000, pp 1–66.
7. DE Frederickson, PG Mantle, WAJ De Milliano. Windborne spread of ergot disease (*Claviceps africana*) in sorghum A-lines in Zimbabwe. Plant Pathol 42:368–377, 1993.
8. R Bandyopadhyay, DE Frederickson, NW McLaren, GN Odvody, MJ Ryley. Ergot: a new disease threat to *Sorghum* in the Americans and Australia. Plant Dis 82: 356–367, 1998.
9. P Tudzynski, KB Tenberge, B Oeser. *Claviceps purpurea*. In: K Kohmoto, US Singh, RP Singh, eds. Pathogenesis and Host Specificity in Plant Diseases: Histopathological, Biochemical, Genetic and Molecular Bases, Vol II—Eukaryotes. New York: Pergamon, Elsevier Science, 1995, pp 161–187.
10. KB Tenberge. Biology and life strategy of the ergot fungi. In: V Kren, L Cvak, eds. Medicinal and Aromatic Plants—Industrial Profiles, Vol 6. Ergot—The Genus *Claviceps*. Amsterdam: Harwood Academic Publishers, 1999, pp 25–56.
11. P Tudzynski. Genetics of *Claviceps purpurea*. In: V Kren, L Cvak, eds. Medicinal and Aromatic Plants—Industrial Profiles, Vol 6. Ergot—The Genus *Claviceps*. Amsterdam: Harwood Academic Publishers, 1999, pp 79–93.
12. A ten Have, KB Tenberge, JAE Benen, P Tudzynski, J Visser, JAL van Kan. The contribution of cell wall degrading enzymes to pathogenesis of fungal plant pathogens. In: F Kempken, ed. The Mycota XI, Application in Agriculture. New York: Springer-Verlag. In press.
13. K Esser, P Tudzynski. Genetics of the ergot fungus *Claviceps purpurea*. I. Proof of a monoecious life-cycle and segregation patterns for mycelial morphology and alkaloid production. Theor Appl Genet 53:145–149, 1978.
14. LR Tulasne. Mémoire sur l'ergot des glumacées. Ann Sci Nat (Partie Botanique) 20:5–56, 1853.
15. C Engelke. Neue Beobachtungen über die Vegetations-Formen des Mutterkornpilzes *Claviceps purpurea* Tulasne. Hedwigia 41:221–222, 1902.
16. H Kirchhoff. Beiträge zur Biologie und Physiologie des Mutterkornpilzes. Centralbl Bakteriol Parasitenk, Abt. II 77:310–369, 1929.

17. RFN Langdon. BR Champs. Insect vectors of *Claviceps purpurea* in Queensland. J Aust Inst Agr Sci 20:24, 1954.
18. DJ Swan, PG Mantle. Parasitic interactions between *Claviceps purpurea* strains in wheat and an acute necrotic host response. Mycol Res 95:807–810, 1991.
19. ES Luttrell. Host–parasite relationships and development of the ergot sclerotium in *Claviceps purpurea*. Can J Bot 58:942–958, 1980.
20. G Hadley. Development of stromata in *Claviceps purpurea*. Trans Br Mycol Soc 51:763–769, 1968.
21. ES Luttrell. The disease cycle and fungus–host relationships in dallisgrass ergot. Phytopathology 67:1461–1468, 1977.
22. A Stoll, A Brack. Über die Entstehung von Sklerotien des Mutterkornpilzes (*Claviceps purpurea*) an den obersten Halmknoten des Roggens. Ber Schweiz Bot Ges 54:252–254, 1944.
23. RW Lewis. Development of conidia and sclerotia of the ergot fungus on inoculated rye seedlings. Phytopathology 46:295–296, 1956.
24. ASt Garay. The germination of ergot conidia as affected by host plant, and the culture of ergot on excised roots and embryos of rye. Physiol Plant 9:350–355, 1956.
25. J Heslop-Harrison, Y Heslop-Harrison. The pollen–stigma interaction in the grasses. 1. Fine-structure and cytochemistry of the stigmas of Hordeum and Secale. Acta Bot Neerl 29:261–276, 1980.
26. J Heslop-Harrison, Y Heslop-Harrison. The pollen–stigma interaction in the grasses. 2. Pollen-tube penetration and the stigma response in Secale. Acta Bot Neerl 30:289–307, 1981.
27. Y Heslop-Harrison, BJ Reger, J Heslop-Harrison. The pollen–stigma interaction in the grasses. 6. The stigma ('silk') of *Zea mays* L. as host to the pollens of *Sorghum bicolor* (L.) Moench and *Pennisetum americanum* (L.) Leeke. Acta Bot Neerl 33:205–227, 1984.
28. Y Heslop-Harrison, J Heslop-Harrison, BJ Reger. The pollen–stigma interaction in the grasses. 7. Pollen-tube guidance and the regulation of tube number in *Zea mays* L. Acta Bot Neerl 34:193–211, 1985.
29. RJ Williams, N Colotelo. Influence of pollen on germination of conidia of *Claviceps purpurea*. Can J Bot 53:83–86, 1975.
30. R You, WA Jensen. Ultrastructural observations of the mature megagametophyte and the fertilization in wheat (*Triticum aestivum*). Can J Bot 63:163–178, 1985.
31. WJ Cherevick. Association of ergot with loose smut of wheat and of barley. Phytopathology 43:461–463, 1953.
32. PG Mantle. An unusual parasitic association between *Claviceps purpurea* and rye. Trans Br Mycol Soc 59:327–330, 1972.
33. NC Carpita, DM Gibeaut. Structural models of primary cell walls in flowering plants: consistency of molecular structure with the physical properties of the walls during growth. Plant J 3:1–30, 1993.
34. KB Tenberge, V Homann, B Oeser, P Tudzynski. Structure and expression of two polygalacturonase genes of *Claviceps purpurea* oriented in tandem and cytological evidence for pectinolytic enzyme activity during infection of rye. Phytopathology 86:1084–1097, 1996.

35. KB Tenberge, V Homann, P Tudzynski. Der Mutterkornpilz *Claviceps purpurea* auf Roggen: Struktur and Funktion zweier Polygalakturonase-Gene in Relation zu Veränderungen von Pektinen in der Wirtszellwand. Phytomedizin 26(3):55–56, 1996.
36. WP Campbell, LE Tyner. Comparison of degree and duration of susceptibility of barley to ergot and loose smut. Phytopathology 49:348–349, 1959.
37. SB Puranik, DE Mathre. Biology and control of ergot on male sterile wheat and barley. Phytopathology 61:1075–1080, 1971.
38. J Willingale, PG Mantle. Stigma constriction in pearl millet, a factor influencing reproduction and disease. Ann Bot 56:109–115, 1985.
39. DE Frederickson, PG Mantle, WAJ De Milliano. Susceptibility to ergot in Zimbabwe of sorghums that remained uninfected in their native climates in Ethiopia and Rwanda. Plant Pathol 43:27–32, 1994.
40. RG Platford, CC Bernier. Resistance to *Claviceps purpurea* in spring and durum wheat. Nature 226:770–770, 1970.
41. MA Flaishman, PE Kolattukudy. Timing of fungal invasion using host's ripening hormone as a signal. Proc Natl Acad Sci USA 91:6579–6583, 1994.
42. DG Parbery. 1996 Trophism and the ecology of fungi associated with plants. Biol Rev 71:473–527, 1994.
43. WP Campbell. Infection of barley by *Claviceps purpurea*. Can J Bot 36:615–619, 1958.
44. BI Shaw, PG Mantle. Host infection by *Claviceps purpurea*. Trans Br Mycol Soc 75:77–90, 1980.
45. KB Tenberge. Infection by the ergot fungus *Claviceps purpurea*: ultrastructure and cytochemistry of a host-pathogen relationship. Fifth International Mycological Congress (IMC 5), August 14–21, 1994, Vancouver, British Columbia, Canada, 1994, p 219.
46. KB Tenberge, P Tudzynski. Der Mutterkornpilz *Claviceps purpurea* auf Roggen: Affinitätscyto-chemische Untersuchungen einer Wirt-Parasit-Beziehung. Phytomedizin 25(3):51–52, 1995.
47. KB Tenberge, B Brockmann, P Tudzynski. Immunogold localization of an extracellular β-1,3-glucanase of the ergot fungus *Claviceps purpurea* during infection of rye. Mycol Res 103:1103–1118, 1999.
48. J Willingale, PG Mantle. Stigmatic constriction in pearl millet following infection by *Claviceps fusiformis*. Physiol Mol Plant Pathol 30:247–257, 1987.
49. NV Raikhel, HI Lee, WF Broekaert. Structure and function of chitin-binding proteins. Annu Rev Plant Physiol Plant Mol Biol 44:591–615, 1993.
50. T Yamagami, G Funatsu. Limited proteolysis and reduction–carboxymethylation of rye seed chitinase-a: role of the chitin-binding domain in its chitinase action. Biosci Biotechnol Biochem 60:1081–1086, 1996.
51. S Giesbert, HB Lepping, KB Tenberge, P Tudzynski. The xylanolytic system of *Claviceps purpurea*: cytological evidence for secretion of xylanases in infected rye tissue and molecular characterization of two xylanase genes. Phytopathology 88: 1020–1030, 1998.
52. RL Mower, JG Hancock. Mechanism of honeydew formation by *Claviceps* species. Can J Bot 53:2826–2834, 1975.

53. A Hambrock. Untersuchungen über pflanzliche Abwehrreaktionen bei der Interaktion von *Claviceps purpurea* (Fr.) Tul. mit Secale cereale L.–Licht- und elektronenmikroskopische sowie biochemische Analysen Ph.D. thesis, Westfälische Wilhelms-Universität Münster, 1996.
54. K Mendgen, M Hahn, H Deising. Morphogenesis and mechanisms of penetration by plant pathogenic fungi. Annu Rev Phytopathol 34:367–386, 1996.
55. J Willingale, PG Mantle. Interactions between *Claviceps purpurea* and *Tilletia caries* in wheat. Trans Br Mycol Soc 89:145–153, 1987.
56. S Osada Kawasoe, L Fucikovsky Zak, ML Ortega Delgado, E Engleman. Study on the germination of the ergot on maize and host parasit interaction *Zea mays* and *Claviceps gigantea*. Agrociencia O 66:57–70, 1986.
57. CW Mims. Using electron microscopy to study plant pathogenic fungi. Mycologia 83:1–19, 1991.
58. M Baum, ES Lagudah, R Appels. Wide crosses in cereals. Annu Rev Plant Physiol Plant Mol Biol 43:117–143, 1992.
59. DE Frederickson, PG Mantle. The path of infection of sorghum by *Claviceps sorghi*. Physiol Mol Plant Pathol 33:221–234, 1988.
60. KB Tenberge, P Tudzynski. Early infection of rye ovaries by *Claviceps purpurea* is inter- and intracellular. BioEng Sondernummer 10(3):22–22, 1994.
61. U Müller, KB Tenberge, B Oeser, P Tudzynski. Cel1, probably encoding a cellobiohydrolase lacking the substrate binding domain, is expressed in the initial infection phase of *Claviceps purpurea* on *Secale cereale*. Mol Plant–Microbe Interact 10: 268–279, 1997.
62. PG Mantle. Emergence and phytopathological properties of a new strain of *Claviceps purpurea* (Fr.) Tul. on rye. Ann Appl Biol 60:353–356, 1967.
63. BI Shaw, PG Mantle. Parasitic differentiation of *Claviceps purpurea*. Trans Br Mycol Soc 75:117–121, 1980.
64. PG Mantle, LJ Nisbet. Differentiation of *Claviceps purpurea* in axenic culture. J Gen Microbiol 93:321–334, 1976.
65. K Corbett, AG Dickerson, PG Mantle. Metabolic studies on *Claviceps purpurea* during parasitic development on rye. J Gen Microbiol 84:39–58, 1974.
66. RA Bassett, EB Chain, K Corbett, AGF Dickerson, PG Mantle. Comparative metabolism of *Claviceps purpurea* in vivo and in vitro. Biochem J 127:3P–4P, 1972.
67. E Ramstad, G Gjerstad. The parasitic growth of *Claviceps purpurea* (Fries) Tulasne on rye and its relation to alkaloid formation. J Am Pharm Assoc Sci Ed 44:741–743, 1955.
68. U Müller. Struktur, Expression und gezielte Inaktivierung von cel1, einem vermutlich Cellobiohydrolase-codierenden Gen von *Claviceps purpurea*. Bibl Mycol 166: 1–100, 1997.
69. P Heidrich, KB Tenberge. The role of cell wall degrading enzymes from *Claviceps purpurea* during infection of rye. 5th European Conference on Fungal Genetics, ECFG5, March 25–29, Arcachon, France, 2000 p 138.
70. B Brockmann, R Smit, P Tudzynski. Characterization of an extracellular β-1,3-glucanase of *Claviceps purpurea*. Physiol Mol Plant Pathol 40:191–201, 1992.
71. C Lamb, RA Dixon. The oxidative burst in plant disease resistance. Annu Rev Plant Physiol Plant Mol Biol 48:251–275, 1997.

72. GL Mandell. Catalase, superoxide dismutase, and virulence of *Staphylococcus aureus*. In vitro und in vivo studies with emphasis on staphylococcae–lencocyte interaction. J Clin Invest 55:561–566, 1975.
73. MA De Groote, UA Ochsner, MU Shiloh, C Nathan, JM McCord, MC Dinauer, SJ Libby, A Vazquez-Torres, Y Xu, FC Fang. Periplasmic superoxide dismutase protects *Salmonella* from products of phagocytc NADPH-oxidase and nitric oxide synthase. Proc Natl Acad Sci USA 94:13997–14001, 1997.
74. AJ Hamilton, MD Holdom. Antioxidant systems in the pathogenic fungi of man and their role in virulence. Med Mycol 37:375–389, 1999.
75. ES Jacobson, ND Jenkins, JM Todd. Relationship between superoxide dismutase and melanin in a pathogenic fungus. Infect Immun 62:4085–4086, 1994.
76. V Garre, KB Tenberge, R Eising. Secretion of a fungal extracellular catalase by *Claviceps purpurea* during infection of rye: putative role in pathogenicity and suppression of host defense. Phytopathology 88:744–753, 1998.
77. V Garre, U Müller, P Tudzynski. Cloning, characterization and targeted disruption of cpcat1, coding for an in planta secreted catalase of *Claviceps purpurea*. Mol Plant–Microbe Interact 11:772–783, 1998.
78. P Tudzynski, K Hölter, T Correia, C Arntz, N Grammel, U Keller. Evidence for an ergot alkaloid gene cluster in *Claviceps purpurea*. Mol Gen Genet 261:133–141, 1999.
79. SJ Stohs, D Bagchi. Oxidative mechanisms in the toxicity of metal ions. Free Radical Biol Med 18:321–336, 1995.
80. MT Carri, F Galliazzo, MR Ciriolo, G Rotilio. Evidence for co-regulation of Cu, Zn superoxide dismutases and metallothionein gene expression in yeast through transcriptional control by copper via the ACE1 factor. FEBS Lett 278:263–266, 1991.
81. EB Gralla, DJ Thiele, P Silar, JS Valentine. ACE1, a copper-dependent transcription factor, activates expression of the yeast copper, zinc superoxide dismutase gene. Proc Natl Acd Sci USA 88:8558–8562, 1991.
82. P Lapinskas, H Ruis, V Culotta. Regulation of *Saccharomyces cerevisiae* catalase gene expression by copper. Curr Genet 24:388–393, 1993.
83. RP Oliver, B Henricot, G Segers. *Cladosporium fulvum*, cause of leaf mould of tomato. In: IW Kronstad, ed. Fungal Pathology. Dordrecht: Kluwer Academic Publishers, 2000, pp 65–92.
84. C Gross, M Keleher, R Vishwanath, PO Brown, DR Winge. Identification of the copper regulon in *Saccharomyces cerevisiae* by DNA microarrays. J Biol Chem 275:32310–32316, 2000.
85. S Labbé, MMO Pena, AR Fernandes, DJ Thiele. A copper sensing transcription factor regulates iron uptake genes in *Schizosaccharomyces pombe*. J Biol Chem 274:36252–36260, 1999.
86. JM De Freitas, A Liba, R Meneghini, J Selverstone Valentine, E Butler Gralla. Yeast lacking Cu-Zn superoxide dismutase show altered iron homeostasis. J Biol Chem 275:11645–11649, 2000.
87. HJ Bussink, A Clark, K Blissett, M Pike, H Tibosch, R Oliver. Reactive oxygen metabolism in *Cladosporium fulvum*: a role for mannitol. 5th European Conference on Fungal Genetics, Arcachon, France, 2000, p 149.

88. JGH Wessels. Developmental regulation of fungal cell wall formation. Annu Rev Phytopathol 32:413–437, 1994.
89. MJ Kershaw, NJ Talbot. Hydrophobins and repellents: proteins with fundamental roles in fungal morphogenesis. Fungal Genet Biol 23:18–23, 1998.
90. C Arntz, P Tudzynski. Identification of genes induced in alkaloid-producing cultures of *Claviceps* sp. Curr Genet 31:357–360, 1997.
91. OMH de Vries, S Moore, C Arntz, JGH Wessels, P Tudzynski. Identification of a tripartite hydrophobin from *Claviceps fusiformis*: a novel type of class II hydrophobin. Eur J Biochem 262:377–385, 1999.
92. KB Tenberge, P Stellamanns, G Plenz, H Robenek. Nonradioactive in situ hybridization for detection of hydrophobin mRNA in the phytopathogenic fungus *Claviceps purpurea* during infection of rye. Eur J Cell Biol 75:265–272, 1998.
93. P Tudzynski, B Tudzynski. Genetics of phytopathology: phytopathogenic fungi: genetic aspects of host–pathogen interaction. Prog Bot 61:118–147, 2000.
94. S Lev, A Sharon, P Hadar, H Ma, BA Horwitz. A mitogen-activated protein kinase of the corn leaf pathogen *Cochliobolus heterostrophus* is involved in conidiation, appressorium formation, and pathogenicity: diverse roles for mitogen-activated protein kinase homologs in foliar pathogens. Proc Natl Acad Sci USA 96:13542–13547, 1999.
95. JR Xu, JE Hamer. MAP kinase and cAMP signaling regulate infection structure formation and pathogenic growth in the rice blast fungus *Magnaporthe grisea*. Genes Dev 10:2696–2706, 1996.
96. L Zheng, M Campbell, J Murphy, S Lam, JR Xu. The BMP1 gene is essential for pathogenicity in the gray mold fungus *Botrytis cinerea*. Mol Plant–Microbe Interact 13:724–732, 2000.
97. JR Xu, CJ Staiger, JE Hamer. Inactivation of the mitogen-activated protein kinase MPS1 in the rice blast fungus prevents penetration of host cells but allows activation of plant defense responses. Proc Natl Acad Sci USA 95:12713–12718, 1998.
98. MA Nelson, S Kang, EL Braun, ME Crawford, PL Dolan, PM Leonard, J Mitchell, AM Armijo, L Bean, E Blueyes, T Cushing, A Errett, M Fleharty, M Gorman, K Judson, R Miller, J Ortega, I Pavlova, J Perea, S Todisco, R Trujillo, J Valentine, A Wells, M Werner-Washburne, S Yazzie, DO Natvig. Expressed sequences from conidial, mycelial and sexual stages of *Neurospora crassa*. Fungal Genet Biol 21: 348–363, 1997.
99. J Keon, A Bailey, J Hargraves. A group of expressed cDNA sequences from the wheat fungal leaf blotch pathogen *Mycosphaerella graminicola* (*Septoria tritici*). Fungal Genet Biol 29:118–113, 2000.
100. JM Görlach, E van der Knaap, JD Walton. Cloning and targeted disruption of mlgl, a gene encoding two of three extracellular mixed-linked glucanases of *Cochliobolus carbonum*. Appl Environ Microbiol 64:385–391, 1998.
101. T Heizler, D Coulson, MT Saenz-Roblens, M Ashburner, J Roote, P Simpson, D Gubb. Genetic and cytogenetic analysis of the 43A-E region containing the segment polarity gene costa and the cellular polarity genes prickle and spiny-legs in *Drosophila melanogaster*. Genetics 135:105–115, 1993.
102. G Janbon, R Magnet, A Arnaud, P Galzy. Cloning and sequencing of the β-glucosi-

dase-encoding gene from *Candida molischiana* strain 35M5N. Gene 165:109–113, 1995.

103. RW Sandrock, D DellaPenna, HD VanEtten. Purification and characterization of β-2-tomatinase, an enzyme involved in the degradation of α-tomatine and isolation of the gene encoding β-2-tomatinase from *Septoria lycopersici*. Mol Plant Microbe Interact 8:960–970, 1995.
104. A Osbourn, P Bowyer, P Lunnes, B Clarke, M Daniels. Fungal pathogens of oat roots and tomato leaves employ closely related enzymes to detoxify different host plant saponins. Mol Plant Microbe Interact 8:971–978, 1995.
105. JP Morrisey, JP Wubben, AE Osbourn. *Stagnopora avenae* secretes multiple enzymes that hydrolyze oat leaf saponins. Mol Plant Microbe Interact 13:1041–1052, 2000.
106. LA Hadwiger, MM Chang, MA Parsons. *Fusarium solani* DNase is a signal for increasing expression on nonhost disease resistance response genes, hypersensitivity, and pisatin production. Mol Plant Microbe Interact 8:871–879, 1995.
107. SJ Kim, D Bernreuther, M Thumm, GK Podila. LB-AUT7, a novel symbiosis-regulated gene from an ectomycorrhizal fungus, *Laccaria bicolor*, is functionally related to vesicular transport and autophagocytosis. J Bacteriol 181:1963–1967, 1999.
108. P van West, S Kamoun, JW van't Klooster, F Govers. Ric1, a *Phytophthora infestans* gene with homology to stress-induced genes. Curr Genet 36:310–315, 1999.
109. P Tudzynski, B Tudzynski. Genetics of plant pathogenic fungi. Prog Bot 59:169–193, 1998.
110. B Oeser, P Heidrich, Ü Mueller, KB Tenberge, P Tudzynski. Polygalacturonase is a pathogenicity factor in the *Claviceps purpurea*/rye interaction. Fungal Genet Biol. In press.
111. S Moore, O deVries, P Tudzynski. The major Cu, Zn SOD of the phytopathogen *Claviceps purpurea* is not essential for pathogenicity. Molec Plant Pathol. In press.
112. G Mey, B Oeser, MH Lebrün, P Tudzynski. The biotrophic, non-appressorium forming grass pathogen *Claviceps purpurea* needs a Fus 3/Pmk 1 homologous MAP kinase for colonization of rye ovarian tissue. Molec Plant-Microbe Interact. In press.

17

Hypovirulence

Helmut Bertrand and
Dipnath Baidyaroy
Michigan State University, East Lansing, Michigan

1 INTRODUCTION

The term "hypovirulence" has been used liberally in the literature to describe the reduced capacity of some variants of a pathogen to either elicit or sustain the normal rate of progression of the corresponding disease in its host. In the context of the phytopathogenic fungi, however, "hypovirulence" has been used more stringently to define states of reduced virulence that are caused by cytoplasmically transmitted genetic factors. Initially, it was believed that all the hypovirulence-causing agents were viruses [1], but now it is evident that other cytoplasmically transmitted factors, namely mutant forms of mitochondrial DNA (mtDNA) and certain mitochondrial plasmids, also can dramatically reduce the pathological aggressiveness of the chestnut blight fungus *Cryphonectria parasitica* [2]. To distinguish the attenuated pathogenic states that are caused by viruses from those that are caused by mtDNA mutations and mitochondrial plasmids, they have been divided into two types and defined as "viral hypovirulence" and "mitochondrial hypovirulence," respectively [2]. Though this distinction facilitates the description of the fundamentally different processes that result in phenomenologically similar phenotypes, it is hardly adequate because some of the double-stranded

RNA (dsRNA) elements that appear to be causatively associated with hypovirulence have been classified as viruses but are located in the mitochondria [3,4].

Most of the mycelia of phytopathogenic fungi consist primarily of branched coenocytic hyphae that extend by apical growth and are not only interconnected with each other through cytoplasmic bridges, but also can fuse with hyphae from other thalli through the formation of anastomoses. Therefore, debilitating cytoplasmic elements that proliferate rapidly, such as viruses, mutant mitochondria, and plasmids, behave like infectious agents because they are transported by cytoplasmic movement within individual hyphae and are communicated from infected hyphae to uninfected hyphae and mycelia through the anastomoses that form at points of contact. Hypovirulence syndromes can therefore be portrayed as debilitating infectious diseases that afflict filamentous fungi, and from this perspective have been used with some success to control fungal pathogens under natural conditions [5,6]. In spite of their capacity to invade natural populations of fungi, none of the cytoplasmic factors that cause hypovirulence is known to exist as an extracellular form that can actively infect healthy cultures of the fungal host.

2 VIRAL HYPOVIRULENCE

Biraghi [7] first recorded the observations that eventually led to the discovery of transmissible hypovirulence by describing several examples of chestnut trees that were surviving in spite of being blighted by *C. parasitica*. Rather than the deeply indented cankers normally found on dying, blight-infested trees, he noted superficial cankers that appeared to be in the process of healing. At that time it was assumed that the trees were resistant to the fungus. Later, French pathologists discovered that these symptoms were not due to a resistant variety of chestnut, but were a consequence of the infection of susceptible trees by altered forms of *C. parasitica* [5,8,9]. The fungal cultures obtained from the surviving trees consisted of two different morphological types: one that appeared normal in culture—i.e., produced an orange pigmentation, sporulated abundantly, and was virulent—and a new form that had significantly reduced pigmentation—i.e., was white, and generated relatively few asexual spores. When reinoculated onto chestnut, the white isolates produced small, superficial cankers that were incapable of killing trees. The margins of the resulting cankers were surrounded by swollen wound periderm and ultimately closed, forming the callused remnant of a lesion [10]. It was this observation that led to the introduction of the term ''hypovirulence'' as a descriptor for the characteristics of the attenuated isolates of *C. parasitica* [8].

The observation that elicited much interest in the hypovirulent strains of *C. parasitica* was that they had a curative effect on diseased chestnut trees when inoculated onto festering cankers on branches and trunks [9]. The application of the hypovirulent strains resulted in the conversion of resident virulent strains into hypovirulent forms [5]. By the criteria originally described by Jinks [11], the observation that virulent strains also could be attenuated in the laboratory through contacts with hypovirulent strains established that this trait was determined by a cytoplasmic genetic factor that could be transmitted through hyphal anastomoses. Subsequently, the natural spread of hypovirulence in Italy resulted in a corresponding suppression of blight to a point where the disease no longer was a serious problem in the cultivation of chestnut in that country [12].

The first clear indication of the physical nature of the cytoplasmic determinant of hypovirulence in *C. parasitica* was provided by the discovery that several hypovirulent isolates contained dsRNA elements that were absent from virulent strains [13]. Subsequently, it was shown that the introduction of this dsRNA element into a virulent strain converted it to the corresponding hypovirulent form [14]. Moreover, the elimination of dsRNA molecules from a hypovirulent strain by treatment of cultured mycelium with cycloheximide restored the virulence phenotype in *C. parasitica* [15]. These and other findings have provided correlative evidence to suggest that the cytoplasmic genetic elements responsible for the induction of the hypovirulence phenotype in *C. parasitica* are cytoplasmically replicating dsRNA molecules. Because of their ''infectious'' character, their capacity to debilitate the host fungus and their similarities with plant and bacterial RNA viruses, these elements, and several other types of fungal dsRNAs that have similar properties, have been designated as viruses [4,16,17].

Double-stranded RNAs have been detected in filamentous fungi belonging to many genera, and some of them have been characterized extensively [for a review, see 17]. However, very few dsRNAs have been implicated definitively in causing hypovirulence in pathogenic fungi, and among these, the best-characterized are the nonlethal, dsRNA mycoviruses of the chestnut blight fungus *C. parasitica* [1,4,18]. Double-stranded RNA elements have also been implicated strongly in the causation of hypovirulence of the Dutch elm disease pathogen *Ophiostoma ulmi* [4] and have been isolated from attenuated strains of *Monosporascus cannonballus* [19], *Rhizoctonia solani* [20], and *Sclerotinia sclerotiorum* [21]. In addition, dsRNA viruses are strongly implicated in the causation of a debilitating lytic disease in the plant-pathogenic fungus *Helminthosporium victoriae* [17,22] and with the appearance of degenerative diseases in some cultivars of edible mushrooms [4,23,24]. However, the most comprehensive insights into virus–fungus reactions that result in the appearance and dissemination of the hypovirulence trait have been derived from studies on the attenuation of *C. parasitica* by a group of dsRNAs identified as hypoviruses (family Hypoviridae) [16].

Since most mycoviruses have genomes composed of dsRNA [4,17], several attempts were made to isolate virus particles from hypovirulent *C. parasitica* strains. However, the hypovirulence-associated elements proved to be unencapsidated but enclosed in membranous vesicles [25–27] that contain an RNA-dependent RNA polymerase activity that probably functions in the replication of the dsRNA [28]. Like mycoviruses, hypovirulence-associated dsRNAs are not infectious in the classical sense; i.e., cell-free preparations are not infectious when applied to fungal hyphae or spheroplasts, and new infection can only occur by hyphal anastomoses [1]. Surveys of hypovirulent isolates of *C. parasitica* revealed considerable variations in the copy number, size, and physical and genetic composition of the associated dsRNAs [14,25,29–32]. Moreover, a single isolate of a given fungus can harbor more than one type of dsRNA [33]. The most common types of dsRNAs from *C. parasitica* have recently been assigned to a new virus family, the Hypoviridae [16]. Members of the Hypoviridae are cytoplasmic and contain a single dsRNA element of 10–13 kb that is phylogenetically most closely related to single-stranded RNAs of the plant virus family Potyviridae. In addition to the full-length dsRNAs, many fungal strains also contain smaller units, most of which appear to have resulted from deletions of internal segments of the viral genome [31,34]. Moreover, in *C. parasitica* [3] and *Ophiostoma novo-ulmi* [35], hypovirulence has also been associated with dsRNA elements that are located in mitochondria and are related to the yeast cytoplasmic T and W dsRNAs rather than the cytosolic hypoviruses. These dsRNAs have been assigned tentatively as belonging to the genus *Mitovirus* of the family Narnaviridae [35].

2.1 The Hypoviruses of *C. parasitica*

Three different types of cytosolic dsRNA hypoviruses have been characterized so far in *C. parasitica* and have been designated as CHV1, CHV2, and CHV3 [36,37] because they share some common characteristics [18,37]. All are large linear molecules without structural proteins and are polyadenylated at their 3′ termini [29,30,38]. Whereas the CHV1 and CHV2 genomes have two open-reading frames (ORFs), CHV3 contains only one. The first member of the genus to be sequenced, CHV1-Ep713, is 12,712 bp long, and its two ORFs, designated as ORF A and ORF B, overlap by a single base pair [1].

In CHV1-Ep713, ORF A encodes two proteins, named p29 and p40, with p29 being a protease which is autocatalytically released from the ORF A polypeptide during translation [39] (Fig. 1). The introduction of a cDNA version of ORF A into *C. parasitica* and subsequent expression of p29 and p40 resulted in some of the morphological traits that are associated with CHV1 infections, including loss of pigmentation, reduced sporulation, and a reduction in laccase activity, but the transgenic strains remained virulent [40]. Deletion of 88% of the p29 ORF partially relieved the suppression of pigment formation and sporulation without any changes in the level of virulence [41]. Analysis of mutant forms of p29 revealed that the manifestation of the p29-mediated symptoms in *C. parasitica*

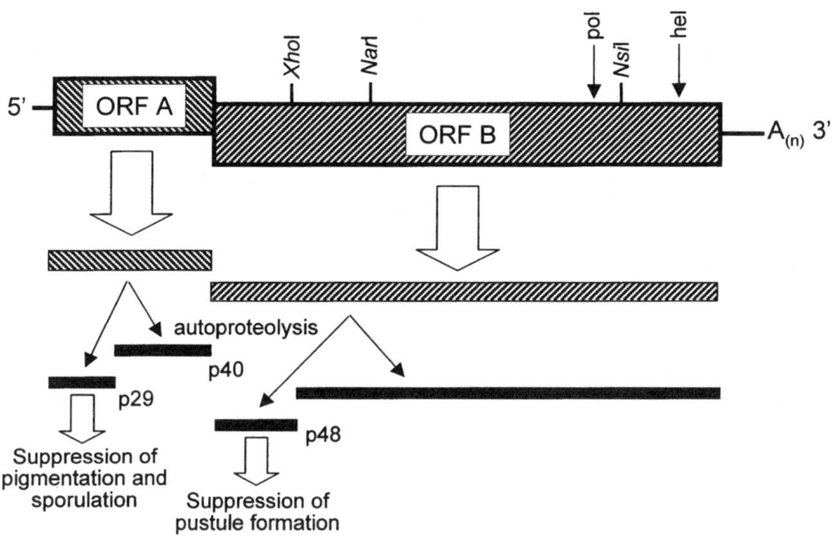

FIGURE 1 Diagrammatic representation of the genome organization of the CHV1-EP713 virus and current view of its basic gene expression strategy [1]. The RNA polymerase and the helicase motifs are designated as pol and hel, respectively. The proteins encoded by the virus are identified as p29, p40, and p48 according to their predicted sizes of 29 kDa, 40 kDa, and 48 kDa, respectively.

are influenced by two cysteine residues that are conserved in the potyvirus-encoded helper component proteinase (HC-Pro), a protein that shares some similarity with p29 [41]. The polyprotein encoded by ORF B includes another protease, p48, which is similar to the p29 protease encoded by ORF A, and a large protein that contains motifs characteristic of RNA-dependent RNA polymerases and RNA helicases [1]. Thus, it can be assumed that the primary function of the ORF-B polyprotein is related to the replication and maintenance of the dsRNA, but it also has been reported that this open reading frame contributes to the virulence-attenuating activity of the virus.

The CHV2 viruses are almost identical to the CHV1 type, except that their ORF A proteins do not undergo autoproteolysis and do not produce active proteases [29,36]. On the other hand, the 9-kb dsRNA of the prototype of the CHV3 group of viruses has a nucleotide sequence that is dissimilar from those of either the CHV1 or the CHV2 dsRNAs. It contains a single ORF which encodes a putative polyprotein with protease, RNA-dependent RNA polymerase, and RNA helicase domains similar to those found in other members of the family Hypoviridae [37].

Transformation of virus-free, virulent *C. parasitica* strains with cDNA copies of CHV1 viruses results in the manifestation of the transmissible hypovirulence phenotype due to the production of full-length transcripts that are expressed and replicated, thus producing complete dsRNA genomes of the hypoviruses [42]. Whereas viral dsRNA genomes are not transmitted sexually, the chromosomally integrated cDNA copies are meiotically stable [43], and different hypovirus strains are known to attenuate their respective hosts to different degrees. To take advantage of these aspects of virus biology for the identification of the roles of the different ORFs in the determination of hypovirulence, chimeric viruses were constructed from infectious cDNA clones of a mildly debilitating (CHV1-Euro7) and a severely debilitating (CHV1-Ep713) hypovirus [44,45]. Chimeric viruses created by swapping of the 5′ untranslated regions or ORF A among the CHV1-EP713 and the CHV1-Euro-7 viruses did not change their respective effects on the host fungus [44]. Similar results were obtained when the 3′ regions (from the *Nsi*I site to the 3′ end in Fig. 1) of the viruses were interchanged. Therefore, the domains that control host morphology and virulence lie upstream of the *Nsi*I site within ORF B. Further mapping of these domains identified multiple sites that are located on either side of the *Nar*I site (Fig. 1) and act independently of each other as determinants of morphological traits of the fungus. Furthermore, it was found that the p48 protein contains a dominant determinant for the suppression of pustules (asexual spore-containing stromal structures) in cankers. However, the results also indicated that p48-mediated suppression of pustule formation depends on interactions between segments of proteins encoded within the ORF B region located between the *Xho*I and the *Nar*I sites (Fig. 1).

The effect of the dsRNA viruses on *C. parasitica* was first analyzed by Powell and Van Alfen [46,47] and later by Kazmierczak et al. [48], who showed that the synthesis of several fungal transcripts, proteins, and secondary metabolites is downregulated by viruses of the CHV1 type. Subsequently, it was suggested that the activities of some of the downregulated components, such as polygalacturonases [49], laccase [50,51], cutinase [52], and oxalic acid [53,54], may be essential virulence factors. Ablation of at least some of these putative virulence factors by disruption of the appropriate genes, however, did not lead to the expected reduction in virulence [49]. Other proteins that have been found to be downregulated in CHV1-infected lines of *C. parasitica* include the cell surface hydrophobin cryparin [55,56] and a precursor protein of a pheromone [57] that is thought to be essential for sexual reproduction [58]. Deletion of the genes encoding the cryparin and the pheromone did not result in a significant decline in the pathogenic potential of the fungus, but did reproduce some of the morphological and fertility traits that are associated with virus infections [58]. The common feature shared by all of the downregulated proteins is that they are translated as preproteins that have conserved signals for processing by serine proteases during their secretion [56,57]. This observation has led to the proposition that the virus might interfere with protein secretion in general by diverting secretory vesicles for its own replication [6].

Studies led by D.L. Nuss [59] on the repression of laccase in *C. parasitica* by dsRNA viruses resulted in the discovery that the transcription of the corresponding gene, *lac1*, is regulated by two opposing pathways: a negative control system that limits transcript accumulation and requires ongoing protein synthesis, and a stimulatory pathway that is dependent on inositol triphosphate and calcium as second messengers [60]. Connected to this discovery was the observation that a hypovirus of the CHV1 type interferes with the transduction of the inositol triphosphate calcium–dependent signal for *lac1* expression [60], and actively perturbs a signal transduction pathway that is essential for the response of the fungus to the availability of nutrients in the immediate environment [61]. Since GTP-binding proteins (G-proteins) are a family of regulatory proteins that play an essential role in the response of eukaryotic cells to many environmental stimuli [62], Choi et al. [63] examined the status of this protein and found that, relative to the virulent control, virus-infected hypovirulent strains were deficient for the α-subunit polypeptides (CPG1) and transcripts from the corresponding gene, *cpg1*. Although the targeted disruption of *cpg1* generated the hypovirulent state, it also resulted in poor growth of the mutant and did not reproduce exactly the complement of morphological and physiological traits that appear in virus-infected hypovirulent strains [59,64]. These observations suggest that the interaction between the CHV1 type of dsRNA elements with the fungal host is complex, and that hypovirulence might be a direct consequence of the repression of the expression of *cpg1* by these viruses. Significantly, disruption of *cpg2*, another gene encoding a G-protein α subunit, did not affect virulence [64], whereas disruption of the G-protein β-subunit gene, *cpgb1*, resulted in a significant reduction in pigmentation, production of aerial mycelia, sporulation, and virulence of the fungus without diminishing the rate of vegetative growth [65]. Finally, the disruption of a gene encoding a different G-protein signal transduction component (*bdm1*) resulted in a phenotype indistinguishable from that of the *cpgb1* β-subunit mutant [66], and the transcription of *cpg1* is repressed in the *cpgb1 bdm1* double mutant. Collectively, these observations strongly support the concept that the CHV1 hypoviruses cause hypovirulence in *C. parasitica* by perturbing signal transduction processes underlying fungal pathogenesis [59].

Very little is known at this time about the physiological causes of the hypovirulence phenotypes that are caused by hypoviruses belonging to groups other than CHV1. Disruption of G-protein signaling is most likely only one of several pathways by which different hypoviruses effect hypovirulence. Since each of the three types of hypoviruses modifies a unique set of morphological traits in *C. parasitica*, it is likely that different groups of viruses debilitate the fungus by perturbing different metabolic of regulatory pathways.

2.2 Control of *C. parasitica* by Hypoviruses

Reports of the successful control of chestnut blight in Europe as a result of natural dissemination (Italy) or artificial application of hypovirulent *C. parasitica* strains (France) [5] stimulated efforts to examine whether transmissible hypovirulence might be effective in controlling blight in North America. Initial reports indicated that the introduction of the European hypovirulent strains indeed could cure cankers incited by virulent North American strains of the pathogen [67,68]. However, subsequent attempts to control chestnut blight in North America by the artificial introduction of natural and genetically engineered strains containing Old World as well as New World viruses, so far, have not resulted in a sustained control of the fungus, even though there was some evidence for the initial introgression of the pertinent hypoviruses and transgenes into some of the indigenous virulent strains [1,18,69–72]. However, sustained biological control of the fungus by native viruses appears to occur naturally at a limited number of specific sites in North America [10,70,72,73]. The reasons for the disappearance of introduced hypovirulence factors from the virulent populations of *C. parasitica* in North America are not obvious at this time. However, factors that are believed to contribute to this process include a complex vegetative incompatibility system that affects the coexistence of genetically diverse nuclei in the same mycelium as well as the efficiency of transmission of cytoplasmic factors through hyphal anastomoses [74–78], and reduced reproductive fitness of the hypovirulent strains [32].

2.3 Hypovirulence and Mitochondrial dsRNAs

In *C. parasitica*, a relatively mild form of hypovirulence is associated with an unencapsidated, 2.7-kb dsRNA element (NB631 dsRNA) that is located in mitochondria and is quite different from the dsRNA molecules of any of the viruses in the family Hypoviridae [3,79]. No ORF of significant size was recognized when the nucleotide sequence of this dsRNA was translated by application of the universal genetic code, but a single and sizable ORF encoding a putative RNA-dependent RNA polymerase was obtained when the translation was carried out by means of the mitochondrial genetic code pertaining to the ascomycetous filamentous fungi (UGA = Trp) [3]. Copurification of the element with a particular subcellular fraction and genetic transmission patterns confirmed that this particular dsRNA is located in the mitochondria, whereas protein-based phylogenetic comparisons indicate that it is related to the T and W cytosolic dsRNAs of yeast rather than the cytosolic hypoviruses of *C. parasitica*. Whereas the cytosolic hypoviruses typically are not transmitted sexually because they cause female sterility, the NB631 dsRNA is transmitted exclusively from the maternal parent in a strain-specific manner to a fraction of the ascospore progeny ranging from 5% to ~50% [79]. The fact that this dsRNA is transmitted from the female parent to all the progeny suggests, but does not prove, that it is not included in all the mitochondria of infected cultures. Recently, the NB631 dsRNA tentatively has been assigned to the genus *Mitovirus* in the Narnaviridiae family, together with related mitochondrial elements from *Rhizoctonia solani* and *Ophiostoma novo-ulmi* [35].

The physiological basis of the attenuation of the chestnut-blight fungus by the NB631 dsRNA remains unclear. Cyanide-insensitive, alternative oxidase activity, which is characteristically induced in *C. parasitica* by chemical agents and mutations that block cytochrome-mediated electron transport [80], is not elevated in NB631 dsRNA-containing strains, indicating that the element does not inhibit the main pathway of cellular respiration or the expression of essential mitochondrial genes [79]. Whether this dsRNA modifies the activity of any other mitochondrion-associated metabolic pathway or affects the expression of nuclear genes by retrograde regulation remains to be established.

At least 10 unencapsidated dsRNA segments are present in the mitochondria of hypovirulent variants of the Dutch elm disease agent *Ophiostoma novo-ulmi* [4,35,81], which have debilitated-growth phenotypes and show symptoms similar to those associated with the senescence syndromes of nonpathogenic filamentous ascomycetes, most notably *Podospora anserina*, *Neurospora crassa*, *N. intermedia*, and *Aspergillus amstelodami* [2]. Not all of these dsRNAs are essential for the manifestation of the hypovirulence trait: when the symptoms are transmitted asexually from diseased to healthy strains of the fungus through hyphal anastomoses, only three of the 10 dsRNA segments appear consistently in all of the converted cultures [4]. Four of the mitochondrial dsRNAs have been sequenced, and the results indicate that they represent four different viruses that are related to each other and to the NB631 mitovirus of *C. parasitica* [35,82].

A possible explanation for the induction of hypovirulence and disease by a mitovirus emerged from the observation that the mitochondria of debilitated, dsRNA-containing mycelia of *O. novo-ulmi* have drastically reduced amounts of cytochrome *c* oxidase relative to the mitochondria of healthy cultures [83]. Moreover, a plasmidlike DNA that is homologous to a small segment of the mitochondrial chromosome accumulates in the mitochondria during the appearance and progression of the diseased state [84]. The plasmidlike DNA is generated by homologous intramolecular recombination between two direct repeats that are located in close proximity to each other within the mitochondrial large-subunit rRNA gene [85], and resembles the small senDNAs that appear during the senescence phase in aging cultures of *Podospora anserina* [2]. When the hypovirulence factor is transferred into virulent strains by transient hyphal contact with diseased cultures, at least some of the dsRNA elements appear to become established rapidly in the recipient mycelium. In contrast, the plasmidlike DNA appears to be eclipsed in the recipients during the initial stages of the conversion process, but reappears in the same form as that which is present in the diseased donor when the symptoms of the disease are fully manifested in the converted cultures [84]. Thus, it is plausible to assume that hypovirulence and disease can be caused by interference of at least one dsRNA segment with the production of cytochrome *c* oxidase [4]. However, on the basis of precedents in other fungi [2,74,80,86], the possibility that the disease factor is a suppressive mutant form of mtDNA that is transmitted in small amounts through hyphal anastomoses and accumulates more slowly than the dsRNA in the recipients cannot be excluded at this time.

3 MITOCHONDRIAL HYPOVIRULENCE

3.1 Mitochondrial DNA Mutations

In 1985, Fulbright [87] reported an incident of hypovirulence in a North American strain of *C. parasitica* that was recovered from a healing canker on a chestnut tree and contained no dsRNA. This strain manifested a reduced growth rate, thin mycelium in regions proximal to the edges of the colonies, reduced sporulation, and ragged colony morphology. These characteristics, together with the extranuclear location of their genetic determinant, prompted the suspicion that this type of hypovirulence might be caused by a mitochondrial DNA (mtDNA) mutation [88]. Indeed, the respiration of this and other dsRNA-free hypovirulent isolates was found to be cyanide (CN) resistant, a trait that is not characteristic of virulent strains or strains that are rendered hypovirulent by hypovirus infections [80,88], but is associated in *Neurospora* with respiratory defects that are caused by mtDNA and nuclear-gene mutations [89,90]. While these observations were consistent with the hypothesis that hypovirulence might be caused by mitochondrial mutations, the initial examination of the dsRNA-free attenuated strains did not reveal abnormal forms of mtDNA. At that time, the effort was thwarted by the fact that the normal progenitor strains that gave rise to the putative mutants could not be identified and the mtDNA of *C. parasitica* proved to be highly variable, making it virtually impossible to distinguish pertinent mutations from optional introns and innocuous insertions and deletions.

To circumvent the difficulties that are inherent in the use of strains that are isolated from trees in fields and forests for the identification of the genetic factors that cause mitochondrial hypovirulence syndromes, Monteiro-Vitorello et al. [80] used a reverse approach: the induction and selection of mutants that showed high levels of CN-resistant respiration in the standard, dsRNA-free, Ep155 laboratory strain of *C. parasitica*. Among the six mutants that were obtained in this manner, two proved to be cytoplasmic on the basis of being capable of transmitting their respiratory defects asexually to wild-type strains by hyphal contact. One of these two mutants, *mit1*, was deficient in cytochromes *a* and *b*, a phenotype that indicates a global defect in mitochondrial gene expression [89]. The other mutant, *mit2*, lacked only cytochrome *a*. At least in *N. crassa*, this phenotype is caused primarily by point mutation in any one of the three mitochondrial genes that code for the three largest subunits of cytochrome *c* oxidase [91,92]. The *mit1* and *mit2* mutants not only proved to be hypovirulent on apples and on chestnut trees, but also transmitted this trait very effectively to a vegetatively compatible, virulent strain by hyphal contact [80]. In other words, the genetic determinant of this type of induced hypovirulence proved to be as ''infectious'' as hypoviruses. Significantly, respiratory defects, hypovirulence, or symptoms of senescence never have appeared spontaneously in the Ep155 strain, even though it has been propagated vegetatively in many laboratories over long periods of time. Thus, the experiment demonstrated that hypovirulence can be associated in *C. parasitica* with extranuclear mutations that have physiological effects similar to those that are caused by well-characterized mtDNA mutations in other ascomycetes [2].

The hypovirulence trait of the *mit1* and *mit2* mutants also was transmitted vertically to all the asexual progeny through conidia [80]. In contrast, a fraction (0–100%, depending on the isolate) of the progeny that are generated from the conidia of strains that are attenuated by the cytosolic dsRNA hypoviruses regain virulence because the virus is lost [1,93]. Furthermore, the hypovirulence phenotype was transmitted to all the sexual progeny when the *mit1* mutant was used as the female parent in crosses [80]. Sexual transmission of the *mit2* attenuating factor, however, was not observed because the mutant was infertile as a female in crosses. As expected of traits that are determined by nonlethal mtDNA mutations, the respiratory defects and hypovirulence phenotype did not reappear in any of the ascospore progeny when the *mit1* and *mit2* mutants were crossed as males with wild-type (virulent) female partners. With respect to its sexual transmission, *mit1* differs markedly from the attenuating hypoviruses, which are not transmitted at all during the sexual process, except in transgenic strains that have cDNA copies of the viral dsRNAs inserted into nuclear chromosomes [1,69]. As mentioned previously, even the attenuating mitochondrial NB631 dsRNA of *C. parasitica* is transmitted sexually from the female parent at best to only half the progeny [79].

An initial comparison of the mtDNAs of the *mit1* and *mit2* mutants with that of their virulent progenitor, Ep155, revealed no gross differences in the length, number, or amount of restriction fragments [80]. However, after having been subcultured asexually five or six times through the transfer of small mycelial plugs over a period of nearly 3 years, both mutants began to show signs of senescence and accumulated a novel 3.2-kb circular species of DNA, resembling those found in mitovirus-infected hypovirulent strains of *O. novo-ulmi* [84] and in senescent cultures of other fungi [2]. The novel senDNA-like molecules proved to be identical in the two *mit* mutants of *C. parasitica*, were located in the mitochondria, and originated from a region of the mtDNA that includes at least a portion of the small-subunit rRNA gene [2,80], as shown in Figure 2. These observations

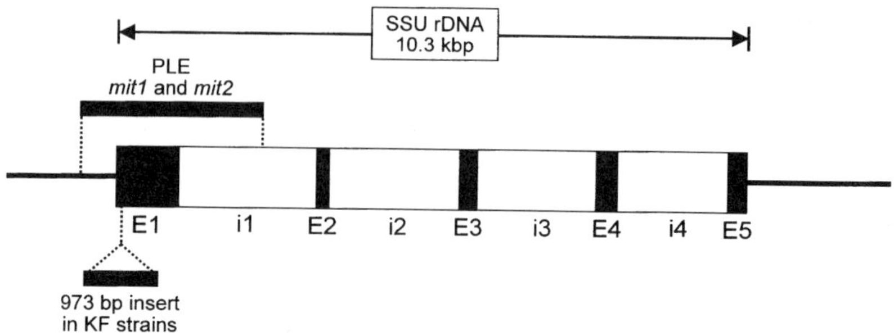

FIGURE 2 Map of the region of the mtDNA of *C. parasitica* that includes the small-subunit ribosomal RNA gene and is involved in rearrangements that appear in induced and field-isolated hypovirulent mitochondrial mutants of *C. parasitica*. Exons are labeled by the letter E and introns are labeled by the letter i, and both are followed by numbers in the order of their occurrence. The segment that is involved in the formation of plasmidlike elements (PLE) in the induced hypovirulent *mit1* and *mit2* mutants and the insert that appears in hypovirulent isolates from the KF site in North America are represented by dark bars labeled accordingly.

suggest that the hypovirulence and respiratory traits of the *mit1* and *mit2* mutants were initially caused by nonlethal suppressive point mutations that had been induced in two different mitochondrial genes, and that senescence and the senDNA-like derivatives of the mtDNA appeared later. It is possible that the *mit* mutations generated metabolic defects, including oxidative stress, that caused a low degree of instability in the mtDNA.

At this point in time, relatively little is known about the emergence, nature, transmission, and persistence of mutant mtDNAs in populations of pathogenic fungi in their natural environments. A few insights have been gained from the analysis of a population of *C. parasitica* that is located in North America in an isolated grove of American chestnut trees, identified henceforth as the KF site. In recent years, trees in this grove have recovered remarkably well from blight because the fungus has been invaded by an ''infectious'' factor that causes hypovirulence [74]. In 1986, there were no signs of tree recovery at the KF site, and no dsRNA viruses, hypovirulence, or senescence was observed in any one of a fairly large number of *C. parasitica* strains from the same location [2]. In 1990, 30% of the isolates from the same site showed some signs of senescence and hypovirulence, and by 1997 the fraction of diseased isolates had increased to 93% and many cultures died after only a few millimeters of growth on complete medium [74]. When several genetically marked, vegetatively compatible and incompatible virulent laboratory strains were grown side by side with diseased isolates from the KF site, almost all were converted to the senescent state and became hypovirulent, indicating that the debilitating genetic factor is transferred through hyphal anastomoses. In every respect, the limits imposed by different *vic* (vegetative incompatibility) genes on the asexual trafficking of the debilitating factor between strains were at least as relaxed as those imposed by the same genes on the asexual transmission of hypoviruses in *C. parasitica* [94]. Sexual transmission of the debilitating factor that has invaded the population of *C. parasitica* at the KF site has not been observed in the laboratory, primarily because the diseased isolates were infertile when used as females in crosses [74] and thus resembled strains infected by hypoviruses. Nevertheless, none of the attenuated or virulent strains that have been recovered from the KF site since 1986 contained detectable amounts of any kind of a dsRNA, but the respiratory activity of the mycelia of all the debilitated and converted strains proved to be cyanide insensitive [74].

When the hypovirulence factor present in the *C. parasitica* population at the KF site was transferred serially among laboratory strains that have RFLP-marked mtDNAs, only an 11.5-kb segment of the 160-kb mitochondrial genome was transmitted in perfect linkage with the hypovirulence trait [74]. Restriction and sequence analyses of this segment of DNA revealed that the mtDNA mutation that causes hypovirulence is a 973-bp nucleotide sequence that has been inserted into the first exon of the mitochondrial small-subunit rRNA gene (Fig. 2). The origin of this insert is unknown, but it is clear that it is not a transposon or an intron, and that it is not derived from any part of the mtDNA that is present in virulent isolates from the KF site or in the virulent Ep155 laboratory strain. Apparently, the mutation interferes with the expression of all the protein-encoding mitochondrial genes at the level of translation because it blocks the assembly of mitochondrial ribosomes. Significantly, small circular DNAs akin to the senDNAs of *P. anserina*, but not derived primarily from any part of the mtDNA, appear in the mitochondria when KF strains become severely diseased (D. Baidyaroy and H. Bertrand, unpublished observation).

3.2 Mitochondrial Plasmids and Hypovirulence

Mitochondrial plasmids occur in virtually all the important species and genera of phytopathogenic fungi [95]. Even though circular and linear mitochondrial plasmids are known to be agents of senescence in the nonpathogenic *Neurospora* spp. [96], very little is known about their effect on the virulence, vigor, or longevity of pathogenic hosts. An exception is a recent report of the attenuation of *C. parasitica* by a 4.2-kb circular plasmid, pCRY1 [75,97]. This plasmid was originally discovered in strains of *C. parasitica* that had been recovered from healing cankers on trees. In the laboratory, however, these isolates only occasionally manifested slightly elevated levels of alternative oxidase activity and at best showed moderately reduced levels of virulence on apples [98]. A study assessing the asexual transmission of pCRY1 among vegetatively compatible and incompatible strains produced several pairs of isogenic plasmid-free and plasmid-containing variants [75]. For three out of four pairs, the plasmid containing culture was markedly attenuated relative to the isogenic plasmid-free, virulent control culture when the mycelia were inoculated into branches of American chestnut trees [75]. Included among the strains that were attenuated by pCRY1 is Ep155, the standard virulent wild type of *C. parasitica*. The lesions produced by the pCRY1-containing strains were as small as or smaller than those produced by strains that are infected by attenuating hypoviruses. There is no indication that pCRY1 causes a respiratory deficiency or debilitating disease in Petri dish cultures of the fungus, and it is not clear why it has an attenuating effect on some strains and not on others.

The pCRY1 plasmid has been found at several, widely separated locations in the eastern United States and in southern Ontario, Canada, but its effect on the pathogenicity of the infected fungal populations in a natural environment has not been assessed [97]. A plasmid that is closely related to pCRY1, named pUG1, exists in a local population of *C. parasitica* in Italy [99], but it remains uncertain whether it moderates the virulence of the fungus. With respect to their structure and genetic organization, pCRY1 and pUG1 are type II circular, double-stranded DNA mitochondrial plasmids [94,97]. The prototypes of this group are the Fiji (pFIJ) and Labelle (pLAB) plasmids of *Neurospora* [95], which appear to have no known effect on the growth, morphological appearance, fertility, or longevity of their hosts. The type II plasmids are distinguished by the fact that they contain a single gene that encodes a unique DNA polymerase [97].

The potential of pCRY1 as an agent for controlling the chestnut blight fungus is accentuated by three aspects of its inheritance: (1) in contacts between strains, it is transmitted horizontally through hyphal anastomoses and across vegetative-incompatibility barriers at least as efficiently as hypoviruses and the *mit1*, *mit2*, and KF mutant [74,75,80,94]; (2) it is transmitted vertically with 100% efficiency through conidia [75]; and (3) in crosses, it is transmitted from the maternal parent to all the ascospore progeny [80,97]. However, pCRY1 and mutant forms of mtDNA have been identified only recently as hypovirulence-causing factors, and their effectiveness as biological control agents under field conditions remains to be established.

REFERENCES

1. DL Nuss. Biological control of chestnut blight: an example of virus-mediated attenuation of fungal pathogenesis. Microbiol Rev 56:561–576, 1992.
2. H Bertrand. Role of mitochondrial DNA in the senescence and hypovirulence of fungi and potential for plant disease control. Annu Rev Phytopathol 38:397–422, 2000.

3. JJ Polashock, BI Hillman. A small mitochondrial double-stranded (ds) RNA element associated with a hypovirulent strain of the chestnut blight fungus and ancestrally related to yeast cytoplasmic T and W dsRNAs. Proc Natl Acad Sci USA 91:8680–8684, 1994.
4. KW Buck. Molecular variability of viruses of fungi. In: P Bridge, Y Couteaudier, JM Clarkson, eds. Molecular Variability of Fungal Pathogens. New York: CAB International, 1998, pp 53–72.
5. J Grente, SS Berthelay-Sauret. Biological control of chestnut blight in France. In: WL MacDonald, FC Cech, J Luchok, C Smith, eds. Proceedings of the American Chestnut Symposium. Morgantown, WV: West Virginia University, 1978, pp 30–34.
6. PM McCabe, P Pfeiffer, NK Van Alfen. The influence of dsRNA viruses on the biology of plant pathogenic fungi. Trends Microbiol 7:377–381, 1999.
7. A Biraghi. Possible active resistance to *Endothia parasitica in Castanea sativa*. In: Reports to 11th Congress of the International Union of Forest Research Organizations. Rome: International Union of Forest Research Organizations, 1953.
8. J Grente. Les formes hypovirulentes d'*Endothia parasitica* les espoirs de lutte contre le chancre du chataignier. C R Hebd Acad Agr France 51:1033–1036, 1965.
9. J Grente, S Berthelay-Sauret. Pathologie vegetale—l'hypoviruence exclusive, phenomene original en pathologie vegetale. C R Acad Sci Paris 268:2347–2350, 1969.
10. DW Fulbright, WH Weidlich, KZ Haufler, CS Thomas, CP Paul. Chestnut blight and recovering American chestnut trees in Michigan. Can J Bot 61:3164–3169, 1983.
11. JL Jinks. Naturally occurring cytoplasmic changes in fungi. C R Travaux Laboratoire Carlsberg, Sér Physiol 26:183–203, 1956.
12. L Mittempergher. The present status of chestnut blight in Italy 1978. In: WL MacDonald, FC Cech, J Luchok, C Smith, eds. Proceedings of the American Chestnut Symposium. Morgantown, WV: West Virginia University, 1978, pp 34–37.
13. PR Day, JA Dodds, JE Elliston, RA Jaynes, SL Anagnostakis. Double-stranded RNA in *Endothia parasitica*. Phytopathology 67:1393–1396, 1977.
14. SL Anagnostakis, PR Day. Hypovirulence conversion in *Endothia parasitica*. Phytopathology 69:1226–1229, 1979.
15. DW Fulbright. Effect of eliminating dsRNA in hypovirulent *Endothia parasitica*. Phytopathology 74:722–724, 1984.
16. BI Hillman, DW Fulbright, DL Nuss, NK Van Alfen. Hypoviridae. In: FA Murphy et al., eds. Virus Taxonomy: Classification and Nomenclature of Viruses. Sixth Report of the International Committee on Taxonomy of Viruses. New York: Springer-Verlag, 1995, pp 261–264.
17. SA Ghabrial. New developments in fungal virology. Adv Virus Res 33:303–388, 1994.
18. DW Fulbright. Chesnut blight and hypovirulence. In: G Stacey, Keen NT, eds. Plant–Microbe Interactions, Vol 4, St. Paul, MN: APS Press, 1999, pp 57–79.
19. JS Batten, KB Scholthof, ME Miller, RD Martyn. cDNA probes for detection of specific dsRNAs from the fungal pathogen, *Monosporascus cannonballus*. J Virol Methods 84:209–215, 2000.
20. DK Lakshman, J Jian, SM Tavantzis. A double-stranded RNA element from a hypovirulent strain of *Rhizoctonia solani* occurs in DNA form and is genetically related

to the pentafunctional AROM protein of the shikimate pathway. Proc Natl Acad Sci USA 95:6425–6429, 1998.

21. GQ Li, DH Jiang, DB Wang, XH Yi, B Zhu, SR Rimmer. Double-standard RNAs associated with the hypovirulence of *Sclerotinia sclerotiorum* strain Ep-1PN. Progr Nat Sci 9:836–841, 1999.
22. S Huang, SA Ghabrial. Organization and expression of the double-standard RNA genome of *Helminthosporium victoriae* 190S virus, a totivirus infecting a plant pathogenic filamentous fungus. Proc Natl Acad Sci USA 93:12541–12546, 1996.
23. TR Van der Lende, EH Duitman, MG Gunnewijk, L Yu, JG Wessels. Functional analysis of dsRNAs (L1, L3, L5, and M2) associated with isometric 34-nm virions of *Agaricus bisporus* (white button mushroom). Virology 217:88–96, 1996.
24. JS Park, YH Kim. Infectious RNA viruses in the edible mushroom *Pleurotus* spp. J Microbiol 34:61–67, 1996.
25. JA Dodds. Association of type 1 viral-like dsRNA with club-shaped particles in hypovirulent strains of *Endothia parasitica*. J Virol 107:1–12, 1980.
26. JR Newhouse, HC Hoch, WL MacDonald. The ultrastructure of *Endothia parasitica*. Comparison of a virulent with a hypovirulent isolate. Can J Bot 61:389–399, 1983.
27. DR Hansen, NK Van Alfen, K Gillies, NA Powell. Naked dsRNA associated with hypovirulence of *Endothia parasitica* is packaged in fungal vesicles. J Gen Virol 66:2605–2614, 1985.
28. T Fahima, P Kazmierczak, DR Hansen, P Pfeiffer, NK Van Alfen. Membrane-associated replication of an unencapsidated double-strand RNA of the fungus, *Cryphonectria parasitica*. Virology 195:81–89, 1993.
29. BI Hillman, Y Tian, PJ Bedker, MP Brown. A North American hypovirulent isolate of the chestnut blight fungus with European isolate-related double-stranded RNA. J Gen Virol 66:2605–2614, 1992.
30. J Tartaglia, CP Paul, DW Fulbright, DL Nuss. Structural properties of double stranded RNAs associated with biological control of chestnut blight fungus. Proc Natl Acad Sci USA 83:9109–9113, 1986.
31. R Shapira, GH Choi, BI Hillman, DL Nuss. The contribution of defective RNAs to the complexity of viral-encoded double-stranded RNA populations present in hypovirulent strains of the chestnut blight fungus *Cryphonectria parasitica*. EMBO J 10: 741–746, 1991.
32. JA Elliston. Characterization of dsRNA-free and dsRNA-containing strains of *Endothia parasitica* in relation to hypovirulence. Phytopathology 75:151–158, 1985.
33. CD Smart, DW Fulbright. Characterization of a strain of *Cryphonectria parasitica* doubly infected with hypovirulence-associated dsRNA viruses. Phytopathology 85: 491–494, 1995.
34. BI Hillman, R Foglia, W Yuan. Satellite and defective RNAs of *Cryphonectria* hypovirus 3–Grand Haven 2, a virus species in the family Hypoviridae with a single open reading frame Virology 276:181–189, 2000.
35. Y Hong, SL Dover, TE Cole, CM Brasier, KW Buck. Multiple mitochondrial viruses in an isolate of the Dutch elm disease fungus *Ophiostoma novo-ulmi*. Virology 258: 118–127, 1999.
36. BI Hillman, BT Halpern, MT Brown. A viral dsRNA element of the chestnut blight fungus with a distinct genetic organization. Virology 201:241–250, 1994.

37. CD Smart, W Yuan, R Foglia, DL Nuss, DW Fulbright, BI Hillman. Cryphonectria hypovirus 3, a virus species in the family hypoviridae with a single open reading frame Virology 265:66–73, 1999.
38. S Hiremath, B L'Hostis, SA Ghabrial, RE Rhoads. Terminal structure of hypovirulence-associated dsRNAs in the chestnut blight fungus *Endothia parasitica*. Nucleic Acids Res 14:9877–9896, 1986.
39. GH Choi, DM Pawlyk, DL Nuss. The autocatalytic protease p29 encoded by a hypovirulence-associated virus of the chestnut blight fungus resembles the potyvirus-encoded protease HC-Pro Virology 183:747–752, 1991.
40. MG Craven, DM Pawlyk, GH Choi, DL Nuss. Papain-like protease p29 as a symptom determinant encoded by a hypovirulence-associated virus of the chestnut blight fungus. J Virol 67:6513–6521, 1993.
41. N Suzuki, B Chen, DL Nuss. Mapping of a hypovirus p29 protease symptom determinant domain with sequence similarity to potyvirus HC-Pro protease. J Virol 73: 9478–84, 1999.
42. GH Choi, DL Nuss. Hypovirulence of chestnut blight fungus conferred by an infectious viral cDNA. Science 257:800–803, 1992.
43. B Chen, GH Choi, DL Nuss. Mitotic stability and nuclear inheritance of integrated viral cDNA in engineered hypovirulent strains of the chestnut-blight. EMBO J 12: 2992–2998, 1993.
44. B Chen, DL Nuss. Infectious cDNA clone of hypovirus CHV1-Euro7: a comparative virology approach to investigate virus-mediated hypovirulence of the chestnut blight fungus *Cryphonectria parasitica*. J Virol 73:985–992, 1999.
45. B Chen, LM Geletka, DL Nuss. Using chimeric hypoviruses to fine-tune the interaction between a pathogenic fungus and its plant host. J Virol 74:7562–7567, 2000.
46. WA Powell, NK Van Alfen. Differential accumulation of poly(A)+RNA between virulent and double-stranded RNA-induced hypovirulent strains of *Cryphonectria* (*Endothia*) *parasitica* Mol Cell Biol 7:3688–3693, 1987.
47. WA Powell Jr, NK Van Alfen. Two nonhomologus viruses of *Cryphonectria* (*Endothia*) *parasitica* reduce accumulation of specific virulence-associated polypeptides. J Bacteriol 169:5324–5326, 1987.
48. P Kazmierczak, P Pfeiffer, L Zhang, NK Van Alfen. Transcriptional repression of specific host genes by the mycovirus *Cryphonectria* hypovirus 1. J Virol 70:1137–1142, 1996.
49. S Gao, GH Choi, L Shain, DL Nuss. Cloning and targeted disruption of *enpg-1*, encoding the major in vitro extracellular endopolygalacturonase of the chestnut blight fungus, *Cryphonectria parasitica*. Appl Environ Microbiol 62:1984–1990, 1996.
50. D Rigling, U Heiniger, UR Hohl. Reduction of laccase activity in dsRNA-containing hypovirulent strains of *Cryphonectria* (*Endothia*) *parasitica*. Phytopathology 79: 219–223, 1989.
51. D Rigling, NK Van Alfen. Upregulation of laccase biosynthesis in the plant-pathogenic fungus *Cryphonectria parasitica* by double-stranded RNA. J Bacteriol 173: 8000–8003, 1991.
52. DA Varley, GK Podila, ST Hiremath. Cutinase in *Cryphonectria parasitica*, the

chestnut blight fungus: suppression of cutinase gene expression in isogenic hypovirulent strains containing double-stranded RNAs. Mol Cell Biol 12:4539–4544, 1992.
53. EA Havir, SL Anagnostakis. Oxalic acid production by virulent but not by hypovirulent strains of *Endothia parasitica*. Physiol Plant Pathol 23:369–376, 1983.
54. A Vannini, CD Smart, DW Fulbright. The comparison of oxalic acid production in vivo and in vitro by virulent and hypovirulent *Cryphonectria* (*Endothia*) *parasitica*. Physiol Mol Plant Pathol 43:443–451, 1993.
55. CE Carpenter, RJ Mueller, P Kazmierczak, L Zhang, DK Villalon, NK Van Alfen. Effect of a virus on accumulation of a tissue-specific cell-surface protein of the fungus *Cryphonectria* (*Endothia*) *parasitica*. Mol Plant Microbe Interact 5:55–61, 1992.
56. L Zhang, D Villalon, Y Sun, P Kazmierczak, NK van Alfen. Virus-associated down regulation of the gene encoding cryparin, an abundant cell-surface protein from the chestnut blight fungus, *Cryphonectria parasitica*. Gene 139:59–64, 1994.
57. L Zhang, RA Baasiri, NK Van Alfen. Viral repression of fungal pheromone precursor gene expression. Mol Cell Biol 18:953–959, 1998.
58. L Zhang, AC Churchill, P Kazmierczak, DH Kim, NK Van Alfen. Hypovirulence-associated traits induced by a mycovirus of *Cryphonectria parasitica* are mimicked by targeted inactivation of a host gene. Mol Cell Biol 13:7782–7792, 1993.
59. DL Nuss. Using hypoviruses to probe and perturb signal transduction processes underlying fungal pathogenesis. Plant Cell 8:1845–1853, 1996.
60. TG Larson, GH Choi, DL Nuss. Regulatory pathways governing modulation of fungal gene expression by a virulence-attenuating mycovirus. EMBO J 11:4539–4548, 1992.
61. TG Larson, DL Nuss. Altered transcriptional response to nutrient availability in hypovirus-infected chestnut blight fungus. EMBO J 13:5616–5623, 1994.
62. AG Gilman. G proteins: transducers of receptor-generated signals. Annu Rev Biochem 56:615–649, 1987.
63. GH Choi, B Chen, DL Nuss. Virus-mediated or transgenic suppression of a G-protein alpha subunit and attenuation of fungal virulence. Proc Natl Acad Sci USA 92: 305–309, 1995.
64. S Gao, DL Nuss. Distinct roles for two G protein alpha subunits in fungal virulence, morphology, and reproduction revealed by targeted gene disruption. Proc Natl Acad Sci USA 93:14122–14127, 1996.
65. S Kasahara, DL Nuss. Targeted disruption of a fungal G-protein beta subunit gene results in increased vegetative growth but reduced virulence. Mol Plant Microbe Interact 10:984–993, 1997.
66. S Kasahara, P Wang, DL Nuss. Identification of bdm-l, a gene involved in G protein beta-subunit function and alpha-subunit accumulation. Proc Natl Acad Sci USA 97: 412–417, 2000.
67. SL Anagnostakis, RA Jaynes. Chestnut blight control: use of hypovirulent cultures. Plant Dis Rep 57:225–226, 1973.
68. NK Van Alfen, RA Jaynes, SL Anagnostakis, PR Day. Chestnut blight: biological control by transmissible hypovirulnece in *Endothia parasitica*. Science 189:890–891, 1975.
69. SL Anagnostakis, B Chen, LM Geletka, DL Nuss. Hypovirus transmission to asco-

spore progeny by field-released transgenic hypovirulent strains of *Cryphonectria parasitica*. Phytopathology 88:598–604, 1998.

70. SL Anagnostakis. Biological control of chestnut blight. Science 215:466–471, 1982.
71. GJ Griffin. Chestnut blight and its control. Hort Rev 8, 291–335, 1986.
72. WL MacDonald, DW Fulbright. Biological control of chestnut blight: use and limitation of transmissible hypovirulence. Plant Dis 75:656–661, 1991.
73. SW Garrod, DW Fulbright, AV Ravenscroft. Dissemination of virulent and hypovirulent forms of a marked strain of *Endothia parasitica* in Michigan. Phytopathology 75:533–538, 1985.
74. D Baidyaroy, DH Huber, DW Fulbright, H Bertrand. Transmissible mitochondrial hypovirulence in a natural population of *Cryphonectria parasitica*. Mol Plant–Microbe Interact 13:88–95, 2000.
75. D Baidyaroy, JM Glynn, H Bertrand. Dynamics of asexual transmission of a mitochondrial plasmid in *Cryphonectria parasitica*. Curr Genet 37:257–267, 2000.
76. SL Anagnostakis, B Hau, J Kranz. Diversity of vegetative compatibility groups of *Cryphonectria parasitica* in Connecticut and Europe. Plant Dis 70:536–538, 1986.
77. P Cortesi, MG Milgroom. Genetics of vegetative incompatibility in *Cryphonectria parasitica*. Appl Environ Microbiol 64:2988–2994, 1998.
78. MG Milgroom, P Cortesi. Analysis of population structure of the chestnut blight fungus based on vegetative incompatibility genotypes. Proc Natl Acad Sci USA 96: 10518–10523, 1999.
79. JJ Polashock, PJ Bedker, BI Hillman. Movement of a small mitochondrial double-stranded RNA element of *Cryphonectria parasitica*: ascospore inheritance and implications for mitochondrial recombination. Mol Gen Genet 256:566–571, 1997.
80. CB Monteiro-Vitorello, JA Bell, DW Fulbright, H Bertrand. A cytoplasmically-transmissible hypovirulence phenotype associated with mitochondrial DNA mutations in the chestnut blight fungus *Cryphonectria parasitica*. Proc Natl Acad Sci USA 92:5935–5939, 1995.
81. HJ Rogers, KW Buck, CM Brasier. Transmission of double-stranded RNA and a disease factor in *Ophiostoma ulmi*. Plant Pathol 35:277–287, 1986.
82. Y Hong, TE Cole, CM Brasier, KW Buck. Evolutionary relationships among putative RNA-dependent RNA polymerases encoded by a mitochondrial virus-like RNA in the Dutch elm disease fungus, *Ophiostoma novo-ulmi*, by other viruses and virus-like RNAs and by the *Arabidopsis* mitochondrial genome. Virology 246:158–169, 1998.
83. HJ Rogers, KW Buck, CM Brasier. A mitochondrial target for double-stranded RNA in diseased isolates of the fungus that causes Dutch elm disease. Nature 329:558–560, 1987.
84. NW Charter, KW Buck, CM Brasier. De-novo generation of mitochondrial DNA plasmids following cytoplasmic transmission of a degenerative disease in *Ophistoma novo-ulmi*. Curr Genet 24:505–514, 1993.
85. SN Abu-Amero, NW Charter, KW Buck, CM Brasier. Nucleotide-sequence analysis indicates that a DNA plasmid in a diseased isolate of *Ophiostoma novo-ulmi* is derived by recombination between two long repeat sequences in the mitochondrial large subunit ribosomal RNA gene. Curr Genet 28:54–59, 1995.

86. EG Diacumakos, L Gamjobst, EL Tatum. A cytoplasmic character in *Neurospora crassa*. The role of nuclei and mitochondria. J Cell Biol 26:427–443, 1965.
87. DW Fulbright. A cytoplasmic hypovirulent strain of *Endothia parasitica* without double-stranded RNA. Phytopathology 75:1328, 1985.
88. N Mahanti, H Bertrand, C Monteiro-Vitorello, DW Fulbright. Elevated mitochondrial alternative oxidase activity in dsRNA-free, hypovirulent isolates of *Cryphonectria parasitica*. Physiol Mol Plant Pathol 42:455–463, 1993.
89. H Bertrand, NA Szakacs, FE Nargang, CA Zagozeski, RA Collins, JC Harrigan. The function of mitochondrial genes in *Neurospora crassa*. Can J Genet Cytol 18: 397–409, 1976.
90. AM Lambowitz, CW Slayman, CL Slayman, WD Bonner Jr. The electron transport components of wild type and *poky* strains of *Neurospora crassa*. J Biol Chem 247: 1536–1545, 1972.
91. EG Lemire, FE Nargang. A missense mutation in the *oxi-3* gene of the [*mi-3*] extranuclear mutant of *Neurospora crassa*. J Biol Chem 261:5610–5615, 1986.
92. EG Lemire, JA Percy, JM Correia, BM Crowther, FE Nargang. Alteration of the cytochrome c oxidase subunit 2 gene in the [*exn-5*] mutant of *Neurospora crassa*. Curr Genet 20:121–127, 1991.
93. MS Melzer, M Dunn, T Zhou, G Boland. Assessment of hypovirulent isolates of *Cryphonectria parasitica* for potential in biological control of chestnut blight. Can J Phytopathol 18:19–28, 1997.
94. Y-C Liu, MG Milgroom. Correlation between hypovirus transmission and the number of vegetative incompatibility (*vic*) genes different among isolates from a natural population of *Cryphonectria parasitica*. Phytopathology 86:79–86, 1998.
95. AJF Griffiths. Natural plasmids of filamentous fungi. Microbiol Rev 59:673–685, 1995.
96. AJF Griffiths. Fungal senescence. Annu Rev Genet 26:351–372, 1992.
97. CB Monteiro-Vitorello, D Baidyaroy, JA Bell, G Hausner, DW Fulbright, H Bertrand. A circular mitochondrial plasmid incites hypovirulence in some strains of *Cryphonectria parasitica*. Curr Genet 37:242–256, 2000.
98. JA Bell, CB Monteiro-Vitorello, G Hausner, DW Fulbright, H Bertrand. Physical and genetic map of the mitochondrial genome of *Cryphonectria parasitica* Ep155. Curr Genet 30:34–43, 1996.
99. E Gobbi, A Carpanelli, G Firrao, R Locci. The *Cryphonectria parasitica* plasmid pUG1 contains a large ORF with motifs characteristic of family B DNA polymerases. Nucleic Acids Res 25:3275–3280, 1997.

18

Pathogenic Development of *Candida albicans*

Daniel Herman
Grand Valley State University, Allendale, Michigan

Richard Calderone
Georgetown University Medical Center, Washington, D.C.

1 INTRODUCTION

Candida albicans is the most common fungal pathogen of humans [1]. In its usual setting, candidiasis occurs as infections of the immunocompromised host. Thus, cutaneous, mucosal, or invasive disease is usually initiated by some predisposing condition, such as cancer chemotherapy, immunosuppression prior to allogeneic transplants, prolonged use of antibacterial antibiotics, indwelling catheters, and T-cell deficiencies. Often predisposing factors occur simultaneously, such as the presence of an indwelling catheter in a patient undergoing prolonged treatment with antibacterial drugs, the result of which is usually an invasive infection. An association with the type of candidiasis and a specific underlying host immune defect has been observed [2]. For example, neutropenia often results in severe invasive disease resulting in high mortality (Fig. 1). A T-cell deficiency (CD_4 T-cell depletion) as occurs in the AIDS patient usually results in mucosal disease (oral, esophagitis, vaginitis), but the organism rarely invades the blood stream and tissues (Fig. 1). While these correlates are often observed, immunity appears to be more complex, especially as it relates to humoral immunity.

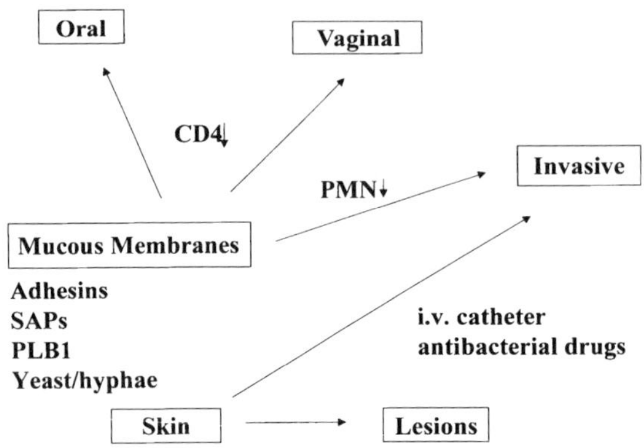

FIGURE 1 The pathogenesis of candidiasis caused by *Candida albicans*. The organism resides as a commensal on the oral, vaginal and gut mucosa. When immunity is immunocompromised, as in AIDS or in patients depleted of PMN cells (neutrophils), the organisms utilize virulence factors (adhesins, SAPs, PLB1 and morphogenesis (yeast/hyphal transition) to establish disease. Other important predisposing conditions exist, including I.V. catheterization and chronic use of antibacterial drugs, which in common trigger bloodborne infection.

Once considered to be nonprotective, more recent data indicate that some antibody(ies) can be associated with protection [3]. Further, in animal models of candidiasis, protection at the systemic level may not protect locally; i.e., vaginal candidiasis can occur even though the host has systemic immunity [4]. In summary, candidiasis exists in many forms and, for protection to occur, a number of immune and nonimmune factors have been shown to be required. Often the protective functions are site specific.

The intent of this chapter is to focus not on the host and the various types of defense mechanisms that have been defined but on the virulence attributes of the organism. At first glance, it would appear that *C. albicans* relies solely on an abrogation of host defenses to cause disease. However, the fact that the organism maintains itself as a commensal in the healthy host indicates that it is able to evade elimination. Further, while disease usually occurs only in the compromised host, animal experiments have shown that specific gene deletions result in avirulence of the organism. It is likely that at least some of these same factors also promote human disease (see discussion of secretory aspartyl proteinases). Thus, *C. albicans* might be considered a highly evolved pathogen that utilizes its own resources to persist in the normal host and invade when host immune deficiencies occur. The virulence factors to be discussed include host recognition through expression of cell surface molecules (adhesins), morphogenesis or a reversible conversion of growth forms, the production of extracellular enzymes, and the switch phenotype (Fig. 1).

The discussion that follows will be presented as a series of events that may occur during invasion by the organism. Thus, colonization (adherence) of mucosal surfaces allows the organism to persist; this stage in pathogenesis is portrayed as an important initial event that is then followed by conversion of unicellular forms (yeast) to a filamentous (hyphae) growth. Intuitively, it would appear that a linear form of growth (hyphae) might be more conducive to invasion of the host, a concept supported by data that indicate agerminative strains of *C. albicans* are avirulent or attenuated in animal models of candidiasis. Further, it would seem that invasion should be accompanied by enzyme production that results in tissue degradation. Thus, secreted aspartyl proteases (SAPs) and phospholipases (PL) have been suggested as factors that augment the invasiveness of the organism. Last but not least, phenotypic switching is viewed as a mechanism whereby the organism can change its phenotype, for example, antigenic changes, resistance to drugs, and changes in tissue affinities, each of which (or collectively) may promote evasion of host defenses. In reality, since the organism is capable of invading a number of diverse body sites, each with its own set of protective factors, it is very likely that the expression of virulence factors in part is site specific. In short, virulence in this organism is multifactoral and invasion will depend on both the attributes of the fungus and the loss/depletion of host defenses.

2 HOST RECOGNITION BY *C. ALBICANS*

We summarize the literature on the adherence of this organism in two sections. First, the earlier, descriptive studies will be examined which emphasize the biochemical characterization of cell surface proteins that confer an adherence phenotype. Second, molecular studies that focus on the identification of genes whose encoded proteins provide adherence properties will be discussed. We use the term adhesin to refer to a cell surface glycoprotein, glycan, or protein produced by *C. albicans* that recognizes host cell ligands.

The view that host recognition by a pathogen might be important to pathogenesis has been suggested for *C. albicans* also [5–8]. During the past two decades, a number of investigators have demonstrated that yeast cells of *C. albicans* as well as other *Candida* spp. adhere to a variety of host cells, including both endothelial and epithelial cells as well as purified ligands of host cells (Table 1). In the latter instance, extracellular matrix proteins such as laminin, fibronectin, and collagen represent some of the ligands that have been suggested as substrates for binding by the organism. King et al. were the first to examine the adherence of *Candida* spp. to human buccal epithelial cells (HBEC) and vaginal cells (HVEC) [9]. They observed that the adherence of *Candida* spp. was hierarchal

Table 1 Confirmed and Putative Adhesins of *Candida albicans*

Adhesin	Feature[a]	Ligand or host cell	Required for virulence[b]	Reference
ALS1	3-domains	endo; epithelial	ND	50
ALA1 *[ALS5]*	3-domains	ECM	ND	49
HWP1	covalent bonding	TG, HBEC	Yes	48
INT1	morphogenesis/adherence	RGD?	Yes	39
FM16	like *Pseduomonas* PAK	Asialosphinoglipids	ND	23–26
EP	lectin	fucosyl residues	ND	17
MTH	immune clearance	splenic macrophages	ND	61
MP60	mannoprotein	iC3b, C3d	Yes	11

[a] *ALS1, ALA1 [ALS5]* = 3-domain structure of putative amino-terminus binding domain, carboxy-terminus with GPI anchor, and tandem repeats located between these two domains; *HWP1* = hyphal cell wall protein 1; *INT1* = required for both morphogenesis and adherence; FM16 = fimbrial protein, similar binding domain to the *Pseudomonas* PAK protein; EP = extracellular polymer with lectin-binding activity; MTH = β-linked mannotetrahexose, promotes binding to splenic macrophages, possibly involved in immune clearance; MP60 = located on cell surface of hyphae but not yeast cells.
[b] Murine hematogenously disseminated model.

in that *C. albicans* adhered to a greater extent than other *Candida* spp. This observation supported to some extent the idea that *C. albicans* was the most virulent of all *Candida* spp. While the studies described above were done in vitro, the first observation that adherence might be associated with the virulence of *C. albicans* was that spontaneous, cerulenin-resistant strains of *C. albicans* not only were less adherent in vitro but were also less virulent in several animal models of candidiasis, including rabbit endocarditis and rat vaginitis [10,11]. Further, these strains were examined for changes in their cell surface protein profile since they were found to be less adherent. The presence of a 60-kDa mannoprotein was observed in wild-type cells but not in the mutant strains [12,13]. In reality, the spontaneous mutants that are less adherent and avirulent are not isogenic with parental cells, meaning that most probably, other mutations exist so that a direct association of adherence with virulence was only speculative. The development of molecular techniques that generate single gene deletions has more recently lead to correlations of adherence and virulence (see below).

2.1 The Adhesins

2.1.1 The Lectin Adhesins

In their studies on adherence, the Douglas laboratory has made the following observations [14–20]: (1) Adherence of the organism to HBEC could be dramatically increased by growing cells in media that contained large amounts of sugar such as galactose or sucrose. (2) Strains of *C. albicans* fell into two categories in regard to adherence properties, based on either the type of lectin or sugar that inhibited adherence. Thus, adherence of most strains was inhibited by a lectin that recognized fucose; the adherence of one other strain was inhibited by a lectin that binds N-acetyl glucosamine (or glucosamine). Importantly, whether inhibition could be observed depended on the order in which the lectin or sugar was incubated with the HBEC or yeast cells. For inhibition to occur, the lectin had to be incubated with the HBEC, whereas each of the sugars mentioned above was inhibitory only if preincubated with yeast cells. From these data a hypothesis was developed that yeast cells possessed lectinlike adhesins that recognized either fucose or N-acetyl glucosamine residues (ligands) of HBEC; in fact, a similar result was observed with HVEC. It should also be pointed out that the lectin that recognizes fucosyl residues has not been fully characterized, and its role in disease progression has not been substantiated.

Brassart et al. [21] defined the structural requirements for ligands of HBEC recognized by *C. albicans*. Oligosaccharides from human milk that inhibited adherence to HBEC shared a terminal fucosyl residue in an α-1,2 linkage with galactose. If the oligosaccharide lacked the terminal fucosyl residue or if the residue was internal, the oligosaccharide did not have inhibitory activity.

When *C. albicans* is grown in high-galactose medium, the surface of yeast cells acquires a fibrillar appearance, and an extracellular polymer (EP) is produced [19]. EP can block the adherence of the organism to HBEC; consequently, this material served as a starting point for the isolation of the adhesin. The adhesin was found to be a mannoprotein since it was bound by Concanavalin A affinity chromatography. However, digestion of the active material with enzymes or alkali that degrade polysaccharide did not result in a loss of adhesin activity. Consequently, the binding site for fucosyl oligosaccharides was thought to be associated with a protein domain, which again is expected if the adhesin is in fact a lectin.

2.1.2 The *Pseudomonas*-like Adhesin

This adhesin has been described extensively [22–26]. The outer surface of yeast cells is composed of thin, filamentous fibrils which can be easily isolated by gentle homogenization. These structures resemble the fimbriae of bacteria; however, unlike fimbriae that are exclusively proteinaceous, the candidal fimbriae are composed of a glycoprotein consisting of ~85% carbohydrate and ~15% protein. The structural subunit of this glycoprotein has a molecular mass of 66 kDa and is referred to as Fm16 (FM = fimbrial) [26]. It appears that the FM 16 glycoprotein recognizes the asialo-GM1 glycosphingolipids of human buccal epithelial cells [24,26]. Interestingly, because the bacterium *Pseudomonas aeruginosa* also binds to the same ligand, a comparison of Fm16 with the pseudomonal pilus protein (PAK) was undertaken. The PAK protein domain that recognizes this ligand is found in the carboxy-terminal disulfide-bonded region of the protein (amino acid residues 128–144). Crossreactivity of antibodies to the respective proteins (or whole cells) of each organism was observed by a number of assays, including whole-cell agglutination and Western blot [22]. Two of the three anti-PAK antibodies reacted with the *C. albicans* Fm16. Both of these antibodies were directed against epitopes that included the ligand-binding domain of the pseudomonal PAK. The third antibody, which did not react with *C. albicans*, was directed against an epitope that was not part of the ligand-binding domain of PAK. Finally, it was determined that an anti-PAK antibody (one of the two mentioned above) could block binding of Fm16 as well as whole cells of *C. albicans* to asialo GM1 and HBEC, respectively [22].

More recently, the delineation of the *C. albicans* Fm16 was investigated. Synthetic peptides corresponding to residues 128–144 and 134–140 of the PAK protein as well as antibodies to the PAK protein were made and used to determine their influence on the binding of *C. albicans* to HBEC [23]. PAK(128–144) inhibited the binding of whole cells of both organisms to immobilized HBEC by 50–55% over control (cells treated with unrelated PAK peptides). PAK(134–140) was about as inhibitory as the larger peptide (PAK128–144). Further, antipeptide antibodies (to PAK128–144 and PAK134–140) also blocked adherence of yeast cells (and *P. aeruginosa*) to HBEC [22]. Thus, it would appear that both the Fm16 and PAK proteins of these diverse pathogens share a similar ligand-binding domain but are otherwise dissimilar.

2.1.3 The Complement Receptors and Extracellular Matrix Adhesins

Hyphal forms of *C. albicans* have been shown to recognize the complement C3 ligands, iC3b and C3d, as well as several extracellular matrix (ECM) proteins [27–37]. Because the integrin family of mammalian proteins binds some of these same ligands, the assumption was put forth that *C. albicans* also expressed an integrinlike protein. This hypothesis has been possibility strengthened by the isolation of *INT1*, a gene thought to encode an integrinlike protein (see below) [38,39]. Initial characterization studies focused on proteins that recognized the complement ligands [27–30]. Binding was lower when yeast forms were used, indicating that either expression of the proteins was reduced in yeast forms or the protein(s) were expressed equally but distributed differently within the cell wall of yeast cells. In fact, other studies indicated that the latter condition existed for the iC3b-C3d-binding protein [40,41]. Reactivity was found in yeast cells, but the location of the protein was below the cell surface [12,41]. Two proteins of 60 and 68 kDa were identified on the basis of their isolation from a C3d/iC3b ligand affinity column [31,40]. These proteins were subsequently referred to as complement-binding proteins or complement receptors (CR), according to the CR2/CR3 like activity of mammalian proteins that recognize C3d or iC3b, respectively. The association of these proteins with an adherence function was established through the identification of nonadhering, avirulent strains of *C. albicans*, which lacked the 60- and 68-kDa proteins [12]. Monoclonal antibodies directed against human integrins have revealed multiple reactive proteins of *C. albicans* homologues in Western blot analyses, but the verification of these proteins as adhesins or virulence factors by gene deletion has not been demonstrated [8].

As stated above, the binding of *C. albicans* to ECM proteins such as fibronectin, fibrinogen, and laminin has been studied extensively, and as with the complement ligands, binding to laminin by hyphal forms of the organism was greater than binding to yeast cells [33]. Preincubation of cells with laminin or with peptides that contain the RGD sequence inhibited the binding of the organism to immobilized fibronectin, indicating that another ECM protein can act as a substrate and that one of the key binding domains of fibronectin involves the RGD sequence [42,43]. The importance of the ECM binding adhesins in pathogenesis was established by in vivo studies which showed that an RGD-containing ligand protected animals when administered prior to infection [37]. Nutritional effects on the expression of fibronectin binding to *C. albicans* have been studied [44–46]. Expression of fibronectin binding proteins required a complex medium such as Sabouraud broth, while a minimal medium such as yeast nitrogen base (YNB) did not support expression [46]. The high activity of Sabouraud broth could be related to the presence of high-molecular-mass components which were not identified. However, the addition of hemoglobin (0.1%) was highly stimulatory for fibronectin binding [46].

The identification of proteins from *C. albicans* that recognize fibronectin has been accomplished using a ligand protection assay [44,45]. Proteins of 55 and 30 kDa were identified that recognized fibronectin (or recombinant fragments) when grown in the presence of hemoglobin. This same assay was used to identify proteins that bind to other ECM proteins such as laminin, fibrinogen, and type IV collagen. A single receptor for these ligands was proposed since binding of each ligand was inhibited by the respective unlabeled ligand with similar IC_{50} values. The 55-kDa protein was purified by affinity chromatography, but sequencing was prohibited since the protein was N-terminally blocked.

In summary, the studies described above indicate that *C. albicans* has several receptors for ECM proteins. Also, several of the ECM proteins may be recognized by the same protein. Thus far, the only gene that encodes a protein with ECM binding activity is the *ALA1* (*ALS5*) (see below).

2.1.4 Adhesin-Encoding Genes

Several genes have been shown to encode proteins with an adherence function including *HWP1*, *ALS1*, and *ALA1* (*ALS5*) as well as the *INT1* mentioned above [38,39,47–50]. All of the encoded proteins have several common features, such as a cell wall localization, an association with the plasma membrane, moderate to extensive glycosylation, and a probable amino-terminal ligand binding domain. Als1p and Als5p are believed to be crosslinked to the cell membrane through a GPI anchor and then relocated to the glucan cell wall polysaccharide. Interestingly, there are as many as seven Als proteins in *C. albicans* reported to date, although functions have not been assigned to all [51–55].

Of significance, strains of *C. albicans* deleted in *HWP1* and *INT1* have been constructed to verify their function in adherence (see below). The standard method for obtaining gene-deleted strains of *C. albicans* is the ''urablaster'' technique [56]. In this procedure, sequential disruption of both alleles (*C. albicans* is diploid) is achieved and, subsequently, a set of parental, single gene–deleted, and double gene–deleted strains are compared phenotypically. This approach is referred to as molecular Koch's Postulates and serves to identify gene function. Observations on each of the genes encoding the proteins mentioned above are provided in the following sections.

HWP1. A developmentally regulated gene, *HWP1* is expressed in germ tubes and true hyphae [47,48,57]. The gene encodes a cell wall mannoprotein (Hwp1), with a cell surface–exposed amino-terminal domain and a carboxy-terminus which is covalently attached to cell wall glucan. Hwp1p is referred to as a nonconventional adhesin since it is thought to form covalent linkages with host cells, as, for example, HBEC [48]. The role of Hwp1 in adherence can be evaluated directly by treating adhering organisms with heat and SDS. A proportion of the total cell population remains adherent, indicating a covalent interaction. All other adhesins reported to date bind to host cell ligands through lectin or hydrophobic interactions, and such interactions are sensitive to heat and SDS. Sequence analysis of Hwp1 has indicated a similarity to substrates of transglutminase (Tgase), an enzyme of stratified squamous epithelium such as HBEC [48]. Tgase has been shown to mediate the formation of an innate host barrier by crosslinking substrate proteins resulting in the formation of an insoluble, cornified envelope through covalent N^{ε}-(γ-glutamyl)lysine isodipeptide bonds. It is postulated that Hwp1 is a substrate for Tgase, and crosslinking of the organism through covalent attachment mediates adherence to HBEC. Proof for the substrate activity of Hwp1 has been demonstrated by the incorporation of (^{14}C) putrescine by a recombinant Hwp1 protein (rHwp1ΔC37) in the presence of purified Tgase2 [48]. No incorporation was observed when Tgase2 inhibitors were added to the assays or when the recombinant protein was not included in the assays. The substrate activity of native Hwp1 was determined by constructing strains of *C. albicans* lacking *HWP1*. The *hwp1/hwp1* mutant was germinated and incubated with Tgase2 and a crosslinking substrate. Parent cells but not the mutant were able to act as a substrate for Tgase2. In regard to the function of Hwp1 as an adhesin, the adherence of the *hwp1/hwp1* mutant to HBEC was reduced by ~80% in stabilized adherence assays (heat- and SDS-resistant adherence) compared to parental and gene-reconstituted strains [48]. Of importance, the *hwp1/hwp1* mutant was less virulent in a murine model of invasive candidiasis and caused less damage to endothelial cells in vitro [47,57]. These data indicate a prominent role for Hwp1 in the biology and virulence of the organism.

INT1. The *INT1* cloning strategy was based on its similarity to vertebrate leukocyte integrins [38]. Primers were designed according to conserved transmembrane sequences of the vertebrate integrin. A PCR product was amplified from genomic DNA of *C. albicans* and used to isolate a cDNA clone whose sequence exhibited an 18% similarity (overall) to the mammalian integrin. Int1 has several putative N-glycosylation sites and is believed to be a transmembrane protein. When expressed in *S. cerevisiae*, *INT1* induced cells to switch from a yeast growth to a filamentous growth pattern [39]. Int1 was observed on the cell surface of transformed cells of *S. cerevisiae*, and such cells were able to adhere to monolayers of human cervical endothelial cells (HeLa) whereas nontransformed cells were unable to adhere [39]. As expression of *INT1* was regulated by the *GAL10* promoter, transformed cells grown in the presence of galactose but not glucose were able to adhere. A polyclonal antibody to the predicted extracellular domain of Int1 inhibited adherence of *S. cerevisiae* cells. To determine if Int1 is involved in adherence of *C. albicans* as well, *int1/int1* strains were constructed by the urablaster procedure described above [39]. Mutant cells were reduced in their adherence to HeLa cells by 39% compared to the parent, indicating that Int1 must be one of several adhesins that recognize host cells. In addition to the affect of the deletion on adherence, the *int1/int1* mutant did not form hyphae on milk-Tween or Spider agar, media that normally induce such a growth pattern. On the other hand, the mutant strain formed normal hyphae on other inducing media, so the protein is not essential for hyphal growth under all conditions. Reintegration of the *INT1* allele in the mutant strain permitted normal hyphal growth on media which failed to induce filamentation in the mutant. As described above for Hwp1, strains deleted of *INT1* were less virulent in a murine model of invasive candidiasis [39]. These data indicate that Int1 is necessary for adherence, morphogenesis, and virulence in *C. albicans*.

ALS *Family of Proteins*. *ALS1* was isolated from a differential screen utilizing a cDNA library constructed from mRNA isolated during the yeast to hyphal conversion [51]. The library was hybridized with yeast (YEPD) and hyphal (RPMI)-specific probes. A clone later identified as an agglutinin like sequence (ALS) because of its significant identity to the *S. cerevisiae* α-agglutinin (see above) was observed when the cDNA library was hybridized with the hyphal probe only [51]. Since that original description of Als1, six other *ALS* genes have been identified [52–55].

Proteins of the Als family of *C. albicans* have a three-domain structure, consisting of a relatively conserved N-terminus which is probably oriented toward the outer surface of the cell, a central domain that is highly glycosylated and composed of tandem repeats of a highly conserved 36 amino acid sequence, and a glycosylated C-terminus that is GPI anchored. Subsequently, the proteins are thought to be covalently attached to cell wall glucan. The N-terminus contains sequence that is similar to the V-domain of the immunoglobulin superfamily of cell recognition proteins; hence, it is proposed (but not proven) that the N-terminus constitutes the ligand-binding domain of the protein. Other members of the Als family (*ALS2*–ALS4) were cloned from a fosmid library with primers designed from the consensus tandem repeat sequence of *ALS1* [55]. *ALS6* and -7 were detected among data of the *C. albicans* genome-sequencing project, and probes developed from unique regions of each were used to screen a genomic library from which both were isolated [52].

Some other general features of the Als family of proteins are also summarized [51–55].

1. Variation exists in the number of tandem repeats for each of the Als proteins, and the C-terminus of the tandem repeats constitutes the least conserved region across this family of proteins. Thus, Als5 appears to have the fewest tandem repeats (five), and the protein may be different enough to represent a subfamily of Als proteins.
2. The *ALS* genes are localized to chromosomes 6 and 3.
3. The N-terminal domain of 330 amino acids is relatively free of glycosylation in comparison to the tandem repeat and carboxy-terminal domains of the proteins. Both putative N- and O-glycosylation sites exist.
4. Hydrophobic sequences are observed at both the N- and C-terminis of each protein and represent consenus sites for GPI addition (C-terminus) and a secretory signal peptide (N-terminus).
5. The cell wall location of Als1 has been verified by the indirect fluorescence utilizing a polyclonal antibody to four 10-mer peptides.
6. Als1 has been detected in a variety of *C. albicans* and *C. stellatoidea* strains.

Functional analysis of the *ALS* gene family has been reported for *ALS1* and *ALS5*. *ALS1* was also cloned from a *C. albicans* library by gain of function experiments in which *S. cerevisiae* was transformed from a nonadherent to an adherent phenotype [50]. The library mentioned above was designed in a single-copy vector (pYesR) that contains the *S. cerevisiae GAL1* promoter. Transformants were selected by their prototrophy and used in adherence assays with endothelial or epithelial (FaDu) cells. Adherence of *S. cerevisiae* was only observed in cells grown in the presence of galactose that were transformed with *C. albicans* sequences. The role of this gene in pathogenesis has not been established. Similarly, expression cloning was used to isolate *ALA1* (*ALS5*) [49]. Nonadherent *S. cerevisiae* was transformed to a phenotype which adhered to ECM-coated magnetic beads. Expression of *ALA1* (*ALS5*) could be varied by using different galactose-inducible promoters that resulted in the expression of low, moderate, or high levels of the protein. Attachment and aggregation of transformed cells to ECM-coated magnetic beads correlated with the strength of the GAL promoter. Subsequent analysis of the *ALA1* (*ALS5*) sequence indicated its homology with the *ALS* family of proteins. The role of this putative adhesin in virulence was not indicated.

Mannan Adhesins. Mannan of *C. albicans* has been suggested as providing an adhesin function. This observation was made from at least two different lines of evidence. First, a mutant of *C. albicans* that was deleted of a protein mannosyltransferase (*PMT1*) gene was shown to have defective hyphal formation, supersensitivity to several antifungal agents, reduced adherence to Caco-2 epithelial cells and avirulence in a systemic candidiasis animal model [58–60]. The enzyme catalyzes the first O-glycosylation step in the mannan synthesis of *S. cerevisiae*, and, likewise, the deletion mutant in *C. albicans* had reduced O-mannosylation activity. As a second line of proof, an antigen purified from a Zymolyase extract by using a monoclonal antibody (mAb 10G) was found to inhibit the attachment of yeast cells to mouse spleen marginal zone macrophages [61,62]. The active fraction was shown to be insensitive to protease but degraded by peroxidate oxidation. From the mannoprotein-rich extract, a fraction containing the purified antigen (10G) was further fractionated to yield the 10G epitope (a mannotetrahexose) that also blocked attachment to similar tissues [61,62]. This epitope may be critical to the immune recognition and may have potential as a candidate vaccine.

3 MORPHOGENESIS

C. albicans and *C. dubliniensis* are the only *Candida* spp. that undergo a conversion from a yeast morphology to a filamentous, septate growth (true hyphae). This process is initiated by the formation of extensions from the yeast cell (germ tubes) followed by apical growth of the germ tube into filaments or hyphae. The term ''dimorphism'' has often been applied to *C. albicans* because of its ability to grow as a unicellular or filamentous organism. In fact, ''polymorphism'' is a more appropriate term since a third growth form, the pseudohypha, is commonly observed both in tissues and in vitro under a variety of growth conditions. Pseudohyphae resemble an intermediate type of growth consisting of deeply constricted elongated cells that remain attached, forming extended filaments. There has been considerable debate concerning which growth form is more virulent—yeast or filaments. In fact, invasion seems to include asynchronous replication of each growth form. Nevertheless, conversion of yeast cells to a filamentous growth is essential for virulence of the organism, at least in the intravenous, systemic candidiasis model in mice. Because of the latter observation, an immense amount of research has been published on the identification of growth phase-specific proteins and the regulation of morphogenesis via signal transduction pathways. Several hyphal-specific proteins have been described, and regulation of morphogenesis is partially understood. Further, the delineation of the morphogenesis signal transduction pathways has advanced considerably. To some extent, study of the conversion process (yeast to hyphae) has been complicated by the numerous factors that trigger this event, including temperature (>30°C is required), pH (better at alkaline pH), the CO_2/O_2 ratio, or the requirement for specific nutrients, such as proline, N-acetylglucosamine, or serum. Morphogenesis is also observed in nutrient-poor media, some of which include milk-Tween, Spider, SLAD (synthetic low ammonium dextrose) agar or in a tissue culture medium such as M199. Thus, there are many inductive conditions, but apparently fewer signal pathways, indicating, perhaps, that receptor-mediated events feed into common pathways.

As described previously, the urablaster technique is used to construct gene-deleted strains to evaluate function [56]. In regard to morphogenesis, the identification of signal transduction and regulatory genes in the morphogenesis pathways have been advanced by the exploitation of the *S. cerevisiae* pseudohyphal phenotype [63–75]. Thus, *C. albicans* homologs have been isolated (but not exclusively) by complementation, over expression and loss of pseudohyphal formation in *S. cerevisiae*. Generally, if one is determining the affect of a deletion on morphogenesis, several media are used for comparison of the gene-deleted strain with parental cells. Also to be included for evaluation are both agar-and liquid-inducing media, since defects on agar media may not be observed in liquid media (see below). The emerging picture is that any mutation that affects the ability of the organism to undergo conversion attenuates its virulence. In the sections below, emphasis will be placed on the morphogenesis regulatory and signal transduction pathways.

3.1 Phase-Specific, Structural Genes

As stated above, developmentally regulated genes that are expressed during filamentous growth include, *ALS3*, *INT1*, and *HWP1* as well as *ECE1*, which encodes a nonessential protein of the cell wall [76], and *HYR1*, encoding a protein whose expression is correlated with the extent of hyphal elongation [63]. Cell cycle regulation is also critical to morphogenesis, but that subject is beyond the focus of this chapter and is appropriately covered in another review [63].

3.2 Regulation of Morphogenesis

Transcriptional regulation of morphogenesis has received considerable attention recently. Thus far, several of these genes have been described, and it appears that they fall into two classes, those that inhibit transcription of genes required for morphogenesis and those that are required for conversion to filamentous growth. *TUP1* and *RBF1* fall into the first category [77,78]. *TUP1* was isolated from a *C. albicans* library using a PCR product as a probe that was obtained with primers to conserved regions of the *S. cerevisiae TUP1*, which in this organism acts as a transcriptional regulator of many genes [77]. The *C. albicans* ORF was 67% identical to the *S. cerevisiae TUP1* and possessed the conserved features typical of DNA binding proteins. Further, the *C. albicans TUP1* complemented a *S. cerevisiae tup1* mutant. A *tup1 C. albicans* mutant was constructed and its phenotype determined. Interestingly, the mutant constitutively formed filaments, most often appearing as pseudohyphae, on all media tested. The *TUP1/tup1* heterozygote grew in a manner intermediate to the parent and mutant strain. Aside from this change, the mutant had a slightly longer doubling time but had a faster growth rate on glycerol-containing media. A phenotypic comparison of the *S. cerevisiae tup1* mutant with that of the *C. albicans tup1* revealed major differences, the most important of which was that invasive pseudohyphal growth of the *S. cerevisiae tup1* was reduced, opposite to the phenotype observed in the *tup1* mutant of *C. albicans*. Thus, it would appear that while *TUP1* homologs in both species are transcriptional regulators, the genes they regulate are different.

RBF1 (RPG-box-binding factor 1), which encodes a putative transcription factor in *C. albicans*, has been characterized [78]. The structure of the protein includes a DNA-binding domain consisting of a putative basic region helix-loop-helix structure located between two glutamine-rich regions (residues 160 and 300). Loss of *RBF1* through gene deletion resulted in a stimulation of filamentous growth on cornmeal agar plates containing 20% serum as well as on all solid media tested. Filamentation in the mutant on solid agar was augmented by conditions that induce hyphal growth except for an alkaline pH. On the other hand, alkaline pH was the most effective condition that amplified filamentation in liquid media.

Genes that are required for morphogenesis include those of two signal transduction pathways that are named according to their specific transcription factors (Fig. 2). Thus, *CPH1* and *EFG1* encode putative DNA-binding proteins both of which are required for morphogenesis [71,72]. *CPH1* is the homolog of the *S. cerevisiae STE12*, which is required for pseudohyphae (invasive growth) as well as mating [72]. Proteins of the Cph1 *C. albicans* pathway include those of the MAP kinase pathway, such as Cst20, Hst7, and Cek1, homologs of the *S. cerevisiae* Ste20, St7, and Kss1, respectively [63].

A second signal transduction pathway has been proposed, Efg1 pathway [71]. Efg1p, which belongs to a family of basic helix-loop-helix transcription factors that have been known to regulate development in fungi, is the homolog of the *PHD1* of *S. cerevisiae*. Epistasis experiments have been critical in aligning signal proteins and, in fact, showing that the Cph1 and Efg1 pathways are distinct and parallel [69]. In addition, data on the phenotypes of knockout strains support

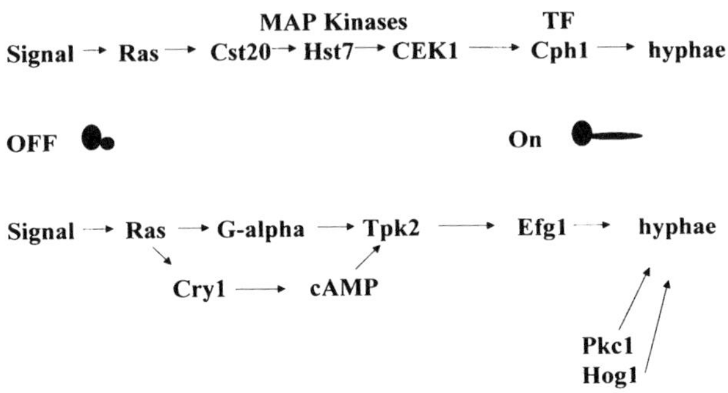

FIGURE 2 Signal transduction pathways that regulate morphogenesis in *Candida albicans*. Four such pathways have been proposed; these are referred to as CPH, EFG, PKC, and HOG1. See text for the description of each pathway.

the notion of distinct pathways. Thus, the individual *cph1* and *efg1* knockouts are unable to form hyphal colonies on specific agar media, but each mutant could germinate in liquid media [72]. On the other hand, the double knockout (*cph1/efg1*) was unable to germinate under any condition tested or kill macrophages in vitro, and was less virulent in a murine model of candidiasis than wild-type or single knockout strains [72]. The Efg1 pathway most likely includes a Ras-cAMP module, although it would appear that Ras may activate both the Efg1 and Cph1 pathways [63]. More recently, protein kinase A (*TPK2*) was shown to be required for hyphal differentiation [73]. Tpk2 is expected to be upstream of Egf1, and this hypothesis has been verified by epistasis experiments in which overexpression of Tpk2 does not rescue the *efg1* phenotype but, interestingly, does rescue the *cek1* phenotype, indicating that while these two pathways are distinct, there may very well be crosstalking in some instances when one of the two pathways is blocked.

In spite of the observations described above on the avirulence of the *cph1/efg1* double mutant, recent data indicate that such a mutant is still able to invade the oral cavity (tongue) of immunosuppressed, gnotobiotic piglets and form filaments [79]. Thus, other transcription factors may very well exist which may be activated by a host signal or by specific growth conditions not yet observed in vitro. In fact, recently, *CZF1*, a putative transcription factor, has been shown to be involved in the regulation of hyphal formation when cells are embedded in a yeast extract-peptone-sucrose (YPS) agar medium [80].

In addition to the Cph1(Ste12) and Efg1 pathways, it is also clear that the cell wall integrity (PKC, protein kinase C) and osmoregulation (HOG1 [high-osmolarity glycerol]) pathways are required for morphogenesis in *C. albicans*

[81–97]. Genes of the PKC pathway have been isolated, and their role in morphogenesis and virulence has been established [81–83].

The *HOG1* pathway of *S. cerevisiae* has been well characterized [86]. Likewise, a putative *C. albicans* homolog of the *S. cerevisiae HOG1* has been identified in *C. albicans*, and *hog1* mutants are unable to invade agar as described above for the *pkc1* mutant, indicating that this gene also is essential for morphogenesis under certain conditions [84]. Additional phenotypes of the *hog1 C. albicans* mutant include, defective budding, resistance to some cell wall inhibitors, and avirulence in a murine model of candidiasis. Proteins upstream of Hog1p have also been identified in *C. albicans*. These proteins belong to the family of two-component histidine kinases (HK) and response regulator (RR) proteins [87–91], including Sln1p (HK), Ypd1p (HK), and Ssk1p (RR). Upregulation of glycerol metabolism by the HOG1 pathway results in the stability of *S. cerevisiae* cells grown under high osmotic conditions. There is some indication that the *C. albicans* homologs, although complementing the *S. cerevisiae sln1* and *ypd1* mutants, may not function similarly in *C. albicans*. The *ssk1* mutant of *C. albicans* is not sensitive to high salt conditions but is unable to germinate in hyphal-inducing agar media. Likewise, *sln1* mutants are only slightly affected by growth in high osmotic media. This pattern of different gene function in homologs of these two organisms is not unusual, and examples have been provided above to suggest this dichotomy (see *TUP1* observations). In addition to the Sln1p, Ypd1p, and Ssk1p proteins described above, there are two other hybrid histidine kinases (referred to as hybrid since each protein has both a histidine kinase and response regulator domain). Both of these genes (*CHK1* and *NIK1*[*COS1*]) have been cloned and their functions determined [88,92–97]. Both are required for morphogenesis, and their deletion also results in avirulence or attenuated virulence in murine candidiasis.

The redundancy in signal transduction pathways for morphogenesis in *C. albicans* is seemingly complex, yet explainable if one supposes that the yeast to hyphal transition is critical to virulence of the organism. As stated earlier, an appropriate generalization is that any mutation that affects morphogenesis results in some degree of attenuation of virulence. For the organism to survive in the host, any loss of a key pathway could result in its elimination. Thus, the organism has evolved mechanisms to coordinate its survival through multiple pathways each of which provides adaptive functions in the host for regulating its invasiveness.

4 INVASIVE ENZYMES OF *C. ALBICANS*

Invasion of tissues by *C. albicans* appears to be accompanied by the production of secreted enzymes that are of two broad types, including phospholipases (PL) and secreted aspartyl proteinases (SAP). As might be expected, there are multiple proteins in each of these groups; in fact, the SAP family is composed of nine proteins. The phospholipases are fewer in number, and in comparison to the SAP family, less is known about the function of this group of proteins in virulence. Below, we describe the PL and SAP proteins and, when possible, discuss their function in virulence.

4.1 Phospholipases

Phospholipases hydrolyze ester linkages of glycerophospholipds. They are of four types—PLA, PLB, PLC, and PLD—and each type has been reported from *C. albicans* [98]. In addition, lysophospholipase-transacylase (LPTA) activity has been identified in PLB1; this enzyme hydrolyzes glycerolphospholipids and transfers a free fatty acid to a lysophospholipid [99]. A variety of assays have been developed to measure phospholipase activity. As expected, there are differences from assay to assay in regard to sensitivity and specificity. The current assay that seems to provide the optimum sensitivity and specificity is a version of an older method that detects PL activity in Sabouraud dextrose agar supplemented with egg yolk [100]. PL-positive strains produce a white halo around colonies. The type of PL can be determined by adding, a specific substrate to the medium [101]. Of the PL, PLB1 has been studied in the greatest detail and its function in virulence determined (see below) [102].

The correlation of PL activity with virulence has been established in two different ways. First, it was observed that isolates differ quantitatively in their production of PL, and that blood isolates produced higher levels of PL than commensal isolates [103,104]. This correlation is statistically higher than for other parameters measured for each isolate, such as adherence, implying that PL is a phenotype of more virulent strains. However, the more direct approach for establishing a cause-and-effect relationship is the construction of specific gene-deleted strains, which will be discussed below.

4.1.1 Plb1

This lipase is a glycoprotein of 84 kDa that appears to have both hydrolase and LPTA activities [102]. The encoding gene was isolated by a PCR-based approach using primers designed from peptide fragments of the purified Plb1 protein. The amino terminus of the protein possesses a stretch of hydrophobic amino acids, which probably represents a signal sequence for secretion, and, in fact, the protein has been detected in culture supernatants. While Plb1 is ~45% homologous to the Plb1 of *S. cerevisiae*, the *C. albicans* Plb1 lacks a GPI anchor, indicating that it may be directly secreted. A *caPLB1* gene-deleted strain has been constructed [102]. Using the standard murine model of hematogenously disseminated candidiasis, the mutant strain produced considerably less Plb1 and was less virulent (40% of the mice died compared to 100% of mice infected with the parent strain). As expected from the survival data, the mutant strain was more rapidly cleared from tissues such as the kidney and liver, while the heterozygote was comparable to the *plb1* null strain in virulence and tissue distribution. The relatively high mortality still observed in mice infected with the *caplb1* mutant suggests that other PL activities may compensate for the loss of Plb1. The *caplb1* mutant was also compared to the parent for in vivo expression of the enzyme in the murine disseminated model mentioned above [102]. Expression of Plb1 was observed by immunoelectron microscopy only from animals infected with parental cells. These data demonstrate that Plb1 is produced during infection of animals and that loss in expression is correlated with reduced virulence.

4.1.2 Other Phospholipases

CaPLB2 was recently cloned from *C. albicans* by a PCR-based approach [105]. *CaPLB2* was similar to *CaPLB1* in a number of aspects, including the lack of a GPI anchor and the presence of a signal peptide. The deduced amino acid sequence of CaPlb2 was 65% identical to that of CaPlb1, and although enzymatic activity was not demonstrated, putative catalytic domains in the protein were found. The role of this enzyme in pathogenesis is unknown. CaPLC activity was detected in cell free extracts by spectrophotometric assays using p-nitrophenylphosphorylcholine (p-NPPC) as a substrate [106]. Activity was higher in hyphal cells than in yeast cells, but whether PLC activity was required for germination was not known. The encoding gene (*CaPLC*) was isolated by PCR-based approaches [106]. The *CaPLC* ORF encodes a putative protein of 1099 amino acids with 16 predicted glycosylation sites and a molecular mass of approximately 124 kDa. Additional amino acid sequence analysis revealed that the protein belongs to the δ-family of phosphoinositide-specific phospholipase C isoenzymes (PI-PLC). *CaPLD* encodes a protein with a putative molecular mass of 196.4 kDa [107]. The protein contains four conserved regions found in a variety of other PLD proteins, such as human, plant, and yeast; these regions are thought to provide a catalytic function [107]. Seven additional domains of unknown function have been identified, five of which are exclusively found in fungi. The other two domains are homologous to mammalian cell PLDs, suggesting that the five fungal-specific domains may provide functions peculiar for fungi. CaPld activity has been recently demonstrated in membrane preparations of the organism but not in a cytosolic extract when fluorescent phosphatidyl choline was used as substrate [108].

4.2 Secreted Aspartyl Proteinases

These enzymes were initially described by Staib [109] and Ruchel and coworkers [110]. The activity was originally referred to as the <u>*C*</u>*andida* <u>a</u>spartyl <u>p</u>roteinase

(CAP) and was described as a protein of 42–45 kDa with a broad substrate specificity that included keratin, dermal collagen, albumin, hemoglobin, immunoglobulin heavy chains, and ECM proteins [110,111]. Most commonly, protease activity is measured in a standard lab medium containing albumin as a substrate. However, the assay can be modified to incorporate other substrates. CAP activity was observed over a broad pH range of 2.0–7.0. The activity of CAP was higher in isolates from patients with vaginitis than from carriers and was detected from vaginal fluids of all women infected with *C. albicans* [112]. CAP activity was also associated with specific phenotypic switching strains; most prominent was the ''opaque'' phenotype that produced CAP while the ''white'' phenotype did not produce CAP in culture [113]. This distinction has been studied in great detail and will be discussed below. In the current nomenclature, CAP corresponds to SAP1.

The SAP family now totals nine proteins, although the possibility exists that others will be added to this list [114]. Cloning of *SAP1* was accomplished using a PCR-based strategy with primers designed from the N-terminus of the purified protein [115]. This PCR product was then used as a probe to screen a genomic library of the organism for hybridizing clones. Similar approaches have been used to clone other members of the SAP family of genes [115]. The ORFs of *SAP1–9* vary from 1173 to 1766 bp, with minimal glycosylation predicted on the basis of the number of potential sites, except for Saps 1, 5, and 8, which do not appear to be glycosylated.

The function of these proteins in disease development has been demonstrated using several experimental approaches, including immunoelectron microscopy, RT-PCR, reporter assays, and the construction of gene deleted strains [116–129]. These observations are summarized in Table 2. There are several conclusions that can be gleaned from these studies:

1. Most of the work has been completed with Saps1–6.
2. All Saps seem to be important for at least one type of candidiasis (or are expressed during disease), while some (Saps1–3) may be essential to the development of disease in several sites.
3. While Saps1–3 seem to be required for the production of vaginitis, Sap2 is the predominant of these three.
4. Saps1 and 3 are predominantly associated with oral disease and infections of the epidermis.
5. Saps1–6 are equally important in disseminated disease.
6. In some forms of candidiasis, there is temporal expression of Sap proteins; i.e., some are associated with early disease development while others are expressed later.
7. Strains deleted in a specific *SAP* compensate by expressing other *SAP* genes. For example, *sap1* deletion resulted in the upregulation of *SAP2*

TABLE 2 *SAP* Genes *of Candidia albicans* and Disease Correlates

In vitro or animal model	Assay[a]	Specific SAP or SAP Group									Ref.
		1–3	1/3	4–6	1	2	3	4	5	6	
Reconstituted HE[b]	RT-PCR				+	+	+			+	116
Reconstituted HE	IEM				+	+	+			+	117
Reconstituted HE	Δ		+	−	+	+	+				118
Human saliva	RT-PCR				+[c]	+	+[c]	+	+	+	119
Mouse disseminated[d]	Reporter					+					120
Mouse disseminated[e]	Δ	+		+							121,122
Mouse peritonitis	Δ	−		+							123
Mouse macrophages	Δ			+							124
Mucin degradation	in vitro					+[f]					125
Rat vaginitis	Δ	+		−		+[g]					126,127
Human oral lesions	IEM	+									128
Reconstituted epidermis	IEM				+[h]	+				+	129

[a] Assays include RT-PCR, IEM (immunoelectron microscopy), or strains deleted of specific *SAP* or multiple *SAP* mutants, indicated as Δ, were used to evaluate gene function or expression.

[b] HE = human epithelial tissue; temporal expression of Sap was observed, Sap1 and 3 (48 h), 6 (54 h), 2 and 8 (60 h).

[c] *SAP4, 5,* and *6* were detected in all subjects; *SAP1* and *3* only in diseased patients; *SAP7* in some patients.

[d] Only *SAP2* studied; esophageal model, *SAP2* not required.

[e] Similar results with a guinea pig model.

[f] Sap2 inhibited mucin-required binding of the organism to HBEC.

[g] Reintegration of *SAP2* in Δ*sap1–3* restored virulence.

[h] *SAP1* and *SAP2* are required for initial invasion, then *SAP8* (not shown), *SAP6*, and *SAP3*.

and 8 expression, while deletion of *SAP1/3* leads to the upregulation of both of these genes as well as *SAP5*.

8. There appears to be good agreement among assays in regard to establishing correlations of *SAP* expression and function in virulence. For example, in the human reconstituted oral epithelium model, IEM analysis revealed that SAP1–3 antigen was found in tissues, and in support of this observation, the Δ*sap1*, *2*, and *3* or the *sap1/3* mutants were the least invasive (required for pathogenesis).

The *SAP2* promoter has been recently used to develop an in vivo expression technology (IVET) with *C. albicans* [130,131]. In this assay, a *SAP2* promoter–*FLP* fusion was used to transform a strain of *C. albicans* that was resistant to mycophenolic acid. *FLP* encodes a site-specific recombinase protein. The gene-encoding mycophenolic acid (MPA) resistance was flanked by direct repeats of the FLP recognition target (*FRT*). Integration of the *SAP2–FLP* fusion into one of the *SAP2* alleles of the resistant strain was accomplished, and a conversion from the resistant to susceptible phenotype was indicative of the promoter activity of *SAP2*. Using this procedure, the authors were able to determine the contribution of *SAP2* to disease development in two animal models of candidiasis. For example, in a disseminated model, *SAP2* promoter activity was not observed in the initial phase of infection but was strongly induced after dissemination into the deep organs. Contrary, in an oropharyngeal model of candidiasis that does not include deep tissue invasion, no *SAP2* promoter activity was observed at any time. These data indicate that *SAP2* is required for deep invasion of tissues but is not apparently needed for mucosal disease, although it appears to be needed for vaginitis to occur. The use of the *SAP2* promoter was also recently employed to control gene expression in *C. dubliniensis* [131].

In summary, both PL (Plb1) and most SAP enzymes are critical to the pathogenesis of candidiasis. In regard to SAP proteins, the data support a site-specific expression. While these observations are interesting, much needs to be done in regard to their exploitation as targets in the development of new antifungals. Interestingly, there are several reports on the activity of the HIV protease inhibitors and their efficacy in reducing candidiasis in the AIDS patient [132,133].

5 PHENOTYPIC SWITCHING

Phenotypic switching, also called high-frequency switching, is another proposed virulence attribute that will be considered. Phenotypic switching is defined as a spontaneous change in colonial morphology that is heritable and occurs at frequencies greater than a spontaneous mutation. Phenotypic switching has been shown to affect virulence factors such as morphogenesis, the expression of secreted aspartyl proteinase genes, and antigenicity [135]. In the following section, we will examine some of the early work describing phenotypic switching as well as more recent studies attempting to elucidate the mechanism of phenotypic switching and its effects on virulence.

5.1 The Rediscovery of Phenotypic Switching

Although atypical colonial morphologies of *C. albicans* have been previously reported [136–138], it was only more recently (1985) that phenotypic switching received serious attention. Pomes et al. [139] demonstrated that low doses of UV light administered to strain 1001 gave rise to rough colonies at a frequency of 3×10^{-3}. Further, although the spontaneous frequency of reversion to the original smooth colonial phenotype was $<9 \times 10^{-4}$, sectored colonies arose at a frequency of 3.5×10^{-3}. Therefore, they were able to demonstrate both heritability and a high frequency of reversion to the wild-type phenotype.

Using the common laboratory strain 3153A, Slutsky et al. [140] were also able to demonstrate high-frequency, reversible switching in *C. albicans*. In this study, *C. albicans* was grown on an amino acid–rich medium that was also limiting for zinc. Although the predominant phenotype was "smooth," alternate morphologies spontaneously arose at a frequency of 1.4×10^{-4}. This frequency could be doubled by exposure to low doses of UV light. The alternate morphologies that appeared were of six specific types—"star," "ring," "irregular wrinkle," "stippled," "hat," and "fuzzy." When cells from a single colony of an alternate morphology were plated, most retained the same switch phenotype. However, a significant minority grew as a different switch phenotype or as the original smooth morphology. These experiments demonstrate that (1) 3153A is capable of high-frequency switching to several different, defined phenotypes, (2) the switch phenotype is heritable, and (3) high-frequency switching can occur between alternate switch phenotypes. It should be noted that the composition of the media and the growth conditions are important in the expression of switch phenotypes [141,142]. Further studies on a variety of strains of *C. albicans* demonstrated that the majority were capable of phenotypic switching [143–145]. However, it was observed that the type of switch phenotypes differed between strains despite identical growth conditions. This led to the supposition that there are multiple switching systems, each giving rise to a particular set of switch phenotypes. Further, it also appears that a given strain of *C. albicans* is capable of expressing just a single switching system.

One of the most-studied switching systems is the white–opaque of *C. albicans* strain WO1 [146]. In this system, the predominant phenotypes are smooth, white colonies (white) or flat, gray colonies (opaque). In addition to these, irregular wrinkle, medusa, and fried egg phenotypes occur at a much lower frequency. However, unlike other switch phenotypes examined, the opaque phase phenotype is markedly different on the cellular level as well. White phase phenotype cells of strain WO1 are round to slightly ovoid and also share a similar size and budding pattern to many common laboratory strains of *C. albicans*. In contrast, opaque cells were elongated or bean shaped and contained twice the cellular volume and mass as white cells. Further examinations using scanning [147] and transmission electron microscopy [148] revealed the presence of cell surface pimples on opaque cells only.

5.2 Effects of Phenotypic Switching on Hyphal Formation

The different colonial morphologies of the switch phenotypes of strains 1001 and 3153A appear to be due to different proportions of budding, pseudohyphal and mycelial growth of each phenotype [139,140]. This suggests that each switch phenotype undergoes morphogenesis using different environmental parameters. The white–opaque system provides an extraordinary example of this phenomenon. As described in an earlier section of the chapter, one of the parameters that influences morphogenesis is the pH of the growth media. *C. albicans* in an acidic environment grows primarily as yeast, while under neutral to basic conditions it will grow in the hyphal form. It was originally noted by Slutsky et al. [146] that although the white phase of strain WO1 could be induced to form hyphae when grown at pH 6.7, 37°C, the opaque phase could not. Further investigation led to the discovery that opaque phase cells could be stimulated to form hyphae when grown in a Sykes-Moore perfusion chamber or on a monolayer of human skin epithelial cells [149]. This suggests that cell–substrate interactions may be important in morphogenesis in opaque phase cells of strain WO1. A surprising finding of this study was that the hyphae produced by opaque cells lacked the cell surface pimples of yeast phase cells. However, when conditions were changed to favor growth as yeast, the cells, which formed from budding hyphae, displayed the unique cellular appearance of opaque phase cells.

5.3 Differential Gene Expression in Switch Phenotypes

The identification of an opaque-specific antigen in strain WO1 [147] as well as differences in carbohydrate assimilation patterns between white and opaque phenotypes [150] supported the idea that there was differential expression of genes in switch phenotypes. Indeed, using strain WO1, white and opaque phase-specific genes have been isolated. Using an opaque cDNA library, Morrow et al. [151] performed a differential hybridization screen and identified an opaque specific gene they named *OP1a*. *OP1a* was identified as a previously cloned secretory aspartyl proteinase, *SAP1*. Subsequently, *SAP3*, another member of the SAP family, was identified as being opaque specific while *SAP2* was shown to be a white-specific gene in strain WO1 [152]. The opaque phase–specific gene *OP4* was also identified using the differential hybridization screen of an opaque cDNA library [153]. The *OP4* gene product did not exhibit homology to any known protein; however, it was shown to be coordinately regulated with *SAP1* in strain WO1. This coordinate regulation occurred despite the fact that the genes are unlinked and reside on different chromosomes.

Screening of a subtracted white cDNA library led to the identification and isolation of the white phase–specific gene *WH11* [154]. *WH11* shows sequence homology to the glucose/lipid-regulated *GLP1* gene of *Saccharomyces cerevisiae*, although it does not appear to share a homologous function in *C. albicans*. Of particular interest is that *WH11* is expressed only in white phase budding cells of WO1 and is not found in white phase hyphal forms. This points to a link between the regulation of *WH11* and morphogenesis.

The basis for the differential expression of genes in the white and opaque phases of strain WO1 appears to lie in the promoter regions of the phase-specific genes. A functional analysis of the promoter of the white phase–specific gene *WH11* led to the identification of two transcription activation domains which control the transcription of *WH11* [155]. Deletion of the distal activation domain (DAD) located between bp 475 and bp 388 resulted in a sixfold decrease in gene transcript while deletion of the proximal activation domain (PAD, bp 307 to bp 230) lead to a 15-fold decrease in transcript. Deletion of both domains resulted in elimination of gene transcription. Gel retardation assays demonstrated that *trans*-acting factors positively regulated transcription of *WH11* through interactions with the DAD and PAD regions in the promoter. A functional analysis of the promoter of the opaque phase–specific gene *OP4* provided similar results [156]. Both DAD and PAD sequences were identified in the promoter region of *OP4*, the deletion of which resulted in the elimination of transcription. Again, *trans*-acting factors were shown to interact with the DAD and PAD elements of the *OP4* promoter to regulate expression in a positive manner.

5.4 Mechanism for Phenotypic Switching

The exact mechanism for phenotypic switching remains unclear. Early observations of changes in karyotype in switch phenotypes led to the suggestion that spontaneous chromosomal rearrangements may be responsible for the generation of switch phenotypes [157–159]. While chromosomal rearrangements are known to occur at relatively high frequencies in *C. albicans*, researchers have been unable to show a cause-and effect relationship between specific changes in karyotype and a particular switch phenotype.

One hypothesis on the control of phenotypic switching involves changes in the state of specific regions of the chromatin. In *Saccharomyces cerevisiae*, genes controlling mating type are located in chromosomal domains where the structure of the chromatin determines whether the genes are active or silent [160]. The *SIR* genes are required to establish the silent state of the mating-type genes

[161]. Once a particular state has been established, it can be passed on to future generations. Therefore, it has been hypothesized that phenotypic switching could be controlled in a similar manner, utilizing active or silent states of chromatin, accounting for the high frequency of switching and the heritability [162].

To test this hypothesis, Perez-Martin et al. [162] cloned the *SIR2* gene of *C. albicans*, a homolog of the *S. cerevisiae* gene. *SIR2* is involved in maintaining the silent state of chromatin in *S. cerevisiae*. Upon deletion of *SIR2* in *C. albicans* strain CAI4, phenotypic switching occurred at frequencies as high as 10^{-1}. Phenotypes observed were similar to those reported by Slutsky et al. [140] in strain 3153A. In addition, a high frequency of karyotypic changes was also observed in the *sir2/sir2* null strain. However, they were unable to correlate a particular phenotype with a specific change in karyotype. Nevertheless, the high frequency of karyotypic changes in the *sir2/sir2* null strain indicates a role for the Sir2 protein in maintaining chromatin structure and stability. It also provides a basis for karyotype changes observed in previous studies.

Morphogenesis was also affected by the *sir2/sir2* null mutation. Null mutants were able to produce hyphae and pseudohyphae under conditions in which the wild-type strain grew only as blastospores. This provides further evidence for a link between phenotypic switching and morphogenesis. Additional proof is provided by the observation that *tup1/tup1* mutants and *tup1/tup1*, *sir2/sir2* double mutants, both of which are locked in the hyphal phase, do not undergo phenotypic switching. This observation implies that not only are morphogenesis and phenotypic switching linked, *TUP1* functions downstream of *SIR2*. Though much work remains to be done, it is clear that *SIR2* plays an important role in controlling phenotypic switching, likely through mechanisms involving chromosome stability and the organization of chromatin structure.

5.5 Phenotypic Switching as a Virulence Factor

To this point, we have described many characteristics of phenotypic switching as well as possible mechanisms for its control. What relevance does phenotypic switching have on the virulence of *C. albicans*? Though a clear picture has yet to emerge, several studies have implicated a role for phenotypic switching in the pathogenesis of *C. albicans*. In studies examining phenotypic switching in vaginal candidiasis [163] and invasive infections [164], switching was evident in isolates recovered from both studies with multiple switch phenotypes being observed. Moreover, Jones et al. [164] reported that 68% of isolates recovered from invasive infections demonstrated switching activity, compared with only 28% of isolates recovered from superficial infections. These studies demonstrate that phenotypic switching is not just a laboratory phenomenon, but occurs in vivo in clinical infections.

Utilizing the white–opaque switching system of *C. albicans* strain WO1, Kvaal et al. [165] examined the virulence of the white and opaque phases in a mouse model of infection. While the white phase was shown to be significantly more virulent in a systemic model of infection, the opaque phase demonstrated an increased ability to colonize the skin in a cutaneous mouse model. As previously mentioned, the opaque phase was shown to differentially express the *SAP1* gene. By placing the *SAP1* gene under the control of the white phase–specific *WH11* promoter, they were able to demonstrate that white phase cells gained an increased capacity to colonize the skin in the cutaneous mouse model. Therefore, phenotypic switching contributed to the ability of *C. albicans* to infect different host niches through differential gene expression.

6 SUMMARY

The development of gene deletion protocols has contributed much to our understanding of virulence of *C. albicans*. Of importance, most of these genes do not appear to be essential for growth of the organism, although they are required for survival in the host. Because of the first feature, these proteins would not appear to be ideal targets for the development of antifungal drugs. However, given the few choices of current drugs, perhaps our immediate goal is to focus upon these proteins as targets, especially the ones that are only found only in *C. albicans* and other fungi. Additionally, the literature is replete with studies on the commensal state of this organism. How is survival achieved on mucosal surfaces and what regulatory events does it use to invade? No doubt some of these questions will be answered within the next few years.

REFERENCES

1. MB Edmond, SE Wallace, DK McClish, MA Pfaller, RN Jones, RP Wenzel. Nosocomial blood stream infections in United States hospitals: a three-year analysis. Clin Infect Dis 29:239–244, 1999.
2. L Cardenas-Freytag, E Cheng, P Mayeux, JE Domer, J Clements. Effectiveness of a vaccine composed of heat-killed *Candida albicans* and a novel mucosal adjuvant, LT (R192G), against systemic candidiasis. Infect Immun 67:826–833, 1999.
3. Y Han, JE Cutler. Antibody response that protects against disseminated candidiasis. Infect Immun 63:2714–2719, 1995.
4. PL Fidel, ME Lynch, DH Conaway, L Tait, JD Sobel. Mice immunized by primary vaginal *Candida albicans* infection develop acquired vaginal mucosal immunity. Infect Immun 63:547–553, 1995.
5. R Calderone, PC Braun. Adherence and receptor relationships of *Candida albicans*. Microbiol Rev 55:1–20, 1991.

6. J Sturtevant, R Calderone. *Candida albicans* adhesins: biochemical aspects and virulence. Iberoam Micol 14:90–97, 1997.
7. P Sundstrom. Adhesins in *Candida albicans*. Curr Opin Microbiol 2:353–357, 1999.
8. WL Chaffin, JL Lopez-Ribot, M Casanova, D Gozalbo, JP Martinez. Cell wall and secreted proteins of *Candida albicans*: identification, function, and expression. Microbiol Mol Biol Rev 62:130–180, 1998.
9. RD King, JC Lee, AL Morris. Adherence of *Candida albicans* and other *Candida* species to mucosal epithelium. Infect Immun 27:667–674, 1980.
10. N Lehrer, E Segal, R Cihlar, RA Calderone. Pathogenesis of *Candida albicans*: studies with a mutant which has reduced ability to adhere. J Med Vet Mycol 24: 127–131, 1986.
11. RA Calderone, R Cihlar, D Lee, K Hoberg, WM Scheld. Yeast adhesion in the pathogenesis of endocarditis due to *Candida albicans*: studies with adherence-negative mutants. J Infect Dis 152:710–715, 1985.
12. MK Ollert, E Wadsworth, RA Calderone. Reduced expression of the functionally active complement receptor for iC3b but not for C3d on an avirulent mutant of *Candida albicans*. Infect Immun 58:909–913, 1990.
13. S Franzke, RA Calderone, K Schaller. Isolation of avirulent clones of *Candida albicans* with reduced ability to recognize the CR2 ligand, C3d. Infect Immun 61: 2662–2669, 1993.
14. IA Critchley, LJ Douglas. Role of glycosides as epithelial cell receptors for *Candida albicans*. J Gen Microbiol 133:637–643, 1987.
15. IA Critchley, LJ Douglas. Isolation and partial characterization of an adhesin from *Candida albicans*. J Gen Microbiol 133:629–636, 1987.
16. LJ Douglas. Adhesion of *Candida* species to epithelial surfaces. Crit Rev Microbiol 15:27–43, 1987.
17. FD Tosh, LJ Douglas. Characterization of a fucoside-binding adhesin of *Candida albicans*. Infect Immun 60:4734–4739, 1992.
18. J McCourtie, LJ Douglas. Relationship between cell surface composition of *Candida albicans* and adherence to acrylic after growth on different carbon sources. Infect Immun 32:1234–1241, 1981.
19. J McCourtie, LJ Douglas. Extracellular polysaccharide of *Candida albicans*: isolation, analysis and role in adhesion. J Gen Microbiol 131:495–503, 1985.
20. J McCourtie, LJ Douglas. Unequal distribution of adhesins within populations of *Candida albicans*. FEMS Microbiol Lett 27:111–115, 1985.
21. D Brassart, A Woltz, M Golliard, J-M Nesser. In vitro inhibition of adhesion of *Candida albicans* clinical isolates to human buccal epithelial cells by Fucα1-2Gal-bearing complex carbohydrates. Infect Immun 59:1605–1613, 1991.
22. KK Lee, L Yu, DL MacDonald, W Paranchych, RS Hodges, RT Irvin. Anti-adhesin antibodies that recognize a receptor-binding motif (adhesintope) inhibit pilus/fimbrial-mediated adherence of *Pseudomonas aeruginosa* and *Candida albicans* to asialo-GM1 receptors and human buccal epithelial cell surface receptors. Can J Microbiol 42:479–486, 1996.
23. L Yu, KK Lee, W Paranchych, RS Hodges, RT Irvin. Use of synthetic peptide to confirm that the *Pseudomonas aeruginosa* PAK pilus adhesin and the *Candida albi-*

cans fimbrial adhesin possess a homologous receptor-binding domain. Mol Microbiol 19:1107–1116, 1996.

24. L Yu, KK Lee, RS Hodges, W Paranchych, RT Irvin. Adherence of *Pseudomonas aeruginosa* and *Candida albicans* to glycosphingolipid (asialo-GM1) receptors is achieved by a conserved receptor-binding domain present on their adhesins. Infect Immun 62:5213–5219, 1994.
25. L Yu, KK Lee, HB Sheth, P Lane-Bell, G Srivastava, O Hindsgaul, W Paranchych, RS Hodges, RT Irvin. Fimbria-mediated adherence of *Candida albicans* to glycosphingolipid receptors on human buccal epithelial cells. Infect Immun 62:2843–2848, 1994.
26. L Yu, KK Lee, K Ens, PC Doig, MR Carpenter, W Staddon, RS Hodges, W Paranchych, RT Irvin. Partial characterization of a *Candida albicans* fimbrial adhesin. Infect Immun 62:2834–2842, 1994.
27. F Heidenreich, MP Dierich. *Candida albicans* and *Candida stellatoidea*, in contrast to other *Candida* species, bind iC3b and C3d but not C3b. Infect Immun 50:598–600, 1985.
28. BJ Gilmore, EM Retsinas, JS Lorenz, MK Hostetter. An iC3b receptor on *Candida albicans*: structure, function and correlates for pathogenicity. J Infect Dis 157:38–46, 1988.
29. RA Calderone, L Linehan, E Wadsworth, A Sandberg. Identification of C3d receptors on *Candida albicans*. Infect Immun 56:252–258, 1988.
30. JE Edwards, TA Gaither, JJ O'Shea, D Rotrosen, TJ Lawley, SA Wright, FF Frank, I Green. Expression of specific binding sites on *Candida albicans* with functional and antigenic characteristics of human complement receptors. J Immunol 137:3577–3583, 1986.
31. L Linehan, E Wadsworth, RA Calderone. *Candida albicans* C3d receptor, isolated by using a monoclonal antibody. Infect Immun 56:1981–1986, 1988.
32. MK Hostetter. Adhesins and ligands involved in the interaction of *Candida* spp with epithelial and endothelial surfaces. Clin Microbiol Rev 7:29–42, 1994.
33. J Bouchara, G Tronchin, V Annaix, R Robert, J-M Senet. Presence of laminin receptors on *Candida albicans*. Infect Immun 58:48–54, 1990.
34. KS Gustafson, GM Vercelloti, JS Lorenz, MK Hostetter. Molecular mimicry in *Candida albicans*. Role of an integrin in adhesion of the yeast to human endothelium. J Clin Invest 87:1896–1902, 1991.
35. KG Skerl, RA Calderone, E Segal, T Sreevalsan, WM Scheld. In vitro binding of *Candida albicans* yeast cells to human fibronectin. Can J Microbiol 30:221–227, 1983.
36. SA Klotz, RL Smith. A fibronectin receptor on *Candida albicans* mediates adherence of the fungus to extracellular matrix proteins. J Infect Dis 163:604–610, 1991.
37. SA Klotz, RL Smith, BW Stewart. Efect of an arginine-glycine-aspartic acid containing peptide on hematogenous candidal infections in rabbits. Antimicrob Agents Chemother 36:132–136, 1992.
38. C Gale, D Finkel, N Tao, M Meinke, M McClellan, J Olson, K Kendrick, M Hostetter. Cloning and expression of a gene encoding an integrin-like protein in *Candida albicans*. Proc Natl Acad Sci USA 93:357–361, 1996.

39. CA Gale, CM Bendel, M McClellan, M Hauser, JM Becker, J Berman, MK Hostetter. Linkage of adhesion, filamentous growth, and virulence in *Candida albicans* to a single gene, *INT1*. Science 279:1355–1357, 1998.
40. MK Ollert, RA Calderone. A monoclonal antibody that defines a surface antigen on *Candida albicans* hyphae cross-reacts with yeast cell protoplasts. Infect Immun 58:625–631, 1990.
41. T Kanbe, R-K Li, E Wadsworth, R. Calderone, JE Cutler. Evidence for expression of the C3d receptor of *Candida albicans* in vitro and in vivo obtained by immunofluorescent and immunoelectron microscopy. Infect Immun 59:1832–1838, 1991.
42. SA Klotz, RC Hein, RL Smith, JB Riuse. The fibronectin adhesin of *Candida albicans*. Infect Immun 62:4679–4681, 1994.
43. SA Klotz, RL Smith. Gelatin fragments block adherence of *Candida albicans* to extracellular matrix proteins. Microbiology 141:2681–2684, 1995.
44. S Yan, RG Rodrigues, DD Roberts. Hemoglobin-induced binding of *Candida albicans* to the cell-binding domain of fibronectin is independent of the Arg-Gly-Asp sequence. Infect Immun 66:1904–1909, 1998.
45. S Yan, RG Rodrigues, D Cahn-Hidalgo, TJ Walsh, DD Roberts. Hemoglobin induces binding of several extracellular matrix proteins to *Candida albicans*. J Biol Chem 273:5638–5644, 1998.
46. S Yan, E Negre, JA Cashel, N Guo, CA Lyman, TJ Walsh, DD Roberts. Specific induction of fibronectin binding activity by hemoglobin in *Candida albicans* grown in defined media. Infect Immun 64:2930–2935, 1996.
47. JF Staab, CA Ferrer, P Sundstrom. Developmental expression of a tandemly repeated, proline- and glutamine-rich amino acid motif on hyphal surfaces on *Candida albicans*. J Biol Chem 271:6298–6305, 1996.
48. JF Staab, SD Bradway, P Fidel, P Sundstrom. Adhesive and mammalian transglutaminase substrate properties of *Candida albicans HWP1*. Science 283:1535–1538, 1999.
49. NK Gaur, SA Klotz, RL Henderson. Overexpression of the *Candida albicans ALA1* gene in *Saccharomyces cerevisiae* results in aggregation following attachment of yeast cells to extracellular matrix proteins, adherence properties similar to those of *Candida albicans*. Infect Immun 67:6040–6047, 1999.
50. Y Fu, G Rieg, WA Fonzi, PH Belanger, JE Edwards, SG Filler. Expression of the *Candida albicans ALS1* in *Saccharomyces cerevisiae* induces adherence to endothelial and epithelial cells. Infect Immun 66:1783–1786, 1998.
51. LL Hoyer, S Scherer, AR Shatzman, GP Livi. *Candida albicans ALS1*: domains related to a *Saccharomyces cerevisiae* sexual agglutinin separated by a repeating motif. Mol Microbiol 15:39–54, 1995.
52. LL Hoyer, JE Hecht. The *ALS6* and *ALS7* genes of *Candida albicans*. Yeast 16: 847–855, 2000.
53. LL Hoyer, TL Payne, M Bell, AM Myers, S Scherer. *Candida albicans ALS3* and insights into the nature of the ALS gene family. Curr Genet 33:451–459, 1998.
54. LL Hoyer, J Clevenger, JE Hecht, EJ Ehrhart, FM Poulet. Detection of Als proteins on the cell wall of *Candida albicans* in murine tissues. Infect Immun 67:4251–4255, 1999.
55. LL Hoyer, TL Payne, JE Hecht. Identification of *Candida albicans ALS2* and *ALS4*

and localization of Als proteins to the fungal cell surface. J Bacteriol 180:5334–5343, 1998.
56. WA Fonzi, MY Irwin. Isogenic strain construction and gene mapping in *Candida albicans*. Genetics 134:717–728, 1993.
57. N Tsuchimorl, LL Sharkey, WA Fonzi, SW French, JE Edwards, SG Filler. Reduced virulence of *HWP1*-deficient mutants of *Candida albicans* and their interactions with host cells. Infect Immun 68:1997–2002, 2000.
58. C Timpel, S Strahl-Bolsonger, K Ziegelbauer, JF Ernst. Multiple functions of Pmt1p-mediated protein O-mannosylation in the fungal pathogen *Candida albicans*. J Biol Chem 273:20837–20846, 1998.
59. ET Buurman, Westwater C, Hube B, Aj Brown, FC Odds, NA Gow. Molecular analysis of CaMnt1p, a mannosyl transferase important for adhesion and virulence of *Candida albicans*. Proc Natl Acad Sci USA 95:7670–7675, 1998.
60. LM Thomson, S Bates, S Yamazaki, M Arisawa, Y Aoki, NA Gow. Functional characterization of the *Candida albicans MNT1* mannosyltransferase expressed heterologously in *Pichia pastoris*. J Biol Chem 275:18933–18938, 2000.
61. R-K Li, JE Cutler. Chemical definition of an epitope/adhesin molecule on *Candida albicans*. J Biol Chem 268:18293–18299, 1993.
62. T Kanbe, JE Culter. Minimum chemical requirements for adhesin activity of the acid-stable part of *Candida albicans* cell wall phosphomannoprotein complex. Infect Immun 66:5812–5818, 1998.
63. AJ Brown, NAR Gow. Regulatory networks controlling *Candida albicans* morphogenesis. Trends Microbiol 7:333–338, 1999.
64. CJ Gimeno, PO Ljungdahl, CA Styles, GR Fink. Unipolar cell division in the yeast *Saccharonyces cerevisiae* leads to filamentous growth: regulation by starvation and RAS. Cell 68:1077–1090, 1992.
65. H Liu, J Kohler, GR Fink. Suppression of hyphal formation in *Candida albicans* by mutation of a *STE12* homolog. Science 266:1723–1726, 1994.
66. JR Kohler, GR Fink. *Candida albicans* strains heterozygous and homozygous for mutations in mitogen-activated protein kinase signaling components have defects in hyphal development. Proc Natl Acad Sci USA 93:13223–13228, 1996.
67. C Csank, C Makris, S Meloche, K Schroppel, M Rollinghoff, D Dignard, D Thomas, M Whiteway. Derepressed hyphal growth and reduced virulence in a VH1 family-related protein phosphatase mutant of the human pathogen *Candida albicans*. Mol Biol Cell 8:2539–2551, 1997.
68. E Leberer, K Ziegelbauer, A Schmidt, D Harcus, D Dignard, J Ash, J Johnson, DY Thomas. Virulence and hyphal formation of *Candida albicans* require the Ste20p-like protein kinase CaCla4p. Curr Biol 7:539–546, 1997.
69. C Csank, K Schroppel, E Leberer, D Harcus, O Mohamed, S Meloche, DY Thomas, M Whiteway. Roles of the *Candida albicans* mitogen-activated protein kinase homolog, Cek1p, in hyphal development and systemic candidiasis. Infect Immun 66: 2713–2721, 1998.
70. E Leberer, D Harcus, ID Broadbent, KL Clark, D Dignard, K Ziegelbauer, A Schmidt, NA Gow, AJ Brown, DY Thomas. Signal transduction through homologs of the Ste20p and Ste7p protein kinases can trigger hyphal formation in the pathogenic fungus *Candida albicans*. Proc Natl Acad Sci USA 93:13217–13222, 1996.

71. VR Stoldt, A Sonneborn, CE Leuker, JF Ernst. Efg1p, an essential regulator of morphogenesis of the human pathogen *Candida albicans*, is a member of a conserved class of bHLH proteins regulating morphogenetic processes in fungi. EMBO J 16:1982–1991, 1997.
72. HJ Lo, JR Kohler, B DiDomenico, D Loebenberg, A Cacciapuoti, GR Fink. Nonfilamentous *Candida albicans* mutants are avirulent. Cell 90:939–949, 1997.
73. A Sonneborn, DP Bockmuhl, M Gerads, K Kurpanek, D Sanglard, JF Ernst. Protein kinase A encoded by *TPK2* regulates dimorphism in *Candida albicans*. Mol Microbiol 35:386–396, 2000.
74. K Malathi, K Ganesan, A Datta. Identification of a putative transcription factor in *Candida albicans* that can complement the mating defect of *Saccharomyces cerevisiae*. J Biol Chem 269:22945–22951, 1994.
75. Q Feng, E Summers, B Guo, G Fink. Ras signaling is required for serum-induced hyphal differentiation in *Candida albicans*. J Bacteriol 181:6339–6346, 1999.
76. CE Birse, MY Irwin, WA Fonzi, PS Sypherd. Cloning and characterization of *ECE1*, a gene expressed in association with cell elongation of the dimorphic pathogen *Candida albicans*. Infect Immun 61:3648–3655, 1993.
77. BR Braun, AD Johnson. Control of filament formation in *Candida albicans* by the transcription repressor *TUP1*. Science 277:10509, 1997.
78. N Ishii, M Yamamoto, F Yoshihara, M Arisawa, Y Aoki. Biochemical characterization of Rbf1p, a putative transcription factor of *Candida albicans*. Microbiology 143:429–435, 1998.
79. P Riggle, KA Andrutis, X Chen, SR Tzipori, C Kumamoto. Invasive lesions containing filamentous forms produced by a *Candida albicans* mutant that is defective in filamentous growth in culture. Infect Immun 67:3649–3652, 1999.
80. DH Brown, AD Giusani, X Chen, CA Kumamoto. Filamentous growth of *Candida albicans* in response to physical environmental clues and its regulation by the unique *CZF1* gene. Mol Microbiol 34:651–662, 1999.
81. F Navarro-Garcia, R Alonso-Monge, H Rico, J Pla, R Sentandreu, C Nombela. A role for the MAP kinase *MKC1* in cell wall construction and morphological transitions in *Candida albicans*. Microbiology 144:411–424, 1998.
82. N Navarro-Garcia, M Sanchez, J Pla, C Nombela. Functional characterization of the *MKC1* gene of *Candida albicans*, which encodes a mitogen-activated protein kinase homolog related to cell integrity. Mol Cell Biol 15:2197–2206, 1995.
83. R Diez-Orejas, G Molero, F Nararro-Garcia, J Pla, C Nombela, M Sanchez-Perez. Reduced virulence of *Candida albicans MKC1* mutants: a role for mitogen activated protein kinase in pathogenesis. Infect Immun 65:833–837, 1997.
84. R Alonso-Monge, F Navarro-Garcia, G Molero, R Diez-Orejas, M Gustin, J Pla, M Sanchez, C Nombela. Role of the mitogen activated protein kinase Hog1p in morphogenesis and virulence of *Candida albicans*. J Bacteriol 181:3058–3068, 1999.
85. C San Jose, R Alonso-Monge, R Perez-Diaz, J Pla, C Nombela. The mitogen-activated protein kinase homolog *HOG1* gene controls glycerol accumulation in the pathogenic fungus *Candida albicans*. J Bacteriol 178:5850–5852, 1996.
86. F Posas, SM Wurgler-Murphy, T Maeda, EA Witten, TC Thai, H Saito. Yeast *HOG1* MAP kinase cascade is regulated by a multistep phosphorelay mecha-

nism in the SLN1-YPD1-SSK1 two-component osmosensor. Cell; 86:865–875, 1996.
87. JF Barrett, JA Hoch. Two-component signal transduction as a target for microbial anti-infective therapy. Antimicrob Agents Chemother 42:1529–1536, 1998.
88. S Nagahashi, T Mio, N Ono, et al. Isolation of *CaSLN1* and *CaN1K1*, the genes for osmosensing histidine kinase homologues, from the pathogenic fungus, *Candida albicans*. Microbiology 144:425–432, 1998.
89. JA Calera, RA Calderone. Identification of a putative response regulator, two-component phosphorelay gene (*CaSSK1*) from *Candida albicans*. Yeast 15:1243–1254, 1999.
90. JA Calera, XJ Zhao, RA Calderone. Defective hyphal formation and avirulence caused by a deletion of the *CSSK1* response regulator gene in *Candida albicans*. Infect Immun 68:518–525, 2000.
91. JA Calera, D Herman, RA Calderone. Identification of *YPD1*, a gene of *Candida albicans* which encodes a two-component phospho-histidine intermediate protein. Yeast 16:1053–1059, 2000.
92. LA Alex, C Korch, CP Selitrennikoff, MI Simon. *COS1*, a two-component histidine kinase that is involved in hyphal development in the opportunistic pathogen, *Candida albicans*. Proc Natl Acad Sci USA 95:3416–3421, 1996.
93. T Srikantha *et al*. The two-component hybrid kinase regulator *CaNIK1* of *Candida albicans*. Microbiology 144:2715–2729, 1998.
94. JA Calera, G Cho G, RA Calderone. Identification of a putative histidine kinase two-component phosphorelay gene (*CaHK1*) in *Candida albicans*. Yeast 14:665–674, 1998.
95. JA Calera, RA Calderone. Flocculation of hyphae is associated with a deletion in the putative *CaHK1* two-component histidine kinase gene from *Candida albicans*. Microbiology 145:1431–1442, 1999.
96. JA Calera, XJ Zhao, M Sheridan, RA Calderone. A virulence of *Candida albicans CaHK1* mutants in a murine model of hematogenously disseminated candidiasis. Infect Immun 67:4280–4284, 1999.
97. T Yamada-Okabe, T Mio, T Ono, Y Kashima, M Matsui, M Arisawa, Y Yamada-Okabe. Roles of three histidine kinase genes in hyphal development and virulence of the pathogenic fungus *Candida albicans*. J Bacteriol 181:7243–7247, 1999.
98. F Muhlschlegel, W Finzi, L Hoyer, T Payne, F Poulet, J Clevenger, J-P Latge, J Calera, A Beauvais, S Paris, M Monod, J Sturtevant, M Ghannoum, Y Nozawa, R Calderone. Molecular mechanisms of virulence in fungus-host interaction for *Aspergillus fumigatus* and *Candida albicans*. Med Mycol 36S:238–248, 1998.
99. F Mirod, Y Banno, MA Ghannoum, AS Ibrahim, S Nakashima, Y Kitajima, GT Cole, Y Nozawa. Purification and characterization of lysophospholipase-transacylase (h-LPTA) from a highly virulent strain of *Candida albicans*. Biochim Biophys Acta 1257:181–188, 1995.
100. MF Price, ID Wilkinson, LO Gentry. Plate method for detection of phospholipase activity in *Candida albicans*. Sabouraudia 20:7–14, 1982.
101. E Habermann, KL Hardt. A sensitive and specific plate assay for the quantitation of phospholipase, Anal Biochem 50:163–173, 1972.
102. S Leidich, AS Ibrahim, Y Fu, A Koul, C Jessup, J Vitullo, W Fonzi, F Mirod,

S Nakashima, Y Nozawa, MA Ghannoum. Cloning and disruption of *caPLB1*, a phospholipase B gene involved in the pathogenicity of *Candida albicans*. J Biol Chem 273:26078–26086, 1998.

103. AS Ibrahim, F Mirod, S Filler, Y Banno, GT Cole, Y Kitajima, JE Edwards, Y Nozawa, MA Ghannoum. Evidence implicating phospholipase as a virulence factor of *Candida albicans*. Infect Immun 63:1993–1998, 1995.
104. K Barrett-Bee, Y Hayes, RG Wilson, JF Ryley. A comparison of phospholipase activity, cellular adherence and pathogenicity of yeast. J Gen Microbiol 131:1217–1221, 1985.
105. Y Sugiyama, S Nakashima, F Mirbod, H Kanoh, Y Kitajima, MA Ghannoum, Y Nozawa. Molecular cloning of a second phospholipase B gene, *caPLB2* from *Candida albicans*. Med Mycol 37:61–67, 1999.
106. De Bennett, CE McCreary, DC Coleman. Genetic evidence of a phospholipase C gene from *Candida albicans*: presence of homologous sequences in *Candida* species other than *Candida albicans* Microbiol 144:55–72, 1998.
107. H Kanoh, S Nakashima, Y Zhao, Y Sugiyama, Y Kitajima, Y Nozawa. Molecular cloning of a gene encoding phospholipase D from the pathogenic and dimorphic yeast, *Candida albicans*. Biochim Biophys Acta 1398:359–364, 1998.
108. N McClain, JW Dolan. Phospholipase D activity is required for dimorphic transition in *Candida albicans*. Microbiol 143:3521–3526, 1997.
109. F Staib. Serum proteins as nitrogen source for yeast-like fungi. Sabouraudia 4:187–193, 1965.
110. R Ruchel. Properties of a purified proteinase from the yeast, *Candida albicans*. Biochim Biophys Acta 659:99–113, 1981.
111. TL Ray, CD Payne. Comparative production and rapid purification of *Candida* proteinase from protein-supplemented cultures. Infect Immun 58:508–514, 1990.
112. A Cassone, F DeBernardis, F Mondello, T Ceddia, L Agatensi. Evidence for a correlation between proteinase secretion and vulvovaginal candidosis. J Infect Dis 156:777–783, 1987.
113. B Morrow, T Srikantha, DR Soll. Transcription of the gene for a pepsinogen, *PEP1*, is regulated by white-opaque switching in *Candida albicans*. Mol Cell Biol 12: 2997–3005, 1992.
114. M Monod, B Hube, D Hess, D Sanglard. Differential regulation of *SAP8* and *SAP9*, which encode two new members of the secreted aspartic proteinase family in *Candida albicans*. Microbiol 144:2731–2737, 1998.
115. B Hube. *Candida albicans* secreted aspartyl proteinases. Curr Top Med Mycol 7: 55–69, 1996.
116. M Schaller, W Schafer, HC Korting, B Hube. Differential expression of secreted aspartyl proteinases in a model of human oral candidosis and in patient samples from the oral cavity. Mol Microbiol 29:605–615, 1998.
117. M Schaller, HC Korting, W Schafer, D Sanglard, B Hube. Investigations on the regulation of secreted aspartyl proteases in a model of oral candidiasis in vivo. Mycoses 41S:69–73, 1998.
118. M Schaller, HC Korting, W Schafer, J Bastert, W Chen, B Hube. Secreted aspartic proteinase (Sap) activity contributes to tissue damage in a model of human oral candidosis. Mol Microbiol 34:169–180, 1999.

119. JR Naglik, G Newport, TC White, LL Fernandes-Naglik, JS Greenspan, SP Sweet, SJ Challacombe, N Agabian. In vivo analysis of secreted aspartyl proteinase expression in human oral candidiasis. Infect Immun 67:2482–2490, 1999.
120. P Staib, M Kretschmar, T Nichterlein, G Kohler, S Michel, H Hof, J Hacker, J Morschhauser. Host-induced, stage-specific virulence gene activation in *Candida albicans* during infection. Mol Microbiol 32:533–546, 1999.
121. B Hube, D Sanglard, FC Odds, D Hess, M Monod, W Schafer, AJ Brown, NA Gow. Disruption of each of the secreted aspartyl proteinase genes, *SAP1*, *SAP2*, and *SAP3* of *Candida albicans* attenuates virulence. Infect Immun 65:3529–3538, 1997.
122. D Sanglard, B Hube, M Monod, FC Odds, NA Gow. A triple deletion of the secreted aspartyl proteinase genes *SAP4*, *SAP5*, *SAP6* of *Candida albicans* causes attenuated virulence. Infect Immun 65:3539–3546, 1997.
123. M Kretschmar, B Hube, T Bertsch, D Sanglard, R Merker, M Schroder, H Hof, T Nichterlein. Germ tubes and proteinase activity contribute to virulence of *Candida albicans* in murine peritonitis. Infect Immun 67:6637–6642, 1999.
124. M Borg–von Zepelin, S Beggah, K Boggian, D Sanglard, M Monod. The expression of the secreted aspartyl proteinases *SAP4* to *SAP6* from *Candida albicans* in murine macrophages. Mol Microbiol 28:543–554, 1998.
125. L de Repentigny, F Aumont, K Bernard, P Belhumeur. Characterization of binding of *Candida albicans* to small intestinal mucin and its role in adherence to mucosal epithelial cells. Infect Immun 68:3172–3179, 2000.
126. F DeBernardis, S Arancia, L Morelli, B Hube, D Sanglard, W Schafer, A Cassone. Evidence that members of the secretory aspartyl proteinase gene family, in particular *SAP2*, are virulence factors for *Candida* vaginitis. J Infect Dis 179:201–208, 1999.
127. F DeBernardis, A Cassone, J Sturtevant, RA Calderone. Expression of *SAP1* and *SAP2* in experimental vaginitis caused by *Candida albicans*. Infect Immun 63: 1887–1892, 1995.
128. M Schaller, B Hube, MW Ollert, W Schafer, M Borg-von Zepelin, E Thoma-Gerber, HC Korting. In vivo expression and localization of *Candida albicans* aspartyl proteinases during oral candidiasis in HIV-infected patients. J Invest Dermatol 112:383–386, 1999.
129. M Schaller, C Schackert, HC Korting, E Januschke, B Hube. Invasion of *Candida albicans* correlates with expression of secreted aspartic proteinases during experimental infection of human epidermis. J Invest Dermatol 114:712–717, 2000.
130. P Staib, S Michel, G Kohler, J Morschhauser. A molecular system for the pathogenic yeast *Candida dubliniensis*. Gene 242:393–398, 2000.
131. P Staib, M Kretschmar, T Nichterlein, H Hof, J Morschhauser. Differential activation of a *Candida albicans* virulence gene family during infection. Proc Natl Acad Sci USA 97:6102–6107, 2000.
132. M Monod, M Borg–von Zepelin, A Telenti, D Sanglard. The inhibition of *Candida albicans* secreted aspartic proteases by three different HIV protease inhibitors. Dermatology 198:412–414, 1999.
133. M Borg–von Zepelin, I Meyer, R Thomssen, R Wurzner, D Sanglard, A Telenti, M Monod. HIV-protease inhibitors reduce cell adherence of *Candida albicans*

strains by inhibition of yeast secreted aspartic proteases. J Invest Dermatol 113: 747–751, 1999.
134. JR Davie, RM Wynn, M Meng, et al. Expression and characterization of branched-chain alpha-ketoacid dehydrogenase kinase from the rat. Is it a histidine–protein kinase? J Biol Chem 270:19861–19867, 1995.
135. DR Soll. High-frequency switching in *Candida albicans*. Clin Microbiol Rev 5: 183–203, 1992.
136. P Negroni. Variacion bacia et tipo R de *Mycotorula albicans*. Rev Soc Argent Biol 11:449–453, 1935.
137. JE MacKinnon. Dissociation in *Candida albicans*. J Infect Dis 66/67:59–77, 1940.
138. J Brown-Thomsen. Variability in *Candida albicans* (Robin) Berkhout. Hereditas 60:335–398, 1968.
139. R Pomes, C Gil, C Nombela. Genetic analysis of *Candida albicans* morphological mutants. J Gen Microbiol 131:2107–2113, 1985.
140. E Slutsky, J Buffo, DR Soll. High frequency switching of colony morphology in *Candida albicans*. Science 230:666–669, 1985.
141. GW Bedell, DR Soll. Effects of low concentrations of zinc on the growth and dimorphism of *Candida albicans*: evidence for zinc-resistant and -sensitive pathways for mycelium formation. Infect Immun 26:348–354, 1979.
142. FC Odds, LA Merson-Davies. Colony variations in *Candida* species. Mycoses 32: 275–282, 1989.
143. DR Soll, CJ Langtimm, J McDowell, J Hicks, R Galask. High-frequency switching in *Candida* strains isolated from vaginitis patients. J Clin Microbiol 25:1611–1622, 1987.
144. DR Soll, B Slutsky, S Mackenzie, C Langtimm, M Staebell. Switching systems in *Candida albicans* and their possible roles in oral candidiasis. In: J Mackenzie, C Squier, E Dabelsteen, eds. Oral Mucosa Diseases: Biology, Etiology and Therapy. Folarg, Denmark: Laegeforeningens, pp 52–59.
145. DR Soll, M Staebell, C Langtimm, M Pfaller, J Hicks, TVG Rao. Multiple *Candida* strains in the course of a single systemic infection. J Clin Microbiol 26:1448–1459, 1988.
146. B Slutsky, M Staebell, J Anderson, L Risen, M Pfaller, DR Soll. ''White–opaque transition'': a second high-frequency switching system in *Candida albicans*. J Bacteriol 169:189–197, 1987.
147. JM Anderson, DR Soll. Unique phenotype of opaque cells in the white–opaque transition of *Candida albicans*. J Bacteriol 169:5579–5588, 1987.
148. JM Anderson, R Mihalik, DR Soll. Ultrastructure and antigenicity of the unique cell and pimple of the *Candida* opaque phenotype. J Bacteriol 172:224–235, 1990.
149. JM Anderson, L Cundiff, B Schnars, M Gao, I Mackenzie, DR Soll. Hypha formation in the white-opaque transition of *Candida albicans*. Infect Immun 57:458–467, 1989.
150. DR Soll. Dimorphism and high frequency switching in *Candida albicans*. In: DR Kirsch, R Kelly, MB Kurtz, eds. The Genetics of *Candida*. Boca Raton, FL: CRC Press, 1990, pp 147–176.
151. B Morrow, T Srikantha, DR Soll. Transcription of the gene for a pepsinogen, *PEP1*,

is regulated by white-opaque switching in *Candida albicans*. Mol Cell Biol 12: 2997–3005, 1992.
152. B Hube, M Monod, DA Schofield, AJP Brown, NAR Gow. Expression of seven members of the gene family encoding secretory aspartyl proteinases in *Candida albicans*. Mol Microbiol 14:87–99, 1994.
153. B Morrow, T Srikantha, J Anderson, DR Soll. Coordinate regulation of two opaque-phase-specific genes during white-opaque switching in *Candida albicans*. Infect Immun 61:1823–1828, 1993.
154. T Srikantha, DR Soll. A white-specific gene in the white-opaque switching system of *Candida albicans*. Gene 131:53–60, 1993.
155. T Srikantha, A Chandrasekhar, DR Soll. Functional analysis of the promoter of the phase-specific *WH11* gene of *Candida albicans*. Mol Cell Biol 15:1797–1805, 1995.
156. SR Lockhart, M Nguyen, T Srikantha, DR Soll. A MADS box protein concensus binding site is necessary and sufficient for activation of the opaque-phase-specific gene *OP4* of *Candida albicans*. J Bacteriol 180:6607–6616, 1998.
157. EP Rutschenko-Bulgac, F Sherman, JB Hicks. Chromosomal rearrangements associated with morphological mutants provide a means for genetic variation of *Candida albicans*. J Bacteriol 173:7436–7442, 1990.
158. T Suzuki, I Kabayashi, T Kanbe, K Tanaka. High frequency variation of colony morphology and chromosome reorganization in the pathogenic yeast *Candida albicans*. J Gen Microbiol 135:425–434, 1989.
159. DR Soll. Gene regulation during high-frequency switching in *Candida albicans*. Microbiology 143:279–288, 1997.
160. S Loo, J Rhine. Silencers and domains of generalized repression. Science 264: 1768–1771, 1994.
161. P Laurenson, J Rhine. Silencers, silencing and heritable transcriptional changes. Microbiol Rev 56:543–560, 1992.
162. J Perez-Martin, JA Uria, AD Johnson. Phenotypic switching in *Candida albicans* is controlled by a *SIR2* gene. EMBO J 18:2580–2592, 1999.
163. DR Soll. High-frequency switching in *Candida albicans* and its relations to vaginal candidiasis. Am J Obstet Gynecol 158:997–1001, 1988.
164. S Jones, G White, PR Hunter. Increased phenotypic switching in strains of *Candida albicans* associated with invasive infections. J Clin Microbiol 132:2869–2870, 1994.
165. C Kvaal, SA Lachke, T Srikantha, K Daniels, J McCoy, DR Soll. Misexpression of the opaque-phase-specific gene *PEP1 (SAP1)* in the white phase of *Candida albicans* confers increased virulence in a mouse model of cutaneous infection. Infect Immun 67:6652–6662, 1999.

19

Cryptococcus neoformans as a Model Fungal Pathogen

Klaus B. Lengeler and Joseph Heitman
Duke University Medical Center, Durham, North Carolina

1 PREFACE

In addition to bacteria, viruses, and parasites, fungi are emerging as an increasingly common threat to human health, primarily as a consequence of immunosuppression caused by AIDS or medical treatments including high-dose steroid therapy or cancer chemotherapy. The basidiomycete *Cryptococcus neoformans* was first identified by Sanfelice in 1894 [1] and shortly thereafter associated with human disease as an opportunistic human fungal pathogen by Busse and Buschke [reviewed in 2]. Over the last two decades this fungal pathogen has emerged as the leading cause of fungal meningitis and infects both immunocompromised and immunocompetent individuals [3–5]. *C. neoformans* has also emerged as one of the model systems for the study of pathobiology at all scientific levels. This chapter introduces the organism and assesses the current status of the most important issues of its pathobiology.

2 INTRODUCTION TO THE ORGANISM

2.1 The Life Cycle of *C. neoformans*

The asexual and sexual cycles of the homobasidiomycetous yeast *C. neoformans* have been defined (Fig. 1) [6–11]. The organism is primarily isolated as stable, haploid yeast cells that divide by budding (Fig. 2A). The cells are spherical to oval in shape, although atypical hyphal forms have been described, even in the early literature [1,12]. One distinct feature of *C. neoformans* that sets this fungus apart from other medically important fungi is the polysaccharide capsule. The capsule gives *C. neoformans* colonies a mucoid appearance and can be easily visualized by staining cells with India ink [13].

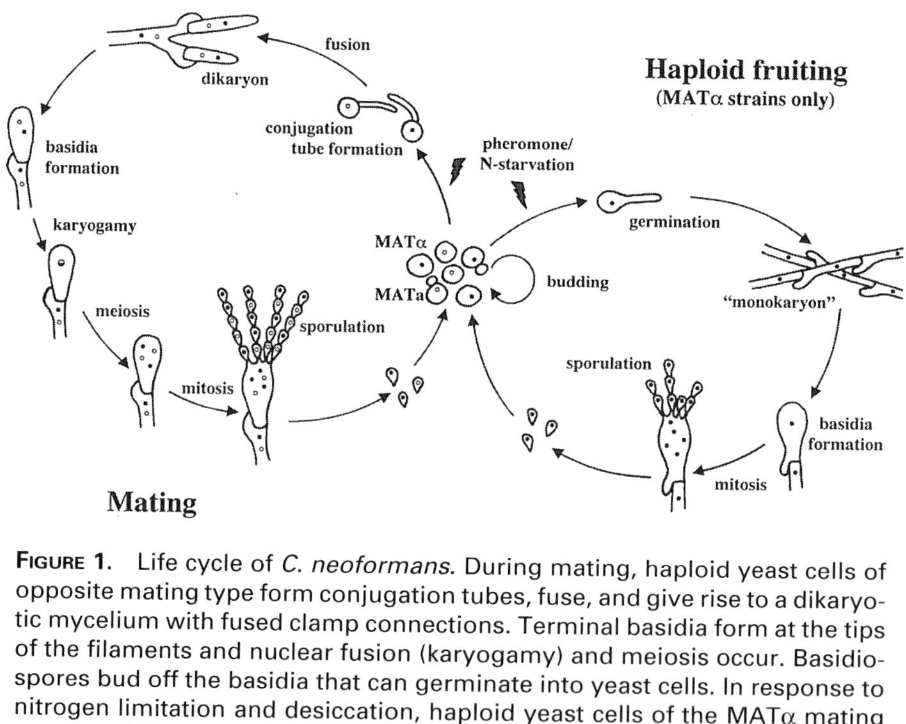

FIGURE 1. Life cycle of *C. neoformans*. During mating, haploid yeast cells of opposite mating type form conjugation tubes, fuse, and give rise to a dikaryotic mycelium with fused clamp connections. Terminal basidia form at the tips of the filaments and nuclear fusion (karyogamy) and meiosis occur. Basidiospores bud off the basidia that can germinate into yeast cells. In response to nitrogen limitation and desiccation, haploid yeast cells of the MATα mating type can differentiate into a monokaryotic mycelium with unfused clamp connections (haploid fruiting). Similar to mating, basidia are formed at the tips of the filaments and basidiospores that are exclusively of MATα mating type are produced. (From Ref. 200.)

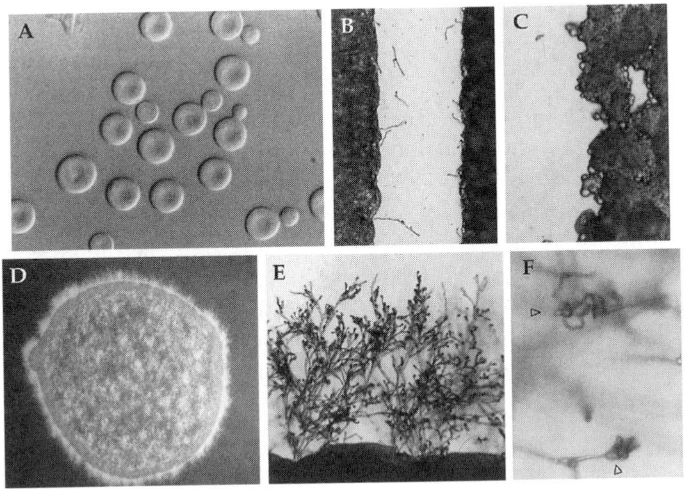

FIGURE 2. Morphological features of *C. neoformans. C. neoformans* is most frequently isolated as a haploid budding yeast (A, 250×). In the presence of the opposite mating type, *C. neoformans* forms conjugation tubes (B). In confrontation assays, cells of opposite mating type are streaked in parallel on nitrogen-deprived medium as close as possible without touching each other. Pheromones diffuse through the agar plate and will induce conjugation tube formation in the mating partner. Panel B shows the conjugation tube formation of MATa (right) and MATα cells (left) in a confrontation assay (25×). In addition to conjugation tubes, MATa cells respond to the opposite mating-type by forming enlarged, round cells (C, 50×). Panel C shows a close-up of MATa cells confronted by MATα cells in a confrontation assay. After cell fusion the perfect state of *C. neoformans, Filobasidiella*, is formed which can readily be observed in an in vitro mating reaction. Mating partners were coincubated on mating promoting V8 solid medium at 24°C. After several days the mating patch shows filament formation predominantly along the periphery (D). Numerous basidia can be found as swollen, refractive structures at the tip of the filaments (E, 25×). After prolonged incubation basidiospores bud off the basidia and form four chains of basidiospores (F; 50×, arrowheads).

2.1.1 The Sexual Cycle

The sexual cycle, first described by Kwon-Chung, involves mating between cells of opposite mating types called MATa and MATα [6–8]. In the presence of a mating partner and under nitrogen limiting conditions, *C. neoformans* first produces thin, filament-like projections called conjugation tubes that are typically

1–1.5 μm in diameter and can extend up to 10–20 cell lengths (Fig. 2B) [10,14,15]. These structures probably share similar functions with the morphological projections known as shmoos that are generated during the mating process of the yeast *Saccharomyces cerevisiae.* The formation of conjugation tubes has also been described for other fungi including *Ustilago maydis* [16], *Ustilago hordei* [17], *Rhodosporidium toruloides* [18], and, interestingly, *Tremella mesenterica* [19] and *Tremella brasiliensis* [20], two close relatives of *C. neoformans.* Conjugation tube formation is induced in response to diffusible pheromones secreted by the mating partner, as shown in confrontation assays and by exposing *C. neoformans* cells to synthetic pheromones [14]. Compared to MATα cells, MATa cells produce fewer mating projections in response to MATα pheromone, but instead produce enlarged, round, and refractive cells (Fig. 2C), which are rarely seen in MATα cells.

2.1.2 Cell Fusion

Following conjugation tube formation, cells of opposite mating type fuse. Whether cell fusion occurs at the tips of the conjugation tubes, along the lengths of the mating projections, or between the tip of a conjugation tube and an enlarged cell of opposite mating type is not known. In addition, it is unknown whether the initial cell fusion event in *C. neoformans* is at all indiscriminate, as occurs in other basidiomycetes in which filaments of identical mating type can fuse but further development is blocked by the absence of a nucleus of different mating type [21].

2.1.3 Dikaryon Formation

After cell fusion, *C. neoformans* elaborates stable, dikaryotic hyphae (Fig. 2D) that are ~3 μm in diameter and share features with the dikaryotic filaments of other basidiomycetes. The filaments are septate and the septa contain a dolipore complex, which is a barrel-shaped structure in the center of the septum [6–8]. Parenthosomes, membraneous structures surrounding the central pore complex, are not formed [22]. Clamp connections, which are structures associated with each septum that ensure proper distribution of both nuclei during mitosis, are formed along the filaments (see Fig. 3). Under proper environmental conditions the tips of these filaments differentiate into rounded, flask-shaped structures called basidia (Fig. 2E). Basidia preferentially form at the tip of filaments, although lateral basidia can be observed. The signals triggering basidium formation in *C. neoformans* are not known.

2.1.4 Nuclear Fusion and Meiosis

Following basidium formation, the two terminal nuclei migrate into the basidium where nuclear fusion (karyogamy) occurs. The signals that inhibit premature nuclear fusion during the initial mating processes and growth of the dikaryotic my-

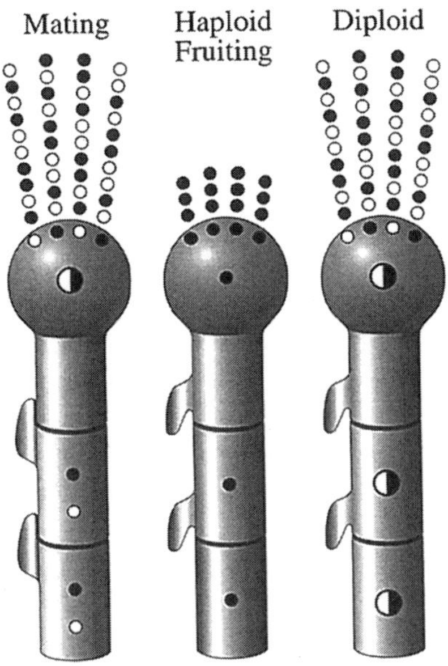

FIGURE 3. Features of *C. neoformans* filaments produced by mating, haploid fruiting, and diploid strains. Several different types of filaments are formed throughout the life cycle of *C. neoformans* that show characteristic features, which can easily be visualized by staining cell wall material with Calcofluor White and nuclei with ethidium bromide. Results of these staining experiments are summarized in the cartoon shown. Filaments produced in mating reactions (left) are dikaryotic showing two unfused nuclei, one of each parental strain, per cell. Clamp connections, structures associated with cell septa, are fused. Filaments produced by haploid fruiting on the other hand are uninucleate and show unfused clamp connections (middle). Finally, filaments produced by diploid strains are morphologically similar to those produced by haploid fruiting. Filaments have unfused clamp connections and cells are uninucleate, but nuclei are generally of larger size in comparison to haploid fruiting (right). Basidiospores isolated from haploid fruiting are genetically identical whereas recombinant basidiospores can be isolated from mating reactions and diploid strains. (From Ref. 11.)

celium are not known. After karyogamy, the diploid nucleus rapidly undergoes meiosis in the apical region of the basidium. In contrast to heterobasidiomycetes, no septa are formed within the basidium. The resulting four haploid nuclei undergo mitosis and each daughter nucleus migrates into a basidiospore that buds off basipetally at four distinct spots from the apex of the basidium [7]. This fixed budding pattern and subsequent rounds of mitosis give rise to four long chains of basidiospores, which can be easily observed under the microscope (Fig. 2F). Dissection of spore chains and analysis of genetic markers revealed that the spores of a single chain are not genetically identical [23]. This indicates that the meiotic nuclei migrate randomly into the budding basidiospores and are not channeled into a fixed spore chain. Finally, the basidiospores germinate to again produce vegetative yeast cells.

2.1.5 Diploid Isolates

The diploid phase of *C. neoformans* is normally transient and most environmental and clinical isolates are haploid. However, rare diploid isolates have been reported based on assays determining cellular DNA content [24–26], the analysis of RAPD markers [27,28], or isozyme analysis [29]. Furthermore, diploid strains have been isolated from basidiospores with a frequency of up to 5–10% following defined genetic crosses [11,30,31]. The mechanism underlying the generation of diploid basidiospores is unknown. Interestingly, atypical basidia have been observed which produce only two spores twice as large as normal, or three spores with one spore having twice the size as the two remaining spores [7]. These larger basidiospores could contain a diploid nucleus.

Stable, well-defined diploid strains should be very useful in the molecular analysis of essential genes. Diploid strains of *C. neoformans* can be transformed like haploid strains using either dominant selective or auxotrophic markers [11]. In addition, basidiospores produced from these diploid strains are viable and haploid. Therefore, it should be feasible to establish the use of diploid strains as a molecular tool similar to tetrad analysis in *S. cerevisiae* in order to analyze genes that are essential in *C. neoformans*. Essential genes can be disrupted in the diploid strain background to produce heterozygous strains. After sporulation and spore dissection haploid progeny can then be tested for linkage between gene disruption and viability of spores.

2.1.6 How Might Diploid Strains Be Generated?

Several mechanisms may result in the generation of diploid basidiospores in *C. neoformans*. First, a random nuclear fusion event at the end of the meiotic process would result in diploid meiotic products. A similar scenario is found in higher plants during ovary development where two haploid nuclei fuse to form the diploid endosperm. In addition to MATa/MATα spores, this postmeiotic nuclear fusion event should also result in the production of MATa/MATa or MATα/MATα diploid basidiospores. Second, following karyogamy an endomitotic event could give rise to a 4n nucleus that would result in diploid spores after completion of meiosis. Several examples of endoreplicative events have been found in nature but the best known is the production of polytene chromosomes in the salivary glands of dipteres such as *Drosophila melanogaster*. Finally, a defect in either the first or second meiotic division that would inhibit the segregation of the chromosomes into separate nuclei could also produce one or two diploid basidiospores. Therefore, factors preventing proper meiosis such as a high degree of genetic variability may dramatically increase the ratio of diploid/aneuploid to haploid spores produced in a defined genetic cross.

2.1.7 Diploid Strains Are Thermally Dimorphic

Diploid strains of *C. neoformans* show an interesting morphological feature: they grow as budding yeasts at 37°C yet spontaneously filament and produce basidia and basidiospores when grown at lower temperatures (24°C; Fig. 4). Using diploid strains heterozygous for different markers it was shown that meiosis occurs in the basidia, giving rise to recombinant, haploid spores [11]. The filaments produced by diploid strains can readily be distinguished from filaments that arise from mating. First, in contrast to the dikaryotic mycelium produced during mating, cells in the diploid mycelium are uninucleate (see Fig. 3). Second, the clamp connections are fused during mating and unfused in the diploid filaments. During mitosis in the dikaryon one nucleus divides along the main cell axis while the second nucleus divides along the axis of the newly formed clamp cell leaving one daughter nucleus in the clamp cell. Fusion of the clamp cell with the postapical cell releases this nucleus resulting in a dikaryotic mother and daughter cell. In contrast, in diploid cells only a single nucleus is present in the filaments and no clamp cell fusion occurs, likely because there are no separate MATa and MATα nuclei to signal to each other and promote fusion. Finally, in contrast to

FIGURE 4. Diploid strains of *C. neoformans* are thermally dimorphic. Diploid strains of *C. neoformans* exhibit a temperature-dependent dimorphic switch. At 37°C diploid *C. neoformans* strains grow as yeast cells (right panel) but switch to a filamentous growth form upon incubation at 24°C (left panel). Strains were grown for several days on synthetic low ammonium galactose (SLAG) medium at the indicated temperatures and the edges of the colony were photographed (25×).

the dikaryon that requires nitrogen starvation for its formation during the mating process, diploid strains filament on nitrogen-rich medium.

It is interesting to note that thermally regulated yeast-to-mycelium transitions have been described for other human pathogenic fungi, including *Blastomyces dermatitidis*, *Histoplasma capsulatum*, and *Paracoccidiodes brasiliensis* [32–35]. These fungi normally grow as saprophytic molds in soil, and cause respiratory and deep tissue infections after spores or conidia are inhaled by susceptible hosts. An increase in temperature results in the formation of the yeast form of these organisms, in contrast to the filamentous form found in soil, and this transition to the yeast form is necessary for virulence.

2.1.8 Serotype AD Strains—Natural Diploid Isolates of *C. neoformans*

C. neoformans can be classified into four predominant serotypes, A, B, C, and D. In addition, less commonly, clinical and environmental isolates have been reported to be of an unusual serotype AD and these isolates were suggested to be diploid [24–29]. In fact, recent analysis of several clinical and environmental isolates confirmed that by FACS analysis serotype AD strains are aneuploid or diploid [25,26,36]. In addition, PCR analysis with primers specific for serotype A or D alleles of several genes revealed both alleles are often present in serotype AD strains [36]. PCR analysis with primers specific to mating type also revealed that serotype AD strains are heterozygous for the mating-type locus [36]. Furthermore, some serotype AD strains exhibited a thermal dimorphism shown to be typical of diploid strains [11]. On the other hand, hybrid strains could easily be isolated following a laboratory cross between a serotype A and D strain. In summary, serotype AD strains of *C. neoformans* are unusual aneuploid or diploid strains that most likely result from matings between serotype A and D strains.

2.1.9 Haploid or Monokaryotic Fruiting

In addition to the two yeast-to-filament transitions that occur during mating or in diploid strains, MATα strains of *C. neoformans* can undergo a third morphological switch called haploid fruiting (Fig. 5). Haploid fruiting was described as early as 1936 by Todd and Herrmann, who noted filamentlike projections from cryptococcal yeast cells recovered from older, dry plates [37]. Subsequently, other researchers also described filaments and basidialike structures independent of mating reactions [3]. Erke suggested that filament production was due to homothallism of some *Cryptococcus* spp. [38]. Kwon-Chung on the other hand explained this morphological transition by self-fertility of heterothallic strains [39]. In 1996 Wickes et al. showed that *C. neoformans* undergoes this morphological transition when grown under nitrogen starvation conditions and severe desiccation [9]. This process, which also occurs in other basidiomycetes, is referred to

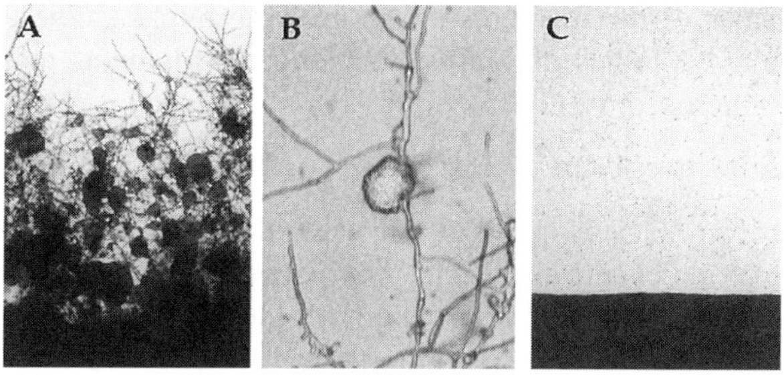

FIGURE 5. Haploid fruiting in *C. neoformans* is linked to MATα mating type. Only MATα cells of *C. neoformans* produce filaments under nutrient-deprived conditions. The MATα strain JEC21 was grown on nutrient-poor filamentation agar (FA) at 24°C. After 10 days of incubation haploid filaments arise from the edge of the colony (A, 10×). Blastospores bud off the filaments at regular intervals that form microcolonies after prolonged incubation (B, 50×), giving the filaments a "beads on a string"–like appearance (A). In contrast, cells of MATa mating type (JEC20) grown under the same conditions do not haploid fruit even after prolonged incubation (C, 10×).

as haploid or monokaryotic fruiting [40,41]. Haploid fruiting in *C. neoformans* was found to be restricted to strains of only the MATα mating type (Fig. 5).

Although haploid fruiting occurs to a limited extent in some MATα strains in response to nitrogen limitation and dessication alone, recent studies reveal that haploid fruiting of MATα strains is dramatically induced by confrontation with MATa cells [42]. This stimulating effect does not require cell–cell contact, and can occur through a dialysis membrane. These findings provide support for the hypothesis that MFa pheromone secreted by MATa cells diffuses and stimulates haploid fruiting of adjacent MATα cells. These findings further suggest that the role of haploid fruiting is linked to the early steps of the sexual cycle, likely to promote the location of distant MATa mating partners by filaments and basidiospores produced by MATα cells.

Several characteristics of the filaments generated by haploid fruiting are distinct from dikaryotic mating hyphae but similar to the filaments produced by diploid strains (Fig. 3). Nuclear staining revealed that cells in haploid filaments are uninucleate although terminal cells may be dikaryotic in some strains [38]. Clamp connections in haploid filaments are unfused. In comparison to mating dikaryons, basidia are less frequently observed in haploid fruiting, and spore chains are generally quite short. In contrast to the recombinant spores that arise from mating or diploid filaments, spores produced by haploid fruiting are genetically identical. A unique feature of haploid fruiting is the production of blastospores that form off the sides of the filaments of the growing hyphae. These spores replicate and form microcolonies of yeast cells along the filaments that can readily be observed by microscopy (Fig. 5).

2.1.10 Are Meiotic Events Involved in Haploid Fruiting?

It is not known if the production of basidiospores during haploid fruiting is strictly mitotic or whether cryptic meiotic events also occur. For example, haploid fruiting could involve the formation of MATα/MATα homozygous diploid nuclei, either prior to filamentation or in the basidia. *S. cerevisiae* strains that are homozygous for the MAT loci (MATa/a or MATα/α) usually do not sporulate, but mutations in the *RME1* gene, which encodes a repressor of meiosis, allow a/a or α/α homozygous strains to sporulate. By analogy, other regulatory mechanisms could allow *C. neoformans* α/α strains to undergo meiosis and produce all haploid MATα spores. Thus, the very restricted haploid fruiting that occurs in wild-type MATα strains after prolonged incubation could result from the formation of MATα/MATα strains. The question of whether meiotic events are involved in the process of haploid fruiting could be assessed by several experiments. First, if meiosis occurs during haploid fruiting, mutations in genes that are required for karyogamy could inhibit either filamentation or sporulation. Similarly, mutations in genes that are involved in proper progression through meiotic events might inhibit sporulation or result in the production of unusual aneuploid strains. Finally, the in vitro generation and analysis of MATα/MATα diploid strains might provide direct proof whether the formation of a diploid nucleus promotes haploid filamentation in *C. neoformans*.

2.2 Taxonomy

2.2.1 *C. neoformans*—A True Basidiomycete

In the late 1960s Shadomy and coworkers made the observation that some *Cryptococcus* strains produced hyphae with clamp connections [43,44]. This observation led to the discovery of the sexual cycle of *C. neoformans* by Kwon-Chung as the filamentous, perfect state (teleomorph) *Filobasidiella neoformans* [6–8]. Morphological examination of these filaments showed for the first time that the genus *F. neoformans* belonged to the Basidiomycota. Dolipores and clamp connections, both structures that are highly distinctive for this phylum of fungi, were identified. In addition, subsequent analysis of small rDNA sequences also confirmed that *F. neoformans* is a basidiomycete [45,46].

The genus *Filobasidiella* contains two different species, *F. neoformans* and *F. depauperata*, which are very different in ecology, biochemistry, and pathogenicity. *F. depauperata* grows exclusively in a filamentous form, and no yeastlike growth form has been identified. Surprisingly, filaments of *F. depauperata* show no clamp connections [47]. Despite these differences, DNA sequence analysis of parts of the rDNA cluster clearly show that *F. depauperata* and *C. neoformans* are closely related [46,48]. Based on morphological studies and 5.8S rDNA analysis, *F. neoformans* was previously thought to be a member of the Filobasidiaceae with its closest relatives being *Filobasidium* and *Cystofilobasidium*. In contrast, subsequent analysis of 16S and 5S rDNA sequences showed that *Filobasidiella* is more closely related to the genera *Tremella*, *Bullera*, and *Trichosporon*, all members of the family Tremellacaeae [45].

Besides *C. neoformans*, the anamorph of *F. neoformans*, there are at least 38 other *Cryptococcus* spp. found in a wide variety of ecological niches. Some species show striking adaptations to extreme environmental conditions such as low temperatures or high salt concentrations [49]. Despite their ability to adapt to various conditions in nature, only a few other *Cryptococcus* spp. have been associated with or linked to human disease. Most cryptococcal species seem to be very susceptible to the elevated growth temperature in the host and to the immune system.

2.2.2 *C. neoformans* Occurs in Three Distinct Varieties

One of the most interesting features of *C. neoformans* is that the organism has evolved into three distinct varieties or species, all of which are pathogenic in humans and which have distinct and interesting physiological properties and unique features of virulence (Table 1) [3]. These different forms of *C. neoformans* are commonly distinguished by differences in capsular polysaccharide antigens that were recognized as early as 1935 and can be detected by the ability of antisera to induce agglutination [50,51]. This morphological reaction serves as the basis for a clinical microbiology test referred to as serotyping. *C. neoformans* occurs in four predominant serotypes—A, B, C, and D [52,53]. As mentioned above, a fifth, less common serotype, AD, represents diploid or aneuploid strains that most likely have arisen from intercrosses between serotype A and D strains [36]. While serotyping detects only minor chemical differences in the capsular polysaccharide antigens, serotype A, serotype D, and the serotype B and C strains are significantly diverged and are classified as distinct varieties [8,54,55]. Serotype A strains are designated *C. neoformans* var. *grubii*, serotype D strains are designated *C. neoformans* var. *neoformans*, and serotype B and C strains are designated *C. neoformans* var. *gattii*.

2.2.3 Genetic Divergence of *C. neoformans* Varieties

DNA sequence analysis indicates that the three varieties are closely related but already show a significant level of sequence divergence, particularly between serotypes A and D compared to B and C. Comparison of the *URA5* gene revealed up to 8% sequence divergence between varieties *gattii* and *neoformans* or *grubii* [56]. On the other hand, variety *grubii* and *neoformans* exhibit up to 5% of sequence divergence for the *URA5* gene [57–59]. Analysis of the 8-kb rDNA cluster

TABLE 1 Characteristics of *C. neoformans* Varieties

	C. neoformans var. *grubii*	*C. neoformans* var. *neoformans*	*C. neoformans* var. *gattii*	References
Perfect state	*Filobasidiella neoformans* var. *neoformans*		*F. neoformans* var. *bacillispora*	6–8,97
Serotype	A	D	B and C	3,52,53,74
Main geographic distribution	worldwide	Europe	tropical and subtropical regions	3,74
Main ecological niche	pigeon droppings hollows of trees		Eucalyptus trees (red and forest river gum)	75–82,84–87
Basidiospore morphology	spherical or cylindrical shape rough spore wall		bacillary smooth spore wall	6–8,96
CGB growth	no	no	yes	93
GCP growth	no	no	yes	94
CDBT growth	no	yes (orange)	yes (blue-green)	95
D-proline assimilation	no	no	yes	89
Creatinine assimilation	repressed by NH_3	repressed by NH_3	not repressed by NH_3	90
Urease inhibited by EDTA	yes	yes	no	92
Laccase inhibition	$(NH_2)_4SO_4$/glutamine	glutamine	glutamine (not serotype B)	91
Average chromosome number	12.1	12.8	13	63,64
Smallest chromosome	~700 kb	~700 kb	400 kb–700 kb	63
G + C content	53%–56%	53%–56%	56%–57%	62

CGB, canavanine–glycine–Bromothymol Blue; GCP, glycine–cycloheximide–Phenol Red; CDBT, creatinine–dextrose–Bromothymol Blue–thymine.

of *C. neoformans* revealed almost no sequence divergence within the rRNA genes of the cluster [60] but significantly reduced sequence identity in the noncoding (ITS) regions (~78.5%) [61]. This observation, and the fact that serotype A and D have a slightly different G + C content (53–56% vs. 56–57% in serotype B and C), may account for the finding that the overall DNA relatedness of the *gattii* variety and *grubii* and *neoformans* varieties is as low as 55–63% [62]. The finding that the rRNA genes in the rDNA cluster are almost 100% identical suggests recent evolutionary divergence. Recent population genetics studies suggest that serotype A and D strains diverged ~20 million years ago, and the serotype B and C variety diverged from the A and D strains even earlier, ~40 million years ago [54]. Serotype B and C strains diverged from a common ancestor most recently, ~10 million years ago.

DNA typing studies revealed significant differences between serotype B and C compared to serotypes A or D. Based on pulsed-field gel electrophoresis, the average chromosome number varies slightly among serotypes, ranging from ~8–12 (serotype A and D) or 11–13 (serotype B and C), respectively [63]. In addition, the smallest chromosome found in serotype B or C strains ranges from 400 kb to 700 kb, whereas the smallest chromosome found in serotype A or D is on average at least 700 kb. Karyotype analysis remains challenging in *C. neoformans* because of the high variability in chromosome patterns found within a single serotype and a relatively high degree of genome instability that can lead to chromosome rearrangements [63,64]. Hybridization with repetitive elements such as CNRE-1 revealed distinct restriction fragment length polymorphism (RFLP) patterns for *C. neoformans* that are similar within a given serotype but are specific to each variety [65,66]. Similar results were obtained from RFLP analysis of mitochondrial DNA [67]. Multilocus enzyme electrophoresis (MEE), a technique to monitor the mobility of enzymes in gel electrophoresis, also revealed differences in enzyme patterns that can be used to identify serotypes of *C. neoformans* strains [29]. In PCR fingerprinting studies by random amplified polymorphic DNA (RAPD), arbitrary single primers are used to identify polymorphic sequence structures of genomic DNA from individual strains. Over the years several synthetic oligonucleotides have been reported that can be used to distinguish among the three varieties of *C. neoformans* based on specific differences in PCR fingerprinting patterns generated by RAPD analysis [68–71]. Most recently, serotype has been established with PCR primers that can distinguish sequence polymorphisms among serotype-specific alleles [36,72,73].

2.2.4 Ecological and Pathological Differences of *C. neoformans* Varieties

Serotype A and D strains are distributed throughout the world [3,74] and are associated with soil contaminated with pigeon droppings [75–78]. More recent ecological studies also associated serotype A strains with several tree species

[79–81]. The most common clinical isolates are the serotype A strains, and in AIDS patients >99% of isolates are serotype A. Serotype D strains are capable of clinical infection and in some regions of the world, such as France, can represent up to 30% of clinical infections [82]. For example, recent studies by Casadevall and coworkers revealed that the proportion of cryptococcal meningitis infections associated with serotype D strains may be higher in New York City than previously appreciated, and is now as much as 12.5% [83]. The serotype B and C variety *gattii* strains are found in tropical or subtropical regions of the world [3,74]. In contrast to varieties *grubii* and *neoformans*, variety *gattii* has not been isolated from bird droppings but is almost exclusively associated with two species of flowering eucalyptus trees, *Eucalyptus camaldulensis* (river red gum) and *Eucalyptus tereticornis* (forest red gum) [84–87]. The association of all *C. neoformans* varieties with trees suggests a primary ecological niche for the species. *C. neoformans* var. *grubii* and *neoformans* primarily infect immunocompromised hosts, including AIDS patients and patients on high-dose steroids or immunosuppressive therapies for organ transplants and cancer chemotherapy. In contrast, *C. neoformans* var. *gattii* predominantly infects immunocompetent hosts and seldom afflicts AIDS patients [88]. While all three varieties infect the central nervous system (CNS), *gattii* strains more commonly cause multiple space-filling lesions in the CNS, hydrocephalus, and papilledema, and higher titers of cryptococcal antigen. While the course of infection with *gattii* strains is often more indolent, neurological deficits are more common following clearing of infection.

2.2.5 Biochemical Differences in *C. neoformans* Varieties

A number of biochemical differences have been described that can be used to distinguish among serotypes in *C. neoformans*. The ability to assimilate various compounds such as D-proline, L-malic, or fumaric acid are slightly different in serotype A and D versus serotype B and C strains [3,5,89]. Other metabolic pathways are present in all serotypes, yet differentially regulated. For example, assimilation of creatinine is repressed by NH_3 in serotype A and D, but not in serotype B or C [90]. Laccase activity is repressed by glutamine in *C. neoformans* serotype A, D, and C strains and is repressed by ammonium sulfate in only serotype A strains [91]. Urease, a key enzyme for the identification of *C. neoformans*, is susceptible to treatment with the ion chelator EDTA in serotype B or C strains but not in serotype A or D strains [92]. Finally, several colorimetric agar tests have been developed to distinguish between varieties of *C. neoformans* based on their ability to assimilate glycine or thymine or their susceptibility to the drug L-canavanine or low concentrations of cyclohexamide [93–95]. The most widely used assay is based on the observation that the majority of serotype A or D strains cannot utilize glycine as the single carbon and nitrogen source and are susceptible to L-canavanine. Therefore, only serotype B and C strains will grow on CGB agar (canavanine–glycine–Bromothymol Blue) and produce blue-green colonies as a result of changes in pH caused by ammonia release [93].

2.2.6 Varieties of *C. neoformans* Show Morphological Specificities but Are Interfertile

Early studies found that basidiospores produced from serotype A or D were spherical to cylindrical with a finely roughened surface visible by electron microscopy. Serotype B and C strains on the other hand produced elongated, bacillus-shaped basidiospores with a smooth appearance. These morphological differences led to the original introduction of *F. neoformans* as the teleomorph of *C. neoformans* var. *grubii* or *neoformans*, and *F. bacillispora* as the perfect state of *C. neoformans* var. *gattii* [6–8]. Classic genetic studies suggest these different serotypes are all interfertile, although viability of the resulting basidiospores is dramatically reduced [96]. Still, the ability to produce viable progeny led to the classification of the serotypes as different varieties of a single species [97]. Interestingly, basidiospores produced from crosses between serotypes A or D and serotypes B and C are a mixed population of A/D and B/C type basidiospores. Unusual hybrid strains can easily be isolated from these intervarietal crosses that are aneuploid or near diploid, suggesting that differences in DNA sequence or genome structure may interfere with proper chromosome segregation during meiosis [36]. The finding that serotype AD isolates are aneuploid or near diploid [36] demonstrates that, at least between serotype A and D strains, intervarietal crosses do occur in nature.

All of the differences listed in the preceeding sections, and recent molecular characterization of mutants whose phenotypes differ significantly between serotypes A and D [98–100], illustrate how these organisms have diverged into distinct varieties and in the future may either evolve into separate species, or be reclassified as different species. In fact, recent population genetic studies suggest that serotypes A, D, and B or C are separate species with only rare recent hybridization and little evidence of recombination between serotypes [54].

3 VIRULENCE OF *CRYPTOCOCCUS NEOFORMANS*

C. neoformans mainly infects the central nervous system, causing a severe meningoencephalitis. Despite the presence of large amounts of fungal organisms, the host inflammatory response is limited. In previous classic studies the virulence of *C. neoformans* has been linked to the ability to produce a characteristic polysaccharide capsule that inhibits phagocytosis and promotes survival in macrophages [101,102]. A second specialized virulence factor involves the synthesis of the pigment melanin by the enzyme laccase. Melanin may serve as an antioxidant to protect fungal cells from oxidative products of macrophages [103]. Other factors linked to virulence include the ability to grow at 37°C [98,104,105], prototrophy [106], and the production of the enzymes urease and phospholipase B [107–109]. In addition, the MATα mating type has been linked to prevalence in the environment and virulence of *C. neoformans* [110,180]. The following section will focus on some of the most important virulence factors illustrating the complexity of virulence in a pathogen like *C. neoformans*.

3.1 The Capsule

Probably the most striking feature of *C. neoformans* that distinguishes this organism from other pathogenic yeasts is the large polysaccharide capsule that surrounds the organism (Fig. 6) [111]. The capsule size can range from 1 to 50 μm depending on strain background and environmental conditions. *C. neoformans* cells normally produce much larger capsules in vivo compared to in vitro even when capsule formation is induced in vitro [112]. The capsule is essential for virulence, and all acapsular mutants studied have been found to be avirulent. No correlation between capsule size and virulence was apparent from studies of nonisogenic clinical isolates of *C. neoformans* [113,114], but recent experiments using congenic *Cryptococcus* strains in animal virulence models strongly suggest that capsule size can be correlated with the pathogenicity of the organism [115,116].

3.1.1 Capsule Composition and Regulation

The majority of the polysaccharide capsule (90%) is glucuronoxylomannan (GXM), a linear chain of α-1,3-linked mannose residues that may be substituted

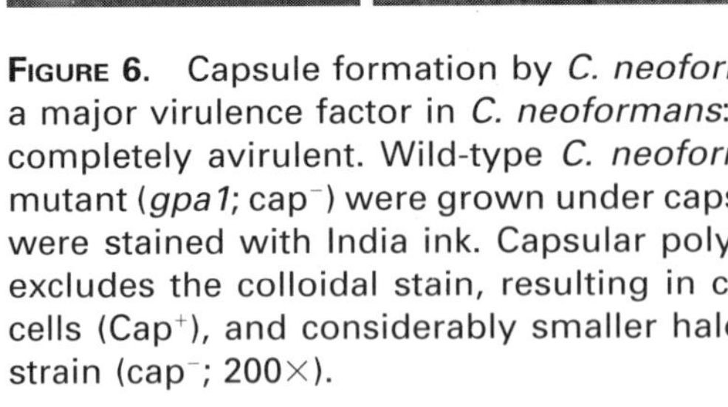

FIGURE 6. Capsule formation by *C. neoformans*. Capsular polysaccharide is a major virulence factor in *C. neoformans*: acapsular mutants are generally completely avirulent. Wild-type *C. neoformans* (Cap$^+$) and a hypocapsular mutant (*gpa1*; cap$^-$) were grown under capsule-inducing conditions and cells were stained with India ink. Capsular polysaccharide surrounding the cells excludes the colloidal stain, resulting in clear halos surrounding the yeast cells (Cap$^+$), and considerably smaller halos with the hypocapsular mutant strain (cap$^-$; 200×).

with singular xylose or β-glucuronyl sidegroups or can be O-acetylated [51,117]. The number and composition of side groups and the degree of acetylation largely define the serotypes of *C. neoformans*. Besides GXM, smaller amounts of galactoxylomannan (GalXM) and mannoproteins are present. The capsular polysaccharide forms a complex network of fibrillar structures that surrounds the cell but, surprisingly, does not appear to be covalently linked to the cell wall [118]. Capsule material can easily be sheared off the yeast cell. Interestingly, capsular polysaccharide added to acapsular strains reassociates on the surface of these cells and restores virulence of these otherwise avirulent strains [119–123]. The capsule is thought to be a very dynamic structure that is constantly changing with the growing cell, and large amounts of capsular material are released into the environment.

Several environmental factors have been identified that regulate capsule production. Whereas high salt concentrations (~1 M) repress capsule production [124], low levels of glucose [125], iron deprivation [126], or physiological CO_2 concentrations [127] dramatically induce capsule production. It is interesting to note that these conditions all mimic the environmental conditions *C. neoformans* encounters in the host. Little is known about the biochemical pathways for capsular polysaccharide synthesis [111]. By complementing acapsular mutants of *C. neoformans*, four genes have been identified that are involved in capsule formation—*CAP59*, *CAP60*, *CAP64*, and *CAP10*. Mutations in any of these genes resulted in acapsular mutant strains that were completely avirulent [101,102,128–130].

3.1.2 Biological Activity of Capsular Polysaccharide

Cell-mediated immunity is thought to be the most important defense mechanism for a host infected with *C. neoformans*. In early stages of infection neutrophils are the primary line of defense. In later stages, monocytes and monocyte-derived macrophages are more important [131,132]. These phagocytotic cells engulf cryptococcal cells and kill them through oxidative burst [133]. The activity of neutrophils during acute inflammations, but especially of monocytes and macrophages during chronic infections with *C. neoformans*, is modulated by various subsets of T-cells via cytokine signaling processes. Therefore, individuals with reduced T-cell numbers, such as AIDS patients or patients receiving immunosuppressive drugs, are highly susceptible to *C. neoformans* infections [134]. The polysaccharide capsule is thought to protect the yeast cells from phagocytosis by activated immune cells and to promote intracellular survival by inhibiting digestion in the phagolysosomes [135–137]. Acapsular strains are phagocytosed by macrophages more rapidly then encapsulated strains and are then destroyed inside the phagolysosome. This seems to be in contradiction to the finding that capsular material activates the complement system and antibody-mediated defense mechanisms as both of these mechanisms are thought to increase the ability of activated immune cells to phagocytose antigens [138]. On the other hand, GXM induces a Th2-type immune response that can repress Th1-type cellular immunity normally activated by *C. neoformans* infection [139]. In addition, the polysaccharide capsule is a physical barrier that prevents complement components from associating with the cell wall and thereby inhibits complement-dependent cell lysis.

Furthermore, large amounts of capsule material are shed into the blood system and suppress the immune system. The capsule material is thought to deplete the immune system of complement components [140]. Several other mechanisms have been proposed for the interference of capsular polysaccharide with the immune system, including the inhibition of leucocyte migration [141,142], dysregulation of cytokine production [143–145], or interference with antigen presentation [146]. It should be noted that the majority of the virulence factors may not have been acquired by the pathogen to protect the organism from the host defense mechanisms but to protect from environmental threats. In this regard, the capsule produced by *C. neoformans* is thought to protect the organism from desiccation or engulfment by amoebae [147]. On the other hand, if *C. neoformans* has an as yet unknown animal host, the capsule may be a true virulence factor.

3.2 Melanin

A second characteristic feature of *C. neoformans* that is correlated with virulence is the ability to produce a dark pigment called melanin (Fig. 7) [148,149]. *C. neoformans* melanin is a very heterogeneous, high-molecular-weight molecule that is synthesized from diphenolic compounds, such as L-dopa, caffeic acid or chlorogenic acid, through the Mason-Raper pathway modified by Ito [150]. In contrast to other melanin-producing organisms, *C. neoformans* is unable to produce melanin de novo from L-tyrosine and therefore is completely dependent on precursors provided by environmental sources. Melanin production can therefore

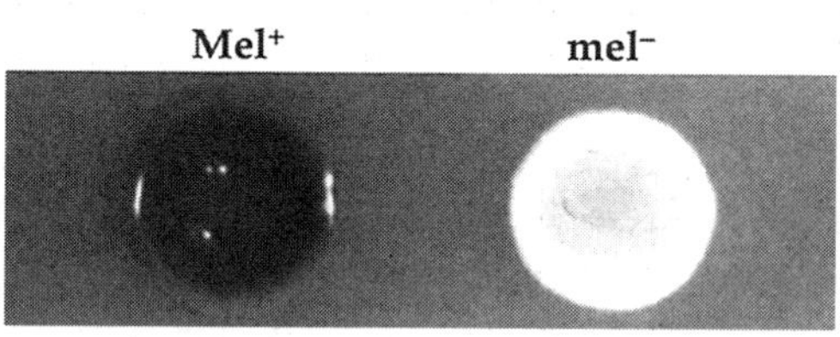

Figure 7. Melanin production by *C. neoformans*. A second important virulence factor of *C. neoformans* is the ability to produce melanin. If provided with diphenolic compounds, which are present in birdseed extract (*Guizotia abyssinica*), wild-type *C. neoformans* cells produce the dark-brown pigment melanin (Mel$^+$). Strains defective in melanin production on the other hand show an albinolike appearance (*pka1*, mel$^-$).

be used in diagnostic tests to identify *C. neoformans* by its ability to convert diphenolic compounds into melanin [151–154]. The actual pigment color can vary depending on the type of chemical precursor provided. Melanin is deposited into the inner layers of the cell wall and may play a role in cell wall integrity [155].

Whether *C. neoformans* produces melanin in vivo has been the subject of much debate. In elegant experiments, Nosanchuk and colleagues recently provided definitive evidence that *C. neoformans* deposits melanin into its cell wall in vivo [156]. In cross sections of brain tissues from patients with AIDS-associated cryptococcal meningoencephalitis, monoclonal antibodies against melanin reacted strongly with the cell wall of fungal yeast cells present in the tissues. Melanin ''ghosts'' can be isolated from melanized cells by treatment with enzymes, denaturants, and acid [157]. Treatment of infected brain tissues with acid resulted in the isolation of particles that strongly reacted with the melanin-binding antibodies and were similar in size and shape to the melanin ghosts previously isolated in in vitro experiments and from rodents infected with the fungal pathogen [157]. Strains that are not able to produce melanin (mel$^-$; Fig. 7) are less virulent [103,158–161]. In addition, strains that were recovered from animals infected with mel$^-$ strains of *C. neoformans* were mostly found to be Mel$^+$ revertants. These findings clearly demonstrate that melanin is important for virulence of *C. neoformans*.

Genetic analysis of mel$^-$ strains identified seven complementation groups involved in pigment production [162]. One of them is the enzyme that catalyzes the rate-limiting step in the biosynthesis of melanin and was found to be a copper-containing phenoloxidase called laccase [163]. The corresponding gene, *CnLAC1*, is induced by various signals: low glucose levels, low temperature, the presence of diphenolic precursors, and stationary growth phase. Interestingly, the brain, the most prominent site of infection, provides almost optimal conditions for melanin production—low glucose levels but high L-dopamine concentrations—and the infectious yeast cells are probably in a growth phase similar to early stationary growth [164]. On the other hand, high body temperature represses *CnLAC1* transcription to some extent, but simultaneously it was found that laccase enzyme activity increases at elevated temperatures [103,152]. As melanin is thought to protect *C. neoformans* cells from oxygen radicals, it is of interest that activity of superoxide dismutase is also increased at elevated temperatures [165].

The biological function of melanin in *C. neoformans* in its environmental niche is not known. Besides a possible role in cell integrity [166], protective functions against oxidative damage induced by UV light [167] or extreme temperature [168] have been proposed. In addition, laccase may enable the organism to degrade lignin and use it as a carbon and nitrogen source [169]. In mammalian hosts melanin most likely protects *C. neoformans* from oxidative damage by neutrophils and macrophages (oxidative burst), which has been demonstrated to be the case in several in vitro studies [160,166,170–172]. In general, strains lacking melanin are much more susceptible to oxygen- or nitrogen-derived oxidants. Melanin might also contribute to virulence by altering the net charge of the cell wall to be more negatively charged [173]. Finally, deposition of melanin in the cell wall may confer increased resistance to amphotericin-B [174].

3.3 Mating Type

Unlike many other basidiomycetes, there are only two mating types in *C. neoformans*: MATa and MATα. The majority of strains isolated from the environment are of the MATα mating type, and virtually all clinical isolates are MATα [110]. In fact, the MATa mating type of *C. neoformans* var. *grubii* has only very recently been identified in a serotype A clinical isolate from Tanzania (strain 125.91) [73]. The finding that only MATα strains undergo haploid fruiting (Fig. 5) reveals a link between mating type and virulence [9], and basidiospores are thought to be the infectious propagule because their size is suitable for deposition into alveoli following inhalation. In addition, MATα strains are more virulent than MATa strains in animal models. It was shown that in a pair of congenic serotype D MATa/MATα strains infection with MATα strains resulted in a more rapid progression to lethal infection than congenic MATa strains [180]. These findings have focused interest on the structure and function of the mating-type loci and their link to the physiology and virulence of this pathogenic fungus.

Moore and Edman implemented a difference cloning method to isolate DNA unique to MATα cells [15]. They identified an ~35-kb region of DNA that is specific to MATα cells and which defines part of the MATα mating-type locus. The mating-type locus was found to contain a mating pheromone precursor gene *MFα1*, which stimulates the production of filaments similar to conjugation tubes when introduced into MATa cells by transformation. A second region of the MATα locus was identified by Wickes and colleagues, who identified a gene encoding a Ste12 transcription factor homolog via its ability to stimulate haploid fruiting of MATα cells [175]. Similar to *MFα1*, the *STE12* gene was found to be unique to MATα cells and was therefore named *STE12α*. Most interestingly, the *STE12α* gene was not contained within the ~35-kb region of the MATα locus that had been previously identified by Moore and Edman, indicating that the MATα locus was significantly larger than previously realized. Further analysis by various laboratories revealed the MATα mating-type locus in *C. neoformans* spans an ~50-kb region. In addition to *MFα1* and *STE12α*, several other MATα mating type–specific genes have been identified within the ~50-kb region including genes that encode for homologs of a putative MAP kinase pheromone response pathway [14,73,99,176]. Similar to the recently identified *Candida albicans* mating type–like loci MTLa and MTLα [177–179], additional genes are also present that are not thought to be involved in mating or virulence.

The MATa mating-type locus has also recently been identified, but much less is known about its structure [73]. Preliminary data, however, suggest that divergent alleles of genes found in the MATα locus are also present in the MATa locus. For example, mating type–specific *STE20a* and *STE20α* genes share only ~67% DNA sequence identity [73]. In addition, the gene order appears to have rearranged between MATa and MATα mating type since their divergence from a common ancestral region of DNA.

The *C. neoformans* mating-type loci are distinct from other known mating-type loci that have been previously characterized from model and pathogenic ascomycetes and basidiomycetes, as discussed elsewhere in this book. The *C. neoformans* MATα locus shares features with the basidiomycete mating-type loci in that it encodes a mating pheromone receptor and multiple mating pheromones. On the other hand, its organization is significantly different, the locus is significantly larger than any previously characterized fungal mating-type locus, and the association of signaling components of the MAP kinase cascade within a mating-type locus is unprecedented. Moreover, the link between the MATα mating type and both virulence and differentiation provides a compelling reason to dissect the evolution, structure, and function of the mating-type loci. Studies on the *C. neoformans* mating-type loci will teach us much about the pathogenesis and life cycle of this human fungal pathogen in particular, and about the evolution and function of mating-type loci in general.

3.4 Growth Under Mammalian Physiological Conditions

To cause disease, pathogens must be able to proliferate in the infected host. Several factors have been found to influence the ability of *C. neoformans* to grow under host physiological conditions. First, and probably most important, is the ability of *C. neoformans* to grow at elevated temperatures. Most other *Cryptococcus* spp. are not able to grow at 37°C and are therefore rarely pathogenic [3,5,49]. In addition, mutant strains that are impaired in their ability to grow at high temperature are less virulent [98,104,105,158]. Similarly, *Candida* spp. that are not able to grow at these temperatures are also less virulent than *C. albicans*, which does grow at elevated temperatures [181]. The *CNA1* gene was identified in *C. neoformans* and found to be indispensable for growth at 37°C [98,104]. *CNA1* encodes the catalytic subunit of the calcium/calmodulin-dependent protein-phosphatase calcineurin, which is involved in signal transduction pathways (see below). Cells that are deleted for the *CNA1* gene, or in which calcineurin enzyme activity was inhibited by cyclosporin A or FK506, grow at 24°C but not at 37°C (Fig. 8). In addition, *cna1* mutant strains were killed by slightly increased pH (7.3–7.4) or increased CO_2 concentrations (5%), and were completely avirulent in animal models. Recent studies reveal that the *CNB1* gene encoding the regulatory subunit of calcineurin is also required for growth at 37°C and virulence [105].

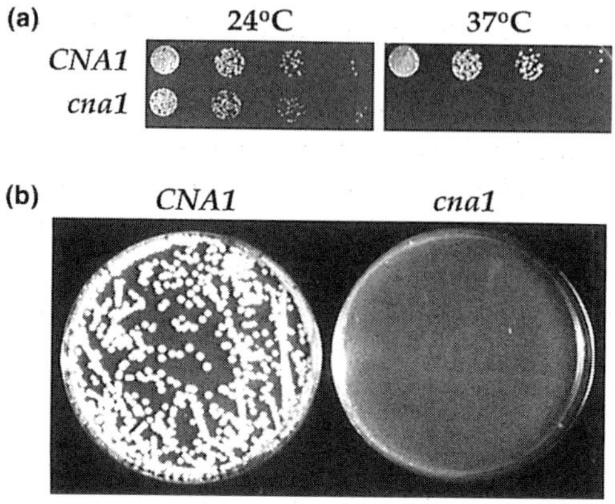

FIGURE 8. Growth at 37°C is dependent on calcineurin A. To cause disease, *C. neoformans* must grow under host physiological conditions. In *C. neoformans*, growth at 37°C is dependent on the calcium/calmodulin-dependent protein phosphatase calcineurin. While wild-type strains are able to grow at both 24°C and 37°C, *cna1* deletion mutants fail to grow at elevated growth temperature (a). Serial dilutions of a wild-type (*CNA1*) and a *cna1* deletion strain, lacking the catalytic subunit of calcineurin, were spotted onto complete medium and incubated at 24°C (left) or 37°C (right). The *cna1* deletion mutant is unable to proliferate inside the host (b) and therefore cannot cause disease. 10^7 cells of a wild-type *C. neoformans* strain (*CNA1*) or a *cna1* disruption mutant were injected into steroid-treated rabbits. After 7 days of infection, 100 μL of cerebral spinal fluid (CSF) was taken from the infected animals, plated on complete medium, and incubated at 24°C. While the wild-type strain proliferates inside the animal, the mutant strain is cleared from the CSF by the immune system.

These findings suggest a role for calcineurin in signal transduction pathways regulating growth at elevated temperatures.

Because of the nutrient-deprived environment inside the host, *C. neoformans* must synthesize nucleotides and amino acids de novo in order to grow in vivo. Mutations in nucleotide biosynthesis (*ade2*, *ura5*) and in several, but not all, amino acid biosynthetic pathways result in avirulent strains [106,182]. This is most likely due to their inability to proliferate inside the mammalian host. Another important trace element that is limiting inside the host is iron. Most iron in the mammalian organism is bound to various molecules (i.e., heme), and therefore the free iron concentration is very low. In addition, most iron is present in the ferric (FeIII) form whereas it has to be in the ferrous (FeII) form to be taken up by *C. neoformans* [183]. Physiological experiments revealed that *C. neoformans* is able to adjust its reducing capacity according to environmental conditions. Three different mechanisms have been proposed that facilitate the reduction of ferric iron: a cell-associated reductase activity, secretion of 3-hydroxyanthranilic acid, and the reducing capacity of melanin [184]. It is interesting to note in this regard that the biosynthesis of capsular material, an additional potential antioxidant (see above), is induced by low iron concentrations [126]. In contrast to many other pathogens, *C. neoformans* has not been found to secrete siderophores but may contain a hydroxamate-dependent iron uptake system [185]. Although a transferrin-like protein has been proposed [186], other iron metabolic enzymes like ferric reductase have not been identified.

3.5 Extracellular Compounds and Enzymes

In addition to capsule, melanin, and the ability to grow at mammalian physiological conditions, several other factors have been identified that influence virulence of *C. neoformans*. *C. neoformans* excretes large amounts of mannitol into the environment, both in liquid cultures and at sites of infection [187–190], and this may in part account for the increase in intracranial pressure associated with heavily infected patients [191]. *C. neoformans* may use mannitol as an energy store and a protectant against stress [187]. Similar to capsular polysaccharide and melanin, mannitol can scavenge host oxidants and may serve to protect the pathogen from an oxidative burst [192]. Strains that are defective in mannitol metabolism are less virulent in animal models [189].

A series of secreted enzymes, including proteinases, esterases, and lipases, have been identified in *C. neoformans* [193]. Many of these enzymes have been associated with virulence in other pathogens. Extracellular protease activity has been identified [194–196] and could contribute to virulence by degrading host tissue or destroying immunologically important proteins at the site of infection. Urease is a nickel metalloenzyme that converts urea into ammonia and carbamate to result in a local increase in pH. Urease is associated with virulence in other pathogens, and *C. neoformans* urease-negative strains have only rarely been found to cause disease. The gene-encoding urease has recently been cloned and urease-deleted strains have been shown to be attenuated for virulence [107]. Histopathological studies on lung tissues of infected mice show a higher inflammatory response in tissues infected with the mutant strain than in a congenic wild-type strain, suggesting a role for urease during primary infection by *C. neoformans*. Finally, extracellular phospholipase activity has been linked to virulence in *C. albicans* [197], and phospholipase activity has also been identified in *C. neoformans* [193]. Levels of enzyme activity were correlated to capsule size and virulence of the organism [108,198]. Strains with higher phospholipase activity on average showed higher virulence. A phospholipase B gene, *PLB1*, has been identified in *C. neoformans*, and virulence studies of *plb1* mutants revealed that phospholipase B activity contributes to virulence in *C. neoformans* [109].

4 SIGNALING PATHWAYS REGULATING DIFFERENTIATION AND PATHOGENICITY

Signal transduction pathways that regulate developmental processes and pathogenicity of *C. neoformans* have been defined (Fig. 9) [199,200]. Work has focused on three major pathways that have been well characterized in other organisms and which have been shown to be important for differentiation and virulence in *C. neoformans.* The first is a pheromone-induced MAP kinase signal transduction pathway that regulates sexual differentiation in *C. neoformans.* The second is the

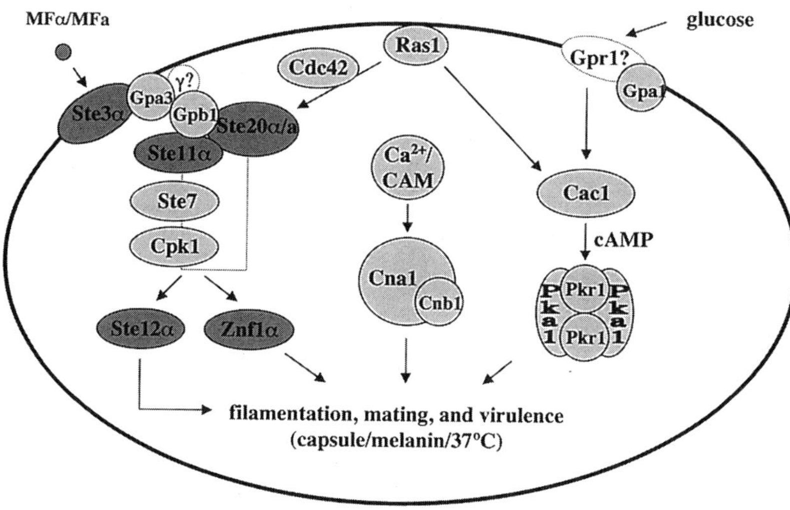

FIGURE 9. Signal transduction pathways in *C. neoformans*. Three signal transduction pathways were found to regulate differentiation and virulence in *C. neoformans* (for details see text). First, a pheromone-activated signaling pathway that contains a highly conserved MAP kinase module (left). Second, a cAMP/PKA-dependent signaling pathway that is regulated by nutritional signals (right). Finally, there is a calcium/calmodulin-dependent signal transduction pathway that is mediated through the calcineurin protein phosphatase complex (middle). Lightly shaded proteins, marked with question marks and dashed lines, have not been identified in *C. neoformans* but are thought to be present by analogy to other organisms. Darkly shaded proteins were found to be mating type specific. (From Ref. 200.)

cAMP-mediated signal transduction pathway that, in response to various environmental stimuli, regulates gene expression through changes in protein kinase A activity. Finally, another signal transduction pathway that is important for virulence in *C. neoformans* is mediated by the Ca^{2+}/calmodulin-dependent protein phosphatase calcineurin. As mentioned above, calcineurin is required for growth at mammalian physiological temperatures and has become of interest as a target for drug development against *C. neoformans*.

4.1 Pheromone-Activated MAP Kinase Cascade in *C. neoformans*

The pheromone response pathway of the ascomycete *Saccharomyces cerevisiae* is to date probably the best-understood MAP kinase signal transduction pathway. Similar to *S. cerevisiae*, sexual differentiation and in part pathogenicity of the basidiomycetous yeast *C. neoformans* are regulated by a pheromone-activated MAP kinase signal transduction pathway, and several homologs of components of the yeast pheromone response pathway have been identified in *C. neoformans*. Interestingly, several components of this signaling pathway are encoded within the mating-type locus and are therefore specific for either MATa or MATα cells [14,15,73,99,175,176]. Most of the work thus far has focused on the MAP kinase signaling pathway of MATα cells because of the higher virulence found for these cells in comparison to MATa cells (see above). MATα mating type–specific components of this pheromone response pathway include three pheromone precursor genes (*MFα1*, *MFα2*, and *MFα3*), a pheromone receptor (*STE3α/CPRα*), a Ste20 protein kinase homolog (*STE20α*), a MAPKK kinase homolog (*STE11α*), and two transcription factors (*STE12α* and *ZNF1α*). MATa mating type–specific alleles of the pheromones and the *STE20* gene have recently been identified [72,73]. In addition, the β-subunit of a heterotrimeric G-protein (*GPB1*) [42] and a MAP kinase homolog (*CPK1*; Davidson and Heitman, submitted) were identified as components of this pheromone activated MAP kinase signaling pathway but were found to be identical in both mating types.

The pheromone precursor gene *MFα1* was the first gene identified as a component of a pheromone response pathway via its ability to induce conjugation tubes and enlarged, round cells after transformation into cells of MATa mating type [14,15]. These morphological transitions are similar to those observed in matings and confrontation assays (see above) and indicate that *MFα1* plays a role in pheromone response and mating in *C. neoformans*. In addition, MFα1 pheromone has been synthesized in an active form and shown to induce conjugation tube formation in MATa cells [14]. Interestingly, the *MFα1* gene is induced in response to nutrient deprivation and in the presence of cells of opposite mating type, and both conditions stimulate mating [14]. Two additional pheromone precursor genes, *MFα2* and *MFα3*, were subsequently identified, and analysis of the effects of all three pheromone genes in mating is in progress [14] (Shen et al., submitted).

Several lines of evidence indicated that the Gβ-subunit Gbp1 is a component of a pheromone response pathway involved in mating of *C. neoformans* [42]. First, overexpression of Gpb1 in MATa and MATα cells induces conjugation tube formation similar to the overexpression of the pheromone gene *MFα1* in MATa cells. Second, mutations in the *GPB1* gene completely abolished mating. Finally, the mating defect exhibited by *gpb1* deletion strains was suppressed by overexpression of Cpk1, a MAP kinase homolog. Cpk1 shows a high similarity to the yeast Fus3/Kss1 kinases of *S. cerevisiae*, which are components of the pheromone response pathway. Therefore, Cpk1 is thought to be a component of a MAP kinase signaling module in *C. neoformans* that is homologous to the pheromone response pathway of *S. cerevisiae*. Surprisingly, the analysis of *gpb1* deletion strains also implied an additional role in haploid filamentation, and mutant strains were unable to haploid fruit in response to nitrogen starvation or Ras1^{Q67L} expression. This finding is in contrast to the ascomycete *S. cerevisiae* where the pheromone sensing G-protein is dispensable for pseudohyphal growth [201], a differentiation process that shares common features with haploid fruiting in *C. neoformans* [202]. Besides its role in mating and haploid fruiting, Gpb1 was required for virulence of *C. neoformans*. Melanin and capsule formation in *gpb1* deletion strains were unaffected, and mutant strains were pathogenic in an animal virulence model [42].

Extensive work has been done on the *STE12α* gene, which encodes a transcription factor similar to Ste12 of *S. cerevisiae*. Disruption mutants have been generated in both serotype A and serotype D strains of *C. neoformans*, but surprisingly, *ste12α* mutant strains showed only subtle mating defects in either serotype [99,100]. In contrast, haploid filamentation was dramatically reduced in both the serotype A and the serotype D background. This finding indicates a role for Ste12α in haploid filamentation but not in mating. Serotype A *ste12α* mutant strains were found to be completely virulent [99]. In serotype D, however, virulence of *ste12α* deletion strains was attenuated and Ste12α may regulate expression of genes associated with virulence [100]. Similar to the results found with the calcineurin *cna1* mutants (see above), this again demonstrates the divergence of serotype A and serotype D strains from a common ancestor ~20 million years ago.

Mating in *C. neoformans* is regulated by a MAP kinase signaling pathway similar to the pheromone response pathway of *S. cerevisiae*, but several interesting questions remain. Several components of this pathway are encoded within the mating-type locus and are therefore specific for MATa or MATα cells [14,15,73,99,175,176]. Investigations are under way to analyze the structures of both mating-type loci and how differences therein contribute to differences in virulence found between MATa and MATα cells. The finding that Ste12α plays a primary role in haploid filamentation but not in mating, and also regulates virulence in the congenic serotype D strains, strongly indicates that other signal transduction cascades work in parallel to the pheromone response pathway to regulate mating. This would be analogous to the yeast *S. cerevisiae* where two different MAP kinase signaling pathways were identified that share several components but are specific for mating or filament formation [203,204]. Specificity for the corresponding developmental process is, in part, achieved at the level of the MAP kinases Fus3 and Kss1.

In addition, a second, as yet unidentified transcription factor might be involved in these signaling processes as was shown for *S. cerevisiae* in which Tec1 acts in conjunction with Ste12 to regulate filament formation [205,206]. Alternatively, an HMG box transcription factor might be involved in mating in *C. neoformans* similar to the Ste11 and Prf1 transcription factors, the main regulators of mating in *Schizosaccharomyces pombe* and *U. maydis* [207,208]. Furthermore, additional signals might be involved during the mating process that are not mediated through a MAP kinase signaling module. In *S. pombe* and *U. maydis* nutritional signals are required to promote mating, which are mediated by a cAMP-dependent signaling pathway [209–211]. As will be discussed below, there is some evidence that there is also crosstalk between the pheromone response pathway and a cAMP-dependent signaling pathway in *C. neoformans*.

4.2 cAMP-Mediated Signaling Cascades in *C. neoformans*

Activation of adenylyl cyclase to produce cAMP regulates several important cellular processes in *C. neoformans*, including capsule production, melanin formation, mating, and virulence [116,212]. Changes in intracellular cAMP concentrations regulate the activity of the cAMP-dependent protein kinase A (PKA), which in turn regulates the expression of specific target genes. Similar to the MAP kinase signaling pathway, the cAMP/PKA signal transduction pathway is highly conserved in eukaryotes. In other organisms it was shown that adenylyl cyclase is in part activated by a highly homologous G-protein α-subunit. A homolog of this protein was identified in *C. neoformans* [213], and a *gpa1* mutant strain was created by homologous recombination using the *ADE2* gene as a selectable marker [116].

Interestingly, *gpa1* mutant strains failed to induce the known virulence factors of melanin synthesis (in response to glucose starvation), capsule production (in response to iron limitation), and mating (in response to nitrogen limitation) [116]. Exogenous cAMP suppressed the *gpa1* mutant phenotypes, restoring mating and melanin and capsule production. This finding showed that, in analogy to other systems, Gpa1 is a component of a cAMP-dependent signal transduction pathway in *C. neoformans*. In addition to the defects in mating, capsule formation, and melanin production, the *gpa1* mutant strain was avirulent in a rabbit model of cryptococcal meningitis, and reconstitution of the strain with the wild-type *GPA1* gene restored virulence. How or whether the various signals converge at Gpa1 is not fully understood, and upstream components have yet to be identified. In the yeast *S. cerevisiae*, the homologous G-protein α-subunit Gpa2 is coupled to a seven-transmembrane domain receptor Gpr1 [214,215]. However, a homolog of Gpr1, which is thought to function as a glucose receptor, has not been identified in *C. neoformans*.

PKA is a tetrameric serine/threonine-specific protein kinase complex formed by two catalytic and two regulatory subunits. Upon binding cAMP, the regulatory subunits release the catalytic subunits in an active form. Recently, genes for the catalytic subunit (*PKA1*) and the regulatory subunit (*PKR1*) have been identified in *C. neoformans* [115]. *pka1* deletion strains have defects similar to the *gpa1* deletion strain; *pka1* mutants are avirulent in a murine model system, do not produce melanin or capsular polysaccharide, and are sterile. In contrast to *gpa1* mutants, the *pka1* mutant phenotypes are not rescued by exogenous cAMP, providing evidence that PKA is the target of cAMP. Most interestingly, mutations in the PKA regulatory subunit result in constitutive PKA activity and signaling. *pkr1* mutants in *C. neoformans* exhibit increased capsule formation under iron limiting conditions and are hypervirulent in animal models.

These findings reveal a prominent role of the cAMP/PKA-mediated signaling pathway in regulating virulence and mating of *C. neoformans*. Several important questions remain. First, how does the organism sense the different signals that require the cAMP/PKA signaling pathway and how are these various stimuli integrated into the pathway? Second, which proteins function upstream and downstream of PKA, and which genes are regulated by this signaling pathway?

4.3 RAS Signaling and Crosstalk Between cAMP and MAP Kinase Signaling Cascades

Ras proteins are small, guanine nucleotide-binding proteins that act as molecular switches by their ability to hydrolyze GTP. Similar to many components of signaling pathways, Ras proteins are highly conserved throughout eukaryotic organisms. In the ascomycete *S. cerevisiae*, two Ras homologs have been identified and these proteins signal changes in the extracellular environment and regulate cell cycle progression by stimulating cAMP production by adenylyl cyclase [216,217].

Recently, *RAS1*, the gene encoding the Ras1 homolog in *C. neoformans*, has been identified and a disruption mutant has been generated in serotype A [218]. While *ras1* mutant strains were able to grow at low temperature (24°C), they exhibited a temperature-sensitive growth defect at elevated temperature (37°C). In addition, *ras1* mutants were defective in mating and agar adherence, and were avirulent in animal virulence models. Exogenous cAMP partially suppressed the mating defect of *ras1* mutants and fully suppressed their agar adherence defect. The *ras1* mutant strain showed no defect in melanin or capsule production, which are regulated by the cAMP/PKA pathway [116]. Epistasis analysis suggests a role for Ras1 upstream of Gpb1 and Ste12 during mating. Overexpression of dominant active Ras1^{Q67L} induced haploid fruiting in wild-type but not in *ste12α* mutant strains [99,218]. The mating defect of *ras1* mutant strains was fully suppressed by overexpressing Gpb1, the G-protein β-subunit that regulates the pheromone response pathway in *C. neoformans*. However, the mating defect exhibited by *gpb1* mutant strains was not suppressed by overexpressing dominant active Ras1^{Q67L}. These findings suggest that Ras1-mediated crosstalk may occur between the pheromone response pathways and the cAMP/PKA-dependent signaling pathway. It is interesting to note that the *C. neoformans* Ste12α protein contains consensus sites for phosphorylation by PKA, which may be a second level of possible crosstalk between the MAP kinase and cAMP/PKA-dependent signaling pathways.

4.4 Calcineurin Regulates Growth at 37°C and Virulence

Calcineurin is a serine/threonine-specific protein phosphatase composed of a catalytic A and a regulatory B subunit [219]. Activity of calcineurin is dependent on the calcium-binding protein calmodulin. Calcineurin is highly conserved in eukaryotes, and the phosphatase has been characterized in several fungi. In the budding yeast *S. cerevisiae*, calcineurin regulates cation homeostasis, cell wall biosynthesis, and recovery from the pheromone-induced cell cycle arrest during mating. In *C. neoformans*, the genes for the catalytic subunit, *CNA1*, and the regulatory subunit, *CNB1*, have been identified, and disruption mutant strains have been characterized in serotypes A and D [98,104,105,220,221]. Both *cna1* and *cnb1* disruption mutants are inviable at elevated temperatures (37°C), and are completely avirulent in animal virulence models. Surprisingly, the functions of calcineurin in serotype A and D strains differ. In serotype A, *cna1* mutant strains are hypersensitive to Na^+ and Li^+, suggesting a role for calcineurin in cation homeostasis similar to that found in the ascomycete *S. cerevisiae* [98]. However, calcineurin was not required for cation homeostasis in serotype D strains [98]. In addition, calcineurin has been found to play a role in filament formation and mating [221].

Because of the temperature-dependent growth defect of calcineurin mutants, it has been suggested that inhibitors of calcineurin may be useful as antifungal compounds. Two macrolides, cyclosporin A (CsA) and FK506, have been identified that are toxic to pathogenic fungi, including *C. neoformans* [104,222–226]. The intracellular receptors for CsA and FK506 are cyclophilin A and FKBP12, respectively. Both drug–protein complexes target calcineurin [227]. CsA and FK506 are potent immunosuppressants, and both compounds exacerbate cryptococcal infection in a rabbit virulence model [228]. Interestingly, analogs of CsA and FK506 that do not have immunosuppressive activity yet retain antifungal activity have been identified [224,225]. Similar to CsA and FK506, these CsA analogs are toxic at 37°C but not at 24°C, consistent with inhibition of calcineurin as their mechanism of action. Mutations on the surface of calcineurin A involved in drug–protein complex binding conferred resistance to both CsA analogs, and mutant strains lacking both cyclophilin A homologs were also drug resistant. Thus, a cyclophilin A–CsA complex targets the calcineurin homolog and inhibits fungal growth. These findings revealed that, although most signaling pathways are highly conserved between eukaryotes, there are sufficient differences that can be exploited in drug discovery to generate antifungal compounds that spare the host immune system.

5 OUTLOOK

In countries with advanced health care systems, highly active antiretroviral therapy (HAART) and fluconazole may reduce the burden of opportunistic fungal infections, including cryptococcosis in AIDS patients. However, several warning signals indicate that we should not be complacent in searching for new approaches to treat and prevent cryptococcal infection:

1. Cryptococcal meningitis continues to be the initial presenting symptom in some AIDS patients.
2. For HIV-infected patients with cryptococcal meningitis, fluconazole suppression is standard therapy, irrespective of antiretroviral therapy, and this is a less than optimal therapeutic approach and is not uniformly applied because of costs and the concern of antifungal drug resistance.
3. Because of the development of antiviral drug resistance and problems with drug regimen intolerance, it is unclear what long-term effect HAART will have on protection from opportunistic infections.
4. Drug-resistant strains of *C. neoformans* do occur [229–232].
5. Besides HIV infection, there is an increasing number of patients immunosuppressed for other reasons (steroids, organ transplantation, chemotherapy), and these populations represent a significant proportion of patients infected with *C. neoformans*.
6. Patients with no known defects in immune function continue to present with cryptococcal meningitis, and may have unknown immune system defects or be infected with hypervirulent strains.

Overall, the annual incidence of cryptococcal meningitis in both AIDS and non-AIDS patients is ~10,000 patients in the U.S. population of ~250 million, so this infection is a serious problem of magnitude comparable to other infections of the central nervous system such as meningococcal meningitis. The incidence of AIDS-associated cryptococcal meningitis is 4–7% in the United States and as high as 30% in Africa [233,234]. These issues illustrate the importance of *C. neoformans* infection and the need to understand the molecular basis of virulence to facilitate the design and implementation of strategies to manage, treat, and prevent infection.

In addition to its clinical significance, *C. neoformans* is an ideal genetic model for studies of fungal development and pathogenesis. The life cycle and sexual cycle of this fungus have been defined [6–11]. This fungus exists primarily as a stable haploid, which allows direct isolation of recessive mutations. The sexual cycle involves mating between cells of the MATα and MATa mating types. Cells of the MATα mating type can also filament and sporulate in response to nitrogen starvation or MATa cells in a process called haploid or monokaryotic fruiting [9,42]. Congenic strains of opposite mating type and with auxotrophic mutations have been developed for genetic studies [15,104,180]. In addition, recent advances in molecular biology approaches make *C. neoformans* an excellent molecular model of fungal pathogenesis. Two different transformation systems have been developed [182,235,236]. Many genes have been disrupted by transformation and homologous recombination [101,103,104,116,220,237]. Moreover, excellent animal model systems have been developed, including the rabbit model of cryptococcal meningitis [238], a murine systemic infection model, and both rat and murine inhalation models [239]. It is now possible to identify a gene of interest, disrupt the gene, reintroduce the wild-type gene, and assess the impact of these genetic manipulations on virulence in vivo.

A public genome sequencing project has been started for *C. neoformans* var. *neoformans* (serotype D strain JEC21) at the Stanford Genome Technology Center (http://baggage.stanford.edu/group/C.neoformans/), Nagasaki University School of Medicine, The Institute for Genome Research (TIGR), and the University of Oklahoma (http://www.genome.ou.edu/cneo.html), and 12× coverage of the genome has already been achieved [240]. This genome project will soon be accompanied by genome sequencing projects for varieties *grubii* and *gattii*. It is anticipated that *C. neoformans* genome sequence data will accelerate research in this area, particularly in the search for genes and studies of their biological function.

Taken together, the accessibility of molecular biology tools, well-developed animal virulence models, and publicly available genome sequence data make *C. neoformans* an outstanding model system to study fungal development in general and fungal pathogenesis in particular.

ACKNOWLEDGMENTS

We thank John Perfect for his advice and guidance, all the members of the Duke mycology group for creating a stimulating environment, and Cristina Cruz, Andy Alspaugh, and Cletus D'Souza for providing images used for figures. Our studies are supported by R01 grants AI39115, AI41937, and AI42159 from NIAID, and P01 program project grant AI44975 to the Duke University Research Mycology Research Unit. Joseph Heitman is a Burroughs Wellcome Fund Scholar in Molecular Pathogenic Mycology and an associate investigator of the Howard Hughes Medical Institute.

REFERENCES

1. F Sanfelice. Contributo alla morfologia e biologia dei blastomiceti che si svilup/pano nei succhi di alcuni frutti. Ann Igien 4:463–495, 1894.
2. M Knoke, G Schwesinger. One hundred years ago: the history of cryptococcosis in Greifswald. Medical mycology in the nineteenth century. Mycoses 37:229–233, 1994.
3. A Casadevall, JR Perfect. *Cryptococcus neoformans*. Washington: ASM Press, 1998, pp 1–541.
4. TG Mitchell, JR Perfect. Cryptococcosis in the era of AIDS—100 years after the discovery of *Cryptococcus neoformans*. Clin Microbiol Rev 8:515–548, 1995.
5. KJ Kwon-Chung, JE Bennett. Cryptococcosis. Medical Mycology, Malvern, PA: Lea & Febiger, 1992, pp 397–446.
6. KJ Kwon-Chung. A new genus, *Filobasidiella*, the perfect state of *Cryptococcus neoformans*. Mycologia 67:1197–1200, 1975.
7. KJ Kwon-Chung. Morphogenesis of *Filobasidiella neoformans*, the sexual state of *Cryptococcus neoformans*. Mycologia 68:821–833, 1976.
8. KJ Kwon-Chung. A new species of *Filobasidiella*, the sexual state of *Cryptococcus neoformans* B and C serotypes. Mycologia 68:943–946, 1976.
9. BL Wickes, ME Mayorga, U Edman, JC Edman. Dimorphism and haploid fruiting in *Cryptococcus neoformans*: association with the α-mating type. Proc Natl Acad Sci USA 93:7327–7331, 1996.
10. JA Alspaugh, RC Davidson, J Heitman. Morphogenesis of *Cryptococcus neoformans*. In: JF Ernst, A Schmidt, eds. Dimorphism in Human Pathogenic and Apathogenic Yeasts, Vol 5. Basel; Karger, 2000, pp 217–238.
11. RA Sia, KB Lengeler, J Heitman. Diploid strains of the pathogenic basidiomycete *Cryptococcus neoformans* are thermally dimorphic. Fungal Genet Biol 29:153–163, 2000.
12. DL Barlow. Primary blasto-mycotic meningitis occurring in a child. Med J Aust 2:302–304, 1923.
13. FD Weidman, W Freeman. India ink in the microscopic study of yeast cells. JAMA 83:1163–1164, 1924.
14. RC Davidson, TDE Moore, AR Odom, J Heitman. Characterization of the MFα pheromone of the human fungal pathogen *Cryptococcus neoformans*. Mol Microbiol 39:1–12, 2000.
15. TDE Moore, JC Edman. The α-mating type locus of *Cryptococcus neoformans* contains a peptide pheromone gene. Mol Cell Biol 13:1962–1970, 1993.
16. T Spellig, E Regenfelder, M Reichmann, F Schauwecker, R Bohlmann, M Urban, M Bölker, J Kämper, R Kahmann. Control of mating and development in *Ustilago maydis*. Antonie van Leeuwenhoek 65:191–197, 1994.

17. AD Martinez-Espinoza, SA Gerhardt, JE Sherwood. Morphological and mutational analysis of mating in *Ustilago hordei*. Exp Mycol 17:200–214, 1993.
18. Y Kamiya, A Sakurai, S Tamura, N Takahashi. Structure of rhodotorucine A, a novel lipopeptide, inducing mating tube formation in *Rhodosporidium toruloides*. Biochem Biophys Res Commun 83:1077–1083, 1978.
19. Y Sakagami, M Yoshida, A Isogai, A Suzuki. Peptidal sex hormones inducing conjugation tube formation in compatible mating-type cells of *Tremella mesenterica*. Science 212:1525–1527, 1981.
20. Y Ishibashi, Y Sakagami, A Isogai, A Suzuki, RJ Bandoni. Isolation of tremerogens A-9291-II, novel sex hormones of *Tremella brasiliensis*. Can J Biochem Cell Biol 61:796–801, 1983.
21. JR Raper. Genetic Sexuality of Higher Fungi. New York: Ronald Press, 1966.
22. KJ Kwon-Chung, TJ Popkin. Ultrastructure of septal complex in *Filobasidiella neoformans* (*Cryptococcus neoformans*). J Bacteriol 126:524–528, 1976.
23. KJ Kwon-Chung. Nuclear genotypes of spore chains in *Filobasidiella neoformans* (*Cryptococcus neoformans*). Mycologia 72:418–422, 1980.
24. K Takeo, R Tanaka, H Taguchi, K Nishimura. Analysis of ploidy and sexual characteristics of natural isolates of *Cryptococcus neoformans*. Can J Microbiol 39:958–963, 1993.
25. R Tanaka, K Nishimura, M Miyaji. Ploidy of serotype AD strains of *Cryptococcus neoformans*. J Med Mycol 40:31–34, 1999.
26. R Tanaka, H Taguchi, K Takeo, M Miyaji, K Nishimura. Determination of ploidy in *Cryptococcus neoformans* by flow cytometry. J Med Vet Mycol 34:299–301, 1996.
27. T Boekhout, AV Belkum. Variability of karyotypes and RAPD types in genetically related strains of *Cryptococcus neoformans*. Curr Genet 32:203–208, 1997.
28. MA Viviani, H Wen, A Roverselli, R Caldarelli-Stefano, M Cogliate, P Ferrante, AM Tortorano. Identification by polymerase chain reaction fingerprinting of *Cryptococcus neoformans* serotype AD. J Med Vet Mycol 35:355–360, 1997.
29. ME Brandt, SL Bragg, RW Pinner. Multilocus enzyme typing of *Cryptococcus neoformans*. J Clin Microbiol 31:2819–2823, 1993.
30. WL Whelan, KJ Kwon-Chung. Genetic complementation in *Cryptococcus neoformans*. J Bacteriol 166:924–929, 1986.
31. CW White, ES Jacobson. Occurrence of diploid strains of *Cryptococcus neoformans*. J Bacteriol 161:1231–1232, 1985.
32. B Maresca, L Carratu, GS Kobayashi. Morphological transition in the human fungal pathogen *Histoplasma capsulatum*. Trends Microbiol 2:110–114, 1994.
33. S Sternberg. The emerging fungal threat. Science 266:1632–1634, 1994.
34. E Brummer, E Castaneda, A Restrepo. Paracoccidioidomycosis: an update. Clin Microbiol Rev 6:89–117, 1993.
35. B Maresca, GS Kobayashi. Dimorphism in *Histoplasma capsulatum* and *Blastomyces dermatitidis*. Contrib Microbiol 5:201–216, 2000.
36. KB Lengeler, GM Cox, J Heitman. Serotype AD strains of *Cryptococcus neoformans* are diploid or aneuploid and are heterozygous at the mating type locus. Infect Immun 69:115–122, 2001.
37. RL Todd, WW Herrmann. The life cycle of the organism causing yeast meningitis. J Bacteriol 32:89–101, 1936.

38. KH Erke. Light microscopy of basidia, basidiospores, and nuclei in spores and hyphae of *Filobasidiella neoformans* (*Cryptococcus neoformans*). J Bacteriol 128: 445–455, 1976.
39. KJ Kwon-Chung. Heterothallism vs. self-fertile isolates of *Filobasidiella neoformans* (*Cryptococcus neoformans*). Proceedings of the Fourth International Conference on Mycoses. Mycoses—Yeast and Yeast-Like Infections. Washington: Pan American Health Organization Scientific Publication No. 356, 1978, pp 204–213.
40. U Stahl, K Esser. Genetics of fruit body production in higher basidiomycetes. I. Monokaryotic fruiting and its correlation with dikaryotic fruiting in *Polyporus ciliatus*. Mol Gen Genet 148:183–197, 1976.
41. K Esser, F Meinhardt. A common genetic control of dikaryotic and monokaryotic fruiting in the basidiomycete *Agrocybe aegerita*. Mol Gen Genet 155:113–115, 1977.
42. P Wang, JR Perfect, J Heitman. The G-protein β subunit GPB1 is required for mating and haploid fruiting in *Cryptococcus neoformans*. Mol Cell Biol 20:352–362, 2000.
43. HJ Shadomy, JP Utz. Preliminary studies on a hypha-forming mutant of *Cryptococcus neoformans*. Mycologica 58:383–390, 1966.
44. HJ Shadomy. Clamp connections in two strains of *Cryptococcus neoformans*. Spectrum Monogr Ser 1:67–72, 1970.
45. A Gargas, PT DePriest, M Grube, A Tehler. Multiple origins of lichen symbioses in fungi suggested by SSU rDNA phylogeny. Science 268:1492–1495, 1995.
46. E Guého, L Improvisi, R Christen, GS Hoog. Phylogenetic relationships of *Cryptococcus neoformans* and some related basidiomycetous yeasts determined from partial large subunit rRNA sequences. Antonie van Leeuwenhoek 63:175–189, 1993.
47. KJ Kwon-Chung, YC Chang, R Bauer, EC Swann, JW Taylor, R Goel. The characteristics that differentiate *Filobasidiella depauperata* from *Filobasidiella neoformans*. In: T Boekhout, A Samson, eds. Heterobasidiomycetes: Systematic and Applied. Studies in Mycology. Baarn, Switzerland: CBS, 1995, pp 67–79.
48. TG Mitchell, TJ White, JW Taylor. Comparison of 5.8S ribosomal DNA sequences among the basidiomycetous yeast genera *Cystofilobasidium*, *Filobasidium* and *Filobasidiella*. J Med Vet Mycol 30:207–218, 1992.
49. JW Fell, A Statzell-Tallman. *Cryptococcus* Vuillemin. In: CP Kurtzman, JW Fell, eds. The Yeasts: A Taxonomic Study. Amsterdam: Elsevier Science, 1999, pp 742–767.
50. RW Benham. Cryptococci—their identification by morphology and by serology. J Infect Dis 57:255–274, 1935.
51. R Cherniak, JB Sundstrom. Polysaccharide antigens of the capsule of *Cryptococcus neoformans*. Infect Immun 62:1507–1512, 1994.
52. EE Evans. The antigenic composition of *Cryptococcus neoformans*. I. A serologic classification by means of the capsular and agglutination reactions. J Immunol 64: 423–430, 1950.
53. DE Wilson, JE Bennett, JW Bailey. Serologic grouping of *Cryptococcus neoformans*. Proc Soc Exp Biol Med 127:820–823, 1968.
54. J Xu, RJ Vilgalys, TG Mitchell. Multiple gene genealogies reveal dispersion and

hybridization in the human pathogenic fungus *Cryptococcus neoformans*. Mol Ecol 9:1471–1482, 2000.
55. SP Franzot, IF Salkin, A Casadevall. *Cryptococcus neoformans* var. *grubii*: separate varietal status for *Cryptococcus neoformans* serotype A isolates. J Clin Microbiol 37:838–840, 1999.
56. A Casadevall, M Fan. *URA5* gene of *Cryptococcus neoformans* var. *gattii*. Evidence for a close phylogenetic relationship between *C. neoformans* var. *gattii* and *C. neoformans* var. *neoformans*. J Gen Appl Microbiol 38:491–495, 1992.
57. A Casadevall, LF Freundlich, L Marsh, MD Scharff. Extensive allelic variation in *Cryptococcus neoformans*. J Clin Microbiol 30:1080–1084, 1992.
58. F Chen, BP Currie, LC Chen, SG Spitzer, ED Spitzer, A Casadevall. Genetic relatedness of *Cryptococcus neoformans* clinical isolates grouped with the repetitive DNA probe CNRE-1. J Clin Microbiol 33:2818–2822, 1995.
59. SP Franzot, JS Hamdan, BP Currie, A Casadevall. Molecular epidemiology of *Cryptococcus neoformans* in Brazil and the United States: evidence for both local genetic differences and a global clonal population structure. J Clin Microbiol 35: 2243–2251, 1997.
60. M Fan, BP Currie, RR Gutell, MA Ragan, A Casadevall. The 16S-like, 5.8S and 23S-like rRNAs of the two varieties of *Cryptococcus neoformans*: sequence, secondary structure, phylogenetic analysis and restriction fragment polymorphisms. J Med Vet Mycol 32:163–180, 1994.
61. M Fan, LC Chen, MA Ragan, RR Gutell, JR Warner, BP Currie, A Casadevall. The 5S rRNA and the rRNA intergenic spacer of the two varieties of *Cryptococcus neoformans*. J Med Vet Mycol 33:215–221, 1995.
62. HS Aulakh, SE Straus, KJ Kwon-Chung. Genetic relatedness of *Filobasidiella neoformans* (*Cryptococcus neoformans*) and *Filobasidiella bacillispora* (*Cryptococcus bacillisporus*) as determined by deoxyribonucleic acid base composition and sequence homology studies. Int J Syst Bacteriol 31:97–103, 1981.
63. BL Wickes, TDE Moore, KJ Kwon-Chung. Comparison of the electrophoretic karyotypes and chromosomal location of ten genes in the two varieties of *Cryptococcus neoformans*. Microbiology 140:543–550, 1994.
64. JR Perfect, BB Magee, PT Magee. Separation of chromosomes of *Cryptococcus neoformans* by pulsed field gel electrophoresis. Infect Immun 57:2624–2627, 1989.
65. SP Franzot, BC Fries, W Cleare, A Casadevall. Genetic relationship between *Cryptococcus neoformans* var. *neoformans* strains of serotypes A and D. J Clin Microbiol 36:2200–2204, 1998.
66. ED Spitzer, SG Spitzer. Use of a dispersed repetitive DNA element to distinguish clinical isolates of *Cryptococcus neoformans*. J Clin Microbiol 30:1094–1097, 1992.
67. A Varma, KJ Kwon-Chung. Restriction fragment polymorphism in mitochondrial DNA of *Cryptococcus neoformans*. J Gen Microbiol 135:3353–3362, 1989.
68. AC Crampin, RC Matthews, D Hall, EGV Evans. PCR fingerprinting *Cryptococcus neoformans* by random amplification of polymorphic DNA. J Med Vet Mycol 31: 463–465, 1993.
69. KA Haynes, DJ Sullivan, DC Coleman, JC Clarke, R Emilianus, C Atkinson, KJ

Cann. Involvement of multiple *Cryptococcus neoformans* strains in a single episode of cryptococcosis and reinfection with novel strains in recurrent infection demonstrated by random amplification of polymorphic DNA and DNA fingerprinting. J Clin Microbiol 33:99–102, 1995.
70. W Meyer, TG Mitchell. Polymerase chain reaction fingerprinting in fungi using single primers specific to minisatellites and simple repetitive DNA sequences: strain variation in *Cryptococcus neoformans*. Electrophoresis 16:1648–1656, 1995.
71. TG Mitchell, RL Sandin, BH Bowman, W Meyer, WG Merz. Molecular mycology: DNA probes and applications of PCR technology. J Med Vet Mycol 32:351–366, 1994.
72. S Chaturvedi, B Rodeghier, J Fan, CM McClelland, BL Wickes, V Chaturvedi. Direct PCR of *Cryptococcus neoformans* MATα and MATa pheromones to determine mating type, ploidy, and variety: a tool for epidemiological and molecular pathogenesis studies. J Clin Microbiol 38:2007–2009, 2000.
73. KB Lengeler, P Wang, GM Cox, JR Perfect, J Heitman. Identification of the MATa mating type locus of *Cryptococcus neoformans* reveals a serotype A MATa strain thought to have been extinct. Proc Natl Acad Sci USA 97:14455–14460, 2000.
74. JE Bennett, KJ Kwon-Chung, D Howard. Epidemiologic differences among serotypes of *Cryptococcus neoformans*. Am J Epidemiol 105:582–586, 1984.
75. CW Emmons. Saprophytic sources of *Cryptococcus neoformans* associated with the pigeon (*Columbia livia*). Am J Hyg 62:227–252, 1955.
76. CW Emmons. Prevalence of *Cryptococcus neoformans* in pigeon habitats. Public Health Rep 75:362–364, 1960.
77. CJ Kao, J Schwarz. The isolation of *Cryptococcus neoformans* from pigeon nests. Am J Clin Pathol 27:652–663, 1957.
78. BM Partridge, HI Winner. *Cryptococcus neoformans* in bird droppings in London. Lancet i:1060–1061, 1965.
79. MS Lazera, FD Pires, L Camillo-Coura, MM Nishikawa, CC Bezerra, L Trilles, B Wanke. Natural habitat of *Cryptococcus neoformans* var. *neoformans* in decaying wood forming hollows in living trees. J Med Vet Mycol 34:127–131, 1996.
80. MS Lazera, MAS Cavalcanti, AT Londero, L Trilles, MM Nishikawa, B Wanke. Possible primary ecological niche of *Cryptococcus neoformans*. Med Mycol 38: 379–383, 2000.
81. H Montenegro, CR Paula. Environmental isolation of *Cryptococcus neoformans* var. *gattii* and *C. neoformans* var. *neoformans* in the city of Sao Paulo, Brazil. Med Mycol 38:385–390, 2000.
82. F Dromer, S Mathoulin, B Dupont, A Laporte. Epidemiology of *Cryptococcosis* in France: a 9-year survey (1985–1993). Clin Infect Dis 23:82–90, 1996.
83. JN Steenbergen, A Casadevall. Prevalence of *Cryptococcus neoformans* var. *neoformans* (serotype D) and *Cryptococcus neoformans* var. *grubii* (serotype A) isolates in New York City. J Clin Microbiol 38:1974–1976, 2000.
84. KJ Kwon-Chung, BL Wickes, L Stockman, GD Roberts, D Ellis, DH Howard. Virulence, serotype, and molecular characteristics of environmental strains of *Cryptococcus neoformans* var. *gattii*. Infect Immun 60:1869–1874, 1992.
85. DH Ellis, TJ Pfeiffer. Ecology, life cycle, and infectious propagule of *Cryptococcus neoformans*. Lancet 336:923–925, 1990.

86. D Ellis, T Pfeiffer. The ecology of *Cryptococcus neoformans*. Eur J Epidemiol 8: 321–325, 1992.
87. TC Sorrell, AG Brownlee, P Ruma, R Malik, TJ Pfeiffer, DH Ellis. Natural environmental sources of *Cryptococcus neoformans* var. *gattii*. J Clin Microbiol 34:1261–1263, 1996.
88. S Chen, T Sorrell, G Nimmo, B Speed, B Currie, D Ellis, D Marriott, T Pfeiffer, D Parr, K Byth. Epidemiology and host- and variety-dependent characteristics of infection due to *Cryptococcus neoformans* in Australia and New Zealand. Clin Infect Dis 31:499–508, 2000.
89. MM Nishikawa, OD Sant'Anna, MS Lazera, B Wanke. Use of D-proline assimilation and CGB medium for screening Brazilian *Cryptococcus neoformans* isolates. J Med Vet Mycol 34:365–366, 1996.
90. I Polacheck, KJ Kwon-Chung. Creatinine metabolism in *Cryptococcus neoformans* and *Cryptococcus bacillisporus*. J Bacteriol 142:15–20, 1980.
91. TA Nurudeen, DG Ahearn. Regulation of melanin production by *Cryptococcus neoformans*. J Clin Microbiol 10:724–729, 1979.
92. KJ Kwon-Chung, BL Wickes, JL Booth, HS Vishniac, JE Bennett. Urease inhibition by EDTA in the two varieties of *Cryptococcus neoformans*. Infect Immun 55: 1751–1754, 1987.
93. KJ Kwon-Chung, I Polacheck, JE Bennett. Improved diagnostic medium for separation of *Cryptococcus neoformans* var. *neoformans* (serotypes A and D) and *Cryptococcus neoformans* var. *gattii* (serotypes B and C). J Clin Microbiol 15:535–537, 1982.
94. IF Salkin, NJ Hurd. New medium for differentiation of *Cryptococcus neoformans* serotype pairs. J Clin Microbiol 15:169–171, 1982.
95. EA Irokanulo, CO Akueshi, AA Makinde. Differentiation of *Cryptococcus neoformans* serotypes A and D using creatinine dextrose Bromothymol Blue thymine medium. Br J Biomed Sci 51:100–103, 1994.
96. AL Rogers, KJ Kwon-Chung, SL Flegler. A scanning electron microscope comparison of basidial structures in *Filobasidiella neoformans* and *Filobasidiella bacillispora*. Sabouraudia 18:85–89, 1980.
97. KJ Kwon-Chung, JE Bennett, JC Rhodes. Taxonomic studies on *Filobasidiella* species and their anamorphs. Antonie van Leeuwenhoek 48:25–38, 1982.
98. MC Cruz, RAL Sia, M Olson, GM Cox, J Heitman. Comparison of the roles of calcineurin in physiology and virulence in serotype D and serotype A strains of *Cryptococcus neoformans*. Infect Immun 68:982–985, 2000.
99. C Yue, LM Cavallo, JA Alspaugh, P Wang, GM Cox, JR Perfect, J Heitman. The *STE12*α homolog is required for haploid filamentation but largely dispensable for mating and virulence in *Cryptococcus neoformans*. Genetics 153:1601–1615, 1999.
100. YC Chang, BL Wickes, GF Miller, LA Penoyer, KJ Kwon-Chung. *Cryptococcus neoformans STE12*α regulates virulence but is not essential for mating. J Exp Med 191:871–882, 2000.
101. YC Chang, KJ Kwon-Chung. Complementation of a capsule-deficient mutation of *Cryptococcus neoformans* restores its virulence. Mol Cell Biol 14:4912–4919, 1994.
102. YC Chang, LA Penoyer, KJ Kwon-Chung. The second capsule gene of *Cryptococ-*

cus neoformans, CAP64, is essential for virulence. Infect Immun 64:1977–1983, 1996.

103. SD Salas, JE Bennett, KJ Kwon-Chung, JR Perfect, PR Williamson. Effect of the laccase gene, *CNLAC1*, on virulence of *Cryptococcus neoformans*. J Exp Med 184: 377–386, 1996.
104. A Odom, S Muir, E Lim, DL Toffaletti, J Perfect, J Heitman. Calcineurin is required for virulence of *Cryptococcus neoformans*. EMBO J 16:2576–2589, 1997.
105. DS Fox, MC Cruz, RAL Sia, H Ke, GM Cox, ME Cardenas, J Heitman. Calcineurin regulatory subunit is essential for virulence and mediates interactions with FKBP12-FK506 in *Cryptococcus neoformans*. Mol Microbiol 39:835–849, 2001.
106. JR Perfect, DL Toffaletti, TH Rude. The gene encoding phosphoribosylaminoimidazole carboxylase (*ADE2*) is essential for growth of *Cryptococcus neoformans* in cerebrospinal fluid. Infect Immun 61:4446–4451, 1993.
107. GM Cox, J Mukherjee, GT Cole, A Casadevall, JR Perfect. Urease as a virulence factor in experimental cryptococcosis. Infect Immun 68:443–448, 2000.
108. SCA Chen, M Muller, JZ Zhou, LC Wright, TC Sorrell. Phospholipase activity in *Cryptococcus neoformans*: a new virulence factor? J Infect Dis 175:414–420, 1997.
109. GM Cox, HC McDade, SCA Chen, SC Tucker, M Gottredsson, LC Wright, TC Sorrell, SD Leidich, A Casadevall, MA Ghannoum, JR Perfect. Extracellular phospholipase activity is a virulence factor for *Cryptococcus neoformans*. Mol Microbiol 39:166–175, 2001.
110. KJ Kwon-Chung, JE Bennett. Distribution of α and a mating types of *Cryptococcus neoformans* among natural and clinical isolates. Am J Epidemiol 108:337–340, 1978.
111. TL Doering. How does *Cryptococcus* get its coat? Trends Microbiol 8:547–553, 2000.
112. F Bergman. Studies on capsule synthesis of *Cryptococcus neoformans*. Sabouraudia 4:23–31, 1965.
113. MA Dykstra, L Friedman, JW Murphy. Capsule size of *Cryptococcus neoformans*: control and relationship to virulence. Infect Immun 16:129–135, 1977.
114. ML Littman, E Tsubura. Effect of degree of encapsulation upon virulence of *Cryptococcus neoformans*. Proc Soc Exp Biol Med 101:773–776, 1959.
115. CA D'Souza, JA Alspaugh, C Yue, T Harashima, GM Cox, JR Perfect, J Heitman. cAMP dependent protein kinase controls mating and virulence of the fungal pathogen *Cryptococcus neoformans*. Mol Cell Biol 21:3179–3191, 2001.
116. JA Alspaugh, JR Perfect, J Heitman. *Cryptococcus neoformans* mating and virulence are regulated by the G-protein α subunit GPA1 and cAMP. Genes Dev 11: 3206–3217, 1997.
117. AK Bhattacharjee, JE Bennett, CP Glaudemans. Capsular polysaccharides of *Cryptococcus neoformans*. Rev Infect Dis 6:619–624, 1984.
118. E Reiss, EH White, R Cherniak, JE Dix. Ultrastructure of acapsular mutant *Cryptococcus neoformans* cap 67 and monosaccharide composition of cell extracts. Mycopathologia 93:45–54, 1986.
119. GS Bulmer, MD Sans. *Cryptococcus neoformans* III. Inhibition of phagocytosis. J Bacteriol 95:5–8, 1968.
120. TR Kozel. Non-encapsulated variant of *Cryptococcus neoformans*. II. Surface re-

ceptors for cryptococcal polysaccharide and their role in inhibition of phagocytosis by polysaccharide. Infect Immun 16:99–106, 1977.

121. TR Kozel, B Highison, CJ Stratton. Localization on encapsulated *Cryptococcus neoformans* of serum components opsonic for phagocytosis by macrophages and neutrophils. Infect Immun 43:574–579, 1984.
122. JM Small, TG Mitchell. Binding of purified and radioiodinated capsular polysaccharides from *Cryptococcus neoformans* serotype A strains to capsule-free mutants. Infect Immun 54:742–750, 1986.
123. JM Small, TG Mitchell. Strain variation in antiphagocytic activity of capsular polysaccharides from *Cryptococcus neoformans* serotype A. Infect Immun 57:3751–3756, 1989.
124. ES Jacobson, MJ Tingler, PL Quynn. Effect of hypertonic solutes upon the polysaccharide capsule in *Cryptococcus neoformans*. Mycoses 32:14–23, 1989.
125. ML Littman. Capsule synthesis by *Cryptococcus neoformans*. Trans NY Acad Sci 20:623–648, 1958.
126. SE Vartivarian, EJ Anaissie, RE Cowart, HA Sprigg, MJ Tingler, ES Jacobson. Regulation of cryptococcal capsular polysaccharide by iron. J Infect Dis 167:186–190, 1993.
127. DL Granger, JR Perfect, DT Durack. Virulence of *Cryptococcus neoformans*: regulation of capsule synthesis by carbon dioxide. J Clin Invest 76:508–516, 1985.
128. YC Chang, KJ Kwon-Chung. Isolation of the third capsule-associated gene, *CAP60*, required for virulence in *Cryptococcus neoformans*. Infect Immun 66:2230–2236, 1998.
129. YC Chang, KJ Kwon-Chung. Isolation, characterization, and localization of a capsule-associated gene, *CAP10*, of *Cryptococcus neoformans*. J Bacteriol 181:5636–5643, 1999.
130. YC Chang, R Cherniak, TR Kozel, DL Granger, LC Morris, LC Weinhold, KJ Kwon-Chung. Structure and biological activities of acapsular *Cryptococcus neoformans* 602 complemented with the *CAP64* gene. Infect Immun 65:1584–1592, 1997.
131. HH Gadebusch. Mechanisms of native and acquired resistance to infection with *Cryptococcus neoformans*. Crit Rev Microbiol 1:311–320, 1972.
132. SM Levitz, HL Mathews, JW Murphy. Direct antimicrobial activity of T cells. Immunol Today 16:387–391, 1995.
133. RD Diamond, RK Root, JE Bennett. Factors influencing killing of *Cryptococcus neoformans* by human leukocytes in vitro. J Infect Dis 125:367–376, 1972.
134. SM Levitz. The ecology of *Cryptococcus neoformans* and the epidemiology of cryptococcosis. Rev Infect Dis 13:1163–1169, 1991.
135. TR Kozel, EC Gotschlich. The capsule of *Cryptococcus neoformans* passively inhibits phagocytosis of the yeast by macrophages. J Immunol 129:1675–1680, 1982.
136. TR Kozel, GS Pfrommer, AS Guerlain, BA Highison, GJ Highison. Role of the capsule in phagocytosis of *Cryptococcus neoformans*. Rev Infect Dis 10(suppl 2):S436–9, 1988.
137. DL Goldman, SC Lee, AJ Mednick, L Montella, A Casadevall. Persistent *Cryptococcus neoformans* pulmonary infection in the rat is associated with intracellular parasitism, decreased inducible nitric oxide synthase expression, and altered anti-

body responsiveness to cryptococcal polysaccharide. Infect Immun 68:832–838, 2000.

138. TR Kozel. Activation of the complement system by pathogenic fungi. Clin Microbiol Rev 9:34–46, 1996.
139. SM Levitz, A Tabuni, SH Nong, DT Golenbock. Effects of interleukin-10 on human peripheral blood mononuclear cell responses to *Cryptococcus neoformans*, *Candida albicans*, and lipopolysaccharide. Infect Immun 64:945–951, 1996.
140. AM Macher, JE Bennett, JE Gadek, MM Frank. Complement depletion in cryptococcal sepsis. J Immunol 120:1686–1690, 1978.
141. E Drouhet. Milestones in the history of *Cryptococcus* and cryptococcosis. J Mycol Med 7:10–27, 1997.
142. ZM Dong, JW Murphy. Effects of the two varieties of *Cryptococcus neoformans* cells and culture filtrate antigens on neutrophil locomotion. Infect Immun 63:2632–2644, 1995.
143. C Retini, A Vecchiarelli, C Monari, C Tascini, F Bistoni, TR Kozel. Capsular polysaccharide of *Cryptococcus neoformans* induces proinflammatory cytokine release by human neutrophils. Infect Immun 64:2897–2903, 1996.
144. A Vecchiarelli, C Retini, D Pietrella, C Monari, C Tascini, T Beccari, TR Kozel. Downregulation by cryptococcal polysaccharide of tumor necrosis factor alpha and interleukin-1 beta secretion from human monocytes. Infect Immun 63:2919–2923, 1995.
145. A Vecchiarelli, C Retini, C Monari, C Tascini, F Bistoni, TR Kozel. Purified capsular polysaccharide of *Cryptococcus neoformans* induces interleukin-10 secretion by human monocytes. Infect Immun 64:2846–2849, 1996.
146. HL Collins, GJ Bancroft. Encapsulation of *Cryptococcus neoformans* impairs antigen-specific T-cell responses. Infect Immun 59:3883–3888, 1991.
147. JB Neilson, MH Ivey, GS Bulmer. *Cryptococcus neoformans*: pseudohyphal forms surviving culture with *Acanthamoeba polyphaga*. Infect Immun 20:262–266, 1978.
148. I Polacheck. The discovery of melanin production in *Cryptococcus neoformans* and its impact on diagnosis and the study of virulence. Zentralbl Bakteriol 276:120–123, 1991.
149. PR Williamson. Laccase and melanin in the pathogenesis of *Cryptococcus neoformans*. Front Biosci 2:e99–e107, 1997.
150. S Ito. Biochemistry and physiology of melanin. In: N Levine, ed. Pigmentation and Pigmentory Disorders. Ann Arbor, MI: CRC Press, 1993, pp 33–59.
151. F Staib. *Cryptococcus neoformans* und *Guizotia abyssinica* (Syn. *G. oleifera*) Farbreaktion für *Cr. neoformans*. Z Hyg 148:466–475, 1962.
152. CE Shaw, L Kapica. Production of diagnostic pigment by phenoloxidase activity of *Cryptococcus neoformans*. Appl Microbiol 24:824–830, 1972.
153. H Korth, G Pulverer. Pigment formation for differentiating *Cryptococcus neoformans* from *Candida albicans*. Appl Microbiol 21:541–542, 1971.
154. AA Strachan, RJ Yu, F Blank. Pigment production of *Cryptococcus neoformans* grown with extracts of *Guizotia abyssinica*. Appl Microbiol 22:478–479, 1971.
155. Y Wang, P Aisen, A Casadevall. Melanin, melanin ''ghosts,'' and melanin composition in *Cryptococcus neoformans*. Infect Immun 64:2420–2424, 1996.

156. JD Nosanchuk, AL Rosas, SC Lee, A Casadevall. Melanisation of *Cryptococcus neoformans* in human brain tissue. Lancet 355:2049–2050, 2000.
157. AL Rosas, JD Nosanchuk, M Feldmesser, GM Cox, HC McDade, A Casadevall. Synthesis of polymerized melanin by *Cryptococcus neoformans* in infected rodents. Infect Immun 68:2845–2853, 2000.
158. KJ Kwon-Chung, I Polacheck, TJ Popkin. Melanin-lacking mutants of *Cryptococcus neoformans* and their virulence for mice. J Bacteriol 150:1414–1421, 1982.
159. KJ Kwon-Chung, JC Rhodes. Encapsulation and melanin formation as indicators of virulence in *Cryptococcus neoformans*. Infect Immun 51:218–223, 1986.
160. I Polacheck, Y Platt, J Aronovitch. Catecholamines and virulence of *Cryptococcus neoformans*. Infect Immun 58:2919–2922, 1990.
161. JC Rhodes, I Polacheck, KJ Kwon-Chung. Phenoloxidase activity and virulence in isogenic strains of *Cryptococcus neoformans*. Infect Immun 36:1175–1184, 1982.
162. H Torres-Guererro, JC Edman. Melanin-deficient mutants of *Cryptococcus neoformans*. J Med Vet Mycol 32:303–313, 1994.
163. PR Williamson. Biochemical and molecular characterization of the diphenol oxidase of *Cryptococcus neoformans*: identification as a laccase. J Bacteriol 176:656–664, 1994.
164. I Polacheck, KJ Kwon-Chung. Melanogenesis in *Cryptococcus neoformans*. J Gen Microbiol 134:1037–1041, 1988.
165. ES Jacobson, ND Jenkins, JM Todd. Relationship between superoxide dismutase and melanin in a pathogenic fungus. Infect Immun 62:4085–4086, 1994.
166. Y Wang, P Aisen, A Casadevall. *Cryptococcus neoformans* melanin and virulence: mechanism of action. Infect Immun 63:3131–3136, 1995.
167. Y Wang, A Casadevall. Decreased susceptibility of melanized *Cryptococcus neoformans* to UV light. Appl Environ Microbiol 60:3864–3866, 1994.
168. AL Rosas, A Casadevall. Melanization affects susceptibility of *Cryptococcus neoformans* to heat and cold. FEMS Microbiol Lett 153:265–272, 1997.
169. Y Kojima, Y Tsukuda, Y Kawai, A Tsukamoto, J Sugiura, M Sakaino, Y Kita. Cloning, sequence analysis, and expression of ligninolytic phenoloxidase genes of the white-rot basidiomycete *Coriolus hirsutus*. J Biol Chem 265:15224–15230, 1990.
170. ES Jacobson, HS Emery. Catecholamine uptake, melanization, and oxygen toxicity in *Cryptococcus neoformans*. J Bacteriol 173:401–403, 1991.
171. ES Jacobson, SB Tinnell. Antioxidant function of fungal melanin. J Bacteriol 175: 7102–7104, 1993.
172. Y Wang, A Casadevall. Susceptibility of melanized and nonmelanized *Cryptococcus neoformans* to nitrogen- and oxygen-derived oxidants. Infect Immun 62:3004–3007, 1994.
173. JD Nosanchuk, A Casadevall. Cellular charge of *Cryptococcus neoformans*: contributions from the capsular polysaccharide, melanin, and monoclonal antibody binding. Infect Immun 65:1836–1841, 1997.
174. Y Wang, A Casadevall. Growth of *Cryptococcus neoformans* in presence of L-dopa decreases its susceptibility to amphotericin B. Antimicrob Agents Chemother 38:2648–2650, 1994.
175. BL Wickes, U Edman, JC Edman. The *Cryptococcus neoformans* *STE12*α gene:

a putative *Saccharomyces cerevisiae STE12* homologue that is mating type specific. Mol Microbiol 26:951–960, 1997.

176. M Karos, YC Chang, CM McClelland, DL Clarke, J Fu, BL Wickes, KJ Kwon-Chung. Mapping of the *Cryptococcus neoformans* MATα locus: presence of mating type–specific mitogen-activated protein kinase cascade homologs. J Bacteriol 182: 6222–6227, 2000.
177. CM Hull, AD Johnson. Identification of a mating type-like locus in the asexual pathogenic yeast *Candida albicans*. Science 285:1271–1275, 1999.
178. CM Hull, RM Raisner, AD Johnson. Evidence for mating of the ''asexual'' yeast *Candida albicans* in a mammalian host. Science 289:307–310, 2000.
179. BB Magee, PT Magee. Induction of mating in *Candida albicans* by construction of MTLa and MTLα strains. Science 289:310–313, 2000.
180. KJ Kwon-Chung, JC Edman, BL Wickes. Genetic association of mating types and virulence in *Cryptococcus neoformans*. Infect Immun 60:602–605, 1992.
181. A Kamiyama, M Niimi, M Tokunaga, H Nakayama. Adansonian study of *Candida albicans*: intraspecific homogeneity excepting *C. stellatoidea* strains. J Med Vet Mycol 27:229–241, 1989.
182. A Varma, JC Edman, KJ Kwon-Chung. Molecular and genetic analysis of *URA5* transformants of *Cryptococcus neoformans*. Infect Immun 60:1101–1108, 1992.
183. ES Jacobson, SE Vartivarian. Iron assimilation in *Cryptococcus neoformans*. J Med Vet Mycol 30:443–450, 1992.
184. KJ Nyhus, AT Wilborn, ES Jacobson. Ferric iron reduction by *Cryptococcus neoformans*. Infect Immun 65:434–438, 1997.
185. ES Jacobson, MJ Petro. Extracellular iron chelation in *Cryptococcus neoformans*. J Med Vet Mycol 25:415–418, 1987.
186. F Tesfa-Selase, RJ Hay. Partial characterization and identification of a transferrin-like molecule of pathogenic yeast *Cryptococcus neoformans*. J Gen Appl Microbiol 42:61–70, 1996.
187. WG Niehaus, T Flynn. Regulation of mannitol biosynthesis and degradation by *Cryptococcus neoformans*. J Bacteriol 176:651–655, 1994.
188. B Wong, JR Perfect, S Beggs, KA Wright. Production of the hexitol D-mannitol by *Cryptococcus neoformans* in vitro and in rabbits with experimental meningitis. Infect Immun 58:1664–1670, 1990.
189. V Chaturvedi, T Flynn, WG Nichaus, B Wong. Stress tolerance and pathogenic potential of a mannitol-mutant of *Cryptococcus neoformans*. Microbiology 142: 937–943, 1996.
190. GM Megson, DA Stevens, JR Hamilton, DW Denning. D-mannitol in cerebrospinal fluid of patients with AIDS and cryptococcal meningitis. J Clin Microbiol 34:218–221, 1996.
191. DW Denning, RW Armstrong, BH Lewis, DA Stevens. Elevated cerebrospinal fluid pressures in patients with cryptococcal meningitis and acquired immunodeficiency syndrome. Am J Med 91:267–272, 1991.
192. V Chaturvedi, B Wong, SL Newman. Oxidative killing of *Cryptococcus neoformans* by human neutrophils. Evidence that fungal mannitol protects by scavenging reactive oxygen intermediates. J Immunol 156:3836–3840, 1996.
193. SC Chen, LC Wright, RT Santangelo, M Muller, VR Moran, PW Kuchel, TC Sor-

rell. Identification of extracellular phospholipase B, lysophospholipase, and acyltransferase produced by *Cryptococcus neoformans*. Infect Immun 65:405–411, 1997.

194. S Aoki, S Ito-Kuwa, K Nakamura, J Kato, K Ninomiya, V Vidotto. Extracellular proteolytic activity of *Cryptococcus neoformans*. Mycopathologia 128:143–150, 1994.
195. CH Brueske. Proteolytic activity of a clinical isolate of *Cryptococcus neoformans*. J Clin Microbiol 23:631–633, 1986.
196. LC Chen, ES Blank, A Casadevall. Extracellular proteinase activity of *Cryptococcus neoformans*. Clin Diagn Lab Immunol 3:570–574, 1996.
197. SD Leidich, AS Ibrahim, Y Fu, A Koul, C Jessup, J Vitullo, W Fonzi, F Mirbod, S Nakashima, Y Nozawa, MA Ghannoum. Cloning and disruption of *caPLB1*, a phospholipase B gene involved in the pathogenicity of *Candida albicans*. J Biol Chem 273:26078–26086, 1998.
198. V Vidotto, A Sinicco, DD Fraia, S Cardaropolis, S Aoki, S Ito-Kuwa. Phospholipase activity in *Cryptococcus neoformans*. Mycopathologia 136:119–123, 1997.
199. P Wang, J Heitman. Signal transduction cascades regulating mating, filamentation, and virulence in *Cryptococcus neoformans*. Curr Opin Microbiol 2:358–362, 1999.
200. KB Lengeler, RC Davidson, C D'Souza, T Harashima, W-C Shen, P Wang, X Pan, M Waugh, J Heitman. Signal transduction cascades regulating fungal development and virulence. Microbiol Mol Biol Rev 64:746–785, 2000.
201. H Liu, CA Styles, GR Fink. Elements of the yeast pheromone response pathway required for filamentous growth of diploids. Science 262:1741–1744, 1993.
202. CJ Gimeno, PO Ljungdahl, CA Styles, GR Fink. Unipolar cell divisions in the yeast *S. cerevisiae* lead to filamentous growth: regulation by starvation and *RAS*. Cell 68:1077–1090, 1992.
203. JG Cook, L Bardwell, J Thorner. Inhibitory and activating functions for MAPK Kss1 in the *S. cerevisiae* filamentous-growth signaling pathway. Nature 390:85–88, 1997.
204. HD Madhani, CA Styles, GR Fink. MAP kinases with distinct inhibitory functions impart signaling specificity during yeast differentiation. Cell 91:673–684, 1997.
205. V Gavrias, A Andrianopoulos, CJ Gimeno, WW Timberlake. *Saccharomyces cerevisiae TEC1* is required for pseudohyphal growth. Mol Microbiol 19:1255–1263, 1996.
206. HD Madhani, GR Fink. Combinatorial control required for the specificity of yeast MAPK signaling. Science 275:1314–1317, 1997.
207. HA Hartmann, R Kahmann, M Bölker. The pheromone response factor coordinates filamentous growth and pathogenicity in *Ustilago maydis*. EMBO J 15:1632–1641, 1996.
208. A Sugimoto, Y Iino, T Maeda, Y Watanabe, M Yamamoto. *Schizosaccharomyces pombe ste11*$^{+}$ encodes a transcription factor with an HMG motif that is a critical regulator of sexual development. Genes Dev 5:1990–1999, 1991.
209. J Hayles, P Nurse. Genetics of the fission yeast *Schizosaccharomyces pombe*. Annu Rev Genet 26:373–402, 1992.
210. N Mochizuki, M Yamamoto. Reduction in the intracellular cAMP level triggers initiation of sexual development in fission yeast. Mol Gen Genet 233:17–24, 1992.

211. R Kahmann, C Basse, M Feldbrugge. Fungal–plant signaling in the *Ustilago maydis*–maize pathosystem. Curr Opin Microbiol 2:647–50, 1999.
212. JA Alspaugh, JR Perfect, J Heitman. Signal transduction pathways regulating differentiation and pathogenicity of *Cryptococcus neoformans*. Fungal Genet Biol 25: 1–14, 1998.
213. T Tolkacheva, P McNamara, E Piekarz, W Courchesne. Cloning of a *Cryptococcus neoformans* gene, *GPA1*, encoding a G-protein α-subunit homolog. Infect Immun 62:2849–2856, 1994.
214. Y Xue, M Batlle, JP Hirsch. *GPR1* encodes a putative G protein–coupled receptor that associates with the Gpa2p Gα subunit and functions in a Ras-independent pathway. EMBO J 17:1996–2007, 1998.
215. MC Lorenz, X Pan, T Harashima, ME Cardenas, Y Xue, JP Hirsch, J Heitman. The G protein–coupled receptor GPR1 is a nutrient sensor that regulates pseudohyphal differentiation in *Saccharomyces cerevisiae.* Genetics 154:609–622, 2000.
216. Y Jiang, C Davis, JR Broach. Efficient transition to growth on fermentable carbon sources in *Saccharomyces cerevisiae* requires signaling through the Ras pathway. EMBO J 17:6942–6951, 1998.
217. T Toda, I Uno, T Ishikawa, S Powers, T Kataoka, D Broek, S Cameron, J Broach, K Matsumoto, M Wigler. In yeast, *RAS* proteins are controlling elements of adenylate cyclase. Cell 40:27–36, 1985.
218. JA Alspaugh, LM Cavallo, JR Perfect, J Heitman. *RAS1* regulates filamentation, mating and growth at high temperature of *Cryptococcus neoformans*. Mol Microbiol 36:352–365, 2000.
219. CS Hemenway, J Heitman. Calcineurin: structure, function, and inhibition. Cell Biochem Biophys 30:115–151, 1999.
220. RC Davidson, MC Cruz, RAL Sia, BM Allen, JA Alspaugh, J Heitman. Gene disruption by biolistic transformation in serotype D strains of *Cryptococcus neoformans*. Fungal Genet Biol 29:38–48, 2000.
221. MC Cruz, DS Fox, J Heitman. Calcineurin is required for hyphal elongation during mating and haploid fruiting in *Cryptococcus neoformans*. EMBO J 20:1020–1032, 2001.
222. TN Kirkland, J Fierer. Cyclosporin A inhibits *Coccidioides immitis* in vitro and in vivo. Antimicrob Agents Chemother 24:921–924, 1983.
223. M Tropschug, IB Barthelmess, W Neupert. Sensitivity to cyclosporin A is mediated by cyclophilin in *Neurospora crassa* and *Saccharomyces cerevisiae*. Nature 342: 953–955, 1989.
224. A Odom, MD Poeta, J Perfect, J Heitman. The immunosuppressant FK506 and its nonimmunosuppressive analog L-685,818 are toxic to *Cryptococcus neoformans* by inhibition of a common target protein. Antimicrob Agents Chemother 41:156–161, 1997.
225. MC Cruz, MD Poeta, P Wang, R Wenger, G Zenke, VFJ Quesniaux, NR Movva, JR Perfect, ME Cardenas, J Heitman. Immunosuppressive and nonimmunosuppressive cyclosporin analogs are toxic to the opportunistic fungal pathogen *Cryptococcus neoformans* via cyclophilin dependent inhibition of calcineurin. Antimicrob Agents Chemother 44:143–149, 2000.
226. T Breuder, CS Hemenway, NR Movva, ME Cardenas, J Heitman. Calcineurin is

essential in cyclosporin A- and FK506-sensitive yeast strains. Proc Natl Acad Sci USA 91:5372–5376, 1994.
227. J Liu, JD Farmer, WS Lane, J Friedman, I Weissman, SL Schreiber. Calcineurin is a common target of cyclophilin–cyclosporin A and FKBP-FK506 complexes. Cell 66:807–815, 1991.
228. JR Perfect, DT Durack. Effects of cyclosporine in experimental cryptococcal meningitis. Infect Immun 50:22–26, 1985.
229. HD Birley, EM Johnson, P McDonald, C Parry, PB Carey, DW Warnock. Azole drug resistance as a cause of clinical relapse in AIDS patients with cryptococal meningitis. Int J STD AIDS 6:353–355, 1995.
230. JP Viard, C Hennequin, N Fortineau, N Pertuiset, C Rothschild, H Zylberberg. Fulminant cryptococcal infections in HIV-infected patients on oral fluconazole. Lancet 346:118, 1995.
231. A Paugam, J Camet-Dupouy, P Blanche, JP Gangneux, C Tourte-Schaefer, D Sicard. Increased fluconazole resistance of *Cryptococcus neoformans* isolated from a patient with AIDS and recurrent meningitis. Clin Infect Dis 19:975–976, 1994.
232. J Berg, CJ Clancy, MH Nguyen. The hidden danger of primary fluconazole prophylaxis for patients with AIDS. Clin Infect Dis 26:186–187, 1998.
233. N Clumeck, J Sonnet, H Taelman, F Mascart-Lemone, MD Bruyere, P Vandeperre, J Dasnoy, L Marcelis, M Lamy, C Jonas, L Eyckmans, H Noel, M Vanhaeverbeek, JP Butzler. Acquired immunodeficiency syndrome in African patients. N Engl J Med 310:492–497, 1984.
234. D Maher, H Mwandumba. Cryptococcal meningitis in Lilongwe and Blentyre, Malawi. J Infect 28:59–64, 1994.
235. DL Toffaletti, TH Rude, SA Johnston, DT Durack, JR Perfect. Gene transfer in *Cryptococcus neoformans* by use of biolistic delivery of DNA. J Bacteriol 175: 1405–1411, 1993.
236. JC Edman. Isolation of telomerelike sequences from *Cryptococcus neoformans* and their use in high-efficiency transformation. Mol Cell Biol 12:2777–2783, 1992.
237. JK Lodge, E Jackson-Machelski, DL Toffaletti, JR Perfect, JI Gordon. Targeted gene replacement demonstrates that myristoyl-CoA: protein *N*-myristoyltransferase is essential for viability of *Cryptococcus neoformans*. Proc Natl Acad Sci USA 91: 12008–12012, 1994.
238. JR Perfect, SDR Lang, DT Durack. Chronic cryptococcal meningitis: a new experimental model in rabbits. Am J Pathol 101:177–194, 1980.
239. D Goldman, Y Cho, M Zhao, A Casadevall, SC Lee. Expression of inducible nitric oxide synthase in rat pulmonary *Cryptococcus neoformans* granulomas. Am J Pathol 148:1275–1282, 1996.
240. J Heitman, A Casadevall, JK Lodge, JR Perfect. The *Cryptococcus neoformans* genome sequencing project. Mycopathologia 148:1–7, 1999.

20

Molecular Mechanisms of Pathogenicity of *Aspergillus fumigatus*

Axel A. Brakhage
University of Hannover, Hannover, Germany

Bernhard Jahn
Johannes Gutenberg-Universität Mainz, Mainz, Germany

1 INTRODUCTION

Fungal pathogens have begun to rival their bacterial counterparts as a significant problem in many settings. Death from fungal infection and bacterial infection are now almost equal in number in neutropenic patients. Besides the rise in occurrence of nosocomial yeast infections, infections with mold pathogens have emerged as an increasing risk faced by patients under sustained immunosuppression. Species of the *Aspergillus* family account for most of these infections and in particular *Aspergillus fumigatus* can be regarded as the primary mold pathogen. The improvement in transplant medicine and the therapy of hematological malignancies is often complicated by the threat of invasive aspergillosis. Specific diagnostics are still limited as are the possibilities of therapeutic intervention, leading to the disappointing fact that invasive aspergillosis is still associated with a high mortality rate that ranges from 30–90%. For example, during the past 15 years, invasive aspergillosis has become the main cause of death in patients with acute leukemia and liver transplantation [1, reviewed in 2]. However, the spectrum of diseases related to *A. fumigatus* is rather heterogeneous and exceeds that of a mere opportunistic pathogen: Colonization with restricted invasiveness is seen in the immunocompetent host, disseminated infections in the immunocompromised patients, and allergic disorders are known for long to be caused by *A. fumigatus* antigens [see articles in 3].

The various aspects of *A. fumigatus*, i.e., virulence factors, diseases caused by the fungus, treatment, diagnosis, etc., were the subject of several reviews [reviewed in 3–5]. Comprehensive information on all aspects of *A. fumigatus* can also be obtained from the *Aspergillus* Web site: www.aspergillus.man.ac.uk/. The identification of virulence factors associated with *A. fumigatus* conidia, in particular the conidial pigment, has seen an increase in knowledge in recent years, and it is this aspect that is considered most in the present review.

2 BIOLOGY AND GENETICS OF *ASPERGILLUS FUMIGATUS*

The genus *Aspergillus* contains >180 species [6], including *A. fumigatus*, a saprobic fungus, associated with decaying organic matter such as compost and hay. *A. fumigatus* plays an essential role in recycling carbon and nitrogen sources [7,8]. Conidia of this fungus can be isolated nearly everywhere, from the winds of the Sahara to the snows of the Antarctic [9]. *A. fumigatus* conidia often account for a surprisingly small fraction of the *Aspergillus* conidia in the environment—e.g., 0.3% of the aerial *Aspergillus* conidia in a particular cancer hospital [10]. In temperate climates, seasonal concentration shifts of airborne *A. fumigatus* conidia have been observed with minimum values of ~5/m^3 in the autumn, and peaks of >30/m^3 in the early spring in an urban environment [10].

A. fumigatus is characterized by gray-green echinulate conidia (spores), 2.5–3 μm in diameter, so that they can reach the lung alveoli. The same may be true for several other *Aspergillus* spp. [11,12; reviewed in 13] (Fig. 1). Conidia are produced in chains basipetally from greenish phialides, 6–8 by 2–3 μm in size. Phialides are flask-shaped and confined to the apical part (Fig. 1). Conidiophores are uniseriate; i.e., there is only a single row of cells (phialides) borne directly on broadly clavate vesicles (20–30 μm in diameter). Metulae are lacking [reviewed in 5]. The first stage in conidia formation is a constriction of an elongated portion of the phialide; the second stage is the development of a septum that separates the conidium from the phialide. Chains of attached conidia evolve as the cutting process continues at the tip of the phialide. The compact mass of conidia around the vesicle is called a conidial head and corresponds to the portion of a brush in which the bristles are set. At its opposite end, the conidiophore arises from a foot cell, an enlarged, thick-walled cell of the segmented mycelium.

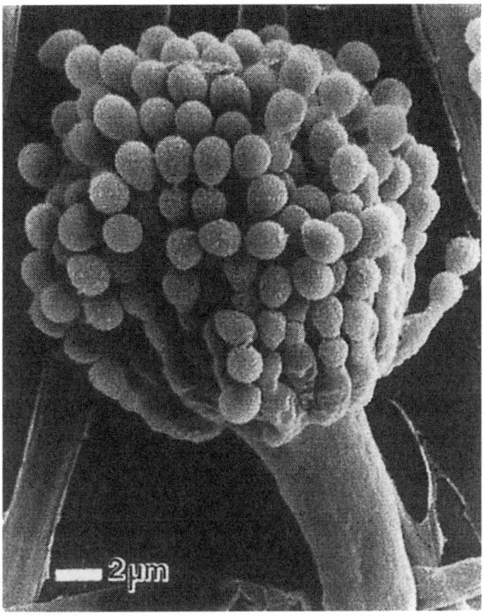

FIGURE 1 Scanning electron microscopy photograph of *A. fumigatus*. (From Ref. 70.)

The conidiophore is borne as a stalk perpendicular to the long axis of the foot cell [9]. *A. fumigatus* has smooth-walled stipes. Hülle cells and sclerotia are unknown [6] (Fig. 1).

A. fumigatus grows quickly on Czapek-Dox agar, but also on minimal agar plates with a carbon source (e.g., glucose) and nitrogen source (e.g., nitrate), and trace elements. *A. fumigatus* can withstand extremes of temperatures and pH. It is a thermophilic species, with growth occurring as high as 55°C and survival maintained at temperatures up to 70°C [reviewed in 5,6]. *A. fumigatus* is morphologically more variable than was originally described by Raper and Fennell [6,14]. In particular clinical isolates can be markedly abnormal and are often more floccose with fewer conidia [6]. These variations have led to the description of several varieties of *A. fumigatus*, including *acolumnaris*, *phialiseptus*, *ellipticus*, and *sclerotiorum*, with the distinctions being based on only slight morphological differences [reviewed in 5].

The existence of a teleomorph of *A. fumigatus* is a matter of debate. Until now, no sexual stage is known for this species. It was proposed that the genus *Neosartorya* might represent the teleomorph. However, DNA-DNA reassociation values, wherein values >92% have been found for strains of *A. fumigatus*, while values <70% have been calculated for *A. fumigatus* and *Neosartorya* species, indicated that the two genera are genetically distinct [15].

The lack of a sexual cycle which would have been useful to exchange genetic markers and has been used extensively for the genetic analysis of other fungi, such as *Aspergillus* (*Emericella*) *nidulans*, makes an analysis based on the presence of a sexual cycle impossible. In the early 1960s [16,17] parasexuality was described for *A. fumigatus*. Its use in classical genetic analysis, however, is only poorly developed. On the other hand, *A. fumigatus* has some physiological characteristics which allow precise molecular analyses. The gray-green conidia of the fungus are uninucleate [18]. The nuclei of *A. fumigatus* are haploid. Both characteristics enable the isolation of clones from conidia and facilitate the isolation of mutants using classical or molecular techniques [19].

Pulsed-field gel electrophoresis has shown the presence of five chromosomal-size DNA bands ranging in size from 1.7 to 4.8 Mb [20]. It is not known whether this is the total number of chromosomes or whether some bands represent double bands which would increase the total number of chromosomes.

3 VIRULENCE FACTORS

3.1 Overview

One of the important questions concerning *A. fumigatus* is the identification of virulence factors. Since *A. fumigatus* is a saprobic fungus, associated with decaying organic matter such as compost and hay [7,8], it is debatable whether *A. fumigatus* has specific virulence factors [21,22]. However, it is becoming more and more obvious that the fungus has certain physiological characteristics that enable it to avoid or suppress the residual immune system in immunocompromised patients, making it an aggressive opportunistic pathogen. In the following, such a physiological characteristic will be called a virulence factor because it seems to distinguish *A. fumigatus* at least from the nonpathogenic species *A. nidulans*. Recently, we and others have identified a gene which encodes a virulence factor. It was designated *pksP* (or alternatively *alb1*) for polyketide synthase involved in pigment biosynthesis [23–26].

Using the strategy to delete a single gene and test such deletion strains in a mouse infection model, a reduction in virulence was observed only for the *pksP* gene (see below). Since *pksP* mutants were still virulent in a mouse infection model, additional factors seem to be required for virulence, supporting the hypothesis [5] that virulence of *A. fumigatus* is a multifactorial process. However, for none of the other genes analyzed until now, such as protease-encoding genes, or a gene encoding an RNAse, was a significant reduction in virulence observed when the respective gene was deleted [reviewed in 5].

Only for genes which are essential for vegetative growth, like *pyrG*- and *pabaA*-encoding proteins involved in the pyrimidine and *p*-aminobenzoic acid biosynthesis, respectively, did their mutation lead to avirulent strains when tested in a mouse infection model [27,28]. However, since these strains are auxotrophic and do not grow on agar plates without supplementation with uracil and *p*-aminobenzoic acid, respectively, these genes cannot be regarded as virulence factors. In the remainder of this chapter, the *pksP* gene and the putative mechanisms by which it might contribute to virulence are discussed.

3.2 Interaction of *A. fumigatus* Conidia with the Host

Because of their abundance in air, *A. fumigatus* conidia are continuously inhaled during routine daily activities. It is estimated that every day humans inhale several hundred *A. fumigatus* conidia [29–31; reviewed in 5]. In immunosuppressed patients, the lung is the site of infection of *A. fumigatus*. In immunocompetent persons, mucociliary clearance and phagocytic defense normally prevent the disease. Alveolar macrophages are the major resident cells of the lung alveoli; along with neutrophils (which are actively recruited during inflammation), they are the major cells in the phagocytosis of *A. fumigatus*. Until now, it was unclear by which mechanism macrophages kill conidia [reviewed in 5]. Moreover, a 100% killing of inhaled conidia by alveolar macrophages has never been reported (Fig. 2). Furthermore, conidia can germinate in monocytes, suggesting an essential role for the second line of phagocytic cells—the neutrophils. It is believed that the primary target of neutrophils is hyphae. However, polymorphonuclear leukocytes (PMN) are also able to kill resting or swollen conidia [5]. Contact between neutrophils and hyphae triggers a respiratory burst, secretion of reactive oxygen intermediates, and degranulation [32–38; reviewed in 5]. A severely depressed immune system, however, provides an opportunity for conidia to germinate and invade lung tissue [reviewed in 13,39] (Fig. 2).

3.3 The *pksP* Gene Encodes a Polyketide Synthase Involved in Conidial Pigment Biosynthesis and Virulence

3.3.1 Identification and Characterization of the *pksP* Gene

It was conceivable that one of the factors that contribute to conidial survival in the human host might be associated with the gray-green conidia because in contrast to other molds, they are able to survive the attack of the residual immune system of immunosuppressed patients, to germinate and invade the tissue [23]. Such a factor could be represented by a diffusible toxin associated with the conidia. This assumption would be in line with several publications reporting that diffusible

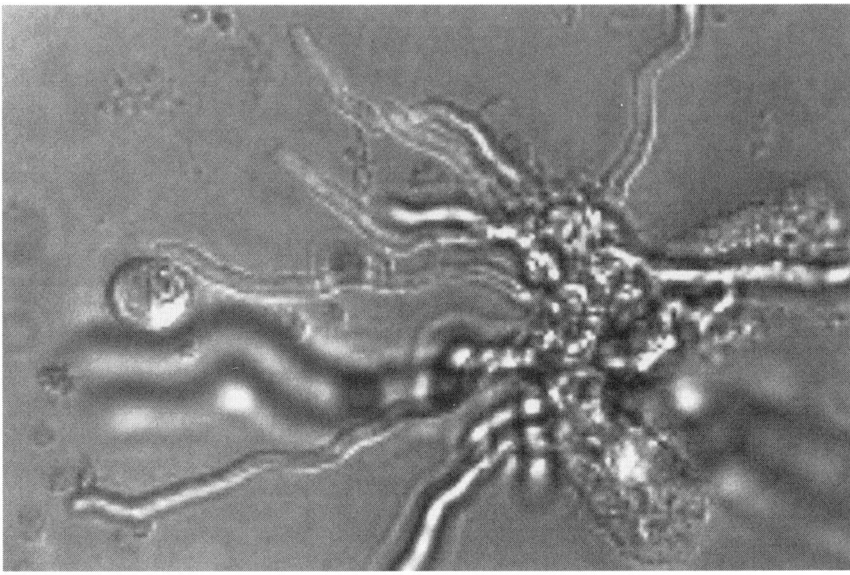

FIGURE 2 Hyphae of *A. fumigatus* grow out of a macrophage.

compounds associated with the fungus have all kinds of toxic effects on immune effector cells [40–55]. Until now, however, only one of these putative compounds, gliotoxin, has been purified and its structure elucidated. Evidence for the necessity of gliotoxin during the infectious process by rigorous genetic analyses, e.g., by deletion of a biosynthesis gene involved, has not been provided. Alternatively, the conidial pigment itself could contribute to resistance against the attack of immune effector cells. The latter reasoning was also based on results obtained by the analysis of other fungi. One example is *Cryptococcus neoformans*, a fungus with a pathogenic yeast phase, which causes life-threatening infections particularly in AIDS patients [56]. It has been shown that production of a melaninlike pigment derived from dihydroxyphenylalanine (DOPA) is a virulence-determining factor in *C. neoformans* [57–60; reviewed in 61]. One mechanism by which pigment might contribute to virulence derives from its ability to confer some resistance to reactive oxygen species (ROS), a major host antimicrobial effector system also active against *Aspergillus* conidia [62–64]. For the dematiacous fungus *Wangiella dermatitidis*, a main causative agent of pheohyphomycosis, a similar biological function of a fungal pigment derived from 1,8-dihydroxynaphthalene (DHN)–melanin was demonstrated [65–67].

To investigate the importance of factors associated with conidia for the virulence of *A. fumigatus* by UV mutagenesis, a mutant of *A. fumigatus* was isolated which lacked the ability to produce the conidial pigment and therefore had pigmentless, white conidia. In an intravenous murine infection model, the virulence of conidia of the white mutant strain was significantly reduced to ~20–50% of that observed with wild-type (wt) conidia. On minimal medium agar plates with glucose as the sole carbon source, at 28°C and 37°C no difference in growth rate of the white mutant strain compared with the wt was observed. Furthermore, germination rates were found to be the same for both strains. However, scanning electron microscopy revealed that wt conidia showed a rough surface covered with ornaments, whereas conidia of the white mutant had a smooth surface showing hardly any ornaments on the conidial wall [23].

Isolation of the white mutant strain enabled us to analyze whether the conidial pigment was of importance as protectant against ROS. Nonpigmented white conidia were 10-fold more susceptible toward the oxidants H_2O_2 and NaOCl than wt conidia, suggesting that the pigment plays an important role as a protectant against ROS [23]. Moreover, white and wt conidia were compared with respect to their capacity to stimulate an oxidative response in human phagocytes and their intracellular survival in human monocytes. The amount of measurable ROS was 10-fold higher when human neutrophils or monocytes were challenged with white conidia compared with wt conidia. In addition, white conidia were more efficiently damaged by human monocytes in vitro than wt conidia [23].

By using a genomic cosmid library with the hygromycin B phosphotransferase gene of *Escherichia coli* as the selectable marker, a gene complementing the white mutant to the wt phenotype was independently isolated by Langfelder et al. [24] and Tsai et al. [26]. Although there were marginal differences in the DNA sequences reported, the two research groups seem to have isolated the same gene. Since its deduced amino acid sequence exhibits a high degree of similarity to polyketide synthases, the gene was designated *pksP* for polyketide synthase involved in pigment biosynthesis (EMBL accession No. Y17317) [24] and *alb1* (Genbank accession No. g3136091) [26]. In the remainder of this chapter the name *pksP* is used. The *pksP* gene has a size of 6660 bp. Its encoded ORF is interrupted by four introns of 47, 47, 55, and 73 bp length (Fig. 3). The deduced PksP protein is composed of 2146 amino acids with a calculated molecular mass of 234.3 kDa [24].

White (*pksP*) mutant strains complemented to the gray-green phenotype of wt conidia using the *pksP* gene also showed a wt ornamentation of the conidial surface. In addition, they triggered the same ROS release by PMN as wt conidia which was 10-fold less than the amount of ROS released by PMN when challenged with *pksP* mutant conidia. In an intravenous murine infection model, the virulence of transformants complemented to the wt was restored and corresponded to that of wt conidia [24]. In addition, the disruption of the *pksP* (*alb1*) gene resulted in the phenotypes of conidia characteristic of white conidia, i.e., white conidia, lack of ornamentation, and reduced virulence of conidia of the disruptant strain in an intravenous murine infection model [26].

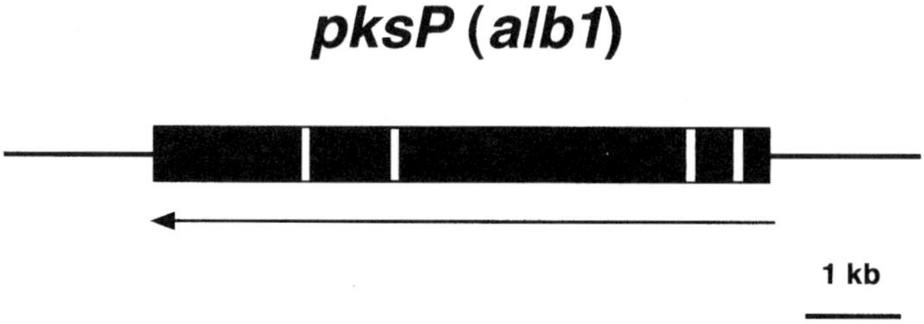

Figure 3 Schematic diagram of the *pksP* gene. The ORF of *pksP* is indicated by an arrow; the four introns are boxed. (From Ref. 24.)

As mentioned above, the deduced amino acid sequence of PksP showed extensive sequence similarities with fungal polyketide synthases, in particular with the WA protein of *A. nidulans* (60.8% identical amino acids) and Pks1 of *Colletotrichum lagenarium*, indicating that *pksP* encodes a polyketide synthase. According to the definition of polyketide synthases [68], the *pksP* gene encodes a typical fungal type I polyketide synthase. PksP shares β-ketotransferase, acyltransferase, acyl carrier protein, and thioesterase motifs with other fungal polyketide synthases. Therefore, *pksP* encodes a polyketide synthase involved in the formation of the *A. fumigatus* conidial pigment [24,26].

3.3.2 The *pksP* Gene Is Part of a Cluster of 1,8-Dihydroxynaphthalene (DHN)–Melanin Biosynthesis Genes

The *pksP* gene is part of a cluster [24,26,69,70] (Fig. 4). The *arp1* gene is located upstream of *pksP*. *Arp1* has a size of 670 bp. The ORF of *arp1* is interrupted by two introns and encodes a protein of 168 amino acids. The *arp1* gene product is very similar to scytalone dehydratase, an enzyme involved in 1,8-dihydroxynaphthalene (DHN)–melanin synthesis in the plant-pathogenic fungi *C. lagenarium* and *Magnaporthe grisea* [71] (Fig. 5). Deletion of *arp1* resulted in reddish-

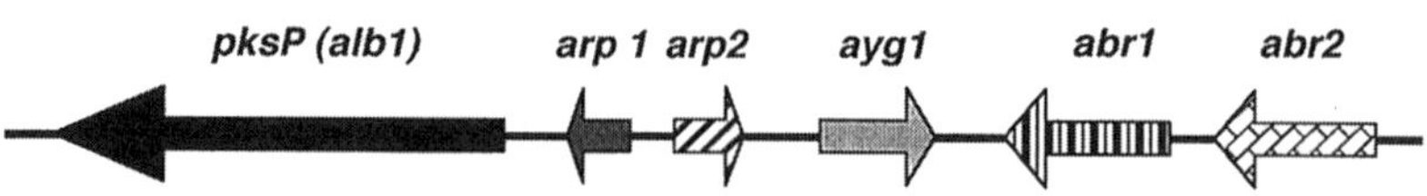

Figure 4 Gene cluster of DHN-melanin biosynthesis according to Tsai et al. (From Ref. 69.)

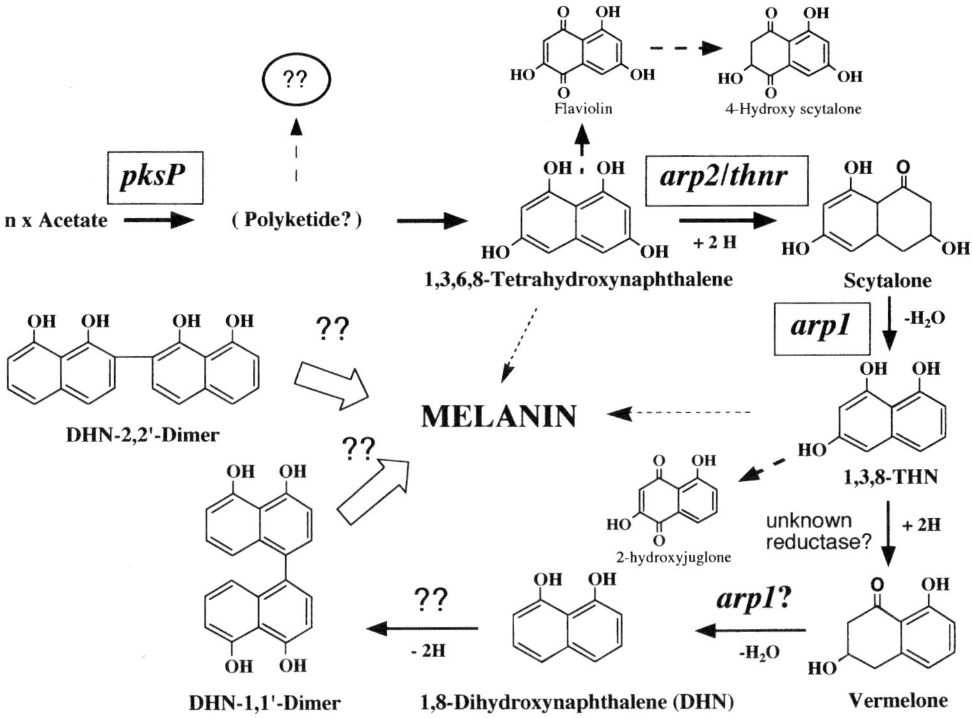

FIGURE 5 DHN-melanin biosynthesis pathway modified on the basis of Butler and Day [61]. The reactions which are assumed to be catalyzed by the products of the *A. fumigatus pksP, arp1,* and *arp2(thnr)* gene are indicated. (From Refs. 69, 70.)

pink conidia, indicating its involvement in the synthesis of the conidial pigment, and increased C3 binding to conidia [71]. For both *arp1* and *pksP* it was shown that the appearance of their mRNA was developmentally regulated; i.e., the respective mRNA of 0.8 kb and 7.0 kb, respectively, was detectable in conidia but not in hyphae [26,69]. Moreover, recent studies using the green fluorescent protein-encoding gene as a reporter suggest that the *pksP* gene might also be expressed in hyphae under certain environmental conditions [71a].

Downstream of *arp1* is a gene located with high similarity to 1,3,6,8-tetrahydroxynapthalene (THN) reductase genes. It was designated *arp2* [69] and *thnR* (Langfelder and Brakhage, Acc. No. Y19033) (Fig. 4). It is very likely that the *arp2* (*thnR*) gene product is also involved in the DHN-melanin pathway since its deletion also led to reddish-pink colonies [69] (Figs. 4, 5). Furthermore, based on similarity comparison, the *arp2* gene appears to encode a hydroxynaphthalene (HN) reductase. This assumption well agrees with the observation that when the wt strain was grown on 8 μg/mL tricyclazole, it produced reddish pink conidia similar to those of the *arp2* deletion strain [69,72]. Tricyclazole is a fungicide which specifically inhibits the HN reductase involved in DHN-melanin biosynthesis in brown and black fungi [72,73]. The transcript of *arp2* has a size of 1.1 kb. The *arp2* gene is not interrupted by introns. It encodes a protein of 273 amino acids [69].

Tsai et al. [69] identified three additional genes as part of the cluster, designated *abr1*, *ayg1*, and *abr2* (Fig. 4). The *abr1* gene is interrupted by two introns. It encodes an ORF of 664 amino acids. Its transcript has a size of 2.2 kb. The *abr2* gene contains six introns and encodes an ORF of 587 amino acids. Its transcript has a size of 2.8 kb. Sequence analysis revealed that *abr1* encodes a putative multicopper oxidase and *abr2* a putative laccase. For *ayg1*, no putative homolog was identified. The *ayg1* gene is interrupted by four introns. It encodes an ORF of 406 amino acids. Its transcript has a size of 1.7 kb [69].

Deletion of each *abr1*, *abr2*, and *ayg1* in *A. fumigatus* resulted in an alteration of the conidial color phenotype; i.e., *abr1* and *abr2* deletion strains produced brown conidia, whereas the *ayg1* deletion produced yellowish-green conidia [69]. These findings indicated that the genes were involved in the biosynthesis of the conidial pigment.

3.3.3 The Conidial Pigment of *A. fumigatus* Is Synthesized via the DHN-Melanin Pathway Which Does Not Seem to Be the Case for the Nonpathogenic *A. nidulans*

In general, including the plant pathogens *C. lagenarium* and *M. grisea*, there are three enzymes conserved among the known DHN-melanin pathways in fungi: a polyketide synthase, a scytalone dehydratase, and an HN reductase [reviewed in 61,74]. The presence of genes encoding homologous proteins in *A. fumigatus* suggests a similarity of melanin biosynthesis in *A. fumigatus* to that of brown and black fungi. However, the final pigments produced by *A. fumigatus* are gray-green. This is supported by the discovery of three additional genes (*abr1*, *abr2*, and *ayg1*), associated with conidial pigment biosynthesis. The DHN-melanin pathways of *A. fumigatus* and black/brown fungi must have diverged at steps prior to the final melanin product; e.g., a laccase was suggested to polymerise and oxidise DHN into melanin based on previous biochemical studies [74] and could well be represented by the ABR2 protein [69]. The presence of the *abr1*, *abr2*, and *ayg1* genes in addition to *pksP*, *arp1*, and *arp2* (*thnr*) suggested that conidial pigment biosynthesis was more complex than the known DHN melanin pathway in brown and black fungi [69]. In addition to the genetic indication for presence of the DHN-melanin biosynthesis in *A. fumigatus*, there are biochemical data [69,72]. As mentioned above, the fungicide tricyclazole specifically inhibits the reductases involved in DHN-melanin biosynthesis in black fungi [73]. Previous studies on the melanin biosynthesis with black and brown fungi showed that blockage of the 1,3,6,8-THN reduction step results in the accumulation of flaviolin. Blockage of the scytalone dehydration step results in accumulation of stable intermediates or branch products, i.e., scytalone, flaviolin, and 4-hydroxy scytalone [reviewed in 61,74] (Fig. 5).

The inhibition of conidial pigmentation of *A. fumigatus* by tricyclazole suggested that the gray-green pigment of *A. fumigatus* was synthesized via a pathway similar to the DHN-melanin pathway of *C. lagenarium* and *M. grisea* [71,72]. Furthermore, Tsai et al. [69] analyzed extracts from deletion mutants of *arp1* and *arp2* by thin-layer chromatography for pigment biosynthetic intermediates. In culture extract of an *arp1* disruptant strain, high levels of scytalone and flaviolin were found, and a detectable amount of 4-hydroxy scytalone. This agrees with the notion that ARP1 is a scytalone dehydratase. Furthermore, evidence was provided that ARP2 is a tricyclazole-sensitive HN reductase which converts 1,3,6,8-THN to scytalone [69]. In *M. grisea*, it was initially suggested that the HN reductase catalyzes the conversion of 1,3,6,8-THN to scytalone as well as the conversion of 1,3,8-THN to vermelone [75]. However, a second reductase present in the DHN-melanin pathway was found in *M. grisea* also catalyzing the reduction of 1,3,6,8-THN to scytalone [76]. A second reductase was also suggested to be present in *C. lagenarium* [77]. In *A. fumigatus*, the presence of a second reductase involved in the pathway which catalyzes the reduction of 1,3,8-THN to vermelone, in addition to the ARP2 reductase, is very likely [69] (Fig. 5). This putative reductase does not seem to be encoded by any of the identified six pigment biosynthesis genes [69].

There are very few publications reporting the isolation of *A. nidulans* from material derived from patients. Since there is no evidence of pathogenicity of *A. nidulans* per se, we regard this fungus as nonpathogenic. Previously, mutants of *A. nidulans* were isolated that also produce white conidia. One of these mutants (WA) was complemented by the *wA* gene of *A. nidulans* [78,79]. Since pigment synthesis in *A. nidulans* involves a polyketide pathway [74], the *wA* gene which codes for a polyketide synthase seems to contribute to conidial pigment biosynthesis [80]. PksP also showed a high degree of similarity to Pks1 of the plant-pathogenic fungus *C. lagenarium*. *pks1* encodes a polyketide synthase which is required for the biosynthesis of DHN-melanin [81]. In *C. lagenarium*, DHN-melanin is essential for the function of the appressorium in penetration of host plants [82]. It is thus regarded as a virulence factor. In agreement with this conclusion, albino strains lacking the *pks1* gene have reduced penetration ability and thus attenuated pathogenicity on the host plant [83].

Interestingly, in contrast to *A. fumigatus*, the *A. nidulans* conidial pigmentation was not inhibited by tricyclazole [74], and in addition, Tsai et al. [71] could not detect any homologs of *arp1* in *A. nidulans* with low-stringency hybridization. Furthermore, in contrast to the *pksP* mutant of *A. fumigatus* which exhibits conidia with a smooth surface, conidia of the corresponding *wA* mutant of *A. nidulans* had still ornaments on the conidial surface [25]. Taken together, these findings suggest dissimilar pathways in *A. fumigatus* and *A. nidulans* pigment synthesis which seems a prerequisite for the virulence of *A. fumigatus*.

Overproduction of PksP in *A. oryzae* implied that its product is a heptaketide naphthopyrone instead of a pentaketide [84], as initially thought by analogy to the products of the *C. lagenarium* Pks1 [85].

3.3.4 Is It the Final Product (Pigment), Intermediates, or Other Metabolites in Whose Synthesis *PksP* Is Involved Which Could Explain Pathogenicity of *A. fumigatus?*

To elucidate the mechanisms by which the pigment or intermediates might protect conidia against the attack of immune effector cells, a series of in vitro experiments was performed. When conidia were tested for their ability to cause ROS production in human monocytes and PMN, wt conidia only led to a small release of ROS [23], which agreed well with previous results reported by Levitz and Diamond [34]. By contrast, conidia of the *pksP* mutant caused a 10-fold higher ROS release by both cell types [23].

There are two possible explanations for the observation that the white conidia induced a 10-fold greater release of ROS when coincubated with PMN and monocytes than the wt conidia. The first explanation was that conidia of both strains induced an identical oxidative burst. However, the resulting ROS were scavenged by the pigment present in wt conidia and thus were not detected by chemiluminescence. Some fungal pigments are known to quench ROS, as has been shown in studies on oxidative damage of *Wangiella dermatitidis* and *Cryptococcus neoformans* [59,66]. This could explain the apparent absence of oxidative burst by PMNs in response to wt conidia of *A. fumigatus* [23]. Thus, ROS might well be released during the interaction of *A. fumigatus* wt conidia with immune effector cells. However, their detection might be impaired owing to immediate quenching of nascent ROS by the conidial pigment. In the case of *A. fumigatus*, this assumption was supported by the observation that *pksP* conidia were about 10-fold more sensitive to both hydrogen peroxide and NaOCl than wt conidia. This indicated that ROS scavenging also occurs in pigmented *A. fumigatus* conidia and thus confers a degree of protection against damage caused by ROS [23].

The second explanation was that *pksP* mutant conidia were better recognized by phagocytes because the conidial surface of the *pksP* mutant conidia markedly differed from that of wt conidia [23]. This could have resulted in a more efficient and stronger activation of phagocytes, which would explain the observed increase in production of ROS when the immune cells were challenged with *pksP* mutant conidia. The latter assumption was also supported by the findings that conidia of a white mutant independently isolated by Tsai et al. [71] exhibited significantly elevated complement component C3 binding capacity compared with wt conidia. Inefficient deposition of complement component C3 on wt *A. fumigatus* conidia has previously been demonstrated [86]. Also, there is a good correlation between binding of C3 molecules per unit conidial surface and pathogenicity; i.e., highly pathogenic *A. fumigatus* and *A. flavus* bound fewer C3 molecules than the nonpathogenic species *A. nidulans* [87]. Therefore, a key question concerned the quenching ability of the *A. fumigatus* pigment and whether this is specific of *A. fumigatus*. Furthermore, since the conidial pigments of *A. fumigatus* and *A. nidulans* were synthesized by different biosynthesis pathways and it can be assumed that their structures are different, it was also interesting to investigate whether the *A. nidulans* pigment is able to quench ROS and to protect conidia against ROS. Therefore, the wt and a mutant of *A. nidulans* producing white conidia (*wA*) which is also characterized by a polyketide synthase defect [78–80] were analyzed. The *A. nidulans* wt conidia showed a similar ornamentation as the *A. fumigatus* wt. Interestingly, in contrast to the *pksP* mutant, *wA* conidia still exhibited the ornamentation characteristic of wt conidia [25].

As noted previously, with both cell types, i.e., human PMN and monocytes, the amounts of ROS detected in response to pigmentless *A. fumigatus pksP* mutant conidia were 10-fold higher than the respective wt conidia [23]. Similar results were obtained with the *A. nidulans* conidia; i.e., the pigmentless *wA* mutant conidia led to a 10-fold increase of ROS release compared with wt conidia. The amounts of ROS released after coincubation of immune effector cells with pigmentless conidia were comparable to those observed after stimulation of the cells with phorbol myristate acetate (PMA). Because the release of ROS was the same for both pigmentless *A. fumigatus* and *A. nidulans* conidia, this suggested that the altered conidial surface of the *A. fumigatus* pigmentless mutant is not responsible for the increased ROS release [25].

To further analyze the ROS quenching ability of conidia, the phagocytes were stimulated with 10 nM PMA in the presence or absence of *A. nidulans* and *A. fumigatus* wt as well as *pksP* and *wA* mutant conidia. When phagocytes were stimulated with PMA in the presence of wt conidia the amounts of ROS detected decreased by 80–90%, compared with the ROS release observed when PMA alone was applied. By contrast, stimulation of both PMN and monocytes by PMA in the presence of *pksP* or *wA* mutant conidia even resulted in an 1.6-fold increase in detectable ROS compared with the stimulation observed with PMA alone, suggesting that the conidia have an additive stimulatory effect on ROS release. These experiments indicated that the pigments of both *A. fumigatus* and *A. nidu-*

lans have ROS quenching ability [25]. Further support of the hypothesis of *Aspergillus* conidia pigment functions as ROS quencher was obtained from experiments on conidial damage by oxidating agents. *A. fumigatus pksP* conidia were significantly more damaged by exposure to H_2O_2 and NaOCl than were pigmented wt conidia [23]. Consistently, exposure to H_2O_2 and NaOCl revealed that *A. nidulans wA* conidia were fivefold more sensitive against these ROS than *A. nidulans* wt conidia. Hence, in both *A. nidulans* and *A. fumigatus* loss of conidial pigmentation was paralleled by an increased susceptibility of conidia toward the oxidating agent [25]. Taken together, the results showed that the release of ROS from human monocytes and PMN in response to pigmentless *A. fumigatus* conidia is not related to the change of the conidial surface, which is associated with the loss of the pigmentation in *A. fumigatus*. Furthermore, although the conidial pigments are apparently chemically different, conidia of both *A. fumigatus* and *A. nidulans* have the capability to scavenge reactive oxygen species. Pigments therefore might contribute to the relative resistance of conidia against neutrophil attack, as described for *A. fumigatus* [34]. Furthermore, the lack of pigmentation in both *A. nidulans* and *A. fumigatus* [25] resulted in an increased susceptibility toward oxidative attack. These findings underline the important role of fungal pigments, present in a variety of human fungal pathogens, as protective agents against oxidant based host defense mechanisms [59,67,88]. However, they do not explain why *A. fumigatus* conidia can be pathogenic whereas this is rarely the case for *A. nidulans* conidia.

3.3.5 PksP-Dependent Reduction of Phagosome–Lysosome Fusion and Intracellular Kill of *Aspergillus fumigatus* Conidia by Human Macrophages

When it is not the ROS quenching ability of the pigment that could explain the predominance of *A. fumigatus* as the infectious *Aspergillus* spp., what else could be a target of conidia-associated compounds whose lack helped explain the reduced virulence of *pksP* mutant conidia?

Survival of conidia and onset of germination in the lung are prerequisite for establishing disease. A protective effect of macrophage transfer against IPA was reported for a murine animal model [89]. However, macrophages from different sources show a limited activity against conidia in vitro [90,91]. Furthermore, in vitro killing of conidia by macrophages is rather slow [91]. The mechanisms underlying the anticonidial activity of macrophages and the relative resistance of conidia against the respective effector mechanisms are not known. Oxygen-dependent effects seem to play a minor role [see above and 5]. Since conidia of the *pksP* mutant showed enhanced susceptibility toward damage by monocytes in vitro [23], these results suggested differences in the intracellular processing of *A. fumigatus* wt conidia compared with nonpigmented *pksP* mutant conidia. One known key element of antimicrobial activity in macrophages is the formation of functional phagolysosomes which contain a large variety of degrading enzymes in an acidic environment. Therefore, the intracellular processing of conidia with respect to phagolysosome formation and acidification following phagocytosis of different *A. fumigatus* conidia was analyzed. It turned out that the presence of a functional *pksP* gene affected intracellular processing of ingested conidia by macrophages [92].

After phagocytosis by human monocyte-derived macrophages (MDM), the percentage of *A. fumigatus* wt conidia residing in an acidic environment was approximately fivefold lower than that observed for nonpigmented *pksP* mutant conidia. It was conceivable that the observed differences in periconidial acidification were due to a direct effect of pigmented conidia on the phagolysosomal acidification. This has been observed for *Histoplasma capsulatum* yeast forms when present in the phagolysosome [93,94]. However, the *A. fumigatus* melanin has no pH-buffering capacity, as demonstrated by comparative titration experiments with pigmented wt and nonpigmented *pksP* conidia, suggesting that the conidial pigment of *A. fumigatus* does not neutralize the acidic pH of phagolysosomes containing wt conidia [92]. Alternatively, the conidial pigment or another PksP-related compound could inhibit the fusion of the conidium-containing phagosome with the lysosome. To determine whether phagolysosome formation was dependent on which conidia were ingested, i.e., wt or *pksP* mutant conidia, the colocalization of LAMP-1, a membrane protein characteristic of the lysosome [95], with phagocytosed conidia was studied by immunofluorescence. The rate of LAMP-1-labeled conidia was 10% for wt conidia and 30% for *pksP* mutant conidia. Hence, the percentage of phagolysosomes containing *pksP* mutant conidia was approximately threefold higher than the proportion of phagolysosomes containing wt conidia, indicating that the lack of periconidial acidification upon phagocytosis of wt conidia resulted from a reduction in the fusion rate of phagosomes with lysosomes rather than from a mechanism acting on mature phagolysosomes [92].

When the intracellular killing of both the wt and *pksP* mutant by MDM was compared, a marked increase in inactivation rates was observed, which was linked to the absence of a functional *pksP* gene. Interestingly, the intracellular kill rate of wt conidia increased to the level observed with *pksP* mutant conidia in the presence of chloroquine, a known enhancer of phagolysosome formation in macrophages [96]. These results indicate that the fusion of a conidium-containing phagosome with a lysosome is a limiting step in the elimination process of intracellular *A. fumigatus* conidia. Furthermore, since bafilomycin A abolished the enhanced killing rates of *pksP* mutant conidia, this notion demonstrates that intracellular elimination of *pksP* mutant conidia, at least in part, depends on a functional lysosomal H^+-ATPase [92]. Taken together, the presence of a functional *pksP* gene in *A. fumigatus* conidia is associated with an inhibition of phagosome–lysosome fusion in human MDM.

This finding might help explain the attenuated virulence of *pksP* mutant strains in a murine animal model and provides a conceptual frame to understand the virulence of *A. fumigatus* [92]. It remains to be shown whether these effects were due to the lack of the conidial pigment and/or intermediates, or the lack of another compound in whose synthesis PksP is involved. We are in favor of the hypothesis that PksP is involved not only in the biosynthesis of the conidial pigment, but also in the biosynthesis of other, so far unknown polyketide derivatives. This also agrees with the observation of Schnitzler et al. [67], who recently reported that shunt products of DHN-melanin deficient *Exophiala* (*Wangiella*) *dermatitidis* strains had no adverse effect on phagocytosis and intracellular killing by human PMN. On the other side, *A. fumigatus* melanin biosynthesis apparently differs from the classical DHN-melanin pathway [see above and 69]. Hence, the possibility still exists that during pigment biosynthesis in *A. fumigatus* compounds toxic to the immune effector cells are synthesized which contribute to the pathogenic potential of the fungus. If, alternatively, PksP were involved in the biosynthesis of an additional polyketide derivative, the lack of such a compound in the *pksP* mutant would explain the differences observed between *pksP* mutant conidia and wt conidia because it is conceivable that such a compound could influence the fusion rate of phagosomes and lysosomes. An example for a dual role of a polyketide synthase is the polyketide derivative 6-demethylchlortetracycline (6-DCT) which is produced by *Streptomyces aureofaciens*. It was reported that strains producing 6-DCT also produce a melaninlike pigment [97]. Interestingly, some of the 6-DCT biosynthesis enzymes are indispensable for the production of the melaninlike pigment, suggesting a dualistic role for these enzymes.

A variety of pathogens have evolved strategies to interfere with the phagosome–lysosome fusion process in order to improve their intracellular survival. These include the inhibition of phagosome–lysosome fusion, as described for *Toxoplasma gondii* and *Legionella pneumophila*, the selective fusion of phagosomes with early lysosomes, as seen with *Mycobacterium tuberculosis*, or the fusion of phagosomes with nonlysosome vesicles, as observed for *Chlamydia* spp. [98,99]. Studies on human mycopathogens have shown that *Cryptococcus neoformans* may survive and replicate within the acidic phagolysosomal vacuole [100]. Furthermore, after phagocytosis of *Histoplasma capsulatum* yeast forms by macrophages, also a selective inhibition of phagosome–lysosome fusion was observed together with an active acidification of the phagosomal pH to 6.5 [94].

The significantly higher fusion rates and the more efficient intracellular killing observed for *A. fumigatus pksP* mutant conidia compared with wt conidia may also contribute to a more effective host response. This could be one factor explaining the attenuated virulence demonstrated in infections with nonpigmented *pksP*-deficient *A. fumigatus* mutant conidia in the murine animal model

[23,24,26]. Furthermore, the effect of PksP dependent compounds on intracellular processing may also contribute to the dominance of *A. fumigatus* as the major human mold pathogen.

4 CONCLUSION

For several human pathogenic fungi, including *Cryptococcus neoformans*, *Sporothrix schenckii*, and *Exophiala* (*Wangiella*) *dermatitidis*, different melanin types were shown to be important for virulence [57–59,66,74,88,101]. Conidia of *A. fumigatus* contain a gray-green pigment. Genetic and biochemical data indicate that the conidial pigment of *A. fumigatus* contains DHN-melanin. The initial reaction of the pigment biosynthesis pathway appears to be catalyzed by a polyketide synthase encoded by the *pksP* gene. Mutation of the *pksP* gene led to pigmentless (white) conidia with a smooth surface. *pksP* mutant conidia showed reduced virulence in a murine infection model. In addition, they were more sensitive against reactive oxygen species and led to the release of 10-fold greater amounts of ROS than gray-green wt conidia. Further studies indicated that the conidia of *A. fumigatus* are able to scavenge ROS. However, although the green pigment of the nonpathogenic *A. nidulans* conidia does not seem to be synthesized via the DHN-melanin pathway, some features of *A. nidulans* conidia compared with *A. fumigatus* conidia were similar: pigmentless (white) conidia of *A. nidulans* were more susceptible against killing by ROS compared with wt conidia, and green wt conidia of *A. nidulans* were also able to scavenge ROS. Taken together, both pigments might act as protective agents against oxidant-based host defense mechanisms and thus contribute to the relative resistance of conidia against neutrophil attack, as described for *A. fumigatus*. However, these findings do not explain why *A. fumigatus* conidia can be pathogenic whereas this is rarely the case for *A. nidulans* conidia. An attractive hypothesis is that besides the pigment the *pksP* gene product of *A. fumigatus* is involved in the production of another compound which is immunosuppressive. This hypothesis is further supported by the notion that the presence of a functional *pksP* gene in *A. fumigatus* conidia is associated with an inhibition of the fusion of phagosomes and lysosomes in human MDM. This finding might help explain the attenuated virulence of *pksP* mutant strains in a murine animal model and provides a conceptual frame to understand the virulence of *A. fumigatus*.

ACKNOWLEDGMENTS

We are grateful to Kim Langfelder for helpful discussions and providing information prior to publication. The work in the laboratory of A. A. B. was supported by the Deutsche Forschungsgemeinschaft.

REFERENCES

1. A Wald, W Leisenring, JA van Burik, RA Bowden. Epidemiology of *Aspergillus* infections in a large cohort of patients undergoing bone marrow transplantation. J Infect Dis 175:1459–1466, 1997.
2. R Kappe, D Rimek. Laboratory diagnosis of *Aspergillus fumigatus*-associated diseases. In: AA Brakhage, B Jahn, A Schmidt, eds. *Aspergillus fumigatus*: Biology, Clinical Aspects and Molecular Approaches to Pathogenicity. Contributions to Microbiology. Basel: Karger, 1999, pp 88–104.
3. AA Brakhage, B Jahn, A Schmidt. *Aspergillus fumigatus*: Biology, Clinical Aspects and Molecular Approaches to Pathogenicity. Contributions to Microbiology, Vol 2. Basel: Karger, 1999.
4. DW Denning. Invasive aspergillosis. Clin Infect Dis 26:781–805, 1998.
5. J-P Latgé. *Aspergillus fumigatus* and aspergillosis. Clin Microbiol Rev 12:310–350, 1999.
6. RA Samson. The genus *Aspergillus* with special regard to the *Aspergillus fumigatus* group. In: AA Brakhage, B Jahn, A Schmidt, eds. *Aspergillus fumigatus*: Biology, Clinical Aspects and Molecular Approaches to Pathogenicity. Contributions to Microbiology. Basel: Karger, 1999, pp 5–20.
7. J Mullins, R Harvey, A Seaton. Sources and incidence of airborne *Aspergillus fumigatus* (Fres.). Clin Allergy 6:209–217, 1976.
8. J-P Debeaupuis, J Sarfati, V Chazalet, J-P Latgé. Genetic diversity among clinical and environmental isolates of *Aspergillus fumigatus*. Infect Immun 65:3080–3085, 1997.
9. AI Braude. The aspergilli. In: AI Braude, CE Davis, J Fierer, eds. Infectious Diseases and Medical Microbiology. 2nd ed. Philadelphia: W.B. Saunders, 1986, pp 592–597.
10. HJ Schmidt, A Blevins, K Sobeck, D Armstrong. *Aspergillus* species from hospital air and from patients. Mycoses 33:539–541, 1991.
11. GS de Hoog, J Guarro. Atlas of Clinical Fungi. Baarn, Netherlands: Centraalbureau voor Schimmelcultures, 1995, pp 136–139.
12. JI Pitt. The current role of *Aspergillus* and *Penicillium* in human and animal health. J Med Vet Mycol 32(suppl 1):17–32, 1994.
13. R Rüchel, U Reichard. Pathogenesis and clinical presentation of Aspergillosis. In: AA Brakhage, B Jahn, A Schmidt, eds. *Aspergillus fumigatus*: Biology, Clinical Aspects and Molecular Approaches to Pathogenicity. Contributions to Microbiology. Basel: Karger, 1999, pp 21–43.
14. KB Raper, DI Fennell. *Aspergillus fumigatus* group. In: KB Raper, DI Fennell, eds. The Genus *Aspergillus*. Baltimore: Williams and Wilkins, 1965, pp 238–268.
15. SW Peterson. *Neosartorya pseudofischeri* sp. nov. and its relationship to other species in *Aspergillus* section Fumigati. Mycol Res 96:547–554, 1992.
16. CM Berg, ED Garber. A genetic analysis of color mutants of *Aspergillus fumigatus*. Genetics 47:1139–1146, 1962.
17. I Strømnæs, ED Garber. Heterocaryosis and the parasexual cycle in *Aspergillus fumigatus*. Genetics 48:653–662, 1962.

18. E Yuill. The number of nuclei in conidia of aspergilli. Trans Br Mycol Soc 33: 324–331, 1950.
19. C d'Enfert, G Weidner, PC Mol, AA Brakhage. Transformation systems of *Aspergillus fumigatus*. In: AA Brakhage, B Jahn, A Schmidt, eds. *Aspergillus fumigatus*: Biology, Clinical Aspects and Molecular Approaches to Pathogenicity. Contributions to Microbiology. Basel: Karger, 1999, pp 149–166.
20. MB Tobin, RB Peery, PL Skatrud. An electrophoretic molecular karyotype of a clinical isolate of *Aspergillus fumigatus* and localization of the MDR-like genes AfuMDR1 and AfuMDR2. Microbiol Infect Dis 29:67–71, 1997.
21. DW Holden, CM Tang, JM Smith. Molecular genetics of *Aspergillus* pathogenicity. Antonie Van Leeuwenhoek 65:251–255, 1994.
22. M Monod, K Jaton-Ogay, U Reichard. *Aspergillus fumigatus* secreted proteases as antigenic molecules and virulence factors. In: AA Brakhage, B Jahn, A Schmidt, eds. *Aspergillus fumigatus*: Biology, Clinical Aspects and Molecular Approaches to Pathogenicity. Contributions to Microbiology. Basel: Karger, 1999, pp 182–192.
23. B Jahn, A Koch, A Schmidt, G Wanner, H Gehringer, S Bhakdi, AA Brakhage. Isolation and characterization of an *Aspergillus fumigatus* mutant strain with pigmentless conidia and reduced virulence. Infect Immun 65:5110–5117, 1997.
24. K Langfelder, B Jahn, H Gehringer, A Schmidt, G Wanner, AA Brakhage. Identification of a polyketide synthase gene (*pksP*) of *Aspergillus fumigatus* involved in conidial pigment biosynthesis and virulence. Med Microbiol Immunol 187:79–89, 1998.
25. B Jahn, F Boukhallouk, J Lotz, K Langfelder, G Wanner, AA Brakhage. Interaction of human phagocytes with pigmentless conidia. Infect Immun 68:3736–3739, 2000.
26. H-F Tsai, CC Yun, RG Washburn, MH Wheeler, KJ Kwon-Chung. The developmentally regulated *alb1* gene of *Aspergillus fumigatus*: its role in modulation of conidial morphology and virulence. J Bacteriol 180:3031–3038, 1998.
27. C d'Enfert, M Diaquin, A Delit, N Wuscher, J-P Debeaupuis, M Huerre, J-P Latgé. Attenuated virulence of uridine-uracil auxotrophs of *Aspergillus fumigatus*. Infect Immun 64:4401–4405, 1996.
28. JS Brown, A Aufauvre-Brown, J Brown, JM Jennings, HN Arst Jr, DW Holden. Signature-tagged and directed mutagenesis identify PABA synthetase as essential for *Aspergillus fumigatus* pathogenicity. Mol Microbiol 36:1371–1380, 2000.
29. JM Goodley, YM Clayton, RJ Hay. Environmental sampling for aspergilli during building construction on a hospital site. J Hosp Infect 26:27–35, 1994.
30. V Chazalet, JP Debeaupuis, J Sarfati, J Lortholary, P Ribaud, P Shah, M Cornet, H Vu Thien, E Gluckman, G Brucker, JP Latgé. Molecular typing of environmental and patient isolates of *Aspergillus fumigatus* from various hospital settings. J Clin Microbiol 36:1494–1500, 1998.
31. DR Hospenthal, KJ Kwon-Chung, JE Bennett. Concentrations of airborne *Aspergillus* compared to the incidence of invasive aspergillosis: lack of correlation. Med Mycol 36:165–168, 1998.
32. RD Diamond, RA Clark. Damage of *Aspergillus fumigatus* and *Rhizopus oryzae* hyphae by oxidative and nonoxidative microbicidal products of human neutrophils in vitro. Infect Immun 38:487–495, 1982.

33. B Jahn, A Stueben, S Bhakdi. Colorimetric susceptibility testing of *Aspergillus fumigatus*: comparison of menandione-augmented 3-(4,5-dimethyl-2-thiazolyl)-2,5-diphenyl-2H-tetrazolium bromide and alamar blue tests. J Clin Microbiol 34: 2039–2041, 1996.
34. SM Levitz, RD Diamond. A rapid colorimetric assay of fungal viability with tetrazolium salt MTT. J Infect Dis 152:938–945, 1985.
35. SM Levitz, TP Farrell. Human neutrophil degranulation stimulated by *Aspergillus fumigatus*. J Leukoc Biol 47:170–175, 1990.
36. T Meshulam, SM Levitz, L Christin, RD Diamond. A simplified new assay for assessment of fungal cell damage with the tetrazolium dye (2,3)-bis-(2-methoxy-4-nitro-5-sulphenyl)-(2H)-tetrazolium-5-carboxyanilide (XTT). J Infect Dis 172: 1153–1156, 1995.
37. A Schaffner, H Douglas, A Braude. Selective protection against conidia by mononuclear and against mycelia by polymorphonuclear phagocytes in resistance to *Aspergillus*. Observations on these two lines of defense in vivo and in vitro with human and mouse phagocytes. J Clin Invest 69:617–631, 1982.
38. A Schaffner, CE Davis, T Schaffner, M Markert, H Douglas, AI Braude. In vitro susceptibility of fungi to killing by neutrophil granulocytes discriminates between primary pathogenicity and opportunism. Clin Invest 78:511–524, 1986.
39. M Ellis. Therapy of *Aspergillus fumigatus*–related diseases. In: AA Brakhage, B Jahn, A Schmidt, eds. *Aspergillus fumigatus*: Biology, Clinical Aspects and Molecular Approaches to Pathogenicity. Contributions to Microbiology. Basel: Karger, 1999, pp 105–129.
40. A Müllbacher, RD Eichner. Immunosuppression in vitro by a metabolite of a human pathogenic fungus. Proc Natl Acad Sci USA 81:3835–3837, 1984.
41. MD Robertson, A Seaton, LJ Milne, JA Raeburn. Resistance of spores of *Aspergillus fumigatus* to ingestion by phagocytic cells. Thorax 42:466–472, 1987.
42. MD Robertson, A Seaton, LJ Milne, JA Raeburn. Suppression of host defences by *Aspergillus fumigatus*. Thorax 42:19–25, 1987.
43. MD Robertson, A Seaton, LJ Milne, JA Raeburn. Inhibition of phagocyte migration and spreading by spore diffusates of *Aspergillus fumigatus*. J Med Vet Mycol 25: 389–396, 1987.
44. R Amitani, G Taylor, EN Elezis, C Llewellyn-Jones, J Mitchell, F Kuze, PJ Cole, R Wilson. Purification and characterization of factors produced by *Aspergillus fumigatus* which affect human ciliated respiratory epithelium. Infect Immun 63: 3266–3271, 1995.
45. WJ Nicholson, J Slight, K Donaldson. Inhibition of the transcription factors NF-kappa B and AP-1 underlies loss of cytokine gene expression in rat alveolar macrophages treated with a diffusible product from the spores of *Aspergillus fumigatus*. Am J Respir Cell Mol Biol 15:88–96, 1996.
46. J Slight, WJ Nicholson, CG Mitchell, N Pouilly, PH Beswick, A Seaton, K Donaldson. Inhibition of the alveolar macrophage oxidative burst by a diffusible component from the surface of the spores of the fungus *Aspergillus fumigatus*. Thorax 51:389–396, 1996.
47. P Sutton, P Waring, A Mullbacher. Exacerbation of invasive aspergillosis by the

immunosuppressive fungal metabolite, gliotoxin. Immunol Cell Biol 74:318–322, 1996.

48. CG Mitchell, J Slight, K Donaldson. Diffusible component from the spore surface of the fungus *Aspergillus fumigatus* which inhibits the macrophage oxidative burst is distinct from gliotoxin and other hyphal toxins. Thorax 52:796–801, 1997.
49. T Murayama, R Amitani, Y Ikegami, R Kawanami, WJ Lee, R Nawada. Effects of *Aspergillus fumigatus* culture filtrate on antifungal activity of human phagocytes in vitro. Thorax 53:975–978, 1998.
50. P Daly, S Verhaegen, M Clynes, K Kavanagh. Culture filtrates of *Aspergillus fumigatus* induce different modes of cell death in human cancer cell lines. Mycopathologia 146:67–74, 1999.
51. RP Hobson. The effects of diffusates from the spores of *Aspergillus fumigatus* and *A. terreus* on human neutrophils, *Naegleria gruberi* and *Acanthamoeba castellanii*. Med Mycol 38:133–141, 2000.
52. A Yamada, T Kataoka, K Nagai. The fungal metabolite gliotoxin: immunosuppressive activity on CTL-mediated cytotoxicity. Immunol Lett 71:27–32, 2000.
53. HL Pahl, B Krauss, K Schulze-Osthoff, T Decker, EB Traenckner, M Vogt, C Myers, T Parks, P Warring, A Muhlbacher, AP Czernilofsky, PA Baeuerle. The immunosuppressive fungal metabolite gliotoxin specifically inhibits transcription factor NF-kappaB. J Exp Med 183:1829–1840, 1996.
54. LS Yoshida, S Abe, S Tsunawaki. Fungal gliotoxin targets the onset of superoxide-generating NADPH oxidase of human neutrophils. Biochem Biophys Res Commun 268:716–723, 2000.
55. GS Bondy, JJ Pestka. Immunomodulation by fungal toxins. J Toxicol Environ Health Part b: Crit Rev 3:109–143, 2000.
56. A Zuger, E Louie, RS Holzman, MS Simberkoff, JJ Rahal. Cryptococcal disease in patients with the acquired immunodeficiency syndrome. Diagnostic features and outcome of treatment. Ann Intern Med 104:234–240, 1986.
57. KJ Kwon-Chung, I Polacheck, TJ Popkin. Melanin-lacking mutants of *Cryptococcus neoformans* and their virulence for mice. J Bacteriol 150:1414–1421, 1982.
58. KJ Kwon-Chung, JC Rhodes. Encapsulation and melanin formation as indicators of virulence in *Cryptococcus neoformans*. Infect Immun 51:218–223, 1986.
59. Y Wang, P Aisen, A Casadevall. *Cryptococcus neoformans* melanin and virulence: mechanism of action. Infect Immun 63:3131–3136, 1995.
60. JA Alspaugh, JR Perfect, J Heitman. *Cryptococcus neoformans* mating and virulence are regulated by the G-protein alpha subunit GPA1 and cAMP. Genes Dev 11:3206–3217, 1997.
61. MJ Butler, AW Day. Fungal melanins: a review. Can J Microbiol 44:1115–1136, 1998.
62. BM Babior. Oxygen-dependent microbial killing by phagocytes (first of two parts). N Engl J Med 298:659–668, 1978.
63. RG Washburn, JI Gallin, JE Bennett. Oxidative killing of *Aspergillus fumigatus* proceeds by parallel myeloperoxidase-dependent and -independent pathways. Infect Immun 55:2088–2092, 1987.
64. Y Wang, A Casadevall. Susceptibility of melanized and nonmelanized *Cryptococ-*

cus neoformans to nitrogen- and oxygen-derived oxidants. Infect Immun 62:3004–3007, 1987.
65. ES Jacobson, SB Tinnell. Antioxidant function of fungal melanin. J Bacteriol 175: 7102–7104, 1993.
66. ES Jacobson, E Hove, HS Emery. Antioxidant function of melanin in black fungi. Infect Immun 63:4944–4945, 1995.
67. N Schnitzler, H Peltroche-Llacsahuanga, N Bestier, J Zundorf, R Lutticken, G Haase. Effect of melanin and carotenoids of *Exophiala* (*Wangiella*) *dermatitidis* on phagocytosis, oxidative burst, and killing by human neutrophils. Infect Immun 67:94–101, 1999.
68. CR Hutchinson, I Fujii. Polyketide synthase gene manipulation: a structure–function approach in engineering novel antibiotics. Annu Rev Microbiol 49:201–238, 1995.
69. H-F Tsai, MH Wheeler, YC Chang, KJ Kwon-Chung. A developmentally regulated gene cluster involved in conidial pigment biosynthesis in *Aspergillus fumigatus*. J Bacteriol 181:6469–6477, 1999.
70. AA Brakhage, K Langfelder, G Wanner, A Schmidt, B Jahn. Pigment biosynthesis and virulence. In: AA Brakhage, B Jahn, A Schmidt, ed. *Aspergillus fumigatus*: Biology, Clinical Aspects and Molecular Approaches to Pathogenicity. Contributions to Microbiology, Vol 2. Basel: Karger, 1999, pp 205–215.
71. H-F Tsai, RG Washburn, YC Chang, KJ Kwon-Chung. *Aspergillus fumigatus arp1* modulates conidial pigmentation and complement deposition. Mol Microbiol 26: 175–183, 1997.
71a. K Langfelder, B Philippe, B Sahn, D-P Latgé, AA Brakhage. Differential expression of the *Aspergillus fumigatus pksP* gene detected in vitro and in vivo with green fluorescent protein. Infect Immun 69:6411–6418, 2001.
72. MH Wheeler, MA Klich. The effects of tricyclazole, pyroquilon, phthalide, and related fungicides on the production of conidial wall pigments by *Penicillium* and *Aspergillus* species. Pestic Biochem Physiol 52:125–136, 1995.
73. MC Tokousbalides, HD Sisler. Site of inhibition by tricyclazole in the melanin biosynthetic pathway of *Verticillium dahliae*. Pestic Biochem Physiol 11:64–73, 1979.
74. MH Wheeler, AA Bell. Melanins and their importance in pathogenic fungi. In: MR McGinnis, ed. Current Topics in Medical Mycology, Vol 2. New York: Springer-Verlag, 1988, pp 338–387.
75. A Vidal-Cros, F Viviani, G Labesse, M Boccara, M Gaudry. Polyhydroxynaphthalene reductase involved in melanin biosynthesis in *Magnaporthe grisea*. Purification, cDNA cloning and sequencing. Eur J Biochem 219:985–992, 1994.
76. JE Thompson, S Fahnestock, L Farrall, DI Liao, B Valent, DB Jordan. The second naphthol reductase of fungal melanin biosynthesis in *Magnaporthe grisea*: tetrahydroxynaphthalene reductase. J Biol Chem 275:34867–34872, 2000.
77. NS Perpetua, Y Kubo, N Yasuda, Y Takano, I Furuswa. Cloning and characterization of a melanin biosynthetic THR1 reductase gene essential for appressorial penetration of *Colletotrichum lagenarium*. Mol Plant Microbe Interact 9:323–329, 1996.

78. ME Mayorga, WE Timberlake. Isolation and molecular characterization of the *Aspergillus nidulans wA* gene. Genetics 126:73–79, 1990.
79. J Tilburn, F Roussel, C Scazzocchio. Insertional inactivation and cloning of the *wA* gene of *Aspergillus nidulans*. Genetics 126:81–90, 1990.
80. ME Mayorga, WE Timberlake. The developmentally regulated *Aspergillus nidulans wA* gene encodes a polypeptide homologous to polyketide and fatty acid synthases. Mol Gen Genet 235:205–212, 1992.
81. Y Takano, Y Kubo, K Shimizu, K Mise, T Okuno, I Furusawa. Structural analysis of *Pks1*, a polyketide synthase gene involved in melanin biosynthesis in *Colletotrichum lagenarium*. Mol Gen Genet 249:162–167, 1995.
82. I Yamaguchi, Y Kubo. Target sites of melanin biosynthesis inhibitors. In: W Koller, ed. Target Sites of Fungicide Action. London: CRC Press, 1992, pp 101–118.
83. Y Kubo, I Furusawa. Melanin biosynthesis: prerequisite for successful invasion of the plant host by appressoria of *Colletotrichum* and *Pyricularia*. In: GT Cole, HC Hoch, eds. The Fungal Spore and Disease Initiation in Plants and Animals. New York: Plenum Publishing, 1991, pp 205–217.
84. A Watanabe, I Fujii, H Tsai, YC Chang, KJ Kwon-Chung, Y Ebizuka. *Aspergillus fumigatus alb1* encodes naphthopyrone synthase when expressed in *Aspergillus oryzae*. FEMS Microbiol Lett 192:39–44, 2000.
85. AA Bell, MH Wheeler. Biosynthesis and functions of fungal melanins. Annu Rev Phytopathol 24:411–451, 1986.
86. TR Kozel, MA Wilson, TP Farrell, SM Levitz. Activation of C3 and binding to *Aspergillus fumigatus* conidia and hyphae. Infect Immun 57:3412–3417, 1989.
87. S Henwick, SV Hetherington, CC Patric. Complement binding to *Aspergillus* conidia correlates with pathogenicity. J Lab Clin Med 122:27–35, 1993.
88. DM Dixon, A Polak, PJ Szaniszlo. Pathogenicity and virulence of wild-type and melanin-deficient *Wangiella dermatitidis*. J Med Vet Mycol 25:97–106, 1987.
89. L de Repentigny, S Petitbois, M Boushira, E Michaliszyn, S Senechal, N Gendron, S Montplaisir. Acquired immunity in experimental murine aspergillosis is mediated by macrophages. Infect Immun 61:3791–3802, 1993.
90. A Schaffner, H Douglas, AI Braude, CE Davis. Killing of *Aspergillus* spores depends on the anatomical source of the macrophage. Infect Immun 42:1109–1115, 1983.
91. B Jahn, A Rampp, C Dick, A Jahn, M Palmer, S Bhakdi. Accumulation of amphotericin B in human macrophages enhances activity against *Aspergillus fumigatus* conidia: quantification of conidial kill at the single-cell level. Antimicrob Agents Chemother 42:2569–2575, 1998.
92. B Jahn, K Langfelder, F Boukhallouk, AA Brakhage. PksP dependent reduction of phagolysosome fusion and intracellular kill of *Aspergillus fumigatus* conidia by human macrophages. Submitted.
93. LG Eissenberg, WE Goldman, PH Schlesinger. *Histoplasma capsulatum* modulates the acidification of phagolysosomes. J Exp Med 177:1605–1611, 1993.
94. JE Strasser, SL Newman, GM Ciraolo, RE Morris, ML Howell, GE Dean. Regulation of the macrophage vacuolar ATPase and phagosome–lysosome fusion by *Histoplasma capsulatum*. J Immunol 162:6148–6154, 1999.

95. JW Chen, W Pan, MP D'Souza, JT August. Lysosome-associated membrane proteins: characterization of LAMP-1 of macrophage P388 and mouse embryo 3T3 cultured cells. Arch Biochem Biophys 239:574–586, 1985.
96. PD Hart, MR Young. Manipulations of the phagosome–lysosome fusion response in cultured macrophages. Enhancement of fusion by chloroquine and other amines. Exp Cell Res 114:486–490, 1978.
97. T Nakano, K Miyake, M Ikeda, T Mizukami, R Katsumata. Mechanism of the incidental production of a melanin-like pigment during 6-demethylchlortetracycline production in *Streptomyces aureofaciens*. Appl Environ Microbiol 66:1400–1404, 2000.
98. BB Finlay, P Cossart. Exploitation of mammalian host cell functions by bacterial pathogens. Science 276:718–725, 1997. (Erratum Science 278:373, 1997.)
99. F Garcia-del Portillo, BB Finlay. The varied lifestyles of intracellular pathogens within eukaryotic vacuolar compartments. Trends Microbiol 3:373–380, 1995.
100. SM Levitz, SH Nong, KF Seetoo, TS Harrison, RA Speizer, ER Simons. *Cryptococcus neoformans* resides in an acidic phagolysosome of human macrophages. Infect Immun 67:885–890, 1999.
101. R Romero-Martinez, M Wheeler, A Guerrero-Plata, G Rico, H Torres-Guerrero. Biosynthesis and functions of melanin in *Sporothrix schenckii*. Infect Immun 68: 3696–3703, 2000.

Index